The Stress Analysis of Cracks Handbook

Third Edition

Hiroshi Tada

Paul C. Paris

George R. Irwin

The American Society of
Mechanical Engineers, New York

Professional Engineering Publishing Limited
London and Bury St Edmunds, UK

International Standard Book Number: 1-86058-304-0

Published by Professional Engineering Publishing, publishers to the Institution of Mechanical Engineers, Northgate Avenue, Bury St. Edmunds, IP32 6BW, UK

A CIP Catalogue record for this book is available from the British Library.

DEDICATION

On 9 October 1998, Dr. George R. Irwin passed away. We, the surviving authors of this book, have lost a great friend and colleague, who deservingly has been called "the Father of Fracture Mechanics." He was our inspiration in continuing to develop new items for the second and third editions of this book. Therefore, we feel compelled to dedicate our effort on this third edition to him, as well as our effort on previous editions. We are proud of the privilage of having worked closely with him.

Paul C. Paris and Hiroshi Tada, May 2000

TABLE OF CONTENTS

Part III
Two-Dimensional Stress Solutions for Various Configurations with Cracks 81

Part IV
Three-Dimensional Cracked Configurations 333

Part V
Crack(s) in a Rod or a Plate by Energy Rate Analysis 415

List of Principal Symbols

Only the principal symbols are listed here. So many symbols are needed that the notation is not necessarily consistent throughout the book. However, most symbols are clearly defined within each solution page and those not included in this list will be readily identified.

In many two-dimensional configurations, forces and moment (P, Q, etc., and M) implicitly designate forces and moment unit thickness. The definition will most often be obvious from the context; otherwise, confirm the definition so that the stress intensity factor, K, yielded has the dimension: force/(length)$^{3/2}$ or stress · (length)$^{1/2}$.

I, II, III (subscripts)	Designations for Mode I, Mode II, and Mode III, respectively
A	Area of crack surface Area of crack opening A crack tip (2-D) or a point on crack front (3-D)
A_n	Coefficients for series expansion of $Z(z)$
A_{ij}	Elastic constants for anisotropic solid
a	Half length of crack in plate or shell Depth of edge crack Length or half length of net ligament (2-D) Radius of circular crack or circular net ligqment Semi-major axis of elliptical crack or elliptical net ligament
a_{ij}	Elastic constants for anisotropic solid
B	A crack tip (2-D) or a point on crack front (3-D) Width of crack surface for rod or beam
$B(B_I, B_{II}, B_{III})$	Strength of Bueckner-type crack-tip singularity

Symbol	Definition
b	Width or half width of strip Depth of bend specimen Radius of cracked disc or circular hole Radius of round bar with circular crack Semi-minor axis of elliptical crack or elliptical net ligament Semi-major axis of elliptical hole or depth of semi-elliptical notch Half width of rectangular hole Coordinate or length-defining position of concentrated or distributed load Half length of second crack (collinear cracks) or of vertical crack (cruciform crack)
C	Elastic compliance Elastic constant: 1 for plane stress, $1/\sqrt{1-\nu^2}$ for plane strain
$C_{ij}(a)$	Elastic compliance of cracked body at ith load point by jth load
c	Coordinate or length defining position of concentrated or distributed load Semi-major or minor axis of elliptical hole or depth of semi-elliptical notch Radial width of annular crack Model crack size for strip yield analysis (actual crack size + plastic zone size) Model net ligament size for strip yield analysis (actual ligament size - plastic zone size) Coordinate of zero crossing of residual stress distribution
D	Diameter of (cracked) disc
$D(x/a)$, $D(\gamma)$, etc.	Configuration functions for crack opening displacement
d	Semi-minor axis of elliptical hole or half width of semi-elliptical notch Distance between center of crack and center of circular hole
E	Young's modulus
E'	Elastic constant = 4 α G : E for plane stress, $E/(1-\nu^2)$ for plane strain
$E(k)$	Complete elliptic integral of the second kind
$E(\varphi, k)$	Elliptic integral of the second kind
e	Eccentricity of crack in strip Base of natural logarithm
F	Concentrated force

$F(a/b)$, $F(\theta)$, $F(\lambda)$ $F_1(a/b)$, $F_2(a/b)$ $F_I(a/b)$, $F_{II}(a/b)$ $F_{III}(a/b)$, etc.	Configuration correction factors for stress intensity factor
$F(\varphi, k)$	Elliptic integral of the first kind
$f_m(s, a)$	Weight function determined from mth loading system
G	Shear modulus
$G(a/b)$, etc.	Alternate forms of configuration correction factors for stress-intensity factor
$G(\theta)$, $G(\lambda)$, etc.	Configuration functions for crack opening area for shells
$\mathcal{G}(\mathcal{G}_I, \mathcal{G}_{II}, \mathcal{G}_{III})$	Crack extension force
H	Distance between parallel cracks
$H(a/b)$, etc.	Configuration functions for crack opening displacement
h	Half depth of beam Height or half height of strip with crack vertical to side edges Half width of strip with crack parallel to side edges Half distance between parallel cracks Half width of rectangular hole Half thickness of uniform wedge Distance from free surface to crack parallel to it in semi-infinite plate
I, I_1, I_2, etc.	Moments of inertia of cross section of rods or beams
$I(\theta)$	Configuration function for crack opening area for cylindrical shell
Im( )	Imaginary part of ( )
J	J-Integral
$K(K_I, K_{II}, K_{III})$	Stress intensity factor
$K_{I\pm a}$, K_{IA}, etc.	K_I at $x = \pm a$, K_I at A, etc.
K_t	Stress concentration factor
$K(k)$	Complete elliptic integral of the first kind
k	Modulus of elliptic integral
k'	Complementary modulus of elliptic integral

L, L', L_1, L_2, etc.	Dimensions defined in solution pages
ℓ	Length of normal for ellipse Total length of crack and hole (or notch) Length of plastic zone for strip yield model Length of crack surface contact
ℓ_1, ℓ_2 ℓ_a, ℓ_b ℓ_α, ℓ_β, etc.	Dimensions defined in solution pages
M	Bending moment In-plane moment per unit thickness
m	Slope of tapered double cantilever specimen
$(\ )_{max}$	Maximum value of ( )
$(\ )_{min}$	Minimum value of ( )
n	Number of radial cracks for star-shaped crack
o	Origin of coordinate system
P	Concentrated load Concentrated load per unit thickness (2-D)
P_i	ith applied load
p	Distributed load (stress) Internal pressure in shell
$\bar{p}$	Line force (force/length)
Q	Concentrated load Concentrated load per unit thickness (2-D)
q	Distributed load (stress)
R	Concentrated load Radius of circular arc crack Radius of circular hole Radius of cylindrical or spherical shell
$Re(\ )$	Real part of ( )
r	First (radial) coordinate of polar $[r, \theta, (z)]$ coordinate system
r_i, r_o	Inner and outer radius, respectively, of thick-walled cylinder

r_p, r_Y	Plastic zone size and size index, respectively $(r_p = 2r_Y)$
$S(a/b)$, etc.	Configuration functions for crack opening area or rotation
s	Span between supports of bend specimen Distance from crack plane to point of concentrated load application Normalized half vertex angle of cracked wedge Dimensionless parameter characterizing geometry of cracked body, e.g., $s = a/(a+b)$
T	Concentrated load per unit thickness (Mode III) Twisting moment
$T(s)$	Distributed traction over surface s
t	Distributed load (stress) Thickness of shell
U	Elastic energy density
U_T	Total strain energy in cracked body
$U(a/b)$, $U_1(a/b)$ $U_2(a/b)$, etc.	Configuration functions for displacement
u	x-direction displacement in (x, y, z) system
u_i	Load point displacement of ith load P_i
u_r, u_θ	Displacement components in (r, θ) system
V	Volume of crack (3-D)
$V(a/b)$, $V_1(a/b)$ $V_2(a/b)$, etc.	Configuration functions for displacement
$V_1(a)$, $V_1(a_0)$ $V_2(a)$, $V_2(a_0)$	Electric potentials
v	y-direction displacement in (x, y, z) system
$v(x, 0)$, $v(r, 0)$	Half crack opening displacements
v_s, v, $v(0, y)$ $v(0, s)$, $v(0, 0, s)$, etc.	Displacements perpendicular to crack plane
W	Width of plate (strip) Interval of periodically repeated collinear cracks
$W(a/b)$, $W_1(a/b)$ $W_2(a/b)$, etc.	Configuration functions for displacement (Mode III)

w	z-direction displacement in (x, y, z) system
X	x-coordinate of point A(X, Y) on circular crack front
x	First coordinate of (x, y, z) Cartesian coordinate system Real part of complex variable $z = x + iy$
x_0	x-coordinate of point of concentrated load application
Y	y-coordinate of point A(X, Y) on circular crack front
y	Second coordinate of (x, y, z) Cartesian coordinate system Imaginary part of complex variable $z = x + iy$
y_0	y-coordinate of point of concentrated load application
$Z(z)\ (Z_I(z), Z_{II}(z), Z_{III}(z))$	Westergaard stress function
$Z'(z), \overline{Z}(z), \overline{\overline{Z}}(z)$	$Z'(z) = dZ(z)/dz$, $Z(z) = d\overline{Z}(z)/dz$, $\overline{Z}(z) = d\overline{\overline{Z}}(z)/dz$
$Z^*(z)$	Complex conjugate of $Z(z)$ (used in Part I)
z	Third coordinate of (x, y, z) or (r, θ, z) coordinate system Complex variable $z = x + iy$
$\bar{z}, z^*$	Complex conjugate of z: $\bar{z} = x - iy$ (z^* is used in Part I)
z_0	Complex coordinate of point of concentrated load application ($z_0 = x_0 + iy_0$)
α	Elastic constant: $(1+\nu)/2$ for plane stress, $1/[2(1-\nu)]$ for plane strain $(= 1/\beta)$ Direction of applied stress Direction of concentrated load (for cracked wedge)
β	Elastic constant $(= 1/\alpha)$ Parameter characterizing relative rigidity of second plate (or material) $(= E_1 h_1/Eh)$
$\Gamma(x)$	Gamma function
γ	Direction of applied stress Exponent specifying distributed load
$\gamma_{xy}, \gamma_{yz}, \gamma_{zx}$	Shear strain components in (x, y, z) coordinate system
$\gamma_{r\theta}, \gamma_{\theta z}, \gamma_{zr}$	Shear strain components in (r, θ, z) coordinate system
Δ	Thickness of uniform wedge

$\Delta\,(\Delta_I,\,\Delta_{II},\,\Delta_{III})$	Load point displacement
Δ_i	Load point displacement at ith load (P_i)
Δ_{crack}	Additional displacement due to the presence of crack
$\Delta_{\text{no crack}}$	Displacement in the absence of crack
Δ_{total}	Total displacement $(= \Delta_{\text{no crack}} + \Delta_{\text{crack}})$
$\delta\,(\delta_I,\,\delta_{II},\,\delta_{III})$	Crack surface relative displacement
$\delta_0,\,\delta_b,\,\delta(x)$, etc.	Crack opening displacements at specific point on crack (Mode I)
ε	Shell parameter: $\varepsilon^2 = (t/R)\sqrt{12(1-\nu^2)}$
$\varepsilon_x,\,\varepsilon_y,\,\varepsilon_z$	Normal strain components in $(x,\,y,\,z)$ coordinate system
$\varepsilon_r,\,\varepsilon_\theta,\,\varepsilon_z$	Normal strain components in $(r,\,\theta,\,z)$ coordinate system
ζ	Complex coordinate taken at crack tip
θ	Second coordinate (polar angle) in $[r,\,\theta,\,(z)]$ coordinate system Parametric angle defining point A on crack front of circular, elliptical, semi-circular (surface), or quarter-circular (corner) crack Half central angle contained by circular arc crack(s) (2-D) or circumferential crack in cylindrical shell Half vertex angle of infinite wedge with symmetrical edge crack Relative rotation at infinity
θ_{crack}	Additional rotation due to the presence of crack
$\theta_{\text{no crack}}$	Rotation in the absence of crack
θ_{total}	Total rotation $(= \theta_{\text{no crack}} + \theta_{\text{crack}})$
$\varkappa$	Elastic constant: $(3-\nu)/(1+\nu)$ for plane stress, $3\text{-}4\nu$ for plane strain $(= 2\beta - 1)$
λ	Ratio of two systems of applied load Shell parameter $(= \theta/\sqrt{t/R} = a\sqrt{Rt})$
ν	Poisson's ratio
$\Pi(\varphi,\,n,\,k)$	Elliptic integral of the third kind

ρ	Notch tip root radius Mass density
σ, σ'	Applied stress
σ_0	Maximum tensile stress in residual stress field
σ_x, σ_y, σ_z	Normal stress components in (x, y, z) coordinate system
σ_r, σ_θ, σ_z	Normal stress components in (r, θ, z) coordinate system
σ_Y, σ_{YP}	Yield strength in tension
τ, τ_ℓ	Applied shear stress in Mode II and Mode III, respectively
τ_Y	Yield strength in shear
τ_{xy}, τ_{yz}, τ_{zx}	Shear stress components in (x, y, z) coordinate system
$\tau_{r\theta}$, $\tau_{\theta z}$, τ_{zr}	Shear stress components in (r, θ, z) coordinate system
Φ	Airy's stress function
$\Phi(a/b)$, $\Phi(\theta)$, $\Phi_1(\theta)$, $\Phi_2(\theta)$, etc.	Configuration functions for rotation
ϕ, $\phi(z)$	Function of complex variable $z = x + iy$
ϕ	Relative rotation at infinity or kink angle $(= \phi_{\text{crack}})$ at cracked section
ϕ_{crack}	Additional rotation (or kink) due to the presence of crack
$\phi_{\text{no crack}}$	Rotation in the absence of crack
ϕ_{total}	Total relative rotation $(= \phi_{\text{no crack}} + \phi_{\text{crack}})$
χ, $\chi(z)$	Function of complex variable $z = x + iy$
ω	Angular velocity of rotating disc Angle defining direction of applied load
$\frac{\partial}{\partial x}, \frac{\partial}{\partial y}, \frac{\partial}{\partial y_0}, \frac{\partial}{\partial s}$ $\frac{\partial^2}{\partial x^2}, \frac{\partial^2}{\partial y^2}, \frac{\partial^2}{\partial x \partial y}$, etc.	Partial differential operators
∇^2	Harmonic operator ($= \frac{\partial^2}{\partial x^2} + \frac{\partial^2}{\partial y^2}$ in (x, y) system)

Foreword

Fracture mechanics was introduced in the 1947-1952 period using the idea that onset of rapid crack extension occurred when the crack extension force became large enough to cause rapid joining of small openings near the leading edge of a crack. The "force" concept used was the rate of loss of stress field energy, G, per unit of new separation area. Unfortunately the usual training in stress analysis of engineers did not provide methods of estimating values of G. However, in the mid-1950s, use of a relatively simple method of crack-stress field analysis, introduced by Westergaard, permitted demonstration that the severity of the enclosing stress field, tending to cause crack extension, could be represented by a stress intensity factor, K.

In addition, values of the force, G, were related to K by the use of equation $G = K^2/E$; where E is Young's modulus. This led to use of toughness measurements in terms of critical values of K necessary for rapid crack extension. This change of concept and nomenclature was of special importance to the understanding and practical use of fracture mechanics by engineers, and led immediately to general acceptance of fracture mechanics. Despite the sound theoretical basis for the force, G, engineers preferred a representation of critical conditions for crack extension in terms of principles of stress analysis with which they were familiar.

The introduction of the K concept was shown to be of special value for studies of fatigue cracking. It was shown that from calibration tests, it was possible to make estimates of the danger of crack growth by small initial cracks due to load fluxuations during periods of use in service. The use of K values for studies of fatigue cracking was followed by the use of K values for studies of corrosion cracking and corrosion fatigue.

In the use of fracture mechanics, estimates of K for potential or real cracks are commonly needed. For this purpose, Tada's Handbook of K Values (renamed The Stress Analysis of Cracks Handbook) for cracks in various structural locations has been widely used. Previously available only in notebook form, this collection of K values has been reviewed and checked carefully.

G. R. Irwin, 1997

PREFACE — THIRD EDITION

The work on this handbook virtually began during the doctoral dissertation of Dr. Hiroshi Tada under the direction of Dr. George R. Irwin during the late 1960s at Lehigh University. In that dissertation, a modest number of new crack-tip stress intensity factor solutions were developed. Upon completion of his degree, Dr. Tada was employed by Del Research Corporation with the primary task of developing material for the Stress Analysis of Cracks Handbook. That led directly to the two earlier editions, with Fracture Proof Design Corporation providing the venue for much of the work on the latter of the two. This third edition has been produced with Dr. Paul C. Paris and Dr. Tada at Washington University, St. Louis, with modest cooperative effort from Dr. Irwin. It seems fitting that this long-term effort to develop such a handbook should finally be published hardbound by a leading engineering society, The American Society of Mechanical Engineers (ASME).

During the 30 years of development of this work, Dr. Tada has continuously devised new solutions, collected others and improved them, and developed fitting formulas and curves to present them in a convenient form for use by practitioners and researchers alike in the field of fracture mechanics. His coauthors herewith recognize his monumental effort in accomplishing that task. The text accompanying the solutions presented in this handbook was the joint task of the three individuals involved, each contributing several sections and editing others. In addition, we acknowledge that the original work related to three of the appendices included contributions from other coauthors: H. Ernst, R. McMeeking, and L. L. Loushin. The involvement of many other individuals through direct assistance, suggestions, corrections, and encouragement throughout the 30 years are also noted and appreciatively acknowledged.

There is a software disk (see pages 676–677) available to purchasers of the third edition which allows rapid numerical computation of some much-used stress intensity factor K formulas for commonly adopted test and crack configurations found in practice. We especially thank Drs. Dilip Dedhia and David Harris for their interest in this book. Incidentally, the new appendix on K values for plates subjected to pinching loads was in fact originally a topic raised by the dissertation of Dr. Harris.

This third edition also adds new appendix sections on the J-Integral, on displacements prescribed on crack surfaces, on plastic zone instability (explaining a potentially interceding "elastic" failure mechanism), on engineering estimates of stress intensity factors, and Mode III plasticity solutions, as well as about 30 new solution pages and modifications of many older solutions.

The objective of this and each edition has been to document all of the important methods and results of elastic stress analysis as may be applied to small-scale yielding fracture mechanics and beyond. The principles and methods are found in the initial text sections and in the many appendices provided. Numerical approaches such as finite element methods, boundary collocation methods, and so on, remain in such a high state of development that discussion of them has been deliberately omitted. However, we have attempted to include all of the relevant and lasting material on elastic analysis as it applies to fracture mechanics and related disciplines.

Hiroshi Tada, Paul C. Paris, George R. Irwin, September 1997

PREFACE — SECOND EDITION

Since the last modification of this handbook in 1975, many new results have been forthcoming that are appropriate to include herein. Over 100 new solutions and other material have been added. Some corrections and modifications for completeness of existing crack stress analysis solutions have also been included in the new edition.

The project of further developing this handbook is ongoing and we hope to offer additional results some time in the future.

We thank the many readers who have offered comments and corrections over the past 10+ years. Further suggestions are welcome.

Hiroshi Tada, Paul C. Paris, October 1985

PREFACE — FIRST EDITION

This text is intended to provide the user with a comprehensive source of formulas and stress analysis information on crack problems. The emphasis is on useful information for treatment of actual problems on crack propagation through fracture mechanics correlation parameters and current fracture criteria, such as K_I approaching K_{Ic} as a plane strain fracture criterion.

The information provided, however, is not limited to that used in current practice, but also embodies other fundamental stress analysis results. For example, where stress functions are known for the complete solution to a crack problem they are either listed or referenced; again, where they are known, functions are listed that may be readily converted to displacements, such as integrals of stress functions.

Each numerical solution and approximation method is accompanied by the author's estimate of the accuracy of the results or the method; moreover, source references are listed in all cases for those users who wish to explore further details.

The information presented is useful only to the degree that it can be understood and properly used. For this reason, descriptive sections of text material are included (a) to define the meaning of the information presented, (b) to indicate and illustrate its conversion to other forms, and (c) to develop methods of applying it to actual cases or problems.

In addition, there are sections of text devoted to (a) the theory and useful methods of compliance calibration analysis; (b) weight function analysis for handling certain cases of arbitrary loading; (c) orthotropic, anisotrpoic, and dynamic effects; and (d) plasticity analysis of crack problems, especially a discussion of the J-Integral methods. Other implications of crack stress analysis (e.g., stress concentrations and notch field equations) and related results (e.g., electric fields in plates with cracks for electrical potential calibration) are given where available.

Obviously, we intended not to limit the material presented to idealized stress analysis results alone, but rather to expand those results where usefulness will be enhanced. For that reason any suggestions by the reader for future additions are welcome.

ACKNOWLEDGMENTS TO THE FIRST EDITION

The able and extensive editorial assistance of Dr. John Srawley and the comments and corrections of several solutions by Professor James R. Rice are most gratefully acknowledged. The efficient secretarial assistance of Christine E. Anders is also noted with thanks. Moreover, the encouragements and comments of numerous collegues has been valuable in expediting the work herein.

Hiroshi Tada, Paul C. Paris, George R. Irwin, June 1973

PART

I

INTRODUCTORY INFORMATION

- Introduction
- Crack-Tip Stress Fields for Linear-Elastic Bodies
- Alternate Expressions for Crack-Tip Elastic Fields
- Slender Notches and Stress Concentrations from Stress Intensity Factors
- Energy Rate Analysis of Crack Extension
- Relationships Between $\mathcal{G}$ and K
- Superposition of $\mathcal{G}$ and K Results
- Meaning of Plane Stress and Plane Strain for Fracture Mechanics Purposes
- Effects of Small-Scale Yielding on Linear-Elastic Fracture Mechanics
- Introduction to Stress Function Methods
- Additivity of Crack Stress Fields and K Values
- Boundary Collocation Method
- Successive Boundary Stress Correction Method
- K Estimates from Finite Element Methods
- Additional Remarks for Part I

INTRODUCTION

Fracture studies of structural elements have been revolutionized in the last 50 years by the analysis of their sensitivity to flaws or cracklike defects. Within these studies an essential ingredient is reasonable and proper stress analysis especially with regard to flaws with high local elevations of stresses from which fractures progress through various crack propagation mechanisms, including corrosion and fatigue cracking.

Full studies of fracture behavior cover both the stress analysis aspects and the material behavior in terms of resistance to the stresses imposed. However, the purpose here is limited to the development of significant stress analysis details and relevant parameters, and to the compilation of available stress analysis results with cracks present insofar as they may be foreseeably related to actual fracture studies.

The redistribution of stress in a body caused by introducing a crack or notch may be solved by methods of linear-elastic stress analysis. Of course the greatest attention should be paid to the high elevation of stresses at or surrounding the crack-tip, which will usually be accompanied by at least some plasticity and other nonlinear effects. Nevertheless linear-elastic stress analysis properly forms the basis of most current fracture analysis, at least for "small scale yielding" where all substantial nonlinearity is confined within a linear-elastic field surrounding the crack-tip. Consequently, the character and significant parameters of linear-elastic crack-tip fields are examined first.

CRACK-TIP STRESS FIELDS FOR LINEAR-ELASTIC BODIES

The surfaces of a crack are the dominating influence on the distribution of stresses near and around the crack-tip, as they are the nearby and stress-free boundaries of the body. Other remote boundaries and loading forces affect only the intensity of the local stress field at the tip.

The stress fields near crack-tips can be divided into three basic types, each associated with a local mode of deformation as illustrated in **Fig. 1**. The opening mode, Mode I, is associated with local displacement in which the crack surfaces move directly apart (symmetric with respect to the x - y and x - z planes). The edge-sliding mode, Mode II, is characterized by displacements in which the crack surfaces slide over one another perpendicular to the leading edge of the crack (symmetric with respect to the x - y plane and skew-symmetric with respect to the x - z plane). Mode III, the tearing mode, finds the crack surfaces sliding with respect to one another parallel to the leading edge (skew-symmetric with respect to the x - y and x - z planes). The superposition of these three modes is sufficient to describe the most general three-dimensional case of local crack-tip deformation and stress fields.

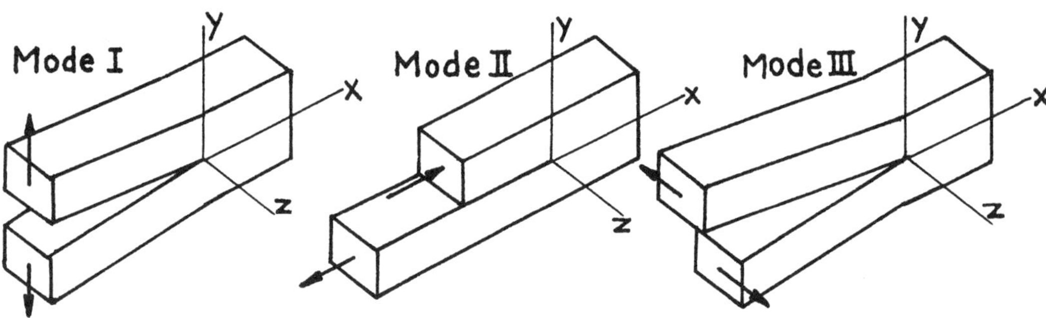

Fig. 1. Basic modes of crack surface displacements.

The most direct approach to determine the stress and displacement fields associated with each mode follows the manner of **Irwin (1957)**, which is based on the method of **Westergaard (1939)**. Modes I and II can be analyzed as two-dimensional plane-extensional problems of the theory of elasticity, which are subdivided as symmetric and skew-symmetric, respectively, with respect to the crack plane. Mode III can be regarded as the two-dimensional pure shear (or torsion) problem. Referring to **Fig. 2** for notation, the resulting stress and displacement fields are given below.

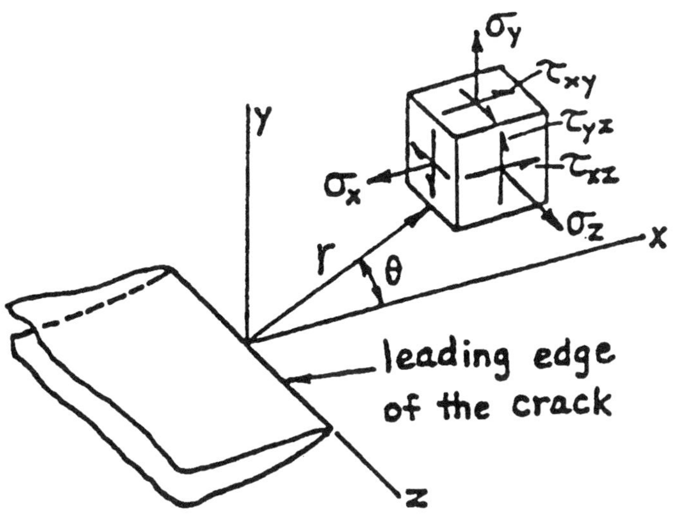

Fig. 2. Coordinates measured from leading edge of a crack and stress components in the crack-tip stress field.

Mode I:

$$\left.\begin{aligned}
\sigma_x &= \frac{K_I}{(2\pi r)^{1/2}}\cos\frac{\theta}{2}\left[1-\sin\frac{\theta}{2}\sin\frac{3\theta}{2}\right]+\sigma_{x0}+O\left(r^{1/2}\right)\\
\sigma_y &= \frac{K_I}{(2\pi r)^{1/2}}\cos\frac{\theta}{2}\left[1+\sin\frac{\theta}{2}\sin\frac{3\theta}{2}\right]+O\left(r^{1/2}\right)\\
\tau_{xy} &= \frac{K_I}{(2\pi r)^{1/2}}\sin\frac{\theta}{2}\cos\frac{\theta}{2}\cos\frac{3\theta}{2}+O\left(r^{1/2}\right)
\end{aligned}\right.$$

and for plane strain (with higher-order terms omitted)

$$\left.\begin{aligned}
&\sigma_z = \nu(\sigma_x+\sigma_y),\ \tau_{xz}=\tau_{yz}=0\\
&u = \frac{K_I}{G}[r/(2\pi)]^{1/2}\cos\frac{\theta}{2}\left[1-2\nu+\sin^2\frac{\theta}{2}\right]\\
&v = \frac{K_I}{G}[r/(2\pi)]^{1/2}\sin\frac{\theta}{2}\left[2-2\nu-\cos^2\frac{\theta}{2}\right]\\
&w = 0
\end{aligned}\right\}\qquad (1)$$

Mode II:

$$\left.\begin{aligned}
\sigma_x &= -\frac{K_{II}}{(2\pi r)^{1/2}}\sin\frac{\theta}{2}\left[2+\cos\frac{\theta}{2}\cos\frac{3\theta}{2}\right]+\sigma_{x0}+O\left(r^{1/2}\right)\\
\sigma_y &= \frac{K_{II}}{(2\pi r)^{1/2}}\sin\frac{\theta}{2}\cos\frac{\theta}{2}\cos\frac{3\theta}{2}+O\left(r^{1/2}\right)\\
\tau_{xy} &= \frac{K_{II}}{(2\pi r)^{1/2}}\cos\frac{\theta}{2}\left[1-\sin\frac{\theta}{2}\sin\frac{3\theta}{2}\right]+O\left(r^{1/2}\right)
\end{aligned}\right.$$

and for plane strain (with higher-order terms omitted) (2)

$$\left.\begin{aligned}
&\sigma_z = \nu(\sigma_x+\sigma_y),\ \tau_{xz}=\tau_{yz}=0\\
&u = \frac{K_{II}}{G}[r/(2\pi)]^{1/2}\sin\frac{\theta}{2}\left[2-2\nu+\cos^2\frac{\theta}{2}\right]\\
&v = \frac{K_{II}}{G}[r/(2\pi)]^{1/2}\cos\frac{\theta}{2}\left[-1+2\nu+\sin^2\frac{\theta}{2}\right]\\
&w = 0
\end{aligned}\right\}$$

Mode III:

$$\left.\begin{aligned}
\tau_{xz} &= -\frac{K_{III}}{(2\pi r)^{1/2}}\sin\frac{\theta}{2}+\tau_{xz0}+O\left(r^{1/2}\right)\\
\tau_{yz} &= \frac{K_{III}}{(2\pi r)^{1/2}}\cos\frac{\theta}{2}+O\left(r^{1/2}\right)\\
\sigma_x &= \sigma_y=\sigma_z=\tau_{xy}=0\\
w &= \frac{K_{III}}{G}[(2r)/\pi]^{1/2}\sin\frac{\theta}{2}\\
u &= v = 0
\end{aligned}\right\} \tag{3}$$

Equations (1) and (2) have been written for the case of plane strain (i.e., $w = 0$) but can be changed to plane stress easily by taking $\sigma_z = 0$ and replacing Poisson's ratio, ν, in the displacements with $\nu/(1+\nu)$.

In **Eqs. (1) – (3)**, higher-order terms such as uniform stresses parallel to cracks, σ_{x0}, and τ_{xz0}, and terms of the order of square root of r, $O(r^{1/2})$, are as indicated. However, normally these terms are omitted since as r becomes small compared with planar dimensions (in the x - y plane) of significance to the stress analysis, these higher-order terms become negligible compared with the leading $1/\sqrt{r}$ term. Therefore these leading terms are the linear-elastic crack-tip stress (and displacement) fields.

The parameters K_I, K_{II}, and K_{III} in these equations are called crack-tip stress (field) intensity factors for the corresponding three modes (**Fig. 1**). Since K_I, K_{II}, and K_{III} are not functions of the coordinates, r and θ, they represent the strength of the stress fields surrounding the crack-tip, as in **Eqs. (1)–(3)**. Alternately, they may be mathematically viewed as the strengths of the $1/\sqrt{r}$ stress singularities at the crack-tip. Their values are determined by the other boundaries of the body and the loads imposed, consequently formulas for their evaluation come from a complete stress analysis of a given configuration and loading.

Physically, K_I, K_{II}, and K_{III} may be regarded as the intensity of load transmittal through the crack-tip region caused by the introduction of a crack into the body of interest. Correspondingly, formulas for K may be regarded as formulas reflecting the redistribution of load paths for transmitting force past a crack. Thus it is plausible to observe that small amounts of plasticity or other nonlinearity at the crack-tip do not seriously further disturb the load redistribution, hence the relevance of K_I, K_{II}, and K_{III} remain.

Similarly, from a physical standpoint, K_I, K_{II}, and K_{III} may be regarded as representing the intensity of the linear-elastic stress distribution surrounding a crack-tip, where small ammounts of nonlinearity at the crack-tip are embedded well within the field and do not significantly disturb it. Thus, a given combination of values of K_I, K_{II}, and K_{III} represents a unique crack-tip stress field environment for small-scale yielding. Because fracture processes of a material may be regarded as "caused" by this surrounding crack-tip stress field environment, the intensity factors K_I, K_{II}, and K_{III} play a large role as fracture correlation parameters in current practice. For this reason, much of the tabulated material to follow includes formulas for K_I, K_{II}, and K_{III} for various configurations and loadings.

Finally, from **Eqs. (1) – (3)** it is significant to note that stress intensity factors have units of

$$\text{(force)} \times \text{(length)}^{-3/2}$$

Moreover, since they are linear factors in linear-elastic stress equations, they must be proportional to the applied loads. Thus it can be observed on a dimensional basis that in addition to the load they must contain other characteristic lengths, such as crack size. This result is a main feature of implying flaw-size effects in fracture, which indeed are observed, and further implies that these size effects can be fully analyzed only if stress analysis effects are included.

NOTE: It is interesting to note that the expressions in the brackets of the displacements u and v in **Eqs. (1)** are identical, that is,

$$\begin{aligned} 1 - 2\nu + \sin^2\frac{\theta}{2} &\equiv 2 - 2\nu - \cos^2\frac{\theta}{2} \\ &= \beta - \cos^2\frac{\theta}{2} \quad \text{(see } \textbf{P. 1.3b and P. 1.3c}) \end{aligned}$$

However, since these distinct expressions have been almost invariably used up to present, they are retained in **Eqs. (1)**. It immediately follows from this identity that the magnitude and direction of the Mode I crack-tip displacement vector $\vec{u} = (u, v)$ are given by

$$|\vec{u}| = \frac{K_I}{G}\sqrt{\frac{r}{2\pi}}\left(\beta - \cos^2\frac{\theta}{2}\right)$$

$$\frac{v}{u} = \tan\frac{\theta}{2}$$

The proportion and direction of the displacement vector in Mode I crack-tip field are schematically presented in **Fig. 3**.

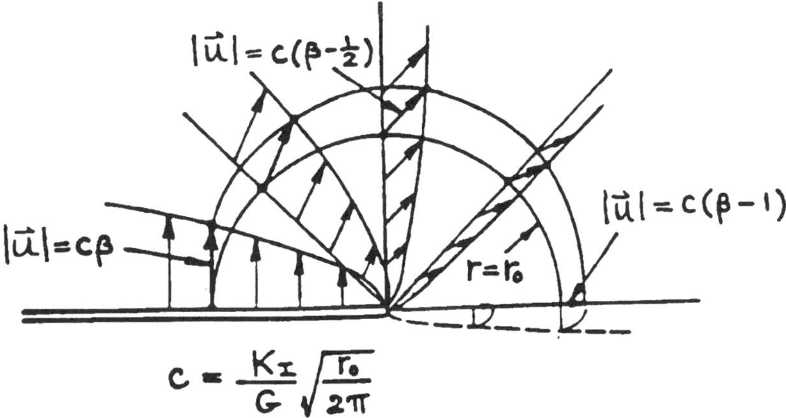

Fig. 3. Mode I displacement field.

ALTERNATE EXPRESSIONS FOR CRACK-TIP ELASTIC FIELD

In Modes I and II, stress and displacement components given by **Eqs. (1) and (2)** are sometimes expressed in alternate forms. These expressions and the corresponding expressions for r - θ components are given below.

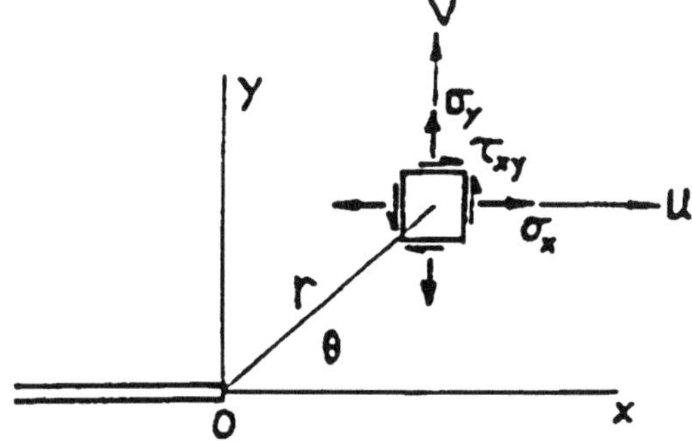

Fig. 4A

(1) x - y Components (Fig. 4A)

Mode I:

$$\begin{Bmatrix} \sigma_x \\ \sigma_y \\ \tau_{xy} \end{Bmatrix} = \frac{K_I}{\sqrt{2\pi r}} \cos\frac{\theta}{2} \begin{Bmatrix} 1 - \sin\frac{\theta}{2}\sin\frac{3\theta}{2} \\ 1 + \sin\frac{\theta}{2}\sin\frac{3\theta}{2} \\ \sin\frac{\theta}{2}\cos\frac{3\theta}{2} \end{Bmatrix} = \frac{K_I}{\sqrt{2\pi r}}\frac{1}{4} \begin{Bmatrix} 3\cos\frac{\theta}{2} + \cos\frac{5\theta}{2} \\ 5\cos\frac{\theta}{2} - \cos\frac{5\theta}{2} \\ -\sin\frac{\theta}{2} + \sin\frac{5\theta}{2} \end{Bmatrix}$$

$$\begin{Bmatrix} u \\ v \end{Bmatrix} = \frac{K_I}{G}\sqrt{\frac{r}{2\pi}} \begin{Bmatrix} \cos\frac{\theta}{2} \\ \sin\frac{\theta}{2} \end{Bmatrix} \left(\beta - \cos^2\frac{\theta}{2}\right) = \frac{K_I}{G}\sqrt{\frac{r}{2\pi}}\frac{1}{4} \begin{Bmatrix} (4\beta - 3)\cos\frac{\theta}{2} - \cos\frac{3\theta}{2} \\ (4\beta - 1)\sin\frac{\theta}{2} - \sin\frac{3\theta}{2} \end{Bmatrix}$$

Mode II:

$$\begin{Bmatrix} \sigma_x \\ \sigma_y \\ \tau_{xy} \end{Bmatrix} = \frac{K_{II}}{\sqrt{2\pi r}} \begin{Bmatrix} -\sin\frac{\theta}{2}\left(2 + \cos\frac{\theta}{2}\cos\frac{3\theta}{2}\right) \\ \sin\frac{\theta}{2}\cos\frac{\theta}{2}\cos\frac{3\theta}{2} \\ \cos\frac{\theta}{2}\left(1 - \sin\frac{\theta}{2}\sin\frac{3\theta}{2}\right) \end{Bmatrix} = \frac{K_{II}}{\sqrt{2\pi r}}\frac{1}{4} \begin{Bmatrix} -5\sin\frac{\theta}{2} - \sin\frac{5\theta}{2} \\ -\sin\frac{\theta}{2} + \sin\frac{5\theta}{2} \\ 3\cos\frac{\theta}{2} + \cos\frac{5\theta}{2} \end{Bmatrix}$$

$$\begin{Bmatrix} u \\ v \end{Bmatrix} = \frac{K_{II}}{G}\sqrt{\frac{r}{2\pi}} \begin{Bmatrix} \sin\frac{\theta}{2}\left(\beta + \cos^2\frac{\theta}{2}\right) \\ -\cos\frac{\theta}{2}\left(\beta - 2 + \cos^2\frac{\theta}{2}\right) \end{Bmatrix} = \frac{K_{II}}{G}\sqrt{\frac{r}{2\pi}}\frac{1}{4} \begin{Bmatrix} (4\beta + 1)\sin\frac{\theta}{2} + \sin\frac{3\theta}{2} \\ -(4\beta - 5)\cos\frac{\theta}{2} - \cos\frac{3\theta}{2} \end{Bmatrix}$$

where $\beta\left(= {}^1/_\alpha\right) = \begin{cases} 2(1-\nu) & \text{plane strain} \\ 2\left(\frac{1}{1+\nu}\right) & \text{plain stress} \end{cases}$

(2) r - θ Components (Fig. 4B)

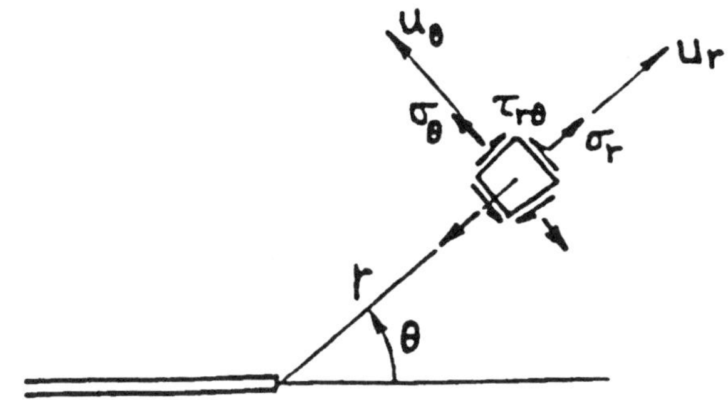

Fig. 4B

Mode I:

$$\begin{Bmatrix} \sigma_r \\ \sigma_\theta \\ \tau_{r\theta} \end{Bmatrix} = \frac{K_I}{\sqrt{2\pi r}}\cos\frac{\theta}{2}\begin{Bmatrix} 1+\sin^2\frac{\theta}{2} \\ \cos^2\frac{\theta}{2} \\ \sin\frac{\theta}{2}\cos\frac{\theta}{2} \end{Bmatrix} = \frac{K_I}{\sqrt{2\pi r}}\cdot\frac{1}{4}\begin{Bmatrix} 5\cos\frac{\theta}{2}-\cos\frac{3\theta}{2} \\ 3\cos\frac{\theta}{2}+\cos\frac{3\theta}{2} \\ \sin\frac{\theta}{2}+\sin\frac{3\theta}{2} \end{Bmatrix}$$

$$\begin{Bmatrix} u_r \\ u_\theta \end{Bmatrix} = \frac{K_I}{G}\sqrt{\frac{r}{2\pi}}\begin{Bmatrix} \cos\frac{\theta}{2} \\ -\sin\frac{\theta}{2} \end{Bmatrix}\left(\beta-\cos^2\frac{\theta}{2}\right) = \frac{K_I}{G}\sqrt{\frac{r}{2\pi}}\frac{1}{4}\begin{Bmatrix} (4\beta-3)\cos\frac{\theta}{2}-\cos\frac{3\theta}{2} \\ -(4\beta-1)\sin\frac{\theta}{2}+\sin\frac{3\theta}{2} \end{Bmatrix}$$

Mode II:

$$\begin{Bmatrix} \sigma_r \\ \sigma_\theta \\ \tau_{r\theta} \end{Bmatrix} = \frac{K_{II}}{\sqrt{2\pi r}}\begin{Bmatrix} \sin\frac{\theta}{2}\left(1-3\sin^2\frac{\theta}{2}\right) \\ -\sin\frac{\theta}{2}\left(3\cos^2\frac{\theta}{2}\right) \\ \cos\frac{\theta}{2}\left(1-3\sin^2\frac{\theta}{2}\right) \end{Bmatrix} = \frac{K_{II}}{\sqrt{2\pi r}}\cdot\frac{1}{4}\begin{Bmatrix} -5\sin\frac{\theta}{2}+3\sin\frac{3\theta}{2} \\ -3\sin\frac{\theta}{2}-3\sin\frac{3\theta}{2} \\ \cos\frac{\theta}{2}+3\cos\frac{3\theta}{2} \end{Bmatrix}$$

$$\begin{Bmatrix} u_r \\ u_\theta \end{Bmatrix} = \frac{K_{II}}{G}\sqrt{\frac{r}{2\pi}}\begin{Bmatrix} -\sin\frac{\theta}{2}\left(\beta-3\cos^2\frac{\theta}{2}\right) \\ -\cos\frac{\theta}{2}\left(\beta+2-3\cos^2\frac{\theta}{2}\right) \end{Bmatrix} = \frac{K_{II}}{G}\sqrt{\frac{r}{2\pi}}\frac{1}{4}\begin{Bmatrix} -(4\beta-3)\sin\frac{\theta}{2}+3\sin\frac{3\theta}{2} \\ -(4\beta-1)\cos\frac{\theta}{2}+3\cos\frac{3\theta}{2} \end{Bmatrix}$$

where $$\beta\left(={}^1/_\alpha\right) = \begin{cases} 2(1-\nu) & \text{plane strain} \\ 2\left(\frac{1}{1+\nu}\right) & \text{plain stress} \end{cases}$$

NOTE: For Mode I, $u_r = u$, $u_\theta = -v$ (displacement is in ${}^\theta/_2$ direction).

SLENDER NOTCHES AND STRESS CONCENTRATIONS FROM STRESS INTENSITY FACTORS

It is worthy to note that crack-tip stress intensity factors, as detailed in **Eqs. (1) – (3)** of the previous section, are fully applicable to the tips of deep slender notches (**Creager 1967**). See **Fig. 5** for location of coordinates. For the region of the notch tip where r' is small compared with other planar (x - y plane) dimensions (except for notch breadth), the stress field becomes ($\rho/2 \leqslant r' < \rho$, small θ')

Mode I:

$$\begin{Bmatrix} \sigma_x \\ \sigma_y \\ \tau_{xy} \end{Bmatrix} = \frac{K_I}{\sqrt{2\pi r'}} \cdot \frac{\rho}{2r'} \begin{Bmatrix} -\cos\frac{3\theta'}{2} \\ \cos\frac{3\theta'}{2} \\ -\sin\frac{3\theta'}{2} \end{Bmatrix} + \frac{K_I}{\sqrt{2\pi r'}} \cos\frac{\theta'}{2} \begin{Bmatrix} 1-\sin\frac{\theta'}{2}\sin\frac{3\theta'}{2} \\ 1+\sin\frac{\theta'}{2}\sin\frac{3\theta'}{2} \\ \sin\frac{\theta'}{2}\cos\frac{3\theta'}{2} \end{Bmatrix} + --- \tag{4}$$

Mode II:

$$\begin{Bmatrix} \sigma_x \\ \sigma_y \\ \tau_{xy} \end{Bmatrix} = \frac{K_{II}}{\sqrt{2\pi r'}} \cdot \frac{\rho}{2r'} \begin{Bmatrix} \sin\frac{3\theta'}{2} \\ -\sin\frac{3\theta'}{2} \\ -\cos\frac{3\theta'}{2} \end{Bmatrix} + \frac{K_{II}}{\sqrt{2\pi r'}} \begin{Bmatrix} -\sin\frac{\theta'}{2}\left(2+\cos\frac{\theta'}{2}\cos\frac{3\theta'}{2}\right) \\ \sin\frac{\theta'}{2}\cos\frac{\theta'}{2}\cos\frac{3\theta'}{2} \\ \cos\frac{\theta'}{2}\left(1-\sin\frac{\theta'}{2}\sin\frac{3\theta'}{2}\right) \end{Bmatrix} + --- \tag{5}$$

Mode III:

$$\begin{Bmatrix} \tau_{xz} \\ \tau_{yz} \end{Bmatrix} = \{O\} + \frac{K_{III}}{\sqrt{2\pi r'}} \begin{Bmatrix} -\sin\frac{\theta'}{2} \\ \cos\frac{\theta'}{2} \end{Bmatrix} + --- \tag{6}$$

Note that by selecting the center of coordinates at the point $\rho/2$ from the notch tip the expansion in **Eqs. (4)–(6)** simply adds an additional term for Modes I and II when directly compared to **Eqs. (1)–(3)**. Moreover, the intensity of the added terms are also given by K_I and K_{II} and these are exactly the same K's as found in **Eqs. (1)–(3)**. Therefore, formulas for stress intensity factors for cracks, as is extensively tabluated in this handbook, are also fully applicable to elastic stress computations for tips of slender notches.

Moreover, the first terms of **Eqs. (4) and (5)** are significant compared to the second only in the region near the end radius of the notch, $r' \to \rho/2$, whereas at greater radius—but still small compared with other planar dimensions—the same crack-tip stress field, as in **Eqs. (1) and (2)**, will dominate. Therefore, with the "disturbance" or "blunting" of a crack, or giving it a finite radius, ρ, the original crack-tip stress fields still surround the crack-tip. This is clear in the linear-elastic analysis case here, and it should be equally clear that a comparably (compared to ρ) sized zone of plasticity and/or other nonlinearity near a crack-tip is probably even less of a disturbance.

For a Mode I-type loading and configuration (i.e., K_{II} and K_{III} zero), for example, **Eq. (4)** may be used to find the stress concentration. In such a case

$$\sigma_{\max} = \sigma_y\Big|_{\substack{r=\rho/2 \\ \theta=0}} = \frac{2K_I}{\sqrt{\pi\rho}} \tag{7}$$

This result is good for slender notches, a practical example of which is "stop-drilled" cracks (a common practice in aircraft maintenance). Applying this result to **Fig. 6**, an elliptical hole through a wide plate where

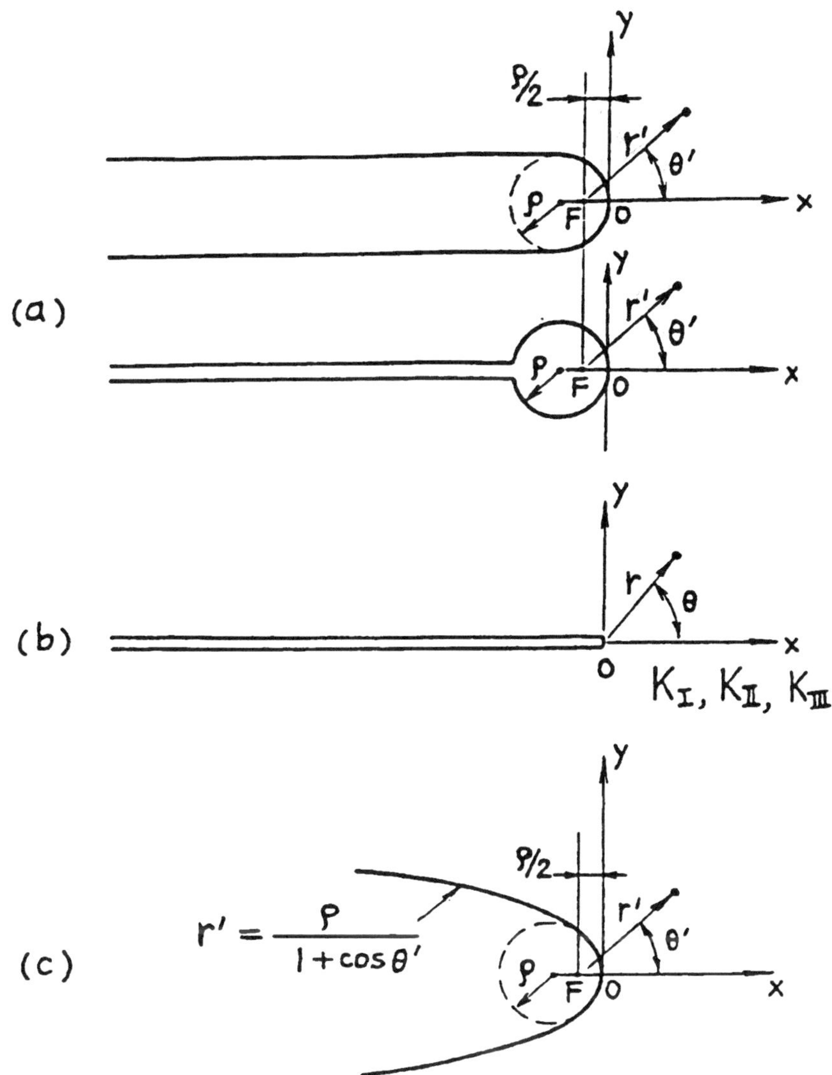

Fig. 5.
(a) Deep slender notches and coordinate system for notch-tip stress fields.
(b) Corresponding crack.
(c) Conic section (parabola) used for the analysis.

the semi-major axis, a, is perpendicular to a remotely applied tension stress, σ, the comparable crack solution (see **page 5.1**) is

$$K_I = \sigma\sqrt{\pi a}\,, \qquad K_{II} = K_{III} = 0 \tag{8}$$

inserting **Eq. (8)** into **Eq. (7)** gives

$$\sigma_{\max} = \sigma\left\{2\sqrt{\frac{a}{\rho}}\right\} \tag{9}$$

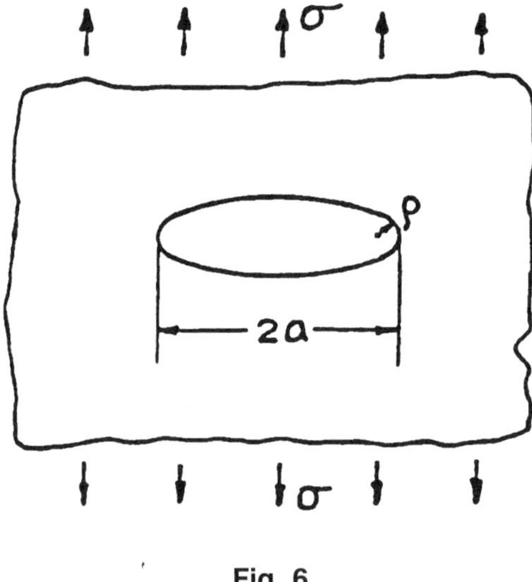

Fig. 6

The result is the well-known stress concentration solution for biaxial tension, σ, for which the comparable K_I solution is also as in **Eq. (8)**. For uniaxial stress, σ, the stress concentration result is

$$\sigma_{\max} = \sigma\left\{1 + 2\sqrt{\frac{a}{\rho}}\right\} \tag{10}$$

For slender notches ($\rho \ll a$), **Eqs. (9) and (10)** are in fact reasonably in agreement. The added term in **Eq. (10)** of one times σ is to be added only in cases of remotely applied uniform uniaxial stress [and is simply comparable to accounting for the additional σ_{xo} term in **Eqs. (1) and (4)**]. Knowing that **Eq. (10)** then follows from **Eq. (9)**, the full stress concentration solution followed from **Eqs. (7) and (8)**. Noting now that these stress concentration solutions, both **Eqs. (9) and (10)**, are, actually not limited to slender ellipses, the power and accuracy of this method is demonstrated.

For further demonstration, for the same elliptical hole in a large plate but loaded by equal and opposite forces (per unit plate thickness), P, on the surface of the hole at the ends of the semi-minor axis, b (see **Fig. 7**), the comparable crack solution (see **page 5.9**) is

$$K_I = \frac{P}{\sqrt{\pi a}} \quad , \quad K_{II} = K_{III} = 0 \tag{11}$$

Combining **Eqs. (7) and (11)** and noting that $\rho a = b^2$ for any ellipse gives

$$\sigma_{\max} = \frac{2P}{\pi b} \tag{12}$$

which again is the complete stress concentration solution, not limited to slender notches (**Savin 1961**).

It is not the intended purpose to present extensive information on stress concentrations here, but simply to illustrate the power of crack stress analysis. Nevertheless, it is evident that close relationships exist between stress concentration analysis and crack analysis. Later, converse to the preceding discussion, it will be pointed out that K formulas can also be derived from stress concentration formulas. For further study in stress concentration theory, **Savin (1961), Neuber (1937) and Peterson (1953)** are recommended as starting points.

In other instances, the role and formulas for crack-tip stress intensity factors K_I, K_{II}, and K_{III} are also preserved; for example, see **Appendix D** for effects of elastic anisotropy. Hence, the most important point of this discussion of notches is to emphasize the generality and scope of crack-tip stress field analysis.

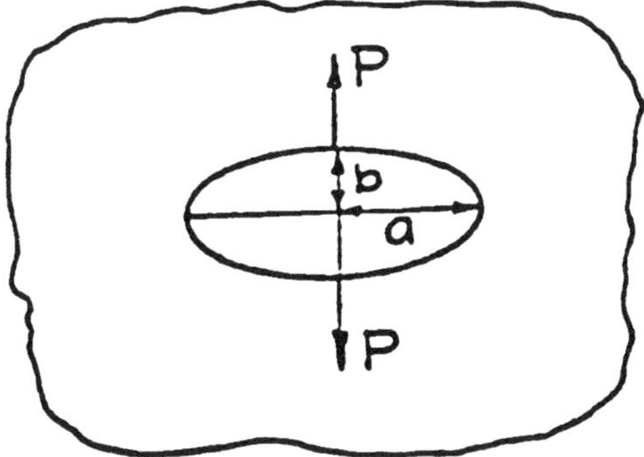

Fig. 7

ENERGY RATE ANALYSIS OF CRACK EXTENSION

Energy rate analysis of the effects of flaws historically preceded crack-tip stress field analysis. The Griffith Theory (**Griffith 1920**) and later modifications (**Irwin 1948, 1952; Orowan 1949**), termed the Griffith–Irwin Theory, made use of this approach. Basically, these methods use an energy balance analysis of crack extension.

The total elastic energy made available per unit increase in crack surface area (one side of the crack surface) is denoted by $\mathcal{G}$ for the linear-elastic case (**Irwin 1957**) (the non-linear counterpart, J, is discussed later; see **Appendices A and J**). Physically, $\mathcal{G}$, may be viewed as the energy made available for the crack extension processes at the crack-tip as a result of the work from displacements of loading forces and/or reductions in strain energy in a body accompanying a unit increase in crack area. Alternatively, $\mathcal{G}$ can be regarded as a "generalized force" based on the potential energy change per unit forward displacement of a unit length of crack front, which results in $\mathcal{G}$ being defined as "the crack extension force."

Following this line of argument, it is not difficult to show that for linear-elastic conditions (**Irwin 1948, 1952; Paris 1957**)

$$\mathcal{G} = -\frac{\partial U_T(\Delta_i, A)}{\partial A} \tag{13}$$

$$\text{and} \qquad \mathcal{G} = +\frac{\partial U_T(P_i, A)}{\partial A} \tag{14}$$

$$\text{and} \qquad \mathcal{G} = \frac{P^2}{2} \cdot \frac{\partial C}{\partial A} \tag{15}$$

where U_T is the total strain energy in a cracked body with a crack area A. U_T is alternately expressed in terms of A and load point displacements, Δ_i, or in terms of A and loads, P_i. In **Eq. (15)**, C is the elastic compliance and the equation is written for a single loading force, but may be generalized for several forces **(Paris 1957)**. Derivations of results such as **Eqs. (13), (14), and (15)** are also available for distributed boundary tractions, and so on **(Bueckner 1958)**. **Equation (15)** and its consequences are also discussed in **Appendix A**.

The $\mathcal{G}$ implied by **Eqs. (13)–(15)** is the average value along a crack front weighted for the extent of crack extension involved for each increment of crack front in the three-dimensional sense. In two-dimensional situations, such as uniform extension of a straight-through crack in a thin plate subject to extension, $\mathcal{G}$ may be viewed as the value of a point quantity along the crack front. Moreover, for certain purposes, the three-dimensional situation may be viewed as being made up of two-dimensional slices to view $\mathcal{G}$ as a point quantity.

The expressions for $\mathcal{G}$, **Eqs. (13)–(15)**, are often useful as tools to compute $\mathcal{G}$, itself, and other quantities. For an example, see **Appendix B**, where a method of computing displacements is developed using **Eq. (14)**. **Equation (13)** will be used as an example as follows.

RELATIONSHIPS BETWEEN $\mathcal{G}$ AND K

If a cracked body is put into a "system-isolated condition," that is, with load point displacements fixed so that no work is done by loading forces, then **Eq. (13)** becomes self-evident. The energy made available for crack extension is the strain energy released by the extension.

Consider that a body with a crack is put into a system-isolated condition for conceptual clarity. Subsequently, presume the crack-tip is elastically pulled closed over a distance α, as illustrated in **Fig. 8**, from (a) to (b). The work done in elastic closure will all go into increasing the total strain energy U_T. Therefore **(Irwin 1957; Paris 1965)**

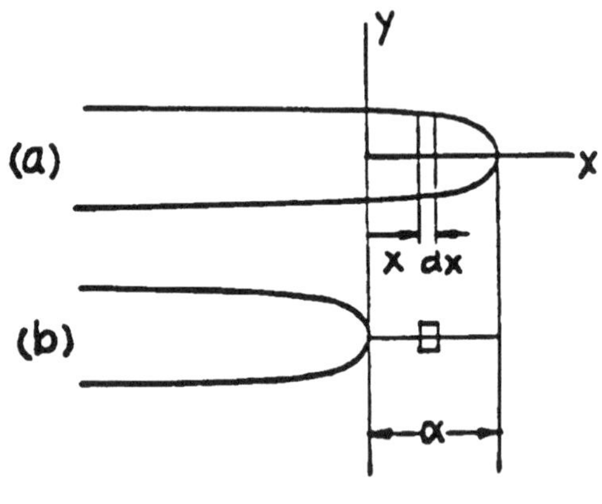

Fig. 8

$$\mathcal{G} = -\left.\frac{\Delta U_T}{\Delta A}\right|_{\text{system isolated}} = \lim_{\alpha \to 0} \frac{2}{\alpha} \int_0^\alpha \left(\frac{\sigma_y v}{2} + \frac{\tau_{yx} u}{2} + \frac{\tau_{yz} w}{2} \right) dx \quad (16)$$

Where σ_y, τ_{yx}, τ_{yz}, and u, v, w are stresses and displacements of the crack surface, respectively, occurring on the portion of the crack surface pulled closed. With the limit $\alpha \to 0$, the stress and displacements may be obtained from the crack-tip fields, **Eqs. (1)–(3)**. The stresses are appropriately obtained with $r = x$, $\theta = 0$, and the displacements with $r = \alpha - x$, $\theta = \pi$. Making these substitutions and integrating **Eq. (16)** leads to (for plane strain)

$$\mathcal{G} = \frac{1-\nu}{2G} K_I^2 + \frac{1-\nu}{2G} K_{II}^2 + \frac{1}{2G} K_{III}^2 \quad (17)$$

It is noted that each term in the integrand of **Eq. (16)** leads to a corresponding term in **Eq. (17)** with no interaction between Modes I, II, and III. Noting $E = 2(1+\nu)G$ and that

$$E' = E \quad \text{(plane stress)}$$
or
$$E' = E/\left(1-\nu^2\right) \quad \text{(plane strain)} \quad (18)$$

The total energy rate, $\mathcal{G}$, may be in general subdivided for each mode

$$\mathcal{G} = \mathcal{G}_I + \mathcal{G}_{II} + \mathcal{G}_{III} \quad (19)$$

where

$$\left.\begin{aligned} \mathcal{G}_I &= K_I^{\,2}\big/E' \\ \mathcal{G}_{II} &= K_{II}^{\,2}\big/E' \\ \mathcal{G}_{III} &= K_{III}^{\,2}\big/2G = \frac{1+\nu}{E}K_{III}^{\,2} \end{aligned}\right\} \tag{20}$$

It should be noted that these relationships are for straight-ahead crack extension, as developed from **Eq. (16)** and **Fig. 8**. Therefore, although they may not suit all physical applications, they form a useful conceptual basis and are suitable for many computations.

Corresponding relationships for anisotropic elastic bodies are noted in **Appendix D**.

SUPERPOSITION OF $\mathcal{G}$ AND K RESULTS

Often in applications, a single cracked body or member is subject to several loading force systems (each system in equilibrium) which can be denoted as systems (1), (2), (3), (4), ..., etc. Because K values are factors containing the load linearly in linear-elastic stresses, superposition applies. That is to say, the total K for all loading systems applied simultaneously is the algebraic sum of K values for each system applied separately. However, because different fields of stress occur for each mode, as noted from **Eqs. (1)–(3)**, the sums must be separated for each mode or

$$\left.\begin{aligned} K_I &= K_{I(1)} + K_{I(2)} + K_{I(3)} + ---- \\ K_{II} &= K_{II(1)} + K_{II(2)} + K_{II(3)} + ---- \\ K_{III} &= K_{III(1)} + K_{III(2)} + K_{III(3)} + ---- \end{aligned}\right\} \tag{21}$$

Using **Eq. (20)** to restate **Eq. (21)** in terms of energy rates

$$\left.\begin{aligned} \mathcal{G}_I &= \left[\mathcal{G}_{I(1)}^{1/2} + \mathcal{G}_{I(2)}^{1/2} + \mathcal{G}_{I(3)}^{1/2} + ----\right]^2 \\ \mathcal{G}_{II} &= \left[\mathcal{G}_{II(1)}^{1/2} + \mathcal{G}_{II(2)}^{1/2} + \mathcal{G}_{II(3)}^{1/2} + ----\right]^2 \\ \mathcal{G}_{III} &= \left[\mathcal{G}_{III(1)}^{1/2} + \mathcal{G}_{III(2)}^{1/2} + \mathcal{G}_{III(3)}^{1/2} + ----\right]^2 \end{aligned}\right\} \tag{22}$$

Equations (21) and (22) along with **Eq. (19)** become the rules for superposition of crack-tip stress intensity factors and energy rates.

It is of interest to note that **Eqs. (21)** imply that Green's function methods may be used for distributed force systems, again separating modes; although here **Eqs. (21)** are stated for discrete systems. The Green's function methods are discussed later (see **pages 1.17** and **1.18**).

MEANING OF PLANE STRESS AND PLANE STRAIN FOR FRACTURE MECHANICS PURPOSES

The mathematical definition of plane strain is that throughout a deformed body

$$\left.\begin{aligned} u &= u(x,y) \\ v &= v(x,y) \\ w &= 0(\text{or constant}) \end{aligned}\right\} \tag{23}$$

An alternate definition of plane strain is

$$\left.\begin{aligned} \varepsilon_z &= \frac{\partial w}{\partial z} = 0 \\ \gamma_{xz} &= \frac{\partial w}{\partial x} + \frac{\partial u}{\partial z} = 0 \\ \gamma_{yz} &= \frac{\partial w}{\partial y} + \frac{\partial v}{\partial z} = 0 \end{aligned}\right\} \tag{24}$$

It can be noted that **Eqs. (24)** follow directly from **Eqs. (23)** or vice-versa. Noting Hooke's laws, yet another definition of plane strain is

$$\left.\begin{aligned} \sigma_z &= \nu(\sigma_x + \sigma_y) \\ \tau_{xz} &= 0 \\ \tau_{yz} &= 0 \end{aligned}\right\} \tag{25}$$

Again **Eqs. (25)** follow from **Eqs. (24)**.

On the other hand, plane stress is mathematically defined as

$$\left.\begin{aligned} \sigma_z &= 0 \\ \tau_{xz} &= 0 \\ \tau_{yz} &= 0 \end{aligned}\right\} \tag{26}$$

where, as before, an alternative definition in terms of strains or displacement derivatives is possible but is not of useful clarity.

The term "generalized plane stress" is applied to cases of deformation where **Eqs. (26)** apply on the average through the thickness of a thin plate subject to extensional forces. Often, when the term "plane stress" is used, "generalized plane stress" is actually implied.

The above definitions of plane stress and plane strain are those used in books and reports on the theory of elasticity, the theory of plasticity, and other such works on solid mechanics in general. However, in fracture mechanics terminology, these terms take on special, more restricted meanings. In fracture mechanics, instead of characterizing stress and strain states throughout a body, special attention is given to the crack-tip and surrounding region. "Plane stress fracture" or "plane strain fracture" have come to mean that the stress and strain conditions within the plastic zone at the crack-tip are plane stress or plane strain, respectively.

Due to the high stress–strain gradients near a crack-tip, the zone of plasticity at the tip is constrained against contraction along the crack front by the elastic material surrounding it, if the plastic zone size is small compared with the length of the crack front. This creates plane strain fracture in the "linear-elastic fracture mechanics" sense, that is, where so-called "small-scale yielding" (compared with x - y planar dimensions) conditions apply. In applications of current linear-elastic fracture mechanics analysis to flat plate test specimens (or structural members) with through-the-thickness cracks, the ratio of the crack-tip plastic zone size to sheet thickness becomes the criterion of plane stress vs. plane strain conditions. Ironically then, the most common applications of fracture mechanics stress analysis to plates with through cracks are situations where stress analysis of elastic portions of the body is properly done using plane stress to characterize conditions, but where frequently the conditions within the crack-tip plastic zone are indeed plane strain. This is called a "plane strain fracture" situation.

In applying **Eqs. (20)** to convert stress intensity factors to energy rates, or vice-versa, along with **Eq. (18)**, a confusion (if not a paradox) arises. If elastic portions of the body are plane stress and the crack-tip region is plane strain, which conditions should be used in the conversion?

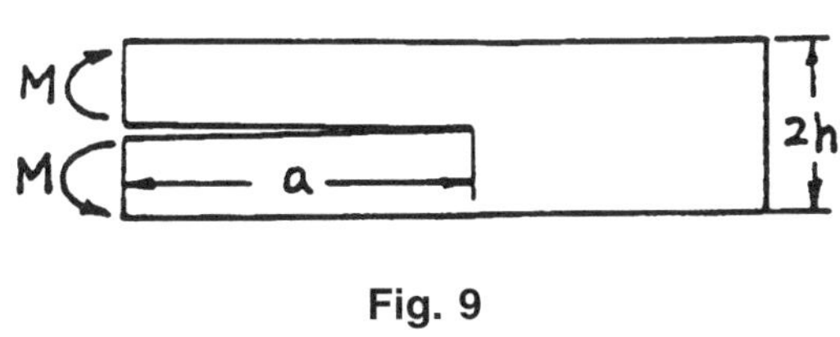

Fig. 9

For example consider a long, slender double cantilever beam configuration of constant thickness, as illustrated in **Fig. 9**. Considering the case of pure moments, M, is instructive **(Rice, 1964)**, since, clearly, adding crack length is directly equivalent to adding material at the center of the cantilever arms. Plane stress conditions are thus present where all of the strain energy is added at the center of the arms. Nevertheless, conditions near the crack-tip can be plane strain, and considering the crack closure derivation of **Eqs. (20)** from **Eq. (16)**, it is also clear that crack-tip stress field conditions apply to **Eqs. (20)** and the conversion should be made using E' for plane strain, **Eq. (18)**. Plasticity at the crack-tip tends to confound this discussion, but not if the surrounding elastic field is plane strain as well as the plastic zone. The complexity of these concepts is due to the attempt to resolve a 3-dimensional situation with 2-dimensional viewpoints, which apparently is unavoidable. Although this example illustrates that E' should be adjusted for crack-tip conditions, for simplicity, it is common practice to use the plane stress value, E, in **Eqs. (20)**. Since ν is normally about 0.3, the resulting error, if any, is less than 5% in computing K formulas. This practice is again mentioned in **Appendix A**, as compliance calibrations require conversions from $\mathcal{G}$ to K.

Finally, for elastic–plastic fracture mechanics, where small-scale yielding at a crack-tip does not apply, plane strain fracture events or tests require only that plane strain conditions exist in the "fracture process zone" at the crack-tip. This fracture process zone is the region in which the immediate crack extension processes such as advanced separations, void growth, and coalescence are taking place. This process zone may be embedded well within the crack-tip plastic zone, near the tip of the crack. Therefore it is evident that for plane strain in the process zone, it is not necessary for the whole plastic zone to be in a state of plane strain. Nevertheless, it is also evident that the immediate region of the crack-tip, that is, the process zone, must be subjected to plane strain. Therefore, blunting of a crack upon loading to a radius (or "crack opening stretch") characterized by J/σ_Y would at least require J/σ_Y to be small compared with specimen thickness, especially for through-the-thickness cracks. (Further discussions of J appear in **Appendix J**.) For this context, J may be viewed as the elastic–plastic analog to $\mathcal{G}$.

In summary, the reader is cautioned that for fracture mechanics purposes, the terms plane stress and plane strain are often used as local definitions of conditions in the crack-tip region. Moreover, the local size of the region involved will depend on the analysis approach used, linear-elastic vs. elastic-plastic, and is consequently subject to future developments in fracture analysis. Nevertheless, even within the most developed context of linear-elastic fracture mechanics confusion still remains, for the example cited of converting $\mathcal{G}$ to K, and other circumstances.

EFFECTS OF SMALL-SCALE YIELDING ON LINEAR-ELASTIC FRACTURE MECHANICS

"Small-scale yielding" means that the nonlinear zone at a crack-tip is small compared with the region in which the elastic crack-tip stress fields, **Eqs. (1)–(3)**, apply. Indeed, the circumstances are that the nonlinear or plastic zone may be regarded as embedded well within a surrounding elastic region. How small the plastic zone must be, compared to other (planar) dimensions, depends on the accuracy desired.

The size of the crack-tip plastic zone may be estimated from **Eq. (1)**, for Mode I, if small-scale yielding is in fact present. In any event, from dimensional considerations of **Eq. (1)**, it is evident that the form for an index of the size, r_Y, of the plastic zone is

$$r_Y = \alpha\left(\frac{K_I}{\sigma_{YP}}\right)^2 \tag{27}$$

where σ_{YP} is a yield strength for the material. The values of α may also be estimated from **Eq. (1)** by taking the stresses ahead of the crack, that is, $\theta = 0$, and computing the point at which yield criterion is first satisfied approaching the crack-tip **(Paris 1957)** and also adjusting for shape of the yield zone, and so on, the most commonly assumed values are **(Irwin 1960b)**

$$\alpha = \begin{cases} \dfrac{1}{2\pi} & (\text{plane stress}) \\ \dfrac{1}{6\pi} & (\text{plane strain}) \end{cases} \tag{28}$$

The plasticity at a crack-tip causes some redistribution of stresses to maintain equilibrium and therefore the full width of the plastic zone, r_P, is estimated at just twice the above results, that is

$$r_P = 2r_Y \tag{29}$$

and it is emphasized that these results, although dimensionally correct, are merely estimates, since work-hardening, large strains, and other obvious influences are ignored.

The redistribution of stress to satisfy equilibrium implies that the center of coordinates (r, θ) for the elastic field, **Eqs. (1)**, is advanced ahead of the real crack-tip into the zone of plasticity **(Paris 1957; Irwin 1960b)**. This correction for the "effective crack size" is often taken as approximately equal to r_Y added to the actual crack size. Using an "effective crack size" in linear-elastic fracture mechanics stress analysis calculation is regarded as sometimes appropriate. However, if very high accuracy is desired, it is appropriate to have r_Y small enough compared to planar dimensions, including crack size, that it may be entirely neglected. On the other hand, for the purpose of examining trends, and for low-accuracy calculations, the "effective crack size" correction has been proposed at times for application to large-scale yield situations. In any event, judgment is required for particular applications.

Nevertheless, provided the scale of yielding is small enough, all of the preceding results and derivations based on linear-elastic theory are, indeed, correct and appropriate to apply to real physical situations. Therefore, this discussion now proceeds based mainly on linear-elastic analysis and methods, but it will also provide other nonlinear analyses and mathematical models as seems appropriate for possible future use (e.g., see **Appendix J**).

INTRODUCTION TO STRESS FUNCTION METHODS

As shown by the preceding discussion, the primary objective of crack stress field analysis is to obtain a characterization of the stress–strain region enclosing the crack-tip, the region within which the progressive separational process occurs. Characterization in terms of K values, assuming linear-elastic behavior, only requires knowledge of stresses and strains close to the crack-tip. However, studies of crack extension often involve displacement measurements at some distance from the crack-tip. Thus solutions of crack problems that permit stress and displacement calculations in the entire stress field are of interest. Solutions of Mode I and Mode II crack problems in closed form are known for a large number of 2-dimensional, linear-elastic problems. The solution procedure uses the stress function approach and, therefore, the stress function method is discussed first. Except for special problems, mainly those of Mode III type, closed-form solutions are strictly applicable only to infinite plate crack problems. Computations of K values and displacements for strip and finite-plate crack problems usually require a numerical approach. In such problems, however, the stress function viewpoint can often enhance the efficiency of numerical methods.

Choosing x,y Cartesian coordinates, the stress equilibrium equations are

$$\frac{\partial \sigma_x}{\partial x} + \frac{\partial \tau_{xy}}{\partial y} = 0, \quad \frac{\partial \tau_{xy}}{\partial x} + \frac{\partial \sigma_y}{\partial y} = 0 \tag{30}$$

These equations are satisfied if we assume

$$\sigma_x = \frac{\partial^2 \Phi}{\partial y^2}, \quad \sigma_y = \frac{\partial^2 \Phi}{\partial x^2}, \quad \tau_{xy} = -\frac{\partial^2 \Phi}{\partial x \partial y} \tag{31}$$

The Hooke's Law equations are

$$\left.\begin{aligned} E\varepsilon_x &= \sigma_x - \nu(\sigma_y + \sigma_z) \\ E\varepsilon_y &= \sigma_y - \nu(\sigma_x + \sigma_z) \\ E\gamma_{xy} &= 2(1+\nu)\tau_{xy} \end{aligned}\right\} \tag{32}$$

where $\sigma_z = 0$ for plane stress and $\sigma_z = \nu(\sigma_x + \sigma_y)$ for plane strain and the identity $E = 2G(1+\nu)$ can be noted.

A convenient equation representing the fact that three strains are defined in terms of derivatives of only two displacements (strain compatibility) is given by

$$\frac{\partial^2 \varepsilon_x}{\partial y^2} + \frac{\partial^2 \varepsilon_y}{\partial x^2} = \frac{\partial^2 \gamma_{xy}}{\partial x \partial y} \tag{33}$$

Substituting **Eq. (31)** into **Eq. (32)**, followed by use of **Eq. (33)**, provides

$$\nabla^2\left(\nabla^2 \Phi\right) = 0 \quad \text{where } \nabla^2 = \frac{\partial^2}{\partial x^2} + \frac{\partial^2}{\partial y^2} \tag{34}$$

Equation (34) is obtained independently of whether plane-stress or plane-strain is assumed in **Eq. (32)**; Φ is termed the Airy stress function.

As a starting point for solving specific problems, **Muskhelishvili (1933)** noted certain analysis advantages were possible if one assumed the solution of **Eq. (34)** was either the real or the imaginary part of

$$F = z^*\phi(z) + \chi(z) \tag{35}$$

where $z = x + iy$ and $z^* = x - iy$. If the problem can be arranged so that the crack of interest occupies a straight segment of the x-axis ($y = 0$), a simpler, one-function approach suggested by **Westergaard (1939)** is often useful. Westergaard discussed several Mode I crack problems that could be solved using

$$\Phi = \mathrm{Re}\left\{\overline{\overline{Z}}(z)\right\} + y\,\mathrm{Im}\left\{\overline{Z}(z)\right\} \tag{36}$$

where

and, for subsequent use,

$$\left.\begin{aligned} \overline{Z} &= \frac{d}{dz}\overline{\overline{Z}} \\ Z &= \frac{d}{dz}\overline{Z},\ Z' = \frac{d}{dz}Z \end{aligned}\right\} \tag{37}$$

From **Eqs. (36) and (31)**

$$\left.\begin{aligned} \sigma_x &= \mathrm{Re}Z - y\,\mathrm{Im}Z' \\ \sigma_y &= \mathrm{Re}Z + y\,\mathrm{Im}Z' \\ \tau_{xy} &= -y\,\mathrm{Re}Z' \end{aligned}\right\} \tag{38}$$

For a straight crack on $y = 0$, a loading symmetry such that $\tau_{xy} = 0$ on $y = 0$,, corresponding to Mode I, is automatically furnished by **Eq. (36)**. The displacements, assuming plane-strain, are given by

$$\left.\begin{aligned} 2Gu &= (1 - 2\nu)\,\mathrm{Re}\overline{Z} - y\,\mathrm{Im}Z \\ 2Gv &= 2(1 - \nu)\,\mathrm{Im}\overline{Z} - y\,\mathrm{Re}Z \end{aligned}\right\} \tag{39}$$

For plane-stress, ν in **Eq. (39)** can be replaced by $\nu/(1 + \nu)$.

In checking the derivation of **Eq. (38)** from **Eq. (36)** and of **Eq. (39)** from integration of **Eq. (32)**, it is helpful to use the Cauchy-Riemann equations. These are

$$\left.\begin{aligned} \mathrm{Re}\left\{f'(z)\right\} &= \frac{\partial}{\partial x}(\mathrm{Re}f) = \frac{\partial}{\partial y}(\mathrm{Im}f) \\ \mathrm{Im}\left\{f'(z)\right\} &= \frac{\partial}{\partial x}(\mathrm{Im}f) = -\frac{\partial}{\partial y}(\mathrm{Re}f) \end{aligned}\right\} \tag{40}$$

In terms of **Eq. (35), Eq. (36)** corresponds to assuming

$$\Phi = \mathrm{Re}F, \quad \chi = \overline{\overline{Z}} - z\phi, \quad \phi = \frac{1}{2}\overline{Z} \tag{41}$$

The solution of one of the crack problems briefly discussed in **Westergaard (1939)** is given by

$$Z(z) = \frac{\sigma}{\sqrt{1-(a/z)^2}} \tag{42}$$

The problem solved with this stress function is the crack problem studied by **Griffith (1920)** with the aid of previous work by **Inglis (1913)**; a central crack of length $2a$, with $\sigma_y = \sigma_x = \sigma$ at distances remote from the crack.

In terms of the vectors

$$z = re^{i\theta}, \quad z-a = r_1 e^{i\theta_1}, \quad z+a = r_2 e^{i\theta_2} \tag{43}$$

Equation (42) can be expressed as

$$Z = \frac{\sigma r}{\sqrt{r_1 r_2}} e^{i\left\{\theta - \frac{\theta_1+\theta_2}{2}\right\}} \tag{44}$$

From differentiation of **Eq. (42)**

$$Z'(z) = \frac{-\sigma a^2}{\left\{z^2 - a^2\right\}^{3/2}} \tag{45}$$

and can be expresses as

$$Z' = \frac{-\sigma a^2}{(r_1 r_2)^{3/2}} e^{-i\frac{3}{2}(\theta_1+\theta_2)} \tag{46}$$

From integration of **Eq. (42)**,

$$\overline{Z}(z) = \sigma\sqrt{z^2 - a^2} \tag{47}$$

and can be expressed as

$$\overline{Z} = \sigma\sqrt{r_1 r_2} e^{i\left(\frac{\theta_1+\theta_2}{2}\right)} \tag{48}$$

The angles in the preceding equations are restricted to the range -π to π (radians). **Equations (44), (46), and (48)** are helpfuul in forming the real and imaginary parts of functions as indicated in the equations for stresses and displacements using the identity $e^{i\phi} = \cos\phi + i\sin\phi$. From these equations it is clear that $\mathrm{Re}Z, y\ \mathrm{Re}Z'$, and y ImZ' are all zero along the line segment occupied by the crack $|x| < a$ and $y = 0$. Thus free boundary conditions along lines of the crack are provided. Remote from the crack, as $|z|$ approaches infinity, y ReZ' and y ImZ' are again zero and Re$Z = \sigma$. Thus the remote stress field is $\sigma_y = \sigma_x = \sigma$ and $\tau_{xy} = 0$.

In the limit of small enough values of r_1/a, taking $r = a, r_2 = 2a, \theta = 0$, and $\theta_2 = 0$, **Eq. (44)** becomes

$$Z = \frac{\sigma\sqrt{a}}{\sqrt{2r_1}} e^{-i\frac{\theta_1}{2}} \tag{49}$$

This relation can be written as

$$Z(\zeta) = K \Big/ \sqrt{2\pi\zeta} \tag{50}$$

$$\text{where} \quad \zeta = r_1 e^{i\theta_1} = z - a \tag{51}$$

$$\text{and} \quad K = K_I = \sigma\sqrt{\pi a} \tag{52}$$

The Mode I stresses and displacements very close to the crack-tip (as shown in the introductory comments) can be derived using **Eq. (50)**, the associated values of Z' and $\overline{Z}$, expressing these in vector form as illustrated above, and substituting real and imaginary parts (as appropriate) into **Eqs. (38) and (39)**.

The single stress function approach of Westergaard is conveniently extended to Mode II crack problems by assuming (**Irwin 1958a**)

$$\Phi = -y\, \mathrm{Re}\overline{Z} \tag{53}$$

In terms of **Eq. (35), Eq. (53)** corresponds to the choices

$$\Phi = \mathrm{Im}F, \quad \chi = -z\phi, \quad \phi = \frac{1}{2}\overline{Z} \tag{54}$$

The stresses are given by

$$\left.\begin{aligned} \sigma_x &= 2\,\mathrm{Im}Z + y\,\mathrm{Re}Z' \\ \sigma_y &= -y\,\mathrm{Re}Z' \\ \tau_{xy} &= \mathrm{Re}Z - y\,\mathrm{Im}Z' \end{aligned}\right\} \tag{55}$$

The displacements (plane strain) are given by

$$\left.\begin{aligned} 2Gu &= 2(1-\nu)\,\mathrm{Im}\overline{Z} + y\,\mathrm{Re}Z \\ 2Gv &= -(1-2\nu)\,\mathrm{Re}\overline{Z} - y\,\mathrm{Im}Z \end{aligned}\right\} \tag{56}$$

The solution of the Mode II counterpart of the Griffith crack problem is obtained by

$$Z(z) = \frac{\tau}{\sqrt{1-(a/z)^2}} \tag{57}$$

The remote stresses are $\sigma_x = \sigma_y = 0, \tau_{xy} = \tau$. The crack-tip stresses and displacements are again provided by **Eq. (50)** with

$$K = K_{II} = \tau\sqrt{\pi a} \tag{58}$$

The use of essentially the same stress function, Z, to solve Mode I and Mode II problems is applicable to many crack stress field problems and can be extended to Mode III by means of the equation

$$Gw = \operatorname{Im}\bar{Z} \tag{59}$$

The stresses are given by

$$\left.\begin{aligned} \tau_{yz} &= \operatorname{Re}Z \\ \tau_{xz} &= \operatorname{Im}Z \end{aligned}\right\} \tag{60}$$

Further use of a Z function, which solves a two-dimensional Mode I crack problem in an isotropic material, in the solution of two-dimensional crack problems (of similar configuration) in orthotropic and anisotropic elastic materials is discussed in **Appendix D**. To provide relations that remain generally valid, it is most convenient to define the three K values as follows [consistent with definitions in **Eqs.(1)–(3)**]:

$$\begin{Bmatrix} K_I \\ K_{II} \\ K_{III} \end{Bmatrix} = \underset{r\to 0}{Limit}\sqrt{2\pi r}\begin{Bmatrix} \sigma_y \\ \tau_{xy} \\ \tau_{yz} \end{Bmatrix} \tag{61}$$

where r is the length of a small vector extending directly forward from the crack-tip.

In the case of Mode I, the invariants used in computing principal stresses are

$$\frac{\sigma_x + \sigma_y}{2} = \operatorname{Re}Z, \quad \tau_{\max} = y\left|Z'\right| \tag{62}$$

For plane strain, the stress field energy density, U, is given by

$$U = \frac{1-2\nu}{2G}(\operatorname{Re}Z)^2 + \frac{1}{2G}\tau^2_{\max} \tag{63}$$

The corresponding relationships for Mode II are

$$\frac{\sigma_x + \sigma_y}{2} = \operatorname{Im}Z, \quad \tau_{\max} = \sqrt{|Z|^2 + y^2|Z'|^2 - 2y\operatorname{Im}(Z^*Z')} \tag{64}$$

$$U = \frac{1-2\nu}{2G}(\operatorname{Im}Z)^2 + \frac{1}{2G}\tau^2_{\max} \tag{65}$$

Williams (1957) called attention to the possibility that studies of U near the leading edge of a crack might be of interest in predicting crack extension behavior. **Irwin (1958a)** noted that the largest tensile stresses at a fixed small distance from a Mode I crack-tip were at 60° to the line of expected crack extension. Either viewpoint predicts a tendency for the location of advance separation to cause roughening of a flat tensile fracture surface. Of course, the subject of these comments pertains to the fracture process zone and a treatment based on stress–strain relations within the crack-tip plastic zone would be more appropriate.

ADDITIVITY OF CRACK STRESS FIELDS AND K VALUES

From the additivity of linear-elastic stress fields and the definitions of K, **Eq. (61)**, several conclusions are evident: (a) the addition of a stress field that does not possess an inverse square root stress singularity at the crack-tip does not alter the value of K for that crack-tip; (b) when each of several superimposed stress fields contributes to the K values, the K values are separately additive for each of these modes; and (c) when several loading configurations are applied to the same crack and the Westergaard Z functions for each are known, the Z functions can be added together, and the stresses and displacements can then be derived from the total Z function using methods discussed in the previous section.

For illustration consider the Z function

$$Z(z) = \frac{P}{\pi(z-b)} \cdot \frac{\sqrt{a^2 - b^2}}{\sqrt{z^2 - a^2}} \tag{66}$$

Using the Mode I value of Φ, **Eq. (36)**, the problem solved is that of a central crack of length, $2a$, opened by a pair of splitting forces, P, acting against the crack surfaces at the position $y = 0$, $x = b$. The value of K at $x = a$ is

$$K = \frac{P}{\sqrt{\pi a}} \sqrt{\frac{a+b}{a-b}} \tag{67}$$

If we add a second pair of equal size splitting forces at $y = 0$, $x = -b$, the total value of Z becomes

$$Z(z) = \frac{2P}{\pi\left(z^2 - b^2\right)} \frac{\sqrt{a^2 - b^2}}{\sqrt{1 - (a/z)^2}} \tag{68}$$

From **Eq. (68)**, the K value at each crack-tip is given by

$$K = \frac{2P}{\sqrt{\pi a}} \cdot \frac{a}{\sqrt{a^2 - b^2}} \tag{69}$$

Simple addition shows that **Eq. (69)** is the sum of **Eq. (67)** plus the same expression after substitution of $-b$ for b inside the radical. Assume next that $P = \sigma db$. From the additivity rule, the stress field for a uniform pressure, σ, acting against the crack surfaces can be derived from the following Z function:

$$Z(z) = \frac{2\sigma}{\pi\sqrt{1 - (a/z)^2}} \int_0^a \frac{\sqrt{a^2 - b^2}\, db}{z^2 - b^2} \tag{70}$$

Performing the integration,

$$Z(z) = \frac{\sigma}{\sqrt{1 - \left(a/z\right)^2}} - \sigma \tag{71}$$

Eq. (71) could have been derived by the alternative method of adding a uniform biaxial compression, $Z = -\sigma$, to the Z value for the Griffith crack (**Eq. 42**). Because the uniform stress field does not alter K, the value of K at each crack-tip is $\sigma\sqrt{\pi a}$, as in **Eq. (52)**. If the result needed is the total K rather than the total Z, a substantial simplification of the computational task can be expected. For example, elementary methods show that

$$\int_0^a \frac{db}{\sqrt{a^2 - b^2}} = \frac{\pi}{2} \tag{72}$$

Use of $P = \sigma db$ in **Eq. (69)** and use of **Eq. (72)** provides $\sigma\sqrt{\pi a}$. The simplicity of this computation can be compared to the integration indicated in **Eq. (70)**.

From the additivity principles just illustrated, it can be seen that the solution of a crack stress field problem can be visualized as a two-step process: (1) solve the stress distribution problem in a manner satisfying the boundary conditions (including applied loads) but with the crack considered absent; (2) add to this stress field a stress field that cancels any stresses acting directly across the crack along the line of the crack. In the case of a crack occupying a segment of the x-axis, the stresses along this segment which must be reduced to zero are σ_y, τ_{xy}, and τ_{yz}. Closed-form solutions of numerous infinite-plate crack problems have been obtained in this way. Because of the analysis simplifications applicable to Mode III, closed-form solutions can be obtained using this method for certain finite plate problems. The two-step approach can be termed a Green's function method when a suitable stress function for local pressures or shears on the crack surfaces is available. A suitable stress function of the Green's function type is one that can be added to the "no crack" stress field without inconsistency with the boundary conditions assumed in the first step of the above method.

BOUNDARY COLLOCATION METHOD

The availability of large, high-speed computers permits a variety of numerical methods that can be used when K cannot be found from a closed-form crack stress field solution. Boundary collocation can be regarded as a relatively simple extension of methods discussed previously.

Assume that the crack occupies a segment of the x-axis with the crack-tip at $z = 0$, and that both of the loads and the shape of the plate containing the crack are symmetrical relative to the x-axis. A simple example would be a long, single-edge-notched tensile specimen with a crack-simulating notch of length, a, open to the left free boundary of the plate, and with uniform tension, σ, applied across the upper and lower boundaries of the plate (parallel to the crack). Let W be the width of the plate and let L be the length. The stress field is of Mode I type and consideration can be given to the use of the stress function Z, where

$$Z(z) = \frac{K}{\sqrt{2\pi z}} + \sum_{n=1}^{N} A_n z^{n-1/2} \tag{73}$$

Using **Eq. (38)**, σ_y and τ_{xy} are zero on $y = 0$ when x is negative. Thus free boundary conditions are exactly satisfied along the line of the crack. In addition, it can be observed that $Ev = 2\ \mathrm{Im}(\bar{Z}) = 0$ on $y = 0$ for positive values of x. Since it is desirable to restrict N to a moderate size, free boundary conditions cannot be exactly satisfied along $x = -a$ and $x = W - a$. However, if the values of A_1, $A_2, \ldots,$ A_N, and K are such that free boundary conditions are nearly satisfied along these lines, the influence of the remaining errors on the stress field close to the crack will be small. A similar consideration applies to the upper and lower boundaries, where the desired boundary conditions, $\sigma_y = \sigma$ and $\tau_{xy} = 0$, can be satisfied only on an average basis. Along the lines $x = -a$ and $x = W - a$, since stress field errors more remote from the crack-tip are of lesser importance, it is convenient to choose the boundary collocation points at y values corresponding to equal spacing of u_1 in the equation $y = a \tan u_1$, where u_1 is an angle measured from the negative branch of the x-axis. A similar method for choosing boundary collocation points on $x = W - a$ might be to use $y = (W - a) \tan u_2$, where u_2 is measured from the positive branch of the x-axis. The preceding methods are continued as a means of selecting boundary collocation points across the line $y = L/2$. Only the specimen half above the x-axis is used because of the symmetry of the problem.

The solution procedure consists of writing the equations for $\tau_{xy} = 0$ and $\sigma_x = 0$ at the points selected along $x = -a$ and $x = W - a$ as well as the equations for $\tau_{xy} = 0$ and $\sigma_y = 1$ (since K is proportional to σ) at points selected along $y = L/2$. If $\frac{1}{2}(N + 1)$ boundary points are selected, the result is a set of equations, linear in terms of the parameters $K, A_1, A_2, \ldots, A_N$, and just sufficient in number to permit determination of each parameter. However, for a given amount of computing time, it has proved most efficient to limit the value of N, select the number of boundary collocation points that is three to four times $\frac{1}{2}(N + 1)$, and use a least squares program to determine the best values for the parameters. Selection of boundary collocation points at sharp corners should be avoided. The outputs needed from the computer are the value of K and (usually) the value of the y-direction displacement at the crack mouth position commonly selected for clip gage crack opening measurements during crack toughness evaluations. A number of calculation refinements can be added. However, only the basic plan of the method is presented here.

SUCCESSIVE BOUNDARY STRESS CORRECTION METHOD

For illustration, consider a straight, two-dimensional crack occupying the segment of the x-axis, $0 \leqslant x \leqslant a$. Assume that the y-axis is the free boundary of a semi-infinite plate and that the stresses remote from the crack are $\sigma_y = \sigma, \sigma_x = 0, \tau_{xy} = 0$. From previous comments, the K is not altered if the remote stresses are all taken as zero and a uniform pressure, σ, is assumed acting inside of the crack. Thus an approximate estimate of K is given by $K \simeq \sigma\sqrt{\pi a}$. An appropriate stress field is provided by the Z function of **Eq. (71)** under the assumption that σ would be added to real calculation of σ_y. Along $x = 0$, the value of τ_{xy} from this Z function is zero and the average of σ_x is zero. Successive additions of stress fields which remove the normal stresses, σ_x, along $x = 0$, and the consequent normal stresses σ_y along the line of the crack, can be visualized as an infinite repetition of additive stress fields (**Irwin 1962a**). Each stress removal along the line of the crack provides a corrective contribution to the value of K. If the σ_x stresses on $x = 0$ from the Z function of **Eq. (71)** are termed $\sigma_x(\sigma)$, the calculation can be compactly summarized by a pair of integral equations as follows:

$$p(y_0) = \sigma_x(\sigma) + \int_0^a q(b) f(b, y_0) db \tag{74}$$

$$q(b) = \int_0^\infty p(y_0) g(b, y_0) dy_0 \tag{75}$$

where

$$\sigma_x(\sigma) = \sigma\left\{\frac{y_0\left(y_0^2+2a^2\right)}{\left(y_0^2+a^2\right)^{3/2}} - 1\right\}$$

$$f(b,y_0) = \frac{2}{\pi}\frac{y_0\sqrt{a^2-b^2}}{\left(y_0^2+b^2\right)\sqrt{a^2+y_0^2}}\left(\frac{2y_0^2}{y_0^2+b^2} - \frac{2a^2+y_0^2}{a^2+y_0^2}\right) \tag{76}$$

$$g(b,y_0) = -\frac{4}{\pi}\frac{by_0^2}{\left(b^2+y_0^2\right)^2} \tag{77}$$

The total K is given by

$$K = \sqrt{\pi a}\left\{\sigma + \frac{2}{\pi}\int_0^a \frac{q(b)db}{\sqrt{a^2-b^2}}\right\} \tag{78}$$

and turns out to be larger than $\sigma\sqrt{\pi a}$ by about one-eighth. For calculation purposes it is convenient to choose $\sigma = 1$ (since K is proportional to σ), $b = a\sin\alpha$, $y_0 = a\tan\beta$, and to use equal intervals of α and β in conducting the necessary numerical integrations. The calculation is started by using of $\sigma_x(\sigma)$ for $p(y_o)$ in **Eq. (75)**. The resulting $q(b)$ is used in **Eq. (74)** to obtain an improved value of $p(y_o)$ for use in **Eq. (75)**. A calculation of K from **Eq. (78)** is made for each new value of $q(b)$ and the computing process is stopped when the change of K becomes smaller than some selected fraction of $\sigma\sqrt{\pi a}$. Similar calculation plans are applicable if the initial "no-crack" stress field is due to thermal stress induced by a uniform rapid cooling along $x = 0$ (**Lachenbruch 1961**).

The use of the preceding method for strip problems is more complex. Even for problems having symmetry relative to the midline of the strip, convenient stress removal functions for the parallel free boundaries of the strip are not available. Nevertheless, a number of problems of this type have been solved (**Tada 1972b**) using a modification of the method. The main advantage of the successive stress removal method is rapid convergence of the total K estimate. In compensation, the programing tasks may be substantially greater than would be encountered using the boundary collocation method.

K ESTIMATES FROM FINITE ELEMENT METHODS*

Although finite element methods are commonly used for practical stress analysis problems, special planning is necessary for efficient use of such methods to determine K for two-dimensional crack problems. For three-dimensional crack problems—for example, a part-through surface crack—even with expert planning, an accuracy of 10% is not easily obtained. The comments given here are mainly limited to two-dimensional problems.

Two kinds of procedures have been used with considerable success. In the first of these, assuming the stress state is caused by externally applied loads, the method is directed toward the computation of the total stress field energy, $U_T = \frac{1}{2}\sum P_i\Delta_i$, for each of a series of crack sizes with the loads, P_i, held constant. In the preceding equation, Δ_i is the displacement parallel to P_i of the local region of load application, If δA is the

* **Historically, this was written for the First Edition in 1973. Progress in numerical methods continues at a rapid pace.**

increase of the severed area corresponding to a crack size increment and if δU_T is the increase of U_T for that increment, K can be obtained from the equation

$$\frac{K^2}{E} = \mathcal{G} = \frac{\delta U_T}{\delta A} = \frac{1}{2}\sum P_i \frac{\delta \Delta i}{\delta A}$$

This procedure simply models an experimental compliance calibration in terms of finite element computations of the load displacements, Δ_i (**Watwood 1969**). The advantage of the procedure is that refinement of the finite element mesh toward very small element sizes near the crack-tip has little influence on the displacement differences, $\delta\Delta_i$, remote from the crack and is therefore not required. In compensation, the computations must include a range of crack sizes and reliable results require careful study of the accuracy of the load displacement differences.

Procedures that use very small elements close to the crack-tip have yielded values of K with an accuracy of better than 2%. Experience is necessary in developing approximate methods of size reduction of the finite elements close to the crack-tip and in choosing the stiffness matrix for this region. Generally, K values have been determined by positioning computed values of an extensional stress at the element centroid; plotting values of stress times the square root of distance from the crack-tip, r, against r; and extrapolating to $r = 0$. A discussion of finite element methods for K determination by **Kobayashi (1973)** suggests that, in terms of computational efficiency, using crack line displacements at nodal points close to the crack-tip along with the displacement equations valid for the crack-tip region possesses advantages over procedures based on the stress equations.

Currently, K estimates from finite element methods are at an intermediate stage of development. Trials of calculations of $\mathcal{G}$ in terms of the J-Integral using finite element methods may provide substantial advantages. In addition, the fact that the stress and displacement patterns at the crack-tip are known and that approximate first-order corrections to these patterns can be estimated has not been fully exploited (**Wilson 1972**). The latter method may be essential as a means for obtaining accurate K estimates for three-dimensional crack problems within reasonable limits of computational expense (**Swedlow 1972**).

ADDITIONAL REMARKS FOR PART I

A. Unified Formulation for In-Plane Two-Dimensional Problems

It is well known that all formulas for in-plane, two-dimensional problems can be expressed in common forms for plane-strain and plane-stress conditions by choosing G and one other elastic constant which is defined in terms of ν according to the condition.

The expression: $G = E/2(1+\nu)$ can be used commonly for both plane-strain and plane-stress conditions. For the second constant, κ defined as follows is frequently used.

$$\kappa = \begin{cases} 3-4\nu & \text{plane strain} \\ \frac{3-\nu}{1+\nu} & \text{plane stress} \end{cases} \tag{79}$$

A choice of a neater combination, however, is $(1-\nu)$ for plane strain and $1/(1+\nu)$ for plane stress. Two constants, α and β $(\beta = 1/\alpha)$, are defined in this handbook as follows for convenience:

$$\beta = \frac{1}{\alpha} = \begin{cases} 2(1-\nu) & \text{plane strain} \\ 2/(1+\nu) & \text{plane stress} \end{cases} \tag{80}$$

A complete, unified formulation in terms of stresses in the x,y Cartesian coordinate system, for example, is summarized as follows.

Equilibrium equations:

$$\left.\begin{aligned}
&\frac{\partial \sigma_x}{\partial x}+\frac{\partial \tau_{xy}}{\partial y}+X=0\\
&\frac{\partial \tau_{xy}}{\partial x}+\frac{\partial \sigma_y}{\partial y}+Y=0
\end{aligned}\right.$$

Compatibility equation:

$$\nabla^2\sigma^* = -\operatorname{div}\vec{F} = -\left(\frac{\partial X}{\partial x}+\frac{\partial Y}{\partial y}\right) \tag{81}$$

Stress-Strain relations:

$$\begin{aligned}
2G\varepsilon_x &= \sigma^* - \sigma_y\\
2G\varepsilon_y &= \sigma^* - \sigma_x\\
2G\gamma_{xy} &= 2\tau_{xy}
\end{aligned}$$

where

$$\begin{aligned}
&\sigma^* = \beta(\sigma_x+\sigma_y)/2\\
&\vec{F}(X,\,Y) = \text{ body force}
\end{aligned}$$

Note that $\sigma_z = \nu(\sigma_x+\sigma_y)$ in plane-strain condition or $\varepsilon_z = -\nu(\sigma_x+\sigma_y)/E$ in plane-stress condition [see **Eqs. (25) and (26)**] is determined simply as a by-product of the analysis.

Various distinct expressions are used throughout this handbook for plane strain and plane stress. All of these pairs of expressions can be unified to single forms with the use of G and α or β. Some examples of unified expressions are

a. **Eq. (79) :** $\kappa = 2\beta - 1$ (79a)

b. **Eq. (39) :**

$$\left.\begin{aligned}
2Gu &= (\beta-1)\,\mathrm{Re}\bar{Z} - y\,\mathrm{Im}\,Z\\
2Gv &= \beta\,\mathrm{Im}\,\bar{Z} - y\,\mathrm{Re}Z
\end{aligned}\right\} \tag{39a}$$

Also refer to **Eq. (1)** and **pages 1.3b and 1.3c**.

c. **Eq. (18) :** $E' = 4\alpha G$ (18a)

d. **Page 16.3 :**

$$\begin{aligned}
v &= \pm\frac{1-\alpha}{G}\sigma h\\
K_I &= 2\sqrt{\alpha(1-\alpha)}\sigma\sqrt{h}
\end{aligned} \tag{82}$$

Also note that $\alpha = 1/2(1-\nu)$ repeatedly appears in solutions for many three-dimensional problems. See, for example, **pages 23.7** and **24.1**.

B. On Completeness of Westergaard Single-Function Method for Analysis of Cracks

The single stress function approach of Westergaard has a certain deficiency that only affects the elastic field in the absence of cracks. For the analysis of the contribution to the elastic field by the presence of cracks, the single-function method is complete.

The deficiency consists in the possible presence of non-zero Z_{II} even in the so-called Mode I field, for example, in the absence of cracks (**Okamura 1976**), as illustrated here in the analysis of a simple example.

The subscripts I and II are used here, as in the solution pages, to represent Mode I and Mode II fields, respectively.

A general Airy stress function Φ in terms of Westergaard stress functions is given, combining **Eqs. (36) and (53)**, by

$$\Phi = \Phi_I + \Phi_{II} = \mathrm{Re}\overline{\overline{Z}}_I + y\,\mathrm{Im}\overline{Z}_I - y\,\mathrm{Re}\overline{Z}_{II} \tag{83}$$

The corresponding stresses and displacements, combining **Eqs. (38) and (55)** and **Eqs. (39) and (56)**, respectively, are given by

$$\begin{Bmatrix} \sigma_x \\ \sigma_y \\ \tau_{xy} \end{Bmatrix} = \begin{Bmatrix} \mathrm{Re}Z_I - y\,\mathrm{Im}Z_I' \\ \mathrm{Re}Z_I + y\,\mathrm{Im}Z_I' \\ -y\,\mathrm{Re}Z_I' \end{Bmatrix} + \begin{Bmatrix} 2\,\mathrm{Im}Z_{II} + y\,\mathrm{Re}Z_{II}' \\ -y\,\mathrm{Re}Z_{II}' \\ \mathrm{Re}Z_{II} - y\,\mathrm{Im}Z_{II}' \end{Bmatrix} \tag{84}$$

$$2G\begin{Bmatrix} u \\ v \end{Bmatrix} = \begin{Bmatrix} (\beta-1)\,\mathrm{Re}\overline{Z}_I - y\,\mathrm{Im}Z_I \\ \beta\,\mathrm{Im}\overline{Z}_I - y\,\mathrm{Re}Z_I \end{Bmatrix} + \begin{Bmatrix} \beta\,\mathrm{Im}\overline{Z}_{II} + y\,\mathrm{Re}Z_{II} \\ -(\beta-1)\,\mathrm{Re}\overline{Z}_{II} - y\,\mathrm{Im}Z_{II} \end{Bmatrix} \tag{85}$$

where $\overline{\overline{Z}}_I(z)$, $\overline{Z}_I(z)$, etc., are abbreviated to $\overline{\overline{Z}}_I$, $\overline{Z}_I$, etc.

Consider, for example, an infinite plate subjected to (a) uniform (biaxial) tension $\sigma_x = \sigma_y = \sigma$ at infinity, **Fig. 10(a)**, and (b) uniaxial tension $\sigma = \sigma$ at infinity, **Fig. 10(b)**. Both elastic fields are symmetric with respect to x-axis and therefore are Mode I fields. Nevertheless, Z_{II} for (b) is not zero, but $Z_{II} = -i(\sigma/2)$. That is, the Westergaard stress functions and, for reference, the Airy stress functions corresponding to **Fig. 10(a) and (b)** are given by

$$\text{(a)}\ Z_I(z) = \sigma,\ \ Z_{II}(z) = 0;\ \ \Phi = \frac{1}{2}\sigma\left(x^2 + y^2\right) \tag{86a}$$

$$\text{(b)}\ Z_I(z) = \sigma,\ \ Z_{II}(z) = -i\frac{\sigma}{2};\ \ \Phi = \frac{1}{2}\sigma x^2 \tag{86b}$$

These Westergaard functions, **Eqs. (86a) and (86b)**, obviously from **Eq. (84)**, yield identical stresses on the x-axis (the presence of nonzero Z_{II} in (b) has no bearing on them), namely,

$$\sigma_y(x, 0) = \sigma,\ \ \tau_{xy}(x, 0) = 0 \tag{87}$$

Therefore, to make a segment $|x| \leqslant a$, $y = 0$ traction-free surfaces, the stress field shown in **Fig. 11** must be superimposed on those of **Fig. 10**. The Westergaard function corresponding to **Fig. 11** has been obtained in the preceding section, that is, **Eq. (71)**.

$$Z_I(z) = \frac{\sigma}{\sqrt{1-(a/z)^2}} - \sigma,\ \ Z_{II}(z) = 0 \tag{71}$$

As readily observed from the analysis, as long as the crack-absent stress distributions to be removed over crack segments are identical, the contributions to the elastic field by the presence of cracks are identical regardless of the difference in the overall elastic fields, and cracks in Mode I fields contribute to Z_I only.

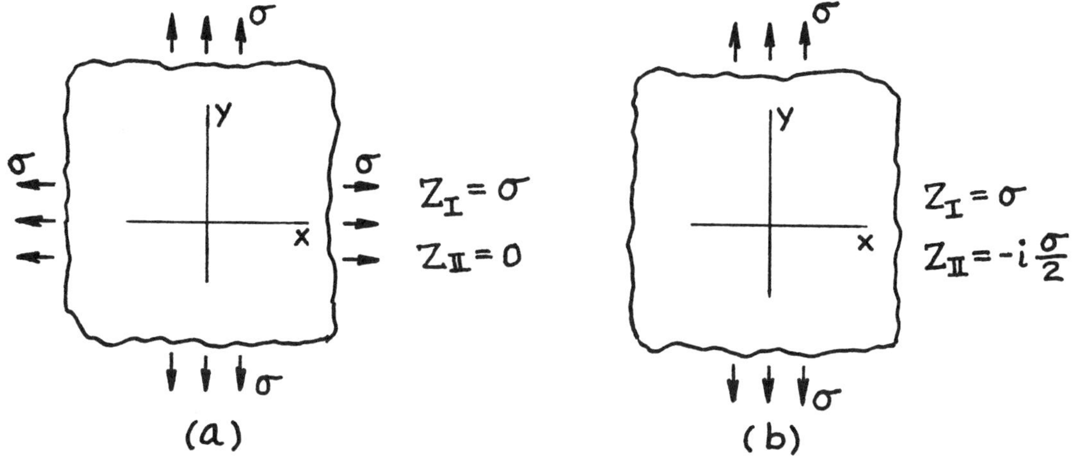

Fig. 10

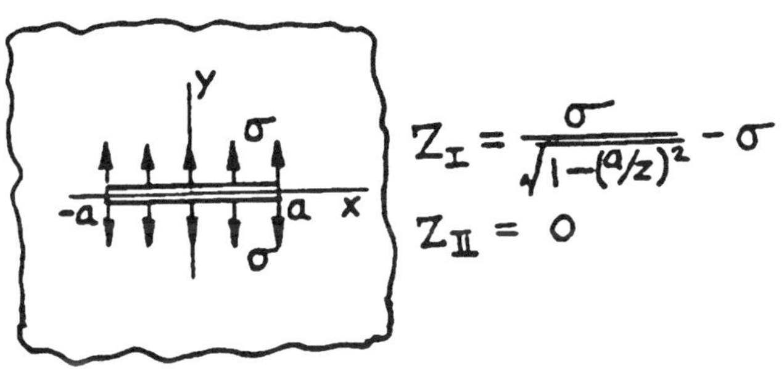

Fig. 11

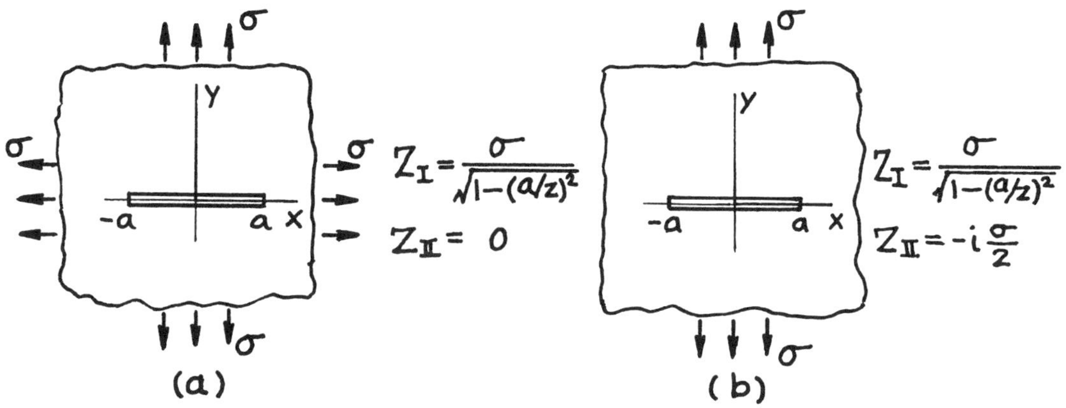

Fig. 12
(Fig. 10) + (Fig. 11)

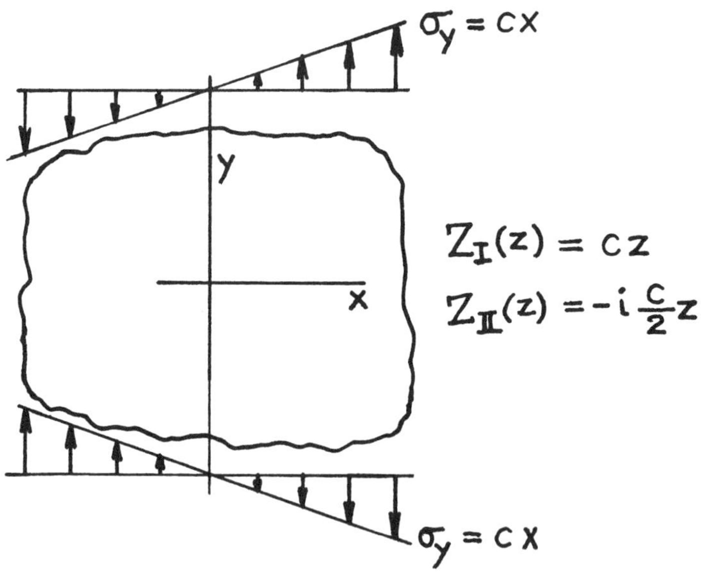

$\sigma_y = cx$
y
x
$Z_I(z) = cz$
$Z_{II}(z) = -i\frac{c}{2}z$
$\sigma_y = cx$

Fig. 13

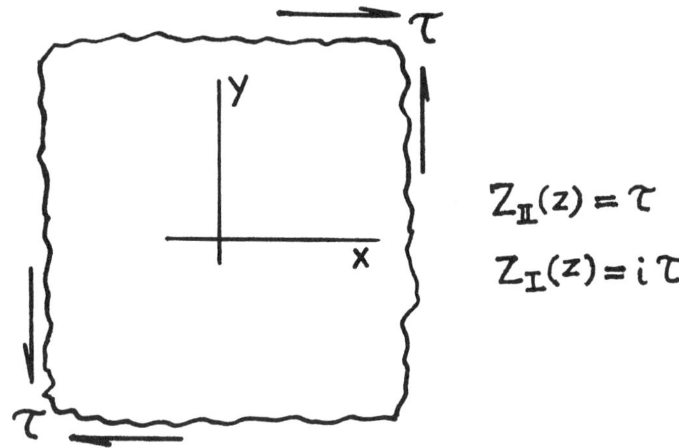

τ
y
x
$Z_{II}(z) = \tau$
$Z_I(z) = i\tau$
τ

Fig. 14

The resultant Westergaard stress functions after the cracks are introduced, **Fig. 12(a) and (b)**, are determined simply by superimposing **Eq. (71)** on **Eqs. (86a) and (86b)**.

$$\text{(a) } Z_I(z) = \frac{\sigma}{\sqrt{1-(a/z)^2}}, \quad Z_{II}(z) = 0 \tag{88a}$$

$$\text{(b) } Z_I(z) = \frac{\sigma}{\sqrt{1-(a/z)^2}}, \quad Z_{II}(z) = -i\frac{\sigma}{2} \tag{88b}$$

However, adjustment for σ_x based on Z_I alone is often made in the manner described in the paragraph following **Eq. (71)** without introducing Z_{II}.

An additional example of nonzero Z_{II} in Mode I field and an example of nonzero Z_I in Mode II field are given next.

(c) An infinite plate subjected to in-plane bending $\sigma_y = cx$ at infinity (**Fig. 13**):

$$Z_I(z) = cz, \; Z_{II}(z) = -i\frac{c}{2}z; \; \Phi = \frac{1}{6}cx^3 \tag{89}$$

(d) An infinite plate subjected to uniform shear $\tau_{xy} = \tau$ at infinity (**Fig. 14**):

$$Z_{II}(z) = \tau, \; Z_I(z) = i\tau; \; \Phi = -\tau xy \tag{90}$$

Note that $Z_{II} = \tau$ alone yields the correct stress field and therefore the correct Airy stress function Φ. The presence of nonzero Z_I may seem trivial. However, although Z_I does not contribute to Φ [$\Phi_I = \text{Re}\bar{\bar{Z}}_I + y\,\text{Im}\bar{Z}_I$ in **Eq. (83)** cancels out], Z_I does contribute to displacements, **Eqs. (85)**, and Z_{II} and Z_I of **Eq. (90)** together yield the correct elastic field (symmetric with respect to $y = \pm x$).

Again, Z_{II} in (c) and Z_I in (d) have no significance in the analysis of cracks.

C. Effect of Surface Interference of Partly Closed Cracks

In Mode I displacement field, when there is a crack-tip with a negative K_I, the crack opening displacement near that tip is also negative and thus the material would "overlap." Such overlapping is physically unacceptable and, consequently, solutions involving negative K_I are not valid by themselves. However, these negative K_I and negative openings can be directly superimposed on the positive values resulting from other applied loads, as long as the resultant K_I remains positive and the surface overlapping is eliminated. In the subsequent solution pages, the effect of crack surface interference is ignored for the most part, and negative K_I and negative opening displacements are given as solutions. These negative values, therefore, must be treated accordingly.

When the surfaces of cracks containing the tips with negative K_I entirely close, the resulting geometric and loading configuration would be obvious, and the analysis can be made in a usual manner for the final crack geometry.

Examples of such situations are:

a. A double-edge cracked strip under in-plane bending (**page 11.5**)
   The resulting configuration is effectively a single-edge cracked strip with a crack on the tension side only, **Fig. 15a (page 2.13)**.

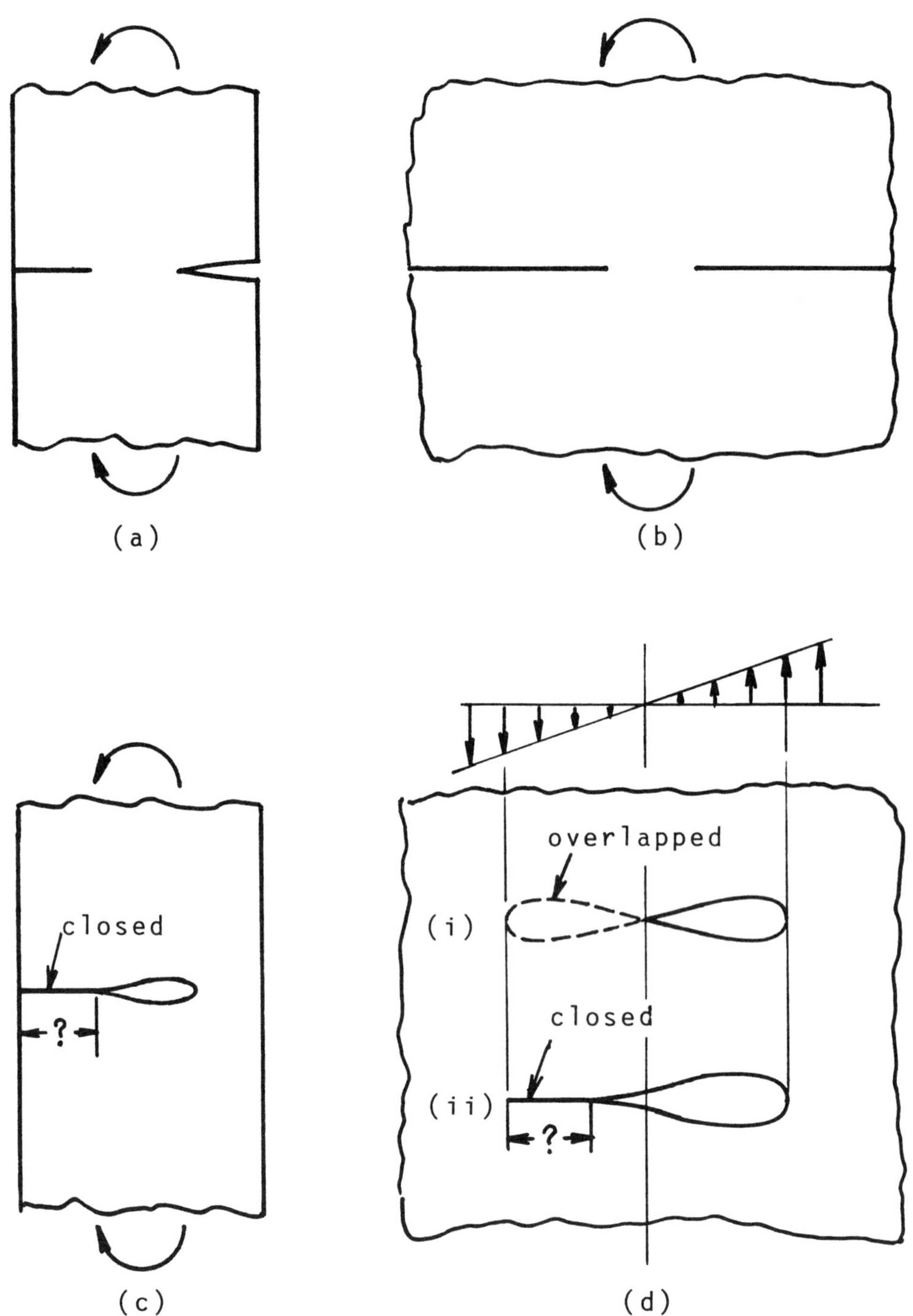

Fig. 15

b. An infinite plate with opposing semi-infinite cracks (leaving a finite net ligament) subjected to a finite in-plane bending moment (**page 4.10**)
   When the crack on the compression side totally closes, the crack on the tension side also loses its significance. The resulting configuration is simply an unloaded plate, **Fig. 15b**.

For these cases, no further discussion is needed here.

When, on the other hand, the crack surfaces only partially close, the interference of closure does affect the resulting geometric configuration and thus the overall elastic field. The geometry of such a crack is not readily knowable beforehand. For example, the length of the closed portion of the crack cannot be determined without analysis.

Examples of such situations are:

c. A strip with a very deep single-edge crack under in-plane bending (**page 2.13** with the bending moment reversed) because the crack-tip is now on the tension side, **Fig. 15(c)**, the crack surfaces on the compression side close smoothly ($K_I = 0$) only partly, and the edge-cracked strip becomes effectively a strip with an internal crack, the resulting geometry of which is not immediately known, **Fig. 15(c)**.
d. An infinite plate with a finite crack subjected to in-plane bending about the symmetric axis of the crack (i.e., linearly varying σ_y at infinity) (**pages 5.18/18a/18b**). (i) in **Fig. 15(d)** is the crack profile of **page 5.18a** where the effect of the crack surface contact is ignored. That is, K_I at the left tip is negative and the left half of the crack surfaces overlap. As discussed earlier, this solution may be directly superimposed on solutions for other loadings when the resulting configurations are physically acceptable ($K_I \geqslant 0$ at the left tip and no surface overlapping). (ii) is the crack profile when the effect of crack surface contact is accounted for (**page 5.18b**, which is to be obtained subsequently). Again, the length of the closed portion and therefore the final geometry of the crack is not known beforehand. Consequently, before determining (by superposition in particular) K_I at the right tip and the crack opening profile, and so on, including the effect of surface interference, the final crack geometry must be known.

For some analyses and discussions on the effect of surface interference of partly closed cracks, see **Seeger (1973), Paris (1975b), Bowie (1976a,b), and Gustafson (1976)**. For additional discussions related to such surface contact, see **Westergaard (1939)** and also **Appendix G** of this handbook.

Next, the effect of crack surface interference is analyzed in detail for example (d). From the analysis of this simple example, various general characteristics of the effect of surface contact, and some features specific to the example, are observed, some of which are well known, or physically or intuitively obvious. The approach is in essence to focus on the portion of the crack that remains open and to remove the stress singularity (K_I) at the point of separation of the closed surfaces by using the usual superposition method.

Let us take, for convenience, the (x, y) coordinate system and $z = x + iy$ with the origin at the midpoint of the crack which remains open, as shown in **Fig. 16**. Other quantities specifying the geometric and loading configurations are also defined in **Fig. 16**. The coordinate system and dimensions associated with the original configuration of the crack are indicated with a subscript "0." The right crack-tips in the two systems, $x = a$ and $x_0 = a_0$, are common, but the position of the left tip, $x = -a$, under loading is not known.

The position of the left tip, that is, the point of separation of surfaces in contact, $x = -a$, is determined from the condition that the resultant K_I vanishes at this point. Refering to **Fig. 16**, the linearly varying stress $\sigma_y = p(x_0/a_0)$ in the original system, line (A), is obtained by superposition of a linearly varying stress $\sigma_y = p'(x/a)$, line (B), and a uniform stress $\sigma_y = p''(p' + p'' = p)$. Therefore, all that is required for the analysis are the solutions found on **pages 5.1/1a and 5.18/18a** applied to the opened portion, $-a \leqslant x \leqslant a$.

The resultant K_I are, by superposition,

$$K_{I x=\pm a} = \pm\left(p'/2\right)\sqrt{\pi a} + p''\sqrt{\pi a} \tag{91}$$

$K_{Ix=-a} = 0$ gives

$$p'' = p'/2 \tag{92a}$$

and correspondingly

$$p' = 2p/3 \tag{92b}$$

$$K_{Ix=+a} = p'\sqrt{\pi a} = (2p/3)\sqrt{\pi a} \tag{92c}$$

In addition, the length of closure

$$\ell = a = 2a_0/3 \tag{92d}$$

is now obvious from the geometry of **Fig. 16**. That is, the left one-third of the original crack closes or, in other words, the one-third of the crack on the compression side remains open $(a_c = a_t = a_0/3)$. These relations are summarized at the bottom of **Fig. 16**.

The complete set of expressions resulting from superposition of **pages 5.1/1a and 5.18/18a** after incorporating the preceding results are now readily obtained. They are summarized in terms of $z = x + iy$ and a in **Fig. 16** as follows:

$$Z_I(z) = \frac{p}{3a}(2z - a)\sqrt{\frac{z+a}{z-a}} \tag{93a}$$

$$\overline{Z}_I(z) = \frac{p}{3a}(z+a)\sqrt{(z+a)(z-a)} \tag{93b}$$

$$K_{Ix=-a} = 0, \; K_{Ix=+a} = \frac{2}{3}p\sqrt{\pi a} \tag{93c}$$

$$\underset{x \leq -a,\; x > a}{\sigma_y(x, 0)} = \frac{p}{3a}(2x - a)\sqrt{\frac{x+a}{x-a}} \tag{93d}$$

$$\underset{|x| \leq a}{2v(x, 0)} = \frac{4pa}{3E'}\left(1 + \frac{x}{a}\right)\sqrt{1 - \left(\frac{x}{a}\right)^2} \tag{93e}$$

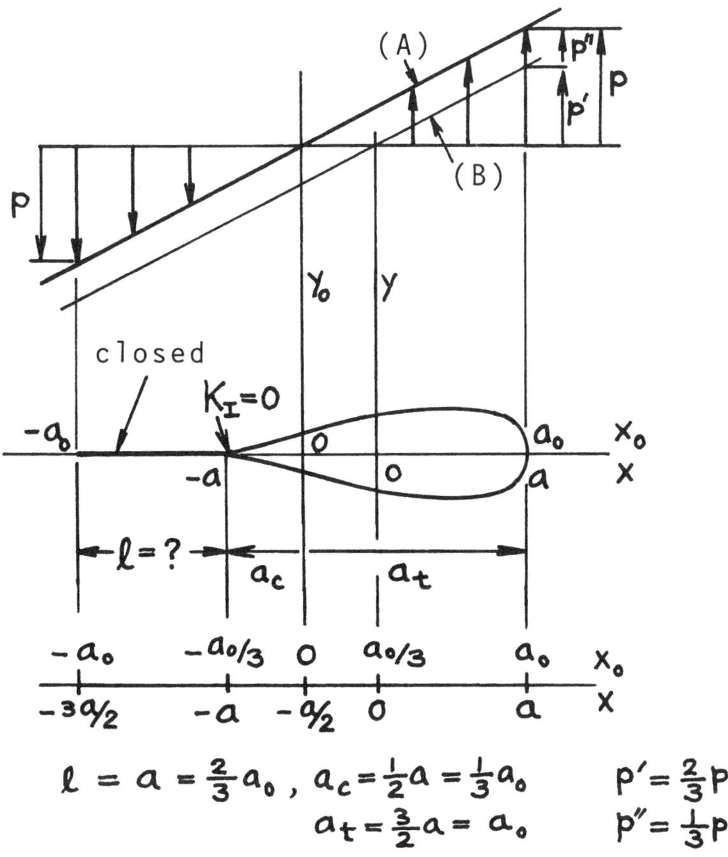

Fig. 16

$$\text{Area of Crack:}\quad A = \frac{2\pi}{3}\frac{pa^2}{E'} \tag{93f}$$

These expressions are converted to those in terms of the original system $z_0 = x_0 + iy_0$ and a_0 by replacing a by $2a_0/3$, **Eq. (92d)**, and z by $z_0 - a/2 = z_0 - a_0/3$, then omitting the subscript "0." The complete expressions corresponding to **Eqs. (93)** and the resulting geometry of the crack are presented on **page 5.18b**. **Page 5.18b** is directly compared with **page 5.18a** for the effect of surface interference. Discussions of the comparison will follow.

The following observations are based on the results of the analysis of the present example. Some of them are general characteristics of surface interference and others are specific features of the present example.

a. The stress variation on the closed portion of the crack near the point of separation of the closed surfaces is in the form of a parabola ($\propto r^{1/2}$), and the separation profile of the surfaces near that point is in the form of a semicubical parabola ($\propto r^{3/2}$), as is well known (**Westergaard 1939**).
b. Once crack surfaces close and remain closed, the presence or absence of cracks beyond the point of separation is immaterial in the subsequent analysis of cracks that remain open.

c. $K_I = 0$ at a tip ensures finite stresses and a smooth separation of the surfaces at that tip. However, $K_I = 0$ by itself does not necessarily require the stresses to be zero at the tip, as observed from the preceding analysis (e.g., **page 5.18b**, "$\sigma_y(x,0)$ - effect of crack" indicated by a dashed line) and the (Dugdale) yield strip model analysis (**pages 30.1 - 32.6**). See f. below for further discussion.

d. The crack surface interference naturally increases K_I at the open end of the crack and the crack opening area. For the present example, K_I is increased to $2(2/3)^{3/2} = 1.089$ times the corresponding value with the surface overlapping permitted, and the crack opening area becomes $4\pi/9 = 1.396$ times that of the opened (right) half. See **pages 5.18a and 5.18b**.

e. A crack-tip located in a compressive region does not necessarily close; that is, K_I remains positive if the other tip remains open $K_I > 0$. For the present example of linearly varying σ_y in **Fig. 16**, when the left tip is located in $x_0 > -a_0/3$, obviously $K_I > 0$, and therefore the left tip located in the compression region $-a_0/3 < x_0 < 0$ will remain open.

f. As discussed in b., the overall solution in terms of $Z(z)$ after cracks are introduced is given (for Mode I) by

$$\left.\begin{aligned} Z_I(z) &= Z_I(z)_{\text{no crack}} + Z_I(z)_{\text{due to crack}} \\ Z_{II}(z) &= Z_{II}(z)_{\text{no crack}} \end{aligned}\right\} \tag{94}$$

where $Z_I(z)_{\text{due to crack}}$ is determined by the integral of **Eq. (70)** or the corresponding integral for other crack configurations. As observed from the example of **Eq. (71)**, the integrals generally result in

$$\left.\begin{aligned} Z_I(z)_{\text{due to crack}} &= Z_I(z)_{\text{function of geometry}} - Z_I(z)_{\text{no crack}} \\ Z_{II}(z)_{\text{due to crack}} &= 0 \end{aligned}\right\} \tag{95}$$

Thus, the resultant field, **Eq. (94)**, is always

$$\left.\begin{aligned} Z_I(z) &= Z_I(z)_{\text{function of crack geometry}} \\ Z_{II}(z) &= Z_{II}(z)_{\text{no crack}} \end{aligned}\right\} \tag{96}$$

$Z_{II}(z)$ may be disregarded here because it has nothing to do with the presence of the cracks.

Therefore, it should be noted that the solution given on **page 5.18** is strictly for the overall situation of **page 5.18a**. That is, for **page 5.18**, $Z_I(z)_{\text{no crack}}(= pz/a)$ and $Z_{II}(z)_{\text{no crack}}(= -ipz/2a)$ [**Eq. (89)** with $c = p/a$] are disregarded.

Similarly, for the present example of **page 5.18b**, the integral of **Eq. (70)** results in

$$\begin{aligned} Z_I(z)_{\text{due to crack}} &= \frac{p}{a}\left(z - \frac{2a}{3}\right)\sqrt{\frac{z + a/3}{z - a}} - \frac{p}{a}z \\ Z_{II}(z)_{\text{due to crack}} &= 0 \end{aligned} \tag{97}$$

and correspondingly

$$\sigma_y(-a/3, 0) = p/3 \neq 0 \tag{98}$$

although K_I at the cuspidal tip, $x = -a/3$, is zero, as discussed in c. above. In addition, it may be interesting to note that

$$\begin{array}{l} \sigma_y(x, 0) > 0 \\ x \leqslant -a/3 \end{array} \tag{99}$$

that is, $\sigma_y(x, 0)$ on the closed portion of the crack, $-a \leqslant x \leqslant -a/3$, and beyond the left tip, $x < -a$, is in tension rather than compression until the remote stress is superimposed. Refer to the distribution of $\sigma_y(x, 0)$, $x \leqslant -a/3$, found on **page 5.18b**.

The final result presented on **page 5.18b** is the superposition of **Eq. (97)** and **Eq. (89)** with $c = p/a$, that is,

$$Z_I(z) = \frac{p}{a}\left(z - \frac{2a}{3}\right)\sqrt{\frac{z + a/3}{z - a}}, \quad Z_{II}(z) = -i\frac{p}{2a}z \tag{100}$$

$Z_{II}(z)$ is not included on **page 5.18b**.

g. It is obvious that cracks having tips with negative K_I totally close, partially close with cuspidal ends, or remain open. It is also obvious, from the analysis of the example, that the final geometry of cracks is unique; that is the positions of the cuspidal ends are uniquely determined as long as the applied load is propotional, regardless of the level of the applied load. The "proportional loading" here is in a sense similar to that used in the theory of plasticity, but less restrictive. It is sufficient if $\sigma_y(x, 0)$ in the absence of cracks is proportional (actually, if it is proportional only on the segments corresponding to the portions of cracks that eventually remain open).For the linearly varying σ_y at infinity (in-plane bending), the proportional loading is realized when the position of zero crossing of σ_y (i.e., the axis of moment) is fixed. Referring to **Fig. 17**, let us assume for simplicity that only one right crack-tip is on the tension side. **Fig. 17a** is obvious; both tips of the right-most crack are in the tensile region. The presence of cracks in the compressive region is immaterial. In **Fig. 17b**, the left tip of the crack is in the compressive region, but the length on the compression side is less than one-third of that on the tension side ($a_c < a_t/3$), and both tips still remain open, as discussed in e. above. The presence of cracks ahead of the left tip is immaterial. When $a_c \geqslant a_t/3$, **Fig. 17c**, a cuspidal end is always formed at $x = -a_t/3$. That is, its position is uniquely determined by the length on the tension side, regardless of the level of the applied load, or the presence or absence of cracks in $x \leqslant -a_t/3$. Only the crack opening profile changes proportionally to the applied load.
h. The analysis of the crack surface interference in three-dimensional configurations will be much more involved and would require a numerical approach.

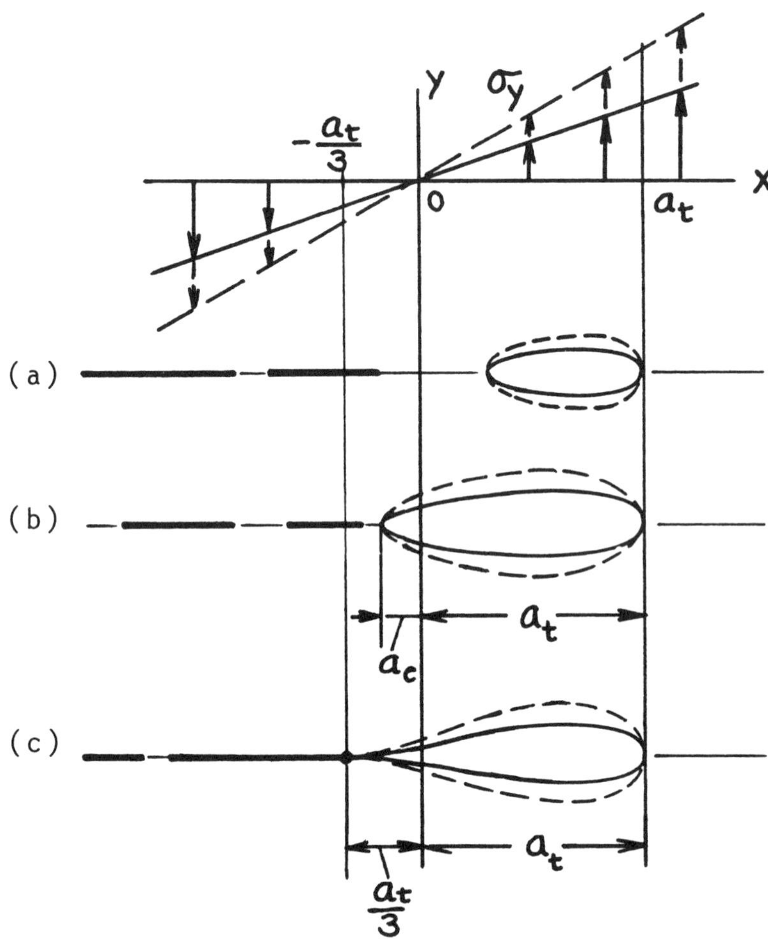

Fig. 17

PART

II

STRESS ANALYSIS RESULTS FOR COMMON TEST SPECIMEN CONFIGURATIONS

- The Center Cracked Test Specimen
- The Double Edge Notch Test Specimen
- The Single Edge Notch Test Specimen
- The Pure Bending Specimen
- The Three-Point Bend Test Specimen
- The Compact Tension Test Specimen
- The Round (Disc-Shaped) Compact Specimen
- The Arc-Shaped (C-Shaped) Specimen
- Other Common Specimen Configurations
- Electrical Potencial Calibraton

THE CENTER CRACKED TEST SPECIMEN

A. Stress Intensity Factor

$$K_I = \sigma\sqrt{\pi a}\, F(a/b)$$

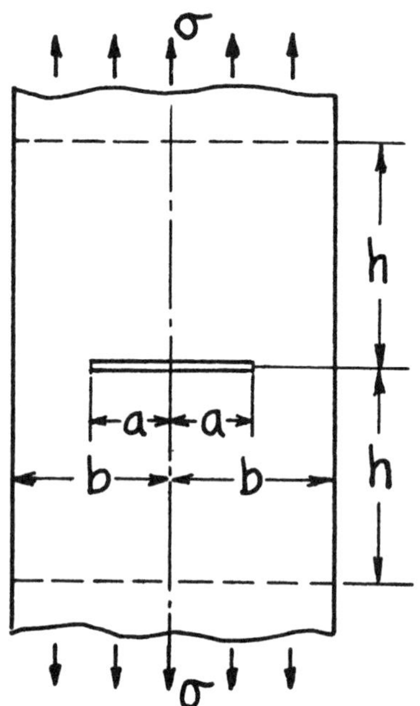

Numerical Values of $F(a/b)$

(Isida 1962, 1965a, b, 1973)

Isida's 36-term power series of $(a/b)^2$ (Laurent series expansion of complex stress potential, 1973) gives practically exact values of $F(a/b)$ up to $a/b = 0.9$. Numerical values of $F(a/b)$ are shown in the following graph and table.

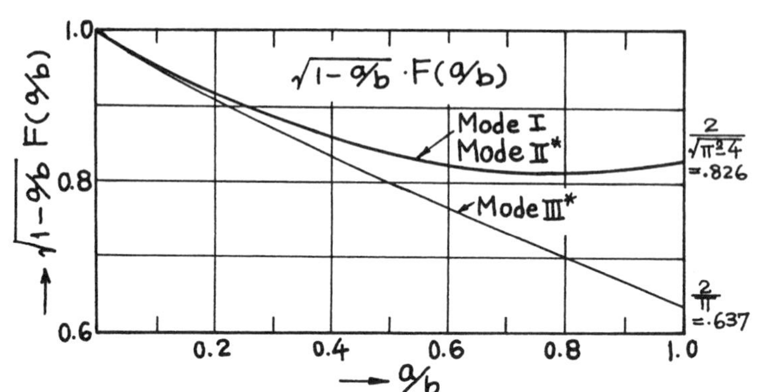

a/b	$F(a/b)$
0.0	1.0000
0.1	1.0060
0.2	1.0246
0.3	1.0577
0.4	1.1094
0.5	1.1867
0.6	1.3033
0.7	1.4882
0.8	1.8160
0.9	2.5776
1.0	$\frac{2}{\sqrt{\pi^2-4}} \Big/ \sqrt{1-a/b}$ **

Exact Limit **(Koiter 1965b)

*See **Note 2**

Empirical Formulas

a. Accuracy
b. Method of derivation, reference

$$F(a/b) = \sqrt{\frac{2b}{\pi a}\tan\frac{\pi a}{2b}}$$

a. Better than 5% for $a/b \leq 0.5$
b. Approximation by periodic crack solution (**Irwin 1957**)

$$F(a/b) = 1 + 0.128(a/b) - 0.288(a/b)^2 + 1.525(a/b)^3$$

a. 0.5% for $a/b \leq 0.7$
b. Least squares fitting to Isida's results (**Brown 1966**)

$$F(a/b) = \sqrt{\sec\frac{\pi a}{2b}}$$

a. 0.3% for $a/b \leq 0.7$, 1% at $a/b = 0.8$
b. Guess based on Isida's results (**Feddersen 1966**)

$$F(a/b) = \frac{1 - 0.5(a/b) + 0.326(a/b)^2}{\sqrt{1 - a/b}}$$

a. 1% for any a/b
b. Asymptotic approximation (**Koiter 1965b**)

$$F(a/b) = \frac{1 - 0.5(a/b) + 0.370(a/b)^2 - 0.044(a/b)^3}{\sqrt{1 - a/b}}$$

a. 0.3% for any a/b
b. Modification of Koiter's formula (**Tada 1973**)

$$F(a/b) = \left\{1 - 0.025(a/b)^2 + 0.06(a/b)^4\right\}\sqrt{\sec\frac{\pi a}{2b}}$$

a. 0.1% for any a/b
b. Modification of Feddersen's formula (**Tada 1973**)

NOTES: 1. Finite height configuration $\left({}^{h}/_{b}\ \textit{finite}\right)$ is given separately. When ${}^{h}/_{b} \geq 3$, the plate is practically regarded as an infinite strip as far as the effects of ${}^{h}/_{b}$ on K are concerned (**Isida 1971a**).

2. For Mode II configuration (II), the correction factor is identical to $F\left({}^{a}/_{b}\right)$ in Mode I.

$$K_{II} = \tau\sqrt{\pi a}\,F\left({}^{a}/_{b}\right)$$

For Mode III configuration (III), the following formula is exact:

$$K_{III} = \tau_{\ell}\sqrt{\pi a}\sqrt{\frac{2b}{\pi a}\tan\frac{\pi a}{2b}}$$

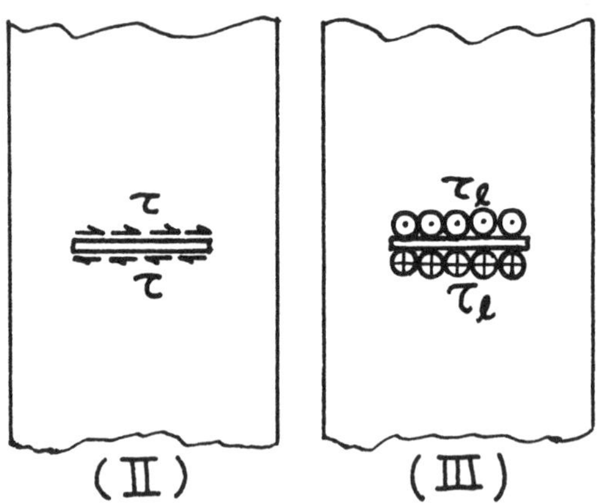

Other Methods and References

1. Compliance Method: **Forman 1964**
2. Fourier Transform - Integral Equation: **Sneddon 1971b**
3. Finite Element Method: **Mendelson 1972**, **Yamamoto 1972**
4. Boundary Collocation Method: **Bowie 1970a**
5. Integral Equations - Successive Stress Relaxation: **Tada 1971, 1972a, b**

(See also **pages 2.24, 2.26, 2.35, 2.36, 7.1, 11.1, 11.2, 11.3, 11.4, 18.1, 18.2, 18.3, 19.4, 20.1, 20.2, 20.3**,etc., for related solutions and corrections for various effects.)

B. Displacements

Crack Opening at Center

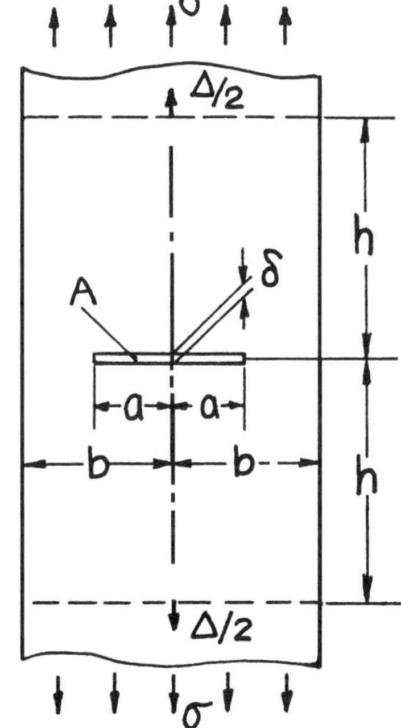

$$\delta = \frac{4\sigma a}{E'} V_1\left(a/b\right)$$

The following formula has better than 0.6% accuracy for any a/b.

$$V_1\left(a/b\right) = -0.071 - 0.535\left(a/b\right) + 0.169\left(a/b\right)^2 - 0.090\left(a/b\right)^3$$

$$+0.020\left(a/b\right)^4 - 1.071 \frac{1}{a/b} \ell n\left(1 - a/b\right)$$

Additional Displacement at Remote Points $(h/b) \geq 3$ Due to Presence of Crack

$$\Delta_{crack} = \Delta_{total} - \Delta_{No\,crack} = \Delta_{total} - \frac{\sigma}{E} \cdot 2h$$

$$\Delta_{crack} = \frac{4\sigma a}{E'} V_2\left(a/b\right)$$

Crack Opening Area

$$A = \Delta_{crack} \cdot (2b)$$

The following formula has better than 0.6% accuracy for any a/b .

$$V_2\left(a/b\right) = -1.071 + 0.250\left(a/b\right) - 0.357\left(a/b\right)^2 + 0.121\left(a/b\right)^3 - 0.047\left(a/b\right)^4 + 0.008\left(a/b\right)^5 - 1.071 \frac{1}{\left(a/b\right)} \ell n\left(1 - a/b\right)$$

Method of Derivation: Paris' equation based on energy principles (**Paris 1957**) (see **Appendix B**)
Reference: **Tada 1973**

NOTES: 1. $E' = E$ for plane stress and $E' = E/(1-\nu^2)$ for plane strain.
2. Uniform pressure σ applied on the crack surfaces results in the same crack opening δ and remote displacement Δ_{crack}.
3. Limiting values of $V_1(a/b)$ and $V_2(a/b)$ at $a/b \to 1$ are exact.

$$V_1(a/b \to 1) = V_2(a/b \to 1) = \frac{2\pi}{\pi^2 - 4} \ell n \frac{1}{1 - a/b}$$

4. For Mode II loading (II), the displacements (II′) are

$$\delta = \frac{4\tau a}{E'} V_1(a/b)$$

$$\Delta = \frac{4\tau a}{E'} V_2(a/b)$$

where V_1 and V_2 are identical to those in Mode I. For Mode III loading (III), the displacements (III′) are given by

$$\delta = \frac{2\tau_\ell a}{G} V_1(a/b)$$

$$V_1(a/b) = \frac{2}{\pi} \frac{1}{a/b} \cosh^{-1}\left(\sec\frac{\pi a}{2b}\right)$$

$$\Delta = \frac{2\tau_\ell a}{G} V_2(a/b)$$

$$V_2(a/b) = \frac{2}{\pi} \frac{1}{a/b} \ell n\left(\sec\frac{\pi a}{2b}\right)$$

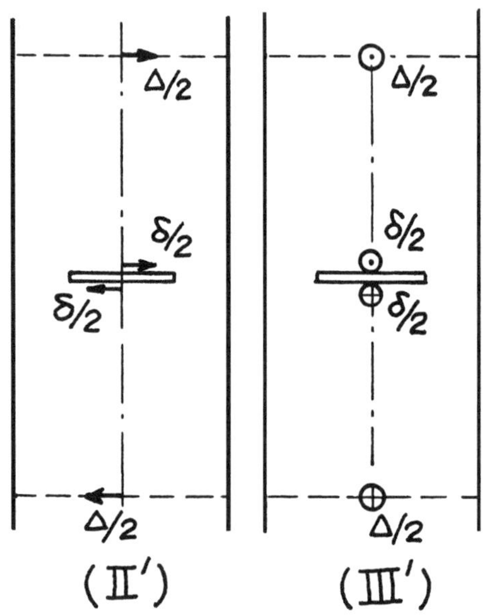

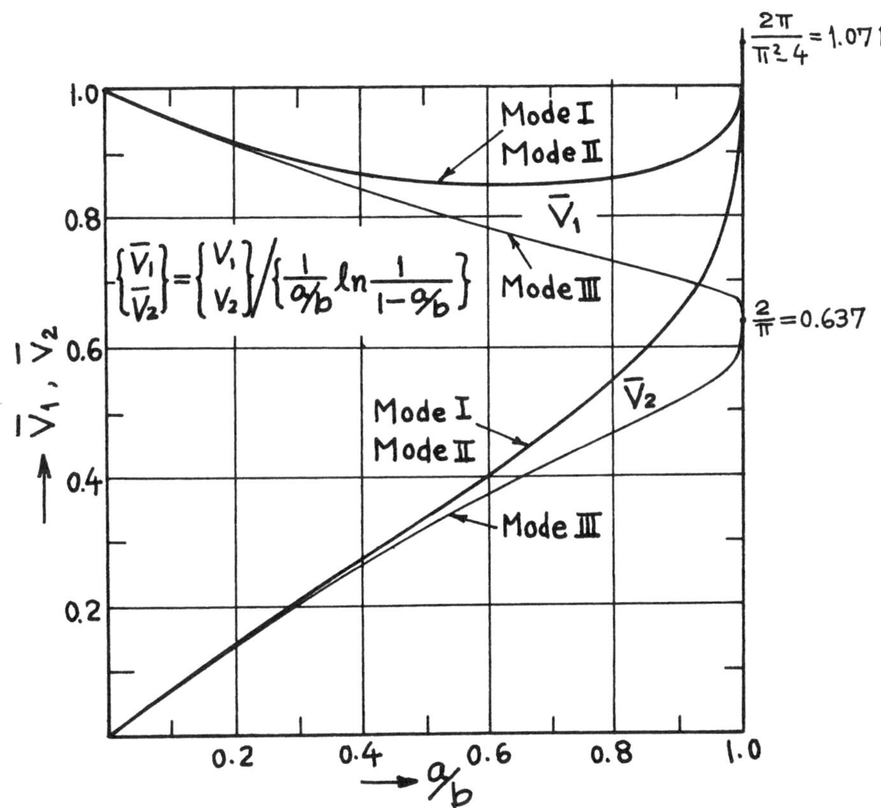

$\frac{2\pi}{\pi^2-4} = 1.071$
1.0
0.8
0.6
0.4
0.2
Mode I
Mode II
$\bar{V}_1$
$\left\{\begin{matrix}\bar{V}_1\\\bar{V}_2\end{matrix}\right\} = \left\{\begin{matrix}V_1\\V_2\end{matrix}\right\} / \left\{\frac{1}{a/b}\ln\frac{1}{1-a/b}\right\}$
Mode III
$\frac{2}{\pi} = 0.637$
$\bar{V}_2$
Mode I
Mode II
Mode III
$\bar{V}_1, \bar{V}_2$
0.2
0.4
0.6
0.8
1.0
a/b

THE DOUBLE EDGE NOTCH TEST SPECIMEN

A. Stress Intensity Factor

$$K_1 = \sigma\sqrt{\pi a}\,F(a/b)$$

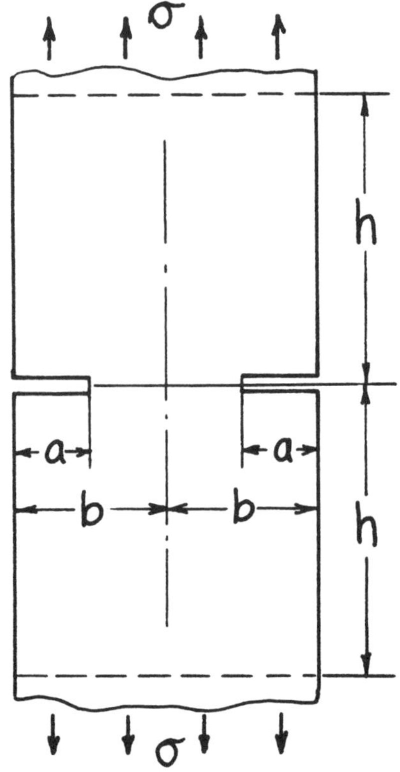

Numerical Values of $F(a/b)$

Bowie's results ($h/b = 3.0$, mapping function method) have 1% accuracy and Yamamoto's results ($h/b = 2.75$, finite element method) have 0.5% accuracy for $0.2 < a/b < 0.9$ (**Bowie 1964a; Yamamoto 1972**).

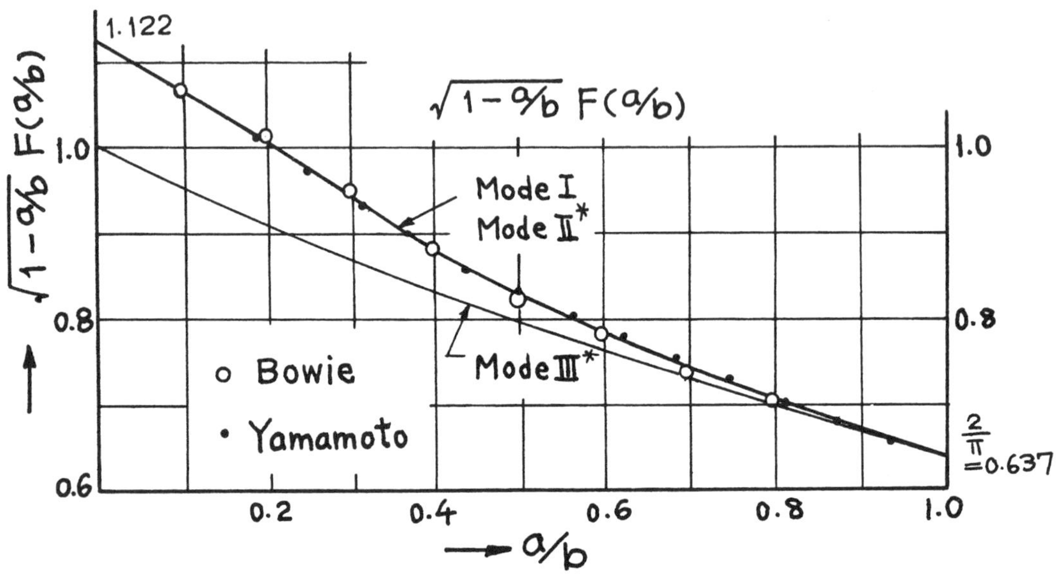

*See **Note 2**

(See also **pages 2.32, 2.33, 2.34, 11.5, 15.1** etc., for corrections and various effects.)

Empirical Formulas
a. Accuracy
b. Method of derivation, reference

$$F\left({}^{a}\!/_{b}\right) = \sqrt{\frac{2b}{\pi a}\tan\frac{\pi a}{2b}}$$

a. Better than 5% for ${}^{a}\!/_{b} > 0.4$
b. Approximation by periodic crack solution (**Irwin 1957**)

$$F\left({}^{a}\!/_{b}\right) = 1.12 + 0.203\left({}^{a}\!/_{b}\right) - 1.197\left({}^{a}\!/_{b}\right)^{2} + 1.930\left({}^{a}\!/_{b}\right)^{3}$$

a. Better than 2% for ${}^{a}\!/_{b} \leq 0.7$
b. Least squares fitting to Bowie's results (**Brown 1966**)

$$F\left({}^{a}\!/_{b}\right) = \frac{1.122 - 0.561\left({}^{a}\!/_{b}\right) - 0.015\left({}^{a}\!/_{b}\right)^{2} + 0.091\left({}^{a}\!/_{b}\right)^{3}}{\sqrt{1 - {}^{a}\!/_{b}}}$$

a. Better than 2% for any ${}^{a}\!/_{b}$
b. Asymptotic approximation (**Benthem 1972**)

$$F\left({}^{a}\!/_{b}\right) = \left(1 + 0.122\cos^{4}\frac{\pi a}{2b}\right)\sqrt{\frac{2b}{\pi a}\tan\frac{\pi a}{2b}}$$

a. 0.5% for any ${}^{a}\!/_{b}$
b. Modification of Irwin's interpolation formula (**Tada 1973**)

$$F\left({}^{a}\!/_{b}\right) = \frac{1.122 - 0.561\left({}^{a}\!/_{b}\right) - 0.205\left({}^{a}\!/_{b}\right)^{2} + 0.471\left({}^{a}\!/_{b}\right)^{3} - 0.190\left({}^{a}\!/_{b}\right)^{4}}{\sqrt{1 - {}^{a}\!/_{b}}}$$

a. 0.5% for any ${}^{a}\!/_{b}$
b. Modification of Benthem's formula (**Tada 1973**)

NOTE: 1. Both $h/b = 3.0$ (Bowie) and $h/b = 2.75$ (Yamamoto) are considered effectively infinite.
2. For Mode II configuration (II), the correction factor is identical to $F(a/b)$ in Mode I.

$$K_{II} = \tau\sqrt{\pi a}\, F(a/b)$$

For Mode III configuration (III), the exact formula is

$$K_{III} = \tau_\ell \sqrt{\pi a}\sqrt{\frac{2b}{\pi a} \tan \frac{\pi a}{2b}}$$

B. Displacements

Crack Opening at Edges

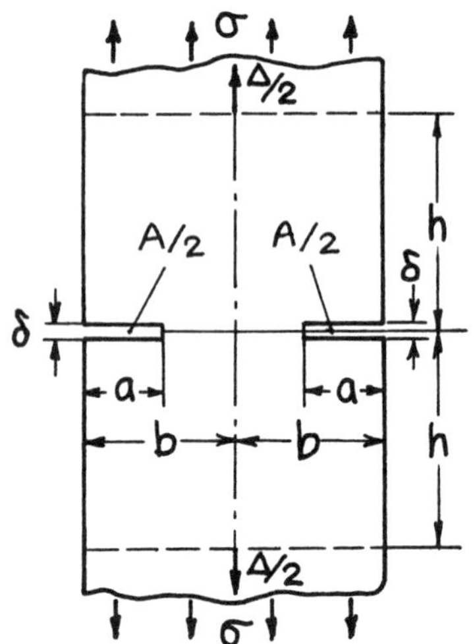

$$\delta = \frac{4\sigma a}{E'} V_1(a/b)$$

The following formula has better than 2% accuracy for any a/b:

$$V_1(a/b) = \frac{1}{\left(\frac{\pi a}{2b}\right)}\left\{0.454\left(\sin\frac{\pi a}{2b}\right) - 0.065\left(\sin\frac{\pi a}{2b}\right)^3 - 0.007\left(\sin\frac{\pi a}{2b}\right)^5 + \cosh^{-1}\left(\sec\frac{\pi a}{2b}\right)\right\}$$

Additional Displacement at Remote Points (h/b >3) Due to Presence of Cracks and Crack Opening Area

$$\Delta_{crack} = \Delta_{total} - \Delta_{no\,crack} = \Delta_{total} - \frac{\sigma}{E} \cdot 2h$$

$$\Delta_{crack} = \frac{4\sigma a}{E'} V_2\left(a/b\right)$$

$$A = \Delta_{crack} \cdot 2b$$

The following formula has better than 1% accuracy for any a/b

$$V_2\left(a/b\right) = \frac{1}{\left(\frac{\pi a}{2b}\right)}\left\{0.0629 - 0.0610\left(\cos\frac{\pi a}{2b}\right)^4 - 0.0019\left(\cos\frac{\pi a}{2b}\right)^8 + \ell n\left(\sec\frac{\pi a}{2b}\right)\right\}$$

Reference: **Tada 1973**

Method of Derivation: Paris' equation based on energy principles (**Paris 1957**) (See **Appendix B**)

NOTE:
1. $E' = E$ for plane stress
   $E' = E/(1-\nu^2)$ for plane strain
2. Uniform pressure σ applied on the crack surfaces results in the same crack opening δ and remote displacement Δ_{crack}.
3. Limiting values of V_1 and V_2 at $a/b \to 0$ and $a/b \to 1$

$$V_1\left(a/b \to 0\right) = 1.454, \quad V_2\left(a/b \to 0\right) = 0$$

$$V_1\left(a/b \to 1\right) = V_2\left(a/b \to 1\right) = \frac{2}{\pi}\ell n \frac{1}{1 - a/b} \quad \text{(exact)}$$

4. For Mode II loading (II), the displacements (II$'$) are

$$\delta = \frac{4\tau a}{E'} V_1\left(a/b\right)$$

$$\Delta = \frac{4\tau a}{E'} V_2\left(a/b\right)$$

where V_1 and V_2 are identical to those in Mode I.
For Mode III loading (III), the displacements (III$'$) are

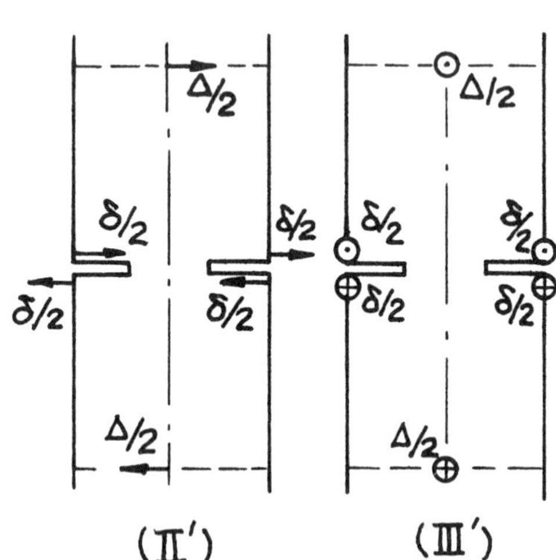

$$\delta = \frac{2\tau_\ell a}{G} V_1\left(a/b\right)$$

$$V_1\left(a/b\right) = \frac{2}{\pi}\frac{1}{a/b}\cosh^{-1}\left(\sec\frac{\pi a}{2b}\right)$$

$$\Delta = \frac{2\tau_\ell a}{G} V_2\left(a/b\right)$$

$$V_2\left(a/b\right) = \frac{2}{\pi}\frac{1}{a/b}\ell n\left(\sec\frac{\pi a}{2b}\right)$$

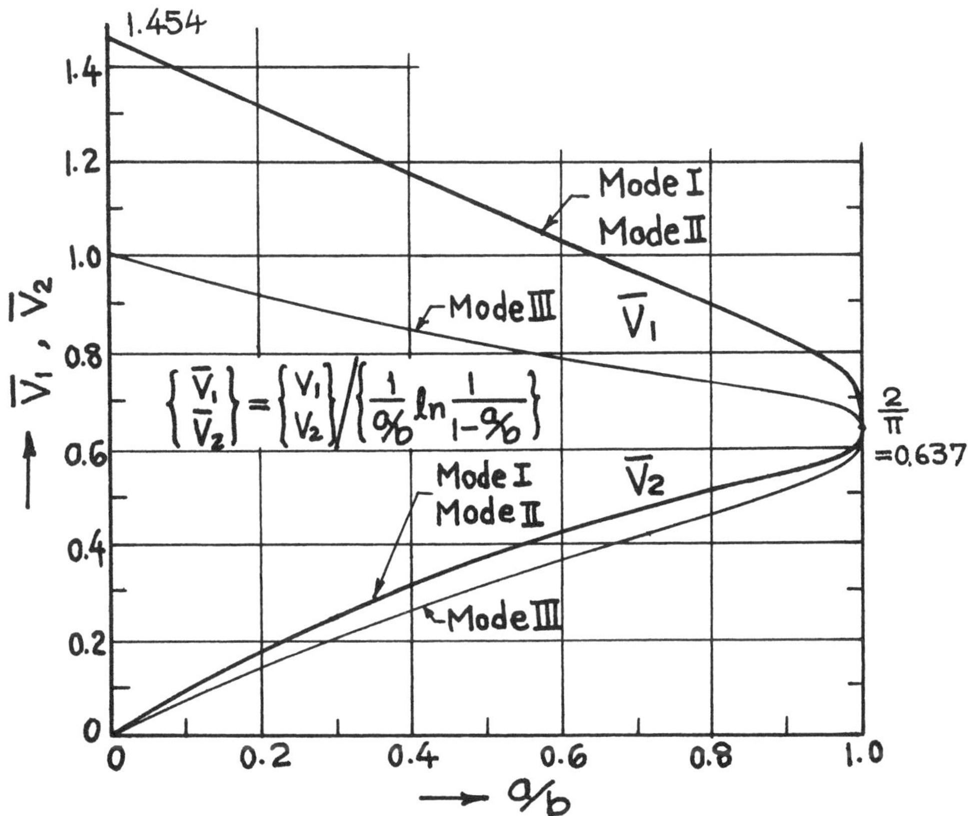
1.454
1.4
1.2
1.0
0.8
0.6
0.4
0.2
0
Mode I
Mode II
Mode III
$\bar{V}_1$
$\bar{V}_1$, $\bar{V}_2$
$\left\{\begin{matrix}\bar{V}_1\\ \bar{V}_2\end{matrix}\right\} = \left\{\begin{matrix}V_1\\ V_2\end{matrix}\right\} \Big/ \left\{\frac{1}{a/b} \ln \frac{1}{1-a/b}\right\}$
$\frac{2}{\pi}$ =0.637
Mode I
Mode II
$\bar{V}_2$
Mode III
0
0.2
0.4
0.6
0.8
1.0
a/b

THE SINGLE EDGE NOTCH TEST SPECIMEN

A. Stress Intensity Factor

$$K_I = \sigma\sqrt{\pi a}\, F(a/b)$$

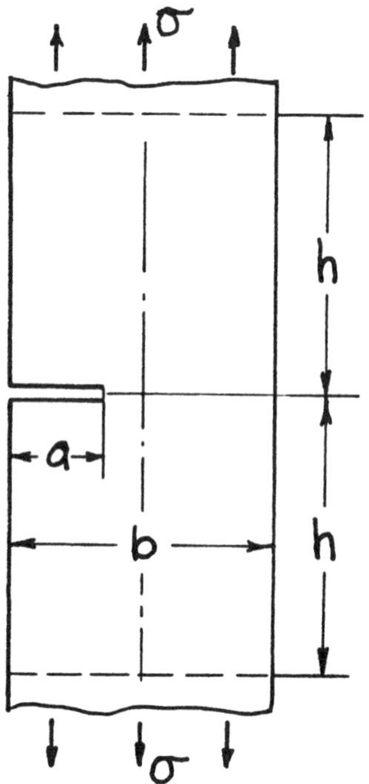

Numerical Values of $F(a/b)$

The curve in the following figure was drawn based on the results having better than 0.5% accuracy.

Methods and References

a. Boundary Collocation Method ($h/b > 0.8$): **Gross 1964**
b. Mapping Function Method ($h/b = 1.53$): **Bowie 1965**
c. Green's Function Method ($h/b > 1.5$): **Emery 1969, 1972**
d. Weight Function Method: **Bueckner 1970, 1971**
e. Asymptotic Approximation: **Benthem 1972**
f. Finite Element Method ($h/b = 2.75, 1.0$): **Yamamoto 1972**

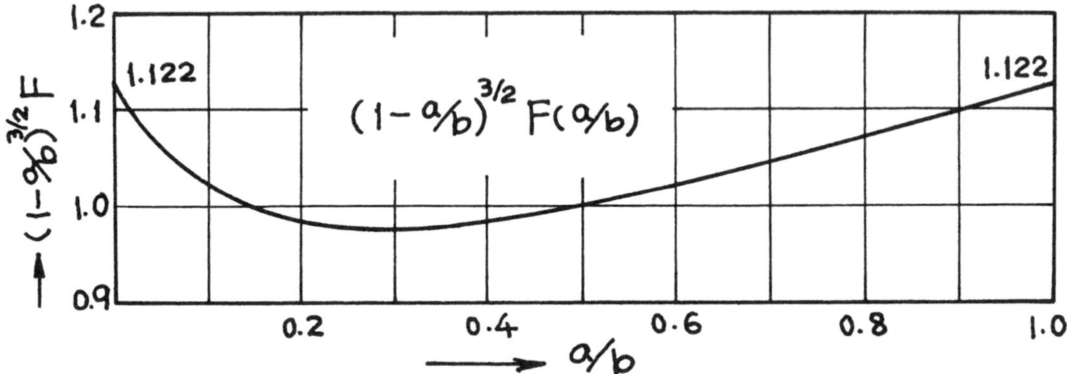

NOTE: 1. Load is applied along the centerline of the strip at the crack location (or uniform pressure on crack surfaces).
2. The effect of h/b is practically negligible for $h/b \geq 1.0$.

(See also **pages 2.13, 2.16, 2.27 to 2.31** etc., for various corrections and effects.)

Empirical Formulas

a. Accuracy
b. Method, reference

$$F(a/b) = 1.122 - 0.231(a/b) + 10.550(a/b)^2 - 21.710(a/b)^3 + 30.382(a/b)^4$$

a. 0.5% for $a/b \leq 0.6$
b. Least squares fitting **(Gross 1964; Brown 1966)**

$$F(a/b) = 0.265(1 - a/b)^4 + \frac{0.857 + 0.265\, a/b}{(1 - a/b)^{3/2}}$$

a. Better than 1% for $a/b < 0.2$, 0.5% for $a/b \geq 0.2$
b. **Tada 1973**

$$F(a/b) = \sqrt{\frac{2b}{\pi a}\tan\frac{\pi a}{2b}} \cdot \frac{0.752 + 2.02(a/b) + 0.37\left(1 - \sin\frac{\pi a}{2b}\right)^3}{\cos\frac{\pi a}{2b}}$$

a. Better than 0.5% for any a/b
b. **Tada 1973**

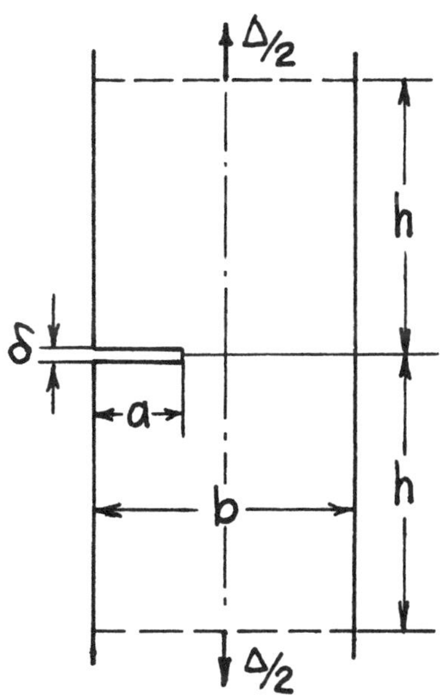

B. Displacements

Crack Opening at Edge

$$\delta = \frac{4\sigma a}{E'} V_1(a/b)$$

Gross' results (**Gross 1967**, Boundary Collocation Method) are expected to have 0.5% accuracy for $0.2 \leq a/b < 0.7$. An empirical formula with 1% accuracy for any a/b is (**Tada 1973**)

$$V_1(a/b) = \frac{1.46 + 3.42\left(1 - \cos\frac{\pi a}{2b}\right)}{\left(\cos\frac{\pi a}{2b}\right)^2}$$

Additional Remote Point ($^{h}/_{b} \geq 1$) Displacement due to Crack (Along the Centerline at the Crack Location)

$$\Delta_{crack} = \Delta_{total} - \Delta_{no\,crack}$$

$$\Delta_{crack} = \frac{4\sigma a}{E'} \cdot V_2\left(^{a}/_{b}\right)$$

The following formula has better than 1% accuracy for any $^{a}/_{b}$:

$$V_2(^{a}/_{b}) = \frac{^{a}/_{b}}{\left(1 - ^{a}/_{b}\right)^2}\left\{0.99 - ^{a}/_{b}\left(1 - ^{a}/_{b}\right)\left(1.3 - 1.2\,^{a}/_{b} + 0.7\left(^{a}/_{b}\right)^2\right)\right\}$$

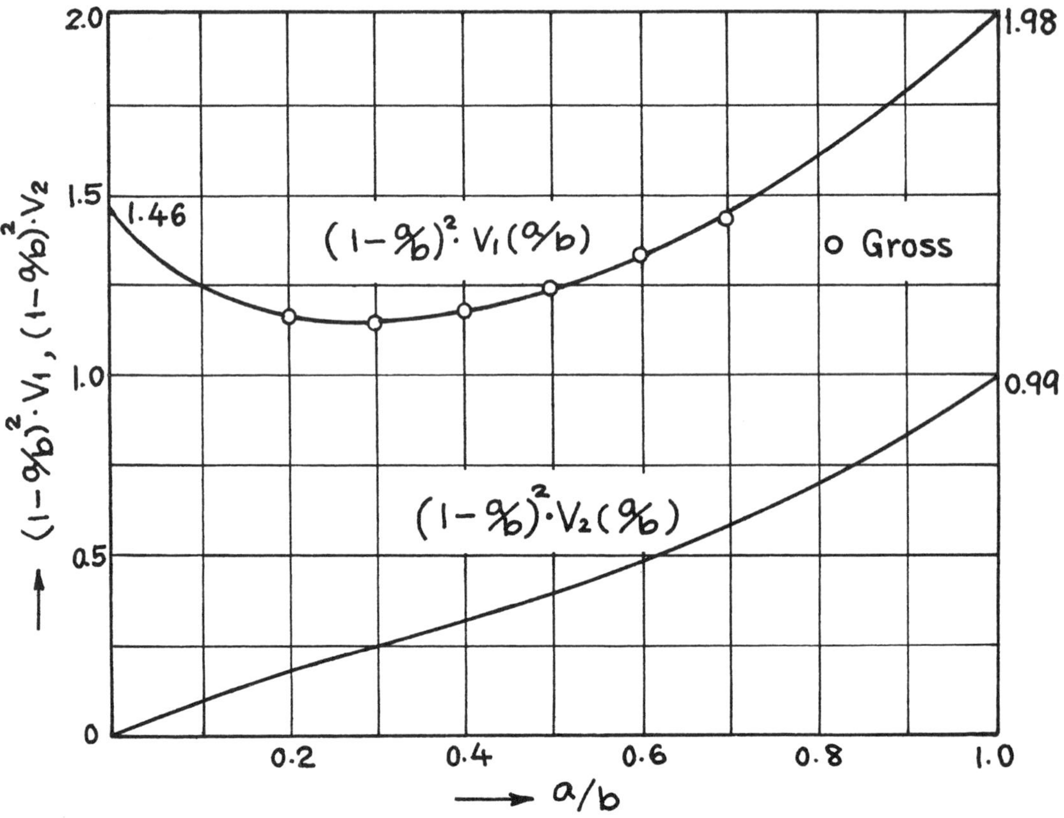

Method: Paris' Equation (**Paris 1957**) (See **Appendix B.**)
Reference: **Tada 1973**

THE PURE BENDING SPECIMEN

A. Stress Intensity Factor

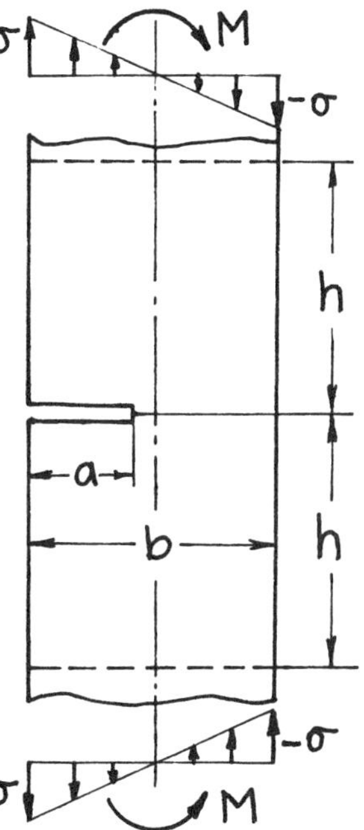

$$\sigma = \frac{6M}{b^2}$$

$$K_I = \sigma\sqrt{\pi a}\, F\left(a/b\right)$$

Numerical Values of $F(a/b)$

The curve in the following figure was drawn based on the results having better than 0.5% accuracy. Also used for four-point bending.

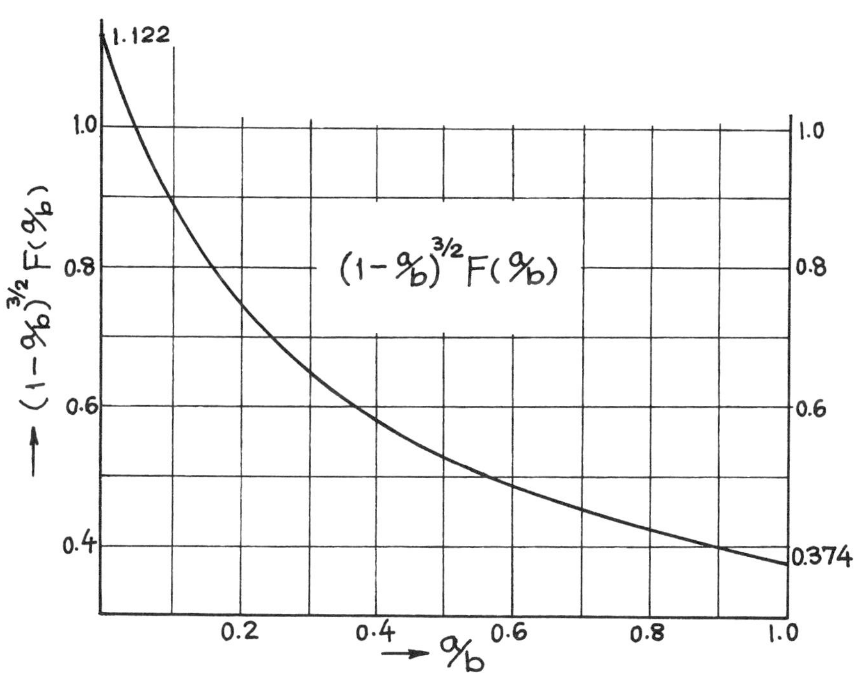

Methods and References

1. Singular Integral Equation, **Bueckner 1960**
2. Boundary Collocation Method $\left(h/b \geq 2\right)$, **Gross 1965a**
3. Weight Function Method, **Bueckner 1970, 1971**
4. Green's Function Method $\left(h/b \geq 1.5\right)$, **Emery 1969**
5. Asymptotic Approximation, **Benthem 1972**

Empirical Formulas
a. Accuracy
b. Method, reference

$$F(a/b) = 1.122 - 1.40(a/b) + 7.33(a/b)^2 - 13.08(a/b)^3 + 14.0(a/b)^4$$

a. 0.2% for $a/b \leq 0.6$
b. Least squares fitting (**Brown 1966**)

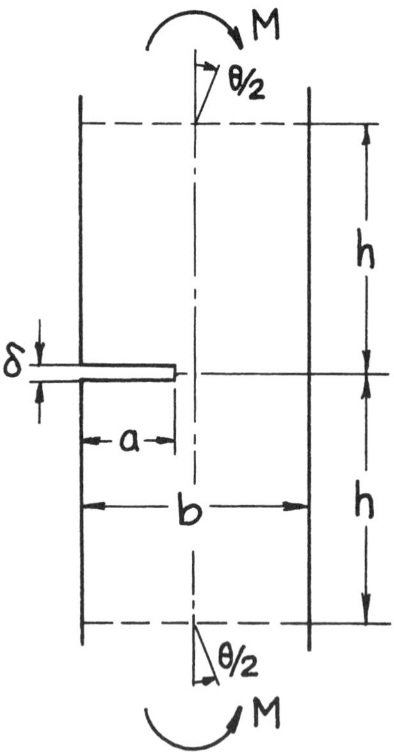

$$F(a/b) = \sqrt{\frac{2b}{\pi a}\tan\frac{\pi a}{2b}}\,\frac{0.923 + 0.199\left(1 - \sin\frac{\pi a}{2b}\right)^4}{\cos\frac{\pi a}{2b}}$$

a. Better than 0.5% for any a/b
b. **Tada 1973**

B. Displacements

Crack Opening at Edge

$$\delta = \frac{4\sigma a}{E'} V(a/b)$$

Gross' results (**Gross 1967**, Boundary Collocation Method) are expected to have 0.5% accuracy for $0.2 \leq a/b \leq 0.7$. An empirical formula with 1% accuracy for any a/b is (**Tada 1973**)

$$V(a/b) = 0.8 - 1.7(a/b) + 2.4(a/b)^2 + \frac{0.66}{(1 - a/b)^2}$$

Additional Remote Point ($h/b > 2$) Displacement (Rotation) Due to Crack

$$\theta_{crack} = \theta_{total} - \theta_{no\,crack}$$

$$\theta_{crack} = \frac{4\sigma}{E'} S(a/b)$$

The following formula has better than 1% accuracy for any a/b.

$$S(a/b) = \left(\frac{a/b}{1 - a/b}\right)^2 \left\{5.93 - 19.69(a/b) + 37.14(a/b)^2 - 35.84(a/b)^3 + 13.12(a/b)^4\right\}$$

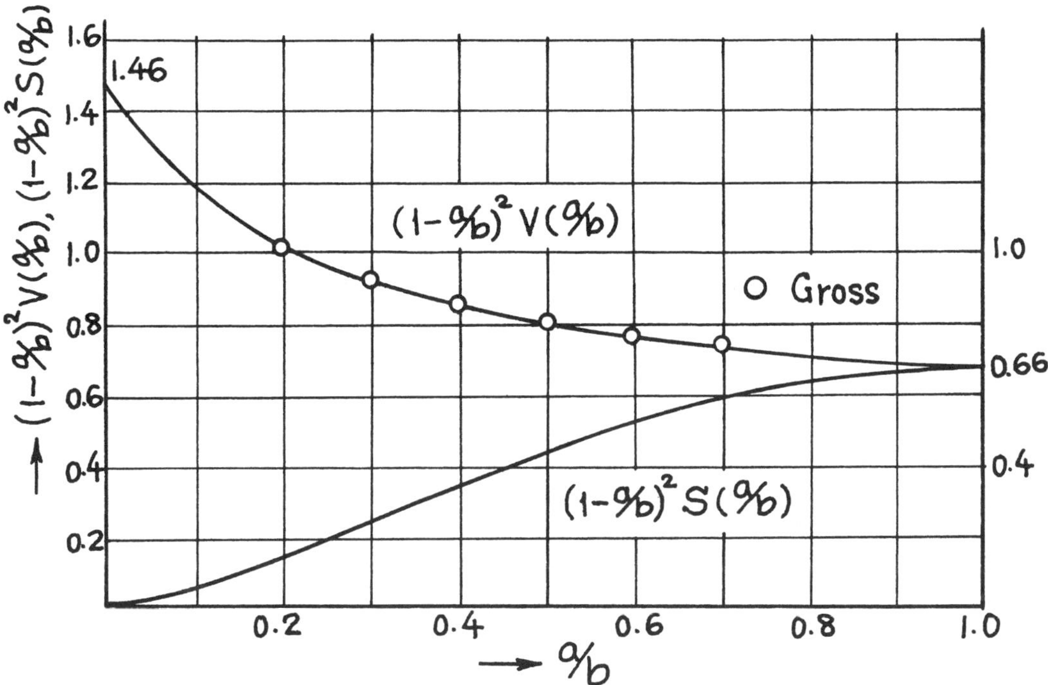

Method: Paris' Equation (**Paris 1957**) (See **Appendix B.**)
Reference: **Tada 1973**
(See also **pages 2.16, 2.27, 9.1** etc., for related solutions.)

THE THREE-POINT BEND TEST SPECIMEN

A. Stress Intensity Factor

$$\sigma = \frac{6M}{b^2} \left(M = \frac{Ps}{4} \right)$$

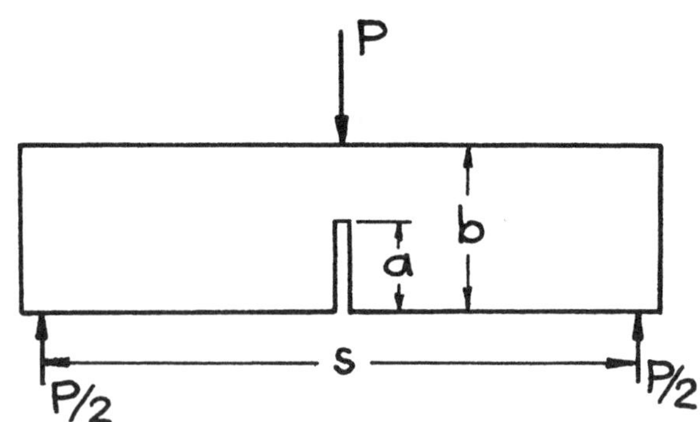

$$K_I = \sigma\sqrt{\pi a}F\left({}^{a}/_{b}\right)$$

Numerical Values of $F({}^{a}/_{b})$

The curves in the following figure have 1% accuracy.

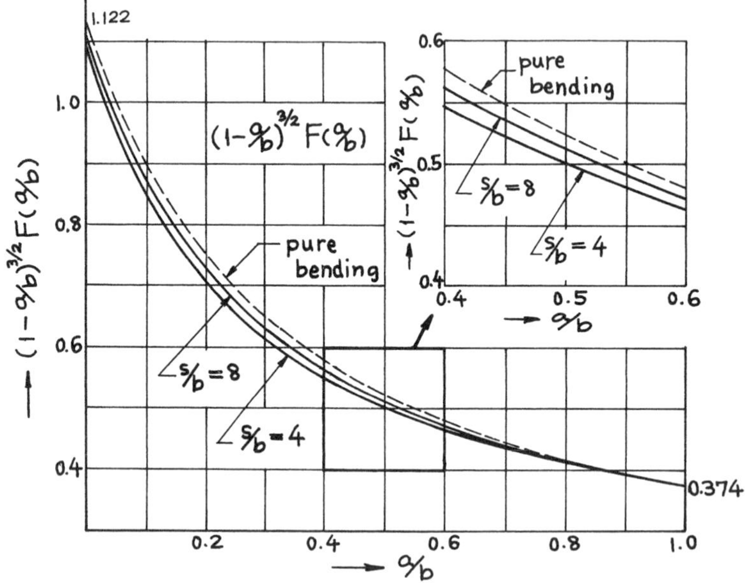

Methods and References

1. Boundary Collocation Method $\left({}^{s}/_{b} = 4, 8\right)$ **(Gross 1965b)**
2. Green's Function Method $\left({}^{s}/_{b} = 3, 8\right)$ **(Emery 1969)**

Empirical Formulas

a. Accuracy
b. Method, reference

For ${}^{s}/_{b} = 4$,

$$F\left({}^{a}/_{b}\right) = \frac{1}{\sqrt{\pi}} \cdot \frac{1.99 - {}^{a}/_{b}(1 - {}^{a}/_{b})\left(2.15 - 3.93\,{}^{a}/_{b} + 2.7\left({}^{a}/_{b}\right)^2\right)}{(1 + 2\,{}^{a}/_{b})(1 - {}^{a}/_{b})^{3/2}}$$

a. 0.5% for any ${}^{a}/_{b}$
b. **Srawley 1976**

For $s/b = 8$,

$$F(a/b) = 1.106 - 1.552(a/b) + 7.71(a/b)^2 - 13.53(a/b)^3 + 14.23(a/b)^4$$

a. 0.2% for $a/b \leq 0.6$
b. Least squares fitting, **Brown 1966**

B. Displacements

Crack Opening at Edge

$$\delta = \frac{4\sigma a}{E'} V_1(a/b)$$

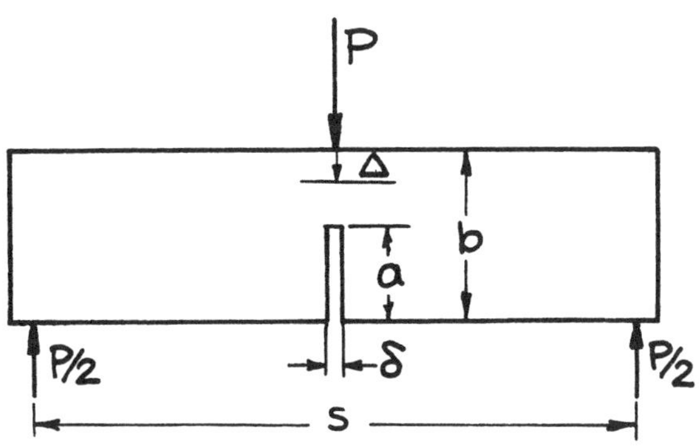

Gross' results (**Gross 1967**, Boundary Collocation Method, $s/b = 4$) are expected to have 0.5% accuracy for $0.2 \leq a/b \leq 0.7$. An empirical formula with 1% accuracy for any a/b for $s/b = 4$ is (**Tada 1973**)

$$V_1(a/b) = 0.76 - 2.28(a/b) + 3.87(a/b)^2 - 2.04(a/b)^3 + \frac{0.66}{(1 - a/b)^2}$$

Additional Load Point Displacement due to Crack

$$\Delta_{crack} = \Delta_{total} - \Delta_{no\ crack}$$

$$\Delta_{crack} = \frac{\sigma}{E'} s V_2(a/b)$$

The following formula has better than 1% accuracy for any a/b ; for $s/b = 4$:

$$V_2(a/b) = \left(\frac{a/b}{1 - a/b}\right)^2 \left\{5.58 - 19.57(a/b) + 36.82(a/b)^2 - 34.94(a/b)^3 + 12.77(a/b)^4\right\}$$

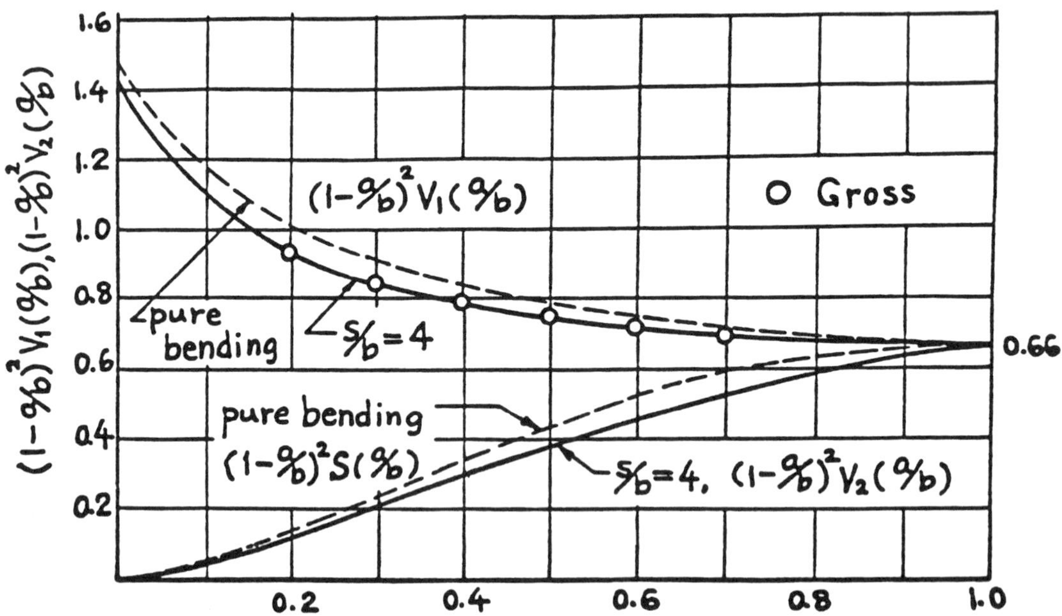

Method: Paris' Equation (**Paris 1957**) (see **Appendix B**)
Reference: **Tada 1973**
(See also **pages 2.13, 2.27, 9.1** etc., for related solutions.)

Note: The curves for $s/b = 8$ are nearly averages between the curves for pure bending and $s/b = 4$. For other cases $(s/b > 4)$, displacements can be estimated by interpolation with fair accuracy.

THE COMPACT TENSION TEST SPECIMEN

$$K_I = \sigma\sqrt{a}\, F_I(a/b, h/b, d/h)$$

where $\sigma = P/b$

or $K_I = \sigma_N \sqrt{b-a}\, F_2(a/b, h/b, d/h)$

where

$$\sigma_N = \sigma_{N\,Tension} + \sigma_{N\,Bending}$$

$$= \frac{P}{b-a} + \frac{6P\left(a + \frac{b-a}{2}\right)}{(b-a)^2}$$

$$= \frac{2P(2b+a)}{(b-a)^2}$$

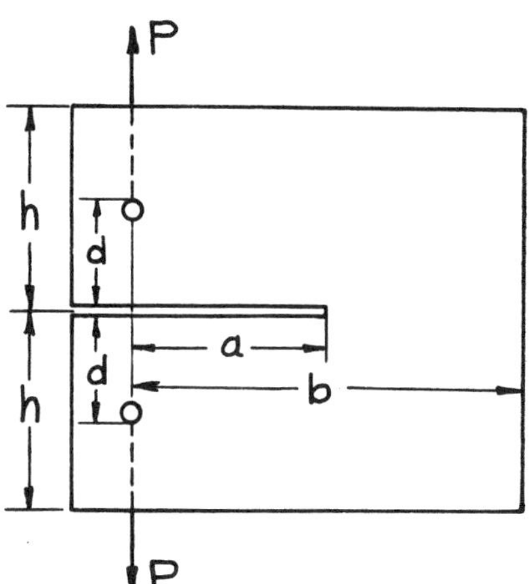

Numerical Values of F_2

$$\left(F_1 = \frac{2(2 + a/b)}{(1 - a/b)^{3/2}} \cdot \frac{1}{\sqrt{a/b}} \cdot F_2\right)$$

The curves in the following figure have better than 1% accuracy.

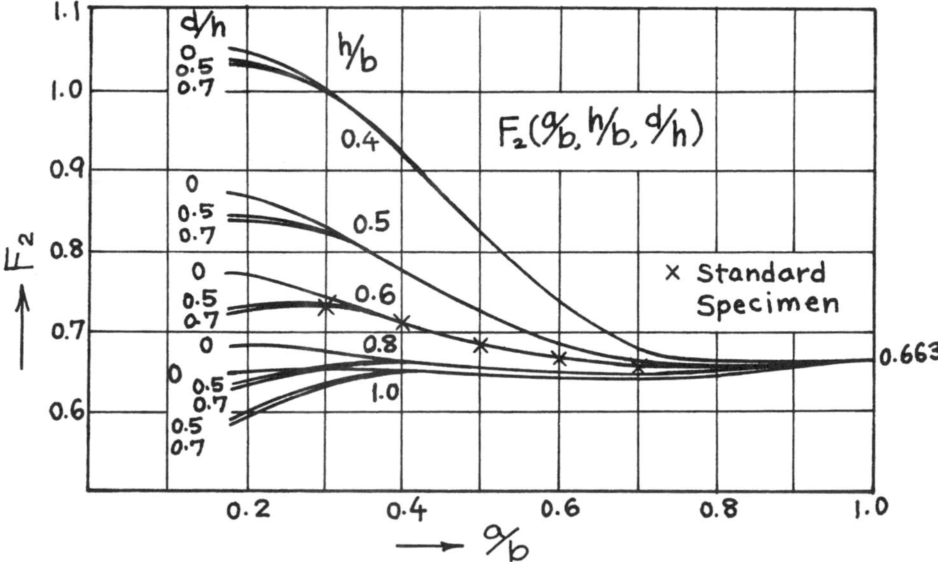

Method: Boundary Collocation Method
References: **Gross 1970; Srawley 1972**

Standard Specimen (ASTM Standard E-399-72)

Standard geometry of compact tension specimen is shown below.

h = 0.6b

h_1 = 0.275b

D = 0.25b

c = 0.25b

(Thickness = b/2)

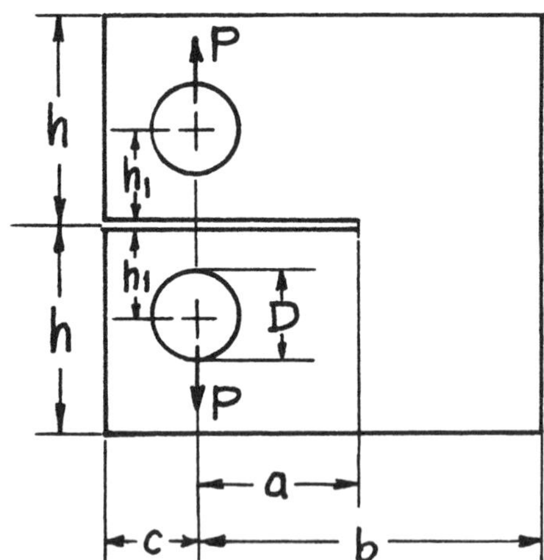

A. Stress Intensity Factor

The F_2 values for the standard specimen are plotted in the previous graph. For the range $0.4 \leq {}^a\!/_b \leq 0.6$, the values of F_1 are plotted.

Note: $F_2 \simeq F_2({}^a\!/_b,\ 0.6,\ 0.7)$

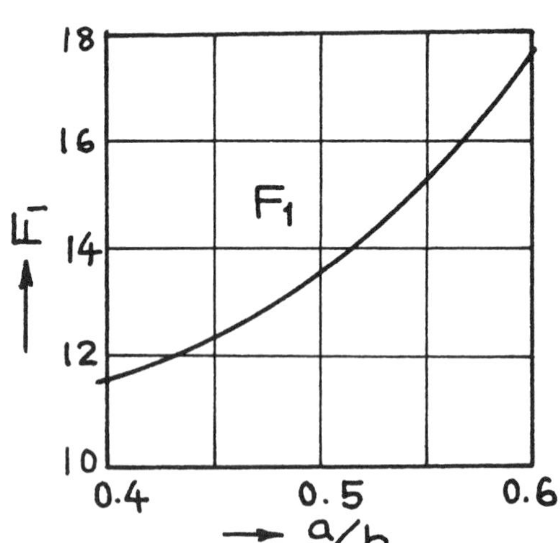

Empirical Formula

For the standard specimen, the following formula has 0.5% accuracy for ${}^a\!/_b > 0.2$ (**Srawley 1976**).

$$F_2({}^a\!/_b) = 0.443 + 2.32({}^a\!/_b) - 6.66({}^a\!/_b)^2 + 7.36({}^a\!/_b)^3 - 2.8({}^a\!/_b)^4$$

B. Displacements

Opening at Crack Edge

$$\delta_1 = \frac{P}{E'} V_1({}^a\!/_b)$$

Opening at Loadline

$$\delta_2 = \frac{P}{E'} V_2({}^a\!/_b)$$

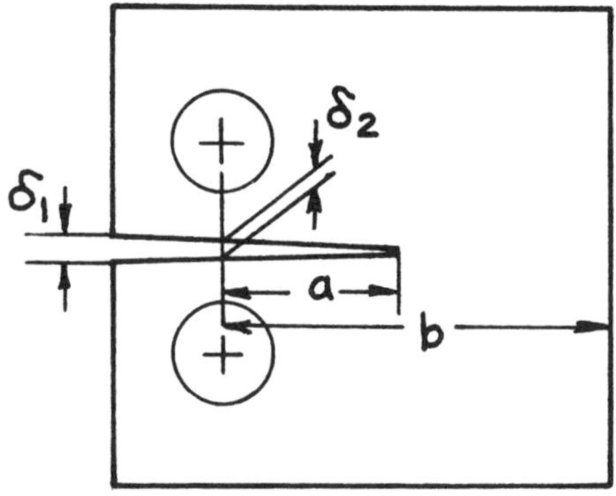

The following formulas for V_1 and V_2 for the standard specimen have 0.5% accuracies for $0.2 \leq {}^{a}/_{b} \leq 0.95$ (**Saxena 1978**).

$$V_1({}^{a}/_{b}) = \left(1 + \frac{0.25}{{}^{a}/_{b}}\right)\left(\frac{1 + {}^{a}/_{b}}{1 - {}^{a}/_{b}}\right)^2 \Big[1.6137 + 12.678({}^{a}/_{b}) - 14.231({}^{a}/_{b})^2 - 16.610({}^{a}/_{b})^3 + 35.050({}^{a}/_{b})^4 - 14.494({}^{a}/_{b})^5\Big]$$

$$V_2({}^{a}/_{b}) = \left(\frac{1 + {}^{a}/_{b}}{1 - {}^{a}/_{b}}\right)^2 \Big[2.1630 + 12.219({}^{a}/_{b}) - 20.065({}^{a}/_{b})^2 - 0.9925({}^{a}/_{b})^3 + 20.609({}^{a}/_{b})^4 - 9.9314({}^{a}/_{b})^5\Big]$$

Method: Boundary Collocation Method

THE ROUND (DISK-SHAPED) COMPACT SPECIMEN

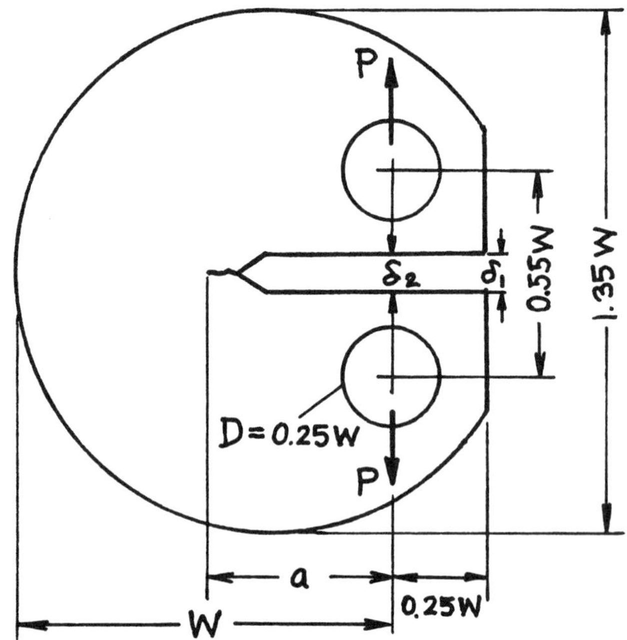

thickness = B

$$A = \frac{a}{W}$$

$$\bar{\sigma} = \frac{P}{WB}$$

A. Stress Intensity Factor

$$K_I = \bar{\sigma}\sqrt{W}\, F(A)$$

$$F(A) = \frac{(2+A)\left(0.76 + 4.8A - 11.58A^2 + 11.43A^3 - 4.08A^4\right)}{(1-A)^{3/2}}$$

B. Displacements

Opening Displacement at Edge

$$\delta_1 = \frac{\bar{\sigma}W}{E'} \cdot V_1(A)$$

$$V_1(A) = \exp\left(1.742 - 0.495A + 14.71A^2 - 22.06A^3 + 14.44A^4\right)$$

Opening Displacement at Load Line

$$\delta_2 = \frac{\bar{\sigma}W}{E'} \cdot V_2(A)$$

$$V_2(A) = \exp\left(0.26 + 5.381A + 2.105A^2 - 8.853A^3 + 9.122A^4\right)$$

Method: Boundary Collocation Method
Accuracy: K_I 0.3% for $0.2 \le A \le 1.0$; δ_1 and δ_2 0.5% for $0.2 \le A \le 0.8$
References: **Newman 1979a, 1981b**

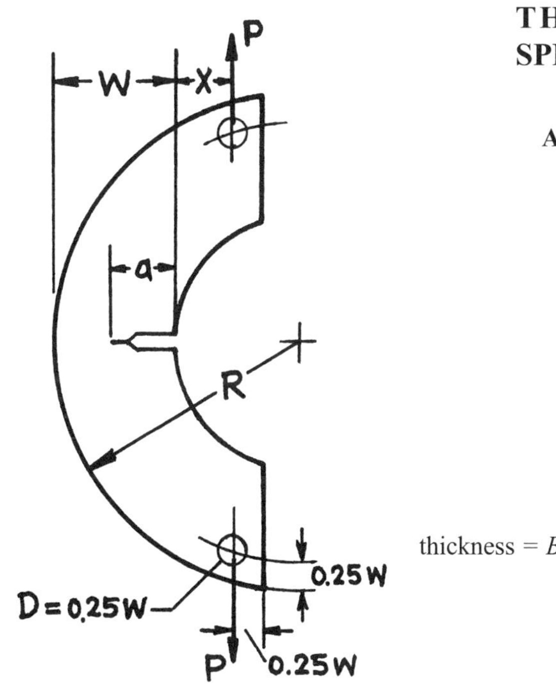

THE ARC-SHAPED (C-SHAPED) SPECIMEN

ASTM E-399 Standard Specimens:

$$X/W = 0 \text{ and } 0.5$$

thickness = B

$$A = \frac{a}{W} \qquad \bar{\sigma} = \frac{P}{WB}$$

$$K_I = \bar{\sigma}\sqrt{\pi a} \cdot f\left(\frac{W}{R}, \frac{X}{W}, A\right) \cdot F(A)$$

$$f\left(\frac{W}{R}, \frac{X}{W}, A\right) = \left(3\frac{X}{W} + 1.9 + 1.1A\right)\left[1 + 0.25(1-A)^2\frac{W}{R}\right]$$

$$F(A) = \frac{3.74 - 6.30A + 6.32A^2 - 2.43A^3}{\sqrt{\pi}(1-A)^{3/2}}$$

Method: Boundary Collocation Method
Accuracy: $X/W = 0$ and 0.5: 1% for $0.45 \le A \le 0.55$; 1.5% for $0.2 \le A \le 1.0$.
$0 \le X/W \le 1.0$: 3% for $0.2 \le A \le 1.0$
Reference: **Kapp 1980**

A. Stress Intensity Factor

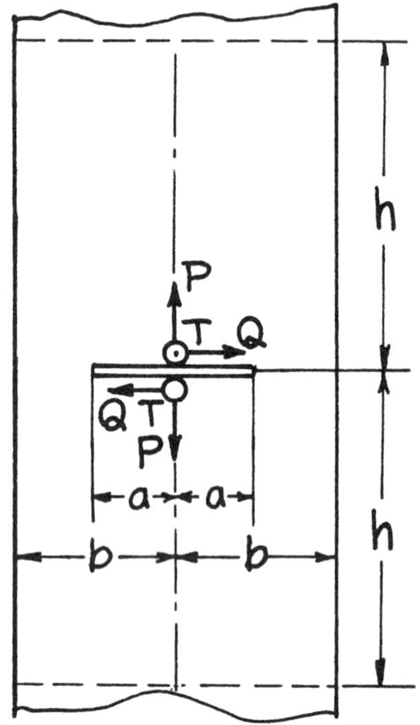

$$\begin{Bmatrix} K_I \\ K_{II} \\ K_{III} \end{Bmatrix} = \frac{1}{\sqrt{\pi a}} \begin{Bmatrix} P \\ Q \\ T \end{Bmatrix} \begin{Bmatrix} F(a/b) \\ F(a/b) \\ F_{III}(a/b) \end{Bmatrix}$$

$$F_{III}(a/b) = \sqrt{\frac{\pi a}{b} \Big/ \sin\frac{\pi a}{b}}$$

Numerical Values of $F(a/b)$

Newman's results based on a method of boundary collocation are expected to have the accuracy of the order of 0.1% for $0.1 \leq a/b \leq 0.8$ (**Newman 1971**, $h/b = 2$).

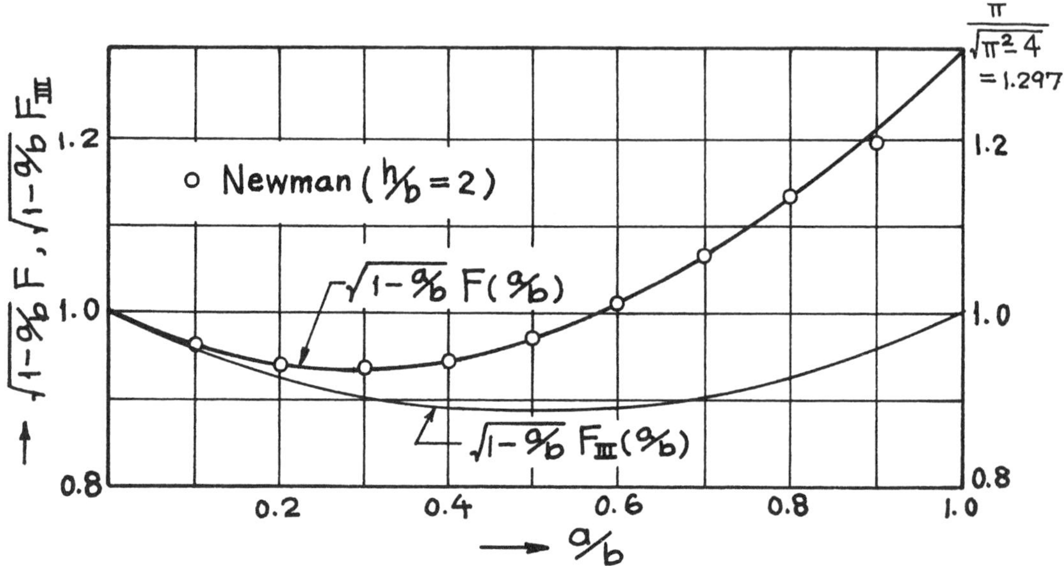

Empirical Formulas

a. Accuracy
b. Method of derivation, reference

$$F(a/b) = \left(1.297 - 0.297\cos\frac{\pi a}{2b}\right)\sqrt{\frac{\pi a}{b}\Big/\sin\frac{\pi a}{b}}$$

a. Better than 1% for any a/b
b. Asymptotic Interpolation, **Tada 1973**

$$F(a/b) = \frac{1 - 0.5(a/b) + 0.957(a/b)^2 - 0.16(a/b)^3}{\sqrt{1 - a/b}}$$

a. Better than 0.3% for any a/b
b. Modified asymptotic formula (**Tada 1973**)

B. Displacements at Remote Points ($h/b > 2$)

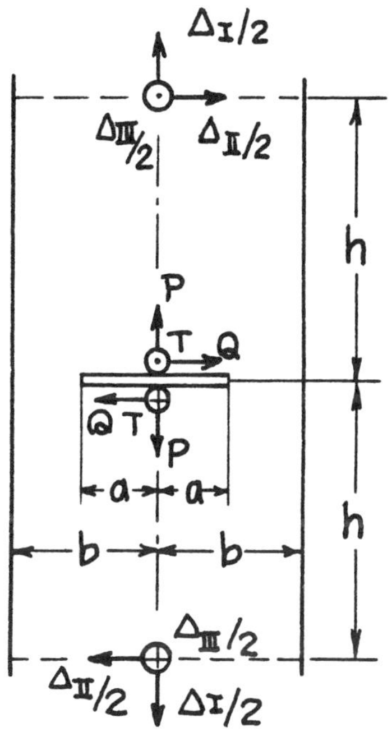

$$\left\{\begin{matrix}\Delta_I \\ \Delta_{II}\end{matrix}\right\} = \frac{2}{E'}\left\{\begin{matrix}P \\ Q\end{matrix}\right\} D(a/b)$$

$$\Delta_{III} = \frac{T}{G}\,\frac{2}{\pi}\cosh^{-1}\left(\sec\frac{\pi a}{2b}\right)$$

The following formula for $D(a/b)$ has better than 0.6% accuracy for any a/b.

$$D(a/b) = -0.071(a/b) - 0.535(a/b)^2 + 0.169(a/b)^3 - 0.090(a/b)^4 + 0.020(a/b)^5 - 1.071\ln(1 - a/b)$$

where $\left(1.071 = \frac{2\pi}{\pi^2 - 4}\right)$

Method: Paris' Equation (**Paris 1957**) (see **Appendix B**)
Reference: **Tada 1973**

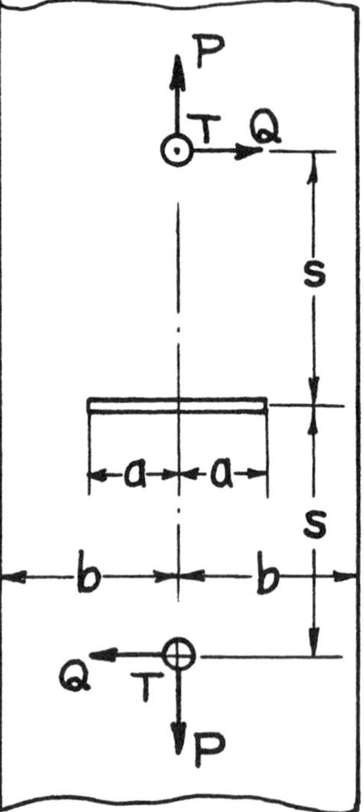

$$\begin{Bmatrix} K_I \\ K_{II} \\ K_{III} \end{Bmatrix} = \frac{1}{\sqrt{2b}} \begin{Bmatrix} P \\ Q \\ T \end{Bmatrix} \begin{Bmatrix} F_I\left({}^a/_b, {}^s/_b\right) \\ F_{II}\left({}^a/_b, {}^s/_b\right) \\ F_{III}\left({}^a/_b, {}^s/_b\right) \end{Bmatrix}$$

$$F_{III}\left({}^a/_b, {}^s/_b\right) = \sqrt{\tan\frac{\pi a}{2b}} \cdot \frac{1}{\sqrt{1 - \left(\dfrac{\cos\frac{\pi a}{2b}}{\cosh\frac{\pi s}{2b}}\right)^2}}$$

$$\begin{Bmatrix} F_I\left({}^a/_b, {}^s/_b\right) \\ F_{II}\left({}^a/_b, {}^s/_b\right) \end{Bmatrix} = f\left({}^a/_b, {}^s/_b\right) \left[1\{\pm\}\alpha \frac{\frac{\pi s}{2b}\tanh\frac{\pi s}{2b}}{\left(\dfrac{\cosh\frac{\pi s}{2b}}{\cosh\frac{\pi a}{2b}}\right)^2 - 1} \right] F_{III}\left({}^a/_b, {}^s/_b\right)$$

$$f\left({}^a/_b, {}^s/_b\right) = 1 + \left\{0.297 - 0.115\left(1 - \operatorname{sech}\frac{\pi s}{2b}\right)\sin\frac{\pi a}{b}\right\}\left(1 - \cos\frac{\pi a}{2b}\right)$$

$$\alpha = \begin{cases} \dfrac{1+\nu}{2} & \text{plane stress} \\ \dfrac{1}{2(1-\nu)} & \text{plane strain} \end{cases}$$

Method: Asymptotic Interpolation
Accuracy: F_{III} exact; F_I, F_{II} better than 1% for any ${}^a/_b$ and ${}^s/_b$
Reference: **Tada 1973**

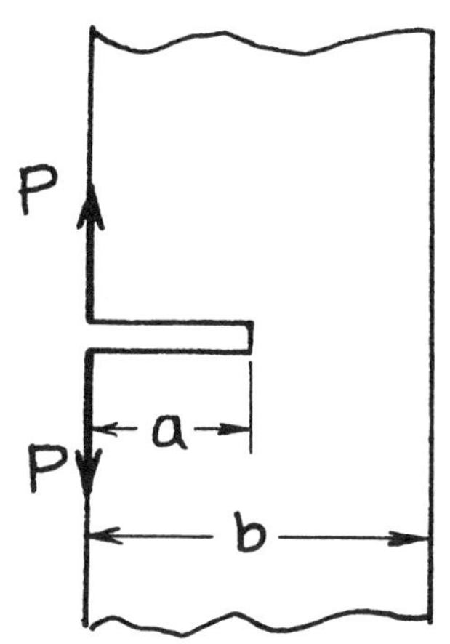

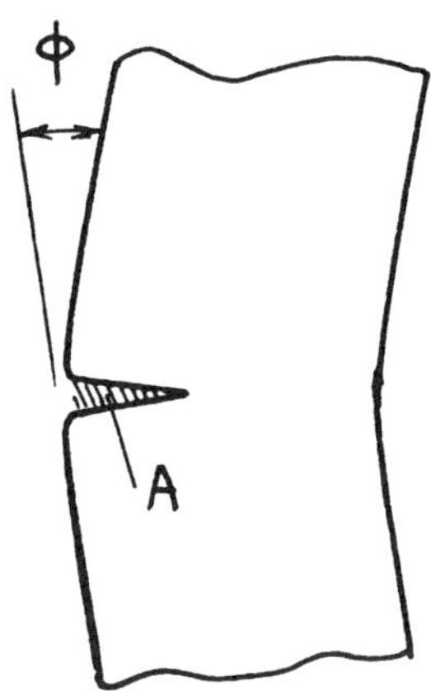

$$K_I = \frac{2P}{\sqrt{\pi a}} F\left(\frac{a}{b}\right)$$

$$F\left(\frac{a}{b}\right) = \frac{.46 + 3.06\frac{a}{b} + .84\left(1 - \frac{a}{b}\right)^5 + .66\left(\frac{a}{b}\right)^2\left(1 - \frac{a}{b}\right)^2}{\left(1 - \frac{a}{b}\right)^{3/2}}$$

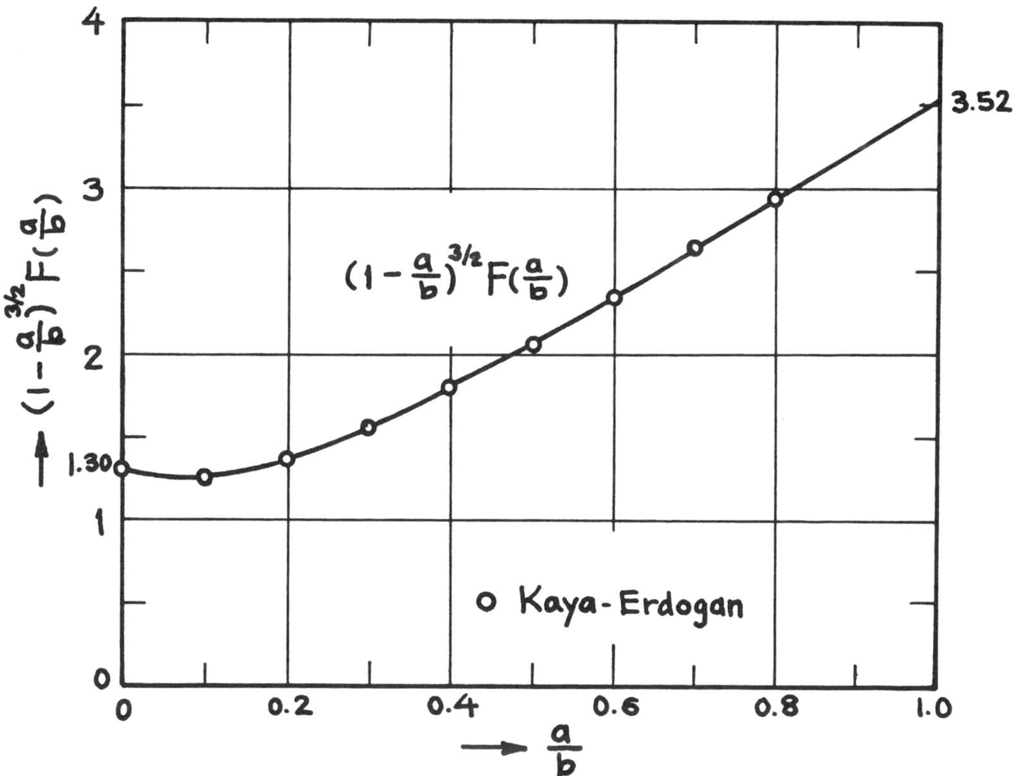

Crack Opening Area

$$A = \frac{4Pa}{E'} \cdot S\left(\frac{a}{b}\right)$$

$$S\left(\frac{a}{b}\right) = \frac{1.46 + 3.42\left(1 - \cos\frac{\pi a}{2b}\right)}{\left(\cos\frac{\pi a}{2b}\right)^2}$$

Rotation due to Crack

$$\phi = \frac{24\,Pa}{E'\,b^2} \cdot \Phi\left(\frac{a}{b}\right)$$

$$\Phi\left(\frac{a}{b}\right) = \frac{.878 - .218\frac{a}{b} + .58\left(1 - \frac{a}{b}\right)^5}{\left(1 - \frac{a}{b}\right)^2}$$

Methods: K Singular Integral Equation; A and ϕ Paris' Equation (see **Appendix B**)
Accuracy: K better than 0.5%; Empirical formula 1%; A and ϕ 1%
References: **Kaya 1980**, **Tada 1985**

$$K_I = \frac{2P}{\sqrt{\pi a}} \cdot \frac{G\left(\frac{c}{a}, \frac{a}{b}\right)}{\left(1 - \frac{a}{b}\right)^{3/2} \sqrt{1 - \left(\frac{c}{a}\right)^2}}$$

$$G\left(\frac{c}{a}, \frac{a}{b}\right) = g_1\left(\frac{a}{b}\right) + g_2\left(\frac{a}{b}\right) \cdot \frac{c}{a} + g_3\left(\frac{a}{b}\right) \cdot \left(\frac{c}{a}\right)^2 + g_4\left(\frac{a}{b}\right) \cdot \left(\frac{c}{a}\right)^3$$

$$g_1\left(\frac{a}{b}\right) = .46 + 3.06\,\frac{a}{b} + .84\left(1 - \frac{a}{b}\right)^5 + .66\left(\frac{a}{b}\right)^2\left(1 - \frac{a}{b}\right)^2$$

$$g_2\left(\frac{a}{b}\right) = -3.52\left(\frac{a}{b}\right)^2$$

$$g_3\left(\frac{a}{b}\right) = 6.17 - 28.22\frac{a}{b} + 34.54\left(\frac{a}{b}\right)^2 - 14.39\left(\frac{a}{b}\right)^3 - \left(1 - \frac{a}{b}\right)^{3/2} - 5.88\left(1 - \frac{a}{b}\right)^5 - 2.64\left(\frac{a}{b}\right)^2\left(1 - \frac{a}{b}\right)^2$$

$$g_4\left(\frac{a}{b}\right) = -6.63 + 25.16\frac{a}{b} - 31.04\left(\frac{a}{b}\right)^2 + 14.41\left(\frac{a}{b}\right)^3 + 2\left(1 - \frac{a}{b}\right)^{3/2} + 5.04\left(1 - \frac{a}{b}\right)^5 + 1.98\left(\frac{a}{b}\right)^2\left(1 - \frac{a}{b}\right)^2$$

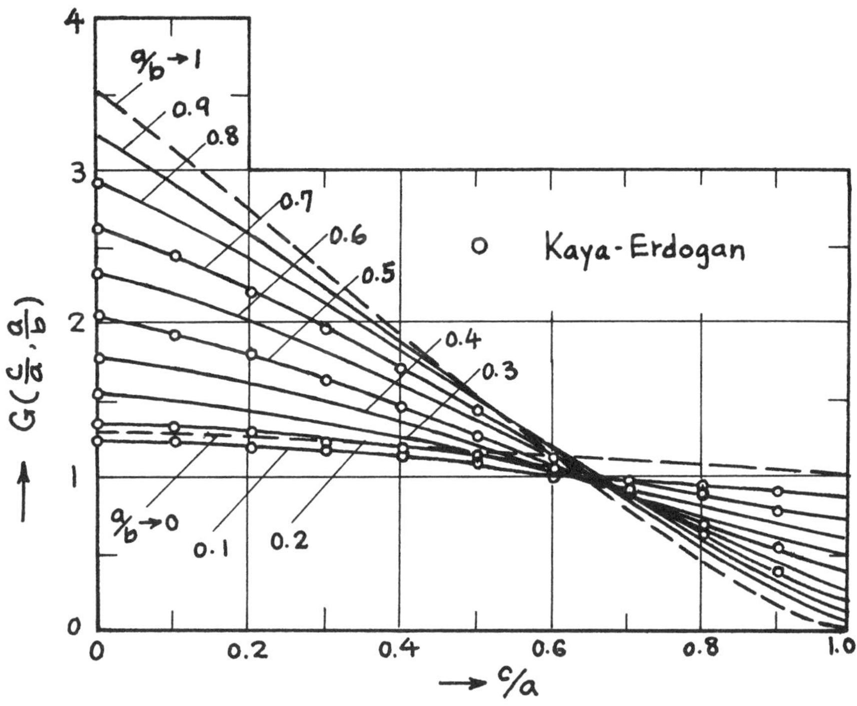

Methods: Singular Integral Equation (Kaya-Erdogan, $a/b = .1, .2, .5, .7$). Estimated by Interpolation for other a/b.

Accuracy: 1% (Curves are based on the empirical formula above)

References: **Kaya 1980**, **Tada 1985**

NOTE: Dashed lines are $G(a/b \rightarrow 0) = 1.3 - .3(c/a)^{5/4}$ and $G(a/b \rightarrow 1) = 3.52(1 \rightarrow c/a)\sqrt{1 - (c/a)^2}$

A. Stress Intensity Factor

$$\begin{Bmatrix} K_{II} \\ K_{III} \end{Bmatrix} = \frac{2}{\sqrt{\pi a}} \begin{Bmatrix} Q \\ T \end{Bmatrix} \begin{Bmatrix} F_{II}(a/b) \\ F_{III}(a/b) \end{Bmatrix}$$

$$F_{II}(a/b) = \frac{1.30 - 0.65(a/b) + 0.37(a/b)^2 + 0.28(a/b)^3}{\sqrt{1 - a/b}}$$

$$F_{III}(a/b) = \sqrt{\frac{\pi a}{b} \Big/ \sin\frac{\pi a}{b}}$$

Method: Asymptotic Interpolation
Accuracy: F_{II} better than 1% for any a/b; F_{III} exact
Reference: **Tada 1973**

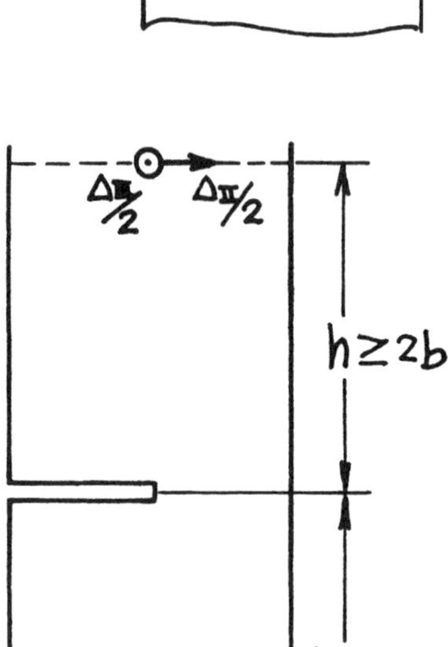

B. Displacements

$$\Delta_{II} = \frac{4Q}{E'} U(a/b)$$

$$\Delta_{III} = \frac{2T}{G} W(a/b)$$

$$U(a/b) = -0.184(a/b) - 0.637(a/b)^2 - 0.129(a/b)^3 + 0.026(a/b)^4 + 0.028(a/b)^5 + 0.008(a/b)^6 - 1.644\ell n(1 - a/b)$$

$$W(a/b) = \frac{2}{\pi}\cosh^{-1}\left(\sec\frac{\pi a}{2b}\right)$$

Method: Paris' Equation (**Paris 1957**) (see **Appendix B**)
Accuracy: U better than 2% for any a/b; W exact
Reference: **Tada 1973**

A. Stress Intensity Factor

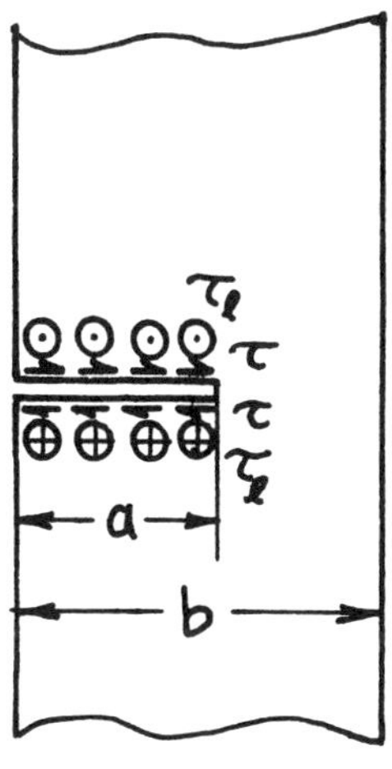

$$\begin{Bmatrix} K_{II} \\ K_{III} \end{Bmatrix} = \begin{Bmatrix} \tau \\ \tau_{\ell} \end{Bmatrix} \sqrt{\pi a} \begin{Bmatrix} F_{II}(a/b) \\ F_{III}(a/b) \end{Bmatrix}$$

$$F_{II}(a/b) = \frac{1.122 - 0.561(a/b) + 0.085(a/b)^2 + 0.180(a/b)^3}{\sqrt{1 - a/b}}$$

$$F_{III}(a/b) = \sqrt{\frac{2b}{\pi a}\tan\frac{\pi a}{2b}}$$

Method: Asymptotic Interpolation
Accuracy: F_{II} better than 2% for any a/b; F_{III} exact
Reference: **Tada 1973**

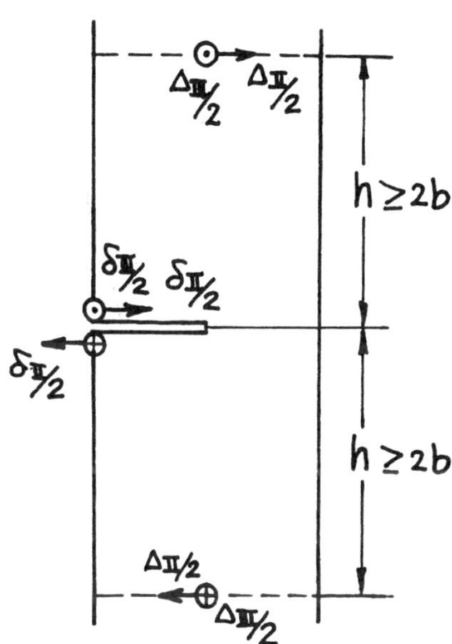

B. Displacements

$$\begin{Bmatrix} \delta_{II} \\ \Delta_{II} \end{Bmatrix} = \frac{4\tau a}{E'} \begin{Bmatrix} U_1(a/b) \\ U_2(a/b) \end{Bmatrix}$$

$$\begin{Bmatrix} \delta_{III} \\ \Delta_{III} \end{Bmatrix} = \frac{2\tau_{\ell} a}{G} \begin{Bmatrix} W_1(a/b) \\ W_2(a/b) \end{Bmatrix}$$

$$U_1(a/b) = -0.184 - 0.637(a/b) - 0.129(a/b)^2 + 0.026(a/b)^3 + 0.028(a/b)^4 + 0.008(a/b)^5 - 1.644(b/a)\,\ell n(1 - a/b)$$

$$U_2(a/b) = 1.46(a/b) - 0.259(a/b)^2 - 0.091(a/b)^3 + 0.052(a/b)^4 - 0.019(a/b)^5 - 0.008(a/b)^6 - 0.518\,\ell n(1 - a/b)$$

$$W_1(a/b) = \frac{2}{\pi}\frac{1}{(a/b)}\cosh^{-1}\left(\sec\frac{\pi a}{2b}\right)$$

$$W_2(a/b) = \frac{2}{\pi}\frac{1}{(a/b)}\ell n\left(\sec\frac{\pi a}{2b}\right)$$

Method: Paris' Equation (**Paris 1957**) (see **Appendix B**)
Accuracy: U_1, U_2 better than 2% for any a/b; W_1, W_2 exact
Reference: **Tada 1973**

A. Stress Intensity Factor

$$\begin{Bmatrix} K_I \\ K_{II} \\ K_{III} \end{Bmatrix} = \frac{2}{\sqrt{\pi a}} \begin{Bmatrix} P \\ Q \\ T \end{Bmatrix} \begin{Bmatrix} F(a/b) \\ F(a/b) \\ F_{III}(a/b) \end{Bmatrix}$$

$$F(a/b) = \frac{1.30 - 0.65(a/b) - 0.10(a/b)^2 + 0.45(a/b)^3}{\sqrt{1 - a/b}}$$

or $$F(a/b) = \left(1 + 0.30\cos^2\frac{\pi a}{2b}\right)\sqrt{\frac{\pi a}{b}\Big/\sin\frac{\pi a}{b}}$$

$$F_{III}(a/b) = \sqrt{\frac{\pi a}{b}\Big/\sin\frac{\pi a}{b}}$$

Method: Asymptotic Interpolation
Accuracy: F better than 2% for any a/b; F_{III} exact
Reference: **Tada 1973**

B. Displacements at Remote Points ($h/b > 2$)

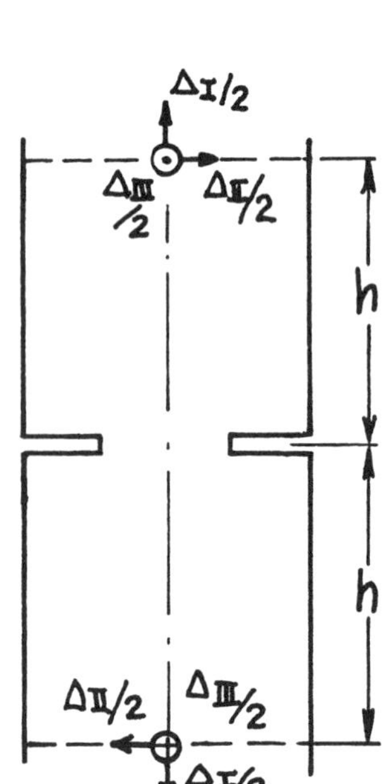

$$\begin{Bmatrix} \Delta_I \\ \Delta_{II} \end{Bmatrix} = \frac{4}{E'} \begin{Bmatrix} P \\ Q \end{Bmatrix} V(a/b)$$

$$\Delta_{III} = \frac{2}{G} T \cdot \frac{2}{\pi} \cosh^{-1}\left(\sec\frac{\pi a}{2b}\right)$$

The following formula for $V(a/b)$ has better than 2% for any a/b.

$$V(a/b) = 0.292\left(\sin\frac{\pi a}{2b}\right) - 0.041\left(\sin\frac{\pi a}{2b}\right)^3 - 0.004\left(\sin\frac{\pi a}{2b}\right)^5 + 0.637(a/b)\cosh^{-1}\left(\sec\frac{\pi a}{2b}\right)$$

Method: Paris' Equation (**Paris 1957**) (see **Appendix B**)
Reference: **Tada 1973**

$$\begin{Bmatrix} K_I \\ K_{II} \\ K_{III} \end{Bmatrix} = \frac{2}{\sqrt{2b}} \begin{Bmatrix} P \\ Q \\ T \end{Bmatrix} \begin{Bmatrix} F(a/b, c/a) \\ F(a/b, c/a) \\ F_{III}(a/b, c/a) \end{Bmatrix}$$

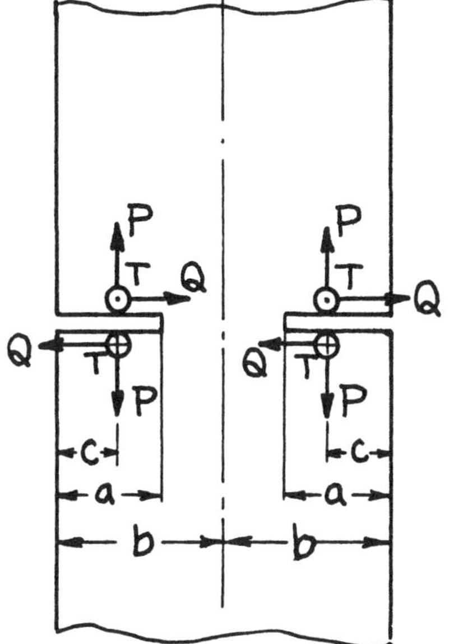

$$F_{III}(a/b, c/a) = \sqrt{\tan\frac{\pi a}{2b}}\;\frac{1}{\sqrt{1-\left(\cos\frac{\pi a}{2b}\Big/\cos\frac{\pi c}{2b}\right)^2}}$$

$$F(a/b, c/a) = \{1 + f(c/a)\cdot g(a/b)\}F_{III}(a/b, c/a)$$

where

$$f(c/a) = 0.3\left\{1-(c/a)^{5/4}\right\}$$

$$g(a/b) = 0.5\left(1-\sin\frac{\pi a}{2b}\right)\left(2+\sin\frac{\pi a}{2b}\right)$$

Method: Asymptotic Interpolation
Accuracy: F_{III} exact; F better than 1% for any a/b and c/a
Reference: **Tada 1985**

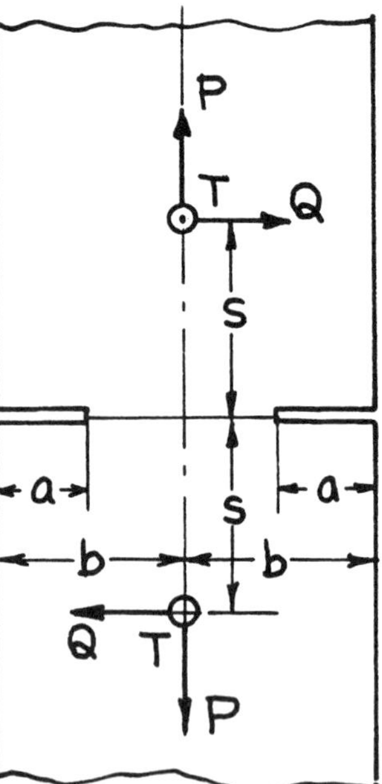

$$\left\{\begin{matrix} K_I \\ K_{II} \\ K_{III} \end{matrix}\right\} = \frac{1}{\sqrt{2b}} \left\{\begin{matrix} P \\ Q \\ T \end{matrix}\right\} \left\{\begin{matrix} F_I(a/b, s/b) \\ F_{II}(a/b, s/b) \\ F_{III}(a/b, s/b) \end{matrix}\right\}$$

$$F_{III}(a/b, s/b) = \sqrt{\tan\frac{\pi a}{2b}} \frac{1}{\sqrt{1 + \left(\dfrac{\cos\frac{\pi a}{2b}}{\sin\frac{\pi s}{2b}}\right)^2}}$$

$$\left\{\begin{matrix} F_I(a/b, s/b) \\ F_{II}(a/b, s/b) \end{matrix}\right\} = \left(1 + 0.122\cos^2\frac{\pi a}{2b}\right) \left[1\{\mp\}\alpha \frac{\frac{\pi s}{2b}\coth\frac{\pi s}{2b}}{1 + \left(\dfrac{\sinh\frac{\pi s}{2b}}{\cos\frac{\pi a}{2b}}\right)^2}\right] F_{III}(a/b, s/b)$$

$$\alpha = \begin{cases} \dfrac{1+\nu}{2} & \text{plane stress} \\ \dfrac{1}{2(1-\nu)} & \text{plane strain} \end{cases}$$

Method: Asymptotic Interpolation
Accuracy: F_I, F_{II} better than 2% for any a/b and s/b; F_{III} exact
Reference: **Tada 1973**

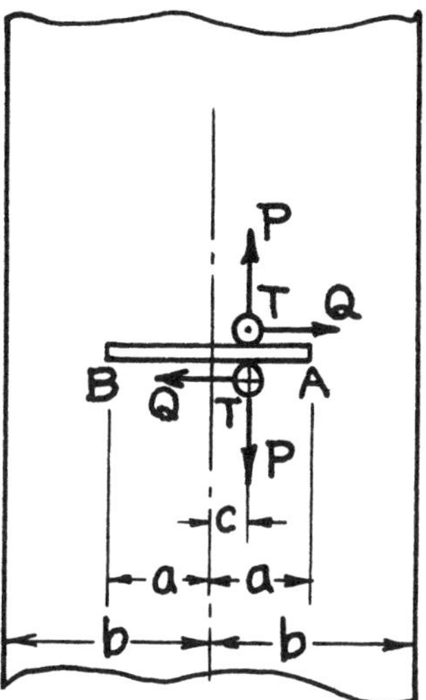

$$\begin{Bmatrix} K_I \\ K_{II} \\ K_{III} \end{Bmatrix}_{\substack{A \\ B}} = \frac{1}{\sqrt{2b}} \begin{Bmatrix} P \\ Q \\ T \end{Bmatrix} \begin{Bmatrix} F(a/b,\, c/a) \\ F(a/b,\, c/a) \\ F_{III}(a/b,\, c/a) \end{Bmatrix}_{\substack{A \\ B}}$$

$$\begin{Bmatrix} F_{III_A}(a/b,\, c/a) \\ F_{III_B}(a/b,\, c/a) \end{Bmatrix} = \sqrt{\tan\frac{\pi a}{2b}} \cdot \frac{1\{\pm\}\left(\dfrac{\sin\frac{\pi c}{2b}}{\sin\frac{\pi a}{2b}}\right)}{\sqrt{1 - \left(\dfrac{\cos\frac{\pi a}{2b}}{\cos\frac{\pi c}{2b}}\right)^2}}$$

$$\{F(a/b,\, c/a)\}_{\substack{A \\ B}} = \left\{1 + 0.297\sqrt{1 - (c/a)^2}\left(1 - \cos\frac{\pi a}{2b}\right)\right\}\{F_{III}(a/b,\, c/a)\}_{\substack{A \\ B}}$$

where $\left(0.297 = \dfrac{\pi}{\sqrt{\pi^2 - 4}} - 1\right)$

Method: Asymptotic Interpolation
Accuracy: F_{III} exact; F better than 1% for any a/b and c/a
Reference: **Tada 1973**

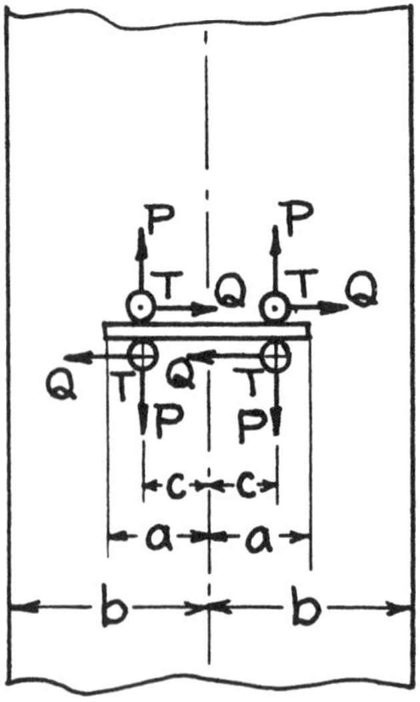

$$\begin{Bmatrix} K_I \\ K_{II} \\ K_{III} \end{Bmatrix} = \frac{2}{\sqrt{2b}} \begin{Bmatrix} P \\ Q \\ T \end{Bmatrix} \begin{Bmatrix} F\left({}^{a}\!/_{b}, {}^{c}\!/_{a}\right) \\ F\left({}^{a}\!/_{b}, {}^{c}\!/_{a}\right) \\ F_{III}\left({}^{a}\!/_{b}, {}^{c}\!/_{a}\right) \end{Bmatrix}$$

$$F_{III}\left({}^{a}\!/_{b}, {}^{c}\!/_{a}\right) = \sqrt{\tan\frac{\pi a}{2b}} \cdot \frac{1}{\sqrt{1 - \left(\dfrac{\cos\frac{\pi a}{2b}}{\cos\frac{\pi c}{2b}}\right)^2}}$$

$$F\left({}^{a}\!/_{b}, {}^{c}\!/_{a}\right) = \left\{1 + 0.297\sqrt{1 - \left({}^{c}\!/_{a}\right)^2}\left(1 - \cos\frac{\pi a}{2b}\right)\right\} F_{III}\left({}^{a}\!/_{b}, {}^{c}\!/_{a}\right)$$

where $$\left(0.297 = \frac{\pi}{\sqrt{\pi^2 - 4}} - 1\right)$$

Method: Asymptotic Interpolation
Accuracy: F_{III} exact; F better than 1% for any ${}^{a}\!/_{b}$ and ${}^{c}\!/_{a}$
Reference: **Tada 1973**

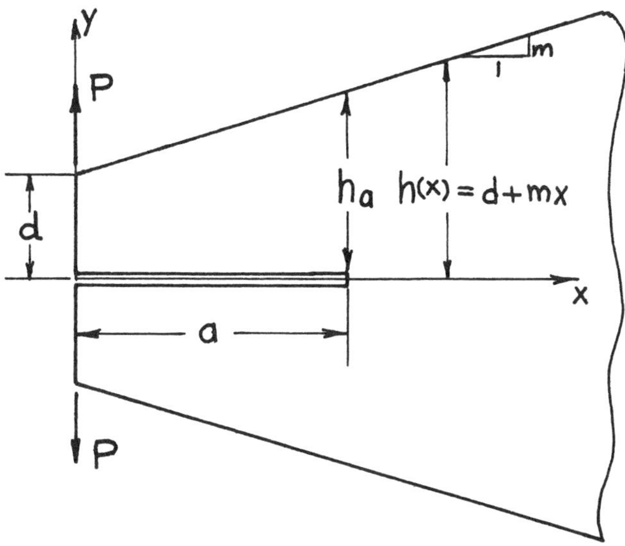

$$h_a = h(a) = d + ma$$

$$K_I = \frac{P}{\sqrt{h_a}} \cdot f(m) \cdot \left(\frac{a}{h_a} + 0.7 \right)$$

$$f(m) = 3.46 - 2.65m + 1.89m^{3/2} \, (m < 0.5)$$

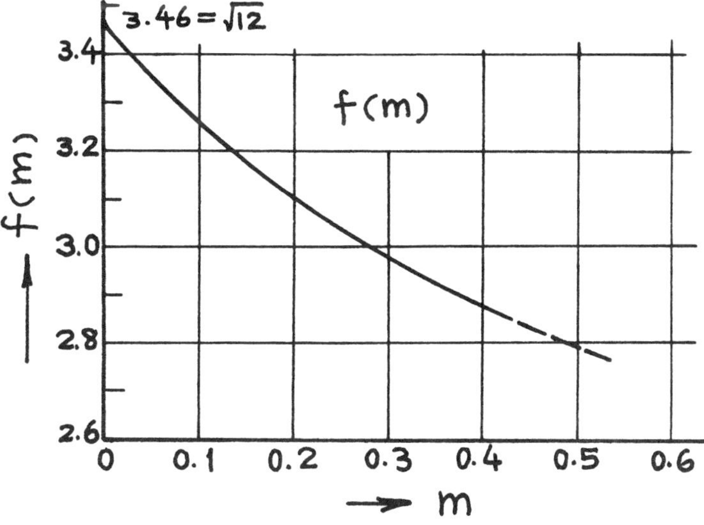

Method: Empirical formula based on the results by Boundary Collocation Method
Accuracy: Order of 1% for $a/h_a > 1$
References: **Gross 1966; Srawley 1967; Tada 2000**

ELECTRICAL POTENTIAL CALIBRATION

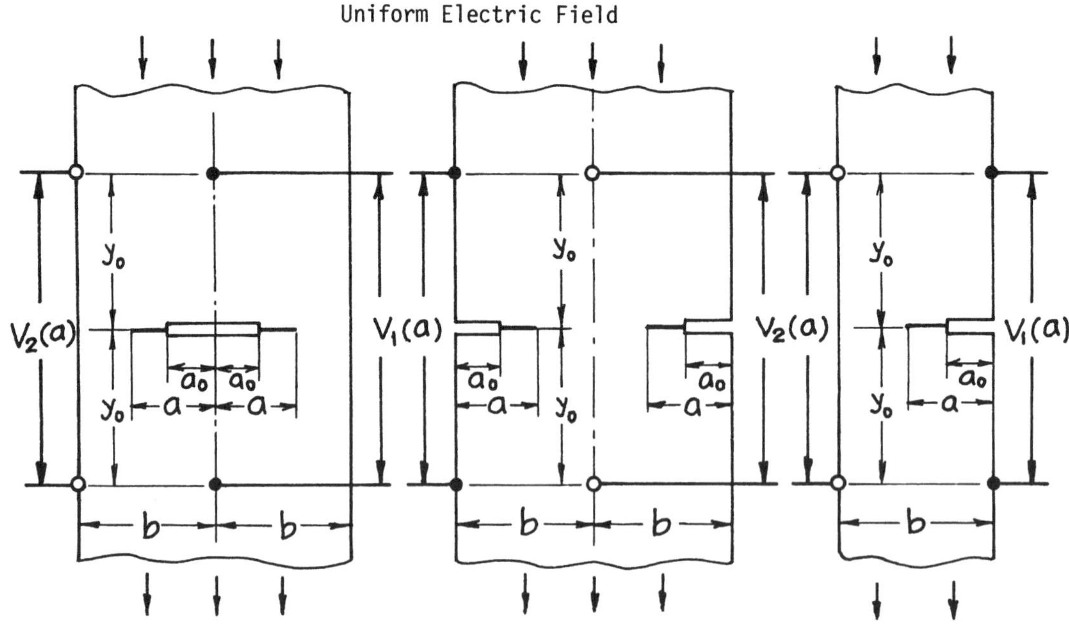

Electric Potentials: $V_1(a)$, $V_2(a)$
Potential Ratios:

$$\frac{V_1(a)}{V_1(a_0)} = \frac{\cosh^{-1}\left(\dfrac{\cosh\frac{\pi y_0}{2b}}{\cos\frac{\pi a}{2b}}\right)}{\cosh^{-1}\left(\dfrac{\cosh\frac{\pi y_0}{2b}}{\cos\frac{\pi a_0}{2b}}\right)}$$

$$\frac{V_2(a)}{V_2(a_0)} = \frac{\sinh^{-1}\left(\dfrac{\sinh\frac{\pi y_0}{2b}}{\cos\frac{\pi a}{2b}}\right)}{\sinh^{-1}\left(\dfrac{\sinh\frac{\pi y_0}{2b}}{\cos\frac{\pi a_0}{2b}}\right)}$$

Method: Conjugate Functions Method
Accuracy: Exact
References: **Johnson 1965; Tada 1973**

PART

III

Two-Dimensional Stress Solutions for Various Configurations with Cracks

- A. Cracks Along a Single Line
- B. Parallel Cracks
- C. Cracks and Holes or Notches
- D. Curved, Angled, Branched, or Radiating Cracks
- E. Cracks in Reinforced Plates

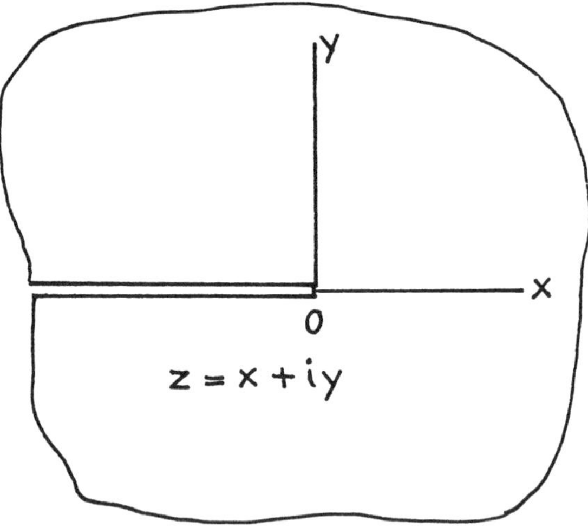

$$\begin{Bmatrix} Z_I(z) \\ Z_{II}(z) \\ Z_{III}(z) \end{Bmatrix} = \begin{Bmatrix} K_I \\ K_{II} \\ K_{III} \end{Bmatrix} \frac{1}{\sqrt{2\pi z}}$$

$$\begin{Bmatrix} \overline{Z}_I(z) \\ \overline{Z}_{II}(z) \\ \overline{Z}_{III}(z) \end{Bmatrix} = \frac{1}{\pi} \begin{Bmatrix} K_I \\ K_{II} \\ K_{III} \end{Bmatrix} \sqrt{2\pi z}$$

Method: Westergaard Stress Function
Accuracy: Exact
References: **Irwin 1958a** (see also **Williams 1957**)

NOTE: These Westergaard stress functions are the solutions for the crack-tip elastic field. That is, **Eqs. (1), (2), and (3)** in the text are directly derived from these functions by use of **Eqs. (38) and (39), Eqs. (55) and (56), and Eqs. (59) and (60),** respectively.

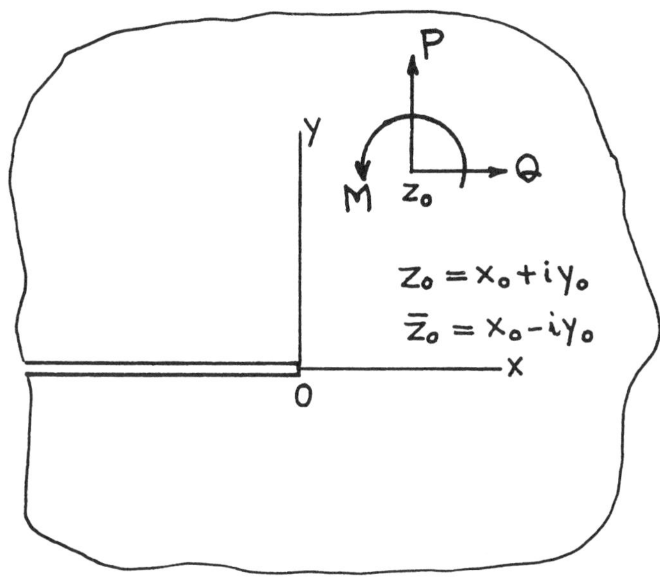

$$K = K_I - iK_{II}$$

$$= \frac{1}{\sqrt{2\pi}} \frac{1}{\kappa + 1} \left\{ (Q + iP) \left(\frac{1}{\sqrt{z_0}} - \kappa \frac{1}{\sqrt{\bar{z}_0}} \right) + \frac{(Q - iP)(\bar{z}_0 - z_0) + i(\kappa + 1)M}{2\,\bar{z}_0 \sqrt{\bar{z}_0}} \right\}$$

where

$$\kappa = \begin{cases} \frac{3-\nu}{1+\nu} & \text{plane stress} \\ 3 - 4\nu & \text{plane strain} \end{cases}$$

Method: Muskhelishvili's Method (Special Case of **page 5.3**)
Accuracy: Exact
References: **Erdogan 1962; Sih 1962a**

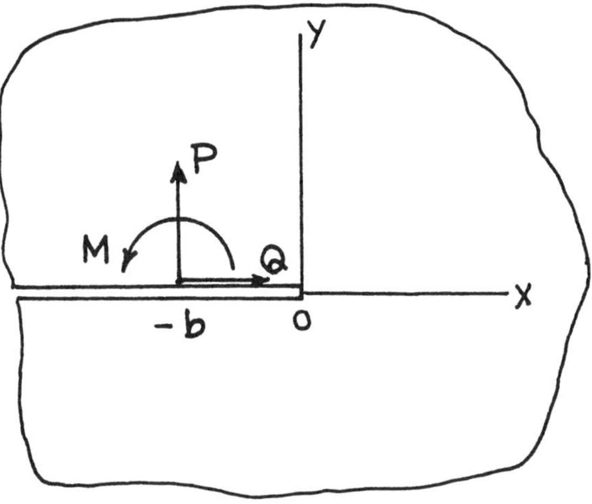

$$K_I = \frac{1}{\sqrt{2\pi b}}\left(P + \frac{M}{2b}\right)$$

$$K_{II} = \frac{1}{\sqrt{2\pi b}} Q$$

Method: Muskhelishvili's Method (Special Case of **page 5.4**)
Accuracy: Exact
Reference: **Erdogan 1962**

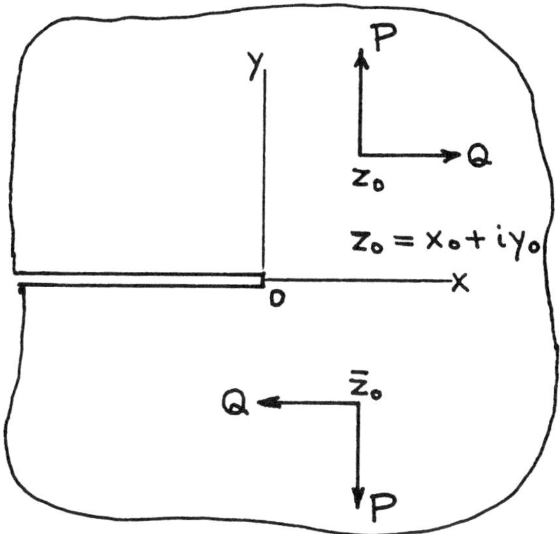

$$z = x + iy$$

$$\begin{Bmatrix} Z_I(z) \\ Z_{II}(z) \end{Bmatrix} = \frac{1}{\pi} \begin{Bmatrix} P \\ Q \end{Bmatrix} \left[1 \begin{Bmatrix} - \\ + \end{Bmatrix} \alpha y_0 \frac{\partial}{\partial y_0} \right] \frac{1}{2} \left(\frac{1}{z - z_0} \sqrt{\frac{-z_0}{z}} + \frac{1}{z - \bar{z}_0} \sqrt{\frac{-\bar{z}_0}{z}} \right)$$

$$\begin{Bmatrix} \overline{Z}_I(z) \\ \overline{Z}_{II}(z) \end{Bmatrix} = \frac{1}{\pi} \begin{Bmatrix} P \\ Q \end{Bmatrix} \left[1 \begin{Bmatrix} - \\ + \end{Bmatrix} \alpha y_0 \frac{\partial}{\partial y_0} \right] \left(\tan^{-1} \sqrt{\frac{z}{-z_0}} + \tan^{-1} \sqrt{\frac{z}{-\bar{z}_0}} \right)$$

$$\begin{Bmatrix} K_I \\ K_{II} \end{Bmatrix} = \frac{1}{\sqrt{2\pi}} \begin{Bmatrix} P \\ Q \end{Bmatrix} \left[1 \begin{Bmatrix} - \\ + \end{Bmatrix} \alpha y_0 \frac{\partial}{\partial y_0} \right] \left\{ \frac{1}{\sqrt{-z_0}} + \frac{1}{\sqrt{-\bar{z}_0}} \right\}$$

where

$$\alpha = \begin{cases} \frac{1}{2}(1 + \nu) & \text{plane stress} \\ \frac{1}{2}\left(\frac{1}{1-\nu}\right) & \text{plane strain} \end{cases}$$

Method: Westergaard Stress Function
Accuracy: Exact
References: **Tada 1972a, 1973**

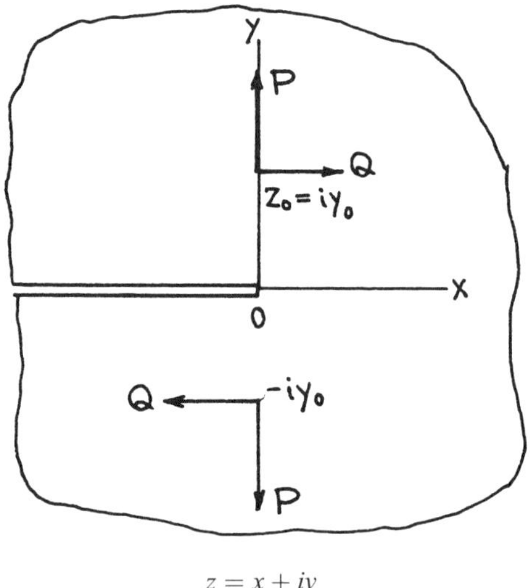

$z = x + iy$

$$\begin{Bmatrix} Z_I(z) \\ Z_{II}(z) \end{Bmatrix} = \frac{1}{\pi} \begin{Bmatrix} P \\ Q \end{Bmatrix} \left[1 \begin{Bmatrix} - \\ + \end{Bmatrix} \alpha y_0 \frac{\partial}{\partial y_0} \right] \frac{1}{\sqrt{2}} \frac{z + y_0}{z^2 + y_0^2} \sqrt{\frac{y_0}{z}}$$

$$\begin{Bmatrix} \overline{Z}_I(z) \\ \overline{Z}_{II}(z) \end{Bmatrix} = \frac{1}{\pi} \begin{Bmatrix} P \\ Q \end{Bmatrix} \left[1 \begin{Bmatrix} - \\ + \end{Bmatrix} \alpha y_0 \frac{\partial}{\partial y_0} \right] \frac{1}{2} \tan^{-1} \frac{\sqrt{2 y_0 z}}{y_0 - z}$$

$$\begin{Bmatrix} K_I \\ K_{II} \end{Bmatrix} = \frac{1}{\sqrt{\pi}} \begin{Bmatrix} P \\ Q \end{Bmatrix} \left[1 \begin{Bmatrix} - \\ + \end{Bmatrix} \alpha y_0 \frac{\partial}{\partial y_0} \right] \frac{1}{\sqrt{y_0}}$$
$$= \begin{Bmatrix} P \\ Q \end{Bmatrix} \left[1 \begin{Bmatrix} + \\ - \end{Bmatrix} \frac{\alpha}{2} \right] \frac{1}{\sqrt{\pi y_0}}$$

$$\underset{x<0}{\operatorname{Im} \begin{Bmatrix} \overline{Z}_I(x) \\ \overline{Z}_{II}(x) \end{Bmatrix}} = \frac{1}{2\pi} \begin{Bmatrix} P \\ Q \end{Bmatrix} \left[1 \begin{Bmatrix} - \\ + \end{Bmatrix} \alpha y_0 \frac{\partial}{\partial y_0} \right] \tanh^{-1} \frac{\sqrt{2 y_0 |x|}}{y_0 - x}$$
$$= \frac{1}{2\pi} \begin{Bmatrix} P \\ Q \end{Bmatrix} \left[\tanh^{-1} \frac{\sqrt{2 y_0 |x|}}{y_0 - x} \begin{Bmatrix} + \\ - \end{Bmatrix} \alpha \frac{x \sqrt{2 y_0 |x|}}{y_0^2 + x^2} \right]$$

where

$$\alpha = \begin{cases} \frac{1}{2}(1+\nu) & \text{plane stress} \\ \frac{1}{2}\left(\frac{1}{1-\nu}\right) & \text{plane strain} \end{cases}$$

Method: Westergaard Stress Function (Special Case of **page 3.4**)
Accuracy: Exact
References: **Tada 1972a, 1973**

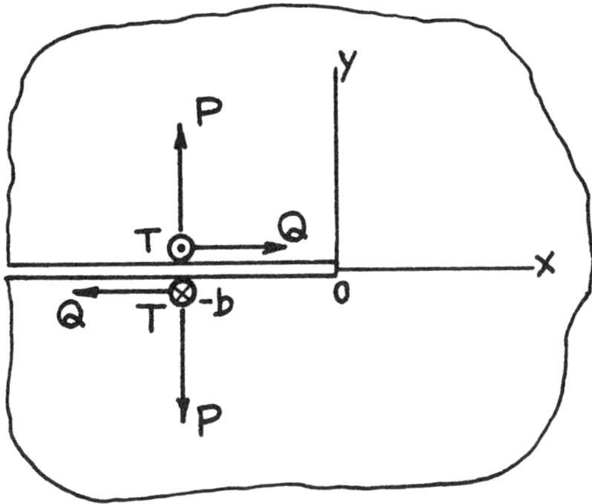

$$z = x + iy$$

$$\begin{Bmatrix} Z_I(z) \\ Z_{II}(z) \\ Z_{III}(z) \end{Bmatrix} = \frac{1}{\pi} \begin{Bmatrix} P \\ Q \\ T \end{Bmatrix} \frac{1}{z+b} \sqrt{\frac{b}{z}}$$

$$\begin{Bmatrix} \overline{Z}_I(z) \\ \overline{Z}_{II}(z) \\ \overline{Z}_{III}(z) \end{Bmatrix} = -\frac{2}{\pi} \begin{Bmatrix} P \\ Q \\ T \end{Bmatrix} \tan^{-1} \sqrt{\frac{b}{z}}$$

$$\begin{Bmatrix} K_I \\ K_{II} \\ K_{III} \end{Bmatrix} = \frac{\sqrt{2}}{\sqrt{\pi b}} \begin{Bmatrix} P \\ Q \\ T \end{Bmatrix}$$

$$\operatorname{Im} \begin{Bmatrix} \overline{Z}_I(x) \\ \overline{Z}_{II}(x) \\ \overline{Z}_{III}(x) \end{Bmatrix}_{x<0} = \frac{1}{\pi} \begin{Bmatrix} P \\ Q \\ T \end{Bmatrix} \ln \left| \frac{\sqrt{|x|} + \sqrt{b}}{\sqrt{|x|} - \sqrt{b}} \right|$$

$$\text{or } = \frac{2}{\pi} \begin{Bmatrix} P \\ Q \\ T \end{Bmatrix} \begin{cases} \tanh^{-1} \sqrt{|x|/b} & (-b < x \leq 0) \\ \coth^{-1} \sqrt{|x|/b} & (x < -b) \end{cases}$$

Methods: Westergaard Stress Function, etc.
Accuracy: Exact
References: **Irwin 1957**, etc.

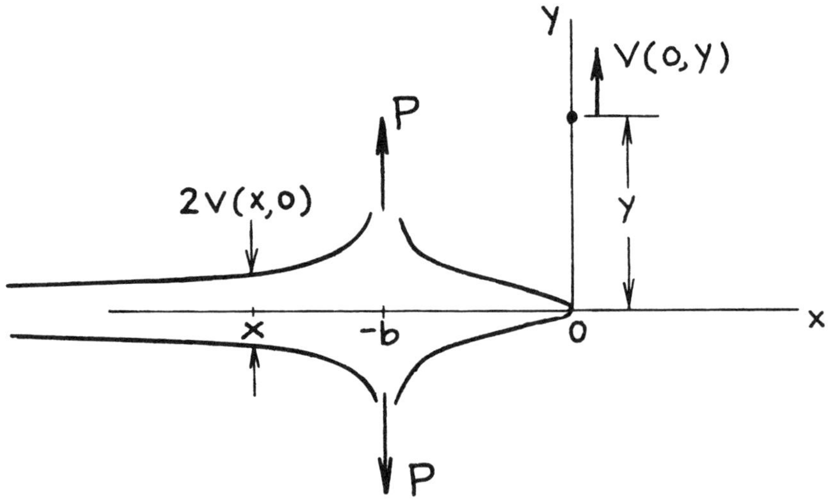

$$K_I = \frac{\sqrt{2}P}{\sqrt{\pi b}}$$

Crack Opening Profile

$$\underset{x \le 0}{2v(x,0)} = \frac{8P}{\pi E'} \begin{cases} \tanh^{-1}\sqrt{\frac{|x|}{b}} & -b < x \le 0 \\ \coth^{-1}\sqrt{\frac{|x|}{b}} & x < -b \end{cases}$$

Vertical Displacement at $(0, y)$

$$v(0,y) = \frac{P}{\pi E'}\left[\tanh^{-1}\frac{\sqrt{2yb}}{y+b} - \alpha\frac{b\sqrt{2yb}}{y^2+b^2}\right]$$

where

$$\alpha = \begin{cases} \frac{1}{2}(1+\nu) & \text{plane stress} \\ \frac{1}{2}\left(\frac{1}{1-\nu}\right) & \text{plane strain} \end{cases}$$

Methods: $v(x,0)$ Westergaard Function (see **page 3.6**); $v(0,y)$ Paris' Equation (see **Appendix B**) or Reciprocity (see **page 3.5**)
Accuracy: Exact
Reference: **Tada 1985**

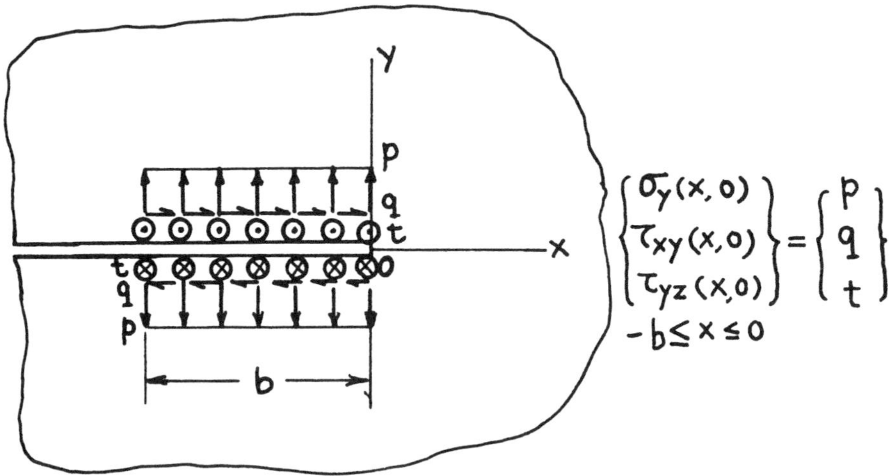

$$z = x + iy$$

$$\begin{Bmatrix} Z_I(z) \\ Z_{II}(z) \\ Z_{III}(z) \end{Bmatrix} = \frac{2}{\pi} \begin{Bmatrix} p \\ q \\ t \end{Bmatrix} \left\{ \sqrt{\frac{b}{z}} - \tan^{-1} \sqrt{\frac{b}{z}} \right\}$$

$$\begin{Bmatrix} \overline{Z}_I(z) \\ \overline{Z}_{II}(z) \\ \overline{Z}_{III}(z) \end{Bmatrix} = \frac{2}{\pi} \begin{Bmatrix} p \\ q \\ t \end{Bmatrix} b \left\{ \sqrt{\frac{z}{b}} - \left(1 + \frac{z}{b}\right) \tan^{-1} \sqrt{\frac{b}{z}} \right\}$$

$$\begin{Bmatrix} K_I \\ K_{II} \\ K_{III} \end{Bmatrix} = \frac{2}{\pi} \begin{Bmatrix} p \\ q \\ t \end{Bmatrix} \sqrt{2\pi b}$$

$$\operatorname{Im} \underset{x<0}{\begin{Bmatrix} \overline{Z}_I(x) \\ \overline{Z}_{II}(x) \\ \overline{Z}_{III}(x) \end{Bmatrix}} = \frac{2}{\pi} \begin{Bmatrix} p \\ q \\ t \end{Bmatrix} b \left\{ \sqrt{\frac{|x|}{b}} + \left(1 + \frac{x}{b}\right) \frac{1}{2} \ln \left| \frac{\sqrt{|x|} + \sqrt{b}}{\sqrt{|x|} - \sqrt{b}} \right| \right\}$$

$$\operatorname{Im} \underset{x=-b}{\begin{Bmatrix} \overline{Z}_I(x) \\ \overline{Z}_{II}(x) \\ \overline{Z}_{III}(x) \end{Bmatrix}} = \frac{2}{\pi} \begin{Bmatrix} p \\ q \\ t \end{Bmatrix} b$$

Methods: Westergaard Stress Function, etc. (Integration of **page 3.6**)
Accuracy: Exact
Reference: **Tada 1973**

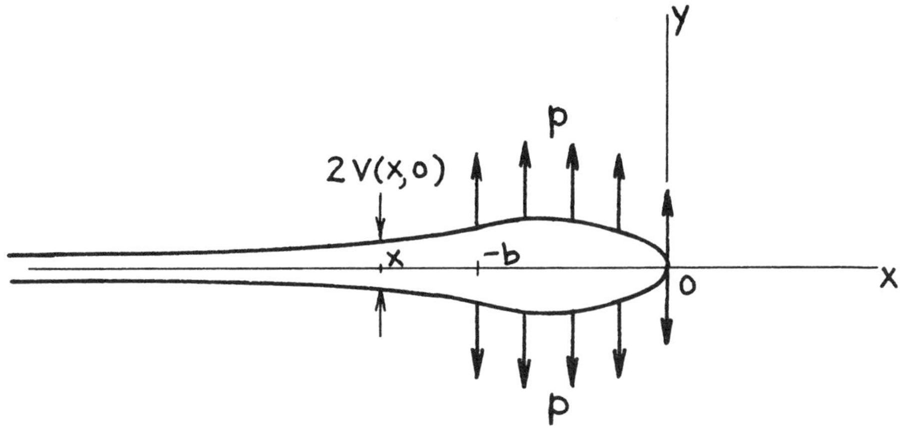

$$K_I = \frac{2\sqrt{2}}{\sqrt{\pi}} p\sqrt{b}$$

Crack Opening Profile

$$\underset{x \leq 0}{2v(x,0)} = \frac{8p}{\pi E'} b \left[\sqrt{\frac{|x|}{b}} + \left(1 + \frac{x}{b}\right) \begin{Bmatrix} \tanh^{-1} \sqrt{\frac{|x|}{b}} & -b < x \leq 0 \\ \coth^{-1} \sqrt{\frac{|x|}{b}} & x < -b \end{Bmatrix} \right]$$

Opening at $(-b, 0)$:

$$2v(-b, 0) = \frac{8p}{\pi E'} b$$

Method: Westergaard Stress Function (see **page 3.7**.)
Accuracy: Exact
Reference: **Tada 1985**

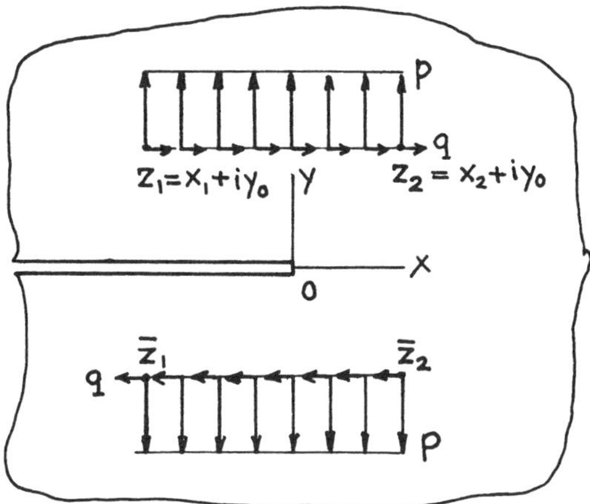

$$z = x + iy$$

$$\begin{Bmatrix} Z_I(z) \\ Z_{II}(z) \end{Bmatrix} = \frac{1}{\pi}\begin{Bmatrix} p \\ q \end{Bmatrix}\left[1\begin{Bmatrix} - \\ + \end{Bmatrix}\alpha y_0 \frac{\partial}{\partial y_0}\right]\left[\left\{\left(\sqrt{-z_2} - \sqrt{-\bar{z}_2}\right) - \left(\sqrt{-z_1} - \sqrt{-\bar{z}_1}\right)\right\}\frac{1}{\sqrt{z}}\right.$$
$$\left. - \left\{\left(\tan^{-1}\sqrt{\frac{-z_2}{z}} - \tan^{-1}\sqrt{\frac{-\bar{z}_2}{z}}\right) - \left(\tan^{-1}\sqrt{\frac{-z_1}{z}} - \tan^{-1}\sqrt{\frac{-\bar{z}_1}{z}}\right)\right\}\right]$$

$$\begin{Bmatrix} \bar{Z}_I(z) \\ \bar{Z}_{II}(z) \end{Bmatrix} = \frac{1}{\pi}\begin{Bmatrix} p \\ q \end{Bmatrix}\left[1\begin{Bmatrix} - \\ + \end{Bmatrix}\alpha y_0 \frac{\partial}{\partial y_0}\right]\left[\left\{\left(\sqrt{-z_2} - \sqrt{-\bar{z}_2}\right) - \left(\sqrt{-z_1} - \sqrt{-\bar{z}_1}\right)\right\}\sqrt{z}\right.$$
$$\left. - \left\{(z - z_2)\tan^{-1}\sqrt{\frac{-z_2}{z}} - (z - \bar{z}_2)\tan^{-1}\sqrt{\frac{-\bar{z}_2}{z}} - (z - z_1)\tan^{-1}\sqrt{\frac{-z_1}{z}} + (z - \bar{z}_1)\tan^{-1}\sqrt{\frac{-\bar{z}_1}{z}}\right\}\right]$$

$$\begin{Bmatrix} K_I \\ K_{II} \end{Bmatrix} = \frac{2}{\sqrt{2\pi}}\begin{Bmatrix} p \\ q \end{Bmatrix}\left[1\begin{Bmatrix} - \\ + \end{Bmatrix}\alpha y_0 \frac{\partial}{\partial y_0}\right]\left\{\left(\sqrt{-z_2} - \sqrt{-\bar{z}_2}\right) - \left(\sqrt{-z_1} - \sqrt{-\bar{z}_1}\right)\right\}$$

where

$$\alpha = \begin{cases} \frac{1}{2}(1 + \nu) & \text{plane stress} \\ \frac{1}{2}\left(\frac{1}{1-\nu}\right) & \text{plane strain} \end{cases}$$

Method: Westergaard Stress Function (Integration of **page 3.4**)
Accuracy: Exact
References: **Tada 1972a, 1973**

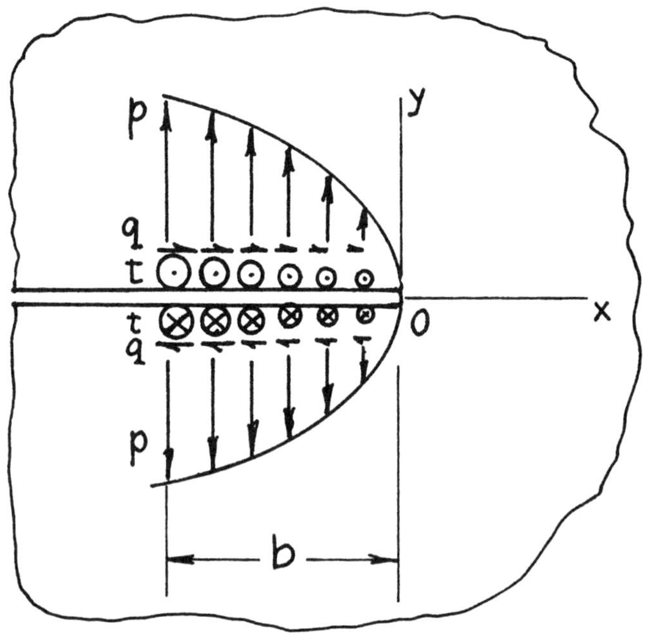

$$\left\{\begin{matrix}\sigma_y(x,0)\\ \tau_{xy}(x,0)\\ \tau_{yz}(x,0)\end{matrix}\right\}_{-b\le x\le 0} = \left\{\begin{matrix}p\\ q\\ t\end{matrix}\right\}\left(\frac{-x}{b}\right)^{\gamma} \quad \left(\gamma > -1/2\right)$$

$$\left\{\begin{matrix}K_I\\ K_{II}\\ K_{III}\end{matrix}\right\} = \frac{2}{\pi}\left\{\begin{matrix}p\\ q\\ t\end{matrix}\right\}\sqrt{2\pi b}\,\frac{1}{2\gamma+1}$$

Method: Integration of **page 3.6**
Accuracy: Exact
Reference: **Tada 1972a, 1974**

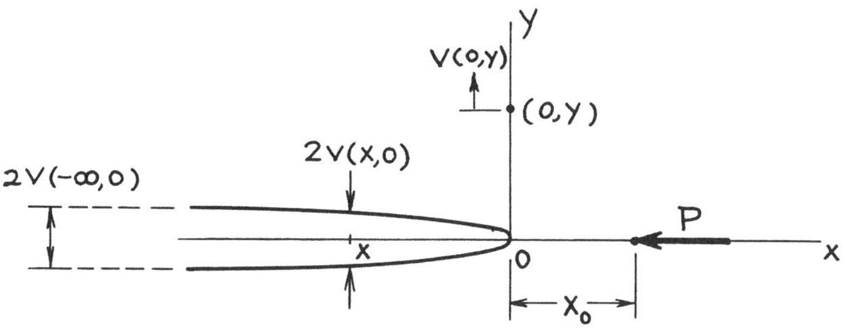

$$z = x + iy$$

$$Z_I(z) = \frac{P}{2\pi}(1-\alpha)\frac{\sqrt{x_0}}{(x_0 - z)\sqrt{z}}$$

$$\overline{Z}_I(z) = \frac{P}{\pi}(1-\alpha)\tanh^{-1}\sqrt{\frac{z}{x_0}}\left(=\frac{P}{2\pi}(1-\alpha)\cosh^{-1}\frac{z+x_0}{z-x_0}\right)$$

$$K_I = \frac{P}{\sqrt{2\pi x_0}}(1-\alpha)$$

Crack Opening Profile:

$$\underset{x\leq 0}{2v(x,0)} = \frac{4P}{\pi E'}(1-\alpha)\tan^{-1}\sqrt{\frac{|x|}{x_0}}$$

Opening at Infinity:

$$2v(-\infty,0) = \frac{2P}{E'}(1-\alpha)$$

Vertical Displacement at $(0, y)$:

$$v(0,y) = \frac{P}{\pi E'}(1-\alpha)\left\{\sin^{-1}\sqrt{\frac{2x_0 y}{x_0^2 + y^2}} - \frac{\alpha}{2}\frac{\sqrt{2x_0 y}(x_0 + y)}{x_0^2 + y^2}\right\}$$

where

$$\alpha = \begin{cases} \frac{1}{2}(1+\nu) & \text{Plane Stress} \\ \frac{1}{2}\left(\frac{1}{1-\nu}\right) & \text{Plane Strain} \end{cases}$$

Method: Integration of **pages 3.6, 3.6a** or Special Case of **page 5.19 or 5.20, or 24.19**
Accuracy: Exact
Reference: **Tada 1985**

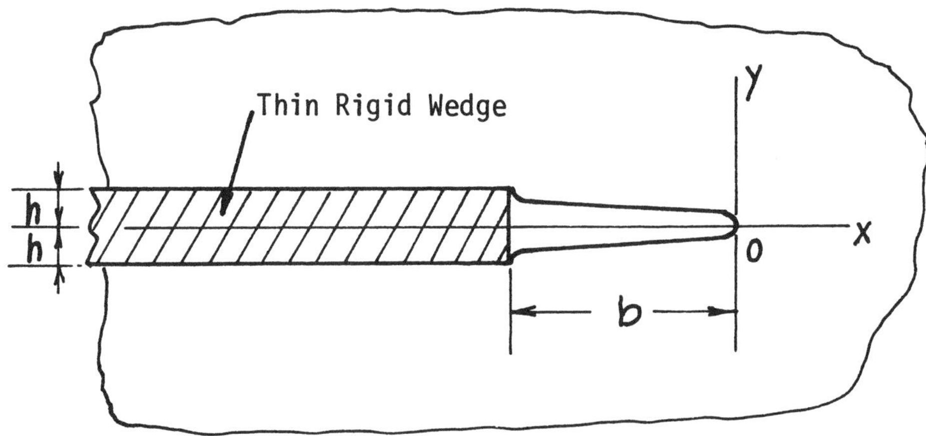

$$\underset{x=0}{K_I} = \frac{E'h}{\sqrt{2\pi b}}$$

$$\underset{x=-b}{K_I} = -\frac{E'h}{\sqrt{2\pi b}}$$

$$\underset{-b<x<0}{2v(x,0)} = \frac{4}{\pi} h \sin^{-1} \sqrt{\frac{-x}{b}}$$

where

$$E' = \begin{cases} E & \text{plane stress} \\ E/(1-\nu^2) & \text{plane strain} \end{cases}$$

Methods: Singular Integral Equation, Westergaard Stress Function or a Special Case of **page 4.15** or **page 5.21**
Accuracy: Exact
References: **Barenblatt 1962, Tada 1985**

The Westergaard Stress Functions are

$$Z_I(z) = \frac{E'h}{2\pi} \cdot \frac{1}{\sqrt{z(z+b)}}$$

$$\overline{Z}_I(z) = \frac{E'h}{2\pi} \sinh^{-1} \sqrt{\frac{z}{b}}$$

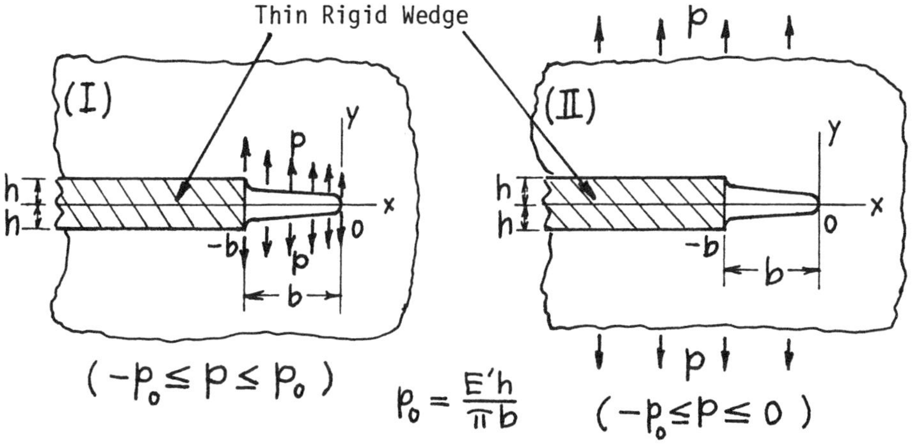

$$\underset{x=0}{K_I} = \frac{E'h}{\sqrt{2\pi b}} + p\sqrt{\pi \cdot b/2}$$

$$\underset{x=-b}{K_I} = -\frac{E'h}{\sqrt{2\pi b}} + p\sqrt{\pi \cdot b/2}$$

$$\underset{-b \leq x \leq 0}{2v(x,0)} = \frac{4}{\pi} h \sin^{-1}\sqrt{\frac{-x}{b}} + \frac{4p}{E'}\sqrt{-x(b-x)}$$

where

$$E' = \begin{cases} E & \text{plane stress} \\ E/(1-\nu^2) & \text{plane strain} \end{cases}$$

Method: Superposition of Solutions of **page 3.11** and **page 5.1**
Accuracy: Exact
References: **Tada 1974**

NOTE: 1. In (I), when $p > p_0$, separation of contact surfaces occurs near $x = -b$, and when $p < -p_0$, crack closure occurs near $x = 0$.
2. In (II), when $p > 0$ (remote tension), separation of contact surfaces occurs for large $-x$, and when $p < -p_0$, crack closure occurs near $x = 0$.

The Westergaard Functions are

$$\left\{ \begin{matrix} Z_I(z) \\ \overline{Z}_I(z) \end{matrix} \right\} = \left\{ \begin{matrix} Z_I(z) \\ \overline{Z}_I(z) \end{matrix} \right\} + \left\{ \begin{matrix} Z_I(z) \\ \overline{Z}_I(z) \end{matrix} \right\}$$

p. 3.12 **p. 3.11** **p. 5.1** (replace σ by p)

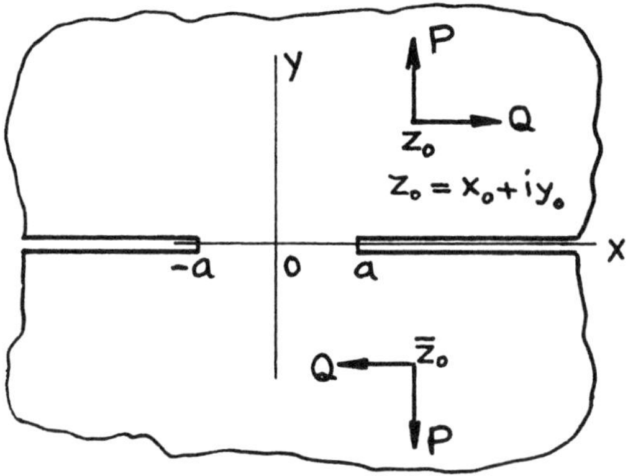

$$z = x + iy$$

$$\left\{ \begin{matrix} Z_I(z) \\ Z_{II}(z) \end{matrix} \right\} = \frac{1}{\pi} \left\{ \begin{matrix} P \\ Q \end{matrix} \right\} \left[1 \left\{ \begin{matrix} - \\ + \end{matrix} \right\} \alpha y_0 \frac{\partial}{\partial y_0} \right] \frac{1}{\sqrt{a^2 - z^2}} \left[\frac{1}{2} \left(\frac{\sqrt{z_0^2 - a^2}}{z_0 - z} - \frac{\sqrt{\bar{z}_0^2 - a^2}}{\bar{z}_0 - z} \right) + \left\{ \begin{matrix} 1 \\ 0 \end{matrix} \right\} \frac{z}{a^2} \left(\sqrt{z_0^2 - a^2} + \sqrt{\bar{z}_0^2 - a^2} \right) \right]$$

$$\left\{ \begin{matrix} \bar{Z}_I(z) \\ \bar{Z}_{II}(z) \end{matrix} \right\} = \frac{1}{\pi} \left\{ \begin{matrix} P \\ Q \end{matrix} \right\} \left[1 \left\{ \begin{matrix} - \\ + \end{matrix} \right\} \alpha y_0 \frac{\partial}{\partial y_0} \right] \left[\frac{1}{2} \left(\sin^{-1} \frac{z_0 z - a^2}{a(z_0 - z)} + \sin^{-1} \frac{\bar{z}_0 z - a^2}{a(\bar{z}_0 - z)} \right) - \left\{ \begin{matrix} 1 \\ 0 \end{matrix} \right\} \frac{\sqrt{a^2 - z^2}}{a^2} \left(\sqrt{z_0^2 - a^2} + \sqrt{\bar{z}_0^2 - a^2} \right) \right]$$

$$K_{I_{\pm a}} = \frac{P}{\sqrt{\pi a}} \left[1 - \alpha y_0 \frac{\partial}{\partial y_0} \right] \left[\frac{1}{2} \left(\sqrt{\frac{z_0 \pm a}{z_0 \mp a}} + \sqrt{\frac{\bar{z}_0 \pm a}{\bar{z}_0 \mp a}} \right) \pm \frac{1}{a} \left(\sqrt{z_0^2 - a^2} + \sqrt{\bar{z}_0^2 - a^2} \right) \right]$$

$$K_{II_{\pm a}} = \frac{Q}{\sqrt{\pi a}} \left[1 + \alpha y_0 \frac{\partial}{\partial y_0} \right] \frac{1}{2} \left(\sqrt{\frac{z_0 \pm a}{z_0 \mp a}} + \sqrt{\frac{\bar{z}_0 \pm a}{\bar{z}_0 \mp a}} \right)$$

where

$$\alpha = \begin{cases} \frac{1}{2}(1 + \nu) & \text{plane stress} \\ \frac{1}{2}\left(\frac{1}{1-\nu}\right) & \text{plane strain} \end{cases}$$

Method: Westergaard Stress Function
Accuracy: Exact
Reference: **Tada 2000**

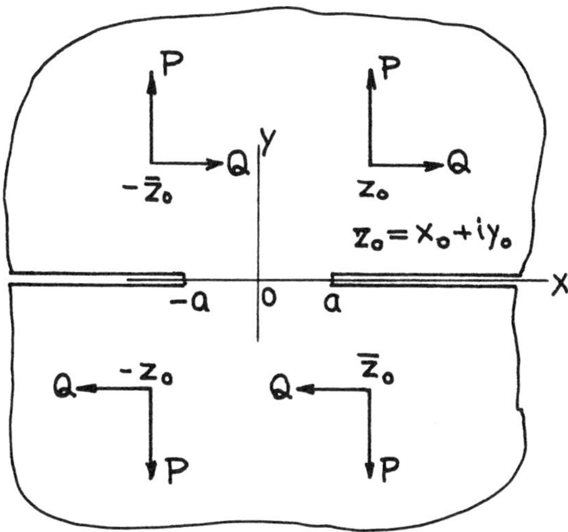

$$z = x + iy$$

$$\begin{Bmatrix} Z_I(z) \\ Z_{II}(z) \end{Bmatrix} = \frac{1}{\pi}\begin{Bmatrix} P \\ Q \end{Bmatrix}\left[1\begin{Bmatrix} - \\ + \end{Bmatrix}\alpha y_0 \frac{\partial}{\partial y_0}\right]\left(\frac{z_0}{z_0^2 - z^2}\sqrt{z_0^2 - a^2} + \frac{\bar{z}_0}{\bar{z}_0^2 - z^2}\sqrt{\bar{z}_0^2 - a^2}\right)\frac{1}{\sqrt{a^2 - z^2}}$$

$$\begin{Bmatrix} \bar{Z}_I(z) \\ \bar{Z}_{II}(z) \end{Bmatrix} = \frac{1}{\pi}\begin{Bmatrix} P \\ Q \end{Bmatrix}\left[1\begin{Bmatrix} - \\ + \end{Bmatrix}\alpha y_0 \frac{\partial}{\partial y_0}\right]\left\{\tan^{-1}\sqrt{\frac{(a/z)^2 - 1}{1 - (a/z_0)^2}} + \tan^{-1}\sqrt{\frac{(a/z)^2 - 1}{1 - (a/\bar{z}_0)^2}}\right\}$$

$$\begin{Bmatrix} K_I \\ K_{II} \end{Bmatrix} = \frac{1}{\sqrt{\pi a}}\begin{Bmatrix} P \\ Q \end{Bmatrix}\left[1\begin{Bmatrix} - \\ + \end{Bmatrix}\alpha y_0 \frac{\partial}{\partial y_0}\right]\left\{\frac{z_0}{\sqrt{z_0^2 - a^2}} + \frac{\bar{z}_0}{\sqrt{\bar{z}_0^2 - a^2}}\right\}$$

where

$$\alpha = \begin{cases} \frac{1}{2}(1 + \nu) & \text{plane stress} \\ \frac{1}{2}\left(\frac{1}{1-\nu}\right) & \text{plane strain} \end{cases}$$

Method: Westergaard Stress Function (Superposition of **page 4.1**)
Accuracy: Exact
References: **Tada 1972a, 1973**

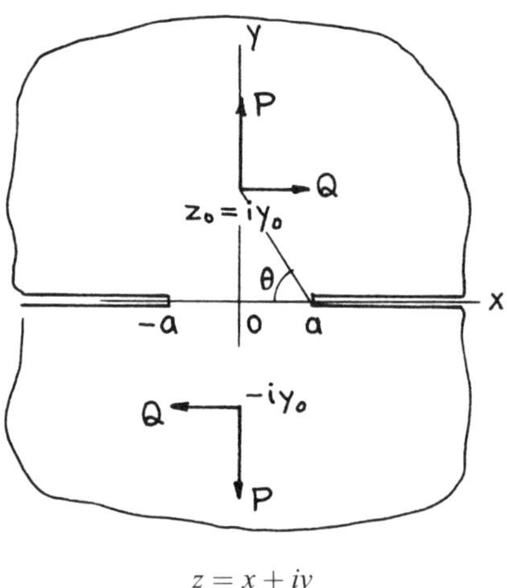

$z = x + iy$

$$\begin{Bmatrix} Z_I(z) \\ Z_{II}(z) \end{Bmatrix} = \frac{1}{\pi}\begin{Bmatrix} P \\ Q \end{Bmatrix}\left[1\begin{Bmatrix} - \\ + \end{Bmatrix}\alpha y_0 \frac{\partial}{\partial y_0}\right]\frac{y_0}{z^2+y_0^2}\sqrt{\frac{a^2+y_0^2}{a^2-z^2}}$$

$$\begin{Bmatrix} \overline{Z}_I(z) \\ \overline{Z}_{II}(z) \end{Bmatrix} = \frac{1}{\pi}\begin{Bmatrix} P \\ Q \end{Bmatrix}\left[1\begin{Bmatrix} - \\ + \end{Bmatrix}\alpha y_0 \frac{\partial}{\partial y_0}\right]\tan^{-1}\sqrt{\frac{(a/z)^2-1}{1+(a/y_0)^2}}$$

$$\begin{Bmatrix} K_I \\ K_{II} \end{Bmatrix} = \frac{1}{\sqrt{\pi a}}\begin{Bmatrix} P \\ Q \end{Bmatrix}\left[1\begin{Bmatrix} - \\ + \end{Bmatrix}\alpha y_0 \frac{\partial}{\partial y_0}\right]\frac{y_0}{\sqrt{a^2+y_0^2}}$$

$$= \frac{1}{\sqrt{\pi a}}\begin{Bmatrix} P \\ Q \end{Bmatrix}\frac{y_0}{\sqrt{a^2+y_0^2}}\left[1\begin{Bmatrix} - \\ + \end{Bmatrix}\alpha\frac{a^2}{a^2+y_0^2}\right]\left(= \frac{1}{\sqrt{\pi a}}\begin{Bmatrix} P \\ Q \end{Bmatrix}\sin\theta\left[1\begin{Bmatrix} - \\ + \end{Bmatrix}\alpha\cos^2\theta\right]\right)$$

$$\underset{|x|\geq a}{\mathrm{Im}\begin{Bmatrix} \overline{Z}_I(x) \\ \overline{Z}_{II}(x) \end{Bmatrix}} = \frac{1}{\pi}\begin{Bmatrix} P \\ Q \end{Bmatrix}\left[1\begin{Bmatrix} - \\ + \end{Bmatrix}\alpha y_0 \frac{\partial}{\partial y_0}\right]\tanh^{-1}\sqrt{\frac{1-(a/x)^2}{1+(a/y_0)^2}}$$

$$= \frac{1}{\pi}\begin{Bmatrix} P \\ Q \end{Bmatrix}\left[\tanh^{-1}\sqrt{\frac{1-(a/x)^2}{1+(a/y_0)^2}}\begin{Bmatrix} - \\ + \end{Bmatrix}\alpha\frac{y_0 x}{x^2+y_0^2}\sqrt{\frac{x^2-a^2}{a^2+y_0^2}}\right]$$

where

$$\alpha = \begin{cases} \frac{1}{2}(1+\nu) & \text{plane stress} \\ \frac{1}{2}\left(\frac{1}{1-\nu}\right) & \text{plane strain} \end{cases}$$

Method: Westergaard Stress Function (Special Case of **page 4.1**)
Accuracy: Exact
References: **Tada 1972a, 1973**

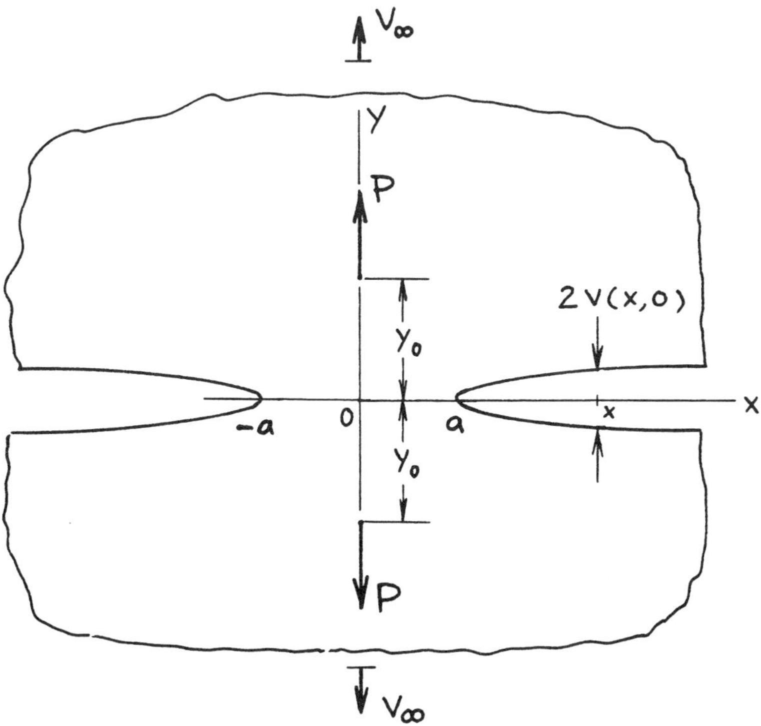

$$K_I = \frac{P}{\sqrt{\pi a}} \frac{y_0}{\sqrt{a^2 + y_0^2}} \left(1 - \alpha \frac{y_0}{\sqrt{a^2 + y_0^2}} \right)$$

Crack Opening Profile:

$$\underset{|x| \geq a}{2v(x, 0)} = \frac{4P}{\pi E'} \left(\tanh^{-1} \sqrt{\frac{1 - (a/x)^2}{1 + (a/y_0)^2}} - \alpha \frac{x^2}{x^2 + y_0^2} \sqrt{\frac{1 - (a/x)^2}{1 + (a/y_0)^2}} \right)$$

Relative Vertical Displacement at Infinity:

$$2v_\infty = \frac{4P}{\pi E'} \left(\sinh^{-1} \frac{y_0}{a} - \alpha \frac{y_0}{\sqrt{a^2 + y_0^2}} \right)$$

$$2v_\infty = 2v(x, 0)_{|x| \to \infty}$$

Methods: $v(x, 0)$ Westergaard Stress Function (see **page 4.3**); v_∞ Paris' Equation (see **Appendix B**) or Reciprocity (see **page 4.9**)
Accuracy: Exact
Reference: **Tada 1985**

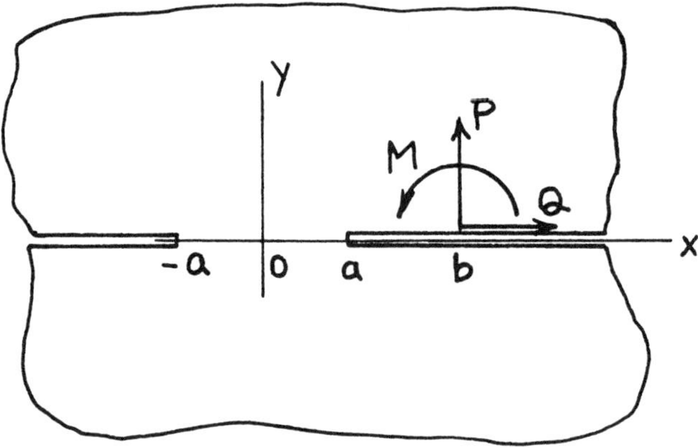

$$K_{I\pm a} = \frac{P}{2\sqrt{\pi a}}\sqrt{\frac{b\pm a}{b\mp a}} \mp \frac{M}{2\sqrt{\pi a}}\frac{a}{(b\mp a)\sqrt{b^2-a^2}}$$

$$K_{II\pm a} = \frac{Q}{2\sqrt{\pi a}}\sqrt{\frac{b\pm a}{b\mp a}}$$

Method: Muskhelishvili's Method (Special Case of **page 6.2**)
Accuracy: Exact
Reference: **Erdogan 1962**

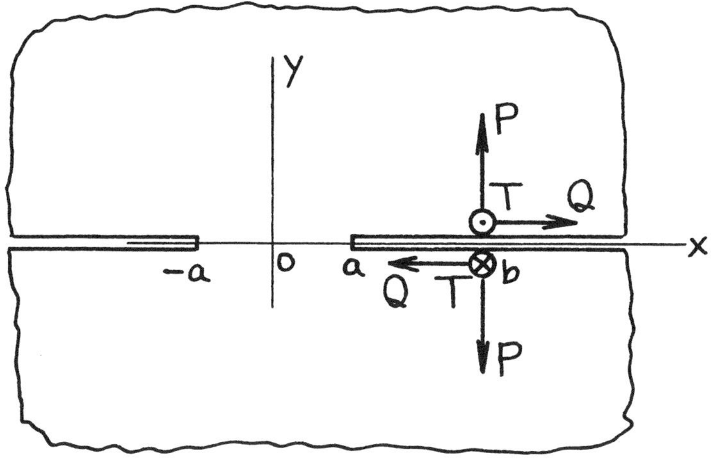

$$z = x + iy$$

$$\begin{Bmatrix} Z_I(z) \\ Z_{II}(z) \\ Z_{III}(z) \end{Bmatrix} = \frac{1}{\pi} \begin{Bmatrix} P \\ Q \\ T \end{Bmatrix} \frac{\sqrt{b^2 - a^2}}{\sqrt{a^2 - z^2}} \left(\frac{1}{b - z} + \frac{2}{a^2} z \begin{Bmatrix} 1 \\ 0 \\ 1 \end{Bmatrix} \right)$$

$$\begin{Bmatrix} \overline{Z}_I(z) \\ \overline{Z}_{II}(z) \\ \overline{Z}_{III}(z) \end{Bmatrix} = \frac{1}{\pi} \begin{Bmatrix} P \\ Q \\ T \end{Bmatrix} \left[\sin^{-1} \frac{bz - a^2}{a(b - z)} - \frac{2}{a^2} \sqrt{b^2 - a^2} \sqrt{a^2 - z^2} \begin{Bmatrix} 1 \\ 0 \\ 1 \end{Bmatrix} \right]$$

$$\begin{Bmatrix} K_I \\ K_{II} \\ K_{III} \end{Bmatrix}_{\pm a} = \frac{1}{\sqrt{\pi a}} \begin{Bmatrix} P \\ Q \\ T \end{Bmatrix} \sqrt{b^2 - a^2} \left(\frac{1}{b \mp a} \pm \frac{2}{a} \begin{Bmatrix} 1 \\ 0 \\ 1 \end{Bmatrix} \right)$$

$$\operatorname{Im} \begin{Bmatrix} \overline{Z}_I(x) \\ \overline{Z}_{II}(x) \\ \overline{Z}_{III}(x) \end{Bmatrix}_{|x| \geq a} = \frac{1}{\pi} \begin{Bmatrix} P \\ Q \\ T \end{Bmatrix} \left[\cosh^{-1} \left| \frac{bx - a^2}{a(b - x)} \right| + 2 \frac{x}{a} \sqrt{\left(\frac{b}{a}\right)^2 - 1} \sqrt{1 - \left(\frac{a}{x}\right)^2} \begin{Bmatrix} 1 \\ 0 \\ 1 \end{Bmatrix} \right]$$

Method: Westergaard Stress Function
Accuracy: Exact
Reference: **Tada 1985**

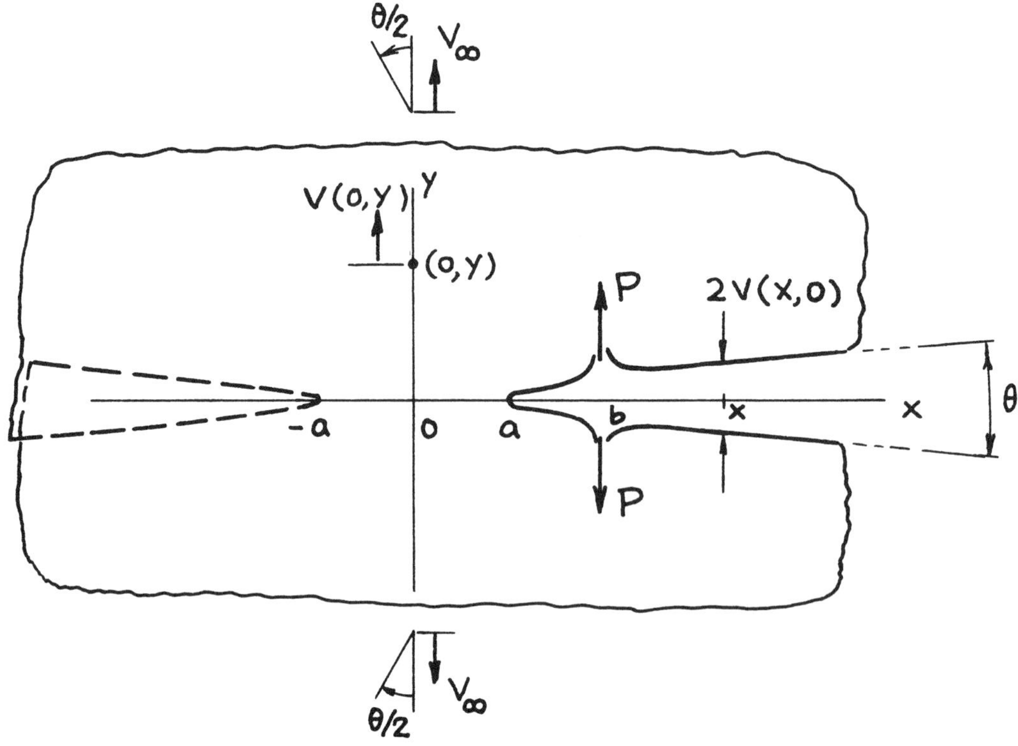

$$K_{I\pm a} = \frac{P}{\sqrt{\pi a}}\sqrt{b^2 - a^2}\left(\frac{1}{b \mp a} \pm \frac{2}{a}\right)$$

Crack Opening Profile:

$$\underset{|x| \geq a}{2v(x,0)} = \frac{4P}{\pi E'}\left[\cosh^{-1}\left|\frac{bx - a^2}{a(b - x)}\right| + \frac{x}{a}\sqrt{\left(\frac{b}{a}\right)^2 - 1}\sqrt{1 - \left(\frac{a}{x}\right)^2}\right]$$

Vertical Displacement at $(0, y)$:

$$v(0,y) = \frac{2P}{\pi E'}\left[\tanh^{-1}\sqrt{\frac{1 - (a/b)^2}{1 + (a/y)^2}} - \alpha\frac{b^2}{b^2 + y^2}\sqrt{\frac{1 - (a/b)^2}{1 + (a/y)^2}}\right]$$

Relative Vertical Displacement along y-Axis at Infinity:

$$2v_\infty = \frac{4P}{\pi E'}\cosh^{-1}\frac{b}{a}$$

Relative Rotation at Infinity:

$$\theta = \frac{8P}{\pi E'}\frac{\sqrt{b^2 - a^2}}{a^2}\left(= \left\{\frac{2v(x,0)}{x}\right\}_{x\to\infty}\right)$$

where

$$\alpha = \begin{cases} \frac{1}{2}(1+\nu) & \text{plane stress} \\ \frac{1}{2}\left(\frac{1}{1-\nu}\right) & \text{plane strain} \end{cases}$$

Methods: v Reciprocity (see **page 4.3a**); θ From **page 4.10a** with $M = P\sqrt{b^2 - a^2}$ (or Paris' Equation — see **Appendix B**)

Accuracy: Exact

Reference: **Tada 1985**

NOTE: 1. Always $K_{I-a} < 0$.
2. No surface interference ($x \leq -a$) was considered.

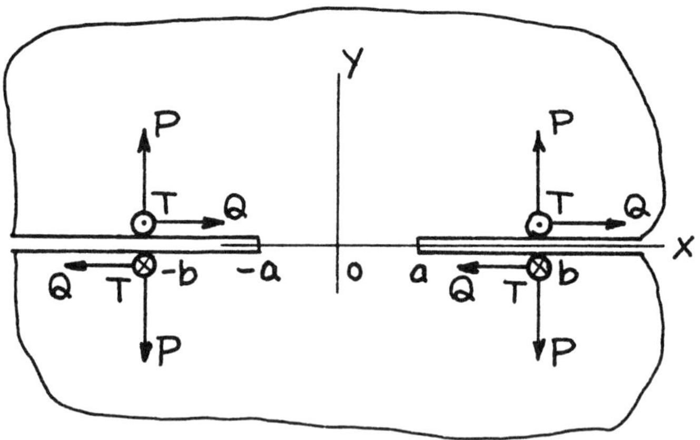

$$z = x + iy$$

$$\left\{ \begin{matrix} Z_I(z) \\ Z_{II}(z) \\ Z_{III}(z) \end{matrix} \right\} = \frac{2}{\pi} \left\{ \begin{matrix} P \\ Q \\ T \end{matrix} \right\} \frac{b\sqrt{b^2 - a^2}}{\left(b^2 - z^2\right)\sqrt{a^2 - z^2}}$$

$$\left\{ \begin{matrix} \overline{Z}_I(z) \\ \overline{Z}_{II}(z) \\ \overline{Z}_{III}(z) \end{matrix} \right\} = \frac{2}{\pi} \left\{ \begin{matrix} P \\ Q \\ T \end{matrix} \right\} \tan^{-1} \sqrt{\frac{1 - (a/b)^2}{(a/z)^2 - 1}}$$

$$\left\{ \begin{matrix} K_I \\ K_{II} \\ K_{III} \end{matrix} \right\} = \frac{2}{\sqrt{\pi a}} \left\{ \begin{matrix} P \\ Q \\ T \end{matrix} \right\} \frac{b}{\sqrt{b^2 - a^2}}$$

$$\operatorname{Im} \underset{|x| \geq a}{\left\{ \begin{matrix} \overline{Z}_I(x) \\ \overline{Z}_{II}(x) \\ \overline{Z}_{III}(x) \end{matrix} \right\}} = \frac{1}{\pi} \left\{ \begin{matrix} P \\ Q \\ T \end{matrix} \right\} \ln \left| \frac{\sqrt{1 - (a/b)^2} + \sqrt{1 - (a/x)^2}}{\sqrt{1 - (a/b)^2} - \sqrt{1 - (a/x)^2}} \right|$$

Methods: Superposition of **page 4.5** (or Special Case of Periodic Cracks, **page 7.7**)
Accuracy: Exact
References: **Erdogan 1962; Tada 1973**

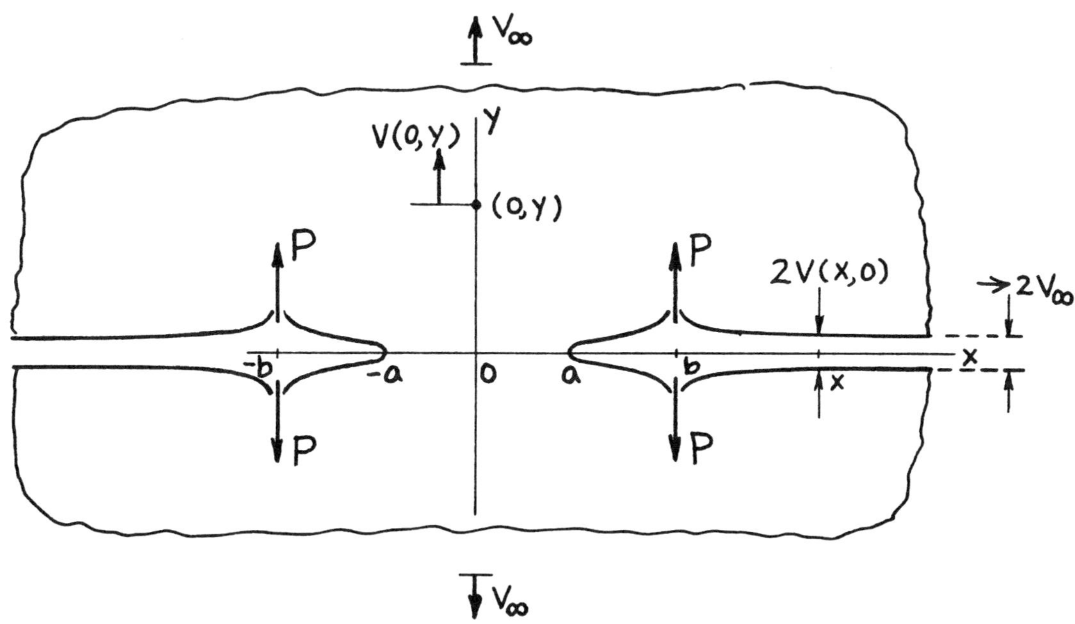

$$K_I = \frac{2P}{\sqrt{\pi a}} \frac{b}{\sqrt{b^2 - a^2}}$$

Crack Opening Profile:

$$\underset{|x| \geq a}{2v(x,0)} = \frac{8P}{\pi E'} \begin{Bmatrix} \tanh^{-1} \\ \coth^{-1} \end{Bmatrix} \sqrt{\frac{1-(a/x)^2}{1-(a/b)^2}} \quad \begin{matrix} a \leq |x| < b \\ |x| > b \end{matrix}$$

Vertical Displacement at $(0, y)$:

$$v(0,y) = \frac{4P}{\pi E'} \left[\tanh^{-1} \sqrt{\frac{1-(a/b)^2}{1+(a/y)^2}} - \alpha \frac{b^2}{b^2+y^2} \sqrt{\frac{1-(a/b)^2}{1+(a/y)^2}} \right]$$

Relative Vertical Displacement at Infinity:

$$2v_\infty = \frac{8P}{\pi E'} \cosh^{-1} \frac{b}{a}$$

$$2v_\infty = \underset{y \to \infty}{2v(0,y)} = \underset{x \to \infty}{2v(x,0)}$$

where

$$\alpha = \begin{cases} \frac{1}{2}(1+\nu) & \text{plane stress} \\ \frac{1}{2}\left(\frac{1}{1-\nu}\right) & \text{plane strain} \end{cases}$$

Method: v Paris' Equation (see **Appendix B**) (or Reciprocity — see **page 4.3a**)
Accuracy: Exact
Reference: **Tada 1985**

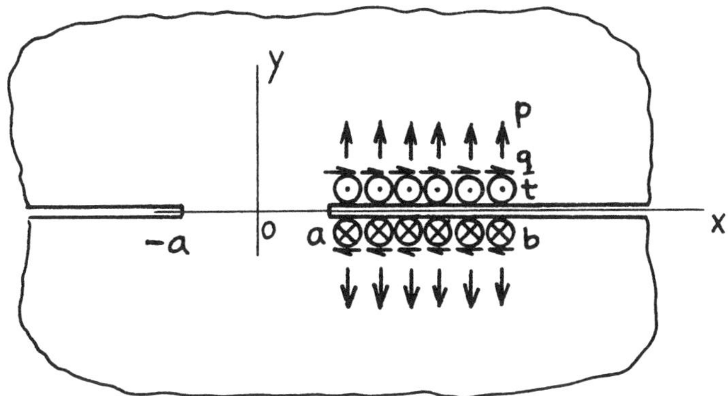

$$z = x + iy$$

$$\begin{Bmatrix} Z_I(z) \\ Z_{II}(z) \\ Z_{III}(z) \end{Bmatrix} = \frac{1}{\pi} \begin{Bmatrix} p \\ q \\ t \end{Bmatrix} \left\{ \frac{\sqrt{b^2 - a^2} + f\left(\frac{b}{a}\right) z}{\sqrt{a^2 - z^2}} - \cos^{-1} \frac{bz - a^2}{a(b - z)} \right\}$$

$$\begin{Bmatrix} \overline{Z}_I(z) \\ \overline{Z}_{II}(z) \\ \overline{Z}_{III}(z) \end{Bmatrix} = \frac{1}{\pi} \begin{Bmatrix} p \\ q \\ t \end{Bmatrix} \left\{ -f\left(\frac{b}{a}\right) \sqrt{a^2 - z^2} + (b - z) \sin^{-1} \frac{bz - a^2}{a(b - z)} \right\}$$

$$\begin{Bmatrix} K_I \\ K_{II} \\ K_{III} \end{Bmatrix}_{\pm a} = \frac{1}{\sqrt{\pi a}} \begin{Bmatrix} p \\ q \\ t \end{Bmatrix} \left\{ \sqrt{b^2 - a^2} \pm a f\left(\frac{b}{a}\right) \right\}$$

$$\operatorname{Im} \begin{Bmatrix} \overline{Z}_I(x) \\ \overline{Z}_{II}(x) \\ \overline{Z}_{III}(x) \end{Bmatrix}_{|x| \geq a} = \frac{1}{\pi} \begin{Bmatrix} p \\ q \\ t \end{Bmatrix} \left\{ f\left(\frac{b}{a}\right) \cdot x \sqrt{1 - \left(\frac{a}{x}\right)^2} + (b - x) \cosh^{-1} \left| \frac{bx - a^2}{a(b - x)} \right| \right\}$$

where
$$f\left(\frac{b}{a}\right) = \begin{Bmatrix} \frac{b}{a} \sqrt{\left(\frac{b}{a}\right)^2 - 1} \\ \cosh^{-1} \frac{b}{a} \\ \frac{b}{a} \sqrt{\left(\frac{b}{a}\right)^2 - 1} \end{Bmatrix}$$

Method: Westergaard Stress Function (Integration of **page 4.5**)
Accuracy: Exact
Reference: **Tada 1985**

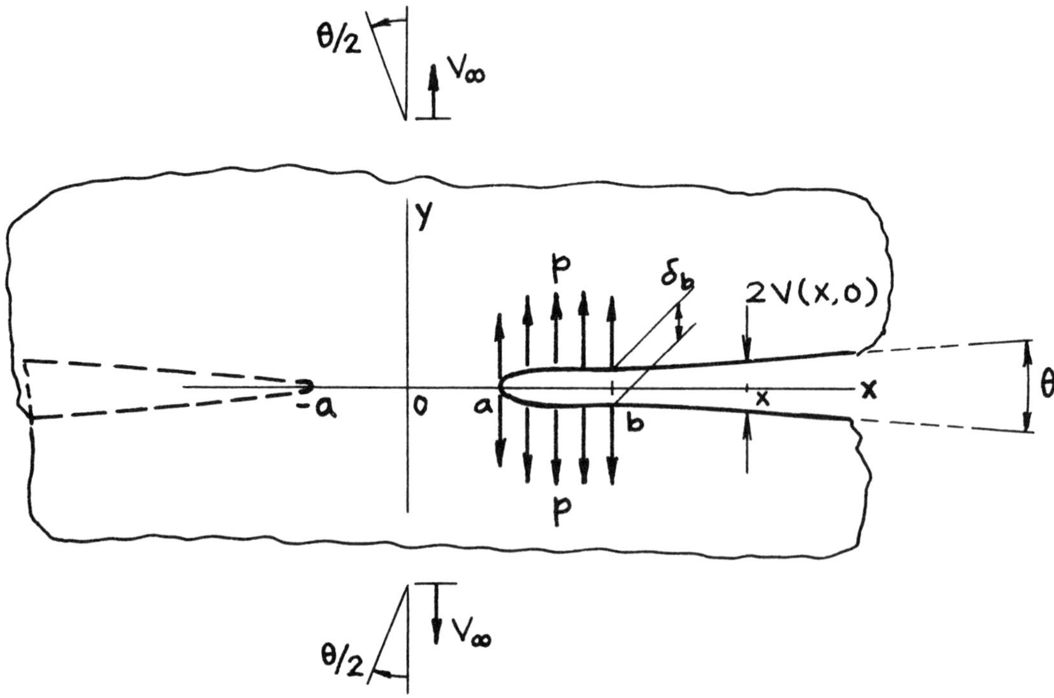

$$K_{I\pm a} = \frac{p}{\sqrt{\pi a}}\sqrt{b^2 - a^2}\left(1 \pm \frac{b}{a}\right)$$

Crack Opening Profile:

$$\underset{|x|\geq a}{2v\,(x,0)} = \frac{4p}{\pi E'}\left\{(b-x)\cosh^{-1}\left|\frac{bx-a^2}{a(b-x)}\right| + b\frac{x}{a}\sqrt{\left(\frac{b}{a}\right)^2 - 1}\sqrt{1-\left(\frac{a}{x}\right)^2}\right\}$$

Opening at $x = b$:

$$\delta_b = 2\,v(b,0) = \frac{4pb}{\pi E'}\left\{\ln\frac{b}{a} + \left(\frac{b}{a}\right)^2 - 1\right\}$$

Relative Vertical Displacement along y-Axis at Infinity:

$$2\,v_\infty = \frac{4pb}{\pi E'}\left\{\cosh^{-1}\frac{b}{a} - \sqrt{1-\left(\frac{a}{b}\right)^2}\right\}$$

Relative Rotation at Infinity:

$$\theta = \frac{4p}{\pi E'}\left\{\frac{b}{a}\sqrt{\left(\frac{b}{a}\right)^2 - 1} - \cosh^{-1}\frac{b}{a}\right\}\left(= \left\{\frac{2\,v(x,0)}{x}\right\}_{x\to\infty}\right)$$

Method: Westergaard Stress Function (Integration of **pages 4.5 and 4.5a,b**)
Accuracy: Exact
Reference: **Tada 1985**

NOTE: 1. Always $K_{I-a} < 0$.
2. No surface interference $(x \leq -a)$ was considered.

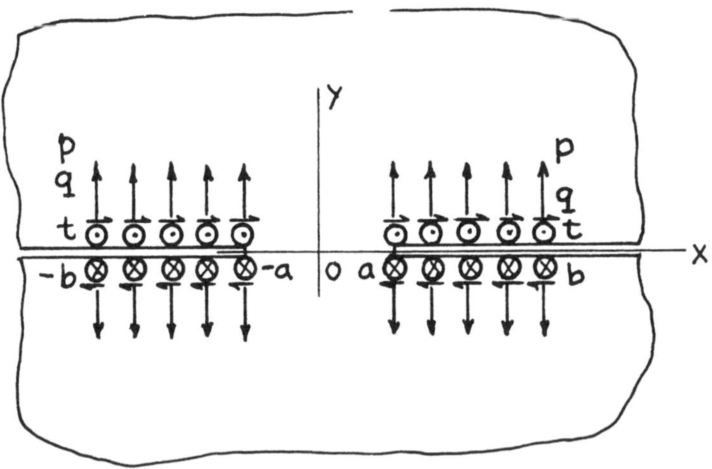

$$z = x + iy$$

$$\begin{Bmatrix} Z_I(z) \\ Z_{II}(z) \\ Z_{III}(z) \end{Bmatrix} = \frac{2}{\pi}\begin{Bmatrix} p \\ q \\ t \end{Bmatrix}\left\{\sqrt{\frac{b^2 - a^2}{a^2 - z^2}} - \tan^{-1}\sqrt{\frac{b^2 - a^2}{a^2 - z^2}}\right\}$$

$$\begin{Bmatrix} \overline{Z}_I(z) \\ \overline{Z}_{II}(z) \\ \overline{Z}_{III}(z) \end{Bmatrix} = \frac{2}{\pi}\begin{Bmatrix} p \\ q \\ t \end{Bmatrix}\left\{b\tan^{-1}\sqrt{\frac{1-(a/b)^2}{(a/z)^2-1}} - z\tan^{-1}\sqrt{\frac{(b/a)^2-1}{1-(z/a)^2}}\right\}$$

$$\begin{Bmatrix} K_I \\ K_{II} \\ K_{III} \end{Bmatrix} = \frac{2}{\sqrt{\pi a}}\begin{Bmatrix} p \\ q \\ t \end{Bmatrix}\sqrt{b^2 - a^2}$$

$$\operatorname{Im}\underset{|x|>a}{\begin{Bmatrix} \overline{Z}_I(x) \\ \overline{Z}_{II}(x) \\ \overline{Z}_{III}(x) \end{Bmatrix}} = \frac{2}{\pi}\begin{Bmatrix} p \\ q \\ t \end{Bmatrix}\left\{b\begin{pmatrix} \tanh^{-1} \\ \coth^{-1} \end{pmatrix}\cdot\sqrt{\frac{1-(a/b)^2}{1-(a/x)^2}} - |x|\begin{pmatrix} \tanh^{-1} \\ \coth^{-1} \end{pmatrix}\sqrt{\frac{(b/a)^2-1}{(x/a)^2-1}}\right\} \quad \begin{pmatrix} |x| > b \\ a \le |x| < b \end{pmatrix}$$

Method: Integration of **page 4.6** or Superposition of **page 4.7**
Accuracy: Exact
Reference: **Tada 1973**

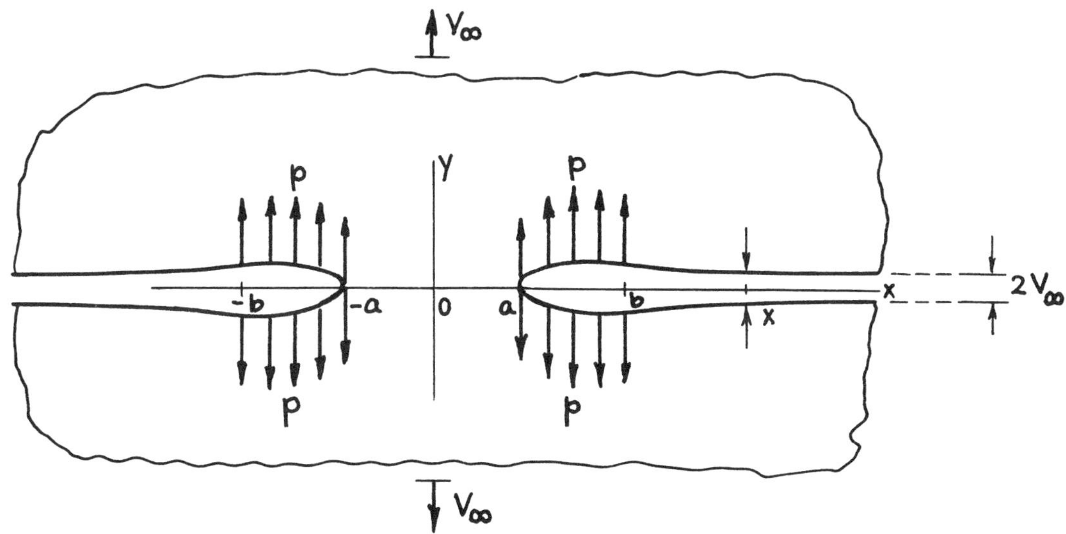

$$K_{I\,\pm a} = \frac{2p}{\sqrt{\pi a}}\sqrt{b^2 - a^2}$$

Crack Opening Profile:

$$\underset{|x| \geq a}{2v(x,0)} = \frac{8p}{\pi E'}\left\{ b\begin{pmatrix}\tanh^{-1}\\ \coth^{-1}\end{pmatrix}\sqrt{\frac{1-(a/x)^2}{1-(a/b)^2}} - |x|\begin{pmatrix}\tanh^{-1}\\ \coth^{-1}\end{pmatrix}\sqrt{\frac{(x/a)^2-1}{(b/a)^2-1}}\right\} \quad \begin{pmatrix} a \leq |x| < b \\ x > b \end{pmatrix}$$

Opening at $|x| = b$:

$$2v(\pm b, 0) = \frac{8pb}{\pi E'}\ell\text{n}\frac{b}{a}$$

Relative Vertical Displacement at Infinity:

$$2v_\infty = \frac{8pb}{\pi E'}\left\{\cosh^{-1}\frac{b}{a} - \sqrt{1-\left(\frac{a}{b}\right)^2}\right\}(= 2v(x \to \infty, 0))$$

Method: Westergaard Stress Function (Integration of **pages 4.6 and 4.6a**)
Accuracy: Exact
Reference: **Tada 1985**

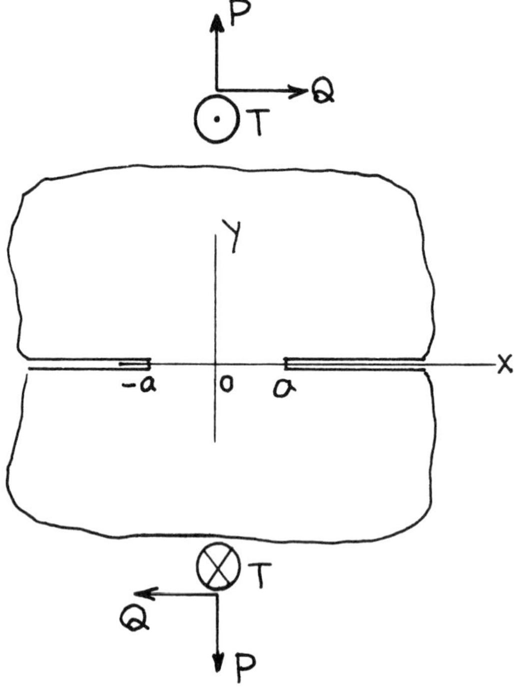

$$z = x + iy$$

$$\begin{Bmatrix} Z_I(z) \\ Z_{II}(z) \\ Z_{III}(z) \end{Bmatrix} = \frac{1}{\pi}\begin{Bmatrix} P \\ Q \\ T \end{Bmatrix} \frac{1}{\sqrt{a^2 - z^2}}$$

$$\begin{Bmatrix} \overline{Z}_I(z) \\ \overline{Z}_{II}(z) \\ \overline{Z}_{III}(z) \end{Bmatrix} = \frac{1}{\pi}\begin{Bmatrix} P \\ Q \\ T \end{Bmatrix} \sin^{-1}\frac{z}{a}$$

$$\begin{Bmatrix} K_I \\ K_{II} \\ K_{III} \end{Bmatrix} = \frac{1}{\sqrt{\pi a}}\begin{Bmatrix} P \\ Q \\ T \end{Bmatrix}$$

$$\mathrm{Im}\begin{Bmatrix} \overline{Z}_I(x) \\ \overline{Z}_{II}(x) \\ \overline{Z}_{III}(x) \end{Bmatrix}_{|x|\geq a} = \frac{1}{\pi}\begin{Bmatrix} P \\ Q \\ T \end{Bmatrix} \cosh^{-1}\frac{x}{a}$$

Methods: Special Case of **page 4.3** or **page 7.1** or by Stress Concentration Factor
Accuracy: Exact
References: **Neuber 1937; Winne 1958; Paris 1960, 1965; Sih 1964**

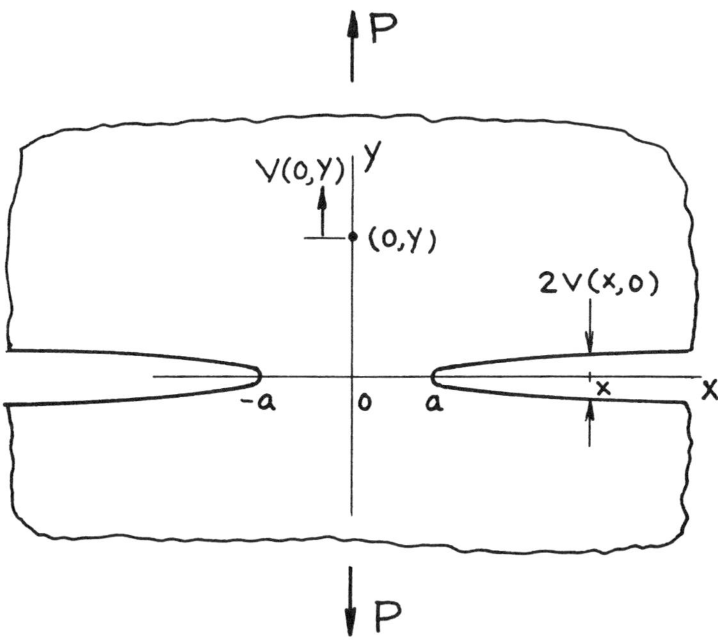

$$K_I = \frac{P}{\sqrt{\pi a}}$$

Crack Opening Profile:

$$\underset{|x| \geq a}{2\, v(x,0)} = \frac{4P}{\pi E'} \cosh^{-1} \frac{x}{a}$$

Vertical Displacement at $(0, y)$:

$$v(0,y) = \frac{2P}{\pi E'} \left(\sinh^{-1} \frac{y}{a} - \alpha \frac{y}{\sqrt{a^2 + y^2}} \right)$$

where

$$\alpha = \begin{cases} \frac{1}{2}(1+\nu) & \text{plane stress} \\ \frac{1}{2}\left(\frac{1}{1-\nu}\right) & \text{plane strain} \end{cases}$$

Methods: v Reciprocity (see **pages 4.3a and 4.6a**) or Paris' Equation (see **Appendix B**)
Accuracy: Exact
Reference: **Tada 1985**

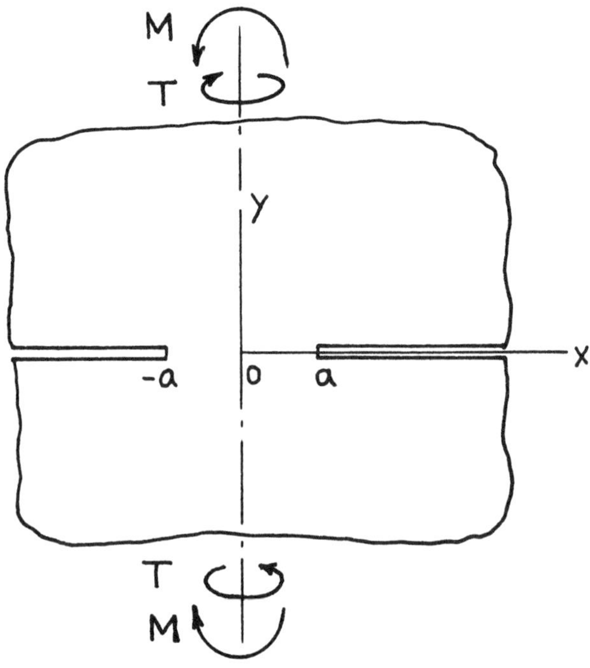

$$z = x + iy$$

$$\left\{ \begin{array}{c} Z_I(z) \\ Z_{III}(z) \end{array} \right\} = \frac{2}{\pi a^2} \left\{ \begin{array}{c} M \\ T \end{array} \right\} \frac{z}{\sqrt{a^2 - z^2}}$$

$$\left\{ \begin{array}{c} \overline{Z}_I(z) \\ \overline{Z}_{III}(z) \end{array} \right\} = \frac{2}{\pi a^2} \left\{ \begin{array}{c} M \\ T \end{array} \right\} \left[-\sqrt{a^2 - z^2} \right]$$

$$\left\{ \begin{array}{c} K_I \\ K_{III} \end{array} \right\}_{\pm a} = \pm \frac{2}{a\sqrt{\pi a}} \left\{ \begin{array}{c} M \\ T \end{array} \right\}$$

$$\operatorname{Im} \left\{ \begin{array}{c} \overline{Z}_I(x) \\ \overline{Z}_{III}(x) \end{array} \right\}_{|x| \geq a} = \frac{2}{\pi a^2} \left\{ \begin{array}{c} M \\ T \end{array} \right\} \cdot x\sqrt{1 - (a/x)^2}$$

Methods: Westergaard Stress Function, etc. or by Stress Concentration Factor
Accuracy: Exact
References: **Neuber 1937; Paris 1960, 1965; Benthem 1972**

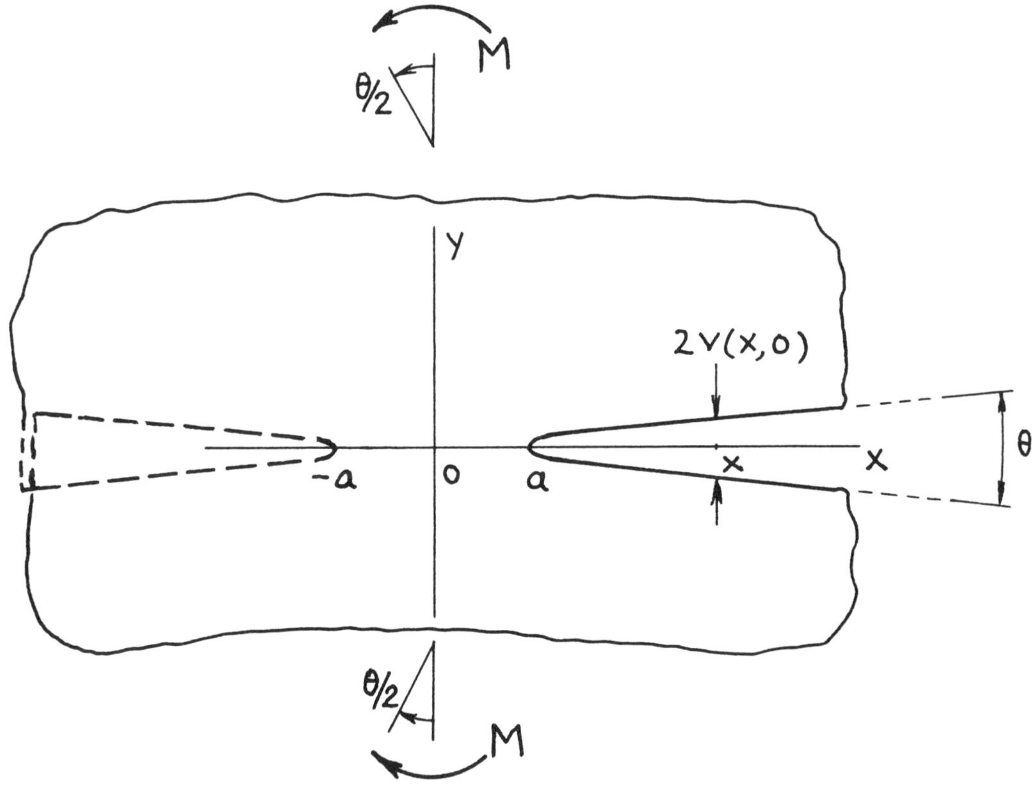

$$K_{I\pm a} = \pm \frac{2}{\sqrt{\pi}} \cdot \frac{M}{a^{3/2}} = \pm \frac{\sqrt{\pi}}{4} E' \theta \sqrt{a}$$

Crack Opening Profile:

$$\underset{|x| \geq a}{2v(x,0)} = \frac{8M}{\pi E'} \frac{x}{a^2} \sqrt{1 - \left(\frac{a}{x}\right)^2}$$

Relative Rotation at Infinity:

$$\theta = \frac{8M}{\pi E' a^2} = \frac{4|K_I|}{\sqrt{\pi} E' \sqrt{a}} \left(= \left\{ \frac{2v(x,0)}{x} \right\}_{x \to \infty} \right)$$

Method: v, θ Paris' Equation (see **Appendix B**)
Accuracy: Exact
Reference: **Tada 1985**

NOTE: No surface interference ($x \leq -a$) was considered.

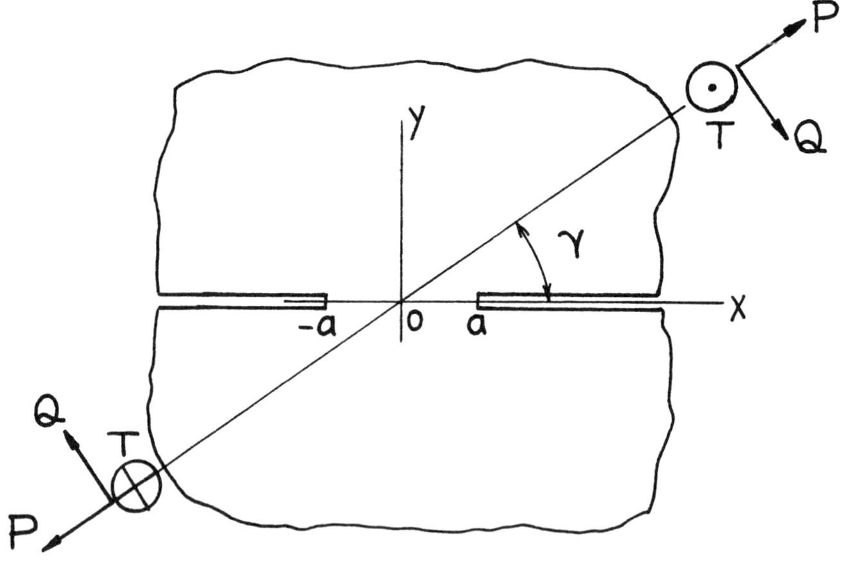

$$z = x + iy$$

$$\begin{Bmatrix} Z_I(z) \\ Z_{II}(z) \\ Z_{III}(z) \end{Bmatrix} = \frac{1}{\pi} \begin{Bmatrix} P\sin\gamma - Q\cos\gamma \\ P\cos\gamma + Q\sin\gamma \\ T \end{Bmatrix} \frac{1}{\sqrt{a^2 - z^2}}$$

$$\begin{Bmatrix} \overline{Z}_I(z) \\ \overline{Z}_{II}(z) \\ \overline{Z}_{III}(z) \end{Bmatrix} = \frac{1}{\pi} \begin{Bmatrix} P\sin\gamma - Q\cos\gamma \\ P\cos\gamma + Q\sin\gamma \\ T \end{Bmatrix} \sin^{-1}\frac{z}{a}$$

$$\begin{Bmatrix} K_I \\ K_{II} \\ K_{III} \end{Bmatrix} = \frac{1}{\sqrt{\pi a}} \begin{Bmatrix} P\sin\gamma - Q\cos\gamma \\ P\cos\gamma + Q\sin\gamma \\ T \end{Bmatrix}$$

$$\operatorname{Im} \begin{Bmatrix} \overline{Z}_I(x) \\ \overline{Z}_{II}(x) \\ \overline{Z}_{III}(x) \end{Bmatrix}_{|x| \geq a} = \frac{1}{\pi} \begin{Bmatrix} P\sin\gamma - Q\cos\gamma \\ P\cos\gamma + Q\sin\gamma \\ T \end{Bmatrix} \cosh^{-1}\frac{x}{a}$$

Method: Superposition of **page 4.9**
Accuracy: Exact
References: **Neuber 1937; Winne 1958; Paris 1960, 1965; Sih 1964**

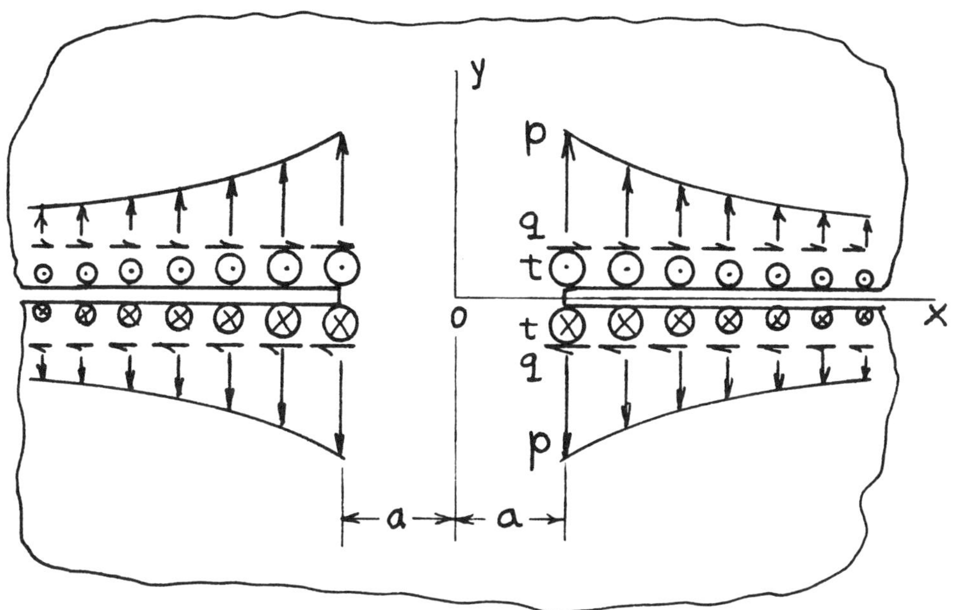

$$\left\{\begin{array}{c}\sigma_y(x,0)\\ \tau_{xy}(x,0)\\ \tau_{yz}(x,0)\end{array}\right\}_{|x|\geq a} = \left\{\begin{array}{c}p\\ q\\ t\end{array}\right\}\left|\frac{a}{x}\right|^{\gamma} \qquad (\gamma > 1)$$

$$\left\{\begin{array}{c}K_I\\ K_{II}\\ K_{III}\end{array}\right\} = \left\{\begin{array}{c}p\\ q\\ t\end{array}\right\}\sqrt{a}\,\frac{\Gamma\left(\frac{\gamma-1}{2}\right)}{\Gamma\left(\frac{\gamma}{2}\right)}$$

where $\Gamma(\gamma) =$ Gamma Function (See **Appendix M**)

Method: Integration of **page 4.6**
Accuracy: Exact
Reference: **Tada 1974**

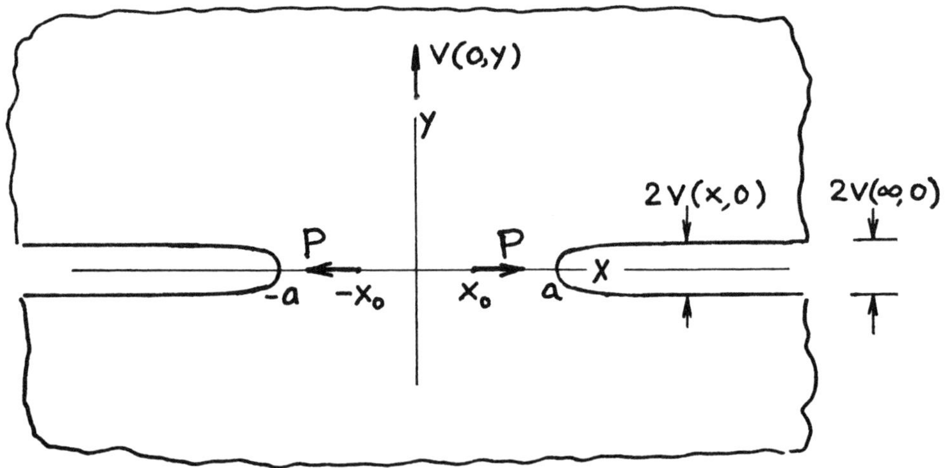

$$K_I = \frac{P(1-\alpha)}{\sqrt{\pi a}} \frac{x_0}{\sqrt{a^2 - x_0^2}}$$

Crack Opening Profile:

$$\underset{|x|\geq a}{2v(x,0)} = \frac{8P(1-\alpha)}{\pi E'} \cos^{-1} \sqrt{\frac{1-(x_0/a)^2}{1-(x_0/x)^2}}$$

$$2v(\infty,0) = \frac{8P(1-\alpha)}{\pi E'} \sin^{-1} \frac{x_0}{a}$$

Displacement at $(0,y)$:

$$v(0,y) = \frac{4P(1-\alpha)}{\pi E'} \left(1 - \alpha y \frac{\partial}{\partial y}\right) \left\{ \sin^{-1} \sqrt{\frac{(y/a)^2 + 1}{(y/b)^2 + 1}} - \tan^{-1} \frac{x_0}{y} \right\}$$

Relative Vertical Displacement at Infinity:

$$2v_\infty = v(0,\infty) - v(0,-\infty) = 2v(\infty,0)$$

where

$$\alpha = \begin{cases} \frac{1}{2}(1+\nu) & \text{plane stress} \\ \frac{1}{2}\left(\frac{1}{1-\nu}\right) & \text{plane strain} \end{cases}$$

Method: Integration of **page 4.6**; Paris' Equation (see **Appendix B**)
Accuracy: Exact
Reference: **Tada 2000**

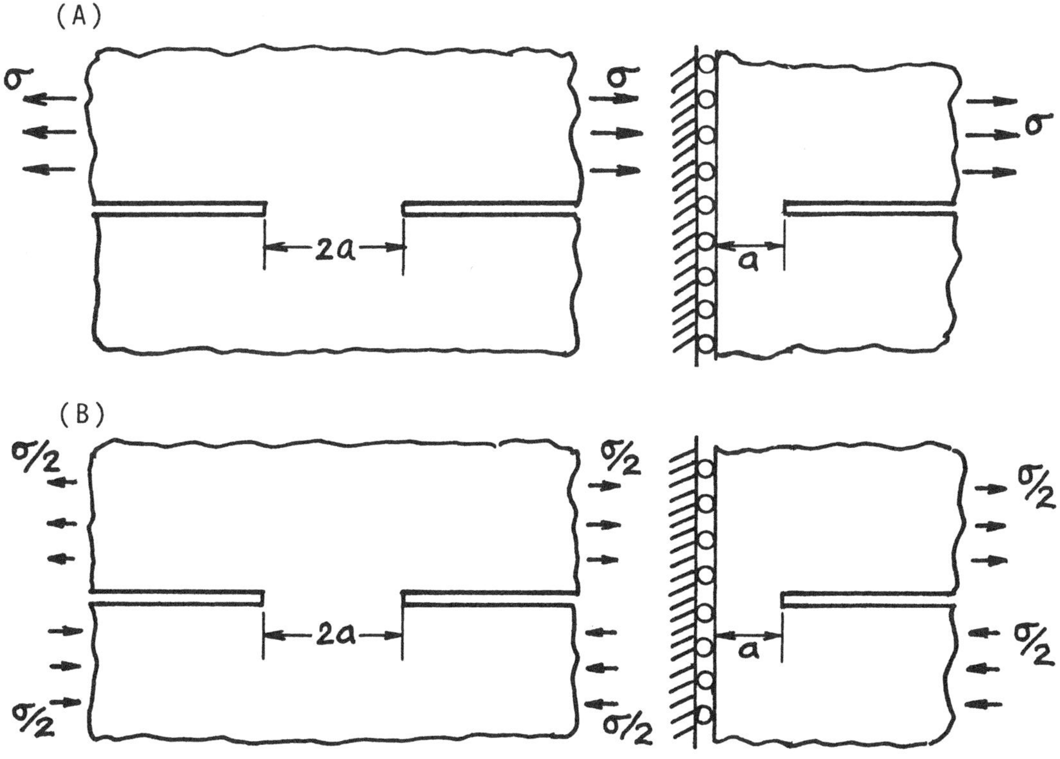

$$K_I = 0$$

$$K_{II} = \frac{1}{4}\sigma\sqrt{\pi a}$$

Method: (A) Special (Limiting) Case of **page 21.1** or **21.3**; (B) Superposition of (A)
Accuracy: Exact
Reference: **Tada 2000**

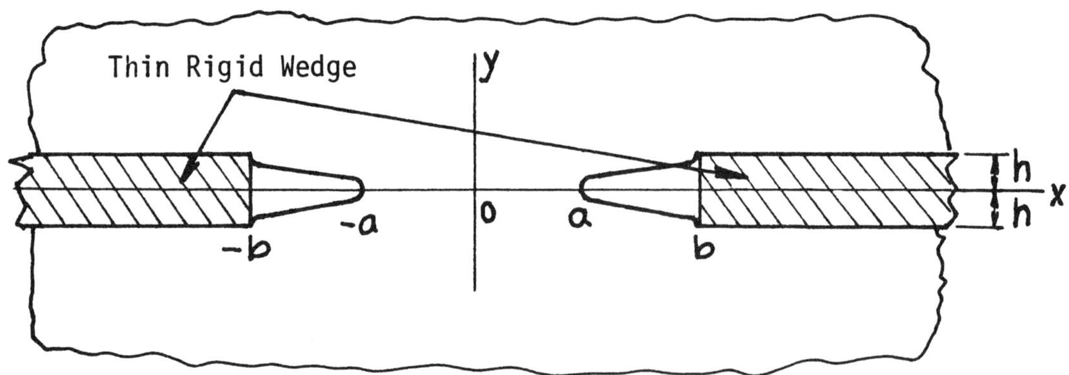

$$K_I\Big|_{x=\pm a} = \frac{E'h}{2kK(k)}\sqrt{\frac{\pi}{a}}$$

$$K_I\Big|_{x=\pm b} = -\frac{E'h}{2kK(k)}\sqrt{\frac{\pi}{b}}$$

$$2v(x,0)\Big|_{a\leq|x|\leq b} = 2h\left\{1 - \frac{F(\varphi, k)}{K(k)}\right\}$$

where

$$E' = \begin{cases} E & \text{plane stress} \\ E/(1-\nu^2) & \text{plane strain} \end{cases}$$

$$k = \sqrt{1-(a/b)^2}$$

$$\varphi = \sin^{-1}\sqrt{\frac{b^2-x^2}{b^2-a^2}}$$

$$F(\varphi, k) = \int_0^{\varphi} \frac{d\varphi}{\sqrt{1 - k^2 \sin^2 \varphi}}$$

$$K(k) = F(\pi/2, k)$$

Method: Westergaard Stress Function (or Negative of **page 5.21**)
Accuracy: Exact
Reference: **Tada 1974**

NOTE: For $K(k)$, see **Appendix L**.

The Westergaard Function is

$$Z_I(z) = -\frac{E'hb}{2K(k)} \cdot \frac{1}{\sqrt{z^2 - a^2}} \frac{1}{\sqrt{z^2 - b^2}}$$

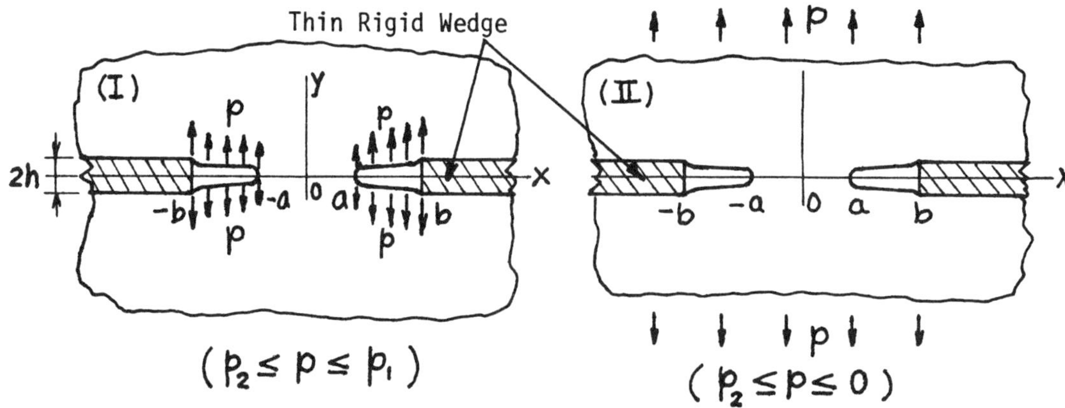

$$p_1 = \frac{E'h}{2b\{K(k) - E(k)\}}$$

$$p_2 = -\frac{E'hb}{2\left\{b^2E(k) - a^2K(k)\right\}}$$

$$K_I\Big|_{x=\pm a} = \left\{\frac{E'h}{2} + p\frac{b^2E(k) - a^2K(k)}{b}\right\}\frac{1}{kK(k)}\sqrt{\frac{\pi}{a}}$$

$$K_I\Big|_{x=\pm b} = \left\{-\frac{E'h}{2} + pb(K(k) - E(k))\right\}\frac{1}{kK(k)}\sqrt{\frac{\pi}{b}}$$

$$\underset{a\leq|x|\leq b}{2v(x,0)} = 2h\left\{1 - \frac{F(\varphi, k)}{K(k)}\right\} + \frac{4pb}{E'}\{K(k)E(\varphi, k) - E(k)F(\varphi, k)\}$$

where

$$E' = \begin{cases} E & \text{plane stress} \\ E/(1-\nu^2) & \text{plane strain} \end{cases}$$

$$k = \sqrt{1 - (a/b)^2}, \quad \varphi = \sin^{-1}\sqrt{\frac{b^2 - x^2}{b^2 - a^2}}$$

$$F(\varphi, k) = \int_0^{\varphi} \frac{d\varphi}{\sqrt{1 - k^2 \sin^2 \varphi}}, \quad K(k) = F(\pi/2, k)$$

$$E(\varphi, k) = \int_0^{\varphi} \sqrt{1 - k^2 \sin^2 \varphi}\, d\varphi, \quad E(k) = E(\pi/2, k)$$

Method: Superposition of **page 4.14** and **page 6.1**
Accuracy: Exact
Reference: **Tada 1974**

NOTE:

1. In (I), when $p > p_1$, separation of contact surfaces occurs near $x = \pm b$, and when $p < p_2$, crack closure occurs near $x = \pm a$.

2. In (II), when $p > 0$ (remote tension), separation of contact surfaces occurs for large $|x|$, and when $p < p_2$, crack closure occurs near $x = \pm a$.

3. For $K(k)$ and $E(k)$, see **Appendix L.**

The Westergaard Function is $Z_I(z) = Z_I(z) + Z_I(z)$
p.4.16 **p.4.15** **p.6.1** (replace σ by p)

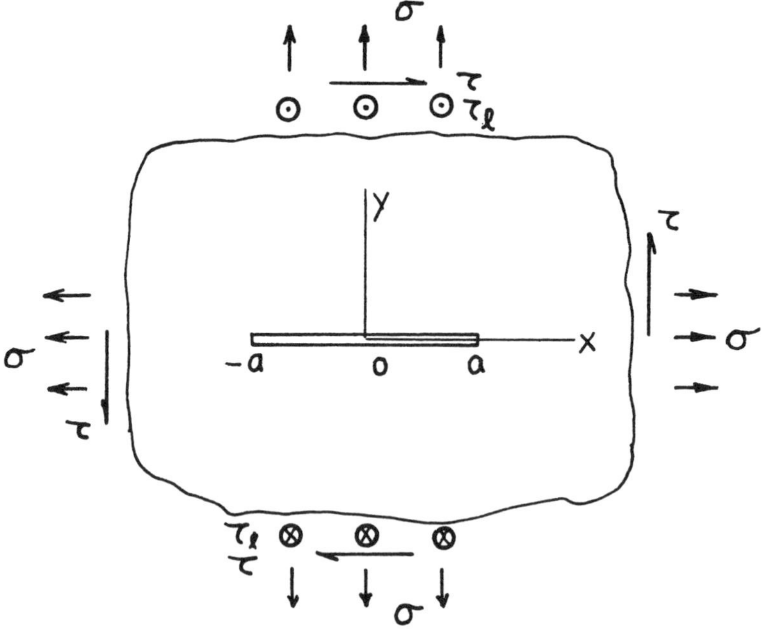

$$z = x + iy$$

$$\begin{Bmatrix} Z_I(z) \\ Z_{II}(z) \\ Z_{III}(z) \end{Bmatrix} = \begin{Bmatrix} \sigma \\ \tau \\ \tau_\ell \end{Bmatrix} \frac{z}{\sqrt{z^2 - a^2}}$$

$$\begin{Bmatrix} \overline{Z}_I(z) \\ \overline{Z}_{II}(z) \\ \overline{Z}_{III}(z) \end{Bmatrix} = \begin{Bmatrix} \sigma \\ \tau \\ \tau_\ell \end{Bmatrix} \sqrt{z^2 - a^2}$$

$$\begin{Bmatrix} K_I \\ K_{II} \\ K_{III} \end{Bmatrix} = \begin{Bmatrix} \sigma \\ \tau \\ \tau_\ell \end{Bmatrix} \sqrt{\pi a}$$

$$\operatorname{Im} \begin{Bmatrix} \overline{Z}_I(x) \\ \overline{Z}_{II}(x) \\ \overline{Z}_{III}(x) \end{Bmatrix}_{|x| \leq a} = \begin{Bmatrix} \sigma \\ \tau \\ \tau_\ell \end{Bmatrix} \sqrt{a^2 - x^2}$$

Methods: Westergaard Stress Function, etc. or by Stress Concentration Factor
Accuracy: Exact
References: **Griffith 1920; Westergaard 1939; Irwin 1957, 1958a; Paris 1965, etc.**

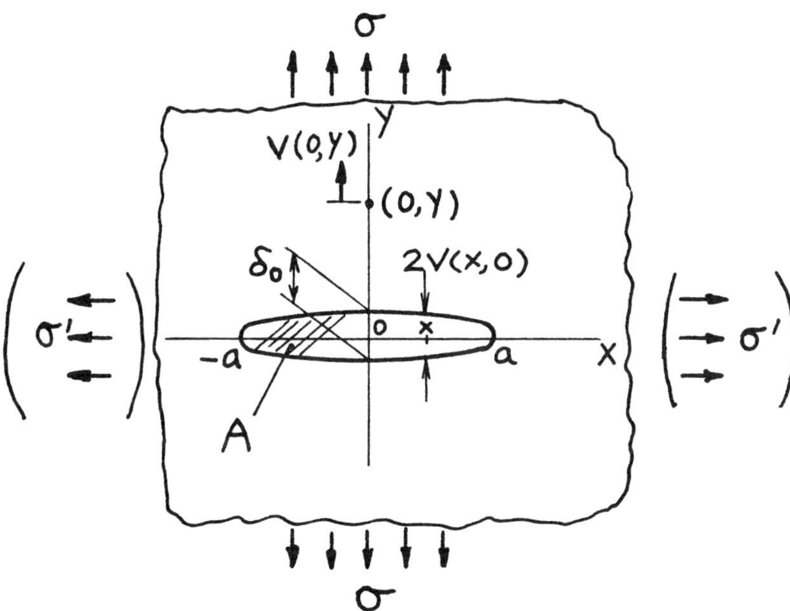

$$K_I = \sigma\sqrt{\pi a}$$

Crack Opening Area:

$$A = \frac{2\sigma\pi a^2}{E'}$$

Crack Opening Profile:

$$\underset{|x|\leq a}{2v(x,0)} = \frac{4\sigma}{E'}\sqrt{a^2 - x^2}$$

Opening at Center:

$$\delta_0 = 2v(0,0) = \frac{4\sigma a}{E'}$$

Additional Vertical Displacement at $(0, y)$ due to Crack:

$$v(0,y) = \frac{2\sigma}{E'}\left(\sqrt{a^2 + y^2} - y\right)\left(1 + \alpha\frac{y}{\sqrt{a^2 + y^2}}\right)$$

where

$$\alpha = \begin{cases} \frac{1}{2}(1+\nu) & \text{plane stress} \\ \frac{1}{2}\left(\frac{1}{1-\nu}\right) & \text{plane strain} \end{cases}$$

Method: Paris' Equation (see **Appendix B**)
Accuracy: Exact
Reference: **Tada 1985**

NOTE: $v(0, y)$ is the displacement at $(0, y)$ when uniform pressure σ is applied on crack surfaces.

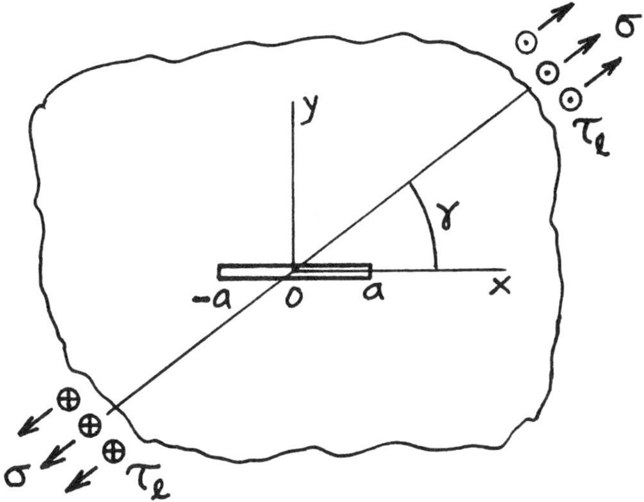

$$\begin{Bmatrix} K_I \\ K_{II} \\ K_{III} \end{Bmatrix} = \begin{Bmatrix} \sigma \sin\gamma \\ \sigma \cos\gamma \\ \tau_\ell \end{Bmatrix} \sin\gamma \cdot \sqrt{\pi a}$$

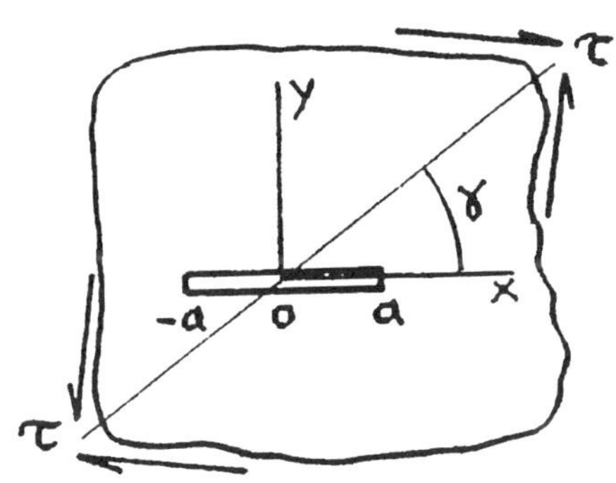

$$\begin{Bmatrix} K_I \\ K_{II} \end{Bmatrix} = \tau \begin{Bmatrix} -\cos 2\gamma \\ \sin 2\gamma \end{Bmatrix} \cdot \sqrt{\pi a}$$

$$K_{II} = 0$$

Method: Superposition
Accuracy: Exact
References: **Paris 1965; Sih 1965a**

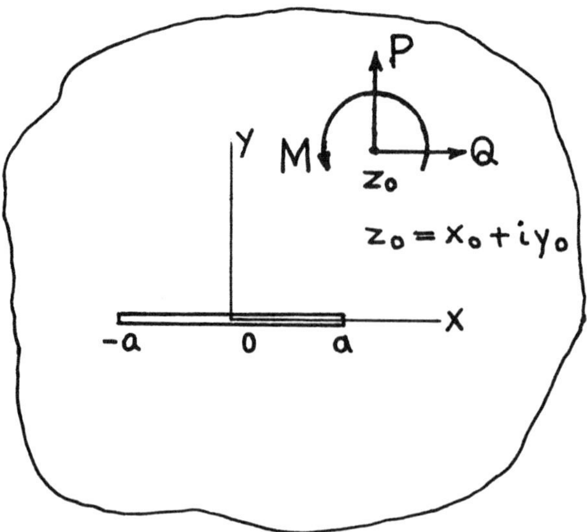

$$K_{+a} = (K_I - iK_{II})_{+a}$$

$$= \frac{1}{2\sqrt{\pi a}} \frac{1}{\kappa+1} \left\{ (Q + iP) \left[\left(\frac{a+z_0}{\sqrt{z_0^2-a^2}} - 1 \right) - \kappa \left(\frac{a+\bar{z}_0}{\sqrt{\bar{z}_0^2-a^2}} - 1 \right) \right] + \frac{a[(Q-iP)(\bar{z}_0 - z_0)+i(1+\kappa)M]}{(\bar{z}_0-a)\sqrt{\bar{z}_0^2-a^2}} \right\}$$

where

$$\kappa = \begin{cases} \frac{3-\nu}{1+\nu} & \text{plane stress} \\ 3 - 4\nu & \text{plane strain} \end{cases}$$

Method: Muskhelishvili's Method
Accuracy: Exact
References: **Erdogan 1962; Sih 1962a**

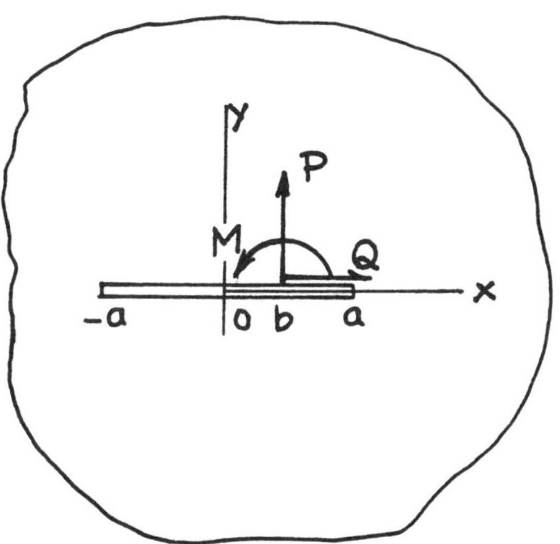

$$K_{I \atop +a} = \frac{1}{2\sqrt{\pi a}} \left\{ P\sqrt{\frac{a+b}{a-b}} + \left(\frac{\kappa-1}{\kappa+1}\right) Q + \frac{Ma}{(a-b)\sqrt{a^2-b^2}} \right\}$$

$$K_{II \atop +a} = \frac{1}{2\sqrt{\pi a}} \left\{ Q\sqrt{\frac{a+b}{a-b}} - \left(\frac{\kappa-1}{\kappa+1}\right) P \right\}$$

where

$$\kappa = \begin{cases} \frac{3-\nu}{1+\nu} & \text{plane stress} \\ 3-4\nu & \text{plane strain} \end{cases}$$

Method: Muskhelishvili's Method
Accuracy: Exact
References: **Erdogan 1962; Sih 1962a**

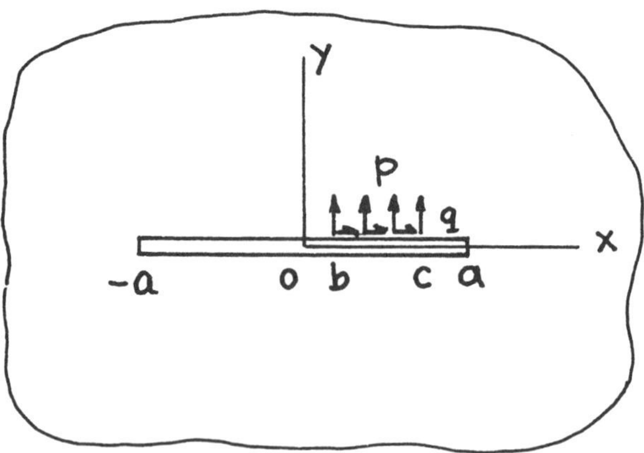

$$K_{I\pm a} = \frac{1}{2}\frac{1}{\sqrt{\pi a}}\left[pa\left(\sin^{-1}\frac{c}{a} - \sin^{-1}\frac{b}{a} \mp \sqrt{1-(c/a)^2} \pm \sqrt{1-(b/a)^2}\right) + \left(\frac{\kappa-1}{\kappa+1}\right)q(c-b)\right]$$

$$K_{II\pm a} = \frac{1}{2}\frac{1}{\sqrt{\pi a}}\left[qa\left(\sin^{-1}\frac{c}{a} - \sin^{-1}\frac{b}{a} \mp \sqrt{1-(c/a)^2} \pm \sqrt{1-(b/a)^2}\right) - \left(\frac{\kappa-1}{\kappa+1}\right)p(c-b)\right]$$

where

$$\kappa = \begin{cases} \frac{3-\nu}{1+\nu} & \text{plane stress} \\ 3-4\nu & \text{plane strain} \end{cases}$$

Method: Muskhelishvili's Method (Integration of **page 5.4**)
Accuracy: Exact
References: **Erdogan 1962; Sih 1962a**

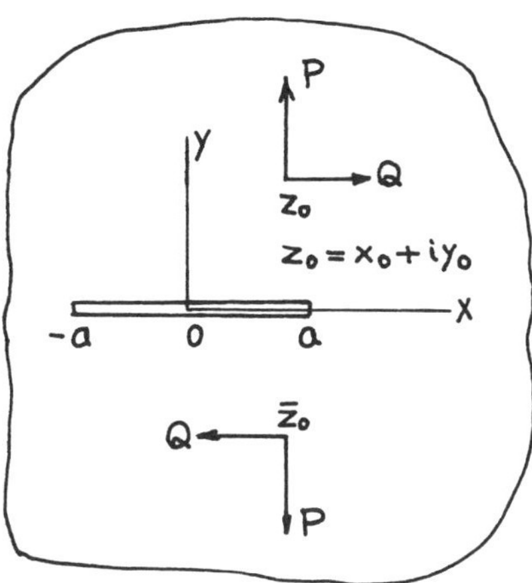

$$z = x + iy$$

$$\left\{\begin{matrix} Z_I(z) \\ Z_{II}(z) \end{matrix}\right\} = \frac{1}{\pi}\left\{\begin{matrix} P \\ Q \end{matrix}\right\}\left[1\left\{\begin{matrix} - \\ + \end{matrix}\right\}\alpha y_0 \frac{\partial}{\partial y_0}\right]\frac{1}{2}\left(\frac{\sqrt{a^2 - z_0^2}}{z - z_0} + \frac{\sqrt{a^2 - \bar{z}_0^2}}{z - \bar{z}_0}\right)\frac{1}{\sqrt{z^2 - a^2}}$$

$$\left\{\begin{matrix} \overline{Z}_I(z) \\ \overline{Z}_{II}(z) \end{matrix}\right\} = \frac{1}{\pi}\left\{\begin{matrix} P \\ Q \end{matrix}\right\}\left[1\left\{\begin{matrix} - \\ + \end{matrix}\right\}\alpha y_0 \frac{\partial}{\partial y_0}\right]\left\{\tan^{-1}\sqrt{\frac{a + z_0}{a - z_0}}\sqrt{\frac{z - a}{z + a}} + \tan^{-1}\sqrt{\frac{a + \bar{z}_0}{a - \bar{z}_0}}\sqrt{\frac{z - a}{z + a}}\right\}$$

$$K_{I\pm a} = \frac{P}{\sqrt{\pi a}}\left[1 - \alpha y_0 \frac{\partial}{\partial y_0}\right]\frac{1}{2}\left\{\sqrt{\frac{a \pm z_0}{a \mp z_0}} + \sqrt{\frac{a \pm \bar{z}_0}{a \mp \bar{z}_0}}\right\}$$

$$K_{II\pm a} = \frac{Q}{\sqrt{\pi a}}\left[1 + \alpha y_0 \frac{\partial}{\partial y_0}\right]\frac{1}{2}\left\{\sqrt{\frac{a \pm z_0}{a \mp z_0}} + \sqrt{\frac{a \pm \bar{z}_0}{a \mp \bar{z}_0}}\right\}$$

where

$$\alpha = \begin{cases} \frac{1}{2}(1 + \nu) & \text{plane stress} \\ \frac{1}{2}\left(\frac{1}{1-\nu}\right) & \text{plane strain} \end{cases}$$

Method: Westergaard Stress Function
Accuracy: Exact
References: **Tada 1972a, 1973**

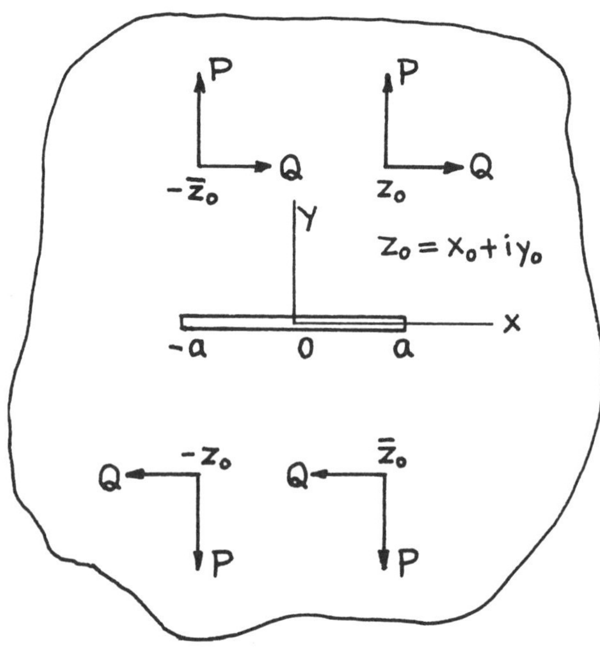

$$z = x + iy$$

$$\begin{Bmatrix} Z_I(z) \\ Z_{II}(z) \end{Bmatrix} = \frac{1}{\pi} \begin{Bmatrix} P \\ Q \end{Bmatrix} \left[1 \begin{Bmatrix} - \\ + \end{Bmatrix} \alpha y_0 \frac{\partial}{\partial y_0} \right] \left(\frac{\sqrt{a^2 - z_0^2}}{z^2 - z_0^2} + \frac{\sqrt{a^2 - \bar{z}_0^2}}{z^2 - \bar{z}_0^2} \right) \frac{1}{\sqrt{1 - (a/z)^2}}$$

$$\begin{Bmatrix} \bar{Z}_I(z) \\ \bar{Z}_{II}(z) \end{Bmatrix} = \frac{1}{\pi} \begin{Bmatrix} P \\ Q \end{Bmatrix} \left[1 \begin{Bmatrix} - \\ + \end{Bmatrix} \alpha y_0 \frac{\partial}{\partial y_0} \right] \left\{ \tan^{-1} \sqrt{\frac{z^2 - a^2}{a^2 - z_0^2}} + \tan^{-1} \sqrt{\frac{z^2 - a^2}{a^2 - \bar{z}_0^2}} \right\}$$

$$\begin{Bmatrix} K_I \\ K_{II} \end{Bmatrix} = \frac{1}{\sqrt{\pi a}} \begin{Bmatrix} P \\ Q \end{Bmatrix} \left[1 \begin{Bmatrix} - \\ + \end{Bmatrix} \alpha y_0 \frac{\partial}{\partial y_0} \right] \left\{ \frac{a}{\sqrt{a^2 - z_0^2}} + \frac{a}{\sqrt{\alpha^2 - \bar{z}_0^2}} \right\}$$

where

$$\alpha = \begin{cases} \frac{1}{2}(1 + \nu) & \text{plane stress} \\ \frac{1}{2}\left(\frac{1}{1-\nu}\right) & \text{plane strain} \end{cases}$$

Method: Westergaard Stress Function (Superposition of **page 5.6**)
Accuracy: Exact
References: **Tada 1972a, 1973**

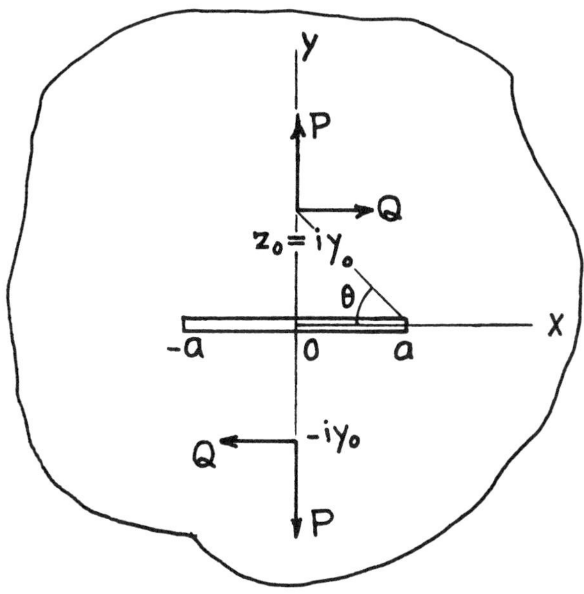

$$z = x + iy$$

$$\left\{\begin{matrix} Z_I(z) \\ Z_{II}(z) \end{matrix}\right\} = \frac{1}{\pi}\left\{\begin{matrix} P \\ Q \end{matrix}\right\}\left[1\left\{\begin{matrix} - \\ + \end{matrix}\right\}\alpha y_0 \frac{\partial}{\partial y_0}\right] \frac{\sqrt{a^2 + y_0^2}}{z^2 + y_0^2} \frac{1}{\sqrt{1 - (a/z)^2}}$$

$$\left\{\begin{matrix} \overline{Z}_I(z) \\ \overline{Z}_{II}(z) \end{matrix}\right\} = \frac{1}{\pi}\left\{\begin{matrix} P \\ Q \end{matrix}\right\}\left[1\left\{\begin{matrix} - \\ + \end{matrix}\right\}\alpha y_0 \frac{\partial}{\partial y_0}\right] \tan^{-1}\sqrt{\frac{z^2 - a^2}{a^2 + y_0^2}}$$

$$\begin{aligned} \left\{\begin{matrix} K_I \\ K_{II} \end{matrix}\right\} &= \frac{1}{\sqrt{\pi a}}\left\{\begin{matrix} P \\ Q \end{matrix}\right\}\left[1\left\{\begin{matrix} - \\ + \end{matrix}\right\}\alpha y_0 \frac{\partial}{\partial y_0}\right] \frac{a}{\sqrt{a^2 + y_0^2}} \\ &= \frac{1}{\sqrt{\pi a}}\left\{\begin{matrix} P \\ Q \end{matrix}\right\} \frac{a}{\sqrt{a^2 + y_0^2}}\left[1\left\{\begin{matrix} + \\ - \end{matrix}\right\}\alpha \frac{y_0^2}{a^2 + y_0^2}\right]\left(= \frac{1}{\sqrt{\pi a}}\left\{\begin{matrix} P \\ Q \end{matrix}\right\}\cos\theta\left[1\left\{\begin{matrix} + \\ - \end{matrix}\right\}\alpha \sin^2\theta\right]\right) \end{aligned}$$

$$\begin{aligned} \operatorname{Im}\left\{\begin{matrix} \overline{Z}_I(x) \\ \overline{Z}_{II}(x) \end{matrix}\right\}_{|x| \le a} &= \frac{1}{\pi}\left\{\begin{matrix} P \\ Q \end{matrix}\right\}\left[1\left\{\begin{matrix} - \\ + \end{matrix}\right\}\alpha y_0 \frac{\partial}{\partial y_0}\right] \tanh^{-1}\sqrt{\frac{a^2 - x^2}{a^2 + y_0^2}} \\ &= \frac{1}{\pi}\left\{\begin{matrix} P \\ Q \end{matrix}\right\}\left[\tanh^{-1}\sqrt{\frac{a^2 - x^2}{a^2 + y_0^2}}\left\{\begin{matrix} + \\ - \end{matrix}\right\}\alpha \frac{y_0^2}{x^2 + y_0^2}\sqrt{\frac{a^2 - x^2}{a^2 + y_0^2}}\right] \end{aligned}$$

where

$$\alpha = \begin{cases} \frac{1}{2}(1+\nu) & \text{plane stress} \\ \frac{1}{2}\left(\frac{1}{1-\nu}\right) & \text{plane strain} \end{cases}$$

Method: Westergaard Stress Function, etc. (Special Case of **page 5.6**)
Accuracy: Exact
References: **Paris 1957; Barenblatt 1962; Irwin 1962a; Tada 1970, 1973**

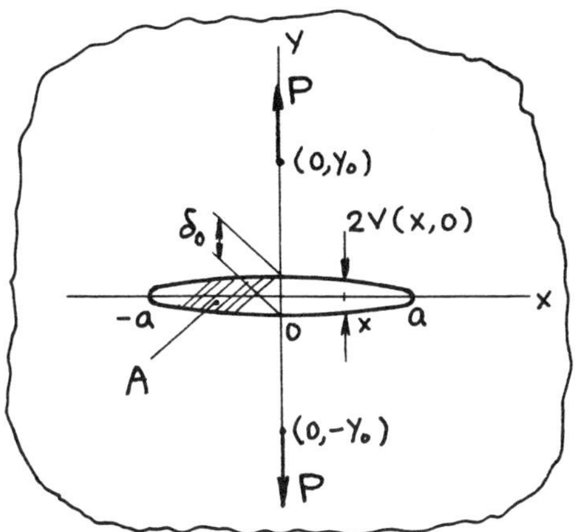

$$K_I = \frac{P}{\sqrt{\pi a}} \frac{a}{\sqrt{a^2 + y_0^2}} \left(1 + \alpha \frac{y_0^2}{a^2 + y_0^2}\right)$$

Crack Opening Area:

$$A = \frac{4P}{E'} \left(\sqrt{a^2 + y_0^2} - y_0\right) \left(1 + \alpha \frac{y_0}{\sqrt{a^2 + y_0^2}}\right)$$

Crack Opening Profile:

$$\underset{|x| \le a}{2v(x, 0)} = \frac{4P}{\pi E'} \left(\tanh^{-1} \sqrt{\frac{a^2 - x^2}{a^2 + y_0^2}} + \alpha \frac{y_0^2}{x^2 + y_0^2} \sqrt{\frac{a^2 - x^2}{a^2 + y_0^2}}\right)$$

Opening at Center:

$$\delta_0 = 2v(0, 0) = \frac{4P}{\pi E'} \left(\sinh^{-1} \frac{a}{y_0} + \alpha \frac{a}{\sqrt{a^2 + y_0^2}}\right)$$

where

$$\alpha = \begin{cases} \frac{1}{2}(1 + \nu) & \text{plane stress} \\ \frac{1}{2}\left(\frac{1}{1-\nu}\right) & \text{plane strain} \end{cases}$$

Methods: A, v Paris' Equation (see **Appendix B**) (or v also from **page 5.8**)
Accuracy: Exact
Reference: **Tada 1985**

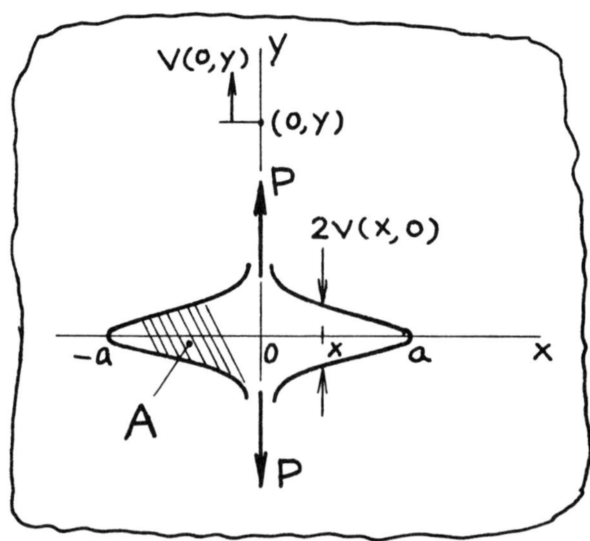

$$z = x + iy$$

$$Z_I(z) = \frac{P}{\pi} \frac{a}{z\sqrt{z^2 - a^2}}$$

$$\overline{Z}_I(z) = \frac{P}{\pi} \cos^{-1} \frac{a}{z}$$

$$K_I = \frac{P}{\sqrt{\pi a}}$$

Crack Opening Area:

$$A = \frac{4Pa}{E'}$$

Crack Opening Profile:

$$\underset{|x| \leq a}{2v(x, 0)} = \frac{4P}{\pi E'} \cosh^{-1} \frac{a}{x}$$

Vertical Displacement at $(0, y)$:

$$v(0,y) = \frac{2P}{\pi E'}\left(\sinh^{-1}\frac{a}{y} + \alpha\frac{a}{\sqrt{a^2+y^2}}\right)$$

where

$$\alpha = \begin{cases} \frac{1}{2}(1+\nu) & \text{plane stress} \\ \frac{1}{2}\left(\frac{1}{1-\nu}\right) & \text{plane strain} \end{cases}$$

Method: Westergaard Stress Function (Special Case of **page 5.8 or 5.10**)
Accuracy: Exact
Reference: **Tada 1985**

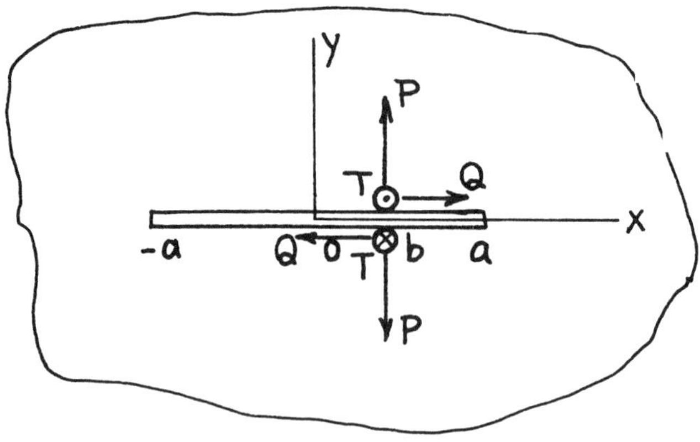

$$z = x + iy$$

$$\begin{Bmatrix} Z_I(z) \\ Z_{II}(z) \\ Z_{III}(z) \end{Bmatrix} = \frac{1}{\pi} \begin{Bmatrix} P \\ Q \\ T \end{Bmatrix} \frac{\sqrt{a^2 - b^2}}{(z-b)\sqrt{z^2 - a^2}}$$

$$\begin{Bmatrix} \overline{Z}_I(z) \\ \overline{Z}_{II}(z) \\ \overline{Z}_{III}(z) \end{Bmatrix} = \frac{1}{\pi} \begin{Bmatrix} P \\ Q \\ T \end{Bmatrix} \sin^{-1} \frac{bz - a^2}{a(z-b)}$$

$$\begin{Bmatrix} K_I \\ K_{II} \\ K_{III} \end{Bmatrix}_{\pm a} = \frac{1}{\sqrt{\pi a}} \begin{Bmatrix} P \\ Q \\ T \end{Bmatrix} \sqrt{\frac{a \pm b}{a \mp b}}$$

$$\mathrm{Im} \begin{Bmatrix} \overline{Z}_I(x) \\ \overline{Z}_{II}(x) \\ \overline{Z}_{III}(x) \end{Bmatrix}_{|x| \le a} = \frac{1}{\pi} \begin{Bmatrix} P \\ Q \\ T \end{Bmatrix} \cosh^{-1} \frac{a^2 - bx}{a|x-b|}$$

Method: Westergaard Stress Function, etc.
Accuracy: Exact
References: **Irwin 1957, 1958a; Erdogan 1962; Sih 1962a, 1964; Paris 1965**

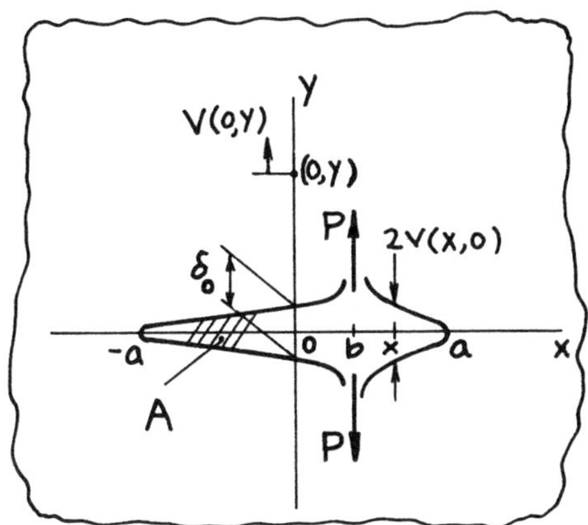

$$K_{I\pm a} = \frac{P}{\sqrt{\pi a}} \frac{\sqrt{a^2 - b^2}}{a \mp b}$$

Crack Opening Area:

$$A = \frac{4P}{E'} \sqrt{a^2 - b^2}$$

Crack Opening Profile:

$$\underset{|x| \le a}{2v(x, 0)} = \frac{4P}{\pi E'} \cosh^{-1} \frac{a^2 - bx}{a|x - b|}$$

Opening at Center:

$$\delta_0 = 2v(0, 0) = \frac{4P}{\pi E'} \cosh^{-1} \frac{a}{b}$$

Vertical Displacement at $(0, y)$:

$$v(0, y) = \frac{2P}{\pi E'} \left(\tanh^{-1} \sqrt{\frac{a^2 - b^2}{a^2 + y^2}} + \alpha \frac{y^2}{b^2 + y^2} \sqrt{\frac{a^2 - b^2}{a^2 + y^2}} \right)$$

where

$$\alpha = \begin{cases} \frac{1}{2}(1 + \nu) & \text{plane stress} \\ \frac{1}{2}\left(\frac{1}{1-\nu}\right) & \text{plane strain} \end{cases}$$

Method: A, v Paris' Equation (see **Appendix B**) or Reciprocity (see **page 5.8b and page 5.9**)
Accuracy: Exact
Reference: **Tada 1985**

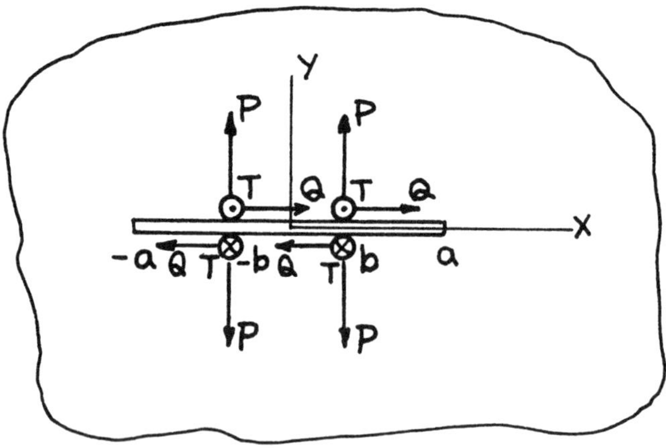

$$z = x + iy$$

$$\begin{Bmatrix} Z_I(z) \\ Z_{II}(z) \\ Z_{III}(z) \end{Bmatrix} = \frac{2}{\pi} \begin{Bmatrix} P \\ Q \\ T \end{Bmatrix} \frac{\sqrt{a^2 - b^2}}{\left(z^2 - b^2\right)\sqrt{1 - \left(a/z\right)^2}}$$

$$\begin{Bmatrix} \overline{Z}_I(z) \\ \overline{Z}_{II}(z) \\ \overline{Z}_{III}(z) \end{Bmatrix} = \frac{2}{\pi} \begin{Bmatrix} P \\ Q \\ T \end{Bmatrix} \tan^{-1} \sqrt{\frac{z^2 - a^2}{a^2 - b^2}}$$

$$\begin{Bmatrix} K_I \\ K_{II} \\ K_{III} \end{Bmatrix} = \frac{2}{\sqrt{\pi a}} \begin{Bmatrix} P \\ Q \\ T \end{Bmatrix} \frac{a}{\sqrt{a^2 - b^2}}$$

$$\operatorname{Im} \underset{|x| \le a}{\begin{Bmatrix} \overline{Z}_I(x) \\ \overline{Z}_{II}(x) \\ \overline{Z}_{III}(x) \end{Bmatrix}} = \frac{1}{\pi} \begin{Bmatrix} P \\ Q \\ T \end{Bmatrix} \ln \left| \frac{\sqrt{a^2 - x^2} + \sqrt{a^2 - b^2}}{\sqrt{a^2 - x^2} - \sqrt{a^2 - b^2}} \right|$$

Method: Westergaard Stress Function, etc. (or Superposition of **page 5.10**)
Accuracy: Exact
References: **Irwin 1957, 1958a; Erdogan 1962; Sih 1962a, 1964; Paris 1965**

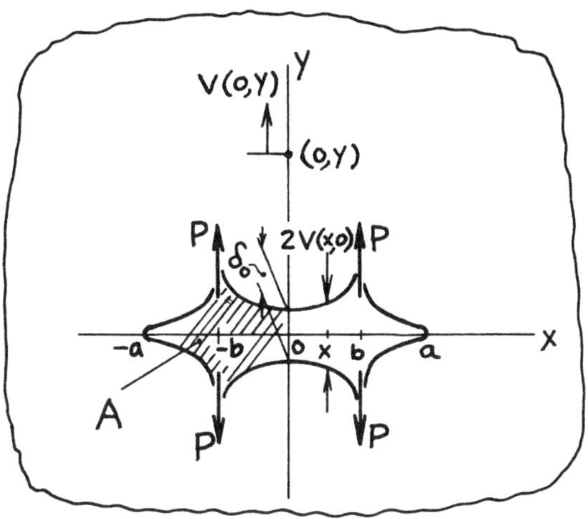

$$K_I = \frac{2P}{\sqrt{\pi a}} \frac{1}{\sqrt{1 - \left(b/a\right)^2}}$$

Crack Opening Area:

$$A = \frac{8P}{E'} \sqrt{a^2 - b^2}$$

Crack Opening Profile:

$$\underset{|x| \le a}{2v(x,0)} = \frac{8P}{\pi E'} \begin{pmatrix} \tanh^{-1} \\ \coth^{-1} \end{pmatrix} \sqrt{\frac{a^2 - b^2}{a^2 - x^2}} \qquad \begin{pmatrix} a \le |x| < b \\ |x| > b \end{pmatrix}$$

Opening at Center:

$$\delta_0 = 2v(0,0) = \frac{8P}{\pi E'} \cosh^{-1} \frac{a}{b}$$

Vertical Displacement at $(0, y)$:

$$v(0,y) = \frac{4P}{\pi E'} \left(\tanh^{-1} \sqrt{\frac{a^2 - b^2}{a^2 + y^2}} + \alpha \frac{y^2}{b^2 + y^2} \sqrt{\frac{a^2 - b^2}{a^2 + y^2}} \right)$$

where

$$\alpha = \begin{cases} \frac{1}{2}(1 + \nu) & \text{plane stress} \\ \frac{1}{2}\left(\frac{1}{1-\nu}\right) & \text{plane strain} \end{cases}$$

Method: Superposition (see **pages 5.10 and 5.10a**)
Accuracy: Exact
Reference: **Tada 1985**

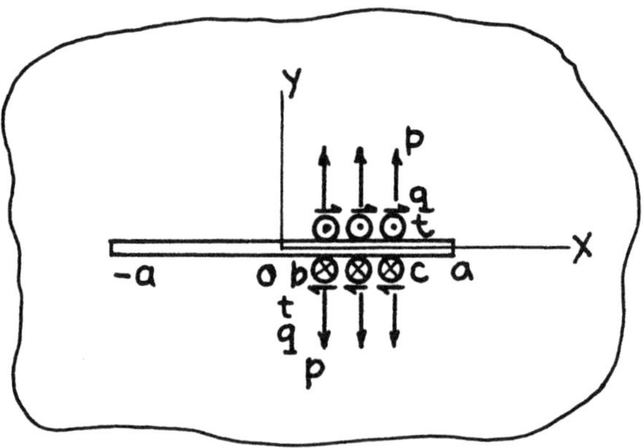

$$z = x + iy$$

$$\begin{Bmatrix} Z_I(z) \\ Z_{II}(z) \\ Z_{III}(z) \end{Bmatrix} = \frac{1}{\pi}\begin{Bmatrix} p \\ q \\ t \end{Bmatrix}\left[\sin^{-1}\frac{a^2 - cz}{a(z-c)} - \sin^{-1}\frac{a^2 - bz}{a(z-b)} + \frac{\sin^{-1}\frac{c}{a} - \sin^{-1}\frac{b}{a}}{\sqrt{1-(a/z)^2}} - \frac{\sqrt{a^2-c^2} - \sqrt{a^2-b^2}}{\sqrt{z^2-a^2}}\right]$$

$$\begin{Bmatrix} \bar{Z}_I(z) \\ \bar{Z}_{II}(z) \\ \bar{Z}_{III}(z) \end{Bmatrix} = \frac{1}{\pi}\begin{Bmatrix} p \\ q \\ t \end{Bmatrix}\left[(z-c)\sin^{-1}\frac{a^2 - cz}{a(z-c)} - (z-b)\sin^{-1}\frac{a^2 - bz}{a(z-b)} + \left(\sin^{-1}\frac{c}{a} - \sin^{-1}\frac{b}{a}\right)\sqrt{z^2-a^2}\right]$$

$$\begin{Bmatrix} K_I \\ K_{II} \\ K_{III} \end{Bmatrix}_{\pm a} = \frac{1}{\pi}\begin{Bmatrix} p \\ q \\ t \end{Bmatrix}\sqrt{\pi a}\left[\sin^{-1}\frac{c}{a} - \sin^{-1}\frac{b}{a} \mp \left(\sqrt{1-(c/a)^2} - \sqrt{1-(b/a)^2}\right)\right]$$

$$\mathrm{Im}\begin{Bmatrix} \bar{Z}_I(x) \\ \bar{Z}_{II}(x) \\ \bar{Z}_{III}(x) \end{Bmatrix}_{|x|\le a} = \frac{1}{\pi}\begin{Bmatrix} p \\ q \\ t \end{Bmatrix}\left[(c-x)\cosh^{-1}\frac{a^2 - cx}{a|x-c|} - (b-x)\cosh^{-1}\frac{a^2 - bx}{a|x-b|} + \left(\sin^{-1}\frac{c}{a} - \sin^{-1}\frac{b}{a}\right)\sqrt{a^2-x^2}\right]$$

Method: Integration of **page 5.10**
Accuracy: Exact
References: **Erdogan 1962; Sih 1962a, 1964; Paris 1965**

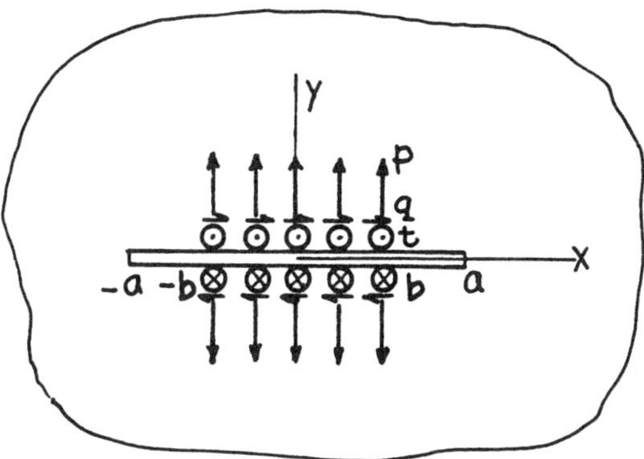

$$z = x + iy$$

$$\begin{Bmatrix} Z_I(z) \\ Z_{II}(z) \\ Z_{III}(z) \end{Bmatrix} = \frac{2}{\pi}\begin{Bmatrix} p \\ q \\ t \end{Bmatrix}\left[\frac{\sin^{-1}\frac{b}{a}}{\sqrt{1-(a/z)^2}} - \tan^{-1}\sqrt{\frac{1-(a/z)^2}{(a/b)^2-1}}\right]$$

$$\begin{Bmatrix} \overline{Z}_I(z) \\ \overline{Z}_{II}(z) \\ \overline{Z}_{III}(z) \end{Bmatrix} = \frac{2}{\pi}\begin{Bmatrix} p \\ q \\ t \end{Bmatrix}\left[\left(\sin^{-1}\frac{b}{a}\right)\sqrt{z^2-a^2} - z\tan^{-1}\sqrt{\frac{1-(a/z)^2}{(a/b)^2-1}} + b\tan^{-1}\sqrt{\frac{(z/a)^2-1}{1-(b/a)^2}}\right]$$

$$\begin{Bmatrix} K_I \\ K_{II} \\ K_{III} \end{Bmatrix} = \frac{2}{\pi}\begin{Bmatrix} p \\ q \\ t \end{Bmatrix}\left(\sin^{-1}\frac{b}{a}\right)\sqrt{\pi a}$$

$$\underset{|x|\le a}{\operatorname{Im}\begin{Bmatrix} \overline{Z}_I(x) \\ \overline{Z}_{II}(x) \\ \overline{Z}_{III}(x) \end{Bmatrix}} = \frac{2}{\pi}\begin{Bmatrix} p \\ q \\ t \end{Bmatrix}\left[\left(\sin^{-1}\frac{b}{a}\right)\sqrt{a^2-x^2} - x\begin{pmatrix} \coth^{-1} \\ \tanh^{-1} \end{pmatrix}\sqrt{\frac{(a/x)^2-1}{(a/b)^2-1}}\right.$$

$$\left. + b\begin{pmatrix} \coth^{-1} \\ \tanh^{-1} \end{pmatrix}\sqrt{\frac{1-(x/a)^2}{1-(b/a)^2}}\right]\begin{pmatrix} |x| < b \\ b < |x| \le a \end{pmatrix}$$

$$\underset{|x|=b}{\operatorname{Im}\begin{Bmatrix} \overline{Z}_I(x) \\ \overline{Z}_{II}(x) \\ \overline{Z}_{III}(x) \end{Bmatrix}} = \frac{2}{\pi}\begin{Bmatrix} p \\ q \\ t \end{Bmatrix}\left[\left(\sin^{-1}\frac{b}{a}\right)\sqrt{a^2-b^2} + b\ln\frac{a}{b}\right]$$

Method: Special Case of **page 5.12**
Accuracy: Exact
References: **Erdogan 1962; Sih 1962a, 1964; Paris 1965**

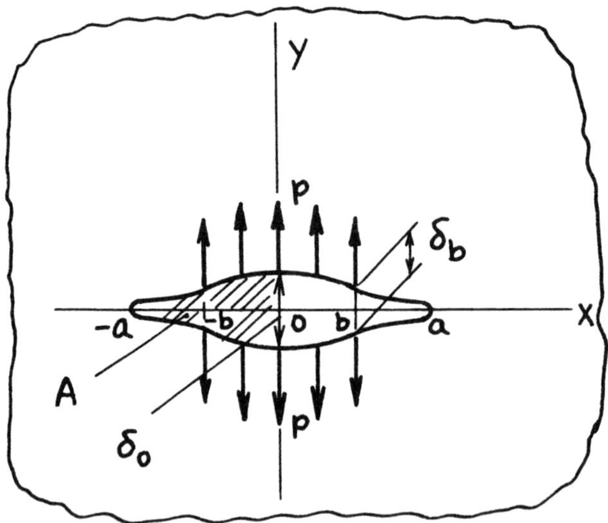

$$K_I = p\sqrt{\pi a} \cdot \frac{2}{\pi} \sin^{-1} \frac{b}{a}$$

Crack Opening Area:

$$A = \frac{2pa^2}{E'} \left\{ \sin^{-1} \frac{b}{a} + \frac{b}{a} \sqrt{1 - \left(\frac{b}{a}\right)^2} \right\}$$

Opening at Center:

$$\delta_0 = 2v(0,0) = \frac{8pa}{\pi E'} \left\{ \sin^{-1} \frac{b}{a} + \frac{b}{a} \cosh^{-1} \frac{a}{b} \right\}$$

Opening at $x = b$:

$$\delta_b = 2v(\pm b, 0) = \frac{8pa}{\pi E'} \left\{ \sqrt{1 - \left(\frac{b}{a}\right)^2} \sin^{-1} \frac{b}{a} - \frac{b}{a} \ell n \frac{b}{a} \right\}$$

Method: A, δ Paris' Equation (see **Appendix B**) or Integration of **pages 5.11 and 5.11a**
Accuracy: Exact
Reference: **Tada 1985**

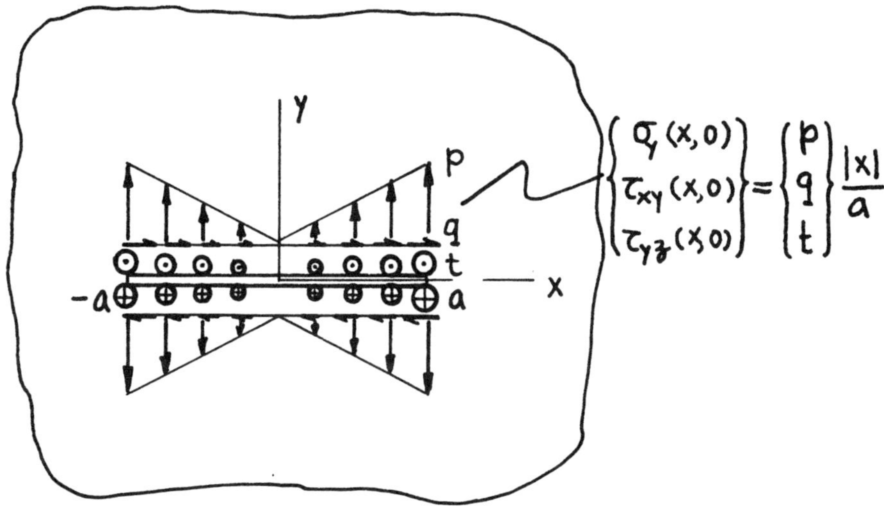

$$z = x + iy$$

$$\begin{Bmatrix} Z_I(z) \\ Z_{II}(z) \\ Z_{III}(z) \end{Bmatrix} = \frac{2}{\pi}\begin{Bmatrix} p \\ q \\ t \end{Bmatrix}\left\{ \frac{1}{\sqrt{1-(a/z)^2}} + \frac{z}{a}\sin^{-1}\frac{a}{z} \right\}$$

$$\begin{Bmatrix} \overline{Z}_I(z) \\ \overline{Z}_{II}(z) \\ \overline{Z}_{III}(z) \end{Bmatrix} = \frac{1}{\pi}\begin{Bmatrix} p \\ q \\ t \end{Bmatrix}\left\{ \sqrt{z^2-a^2} + \frac{z^2}{a}\sin^{-1}\frac{a}{z} \right\}$$

$$\begin{Bmatrix} K_I \\ K_{II} \\ K_{III} \end{Bmatrix} = \frac{2}{\pi}\begin{Bmatrix} p \\ q \\ t \end{Bmatrix}\sqrt{\pi a}$$

$$\operatorname{Im}\underset{|x|\leq a}{\begin{Bmatrix} \overline{Z}_I(x) \\ \overline{Z}_{II}(x) \\ \overline{Z}_{III}(x) \end{Bmatrix}} = \frac{1}{\pi}\begin{Bmatrix} p \\ q \\ t \end{Bmatrix} a\left\{ \sqrt{1-(x/a)^2} + \left(\frac{x}{a}\right)^2\cosh^{-1}\frac{a}{x} \right\}$$

Method: Integration of **page 5.11**
Accuracy: Exact
Reference: **Tada 1973**

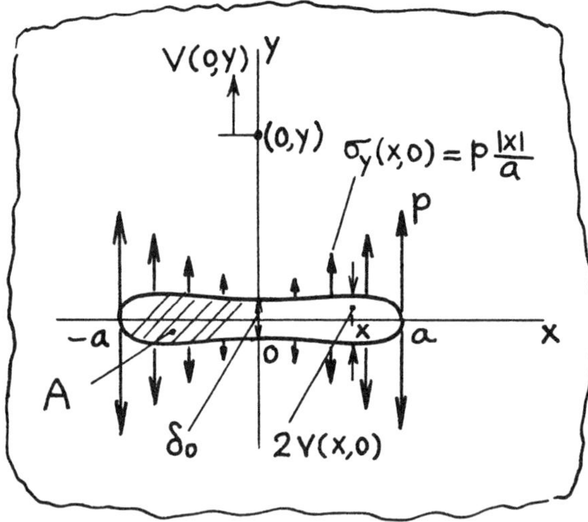

$$K_I = \frac{2}{\pi} p\sqrt{\pi a}$$

Crack Opening Area:

$$A = \frac{8}{3} \cdot \frac{pa^2}{E'}$$

Crack Opening Profile:

$$\underset{|x| \leq a}{2v(x,0)} = \frac{4pa}{\pi E'}\left\{\sqrt{1-\left(\frac{x}{a}\right)^2} + \left(\frac{x}{a}\right)^2 \cosh^{-1}\frac{a}{x}\right\}$$

Opening at Center:

$$\delta_0 = 2v(0,0) = \frac{4pa}{\pi E'}$$

Vertical Displacement at $(0, y)$:

$$v(0,y) = \frac{2p}{\pi E'}\, y\left\{\sqrt{1+\left(\frac{a}{y}\right)^2} - \frac{y}{a}\sinh^{-1}\frac{a}{y} - 2\alpha\left(\frac{1}{\sqrt{1+\left(\frac{a}{y}\right)^2}} - \frac{y}{a}\sinh^{-1}\frac{a}{y}\right)\right\}$$

where

$$\alpha = \begin{cases} \frac{1}{2}(1+\nu) & \text{plane stress} \\ \frac{1}{2}\left(\frac{1}{1-\nu}\right) & \text{plane strain} \end{cases}$$

Method: Integration of **pages 5.11 and 5.11a** or Paris' Equation (see **Appendix B**)
Accuracy: Exact
Reference: **Tada 1985**

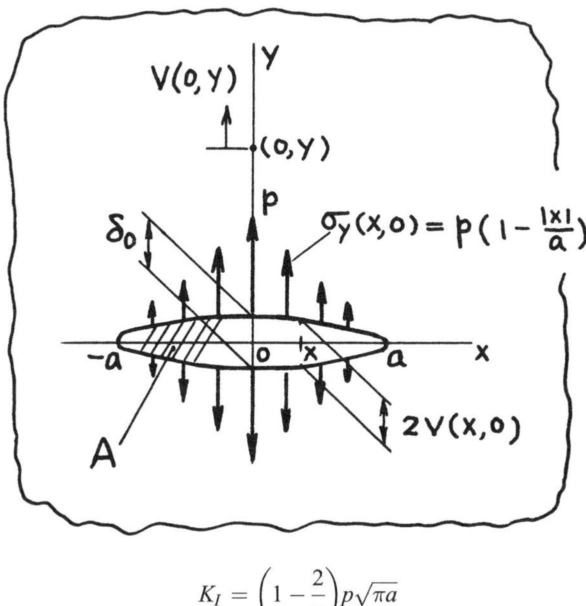

$$K_I = \left(1 - \frac{2}{\pi}\right) p\sqrt{\pi a}$$

Crack Opening Area:

$$A = 2\left(\pi - \frac{4}{3}\right)\frac{pa^2}{E'}$$

Crack Opening Profile:

$$\underset{|x| \leq a}{2v(x,0)} = \frac{4pa}{E'}\left\{\left(1 - \frac{1}{\pi}\right)\sqrt{1 - \left(\frac{x}{a}\right)^2} - \frac{1}{\pi}\left(\frac{x}{a}\right)^2 \cosh^{-1}\frac{a}{x}\right\}$$

Opening at Center:

$$\delta_0 = 2v(0,0) = \frac{4pa}{E'}\left(1 - \frac{1}{\pi}\right)$$

Vertical Displacement at $(0, y)$:

$$v(0,y) = \frac{2p}{E'}\left(1 - \alpha y\frac{\partial}{\partial y}\right)\left\{\left(1 - \frac{1}{\pi}\right)\sqrt{a^2 + y^2} - y\left(1 - \frac{1}{\pi}\frac{y}{a}\sinh^{-1}\frac{a}{y}\right)\right\}$$

where

$$\alpha = \begin{cases} \frac{1}{2}(1+\nu) & \text{plane stress} \\ \frac{1}{2}\left(\frac{1}{1-\nu}\right) & \text{plane strain} \end{cases}$$

Method: Superposition of **pages 5.1, 5.1a and 5.14, 5.14a**
Accuracy: Exact
Reference: **Tada 1985**

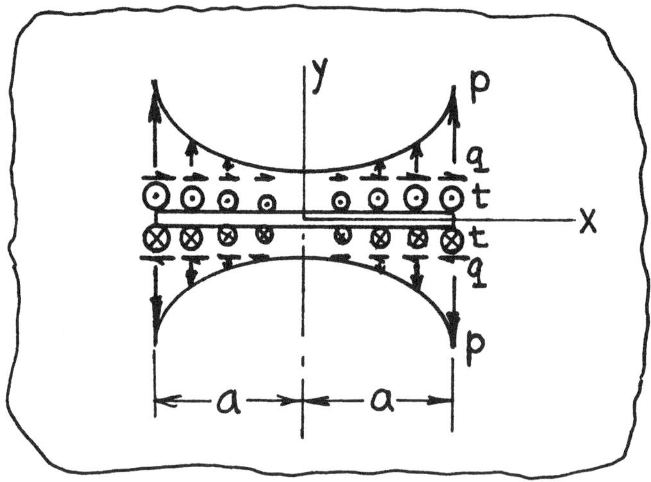

$$\left\{\begin{array}{c}\sigma_y(x,0)\\ \tau_{xy}(x,0)\\ \tau_{yz}(x,0)\end{array}\right\}_{|x|\le a} = \left\{\begin{array}{c}p\\ q\\ t\end{array}\right\}\left(\frac{|x|}{a}\right)^{\gamma} \qquad (\gamma > -1)$$

$$\left\{\begin{array}{c}K_I\\ K_{II}\\ K_{III}\end{array}\right\} = \left\{\begin{array}{c}p\\ q\\ t\end{array}\right\}\sqrt{a}\frac{\Gamma\left(\frac{\gamma+1}{2}\right)}{\Gamma\left(\frac{\gamma}{2}+1\right)}$$

where $\Gamma(\gamma) =$ Gamma Function (See **Appendix M**)

Methods: Fourier Transform (Sneddon); Integration of **page 5.11** (Tada)
Accuracy: Exact
References: **Sneddon 1951; Tada 1974**

NOTE: For special cases of $\gamma = 0$ and $\gamma = 1$, see **page 5.1** and **page 5.14**, respectively.

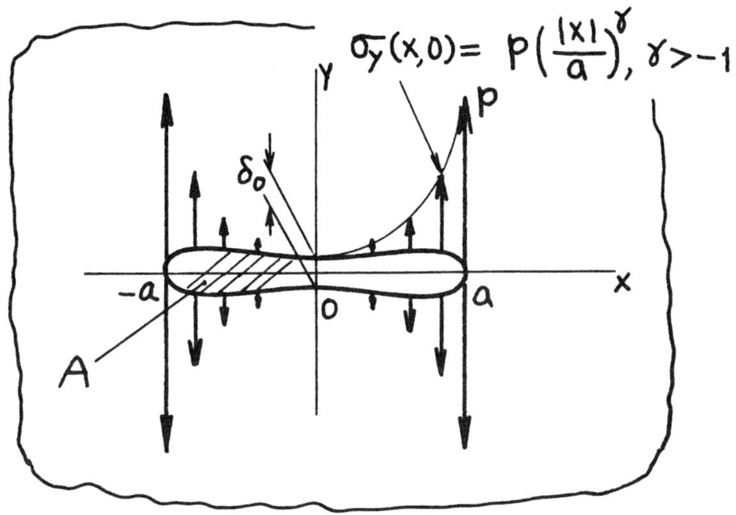

$$K_I = p\sqrt{a}\frac{\Gamma\left(\frac{\gamma+1}{2}\right)}{\Gamma\left(\frac{\gamma}{2}+1\right)}$$

Crack Opening Area:

$$A = \frac{2pa^2}{E'}\sqrt{\pi}\frac{\Gamma\left(\frac{\gamma+1}{2}\right)}{\Gamma\left(\frac{\gamma}{2}+2\right)}$$

Opening at Center:

$$\delta_0 = 2v(0,0) = \frac{4pa}{E'}\cdot\frac{1}{\sqrt{\pi}}\cdot\frac{1}{\gamma+1}\frac{\Gamma\left(\frac{\gamma+1}{2}\right)}{\Gamma\left(\frac{\gamma}{2}+1\right)}$$

where

$\Gamma(\gamma)$ = Gamma Function (See **Appendix M**)

Method: Integration of **pages 5.11 and 5.11a** or Paris' Equation (see **Appendix B**)
Accuracy: Exact
Reference: **Tada 1985**

NOTE: For special cases $\gamma = 0$ and $\gamma = 1$, see **pages 5.1, 5.1a, and 5.14, 5.14a.**

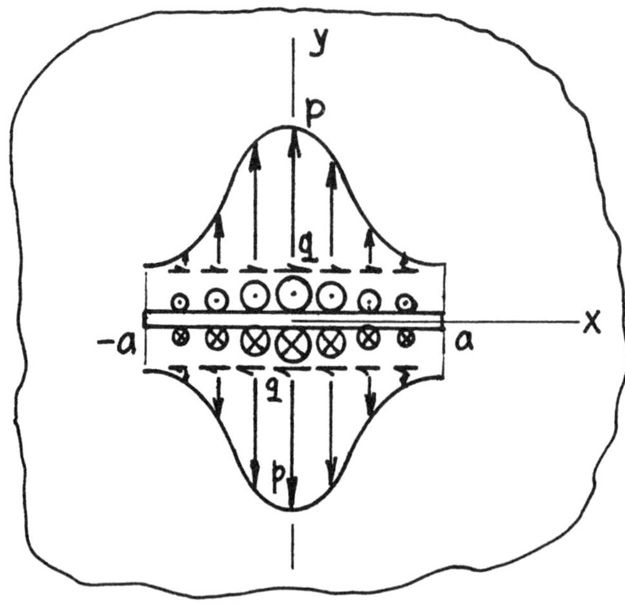

$$\left\{\begin{matrix}\sigma_y(x,0)\\ \tau_{xy}(x,0)\\ \tau_{yz}(x,0)\end{matrix}\right\}_{|x|\le a} = \left\{\begin{matrix}p\\ q\\ t\end{matrix}\right\}\left\{\sqrt{1-\left(\frac{x}{a}\right)^2}\right\}^{\gamma} \quad (\gamma > -1)$$

$$\left\{\begin{matrix}K_I\\ K_{II}\\ K_{III}\end{matrix}\right\} = \left\{\begin{matrix}p\\ q\\ t\end{matrix}\right\}\sqrt{a}\frac{\Gamma\left(\frac{\gamma+1}{2}\right)}{\Gamma\left(\frac{\gamma}{2}+1\right)}$$

where $\Gamma(\gamma)$Gamma Function (See **Appendix M**)

Method: Integration of **page 5.11**
Accuracy: Exact
Reference: **Tada 1974**

NOTE: For special case of $\gamma = 0$, see **page 5.1.**

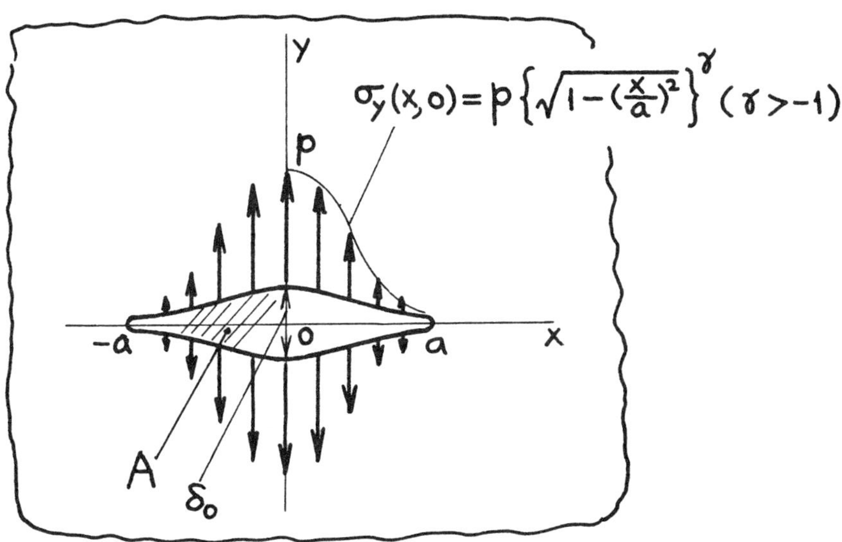

$$K_I = p\sqrt{a}\frac{\Gamma\left(\frac{\gamma+1}{2}\right)}{\Gamma\left(\frac{\gamma}{2}+1\right)}$$

Crack Opening Area:

$$A = \frac{4pa^2}{E'}\sqrt{\pi}\frac{\Gamma\left(\frac{\gamma+3}{2}\right)}{\Gamma\left(\frac{\gamma}{2}+2\right)}$$

Opening at Center:

$$\delta_0 = 2v(0,0) = \frac{4pa}{E'}V(\gamma)$$

where $\Gamma(\gamma)$ = Gamma Function (see **Appendix M**)

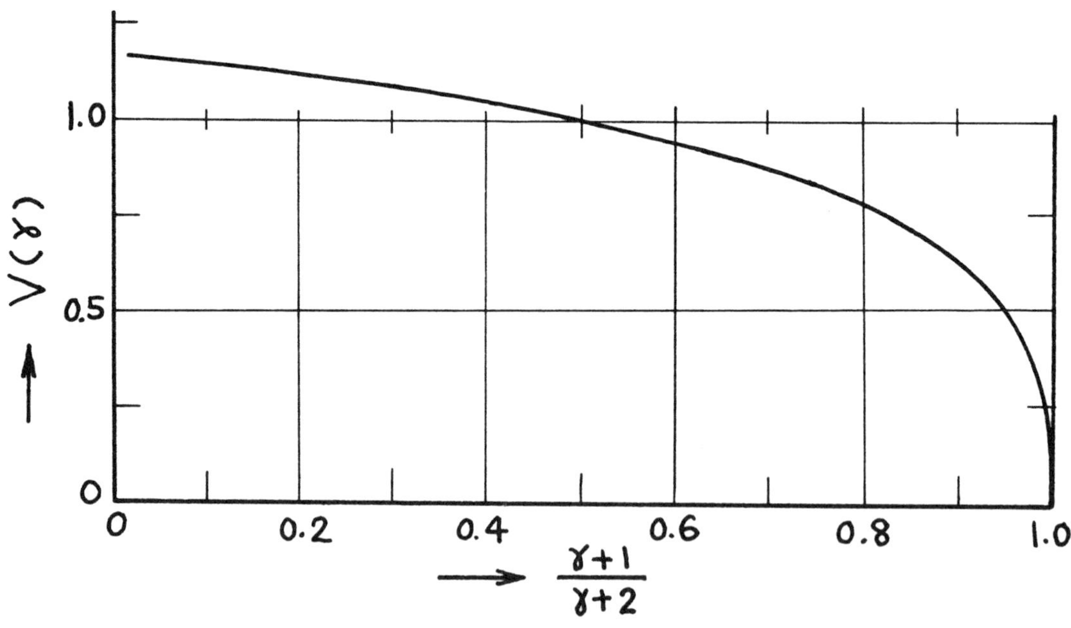

Method: Integration of **pages 5.11 and 5.11a**
Accuracy: A Exact; $V(\gamma)$ curve is based on accurate numerical values.
Reference: **Tada 1985**

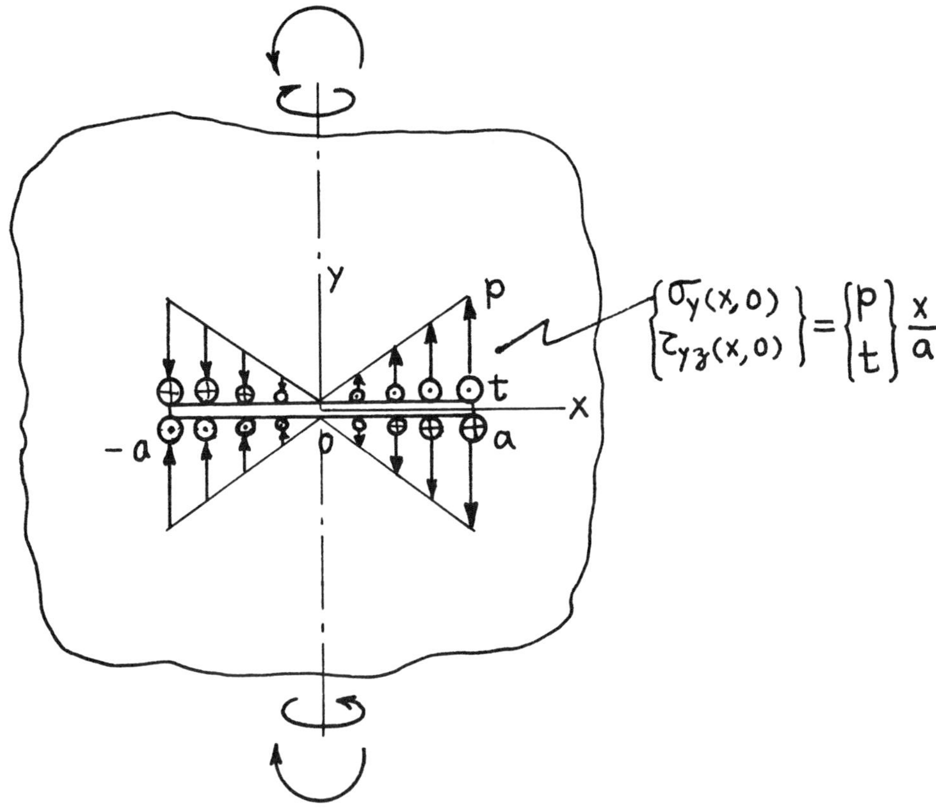

$$z = x + iy$$

$$\begin{Bmatrix} Z_I(z) \\ Z_{III}(z) \end{Bmatrix} = \frac{1}{2a} \begin{Bmatrix} p \\ t \end{Bmatrix} \frac{2z^2 - a^2}{\sqrt{z^2 - a^2}}$$

$$\begin{Bmatrix} \overline{Z}_I(z) \\ \overline{Z}_{III}(z) \end{Bmatrix} = \frac{1}{2a} \begin{Bmatrix} p \\ t \end{Bmatrix} \left[z\sqrt{z^2 - a^2} \right]$$

$$\begin{Bmatrix} K_I \\ K_{III} \end{Bmatrix}_{\pm a} = \pm \frac{1}{2} \begin{Bmatrix} p \\ t \end{Bmatrix} \sqrt{\pi a}$$

$$\operatorname{Im} \begin{Bmatrix} \overline{Z}_I(x) \\ \overline{Z}_{III}(x) \end{Bmatrix}_{|x| \le a} = \frac{1}{2a} \begin{Bmatrix} p \\ t \end{Bmatrix} \left[x\sqrt{a^2 - x^2} \right]$$

Method: Integration of **page 5.11** or by Stress Concentration Factor
Accuracy: Exact
References: **Neuber 1937; Benthem 1972**

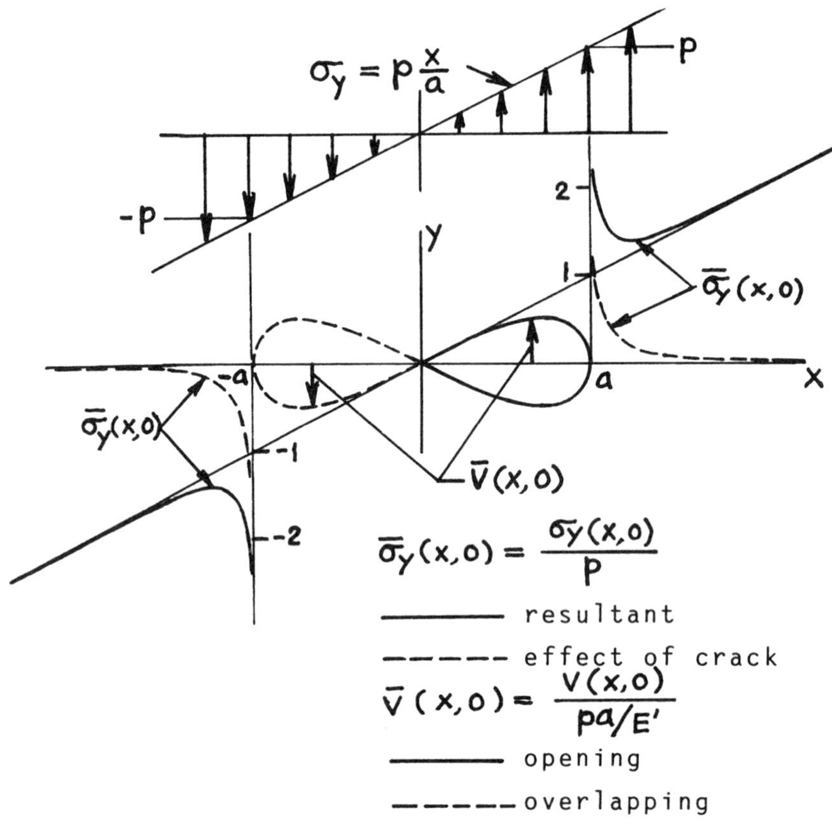

$$K_{I\pm a} = \pm\frac{1}{2}p\sqrt{\pi a}$$

$$\underset{|x|>a}{\sigma_y(x,0)} = p\frac{|x|}{x}\frac{\left(\frac{x}{a}\right)^2 - \frac{1}{2}}{\sqrt{\left(\frac{x}{a}\right)^2 - 1}}$$

$$\underset{|x|\le a}{2v(x,0)} = \frac{2pa}{E'}\cdot\left(\frac{x}{a}\right)\sqrt{1-\left(\frac{x}{a}\right)^2}$$

Crack Opening Area of Right Half:

$$A_{x>0} = \frac{2pa^2}{3E'}$$

Method: Westergaard Stress Functions
Accuracy: Exact
Reference: **Tada 2000**

NOTE: Crack surface interference was ignored. See **page 5.18b** for the effect of surface interference.

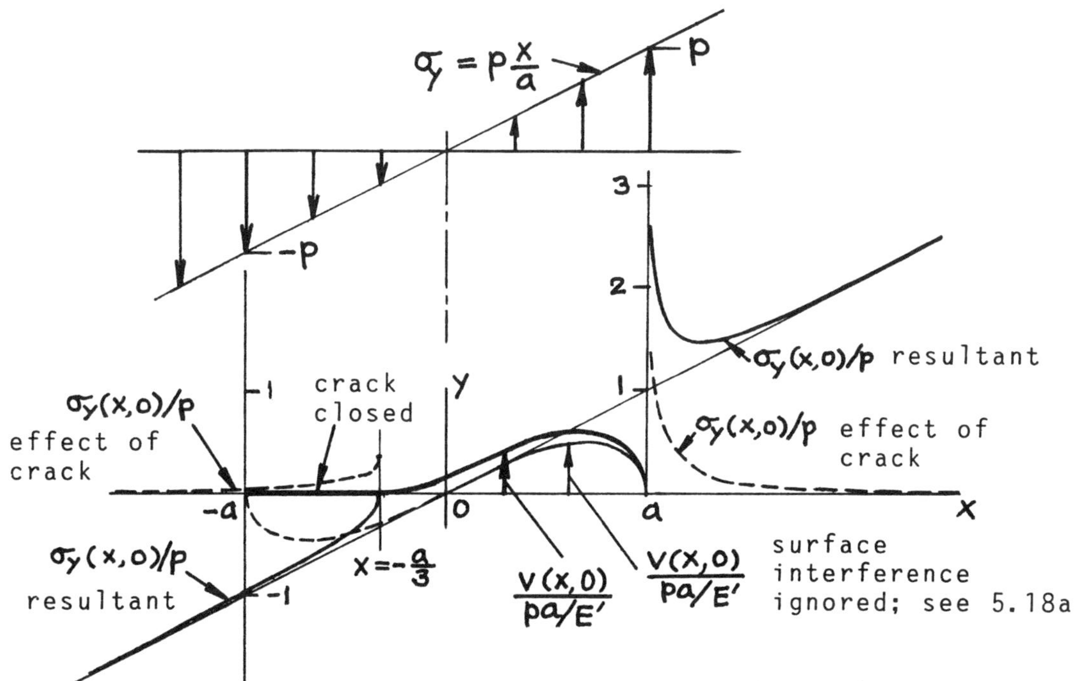

Crack closes from $x = -a$ to $x = -{}^a\!/_3$.

$$Z_I(z) = \frac{p}{a}\left(z - \frac{2a}{3}\right)\sqrt{\frac{z + {}^a\!/_3}{z - a}}$$

$$\overline{Z}_I(z) = \frac{p}{2a}\left(z + \frac{a}{3}\right)\sqrt{(z + {}^a\!/_3)(z - a)}$$

$$K_{I,\,x=a} = \left(\frac{2}{3}\right)^{3/2} p\sqrt{\pi a} = 0.5443p\sqrt{\pi a}; \quad K_{I,\,x=-a/3} = 0$$

$$\underset{x<-a/3,\,x>a}{\sigma_y(x,\ 0)} = \frac{p}{a}\left(x - \frac{2a}{3}\right)\sqrt{\frac{x + {}^a\!/_3}{x - a}}$$

$$\underset{-a/3\le x\le a}{2v(x,\ 0)} = \frac{2p}{E'}\left(x + \frac{a}{3}\right)\sqrt{\left(x + \frac{a}{3}\right)(a - x)}$$

Crack Opening Area:

$$A = \left(\frac{2}{3}\right)^3 \frac{\pi p a^2}{E'} = 0.9308 \frac{p a^2}{E'}$$

Method: Superposition of **pages 5.1/1a and 5.18/18a**
Accuracy: Exact
References: **Seeger 1973; Tada 2000**

NOTE: Compare with **page 5.18a** for the effect of crack surface interference.

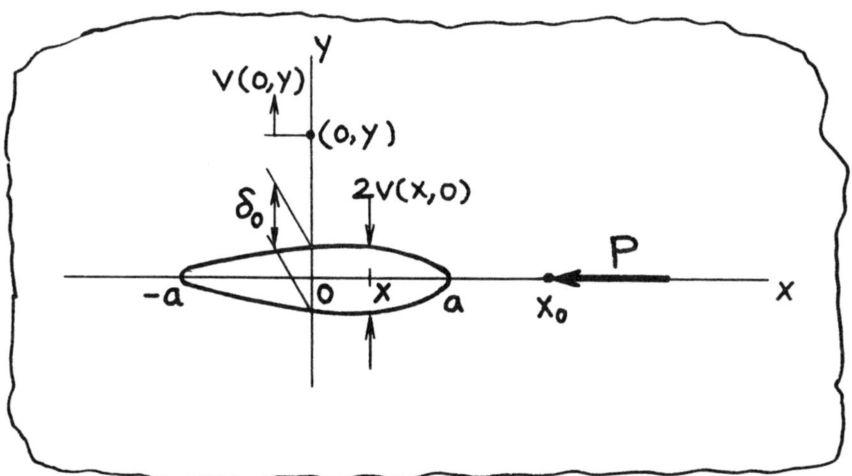

$$K_{I+a} = \frac{P}{2\sqrt{\pi a}}(1-\alpha)\left(\sqrt{\frac{x_0+a}{x_0-a}}-1\right)$$

$$K_{I-a} = \frac{P}{2\sqrt{\pi a}}(1-\alpha)\left(1-\sqrt{\frac{x_0-a}{x_0+a}}\right)$$

Crack Opening Area:

$$A = \frac{2(1-\alpha)}{E'}Pa\left\{\frac{x_0}{a}-\sqrt{\left(\frac{x_0}{a}\right)^2-1}\right\}$$

Crack Opening Profile:

$$\underset{|x|\le a}{2v(x,0)} = \frac{2(1-\alpha)}{\pi E'}P\left\{\sin^{-1}\frac{1-\frac{x_0}{a}\frac{x}{a}}{\frac{x_0}{a}-\frac{x}{a}}+\sin^{-1}\frac{x}{a}\right\}$$

Opening at Center:

$$\delta_0 = 2v(0,0) = \frac{2(1-\alpha)}{\pi E'}P\ \sin^{-1}\frac{a}{x_0}$$

Vertical Displacement at $(0, y)$:

$$v(0,y) = \frac{1-\alpha}{\pi E'} P\left(1 - \alpha y \frac{\partial}{\partial y}\right) \left\{ \cos^{-1}\sqrt{\frac{x_0^2 - a^2}{x_0^2 + y^2}} - \cos^{-1}\frac{x_0}{\sqrt{x_0^2 + y^2}} \right\}$$

where

$$\alpha = \begin{cases} \frac{1}{2}(1+\nu) & \text{plane stress} \\ \frac{1}{2}\left(\frac{1}{1-\nu}\right) & \text{plane strain} \end{cases}$$

Method: Integration of **pages 5.9, 5.9a**; A, δ_0, One half of **page 5.20**
Accuracy: Exact
Reference: **Tada 1985**

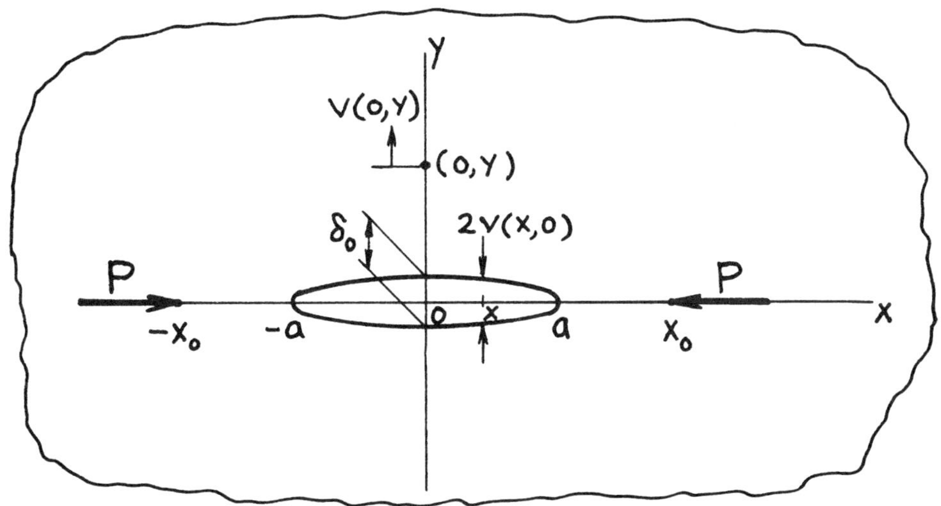

$$K_I = \frac{P}{\sqrt{\pi a}}(1-\alpha)\frac{a}{\sqrt{x_0^2 - a^2}}$$

Crack Opening Area:

$$A = \frac{4(1-\alpha)}{E'}Pa\left\{\frac{x_0}{a} - \sqrt{\left(\frac{x_0}{a}\right)^2 - 1}\right\}$$

Crack Opening Profile:

$$\underset{|x|\leq a}{2v(x,0)} = \frac{4(1-\alpha)}{\pi E'}P\cos^{-1}\sqrt{\frac{x_0^2 - a^2}{x_0^2 - x^2}}$$

Opening at Center:

$$\delta_0 = 2v(0,0) = \frac{4(1-\alpha)}{\pi E'}P\sin^{-1}\frac{a}{x_0}$$

Displacement at $(0,y)$:

$$v(0,y) = \frac{2(1-\alpha)}{\pi E'}P\left(1 - \alpha y\frac{\partial}{\partial y}\right)\left\{\cos^{-1}\sqrt{\frac{x_0^2 - a^2}{x_0^2 + y^2}} - \cos^{-1}\frac{x_0}{\sqrt{x_0^2 + y^2}}\right\}$$

where

$$\alpha = \begin{cases} \frac{1}{2}(1+\nu) & \text{plane stress} \\ \frac{1}{2}\left(\frac{1}{1-\nu}\right) & \text{plane strain} \end{cases}$$

Methods: Superposition of **page 5.19**; Paris' Equation for A, v (see **Appendix B**)
Accuracy: Exact
Reference: **Tada 1985**

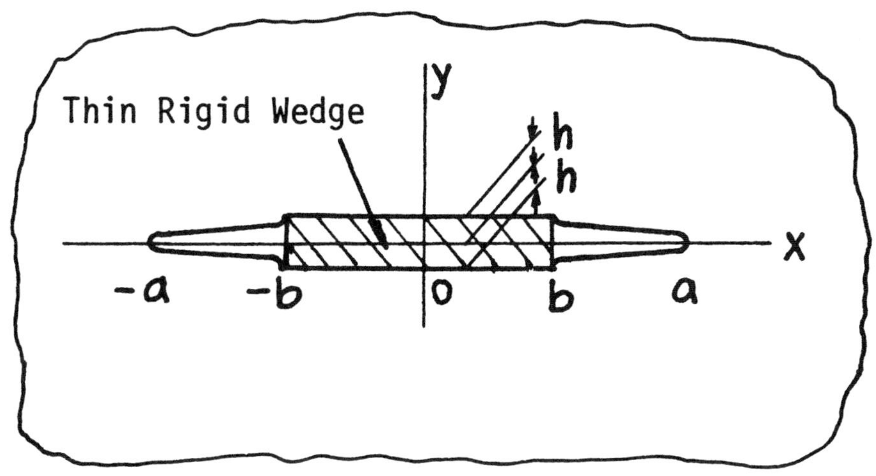

$$K_{I\pm a} = \frac{E'h}{2k\,K(k)}\sqrt{\frac{\pi}{a}}$$

$$K_{I\pm b} = -\frac{E'h}{2k\,K(k)}\sqrt{\frac{\pi}{b}}$$

$$\underset{b\le|x|\le a}{2v(x,0)} = 2h\cdot\frac{F(\varphi,k)}{K(k)}$$

where

$$E' = \begin{cases} E & \text{plane stress} \\ E\Big/\left(1-\nu^2\right) & \text{plane strain} \end{cases}$$

$$k = \sqrt{1-\left(b/a\right)^2}$$

$$\varphi = \sin^{-1}\sqrt{\frac{a^2-x^2}{a^2-b^2}}$$

$$F(\varphi,k) = \int_0^{\varphi}\frac{d\varphi}{\sqrt{1-k^2\sin^2\varphi}}$$

$$K(k) = F\left(\pi/2,k\right)$$

Methods: Muskhelishvili's Method (Markuzon); Triple Integral Equations (Tweed)
Accuracy: Exact
References: **Markuzon 1961; Tweed 1970; Tada 1974**

NOTE: For $K(k)$, see **Appendix L.**

The Westergaard Function is

$$Z_I(z) = \frac{E'ha}{2K(k)} \cdot \frac{1}{\sqrt{z^2 - a^2}} \cdot \frac{1}{\sqrt{z^2 - b^2}}$$

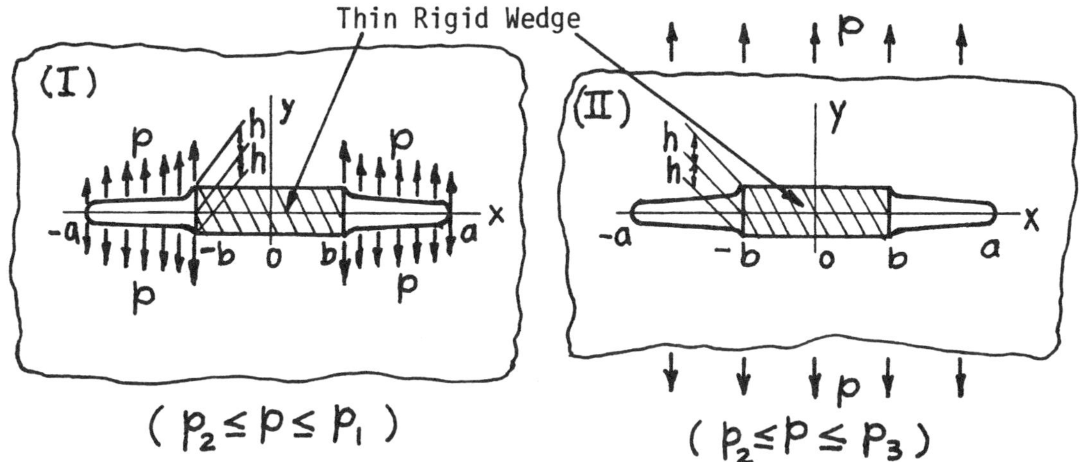

$$p_1 = \frac{E'ha}{2\left\{a^2 E(k) - b^2 K(k)\right\}}$$

$$p_2 = -\frac{E'h}{2a\{K(k) - E(k)\}}$$

$$p_3 = \frac{E'h}{2bK(k)}$$

$$\underset{x=\pm a}{K_I} = \left\{\frac{E'h}{2} + pa(K(k) - E(k))\right\} \frac{1}{k\,K(k)} \sqrt{\frac{\pi}{a}}$$

$$\underset{x=\pm b}{K_I} = \left\{-\frac{E'h}{2} + p\frac{a^2 E(k) - b^2 K(k)}{a}\right\} \frac{1}{k\,K(k)} \sqrt{\frac{\pi}{b}}$$

$$\underset{b\le|x|\le a}{2v(x,0)} = 2h \cdot \frac{F(\varphi,k)}{K(k)} + \frac{4pa}{E'}\{K(k)E(\varphi,k) - E(k)F(\varphi,k)\}$$

where

$$E' = \begin{cases} E & \text{plane stress} \\ E/\left(1-\nu^2\right) & \text{plane strain} \end{cases}$$

$$k = \sqrt{1 - \left({}^{b}\!/_{a}\right)^2}, \quad \varphi = \sin^{-1}\sqrt{\frac{a^2 - x^2}{a^2 - b^2}}$$

$$F(\varphi, k) = \int_0^{\varphi} \frac{d\varphi}{\sqrt{1 - k^2 \sin^2 \varphi}}, \quad K(k) = F\left({}^{\pi}\!/_{2}, k\right)$$

$$E(\varphi, k) = \int_0^{\varphi} \sqrt{1 - k^2 \sin^2 \varphi}\, d\varphi, \quad E(k) = E\left({}^{\pi}\!/_{2}, k\right)$$

Methods: Triple Integral Equations or Superposition of **page 5.21** and **page 6.1**
Accuracy: Exact
References: **Tweed 1970; Tada 1974**

NOTE: 1. In (I), when $p > p_1$, separation of contact surfaces occurs near $x = \pm b$, and when $p < p_2$, crack closure occurs near $x = \pm a$.
2. In (II), when $p > p_3$, separation of contact surfaces occurs near $x = 0$, and when $p < p_2$, crack closure occurs near $x = \pm a$.
3. For $K(k)$ and $E(k)$, see **Appendix L**

The Westergaard Function is $\underset{\textbf{p. 5.22}}{Z(z)} = \underset{\textbf{p. 5.21}}{Z(z)} + \underset{\textbf{p. 6.1}}{Z(z)}$

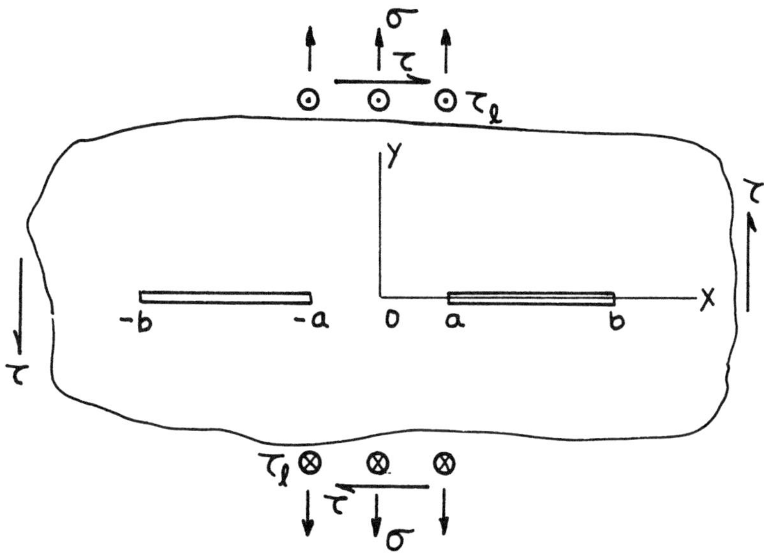

$$\begin{Bmatrix} K_I \\ K_{II} \\ K_{III} \end{Bmatrix}_{\pm a} = \begin{Bmatrix} \sigma \\ \tau \\ \tau_\ell \end{Bmatrix} \sqrt{\pi a} \frac{b^2 \frac{E(k)}{K(k)} - a^2}{a\sqrt{b^2 - a^2}}$$

$$\begin{Bmatrix} K_I \\ K_{II} \\ K_{III} \end{Bmatrix}_{\pm b} = \begin{Bmatrix} \sigma \\ \tau \\ \tau_\ell \end{Bmatrix} \sqrt{\pi b} \frac{1}{k} \left\{ 1 - \frac{E(k)}{K(k)} \right\}$$

where
$$k = \sqrt{1 - a^2/b^2}$$

$$K(k) = \int_0^{\pi/2} \frac{d\varphi}{\sqrt{1 - k^2 \sin^2 \varphi}}$$

$$E(k) = \int_0^{\pi/2} \sqrt{1 - k^2 \sin^2 \varphi}\, d\varphi$$

(See **Appendix L** for values of $K(k)$ and $E(k)$)
Method: Muskhelishvili's Method
Accuracy: Exact
References: **Barenblatt 1962, Erdogan 1962, Sih 1964**

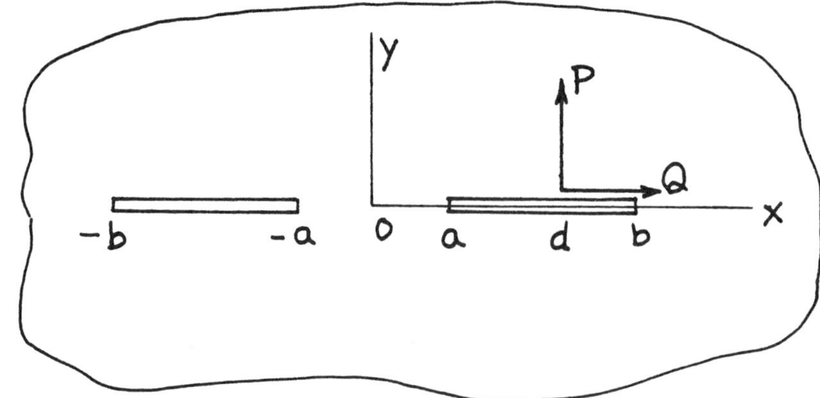

$$K_{I\pm b} = \frac{1}{2\sqrt{\pi b}\sqrt{b^2 - a^2}}\left\{P\sqrt{d^2 - a^2}\sqrt{\frac{b\pm d}{b\mp d}} + \left(\frac{\kappa - 1}{\kappa + 1}\right)\left(1\pm\frac{\pi}{2K(k)}\right)Qb\mp\frac{Pb}{K(k)}[E(k)F(\theta,k) - K(k)E(\theta,k)]\right\}$$

$$K_{II\pm b} = \frac{1}{2\sqrt{\pi b}\sqrt{b^2 - a^2}}\left\{Q\sqrt{d^2 - a^2}\sqrt{\frac{b\pm d}{b\mp d}} - \left(\frac{\kappa - 1}{\kappa + 1}\right)\left(1\pm\frac{\pi}{2K(k)}\right)Pb\mp\frac{Qb}{K(k)}[E(k)F(\theta,k) - K(k)E(\theta,k)]\right\}$$

$$K_{I\pm a} = \frac{1}{2\sqrt{\pi a}\sqrt{b^2 - a^2}}\left\{P\sqrt{b^2 - d^2}\sqrt{\frac{d\pm a}{d\mp a}} - \left(\frac{\kappa - 1}{\kappa + 1}\right)Q\left(a\pm\frac{\pi}{2K(k)}b\right)\pm\frac{Pb}{K(k)}[E(k)F(\theta,k) - K(k)E(\theta,k)]\right\}$$

$$K_{II\pm a} = \frac{1}{2\sqrt{\pi a}\sqrt{b^2 - a^2}}\left\{Q\sqrt{b^2 - d^2}\sqrt{\frac{d\pm a}{d\mp a}} + \left(\frac{\kappa - 1}{\kappa + 1}\right)P\left(a\pm\frac{\pi}{2K(k)}b\right)\pm\frac{Qb}{K(k)}[E(k)F(\theta,k) - K(k)E(\theta,k)]\right\}$$

where

$$k = \sqrt{1 - a^2/b^2}, \quad \theta = \sqrt{\frac{b^2 - d^2}{b^2 - a^2}}$$

$$K(k) = \int_0^{\pi/2} \frac{d\varphi}{\sqrt{1 - k^2\sin^2\varphi}}, \quad E(k) = \int_0^{\pi/2} \sqrt{1 - k^2\sin^2\varphi}\ \ d\varphi$$

$$F(\theta,k) = \int_0^{\theta} \frac{d\varphi}{\sqrt{1 - k^2\sin^2\varphi}}, \quad E(\theta,k) = \int_0^{\theta} \sqrt{1 - k^2\sin^2\varphi}\ \ d\varphi$$

Method: Muskhelishvili's Method (See **Appendix L** for tables of $K(k)$ and $E(k)$)
Accuracy: Exact
References: **Barenblatt 1962, Erdogan 1962**

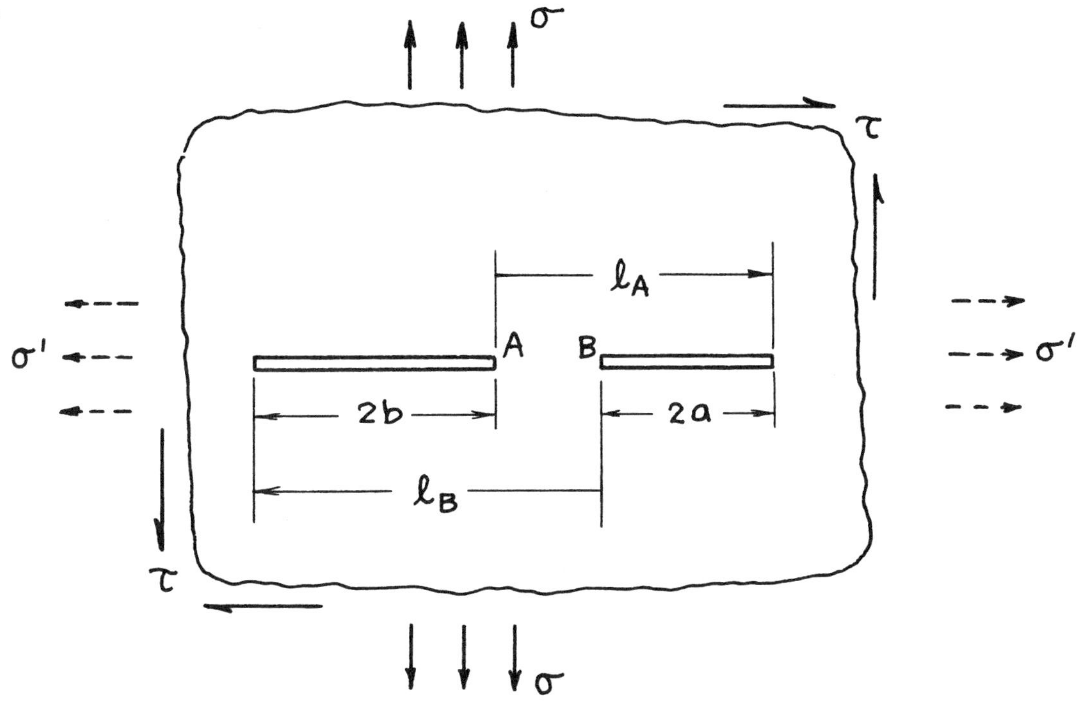

$$\left\{ \begin{matrix} K_I \\ K_{II} \end{matrix} \right\}_A = \left\{ \begin{matrix} \sigma \\ \tau \end{matrix} \right\} \sqrt{\pi b} \frac{1}{\sqrt{1-\alpha_A}} \left[1 - \frac{1}{\alpha_B} \left\{ 1 - \frac{E(k)}{K(k)} \right\} \right]$$

$$\left\{ \begin{matrix} K_I \\ K_{II} \end{matrix} \right\}_B = \left\{ \begin{matrix} \sigma \\ \tau \end{matrix} \right\} \sqrt{\pi a} \frac{1}{\sqrt{1-\alpha_B}} \left[1 - \frac{1}{\alpha_A} \left\{ 1 - \frac{E(k)}{K(k)} \right\} \right]$$

where

$$\alpha_A = \frac{2a}{\ell_A}, \quad \alpha_B = \frac{2b}{\ell_B}$$

$$k = \sqrt{\alpha_A \alpha_B}$$

$$K(k) = \int_0^{\pi/2} \frac{d\varphi}{\sqrt{1 - k^2 \sin^2 \varphi}}, \quad E(k) = \int_0^{\pi/2} \sqrt{1 - k^2 \sin^2 \varphi} \ \ d\varphi$$

Method: Complex Potentials
Accuracy: Exact
References: **Yokobori 1965 (or Kamei 1974), Isida 1973**

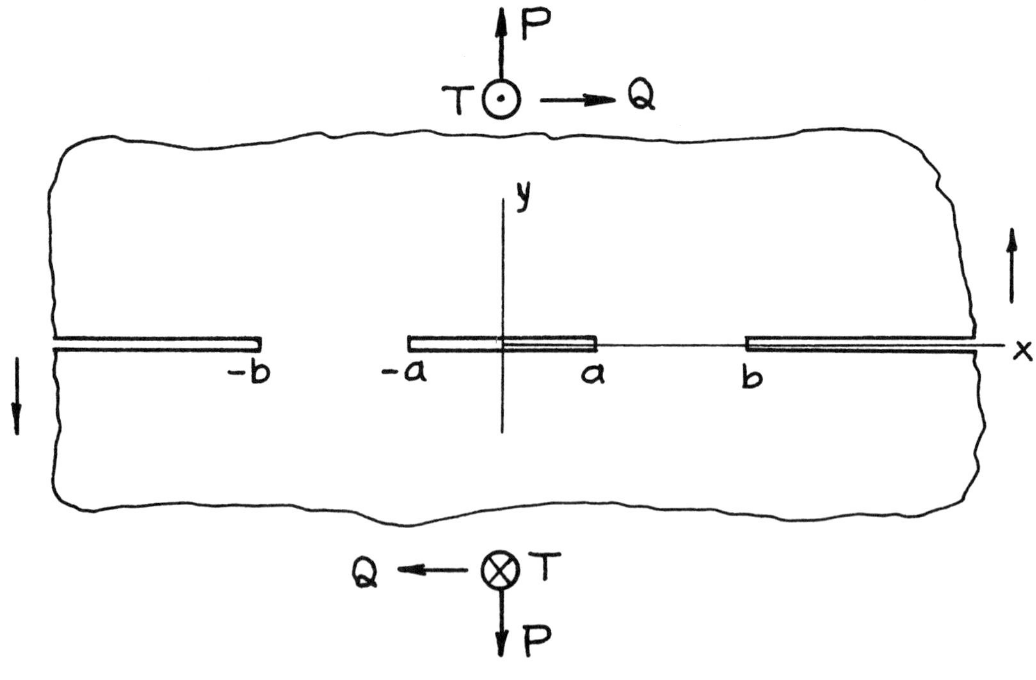

$$\left\{\begin{array}{l} K_I \\ K_{II} \\ K_{III} \end{array}\right\}_{\pm a} = \frac{1}{\sqrt{\pi a}} \left\{\begin{array}{l} P \\ Q \\ T \end{array}\right\} \frac{a}{\sqrt{b^2 - a^2}}$$

$$\left\{\begin{array}{l} K_I \\ K_{II} \\ K_{III} \end{array}\right\}_{\pm b} = \frac{1}{\sqrt{\pi b}} \left\{\begin{array}{l} P \\ Q \\ T \end{array}\right\} \frac{b}{\sqrt{b^2 - a^2}}$$

Methods: Westergaard Stress Function or from Solution of Punch Problem
Accuracy: Exact
References: **Galin 1953, Tada 1974**

The Westergaard Stress Function is

$$Z(z) = \frac{P}{\pi} \frac{z}{\sqrt{z^2 - a^2}\sqrt{b^2 - z^2}}$$

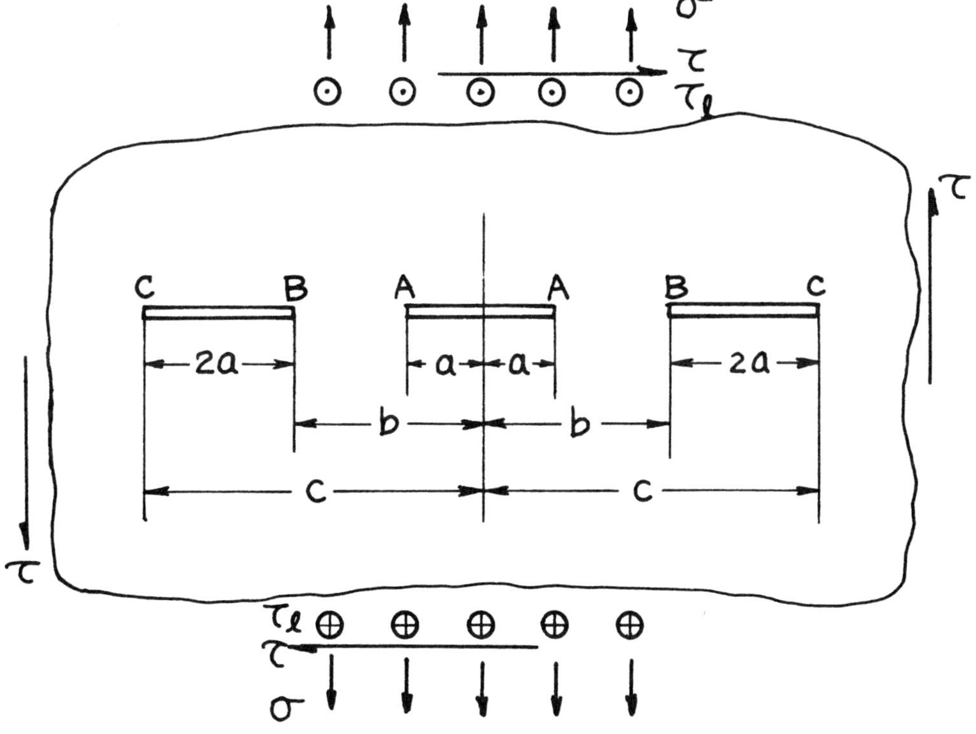

$$\begin{Bmatrix} K_I \\ K_{II} \\ K_{III} \end{Bmatrix}_A = \begin{Bmatrix} \sigma \\ \tau \\ \tau_\ell \end{Bmatrix} \sqrt{\pi a} \sqrt{\frac{c^2 - a^2}{b^2 - a^2}} \frac{E(k)}{K(k)}$$

$$\begin{Bmatrix} K_I \\ K_{II} \\ K_{III} \end{Bmatrix}_B = \begin{Bmatrix} \sigma \\ \tau \\ \tau_\ell \end{Bmatrix} \sqrt{\pi b} \sqrt{\frac{b^2 - a^2}{c^2 - b^2}} \left\{ \frac{c^2 - a^2}{b^2 - a^2} \frac{E(k)}{K(k)} - 1 \right\}$$

$$\begin{Bmatrix} K_I \\ K_{II} \\ K_{III} \end{Bmatrix}_C = \begin{Bmatrix} \sigma \\ \tau \\ \tau_\ell \end{Bmatrix} \sqrt{\pi c} \sqrt{\frac{c^2 - a^2}{c^2 - b^2}} \left\{ 1 - \frac{E(k)}{K(k)} \right\}$$

$$k^2 = \left(c^2 - b^2\right) \Big/ \left(c^2 - a^2\right)$$

$$K(k) = \int_0^{\pi/2} \frac{d\varphi}{\sqrt{1 - k^2 \sin^2 \varphi}}, \quad E(k) = \int_0^{\pi/2} \sqrt{1 - k^2 \sin^2 \varphi} \; d\varphi$$

(See **Appendix L** for values of $K(k)$ and $E(k)$.)
Method: Muskhelishvili's Method
Accuracy: Exact
Reference: **Sih 1964**

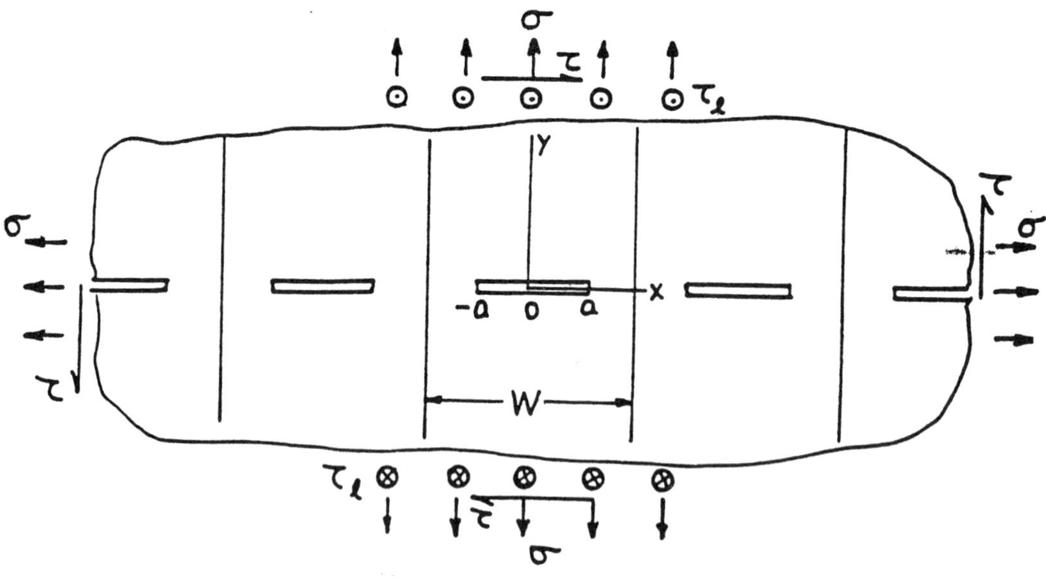

$$z = x + iy$$

$$\begin{Bmatrix} Z_I(z) \\ Z_{II}(z) \\ Z_{III}(z) \end{Bmatrix} = \begin{Bmatrix} \sigma \\ \tau \\ \tau_\ell \end{Bmatrix} \frac{1}{\sqrt{1 - \left(\sin\frac{\pi a}{W} / \sin\frac{\pi z}{W}\right)^2}}$$

$$\begin{Bmatrix} \overline{Z}_I(z) \\ \overline{Z}_{II}(z) \\ \overline{Z}_{III}(z) \end{Bmatrix} = \frac{1}{\pi} \begin{Bmatrix} \sigma \\ \tau \\ \tau_\ell \end{Bmatrix} W \cos^{-1}\left(\frac{\cos\frac{\pi z}{W}}{\cos\frac{\pi a}{W}}\right)$$

$$\begin{Bmatrix} K_I \\ K_{II} \\ K_{III} \end{Bmatrix} = \begin{Bmatrix} \sigma \\ \tau \\ \tau_\ell \end{Bmatrix} \sqrt{W \tan\frac{\pi a}{W}}$$

$$\operatorname{Im} \begin{Bmatrix} \overline{Z}_I(x) \\ \overline{Z}_{II}(x) \\ \overline{Z}_{III}(x) \end{Bmatrix}_{|x| \le a} = \frac{1}{\pi} \begin{Bmatrix} \sigma \\ \tau \\ \tau_\ell \end{Bmatrix} W \cosh^{-1}\left(\frac{\cos\frac{\pi x}{W}}{\cos\frac{\pi a}{W}}\right)$$

Method: Westergaard Stress Function, etc. (Special Case of **page 7.9**)
Accuracy: Exact
References: **Irwin 1957; Koiter 1959**

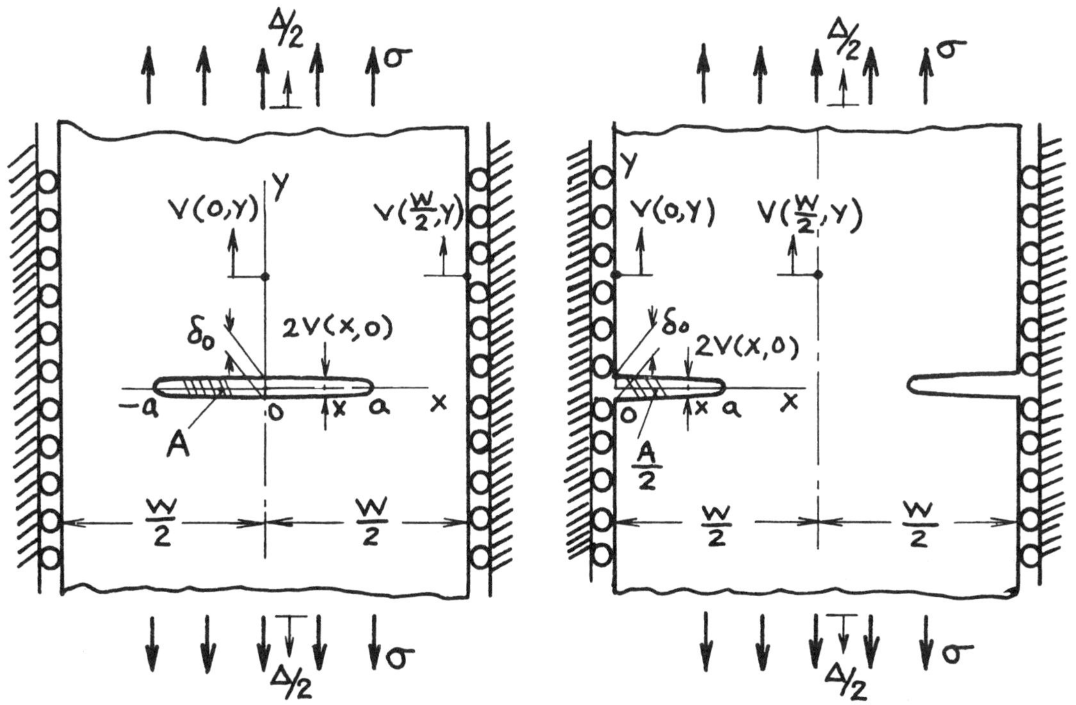

$$K_I = \sigma\sqrt{W \tan\frac{\pi a}{W}}$$

Crack Opening Area:

$$A = \frac{4\sigma W^2}{\pi E'}\,\ell\,\mathrm{n}\left(\sec\frac{\pi a}{W}\right)$$

Additional Relative Displacement at Infinity due to Crack:

$$\Delta_{crack} = A/W = \frac{4\sigma W}{\pi E'}\,\ell\,\mathrm{n}\left(\sec\frac{\pi a}{W}\right) \quad (\Delta = \Delta_{no\ crack} + \Delta_{crack})$$

Crack Opening Profile:

$$\underset{|x|\le a}{2v(x,\,0)} = \frac{4\sigma W}{\pi E'}\cosh^{-1}\left(\frac{\cos\frac{\pi x}{W}}{\cos\frac{\pi a}{W}}\right)$$

Opening at Center or Edge:

$$\delta_0 = 2v(0,\,0) = \frac{4\sigma W}{\pi E'}\cosh^{-1}\left(\sec\frac{\pi a}{W}\right)$$

Additional Vertical Displacements at $(0, y)$ and $({}^{W}\!/_{2}, y)$ due to Crack:

$$v(0,\ y) = \frac{2\sigma W}{\pi E'}\left(1 - \alpha y\frac{\partial}{\partial y}\right)\left\{\cosh^{-1}\left(\frac{\cosh\frac{\pi y}{W}}{\cos\frac{\pi a}{W}}\right) - \frac{\pi y}{W}\right\}$$

$$v\left(\frac{W}{2},\ y\right) = \frac{2\sigma W}{\pi E'}\left(1 - \alpha y\frac{\partial}{\partial y}\right)\left\{\sinh^{-1}\left(\frac{\sinh\frac{\pi y}{W}}{\cos\frac{\pi a}{W}}\right) - \frac{\pi y}{W}\right\}$$

where

$$\alpha = \begin{cases} \frac{1}{2}(1+\nu) & \text{plane stress} \\ \frac{1}{2}\left(\frac{1}{1-\nu}\right) & \text{plane strain} \end{cases}$$

Methods: Westergaard Stress Function, Paris' Equation (see **Appendix B**)
Accuracy: Exact
Reference: **Tada 1985**

NOTE: $\Delta_{crack}/2$, $v(0,\ y)$ and $v({}^{W}\!/_{2},\ y)$ are the vertical displacements at $y = \infty$, $(0,\ y)$ and $({}^{W}\!/_{2},\ y)$, respectively, when uniform pressure σ is applied on the crack surfaces.

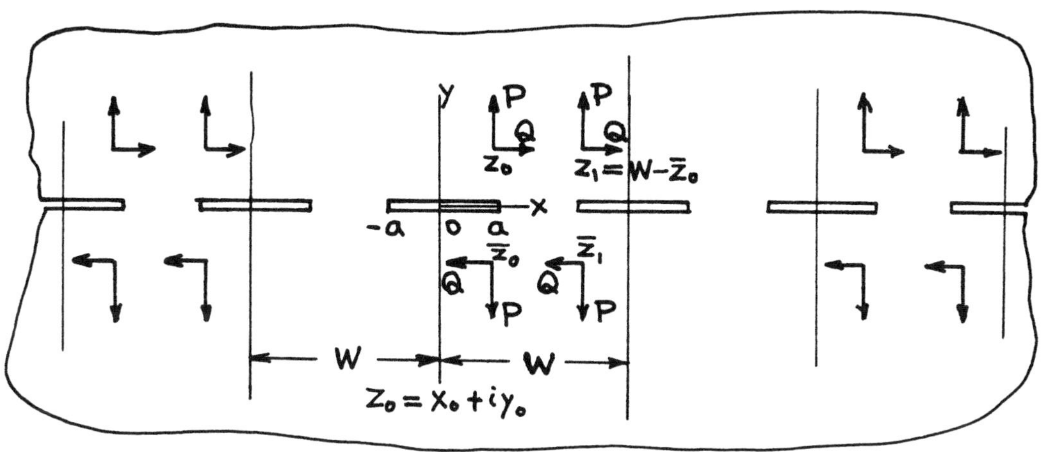

$$z = x + iy$$

$$\begin{Bmatrix} Z_I(z) \\ Z_{II}(z) \end{Bmatrix} = \frac{1}{2W}\begin{Bmatrix} P \\ Q \end{Bmatrix}\left[1\begin{Bmatrix} - \\ + \end{Bmatrix}\alpha y_0 \frac{\partial}{\partial y_0}\right]\left[\left\{\frac{\cos\frac{\pi z_0}{W}\sqrt{\left(\sin\frac{\pi a}{W}\right)^2 - \left(\sin\frac{\pi z_0}{W}\right)^2}}{\sin\frac{\pi z}{W} - \sin\frac{\pi z_0}{W}} + \frac{\cos\frac{\pi \bar{z}_0}{W}\sqrt{\left(\sin\frac{\pi a}{W}\right)^2 - \left(\sin\frac{\pi \bar{z}_0}{W}\right)^2}}{\sin\frac{\pi z}{W} - \sin\frac{\pi \bar{z}_0}{W}}\right.\right.$$

$$\left.\left. + i\left(\cos\frac{\pi z_0}{W} - \cos\frac{\pi \bar{z}_0}{W}\right)\right\}\frac{1}{\sqrt{\left(\sin\frac{\pi z}{W}\right)^2 - \left(\sin\frac{\pi a}{W}\right)^2}}\right]$$

$$\begin{Bmatrix} K_I \\ K_{II} \end{Bmatrix}_{\pm a} = \frac{1}{2W}\begin{Bmatrix} P \\ Q \end{Bmatrix}\frac{\sqrt{W\tan\frac{\pi a}{W}}}{\sin\frac{\pi a}{W}}\left[1\begin{Bmatrix} - \\ + \end{Bmatrix}\alpha y_0 \frac{\partial}{\partial y_0}\right]\left\{\cos\frac{\pi z_0}{W}\sqrt{\frac{\sin\frac{\pi a}{W} \pm \sin\frac{\pi z_0}{W}}{\sin\frac{\pi a}{W} \mp \sin\frac{\pi z_0}{W}}}\right.$$

$$\left. + \cos\frac{\pi \bar{z}_0}{W}\sqrt{\frac{\sin\frac{\pi a}{W} \pm \sin\frac{\pi \bar{z}_0}{W}}{\sin\frac{\pi a}{W} \mp \sin\frac{\pi \bar{z}_0}{W}}} + i\left(\cos\frac{\pi z_0}{W} - \cos\frac{\pi \bar{z}_0}{W}\right)\right\}$$

where

$$\alpha = \begin{cases} \frac{1}{2}(1+\nu) & \text{plane stress} \\ \frac{1}{2}\left(\frac{1}{1-\nu}\right) & \text{plane strain} \end{cases}$$

Method: Westergaard Stress Function
Accuracy: Exact
References: **Tada 1972a, 1973**

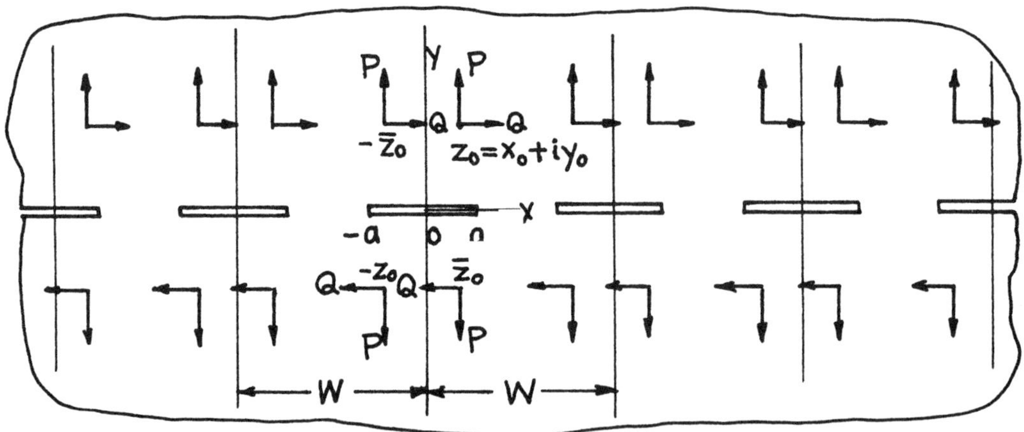

$$z = x + iy$$

$$\left\{\begin{matrix} Z_I(z) \\ Z_{II}(z) \end{matrix}\right\} = \frac{1}{W}\left\{\begin{matrix} P \\ Q \end{matrix}\right\}\left[1\left\{\begin{matrix} - \\ + \end{matrix}\right\}\alpha y_0 \frac{\partial}{\partial y_0}\right]\left[\left\{\frac{\cos\frac{\pi z_0}{W}\sqrt{\left(\sin\frac{\pi a}{W}\right)^2-\left(\sin\frac{\pi z_0}{W}\right)^2}}{\left(\sin\frac{\pi z}{W}\right)^2-\left(\sin\frac{\pi z_0}{W}\right)^2}\right.\right.$$

$$\left.\left.+\frac{\cos\frac{\pi \bar{z}_0}{W}\sqrt{\left(\sin\frac{\pi a}{W}\right)^2-\left(\sin\frac{\pi \bar{z}_0}{W}\right)^2}}{\left(\sin\frac{\pi z}{W}\right)^2-\left(\sin\frac{\pi \bar{z}_0}{W}\right)^2}\right\}\frac{1}{\sqrt{\left(\sin\frac{\pi z}{W}\right)^2-\left(\sin\frac{\pi a}{W}\right)^2}}\right]$$

$$\left\{\begin{matrix} \bar{Z}_I(z) \\ \bar{Z}_{II}(z) \end{matrix}\right\} = \frac{1}{\pi}\left\{\begin{matrix} P \\ Q \end{matrix}\right\}\left[1\left\{\begin{matrix} - \\ + \end{matrix}\right\}\alpha y_0 \frac{\partial}{\partial y_0}\right]\left[-\tan^{-1}\sqrt{\frac{\left(\cos\frac{\pi a}{W}/\cos\frac{\pi z}{W}\right)^2-1}{1-\left(\cos\frac{\pi a}{W}/\cos\frac{\pi z_0}{W}\right)^2}}-\tan^{-1}\sqrt{\frac{\left(\cos\frac{\pi a}{W}/\cos\frac{\pi z}{W}\right)^2-1}{1-\left(\cos\frac{\pi a}{W}/\cos\frac{\pi \bar{z}_0}{W}\right)^2}}\right.$$

$$\left\{\begin{matrix} K_I \\ K_{II} \end{matrix}\right\} = \frac{1}{W}\left\{\begin{matrix} P \\ Q \end{matrix}\right\}\sqrt{W\tan\frac{\pi a}{W}}\left[1\left\{\begin{matrix} - \\ + \end{matrix}\right\}\alpha y_0 \frac{\partial}{\partial y_0}\right]\left\{\frac{\cos\frac{\pi z_0}{W}}{\sqrt{\left(\sin\frac{\pi a}{W}\right)^2-\left(\sin\frac{\pi z_0}{W}\right)^2}}+\frac{\cos\frac{\pi \bar{z}_0}{W}}{\sqrt{\left(\sin\frac{\pi a}{W}\right)^2-\left(\sin\frac{\pi \bar{z}_0}{W}\right)^2}}\right\}$$

where

$$\alpha = \begin{cases} \frac{1}{2}(1+\nu) & \text{plane stress} \\ \frac{1}{2}\left(\frac{1}{1-\nu}\right) & \text{plane strain} \end{cases}$$

Method: Westergaard Stress Function (Superposition of **page 7.2**)
Accuracy: Exact
References: **Tada 1972a, 1973**

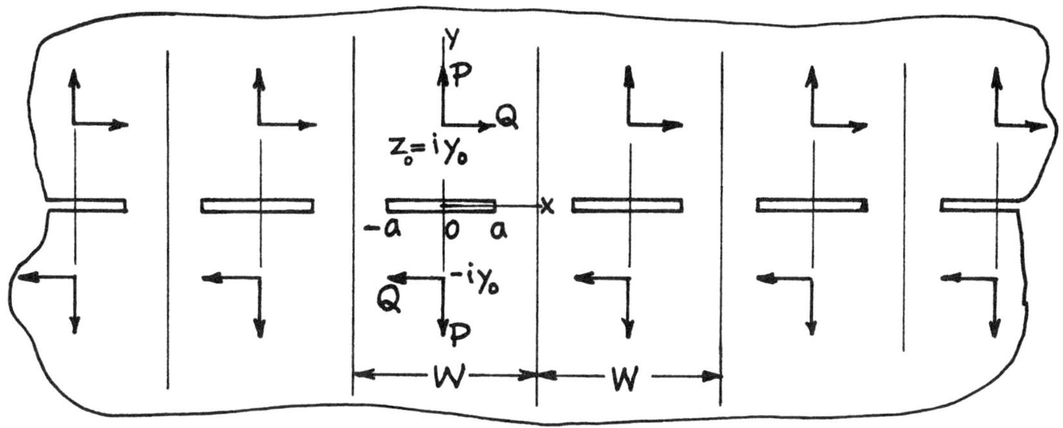

$$z = x + iy$$

$$\begin{Bmatrix} Z_I(z) \\ Z_{II}(z) \end{Bmatrix} = \frac{1}{W}\begin{Bmatrix} P \\ Q \end{Bmatrix}\left[1\begin{Bmatrix} - \\ + \end{Bmatrix}\alpha y_0 \frac{\partial}{\partial y_0}\right]\cosh\frac{\pi y_0}{W}\left[\frac{\sqrt{\left(\sin\frac{\pi a}{W}\right)^2+\left(\sinh\frac{\pi y_0}{W}\right)^2}}{\left(\sin\frac{\pi z}{W}\right)^2+\left(\sinh\frac{\pi y_0}{W}\right)^2}\;\frac{1}{\sqrt{1-\left(\sin\frac{\pi a}{W}/\sin\frac{\pi z}{W}\right)^2}}\right]$$

$$\begin{Bmatrix} \overline{Z}_I(z) \\ \overline{Z}_{II}(z) \end{Bmatrix} = \frac{1}{\pi}\begin{Bmatrix} P \\ Q \end{Bmatrix}\left[1\begin{Bmatrix} - \\ + \end{Bmatrix}\alpha y_0 \frac{\partial}{\partial y_0}\right]\left[-\tan^{-1}\sqrt{\frac{\left(\cos\frac{\pi a}{W}/\cos\frac{\pi z}{W}\right)^2-1}{1-\left(\cos\frac{\pi a}{W}/\cosh\frac{\pi y_0}{W}\right)^2}}\right]$$

$$\begin{Bmatrix} K_I \\ K_{II} \end{Bmatrix} = \frac{1}{W}\begin{Bmatrix} P \\ Q \end{Bmatrix}\sqrt{W\tan\frac{\pi a}{W}}\left[1\begin{Bmatrix} - \\ + \end{Bmatrix}\alpha y_0 \frac{\partial}{\partial y_0}\right]\frac{\cosh\frac{\pi y_0}{W}}{\sqrt{\left(\sin\frac{\pi a}{W}\right)^2+\left(\sinh\frac{\pi y_0}{W}\right)^2}}$$

$$= \frac{1}{W}\begin{Bmatrix} P \\ Q \end{Bmatrix}\sqrt{W\tan\frac{\pi a}{W}}\left[\frac{\cosh\frac{\pi y_0}{W}}{\sqrt{\left(\sin\frac{\pi a}{W}\right)^2+\left(\sinh\frac{\pi y_0}{W}\right)^2}}\begin{Bmatrix} + \\ - \end{Bmatrix}\alpha\frac{\pi y_0}{W}\frac{\sinh\frac{\pi y_0}{W}\left(\cos\frac{\pi a}{W}\right)^2}{\left\{\left(\sin\frac{\pi a}{W}\right)^2+\left(\sinh\frac{\pi y_0}{W}\right)^2\right\}^{3/2}}\right]$$

$$\operatorname{Im}\underset{|x|\le a}{\begin{Bmatrix} \overline{Z}_I(x) \\ \overline{Z}_{II}(x) \end{Bmatrix}} = \frac{1}{\pi}\begin{Bmatrix} P \\ Q \end{Bmatrix}\left[1\begin{Bmatrix} - \\ + \end{Bmatrix}\alpha y_0 \frac{\partial}{\partial y_0}\right]\tanh^{-1}\sqrt{\frac{1-\left(\cos\frac{\pi a}{W}/\cos\frac{\pi x}{W}\right)^2}{1-\left(\cos\frac{\pi a}{W}/\cosh\frac{\pi y_0}{W}\right)^2}}$$

where

$$\alpha = \begin{cases} \frac{1}{2}(1+\nu) & \text{plane stress} \\ \frac{1}{2}\left(\frac{1}{1-\nu}\right) & \text{plane strain} \end{cases}$$

Method: Westergaard Stress Function (Special Case of **page 7.2 or 7.3**)
Accuracy: Exact
References: **Tada 1970, 1973**

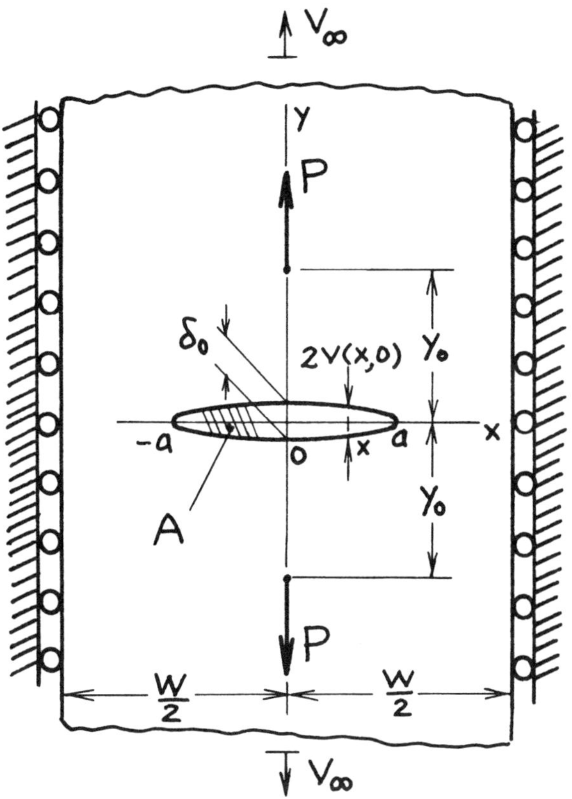

$$K_I = \frac{P}{\sqrt{W}} \sqrt{\tan \frac{\pi a}{W}} \left(1 - \alpha y_0 \frac{\partial}{\partial y_0} \right) \frac{\cosh \frac{\pi y_0}{W}}{\sqrt{\left(\sin \frac{\pi a}{W} \right)^2 + \left(\sinh \frac{\pi y_0}{W} \right)^2}}$$

Crack Opening Area:

$$A = \frac{8PW}{\pi E'} \left(1 - \alpha y_0 \frac{\partial}{\partial y_0} \right) \left\{ \cosh^{-1} \left(\frac{\cosh \frac{\pi y_0}{W}}{\cos \frac{\pi a}{W}} \right) - \frac{\pi y_0}{W} \right\}$$

Relative Displacement at Infinity:

$$2v_\infty = A/W$$

Crack Opening Profile:

$$\underset{|x| \le a}{2v(x,0)} = \frac{4P}{\pi E'} \left(1 - \alpha y_0 \frac{\partial}{\partial y_0} \right) \tanh^{-1} \sqrt{\frac{1 - \left(\cos \frac{\pi a}{W} / \cos \frac{\pi x}{W} \right)^2}{1 - \left(\cos \frac{\pi a}{W} / \cosh \frac{\pi y_0}{W} \right)^2}}$$

Crack Opening at Center:

$$\delta_0 = 2v(0,0) = \frac{4P}{\pi E'}\left(1 - \alpha y_0 \frac{\partial}{\partial y_0}\right)\tanh^{-1}\left\{\frac{\sin\frac{\pi a}{W}}{\sqrt{1-\left(\cos\frac{\pi a}{W} / \cosh\frac{\pi y_0}{W}\right)^2}}\right\}$$

where
$$\alpha = \begin{cases} \frac{1}{2}(1+\nu) & \text{plane stress} \\ \frac{1}{2}\left(\frac{1}{1-\nu}\right) & \text{plane strain} \end{cases}$$

Methods: Westergaard Stress Function, Reciprocity (see **page 7.1a**)
Accuracy: Exact
Reference: **Tada 1985**

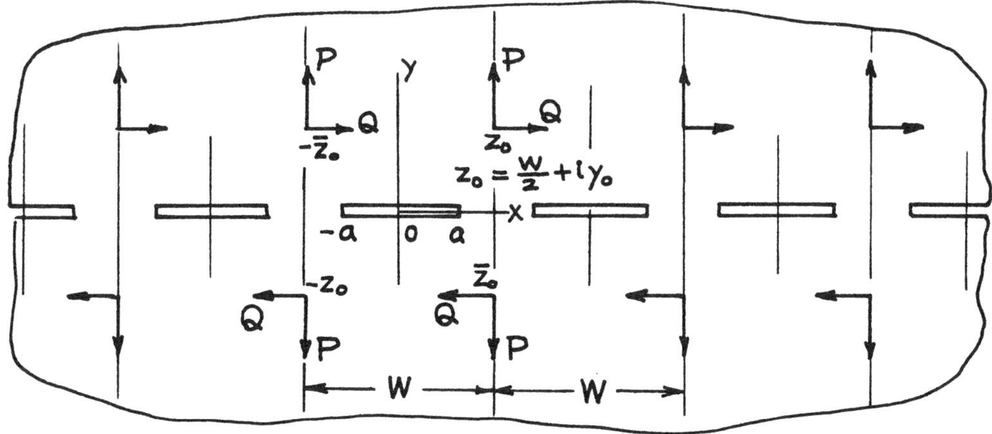

$$z = x + iy$$

$$\left\{ \begin{array}{c} Z_I(z) \\ Z_{II}(z) \end{array} \right\} = \frac{1}{W} \left\{ \begin{array}{c} P \\ Q \end{array} \right\} \left[1 \left\{ \begin{array}{c} - \\ + \end{array} \right\} \alpha y_0 \frac{\partial}{\partial y_0} \right] \sinh \frac{\pi y_0}{W} \left[\frac{\sqrt{\left(\cos \frac{\pi a}{W}\right)^2 + \left(\sinh \frac{\pi y_0}{W}\right)^2}}{\left(\cos \frac{\pi z}{W}\right)^2 + \left(\sinh \frac{\pi y_0}{W}\right)^2} \frac{1}{\sqrt{1 - \left(\sin \frac{\pi a}{W} / \sin \frac{\pi z}{W}\right)^2}} \right]$$

$$\left\{ \begin{array}{c} \bar{Z}_I(z) \\ \bar{Z}_{II}(z) \end{array} \right\} = \frac{1}{\pi} \left\{ \begin{array}{c} P \\ Q \end{array} \right\} \left[1 \left\{ \begin{array}{c} - \\ + \end{array} \right\} \alpha y_0 \frac{\partial}{\partial y_0} \right] \left[-\tan^{-1} \sqrt{\frac{\left(\cos \frac{\pi a}{W} / \cos \frac{\pi z}{W}\right)^2 - 1}{1 + \left(\cos \frac{\pi a}{W} / \cosh \frac{\pi y_0}{W}\right)^2}} - \tan^{-1} \left(\tan \frac{\pi z}{W} \tanh \frac{\pi y_0}{W} \right) \right]$$

$$\left\{ \begin{array}{c} K_I \\ K_{II} \end{array} \right\} = \frac{1}{W} \left\{ \begin{array}{c} P \\ Q \end{array} \right\} \sqrt{W \tan \frac{\pi a}{W}} \left[1 \left\{ \begin{array}{c} - \\ + \end{array} \right\} \alpha y_0 \frac{\partial}{\partial y_0} \right] \frac{\sinh \frac{\pi y_0}{W}}{\sqrt{\left(\cos \frac{\pi a}{W}\right)^2 + \left(\sinh \frac{\pi y_0}{W}\right)^2}}$$

$$= \frac{1}{W} \left\{ \begin{array}{c} P \\ Q \end{array} \right\} \sqrt{W \tan \frac{\pi a}{W}} \left[\frac{\sinh \frac{\pi y_0}{W}}{\left(\cos \frac{\pi a}{W}\right)^2 + \left(\sinh \frac{\pi y_0}{W}\right)^2} \left\{ \begin{array}{c} - \\ + \end{array} \right\} \alpha \frac{\pi y_0}{W} \frac{\cos \frac{\pi y_0}{W} \left(\cos \frac{\pi a}{W}\right)^2}{\left\{ \left(\cos \frac{\pi a}{W}\right)^2 + \left(\sinh \frac{\pi y_0}{W}\right)^2 \right\}^{3/2}} \right]$$

$$\operatorname{Im} \left\{ \begin{array}{c} \bar{Z}_I(x) \\ \bar{Z}_{II}(x) \end{array} \right\}_{|x| \le a} = \frac{1}{\pi} \left\{ \begin{array}{c} P \\ Q \end{array} \right\} \left[1 \left\{ \begin{array}{c} - \\ + \end{array} \right\} \alpha y_0 \frac{\partial}{\partial y_0} \right] \tanh^{-1} \sqrt{\frac{1 - \left(\cos \frac{\pi a}{W} / \cos \frac{\pi x}{W}\right)^2}{1 + \left(\cos \frac{\pi a}{W} / \sinh \frac{\pi y_0}{W}\right)^2}}$$

where

$$\alpha = \begin{cases} \frac{1}{2}(1+\nu) & \text{plane stress} \\ \frac{1}{2}\left(\frac{1}{1-\nu}\right) & \text{plane strain} \end{cases}$$

Method: Westergaard Stress Function (Special Case of **page 7.3**)
Accuracy: Exact
References: **Tada 1970, 1973**

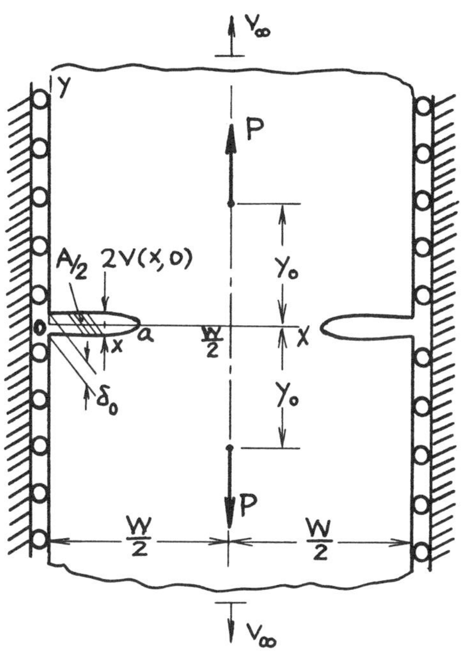

$$K_I = \frac{P}{\sqrt{W}}\sqrt{\tan\frac{\pi a}{W}}\left(1 - \alpha y_0\frac{\partial}{\partial y_0}\right)\frac{\sinh\frac{\pi y_0}{W}}{\sqrt{\left(\cos\frac{\pi a}{W}\right)^2 + \left(\sinh\frac{\pi y_0}{W}\right)^2}}$$

Crack Opening Area:

$$A = \frac{8PW}{\pi E'}\left(1 - \alpha y_0\frac{\partial}{\partial y_0}\right)\left\{\sinh^{-1}\left(\frac{\sinh\frac{\pi y_0}{W}}{\cos\frac{\pi a}{W}}\right) - \frac{\pi y_0}{W}\right\}$$

Relative Displacement at Infinity:

$$2\,v_\infty = {}^{A}/{W}$$

Crack Opening Profile:

$$2\,v(x,\,0) = \frac{4P}{\pi E'}\left(1 - \alpha y_0\frac{\partial}{\partial y_0}\right)\tanh^{-1}\sqrt{\frac{1 - \left(\cos\frac{\pi a}{W}\Big/\cos\frac{\pi x}{W}\right)^2}{1 + \left(\cos\frac{\pi a}{W}\Big/\sinh\frac{\pi y_0}{W}\right)^2}}$$

Opening at Edge:

$$\delta_0 = 2\,v(0,\,0) = \frac{4P}{\pi E'}\left(1 - \alpha y_0\frac{\partial}{\partial y_0}\right)\tanh^{-1}\left\{\frac{\sin\frac{\pi a}{W}}{\sqrt{1 + \left(\cos\frac{\pi a}{W}\Big/\sinh\frac{\pi y_0}{W}\right)^2}}\right\}$$

where

$$\alpha = \begin{cases} \frac{1}{2}(1+\nu) & \text{plane stress} \\ \frac{1}{2}\left(\frac{1}{1-\nu}\right) & \text{plane strain} \end{cases}$$

Methods: Westergaard Stress Function, Reciprocity (see **page 7.1a**)
Accuracy: Exact
Reference: **Tada 1985**

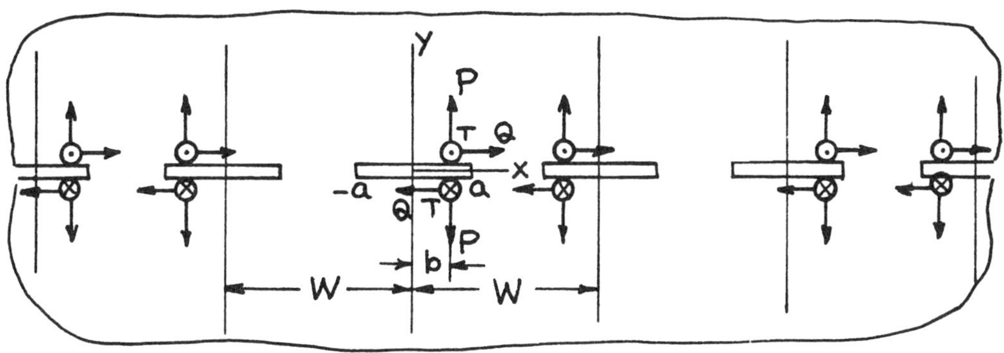

$$z = x + iy$$

$$\begin{Bmatrix} Z_I(z) \\ Z_{II}(z) \\ Z_{III}(z) \end{Bmatrix} = \frac{1}{W} \begin{Bmatrix} P \\ Q \\ T \end{Bmatrix} \frac{\cos\frac{\pi b}{W}\sqrt{\left(\sin\frac{\pi a}{W}\right)^2 - \left(\sin\frac{\pi b}{W}\right)^2}}{\left(\sin\frac{\pi z}{W} - \sin\frac{\pi b}{W}\right)\sqrt{\left(\sin\frac{\pi z}{W}\right)^2 - \left(\sin\frac{\pi a}{W}\right)^2}}$$

$$\begin{Bmatrix} \bar{Z}_I(z) \\ \bar{Z}_{II}(z) \\ \bar{Z}_{III}(z) \end{Bmatrix} = \frac{1}{\pi} \begin{Bmatrix} P \\ Q \\ T \end{Bmatrix} \left[\tan^{-1} \sqrt{\frac{\left(\cos\frac{\pi a}{W} / \cos\frac{\pi z}{W}\right)^2 - 1}{1 - \left(\cos\frac{\pi a}{W} / \cos\frac{\pi b}{W}\right)^2}} \right.$$

$$\left. + i \cot\frac{\pi b}{W} \sqrt{1 - \left(\sin\frac{\pi b}{W} / \sin\frac{\pi a}{W}\right)^2}\, \Pi_c \left(\sin^{-1}\left(\frac{\sin\frac{\pi z}{W}}{\sin\frac{\pi a}{W}}\right), \left(\frac{\sin\frac{\pi a}{W}}{\sin\frac{\pi b}{W}}\right)^2, \sin\frac{\pi a}{W} \right) \right]$$

$$\begin{Bmatrix} K_I \\ K_{II} \\ K_{III} \end{Bmatrix}_{\pm a} = \frac{1}{W} \begin{Bmatrix} P \\ Q \\ T \end{Bmatrix} \sqrt{W \tan\frac{\pi a}{W}}\, \frac{\cos\frac{\pi b}{W}}{\sin\frac{\pi a}{W}} \sqrt{\frac{\sin\frac{\pi a}{W} \pm \sin\frac{\pi b}{W}}{\sin\frac{\pi a}{W} \mp \sin\frac{\pi b}{W}}}$$

$$\text{Im}\left\{\begin{matrix} \overline{Z}_I(x) \\ \overline{Z}_{II}(x) \\ \overline{Z}_{III}(x) \end{matrix}\right\}_{|x|\leq a} = \frac{1}{\pi}\left\{\begin{matrix} P \\ Q \\ T \end{matrix}\right\}\left[\begin{pmatrix} \tanh^{-1} \\ \coth^{-1} \end{pmatrix}\sqrt{\frac{1-\left(\cos\frac{\pi a}{W}/\cos\frac{\pi b}{W}\right)^2}{1-\left(\cos\frac{\pi a}{W}/\cos\frac{\pi x}{W}\right)^2}} \quad \begin{pmatrix} -a<x<b \\ b<x<a \end{pmatrix}\right.$$

$$\left. + \cot\frac{\pi b}{W}\sqrt{1-\left(\sin\frac{\pi b}{W}/\sin\frac{\pi a}{W}\right)^2}\,\Pi\left(\sin^{-1}\left(\frac{\sin\frac{\pi x}{W}}{\sin\frac{\pi a}{W}}\right), \left(\frac{\sin\frac{\pi a}{W}}{\sin\frac{\pi b}{W}}\right)^2, \sin\frac{\pi a}{W}\right)\right]$$

where

$$\Pi_c(\varphi, n, k) = \int \frac{d\varphi}{\left(1-n\sin^2\varphi\right)\sqrt{1-k^2\sin^2\varphi}}$$

$$\Pi(\varphi, n, k) = \int_0^{\varphi} \frac{d\varphi}{\left(1-n\sin^2\varphi\right)\sqrt{1-k^2\sin^2\varphi}}$$

Method: Westergaard Stress Function
Accuracy: Exact
References: **Irwin 1957; Tada 1973**

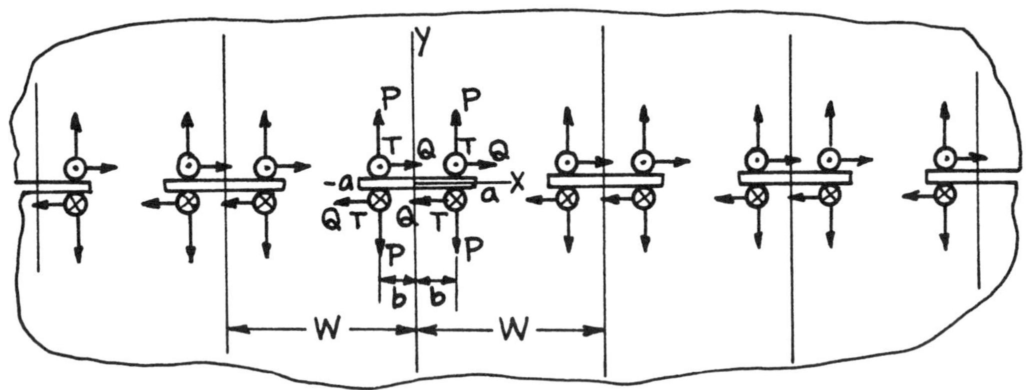

$$z = x + iy$$

$$\left\{\begin{matrix} Z_I(z) \\ Z_{II}(z) \\ Z_{III}(z) \end{matrix}\right\} = \frac{2}{W}\left\{\begin{matrix} P \\ Q \\ T \end{matrix}\right\} \frac{\cos\frac{\pi b}{W}\sqrt{\left(\sin\frac{\pi a}{W}\right)^2 - \left(\sin\frac{\pi b}{W}\right)^2}}{\left\{\left(\sin\frac{\pi z}{W}\right)^2 - \left(\sin\frac{\pi b}{W}\right)^2\right\}\sqrt{1-\left(\sin\frac{\pi a}{W}/\sin\frac{\pi z}{W}\right)^2}}$$

$$\left\{\begin{matrix} \overline{Z}_I(z) \\ \overline{Z}_{II}(z) \\ \overline{Z}_{III}(z) \end{matrix}\right\} = \frac{2}{\pi}\left\{\begin{matrix} P \\ Q \\ T \end{matrix}\right\} \tan^{-1}\sqrt{\frac{\left(\cos\frac{\pi a}{W}/\cos\frac{\pi z}{W}\right)^2 - 1}{1-\left(\cos\frac{\pi a}{W}/\cos\frac{\pi b}{W}\right)^2}}$$

$$\left\{\begin{matrix} K_I \\ K_{II} \\ K_{III} \end{matrix}\right\} = \frac{2}{W}\left\{\begin{matrix} P \\ Q \\ T \end{matrix}\right\} \sqrt{W\tan\frac{\pi a}{W}} \frac{\cos\frac{\pi b}{W}}{\sqrt{\left(\sin\frac{\pi a}{W}\right)^2 - \left(\sin\frac{\pi b}{W}\right)^2}}$$

$$\operatorname{Im}\left\{\begin{matrix} \overline{Z}_I(x) \\ \overline{Z}_{II}(x) \\ \overline{Z}_{III}(x) \end{matrix}\right\}_{|x|\le a} = \frac{1}{\pi}\left\{\begin{matrix} P \\ Q \\ T \end{matrix}\right\} \ln\left|\frac{\sqrt{1-\left(\cos\frac{\pi a}{W}/\cos\frac{\pi x}{W}\right)^2} + \sqrt{1-\left(\cos\frac{\pi a}{W}/\cos\frac{\pi b}{W}\right)^2}}{\sqrt{1-\left(\cos\frac{\pi a}{W}/\cos\frac{\pi x}{W}\right)^2} - \sqrt{1-\left(\cos\frac{\pi a}{W}/\cos\frac{\pi b}{W}\right)^2}}\right|$$

Method: Westergaard Stress Function
Accuracy: Exact
References: **Irwin 1957, 1958a**

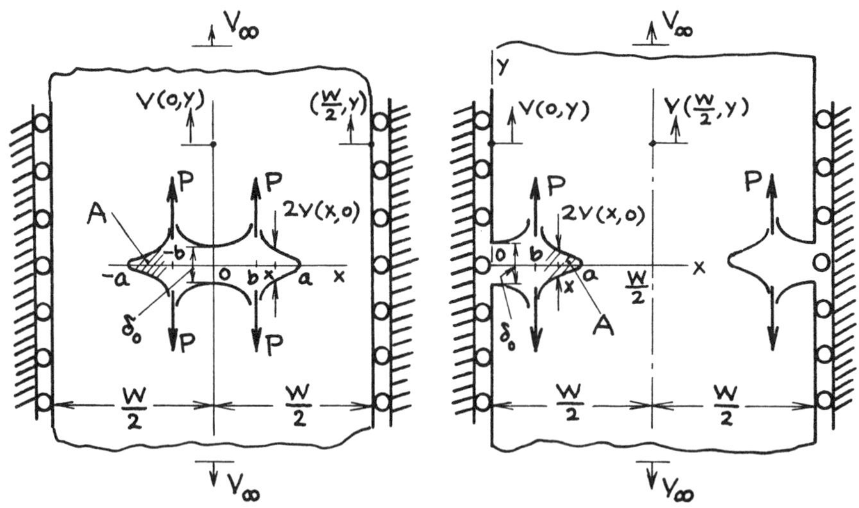

$$K_I = \frac{2P}{\sqrt{W}}\sqrt{\tan\frac{\pi a}{W}}\,\frac{\cos\frac{\pi b}{W}}{\sqrt{\left(\sin\frac{\pi a}{W}\right)^2 - \left(\sin\frac{\pi b}{W}\right)^2}}$$

Crack Opening Area:

$$A = \frac{8PW}{\pi E'}\cosh^{-1}\left(\cos\frac{\pi b}{W} \Big/ \cos\frac{\pi a}{W}\right)$$

Relative Displacement at Infinity:

$$2v_\infty = A/W$$

Crack Opening Profile:

$$\underset{|x|\le a}{2v(x,0)} = \frac{8P}{\pi E'}\begin{pmatrix}\tanh^{-1}\\ \coth^{-1}\end{pmatrix}\sqrt{\frac{1-\left(\cos\frac{\pi a}{W} \big/ \cos\frac{\pi b}{W}\right)^2}{1-\left(\cos\frac{\pi a}{W} \big/ \cos\frac{\pi x}{W}\right)^2}}\begin{pmatrix}|x|\le b\\ b<|x|\le a\end{pmatrix}$$

Opening at Center or Edge:

$$\delta_0 = 2v(0,0) = \frac{8P}{\pi E'}\tanh^{-1}\frac{\sqrt{1-\left(\cos\frac{\pi a}{W} \big/ \cos\frac{\pi b}{W}\right)^2}}{\sin\frac{\pi a}{W}}$$

Vertical Displacement at $(0, y)$:

$$v(0,y) = \frac{4P}{\pi E'}\left(1 - \alpha y \frac{\partial}{\partial y}\right) \tanh^{-1} \sqrt{\frac{1 - \left(\cos\frac{\pi a}{W} / \cos\frac{\pi b}{W}\right)^2}{1 - \left(\cos\frac{\pi a}{W} / \cosh\frac{\pi y}{W}\right)^2}}$$

Vertical Displacement at $(W/2, y)$:

$$v\left(\frac{W}{2}, y\right) = \frac{4P}{\pi E'}\left(1 - \alpha y \frac{\partial}{\partial y}\right) \tanh^{-1} \sqrt{\frac{1 - \left(\cos\frac{\pi a}{W} / \cos\frac{\pi b}{W}\right)^2}{1 + \left(\cos\frac{\pi a}{W} / \sinh\frac{\pi y}{W}\right)^2}}$$

where

$$\alpha = \begin{cases} \frac{1}{2}(1 + \nu) & \text{plane stress} \\ \frac{1}{2}\left(\frac{1}{1-\nu}\right) & \text{plane strain} \end{cases}$$

Methods: Westergaard Stress Function, Reciprocity (see **pages 7.1, 7.1a, 7.4a, 7.5a**)
Accuracy: Exact
Reference: **Tada 1985**

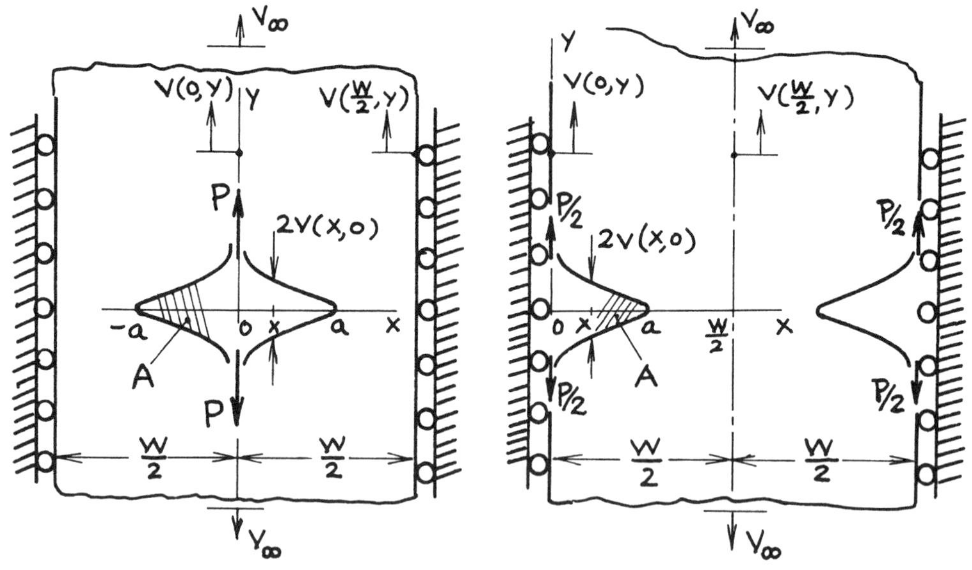

$$K_I = \frac{P}{\sqrt{\pi a}} \sqrt{\frac{2\pi a}{W} \Big/ \sin \frac{2\pi a}{W}}$$

Crack Opening Area:

$$A = \frac{4PW}{\pi E'} \cosh^{-1}\left(\sec \frac{\pi a}{W}\right)$$

Relative Displacement at Infinity:

$$2v_\infty = A/W = \frac{4P}{\pi E'} \cosh^{-1}\left(\sec \frac{\pi a}{W}\right)$$

Crack Opening Profile:

$$\underset{|x| \leq a}{2v(x,0)} = \frac{4P}{\pi E'} \tanh^{-1}\left\{\frac{\sqrt{1 - \left(\cos \frac{\pi a}{W} \Big/ \cos \frac{\pi x}{W}\right)^2}}{\sin \frac{\pi a}{W}}\right\}$$

Vertical Displacements at $(0, y)$ and $(W/2, y)$:

$$v(0,y) = \frac{2P}{\pi E'}\left(1 - \alpha y \frac{\partial}{\partial y}\right) \tanh^{-1}\left\{\frac{\sin \frac{\pi a}{W}}{\sqrt{1 - \left(\cos \frac{\pi a}{W} \Big/ \cosh \frac{\pi y}{W}\right)^2}}\right\}$$

$$v\left(\frac{W}{2}, y\right) = \frac{2P}{\pi E'}\left(1 - \alpha y \frac{\partial}{\partial y}\right) \tanh^{-1}\left\{\frac{\sin\frac{\pi a}{W}}{\sqrt{1 + \left(\cos\frac{\pi a}{W} / \sinh\frac{\pi y}{W}\right)^2}}\right\}$$

where

$$\alpha = \begin{cases} \frac{1}{2}(1+\nu) & \text{plane stress} \\ \frac{1}{2}\left(\frac{1}{1-\nu}\right) & \text{plane strain} \end{cases}$$

Method: Special Case of **page 7.7a**
Accuracy: Exact
Reference: **Tada 1985**

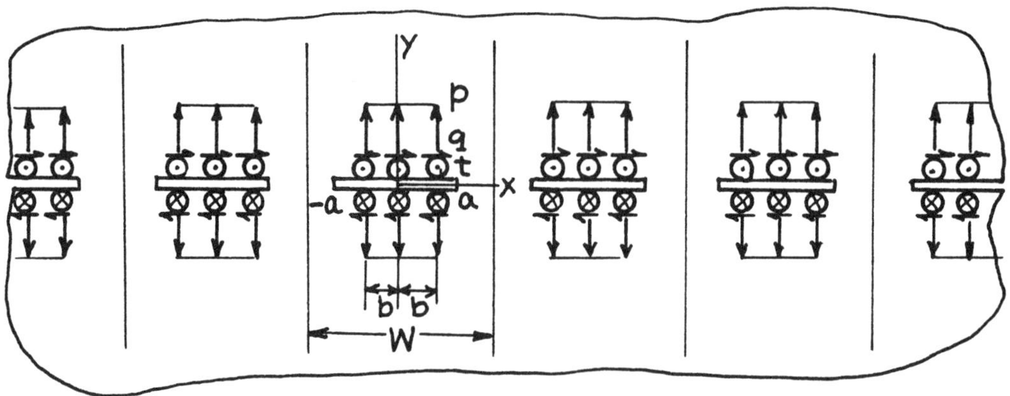

$$z = x + iy$$

$$\begin{Bmatrix} Z_I(z) \\ Z_{II}(z) \\ Z_{III}(z) \end{Bmatrix} = \frac{2}{\pi}\begin{Bmatrix} p \\ q \\ t \end{Bmatrix}\left[\frac{\sin^{-1}\left(\sin\frac{\pi b}{W} / \sin\frac{\pi a}{W}\right)}{\sqrt{1-\left(\sin\frac{\pi a}{W} / \sin\frac{\pi z}{W}\right)^2}} + \tan^{-1}\sqrt{\frac{\left(\sin\frac{\pi a}{W} / \sin\frac{\pi b}{W}\right)^2 - 1}{1-\left(\sin\frac{\pi a}{W} / \sin\frac{\pi z}{W}\right)^2}}\right]$$

$$\begin{Bmatrix} \overline{Z}_I(z) \\ \overline{Z}_{II}(z) \\ \overline{Z}_{III}(z) \end{Bmatrix} = \frac{2}{\pi}\begin{Bmatrix} p \\ q \\ t \end{Bmatrix}\int_0^b \left\{ -\tan^{-1}\sqrt{\frac{\left(\cos\frac{\pi a}{W} / \cos\frac{\pi z}{W}\right)^2 - 1}{1-\left(\cos\frac{\pi a}{W} / \cos\frac{\pi x_0}{W}\right)^2}} \right\} dx_0$$

$$\begin{Bmatrix} K_I \\ K_{II} \\ K_{III} \end{Bmatrix} = \frac{2}{\pi}\begin{Bmatrix} p \\ q \\ t \end{Bmatrix}\sqrt{W\tan\frac{\pi a}{W}}\cdot\sin^{-1}\left(\frac{\sin\frac{\pi b}{W}}{\sin\frac{\pi a}{W}}\right)$$

$$\operatorname{Im}\begin{Bmatrix} \overline{Z}_I(x) \\ \overline{Z}_{II}(x) \\ \overline{Z}_{III}(x) \end{Bmatrix}_{|x|\le a} = \frac{1}{\pi}\begin{Bmatrix} p \\ q \\ t \end{Bmatrix}\int_0^b \ln\left|\frac{\sqrt{1-\left(\cos\frac{\pi a}{W} / \cos\frac{\pi x}{W}\right)^2} + \sqrt{1-\left(\cos\frac{\pi a}{W} / \cos\frac{\pi x_0}{W}\right)^2}}{\sqrt{1-\left(\cos\frac{\pi a}{W} / \cos\frac{\pi x}{W}\right)^2} - \sqrt{1-\left(\cos\frac{\pi a}{W} / \cos\frac{\pi x_0}{W}\right)^2}}\right| dx_0$$

Method: Westergaard Stress Function (Integration of **page 7.6** or **page 7.7**)
Accuracy: Exact
References: **Irwin 1957; Tada 1973**

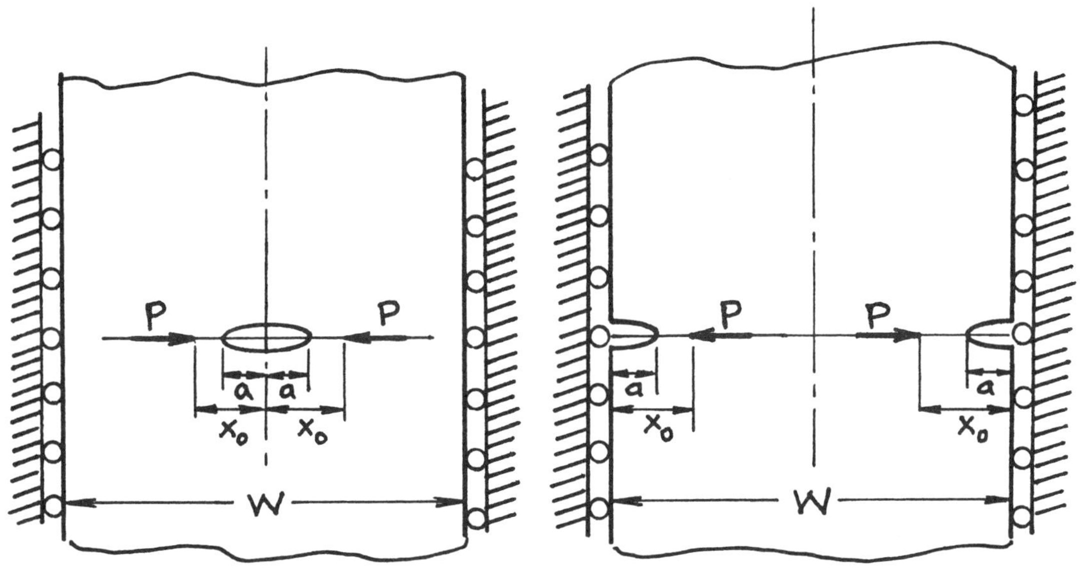

$$K_I = \frac{P(1-\alpha)}{\sqrt{W}} \sqrt{\tan\frac{\pi a}{W}} \frac{\cos\frac{\pi x_0}{W}}{\sqrt{\left(\sin\frac{\pi x_0}{W}\right)^2 - \left(\sin\frac{\pi a}{W}\right)^2}}$$

where

$$\alpha = \begin{cases} \frac{1}{2}(1+\nu) & \text{plane stress} \\ \frac{1}{2}\left(\frac{1}{1-\nu}\right) & \text{plane strain} \end{cases}$$

Method: Integration of **page 7.7**
Accuracy: Exact
Reference: **Tada 2000**

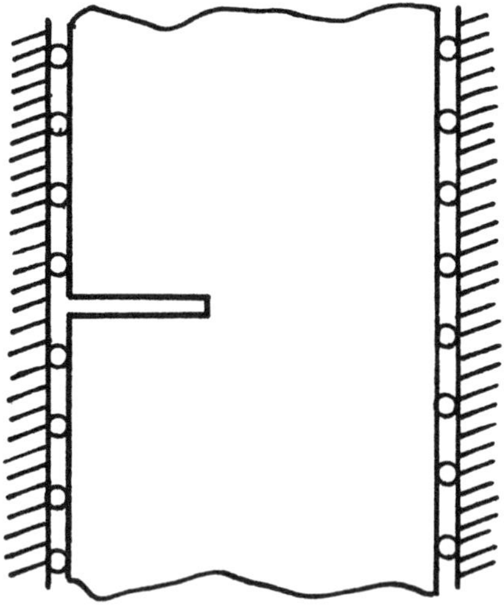

Equivalent to Periodic Cracks with Symmetric Loadings

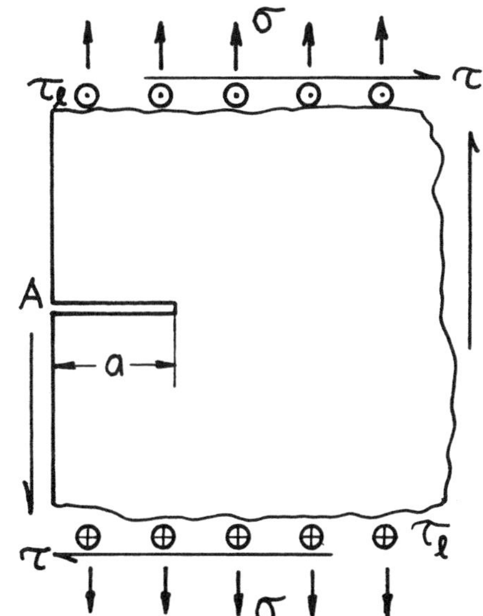

Stress Intensity Factors

$$\begin{Bmatrix} K_I \\ K_{II} \end{Bmatrix} = 1.1215 \begin{pmatrix} \sigma \\ \tau \end{pmatrix} \sqrt{\pi a}$$

$$K_{III} = \tau_\ell \sqrt{\pi a}$$

Displacements at A

$$\begin{Bmatrix} 2v \\ 2u \end{Bmatrix} = 1.454 \cdot \frac{4}{E'} \begin{Bmatrix} \sigma \\ \tau \end{Bmatrix} a = \frac{5.816}{E'} \begin{Bmatrix} \sigma \\ \tau \end{Bmatrix} a$$

$$2w = \frac{2\tau_\ell a}{G}$$

Methods and References: K_I Integral Transform-Singular Integral Equation (**Wigglesworth 1957; Koiter 1965a, Bueckner 1966; Sneddon 1971a; Benthem 1972**); Successive Stress Relaxation-Integral Equation (**Irwin 1958b, 1960a; Lachenbruch 1961; Nishitani 1971a; Tada 1972a)**

K_{II} From similarity of free boundary corrections between Mode I and Mode II (**Tada 1973**)

K_{III} Westergaard Stress Function, etc.

Accuracy: K_I, K_{II} Within one unit of last digit

K_{III} Exact

Displacements were derived by Paris' equation (**Paris 1957**) (see **Appendix B**)

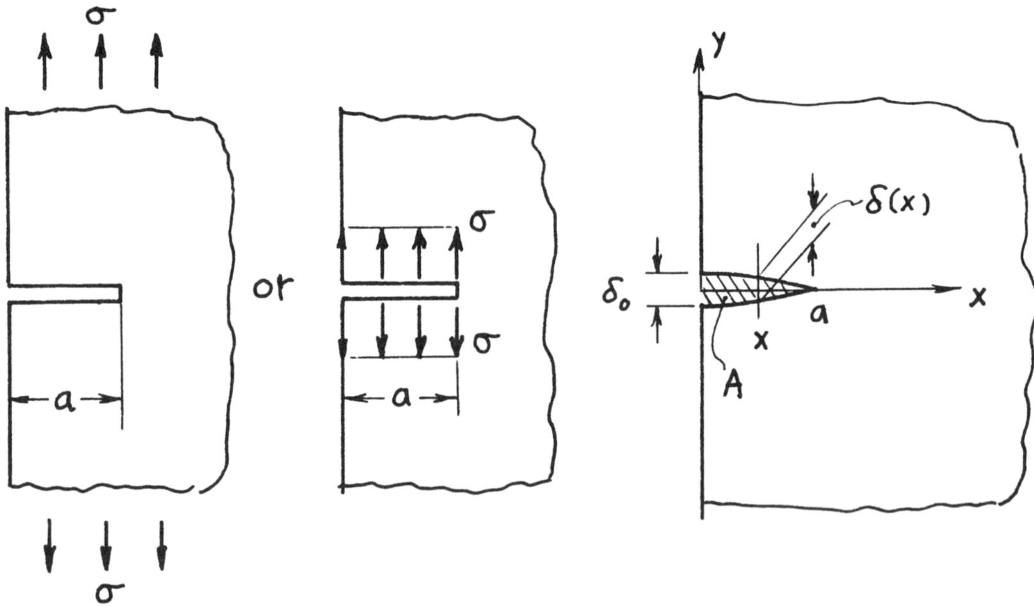

$$K_I = 1.1215\,\sigma\sqrt{\pi a}$$

Crack Opening Area:

$$A = 1.258\,\frac{\sigma\pi a^2}{E'}$$

Crack Profile:

$$\delta(x) = \frac{4\sigma}{E'}\sqrt{a^2 - x^2}\,D\left(\frac{x}{a}\right)$$

$$D\left(\frac{x}{a}\right) = 1.454 - .727\frac{x}{a} + .618\left(\frac{x}{a}\right)^2 - .224\left(\frac{x}{a}\right)^3$$

$$\delta_0 = \delta(0) = 1.454\frac{4\sigma a}{E'}$$

Method: A and δ Paris' Equation (see **Appendix B**)
Accuracy: A and δ_0 Within one unit of last digit; $\delta(x)$ 1%
Reference: **Tada 1985**

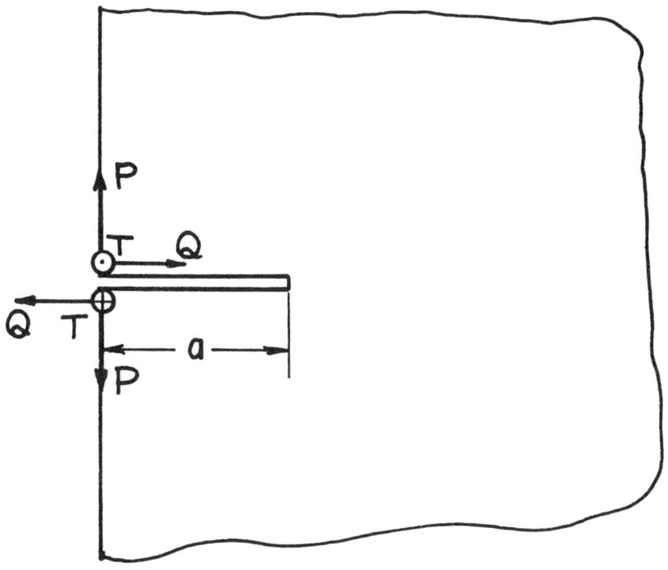

$$K_I = 1.297 \cdot \frac{2P}{\sqrt{\pi a}}$$

$$K_{II} = 1.297 \cdot \frac{2Q}{\sqrt{\pi a}} \qquad \left(\frac{\pi}{\sqrt{\pi^2 - 4}} = 1.297\right)$$

$$K_{III} = \frac{2T}{\sqrt{\pi a}}$$

Methods: K_I, K_{II} Combined Method of Integral Transform and Paris' Equation (see **Appendix B**);
K_{III} Westergaard Stress Function, etc.

Accuracy: Exact

References: **Ouchterlony 1975; Tada 1973, 1985**

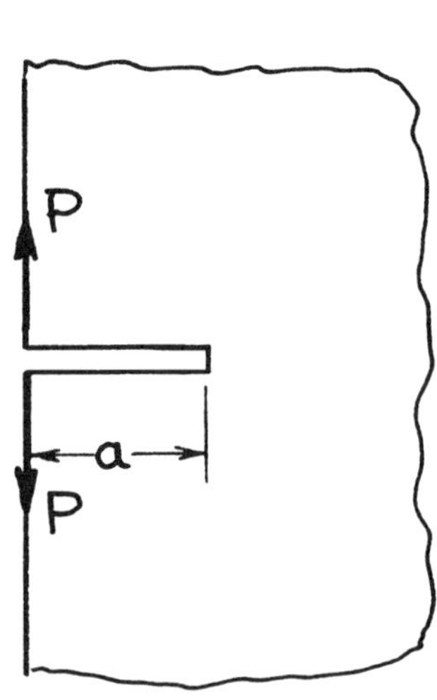

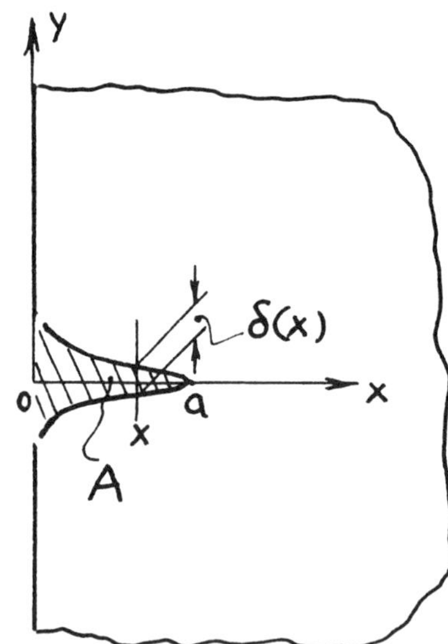

$$K_I = 1.297 \frac{2P}{\sqrt{\pi a}}$$

Crack Opening Area:

$$A = 1.454 \frac{4Pa}{E'}$$

Crack Profile:

$$\delta(x) = \frac{8P}{\pi E'} \cosh^{-1} \frac{a}{x} \cdot D\left(\frac{x}{a}\right)$$

$$D\left(\frac{x}{a}\right) = 1.681 - .384\left(\frac{x}{a}\right)^{.38}$$

Method: A and δ Paris' Equation (see **Appendix B**)
Accuracy: A Within one unit of last digit; δ 1%
Reference: **Tada 1985**

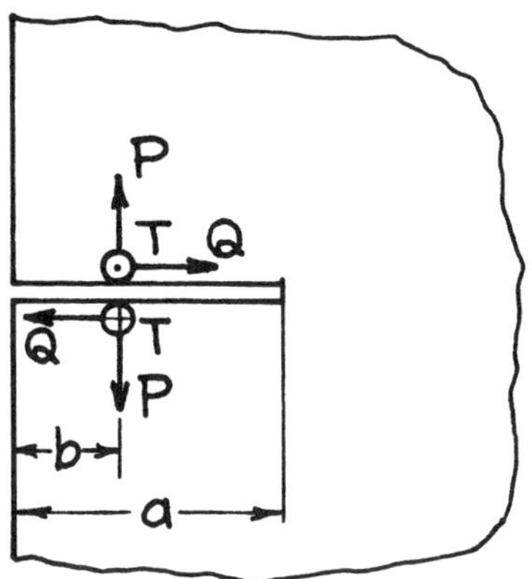

$$\begin{Bmatrix} K_I \\ K_{II} \\ K_{III} \end{Bmatrix} = \frac{2}{\sqrt{\pi a}} \begin{Bmatrix} P \\ Q \\ T \end{Bmatrix} \frac{1}{\sqrt{1-\left({}^{b}\!/_{a}\right)^{2}}} \begin{Bmatrix} F\left({}^{b}\!/_{a}\right) \\ F\left({}^{b}\!/_{a}\right) \\ 1 \end{Bmatrix}$$

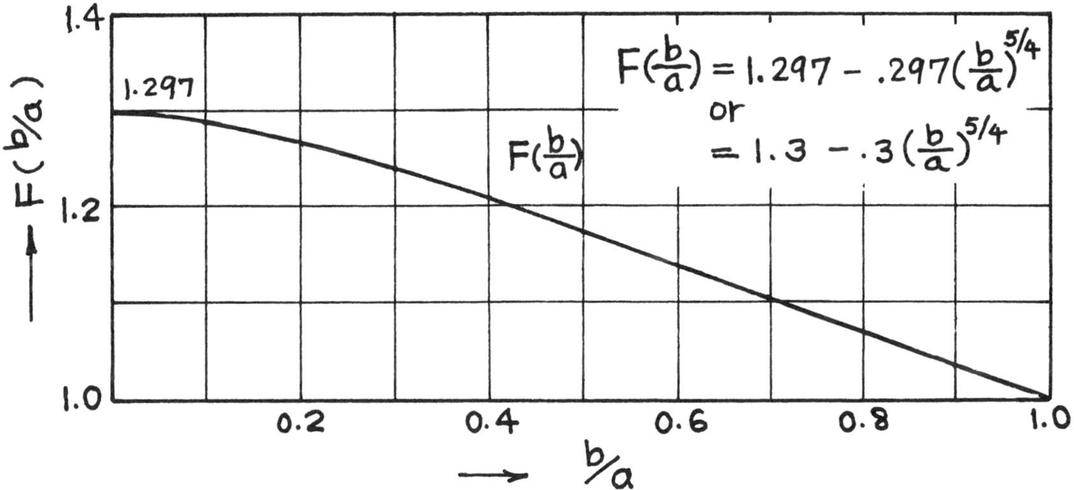

Methods: K_I, K_{II} Alternating Method (Successive Stress Adjustment)
K_{III} Westergaard Stress Function, etc.

Accuracy: K_I, K_{II} 0.5%
K_{III} Exact

References: **Hartranft 1973; Tada 1973, 1985**

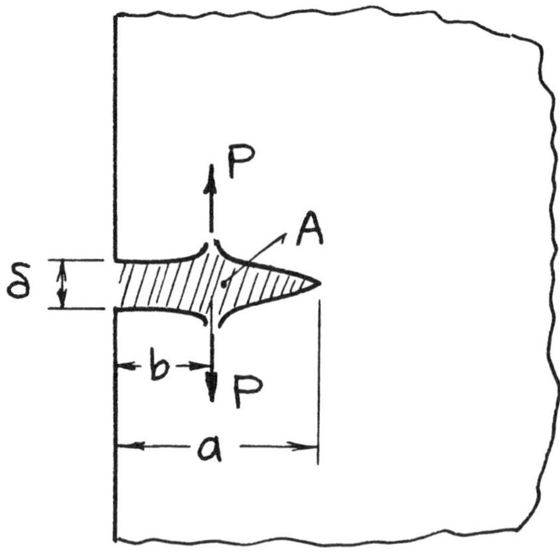

$$K_I = \frac{2P}{\sqrt{\pi a}} F\left(\frac{b}{a}\right)$$

$$F\left(\frac{b}{a}\right) = \frac{1.3 - .3\left(\frac{b}{a}\right)^{5/4}}{\sqrt{1 - \left(\frac{b}{a}\right)^2}}$$

Crack Opening Area:

$$A = \frac{4P}{E'}\sqrt{a^2 - b^2} \cdot G\left(\frac{b}{a}\right)$$

$$G\left(\frac{b}{a}\right) = 1.454 - .727\frac{b}{a} + .618\left(\frac{b}{a}\right)^2 - .224\left(\frac{b}{a}\right)^3$$

Opening at Edge:

$$\delta = \frac{8P}{\pi E'}\cosh^{-1}\frac{a}{b} \cdot H\left(\frac{b}{a}\right)$$

$$H\left(\frac{b}{a}\right) = 1.681 - .384\left(\frac{b}{a}\right)^{.38}$$

Method: A and δ Paris' Equation (see **Appendix B**)
Accuracy: A and δ 1%
Reference: **Tada 1985**

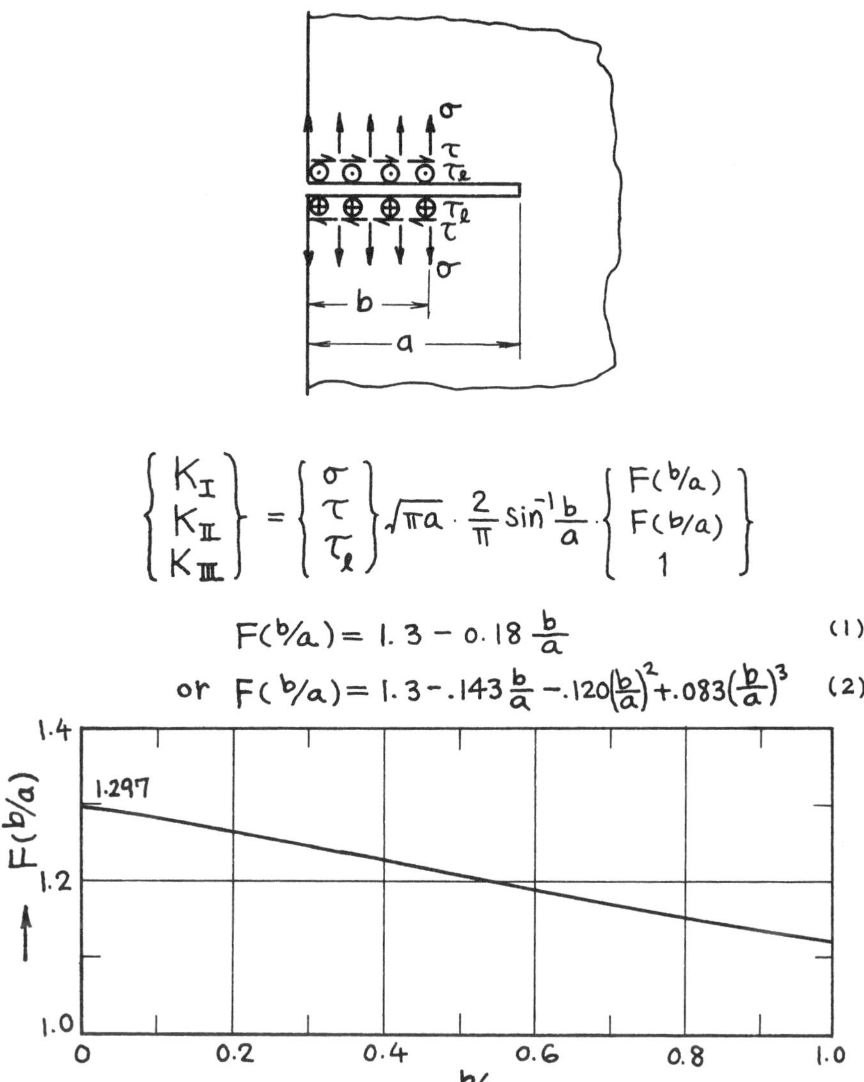

Methods: K_I, K_{II} Alternating Method or Integration of **page 8.3;**
K_{III} Westergaard Stress Function, etc.
Accuracy: Empirical Formula (1) 0.5%, (2) 0.2%; K_{III} Exact
References: **Hartranft 1973; Tada 1985**

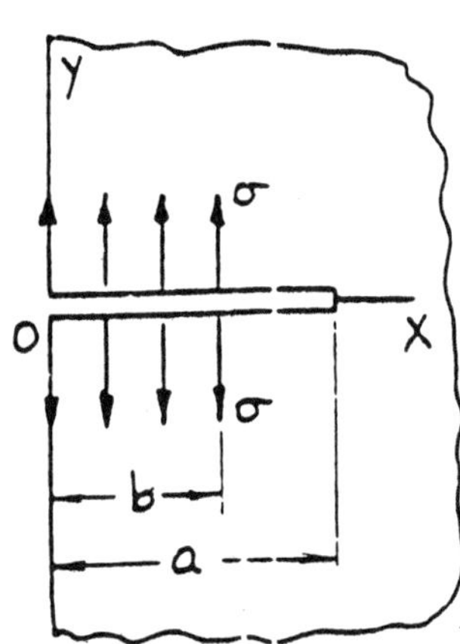

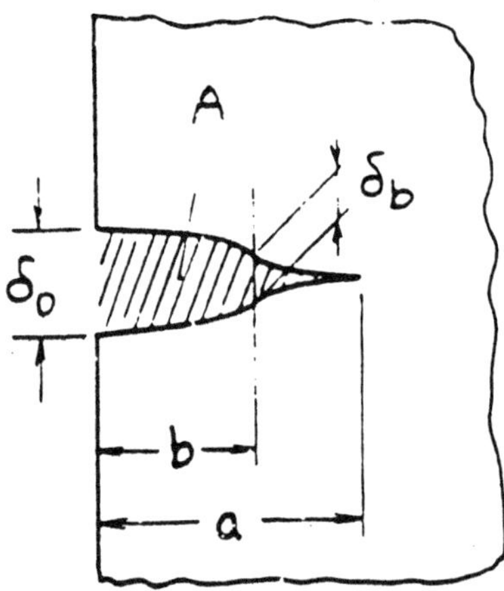

$$K_I = \sigma\sqrt{\pi a}\cdot\frac{2}{\pi}\sin^{-1}\frac{b}{a}\cdot F\left(\frac{b}{a}\right)$$

$$F\left(\frac{b}{a}\right) = 1.3 - .18\frac{b}{a}$$

or

$$F\left(\frac{b}{a}\right) = 1.3 - .143\frac{b}{a} - .120\left(\frac{b}{a}\right)^2 + .083\left(\frac{b}{a}\right)^3$$

Crack Opening Area:

$$A = \frac{\sigma\pi a^2}{E'}\cdot\frac{2}{\pi}\left(\sin^{-1}\frac{b}{a} + \frac{b}{a}\sqrt{1-\left(\frac{b}{a}\right)^2}\right)\cdot G\left(\frac{b}{a}\right)$$

or

$$= \frac{\sigma\pi a^2}{E'}\cdot\overline{G}\left(\frac{b}{a}\right)$$

$$G\left(\frac{b}{a}\right) = 1.258 + .196\left(1-\frac{b}{a}\right)^{5/2}$$

$$\overline{G}\left(\frac{b}{a}\right) = \left(1.454 - .402\frac{b}{a}\right)\cdot\frac{2}{\pi}\sin^{-1}\frac{b}{a} + .926\,\frac{b}{a}\sqrt{1-\left(\frac{b}{a}\right)^2} + \left(\frac{b}{a}\right)^2\left(.206 - .257\cosh^{-1}\frac{a}{b}\right)$$

Opening at Edge:

$$\delta_0 = \frac{4\sigma a}{E'} \cdot \frac{2}{\pi}\left(\sin^{-1}\frac{b}{a} + \frac{b}{a}\cosh^{-1}\frac{a}{b}\right) \cdot H_0\left(\frac{b}{a}\right)$$

or

$$= \frac{4\sigma a}{E'} \cdot \overline{H}_0\left(\frac{b}{a}\right)$$

$$H_0\left(\frac{b}{a}\right) = 1.681 - .227\frac{b}{a}\left[1 + \left(1 - \frac{b}{a}\right)\left(\frac{b}{a}\right)^{-3/4}\right]$$

$$\overline{H}_0\left(\frac{b}{a}\right) = 1.681\frac{2}{\pi}\left(\sin^{-1}\frac{b}{a} + \frac{b}{a}\cosh^{-1}\frac{a}{b}\right) - .392\frac{b}{a} + .140\left(\frac{b}{a}\right)^2 + .025\left(\frac{b}{a}\right)^3$$

Opening at $x = b$:

$$\delta_b = \frac{4\sigma a}{E'} \cdot \frac{2}{\pi}\left[\sqrt{1 - \left(\frac{b}{a}\right)^2}\sin^{-1}\frac{b}{a} - \frac{b}{a}\ell n\frac{b}{a}\right] \cdot H_b\left(\frac{b}{a}\right)$$

or

$$= \frac{4\sigma a}{E'} \cdot \overline{H}_b\left(\frac{b}{a}\right)$$

$$H_b\left(\frac{b}{a}\right) = 1.681\left[1 - .215\left(\frac{b}{a}\right)^{.320}\left(1 - \frac{b}{a}\right)^{.565}\right]$$

$$\overline{H}_b\left(\frac{b}{a}\right) = 1.681\frac{2}{\pi}\left[\sqrt{1 - \left(\frac{b}{a}\right)^2}\sin^{-1}\frac{b}{a} - \frac{b}{a}\ell n\frac{b}{a}\right]$$
$$- \frac{b}{a}\left(1 - \frac{b}{a}\right)\left[.382 + .143\left(\frac{b}{a}\right)^2\left(1 - \frac{b}{a}\right)\right]$$

Method: A and δ Paris' Equation (see **Appendix B**) or Integration of **page 8.3a**
Accuracy: G 2%, $\overline{G}$ 0.5%; H_0 1%, $\overline{H}_0$ 0.5%; H_b and $\overline{H}_b$ 1%
Reference: **Tada 1985**

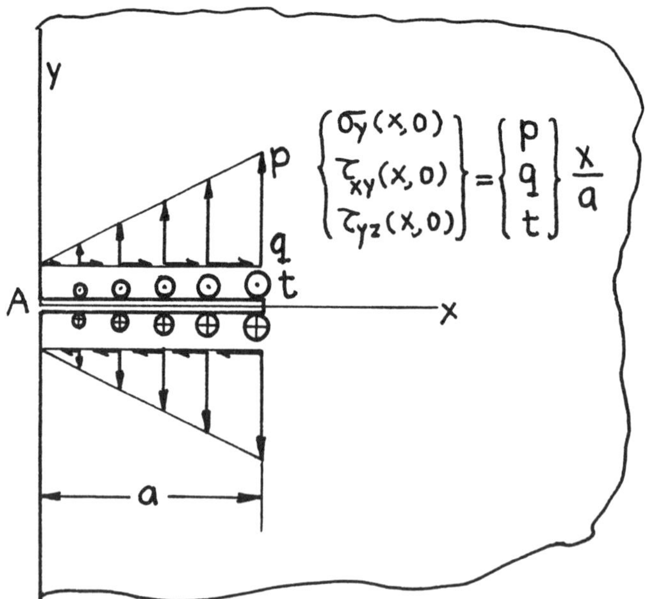

Stress Intensity Factors

$$\begin{Bmatrix} K_I \\ K_{II} \end{Bmatrix} = 0.683 \begin{Bmatrix} p \\ q \end{Bmatrix} \sqrt{\pi a} = 1.072 \frac{2}{\pi} \begin{Bmatrix} p \\ q \end{Bmatrix} \sqrt{\pi a}$$

$$K_{III} = \frac{2}{\pi} t \sqrt{\pi a}$$

Displacements at A

$$\begin{Bmatrix} 2v \\ 2u \end{Bmatrix} = \frac{1.770}{E'} \begin{Bmatrix} p \\ q \end{Bmatrix} a = 1.390 \cdot \frac{4}{E'} \cdot \frac{1}{\pi} \begin{Bmatrix} p \\ q \end{Bmatrix} a$$

$$2w = \frac{2}{G} \cdot \frac{1}{\pi} ta$$

Method: Superposition of **page 8.1** and **page 8.6**
Accuracy: K_I, K_{II} 0.2%
K_{III} Exact
Reference: **Tada 1973**
Displacements were derived by Paris' equation (See **Appendix B**).

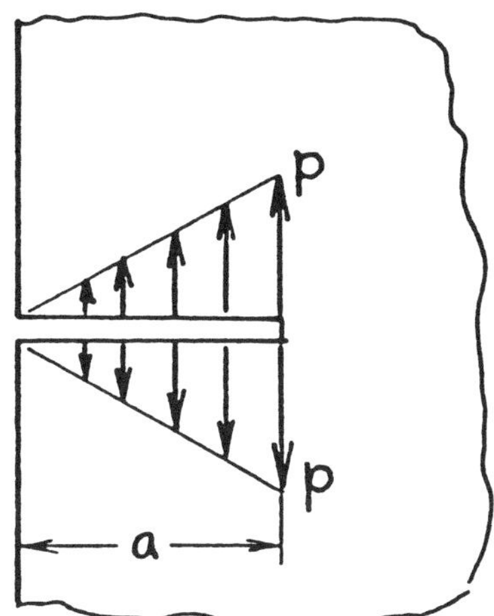

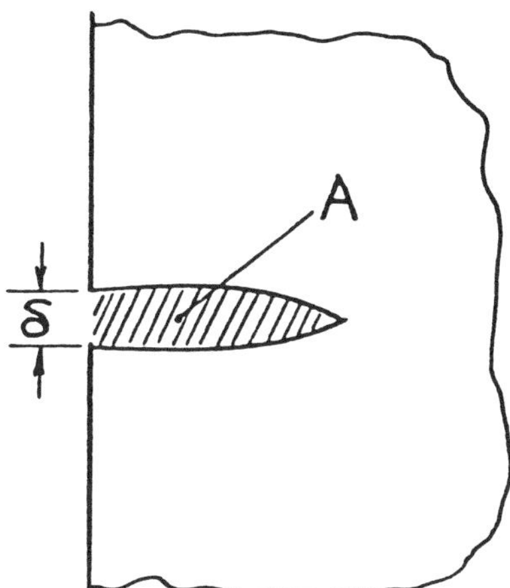

$$K_I = \left(\frac{2}{\pi}p\sqrt{\pi a}\right)(1.072) = .683p\sqrt{\pi a}$$

Crack Opening Area:

$$A = \left(\frac{4pa^2}{3E'}\right)(1.202) = 1.603\frac{pa^2}{E'}$$

Opening at Edge:

$$\delta = \left(\frac{4pa}{\pi E'}\right)(1.390) = 1.770\frac{pa}{E'}$$

Method: A and δ Paris' Equation (see **Appendix B**) or Integration of **page 8.3a**
Accuracy: Within one unit of last digit
Reference: **Tada 1985**

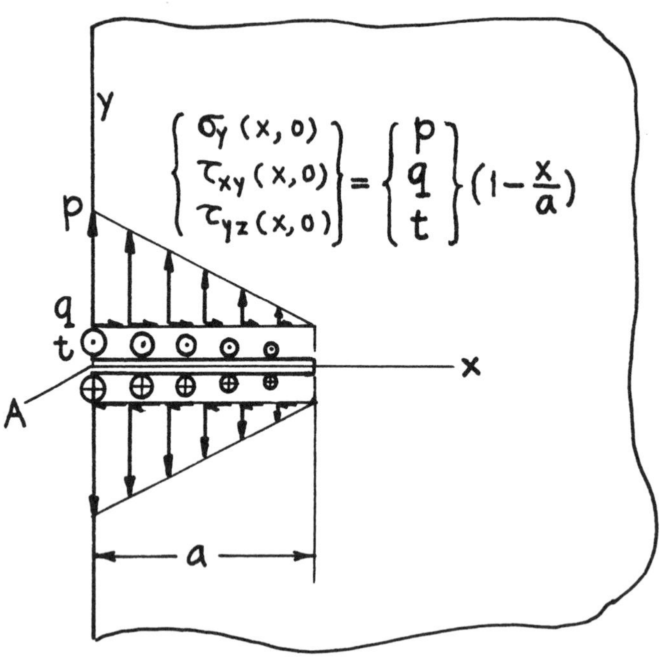

Stress Intensity Factors

$$\begin{Bmatrix} K_I \\ K_{II} \end{Bmatrix} = 0.439 \begin{Bmatrix} p \\ q \end{Bmatrix} \sqrt{\pi a} = 1.208 \left(1 - \frac{2}{\pi}\right) \begin{Bmatrix} p \\ q \end{Bmatrix} \sqrt{\pi a}$$

$$K_{III} = \left(1 - \frac{2}{\pi}\right) t \sqrt{\pi a}$$

Displacements at A

$$\begin{Bmatrix} 2v \\ 2u \end{Bmatrix} = \frac{4.046}{E'} \begin{Bmatrix} p \\ q \end{Bmatrix} a = 1.484 \cdot \frac{4}{E'} \left(1 - \frac{1}{\pi}\right) \begin{Bmatrix} p \\ q \end{Bmatrix} a$$

$$2w = \frac{2}{G} \left(1 - \frac{1}{\pi}\right) ta$$

Methods: K_I Integral Transform (Benthem)
K_{II} From Similarity of Free Boundary Corrections (Tada)
K_{III} Westergaard Stress Function, etc., Stress Concentration Factor (Neuber)
Accuracy: K_I, K_{II} 0.2%
K_{III} Exact
References: **Neuber 1937; Benthem 1972; Tada 1973**
Displacements were derived by Superposition of **page 8.1** and **page 8.5.**

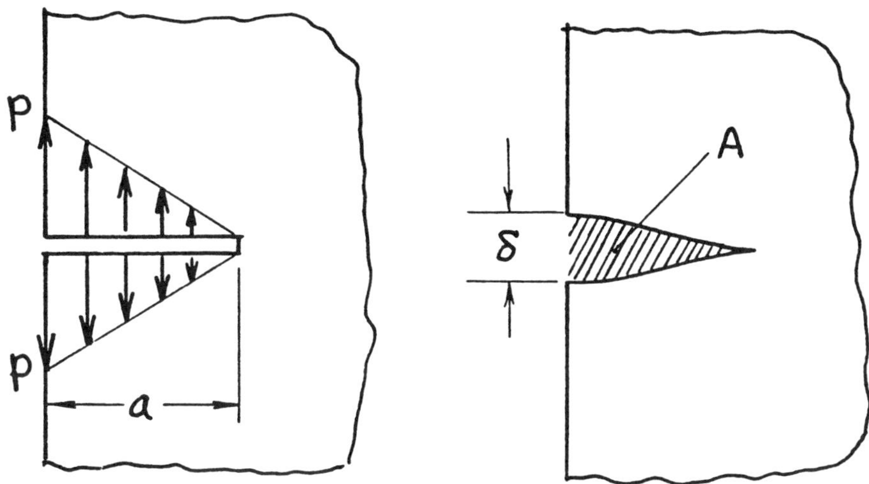

$$K_I = .439 p\sqrt{\pi a} = \left(1 - \frac{2}{\pi}\right) p\sqrt{\pi a} \quad (1.208)$$

Crack Opening Area:

$$A = 2.349 \frac{pa^2}{E'} = \left(\pi - \frac{4}{3}\right) \frac{pa^2}{E'} \quad (1.299)$$

Opening at Edge:

$$\delta = 4.046 \frac{pa}{E'} = 4\left(1 - \frac{1}{\pi}\right) \frac{pa}{E'} \quad (1.484)$$

Method: A and δ Superposition of **page 8.1a** and **page 8.5a**
Accuracy: Within one unit of last digit
Reference: **Tada 1985**

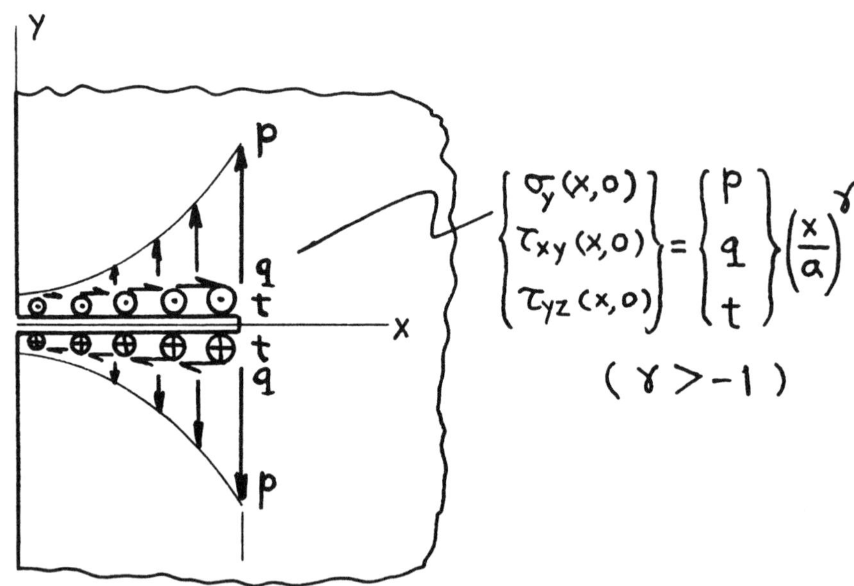

$$\begin{Bmatrix} K_I \\ K_{II} \\ K_{III} \end{Bmatrix} = \begin{Bmatrix} p \\ q \\ t \end{Bmatrix} \sqrt{a} \frac{\Gamma\left(\frac{\gamma+1}{2}\right)}{\Gamma\left(\frac{\gamma}{2}+1\right)} \cdot \begin{Bmatrix} F(\gamma) \\ F(\gamma) \\ 1 \end{Bmatrix}$$

$$F(\gamma) = 1 + \frac{.188\gamma + .488}{(\gamma+2)^2}$$

$\Gamma(\gamma)$ = Gamma Function (See **Appendix M**)

Methods: Integral Equation (Stallybrass), Alternating Method (Hartranft), Integration of **page 8.3**
Accuracy: K_I and K_{II} 0.5%
K_{III} Exact
References: **Stallybrass 1970; Hartranft 1973; Tada 1973, 1985**

NOTE: For special cases $\gamma = 0$ and $\gamma = 1$, see **page 8.1** and **page 8.5**, respectively. γ does not have to be an integer; it can be any real value $\gamma > -1$.

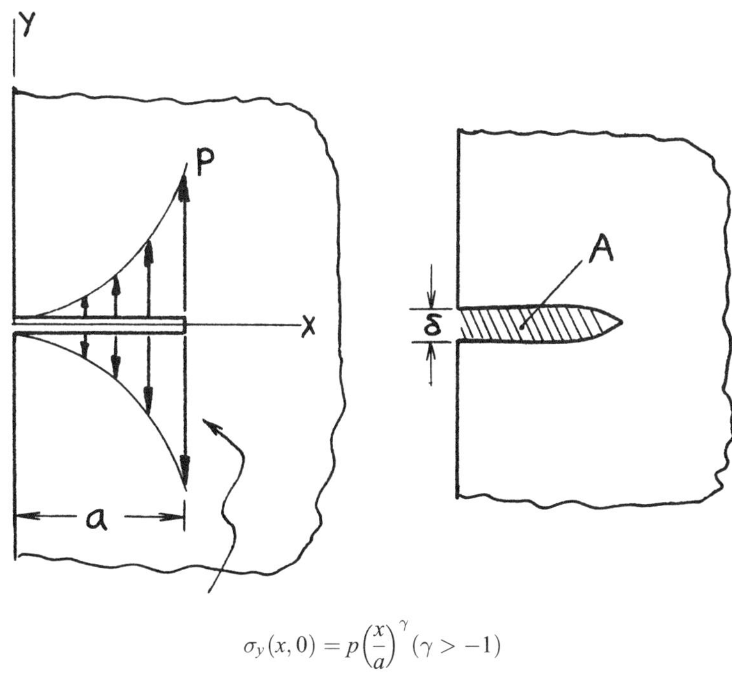

$$\sigma_y(x,0) = p\left(\frac{x}{a}\right)^{\gamma} (\gamma > -1)$$

$$K_I = p\sqrt{a}\,\frac{\Gamma\left(\frac{\gamma+1}{2}\right)}{\Gamma\left(\frac{\gamma}{2}+1\right)}\cdot F(\gamma)$$

Crack Opening Area:

$$A = \frac{pa^2}{E'}\sqrt{\pi}\,\frac{\Gamma\left(\frac{\gamma+1}{2}\right)}{\Gamma\left(\frac{\gamma}{2}+2\right)}\cdot\{1.1215\cdot F(\gamma)\}$$

Opening at Edge:

$$\delta = \frac{4pa}{E'}\cdot\frac{1}{\sqrt{\pi}}\cdot\frac{\Gamma\left(\frac{\gamma+1}{2}\right)}{(\gamma+1)\Gamma\left(\frac{\gamma}{2}+1\right)}\cdot\{1.2967\cdot F(\gamma)\}$$

$$F(\gamma) = 1 + \frac{.188\gamma + .488}{(\gamma+2)^2}$$

$\Gamma(\gamma)$ = Gamma Function (See **Appendix M**)

Method: A and δ Paris' Equation (see **Appendix B**) or Integration of **page 8.3a**
Accuracy: 0.5%
Reference: **Tada 1985**

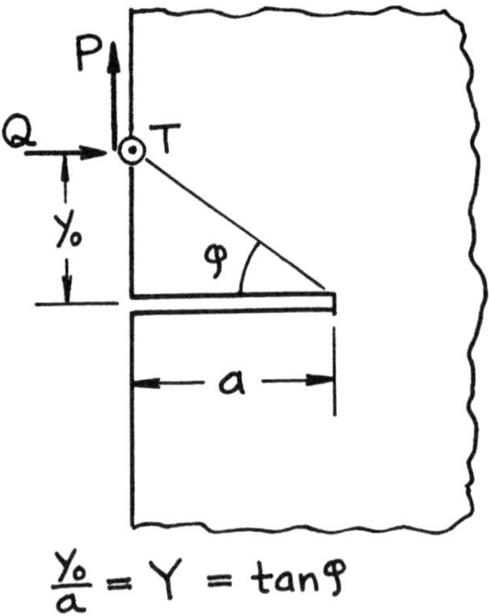

$$\begin{Bmatrix} K_I \\ K_{II} \end{Bmatrix} = \frac{1}{\sqrt{\pi a}} \begin{Bmatrix} P \cdot F_1(Y) \\ Q \cdot F_3(Y) \end{Bmatrix} - \frac{1}{\sqrt{\pi a}} \cdot \frac{2}{\pi} \begin{Bmatrix} Q \\ P \end{Bmatrix} \cdot F_2(Y)$$

$$K_{III} = \frac{T}{\sqrt{\pi a}} \cdot \frac{1}{\sqrt{1+Y^2}} = \frac{T}{\sqrt{\pi a}} \cos\varphi$$

where

$$F_1(Y) = \frac{1+2Y^2}{\left(1+Y^2\right)^{3/2}} \left[1.3 - .3\left(\frac{Y}{\sqrt{1+Y^2}}\right)^{5/4} \left\{.665 - .267\left(\frac{Y}{\sqrt{1+Y^2}}\right)^{5/4} \left(\frac{Y}{\sqrt{1+Y^2}} - .73\right)\right\}\right]$$

$$F_2(Y) = \left[\frac{1}{1+Y^2} + \frac{Y^2}{\left(1+Y^2\right)^{3/2}} \tanh^{-1}\left(\frac{1}{\sqrt{1+Y^2}}\right)\right]\left[1.3 - .375\frac{Y}{\sqrt{1+Y^2}}\left(1 - .4\frac{Y}{\sqrt{1+Y^2}}\right)\right]$$

$$F_3(Y) = \frac{1}{\left(1+Y^2\right)^{3/2}}\left[1.3 - .75\frac{Y}{\sqrt{1+Y^2}}\left(1 - 1.184\frac{Y}{\sqrt{1+Y^2}} + .512\frac{Y^2}{1+Y^2}\right)\right]$$

Method: Integration of **page 8.3**
Accuracy: Formulas $F_1(Y)$, $F_2(Y)$, and $F_3(Y)$ 0.5%
K_{III} Exact
References: **Tada 1985, 2000**

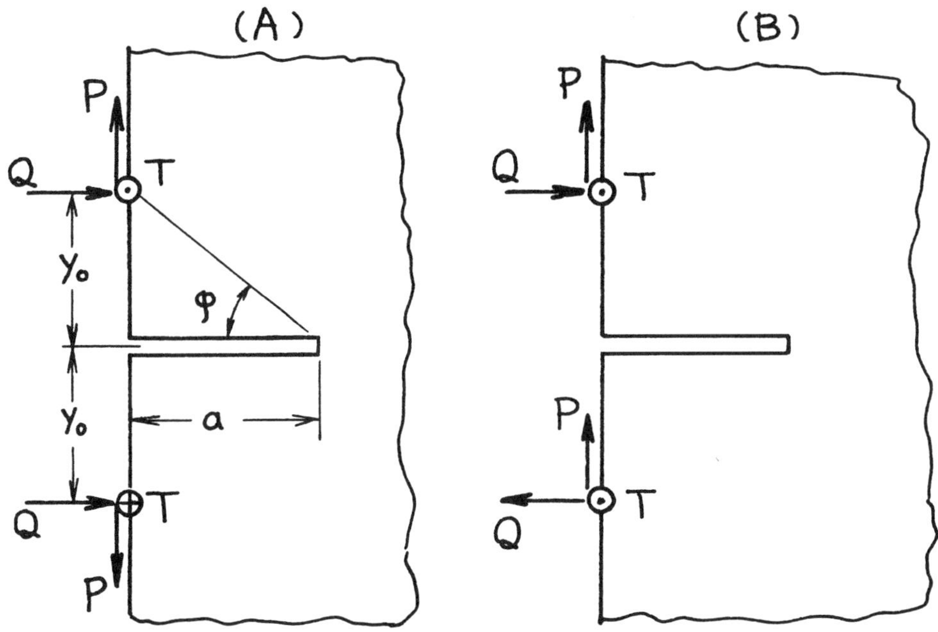

$$\frac{y_0}{a} = Y = \tan\varphi$$

$$(A) \qquad K_I = \frac{2}{\sqrt{\pi a}} P \cdot F_1(Y) - \frac{2}{\sqrt{\pi a}} \left\{ \frac{2}{\pi} Q \right\} \cdot F_2(Y)$$

$$K_{II} = 0$$

$$K_{III} = \frac{2}{\sqrt{\pi a}} T \cdot \frac{1}{\sqrt{1+Y^2}} = \frac{2}{\sqrt{\pi a}} T \cdot \cos\varphi$$

$$(B) \qquad K_I = 0$$

$$K_{II} = \frac{2}{\sqrt{\pi a}} Q \cdot F_3(Y) - \frac{2}{\sqrt{\pi a}} \left\{ \frac{2}{\pi} P \right\} \cdot F_2(Y)$$

$$K_{III} = 0$$

For $F_1(Y)$, $F_2(Y)$, and $F_3(Y)$, see **page 8.8**
Method: Superposition of **page 8.8**
Accuracy: Formulas $F_1(Y)$, $F_2(Y)$, and $F_3(Y)$ 0.5%
K_{II} and K_{III} in **(A)** and K_I and K_{III} in **(B)** are exact.
Reference: **Tada 1985, 2000**

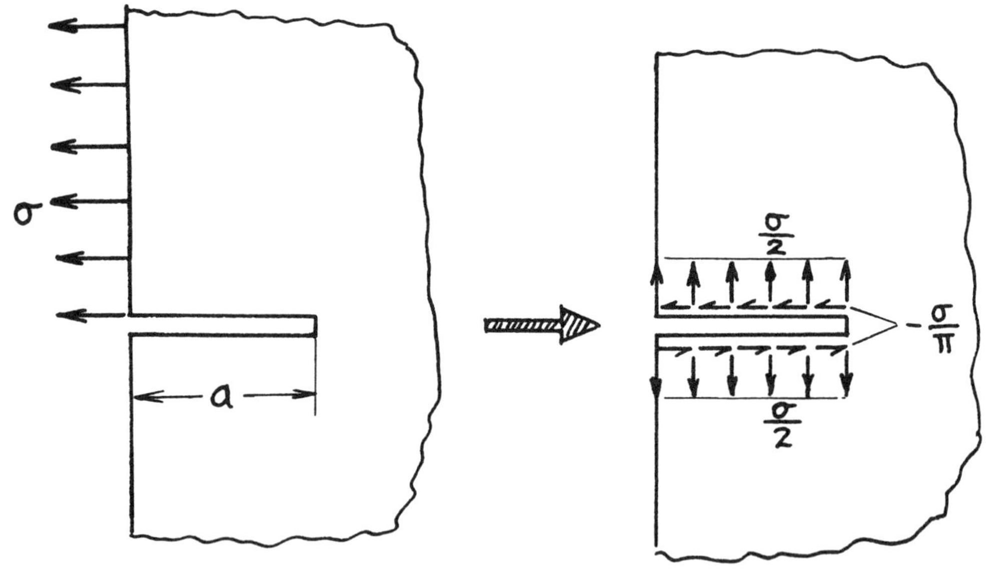

$$K_I = \left(\frac{\sigma}{2}\right)\sqrt{\pi a}(1.1215) = .561\sigma\sqrt{\pi a}$$

$$K_{II} = \left(-\frac{\sigma}{\pi}\right)\sqrt{\pi a}(1.1215) = -.367\sigma\sqrt{\pi a}$$

Method: Direct use of **page 8.1** or Integration of **page 8.8**
Accuracy: The value 1.1215 is accurate within one unit of last digit.
References: **Tada 1985**

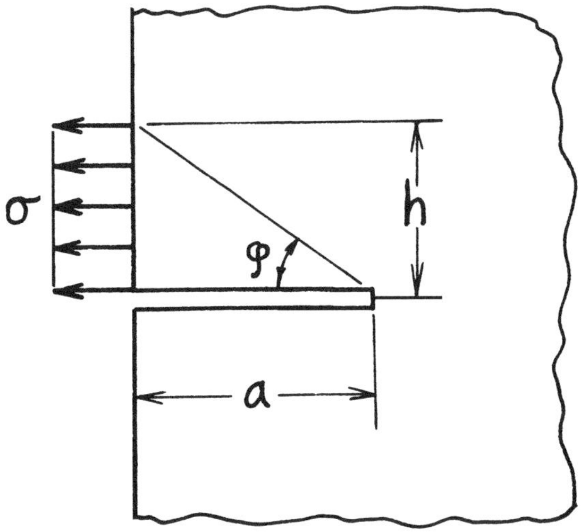

$$s = \frac{\varphi}{(\pi/2)} = \frac{2}{\pi}\tan^{-1}\frac{h}{a}$$

$$K_I = \left(\frac{1}{2}\sigma\right)\sqrt{\pi a}\cdot F_I(s)$$

$$K_{II} = \left(-\frac{1}{\pi}\sigma\right)\sqrt{\pi a}\cdot F_{II}(s)$$

$$F_I(s) = s\left[1.122 - (1-s)\left\{.296 + .25s^{3/4}(.75 - s)\right\}\right]$$

$$F_{II}(s) = s\left[1.122 + (1-s)\left(.915 + .26s^{3/2}\right)\right]$$

Method: Integration of **page 8.3** or **page 8.8**
Accuracy: 0.5%
References: **Tada 1985**

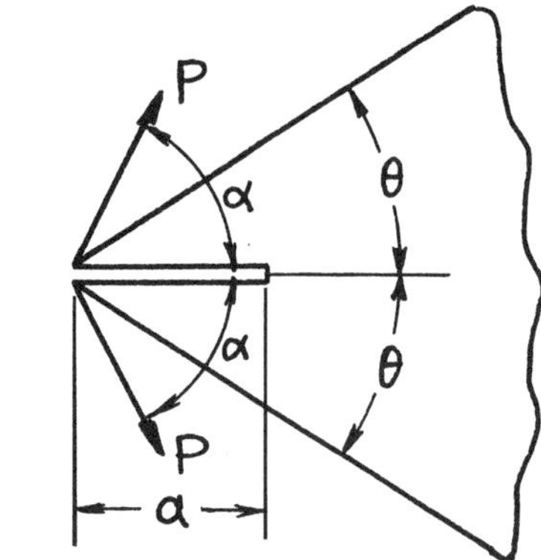

$$K_I = \frac{P}{\sqrt{a}}\left(\sin\alpha - \cos\alpha \frac{2\sin^2\theta}{2\theta + \sin 2\theta}\right)\sqrt{\frac{2\theta + \sin 2\theta}{\theta^2 - \sin^2\theta}}$$

$$K_I \gtreqless 0 : \alpha \gtreqless \tan^{-1}\left(\frac{2\sin^2\theta}{2\theta + \sin 2\theta}\right)$$

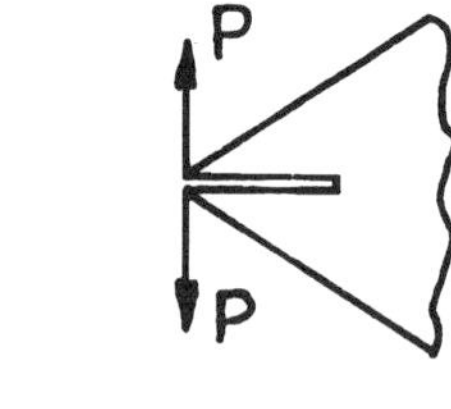

$$\alpha = \frac{\pi}{2}$$

$$K_I = \frac{P}{\sqrt{a}} \cdot \sqrt{\frac{2\theta + \sin 2\theta}{\theta^2 - \sin^2\theta}}$$

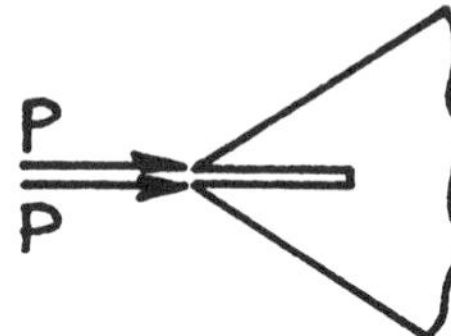

$$\alpha = 0$$

$$K_I = \frac{P}{\sqrt{a}} \cdot \frac{-2\sin^2\theta}{\sqrt{(2\theta + \sin 2\theta)\left(\theta^2 - \sin^2\theta\right)}}$$

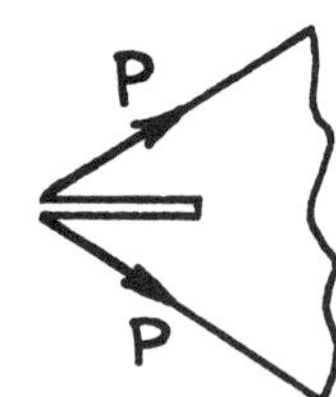

$$\alpha = \theta$$

$$K_I = \frac{P}{\sqrt{a}} \cdot \frac{2\theta \cdot \sin\theta}{\sqrt{(2\theta + \sin 2\theta)\left(\theta^2 - \sin^2\theta\right)}}$$

Method: Integral Transform
Accuracy: Exact
Reference: **Ouchterlony 1975**

NOTE: For special cases of $s = 1/2$ and $s = 1(\alpha = \pi/2)$, see **pages 8.2** and **3.6**, respectively.

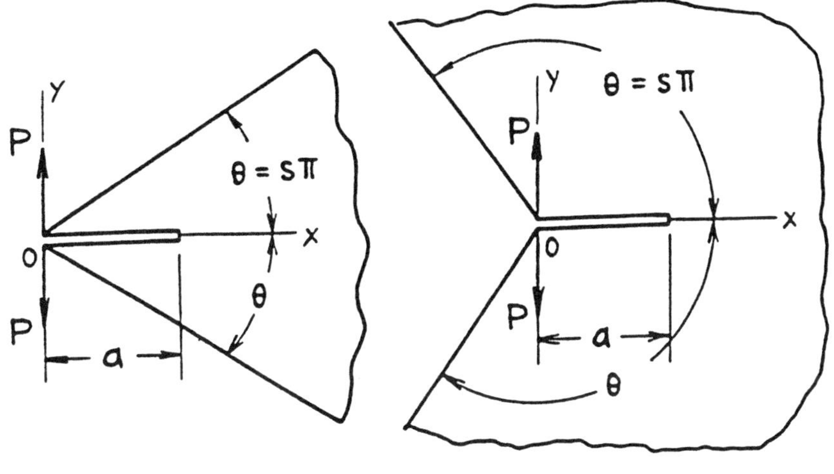

$s = \theta/\pi$

$$K_I = \frac{P}{\sqrt{a}} F(s)$$

Crack Opening Area:

$$A = \frac{2Pa}{E'} \sqrt{\pi} F(s) \cdot G(s)$$

Crack Profile:

$$v(x,0) = \frac{4P}{\pi E'} \sqrt{\pi} F(s) \cdot H\left(\frac{x}{a}, s\right)$$

where

$$F(s) = \sqrt{\frac{2\pi s + \sin 2\pi s}{(\pi s)^2 - (\sin \pi s)^2}} = \sqrt{\frac{2\theta + \sin 2\theta}{\theta^2 - (\sin\theta)^2}}$$

$$G(s) = \frac{.1755 + .219s + .385s^2 + .120s^3}{s^{3/2}}$$

$$H\left(\frac{x}{a}, s\right) = \frac{1}{\sqrt{2}}\left[f(s)\tanh^{-1}\sqrt{1-\frac{x}{a}} + g(s)\cdot\sqrt{1-\frac{x}{a}} + h(s)\cdot\frac{1}{3}\left(2+\frac{x}{a}\right)\sqrt{1-\frac{x}{a}}\right]$$

For $f(s)$, $g(s)$ and $h(s)$, see **page 8.17**

Methods: K_I Combined method of Integral Transform and **Appendix B**
A and v Paris' Equation (see **Appendix B**) and Interpolation

Accuracy: K_I Exact; A and v 1%

References: **Ouchterlony 1975; Tada 1985**

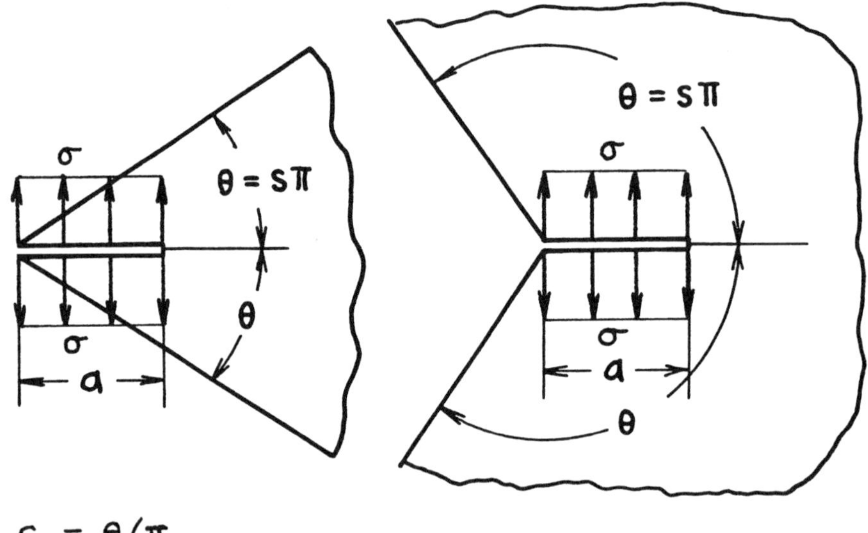

$$K_I = \sigma\sqrt{\pi a} \cdot F(s)$$

Crack Opening Area:

$$A = \frac{\sigma \pi a^2}{E'} \{F(s)\}^2$$

Opening at Edge:

$$\delta = \frac{4\sigma a}{E'} F(s) \cdot G(s)$$

where

$$F(s) = \frac{\cdot 1755 + \cdot 219s + \cdot 385s^2 + \cdot 120s^3}{s^{3/2}}$$

$$G(s) = \frac{\sqrt{\pi}}{2}\sqrt{\frac{2\pi s + \sin 2\pi s}{(\pi s)^2 - (\sin \pi s)^2}}$$

Methods: K_I Integral Transform and Integral Equation (**Doran** $s \geq \cdot 2$), Beam Theory ($s \to 0$), and Interpolation ($0 < s < .2$)
A and δ Paris' Equation (see **Appendix B**)
Accuracy: K_I and δ Better than 1%; A Better than 2%
References: **Doran 1969; Tada 1973, 1985**

NOTE: For special cases $s = 1/2$ and $s = 1$, see also **page 8.1** and **page 3.7**, respectively.

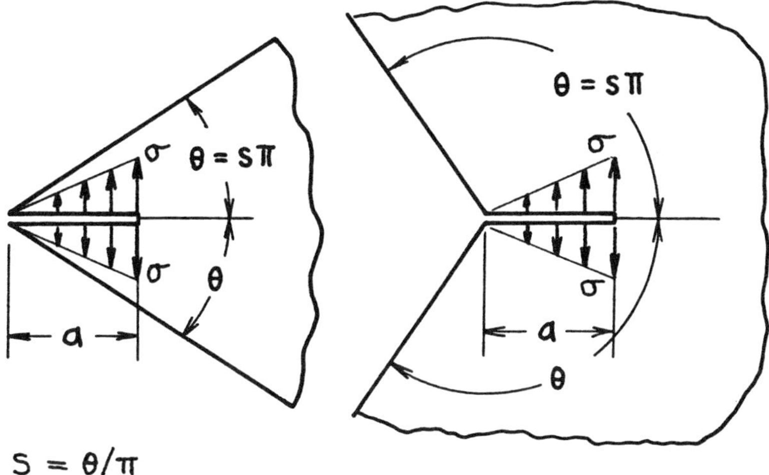

$$K_I = \sigma\sqrt{\pi a} \cdot F(s)$$

Crack Opening Area:

$$A = \frac{\sigma \pi a^2}{E'} \cdot \frac{2}{3} F(s) \cdot G(s)$$

Opening at Edge:

$$\delta = \frac{4\sigma a}{E'} \cdot F(s) \cdot H(s)$$

where

$$F(s) = \frac{.0585 + .196s + .333s^2 + .013s^3}{s^{3/2}}$$

$$G(s) = \frac{.1755 + .219s + .385s^2 + .120s^3}{s^{3/2}}$$

$$H(s) = \frac{\sqrt{\pi}}{4}\sqrt{\frac{2\pi s + \sin 2\pi s}{(\pi s)^2 - (\sin \pi s)^2}}$$

Methods: K_I Asymptotic Interpolation; A and δ Paris' Equation (see **Appendix B**)
Accuracy: K_I and δ Better than 1%; A Better than 2%
Reference: **Tada 1985**

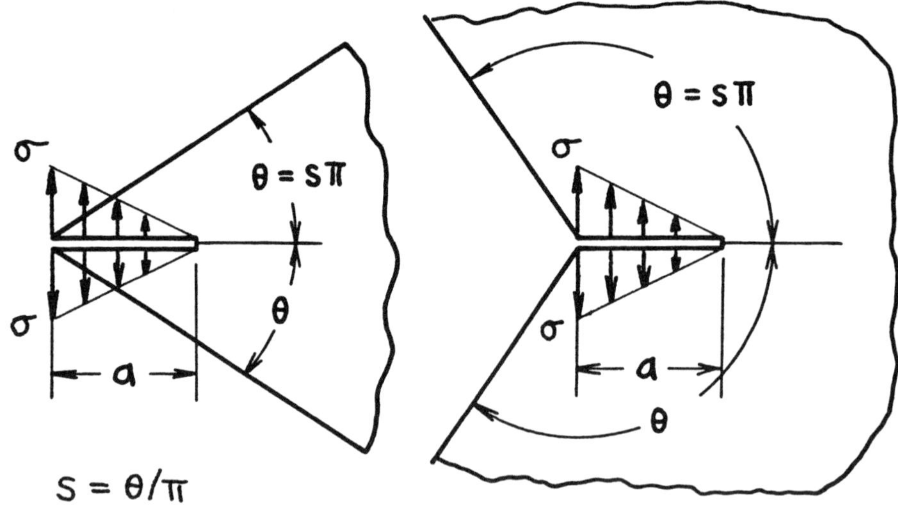

$$K_I = \sigma\sqrt{\pi a} \cdot F(s)$$

Crack Opening Area:

$$A = \frac{\sigma \pi a^2}{E'} \cdot G(s) \cdot H(s)$$

Opening at Edge:

$$\delta = \frac{4\sigma a}{E'} \cdot I(s) \cdot J(s)$$

where

$$F(s) = \frac{.117 + .023s + .052s^2 + .107s^3}{s^{3/2}}$$

$$G(s) = \frac{.1755 + .219s + .385s^2 + .120s^3}{s^{3/2}}$$

$$H(s) = \frac{.1365 + .088s + .163s^2 + .111s^3}{s^{3/2}}$$

$$I(s) = \frac{.1463 + .121s + .219s^2 + .114s^3}{s^{3/2}}$$

$$J(s) = \frac{\sqrt{\pi}}{2}\sqrt{\frac{2\pi s + \sin 2\pi s}{(\pi s)^2 - (\sin \pi s)^2}}$$

Method: Superposition of **page 8.14** and **page 8.15**
Accuracy: K_I and δ Better than 1%; A Better than 2%
Reference: **Tada 1985**

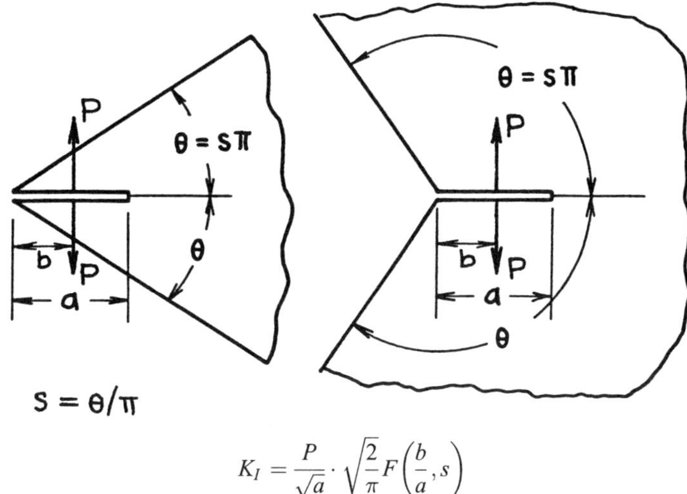

$$K_I = \frac{P}{\sqrt{a}} \cdot \sqrt{\frac{2}{\pi}} F\left(\frac{b}{a}, s\right)$$

Crack Opening Area:

$$A = \frac{4Pa}{E'} \cdot F_1(s) \cdot G\left(\frac{b}{a}, s\right)$$

Opening at Edge:

$$\delta = \frac{4P}{\pi E'} \sqrt{2} F_1(s) \cdot H\left(\frac{b}{a}, s\right)$$

where

$$F_1(s) = \frac{.1755 + .219s + .385s^2 + .120s^3}{s^{3/2}}$$

$$F\left(\frac{b}{a}, s\right) = \frac{f(s) + g(s)\frac{b}{a} + h(s)\left(\frac{b}{a}\right)^2}{\sqrt{1 - \frac{b}{a}}}$$

$$G\left(\frac{b}{a}, s\right) = \frac{1}{\sqrt{2}}\left[f(s)\left(\sqrt{1 - \frac{b}{a}} + \frac{b}{a}\tanh^{-1}\sqrt{1 - \frac{b}{a}}\right) + g(s) \cdot 2\frac{b}{a}\tanh^{-1}\sqrt{1 - \frac{b}{a}} + h(s) \cdot 2\frac{b}{a}\sqrt{1 - \frac{b}{a}}\right]$$

$$H\left(\frac{b}{a}, s\right) = f(s)\tanh^{-1}\sqrt{1 - \frac{b}{a}} + g(s)\sqrt{1 - \frac{b}{a}} + h(s) \cdot \frac{1}{3}\left(2 + \frac{b}{a}\right)\sqrt{1 - \frac{b}{a}}$$

$$f(s) = \sqrt{\frac{\pi}{2}}\sqrt{\frac{2\pi s + \sin 2\pi s}{(\pi s)^2 - (\sin \pi s)^2}}$$

$$g(s) = -1 - 3f(s) + f_1(s)$$

$$h(s) = 2 + 2f(s) - f_1(s)$$

$$f_1(s) = \left(1.103 + 3.615s^2 - .718s^3\right)/s^{3/2}$$

Method: Estimated by Interpolation
Accuracy: K_I and δ 2%; A 3%
Reference: **Tada 1985**

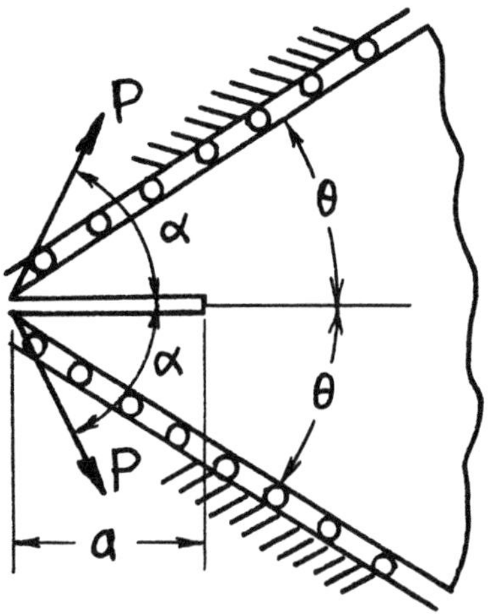

$$K_I = \frac{2P}{\sqrt{a}} \cdot \frac{\cos(\theta - \alpha)}{\sqrt{2\theta + \sin 2\theta}}$$

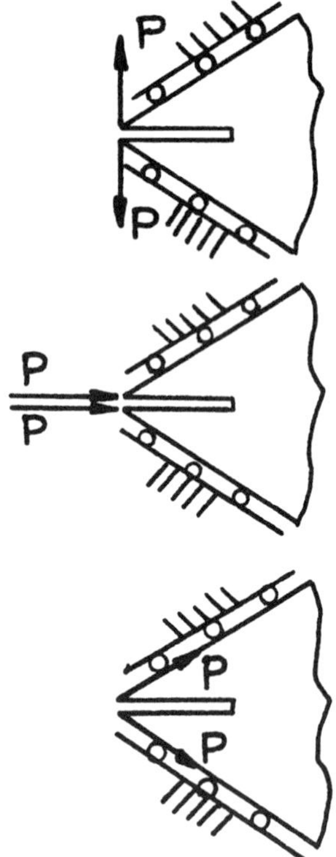

$$\alpha = \frac{\pi}{2}$$

$$K_I = \frac{2P}{\sqrt{a}} \cdot \frac{\sin\theta}{\sqrt{2\theta + \sin 2\theta}}$$

$$\alpha = 0$$

$$K_I = \frac{2P}{\sqrt{a}} \cdot \frac{\cos\theta}{\sqrt{2\theta + \sin 2\theta}}$$

$$\alpha = \theta$$

$$K_I = \frac{2P}{\sqrt{a}} \cdot \frac{1}{\sqrt{2\theta + \sin 2\theta}}$$

Method: Integral Transform
Accuracy: Exact
Reference: **Ouchterlony 1975**

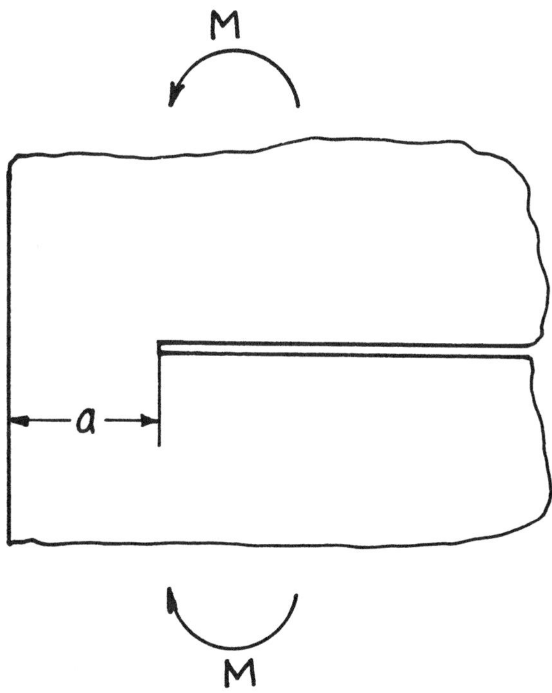

Stress Intensity Factors

$$K_I = 3.975 \frac{M}{a\sqrt{a}} = \frac{E'\theta}{3.975} \cdot \sqrt{a}$$

$$K_{II} = K_{III} = 0$$

Displacement (Relative Rotation at Infinity)

$$\theta = \frac{15.8}{E'} \cdot \frac{M}{a^2}$$

Methods: Integral Transform (Benthem), Extrapolation from the Results by Boundary Collocation Method (Wilson)
Accuracy: Within 0.1%
References: **Wilson 1969, 1970; Benthem 1972**

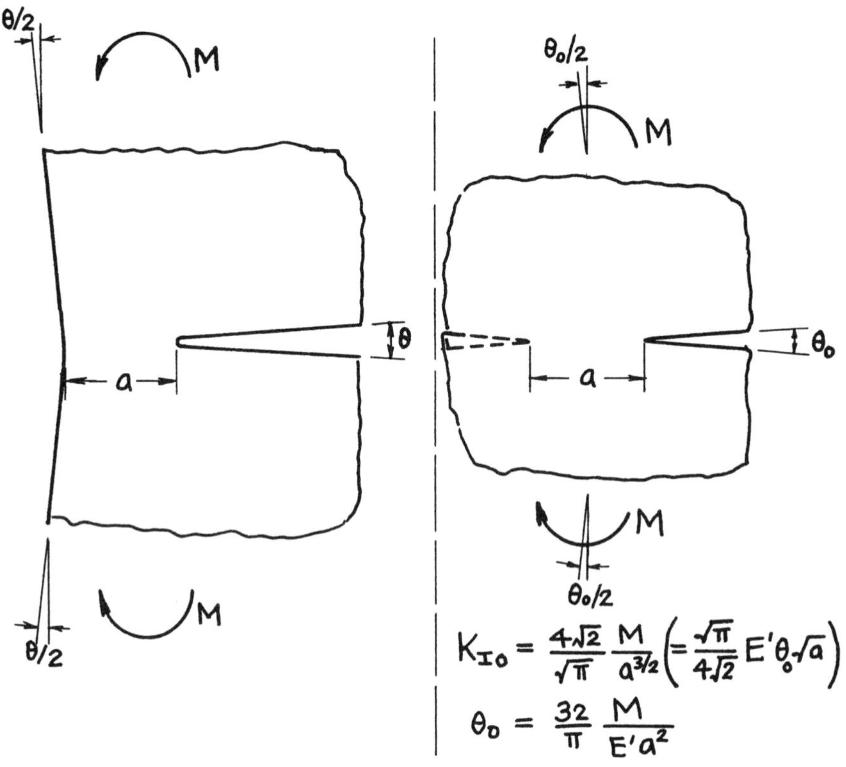

$$K_I = 3.975 \frac{M}{a^{3/2}} = 1.245 K_{I0}$$

Relative Rotation at Infinity:

$$\theta = 15.80 \frac{M}{E' a^2} = 1.551 \theta_0 \left(= 1.245^2 \theta_0 \right)$$

When θ is prescribed:

$$K_I = \frac{1}{3.975} E' \theta \sqrt{a} = \frac{1}{1.245} \left(\frac{\sqrt{\pi}}{4\sqrt{2}} E' \theta a \right) = \frac{K_{I0}}{1.245}$$

Method: (Comparison of **page 9.1** and **pages 4.10, 4.10a**)
Accuracy: Within one unit of last digit
Reference: **Tada 1985**

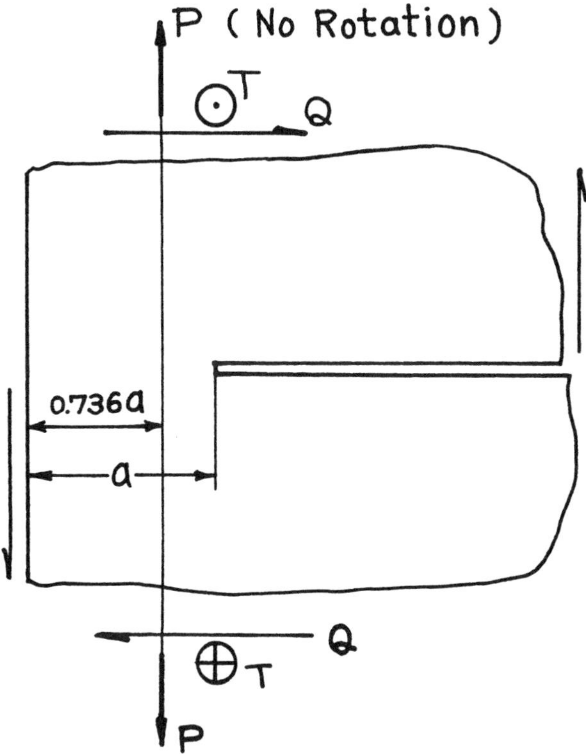

$$K_I = 1.297 \cdot \frac{2P}{\sqrt{\pi a}} \qquad \left(\frac{\pi}{\sqrt{\pi^2 - 4}} = 1.297 \right)$$

$$K_{II} = 1.297 \cdot \frac{2Q}{\sqrt{\pi a}}$$

$$K_{III} = \frac{2T}{\sqrt{\pi a}}$$

Methods: K_I Integral Transform (Benthem, Stallybrass), K_{II} Integral Equation (Tada), Boundary Collocation Method (Wilson), K_{III} Westergaard Stress Function, etc.
Accuracy: K_I 0.1%; K_{II}, K_{III} Exact
References: **Koiter 1965b; Wilson 1970; Stallybrass 1971; Benthem 1972; Tada 1973**

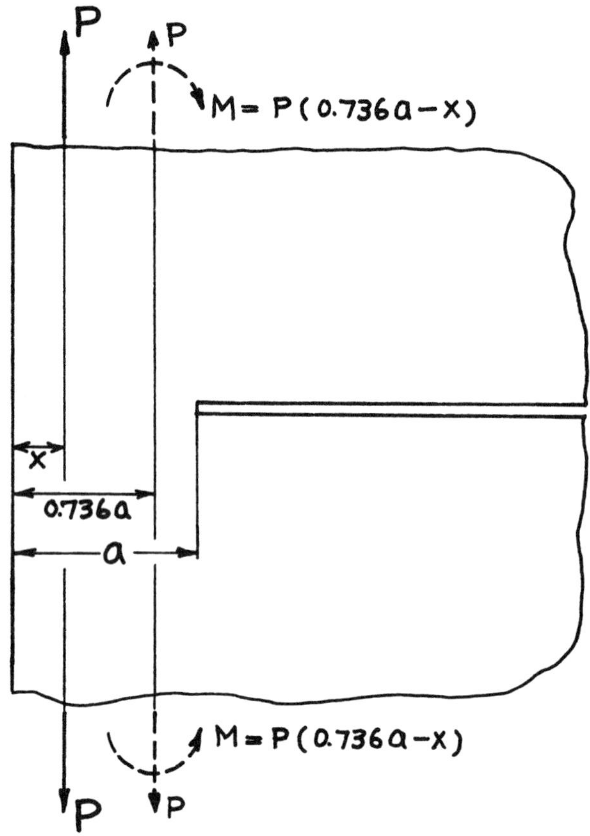

$$K_I = 3.522\left(\frac{x}{a} - 0.368\right)\frac{2P}{\sqrt{\pi a}}$$

$$K_{II} = K_{III} = 0$$

Method: Superposition of **page 9.1** and **page 9.2**
Accuracy: 0.1%
References: **Benthem 1972; Tada 1973**

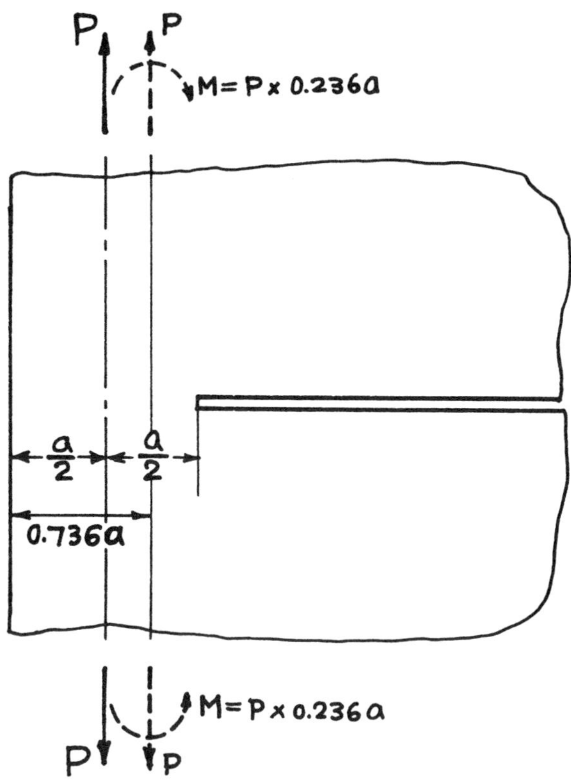

$$K_I = 0.464 \cdot \frac{2P}{\sqrt{\pi a}}$$

$$K_{II} = K_{III} = 0$$

Method: Superposition of **page 9.1** and **page 9.2**
Accuracy: 0.2%
References: **Benthem 1972; Tada 1973**

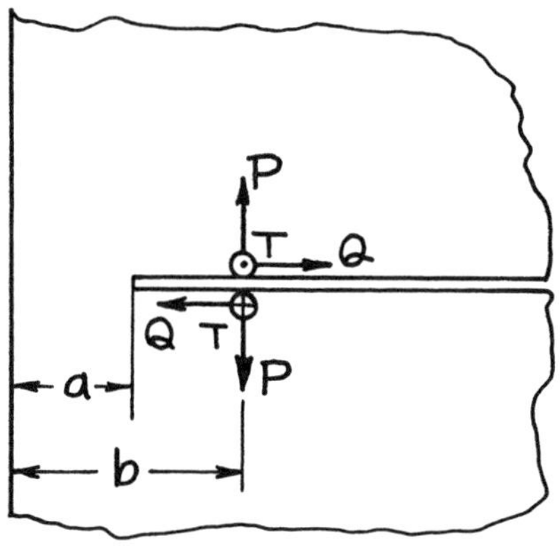

$$s = 1 - a/b$$

$$K_I = \frac{2P}{\sqrt{\pi a}} \cdot \frac{1}{\sqrt{1-\left(a/b\right)^2}} \cdot F(s) + 3.975 \cdot \frac{P(b-0.736a)}{a\sqrt{a}}$$

$$= \frac{2P}{\sqrt{\pi a}} \left\{ \frac{1}{\sqrt{1-\left(a/b\right)^2}} \cdot F(s) + 3.52\left(\frac{b}{a} - 0.736\right) \right\}$$

$$K_{II} = \frac{2Q}{\sqrt{\pi a}} \cdot \frac{1}{\sqrt{1-\left(a/b\right)^2}} \cdot F(s)$$

$$K_{III} = \frac{2T}{\sqrt{\pi a}} \cdot \frac{1}{\sqrt{1-\left(a/b\right)^2}}$$

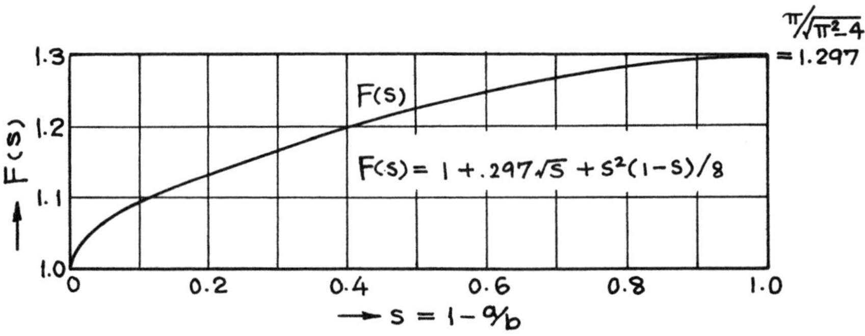

Method: K_I, K_{II} Successive Stress Relaxation - Integral Equations; K_{III} Westergaard Stress Function, etc.
Accuracy: K_I, K_{II} Estimated at 1%; K_{III} Exact
References: **Tada 1972a, 1973**

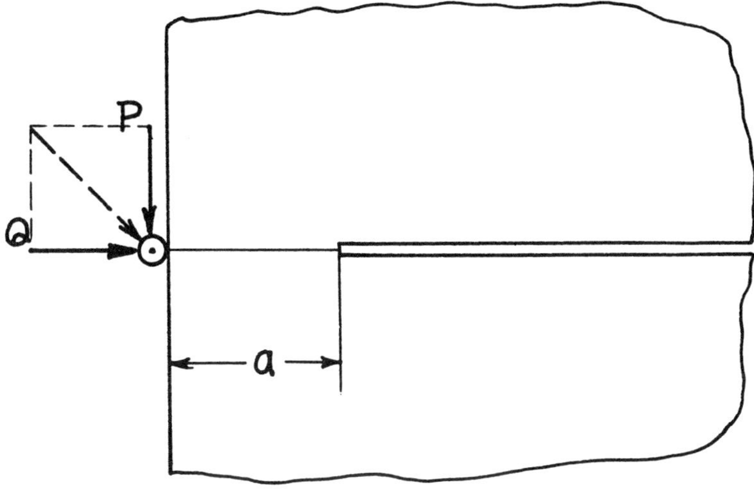

$$K_I = K_{II} = K_{III} = 0$$

Method: Special case of **page 9.6a**
Accuracy: Exact
Reference: **Tada 1973**

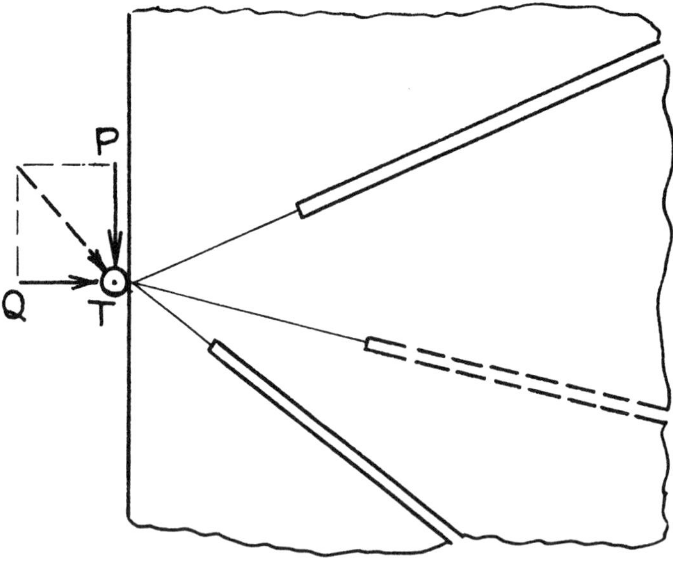

$$K_I = K_{II} = K_{III} = 0$$

Methods: Westergaard Stress Function, etc. (Simple Radial Stresses for P and Q)
Accuracy: Exact
Reference: **Tada 1985**

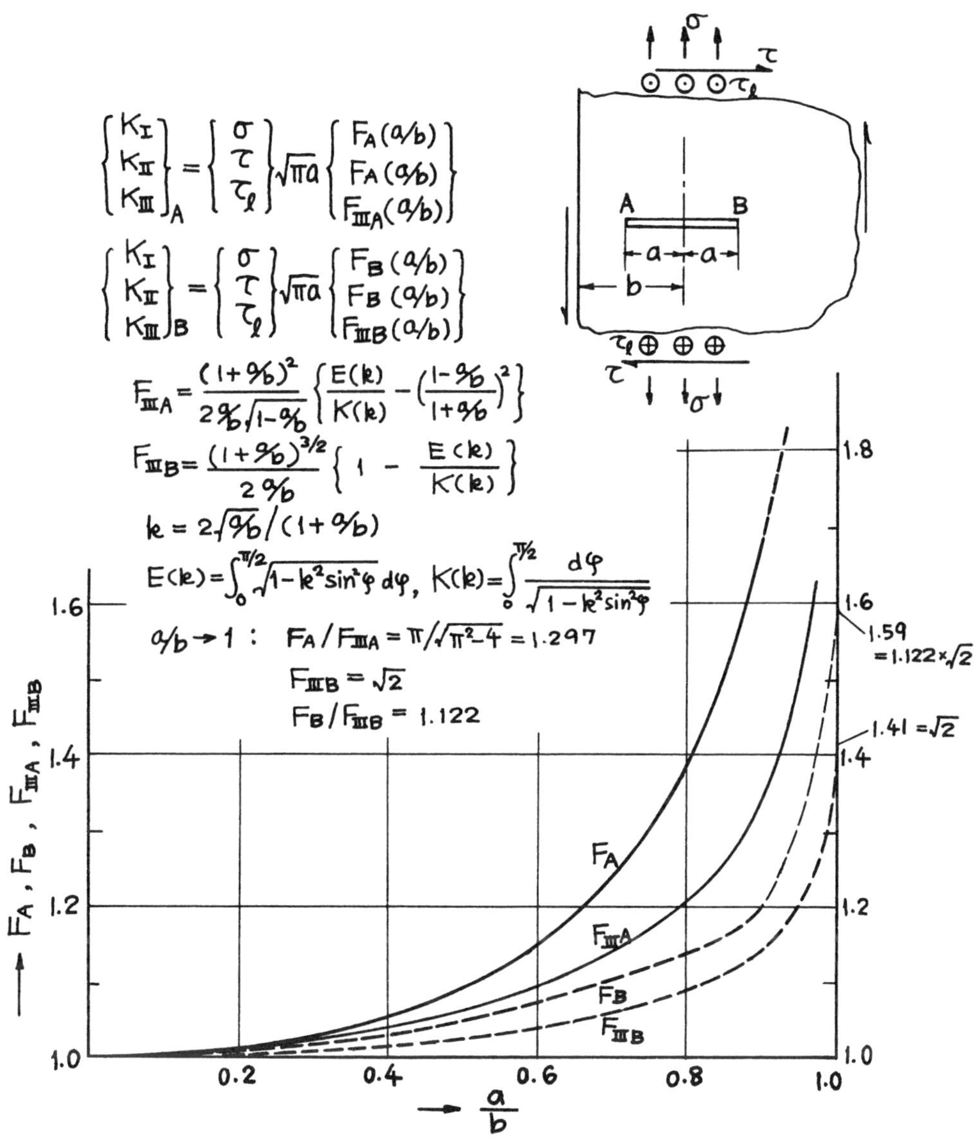

Methods: K_I Expansions of Complex Stress Potentials (Isida)
K_{II} From similarity of free boundary corrections (Tada)
K_{III} Muskhelishvili Method, etc. (**page 6.1**)

Accuracy: K_I, K_{II} 1%
K_{III} Exact

References: K_I **Isida 1965a, 1970a**; K_{II} **Tada 1973**; K_{III} **Erdogan 1962; Sih 1964**

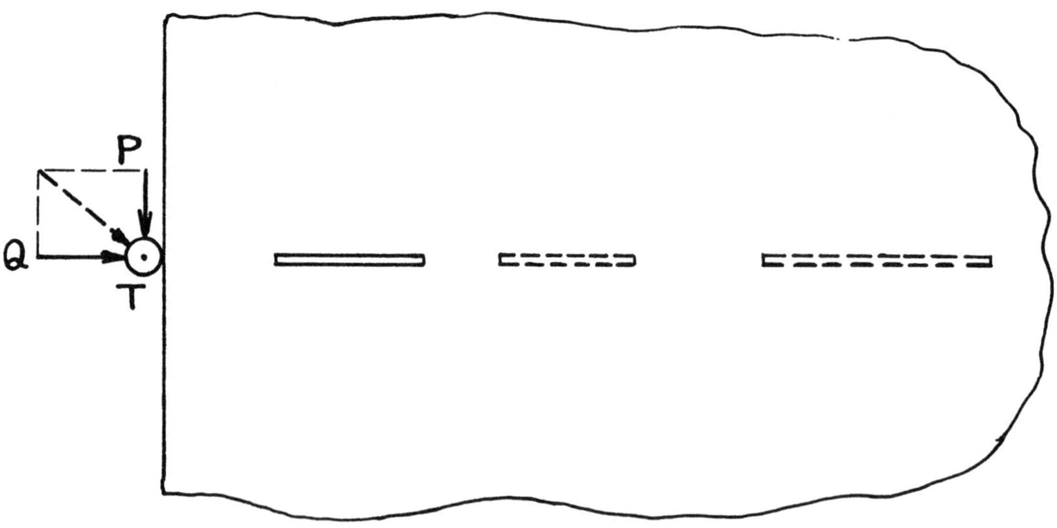

$$K_I = K_{II} = K_{III} = 0$$

Method: Special case of **page 10.2a**
Accuracy: Exact
Reference: **Tada 1973**

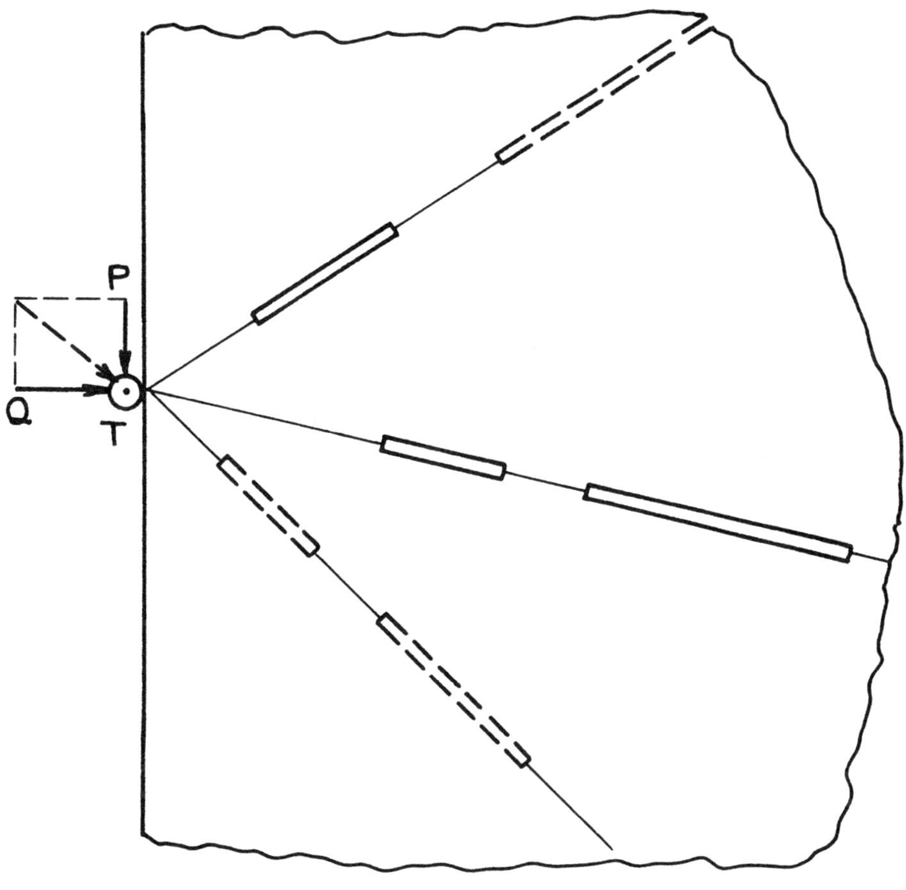

$$K_I = K_{II} = K_{III} = 0$$

Methods: Westergaard Stress Function, etc. (Simple Radial Stresses for P and Q)
Accuracy: Exact
Reference: **Tada 1985**

$$K_{I,A} = \sigma\sqrt{\pi a} \cdot F_A\left({}^{a}\!/_{b}\right)$$

$$K_{I,B} = \sigma\sqrt{\pi a} \cdot F_B\left({}^{a}\!/_{b}\right)$$

$$F_A\left({}^{a}\!/_{b}\right) = 1 - .175\left({}^{a}\!/_{b}\right)^2 - .245\left({}^{a}\!/_{b}\right)^3 \quad \left({}^{a}\!/_{b} \leq .8\right)$$

$$F_B\left({}^{a}\!/_{b}\right) = 1 - .145\left({}^{a}\!/_{b}\right)^2 \quad \left({}^{a}\!/_{b} \leq .9\right)$$

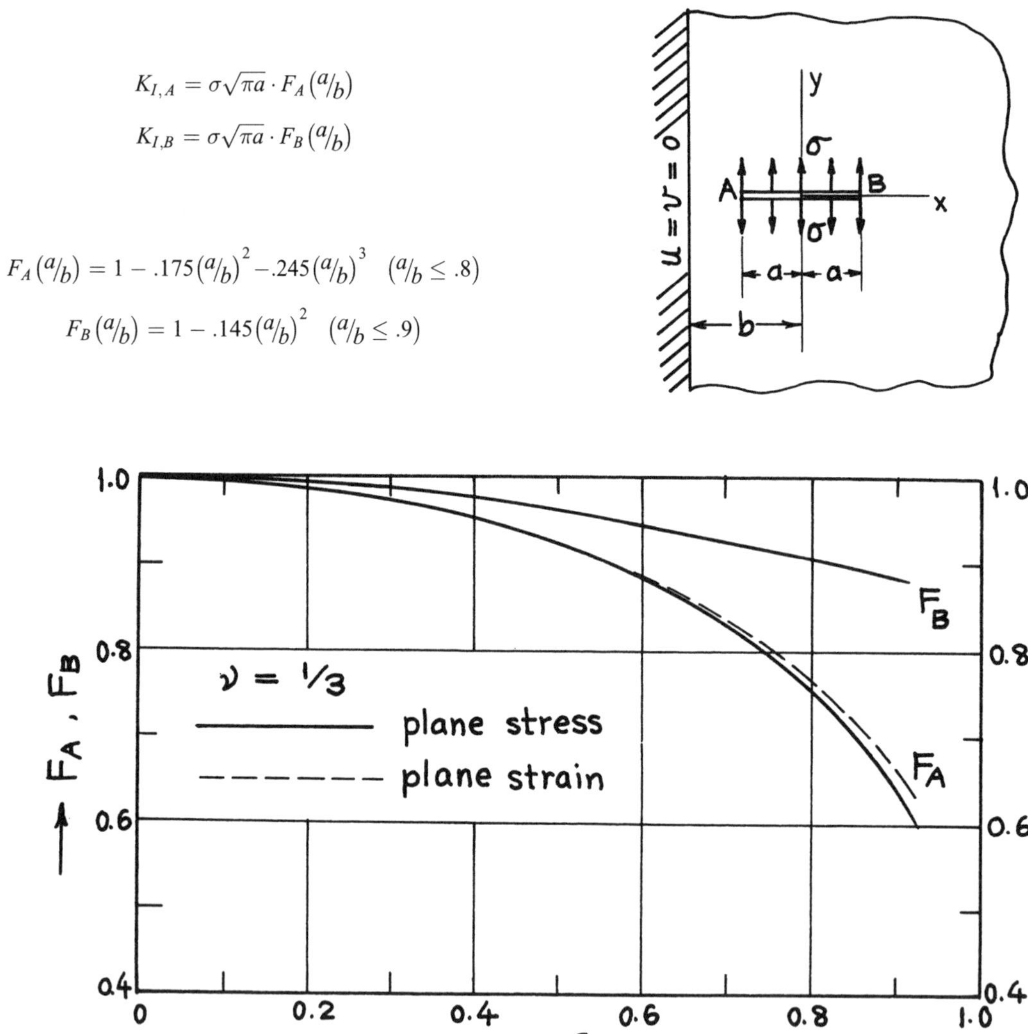

Method: Series Expansions of Complex Potentials

Accuracy: Curves are based on numerical values with 0.1% accuracy

Formulas: F_A 1% for ${}^{a}\!/_{b} < 0.8$; F_B 1% for ${}^{a}\!/_{b} < 0.9$

References: **Isida 1970a; Tada 1985**

$$K_I = \sigma\sqrt{\pi a} \cdot F_I(s)$$
$$K_{II} = \sigma\sqrt{\pi a} \cdot F_{II}(s)$$

$$s = \frac{a}{a+h}$$

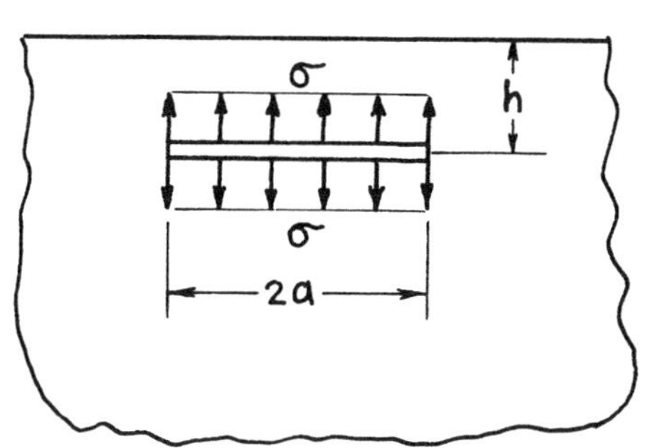

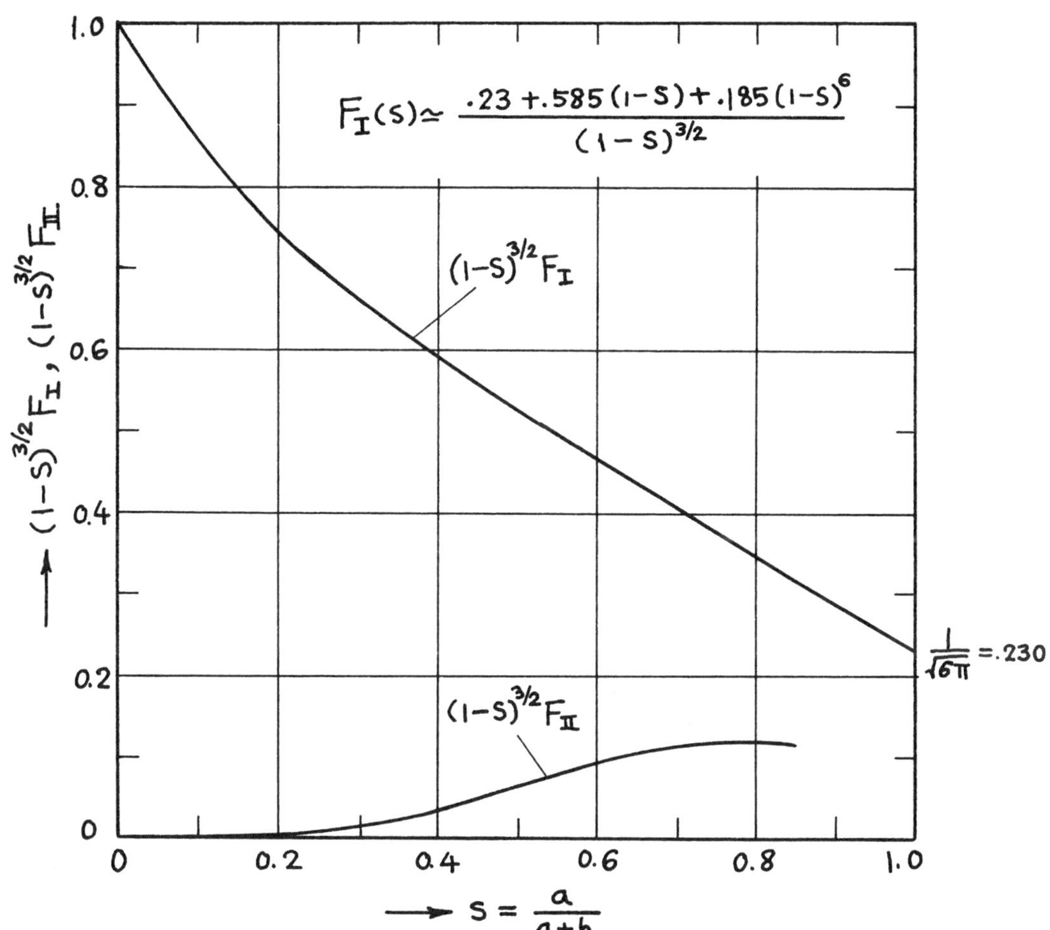

Methods: Singular Integral Equations, Beam Theory ($s \rightarrow 1$)
Accuracy: Better than 1%
References: **Erdogan 1973, Tada 1985**

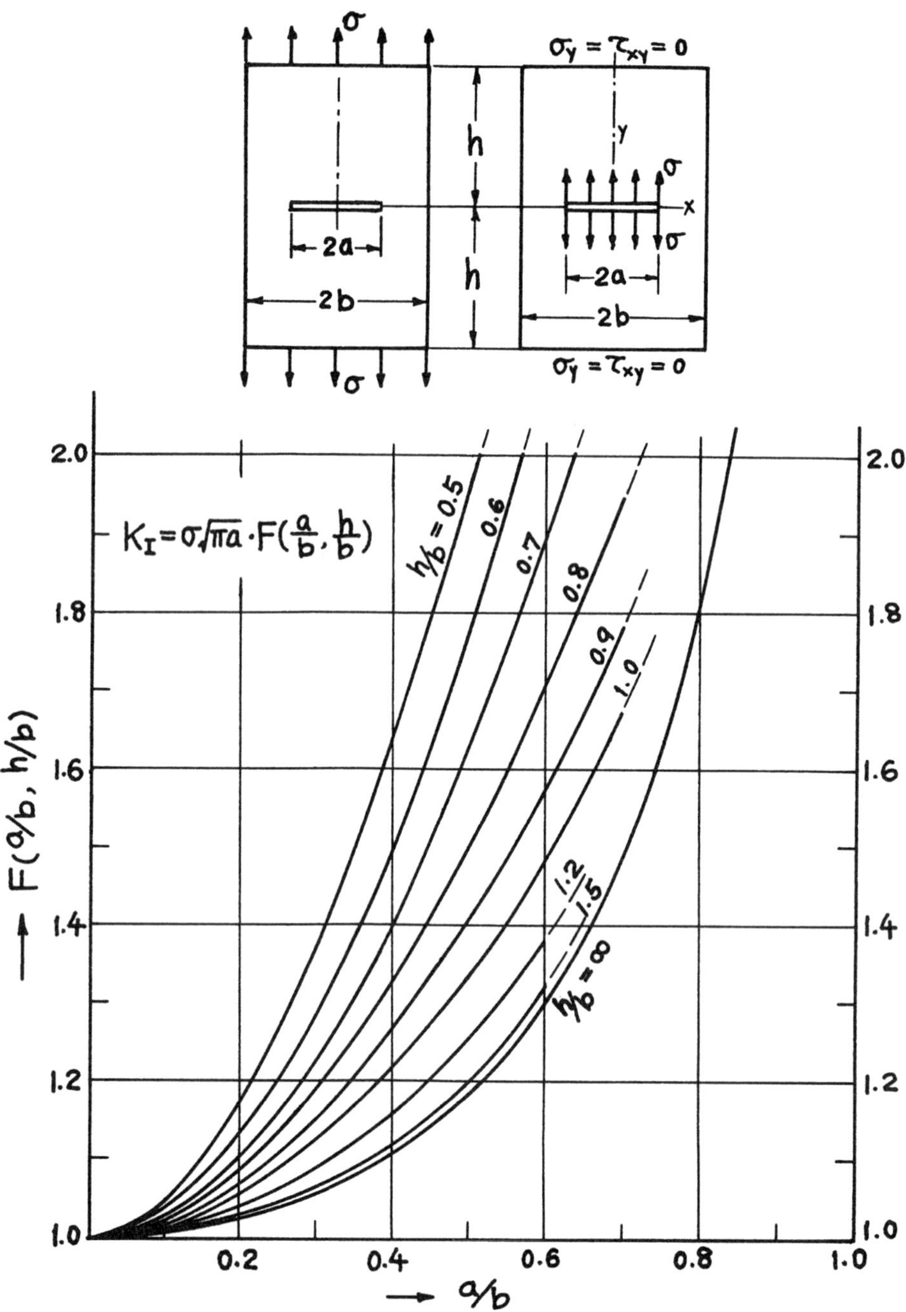

Methods: Expansions of Complex Stress Potentials Combined with a Boundary Collocation Method (Isida), Boundary Collocation Method (Kobayashi)

Accuracy: Better than 1%

References: **Kobayashi 1964; Isida 1971a,b**

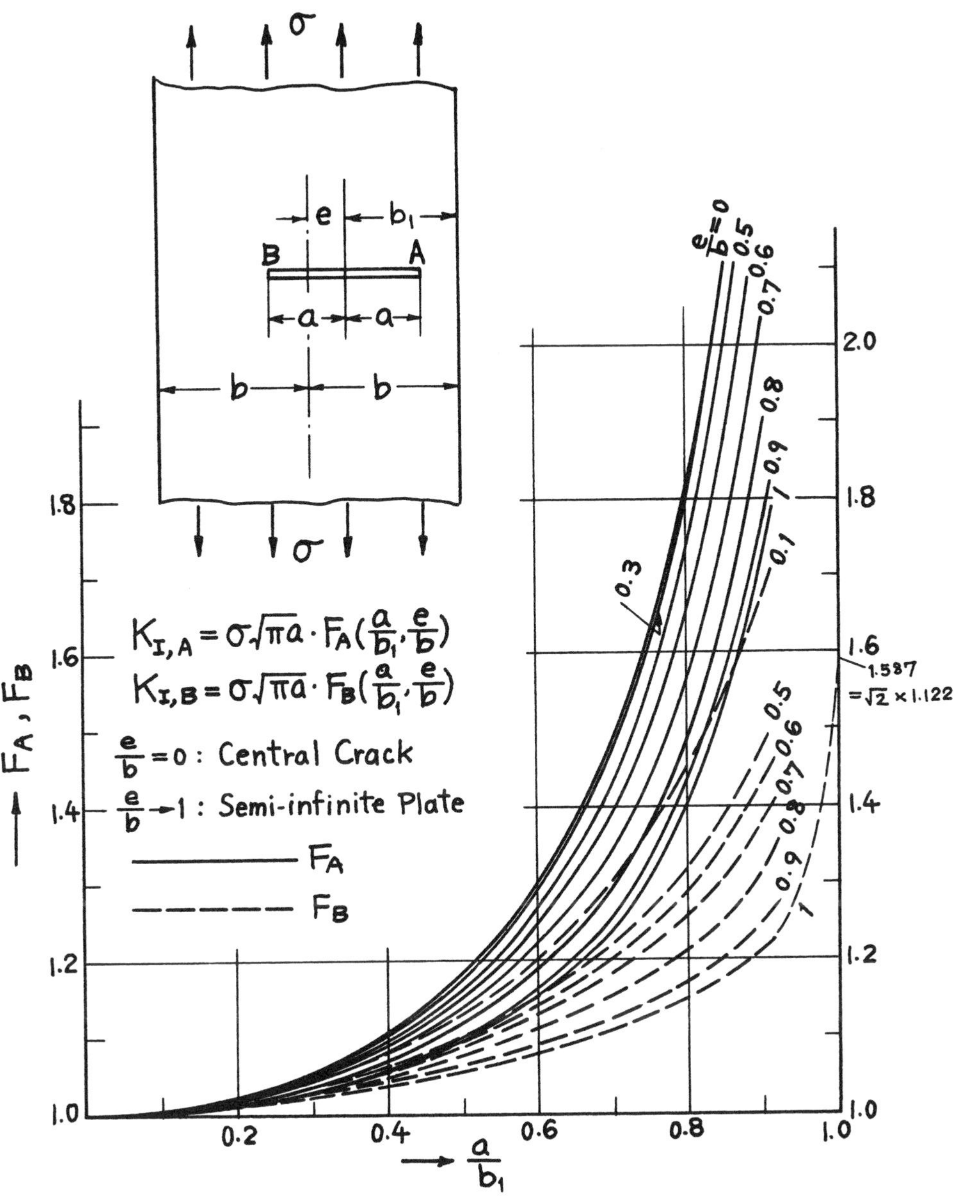

Method: Laurent's Expansions of Complex Stress Potentials
Accuracy: Better than 1%
Reference: **Isida 1965a**

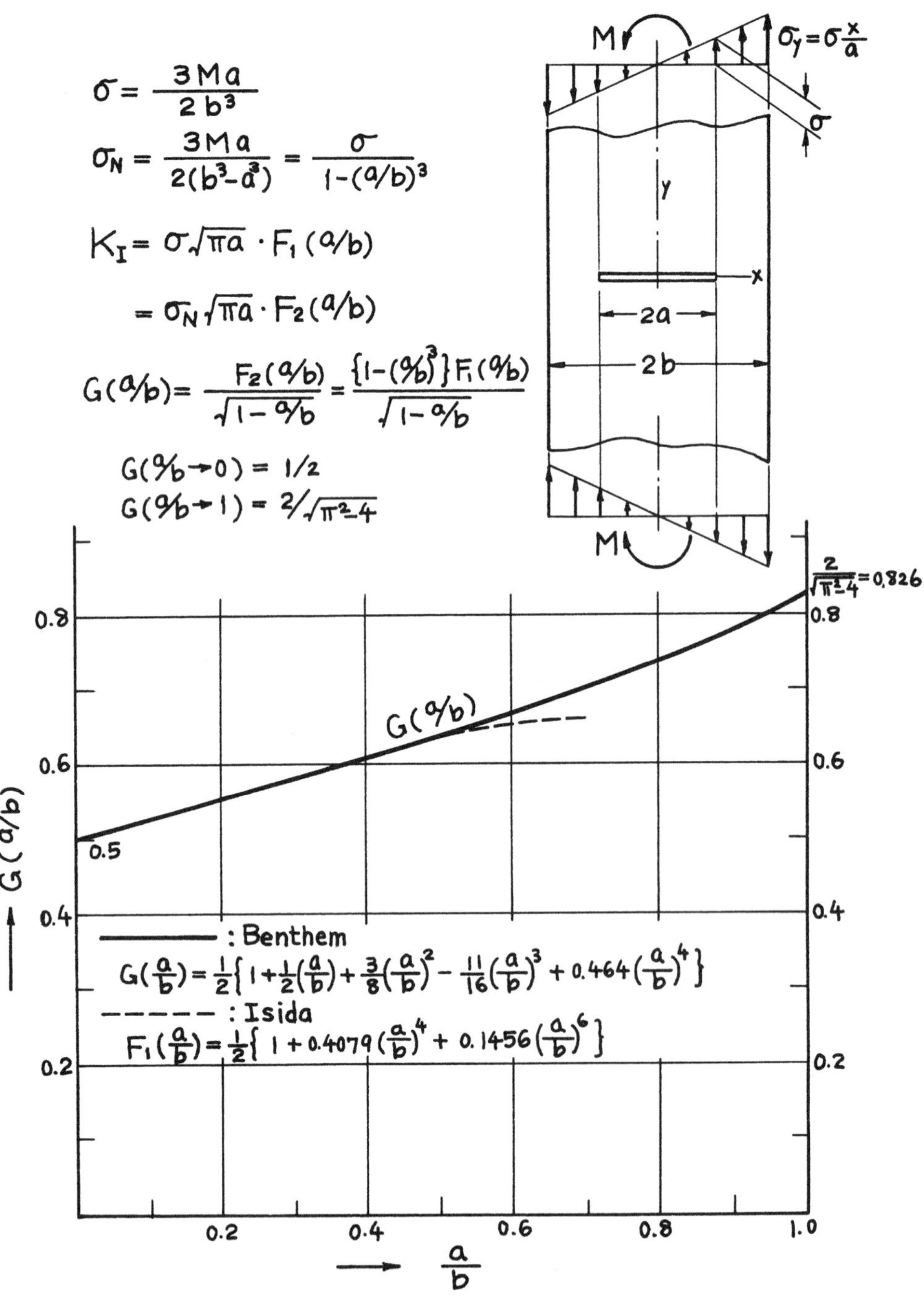

Method: From Stress Concentration Factor for an Elliptical Hole (Isida), Asymptotic Approximation (Benthem)

Accuracy: Better than 1% [$G(a/b)$ formula by Benthem]

References: **Isida 1956; Benthem 1972**

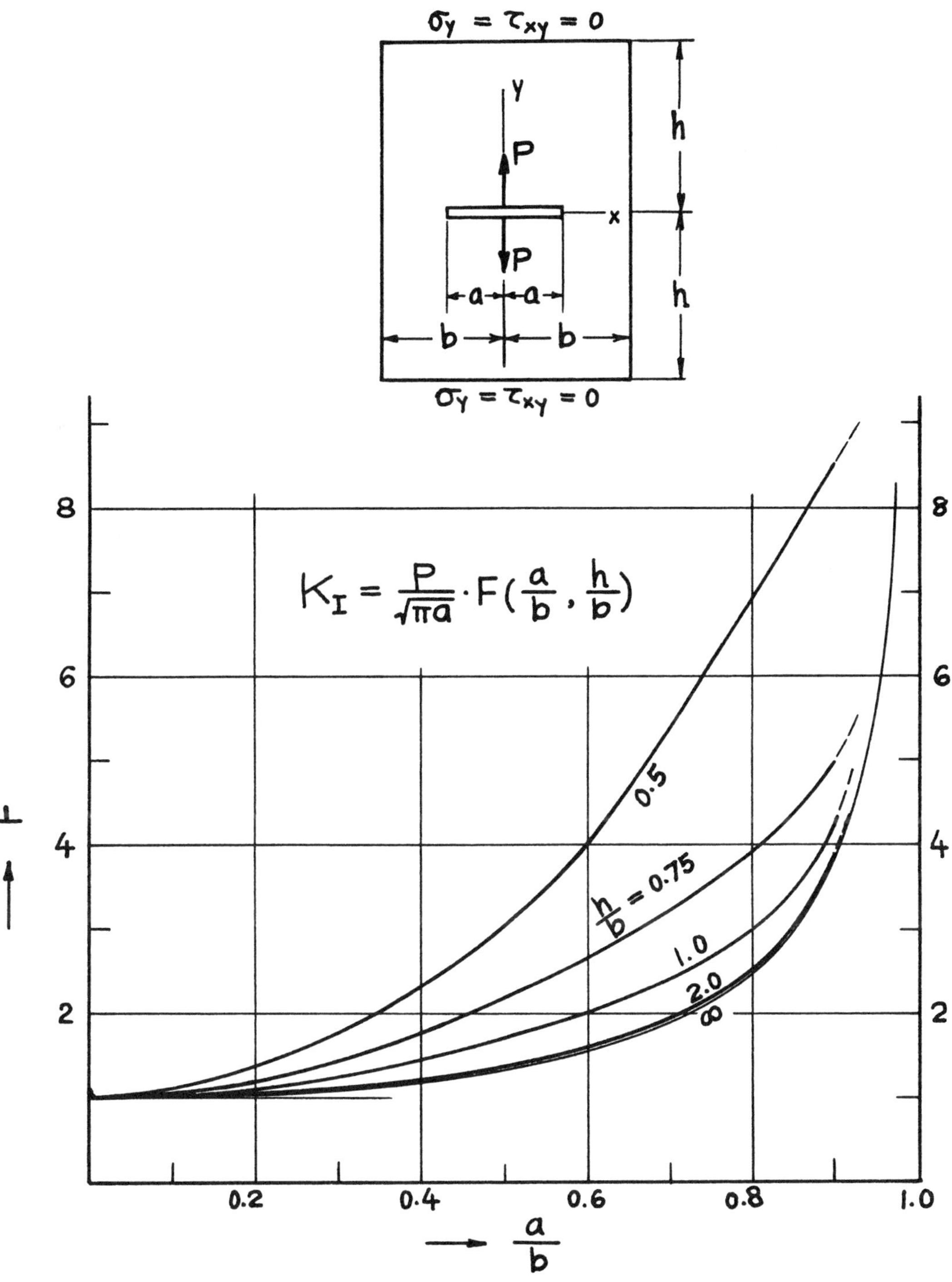

Method: Boundary Collocation Method
Accuracy: Curves were drawn based on results having better than 0.1% accuracy.
Reference: **Newman 1971**

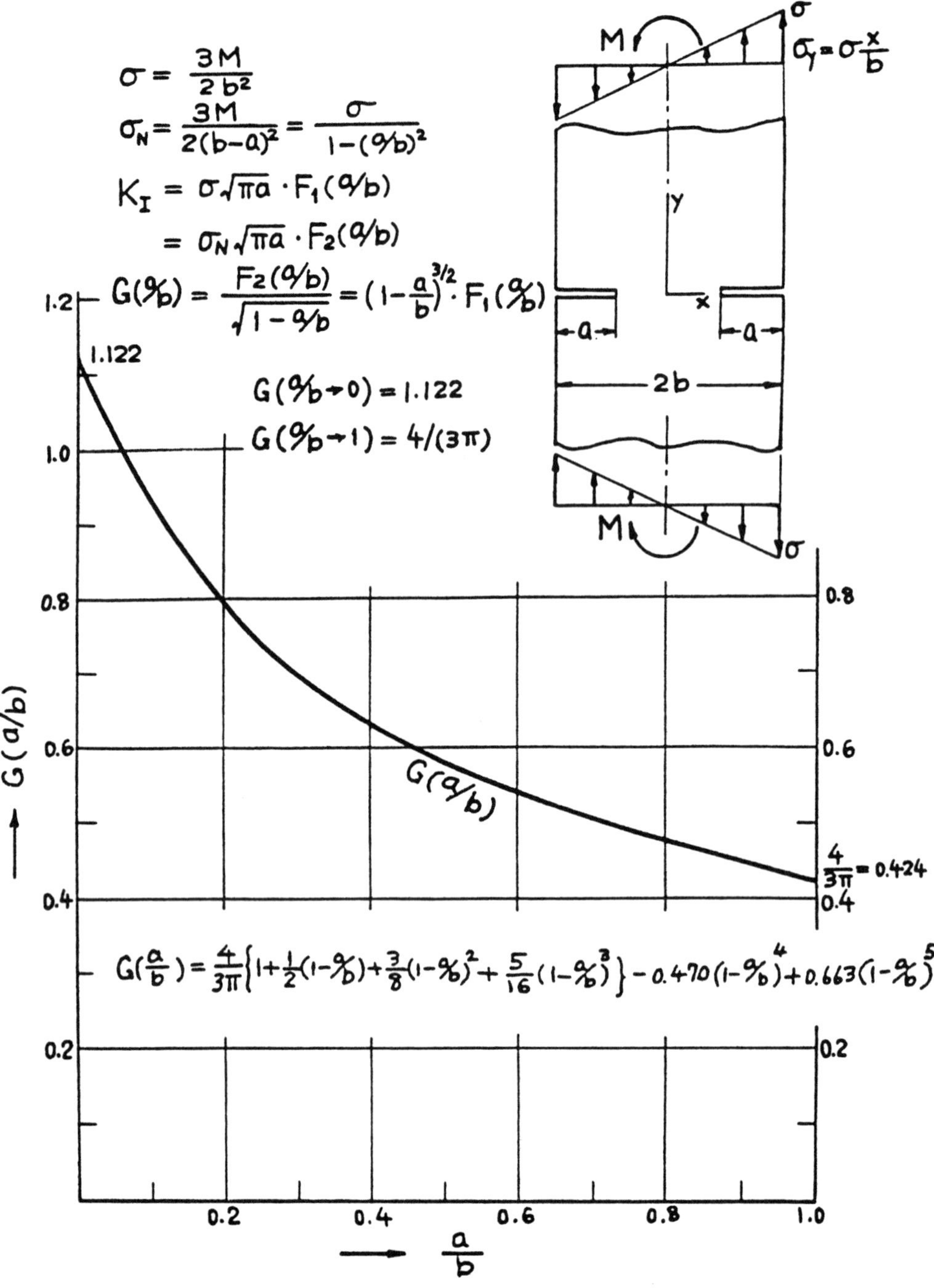

Method: Asymptotic Approximation
Accuracy: Better than 1%
Reference: **Benthem 1972**

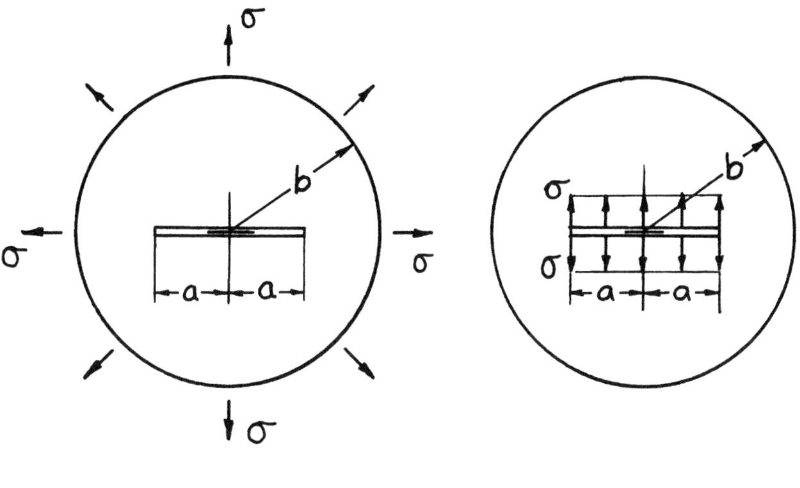

$$K_I = \sigma\sqrt{\pi a} \cdot F(a/b)$$

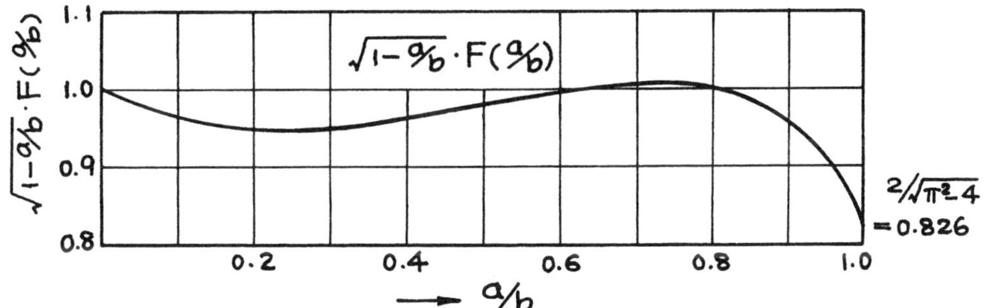

Methods: Mapping Collocation Technique (Bowie), Integral Transform (Tweed)
Accuracy: Curve was drawn based on the values having four significant figures for $0 \leq a/b \leq 0.9$.
References: **Bowie 1970b; Tweed 1972a**

NOTE: For eccentrically located radial internal cracks, see **Tweed 1972a**.

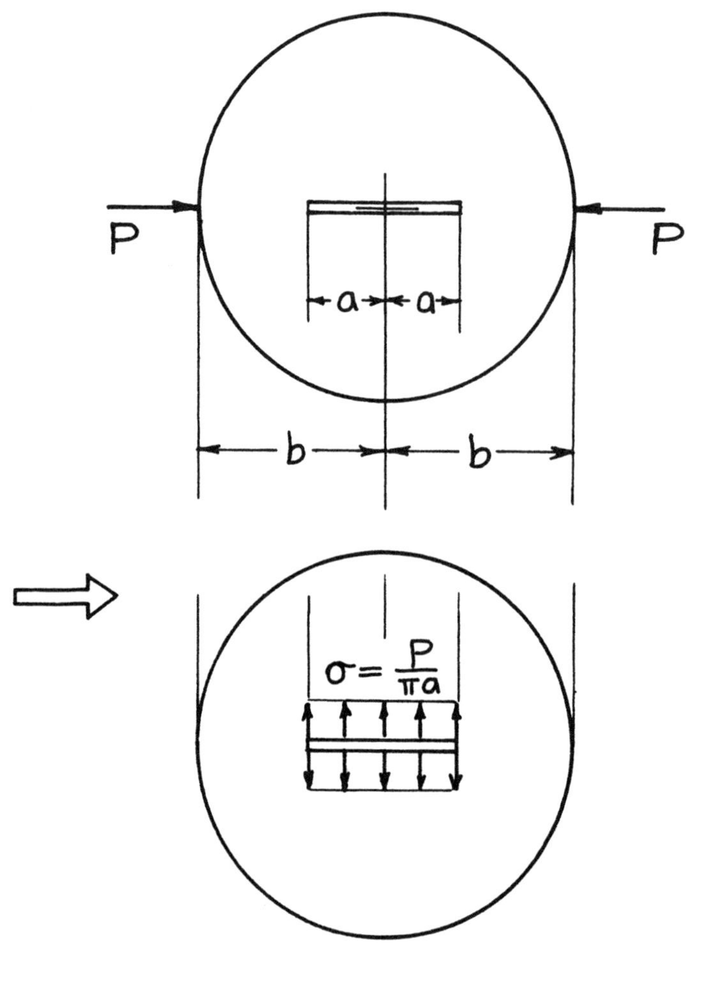

$$K_I = \frac{P}{\pi b}\sqrt{\pi a} \cdot F(a/b)$$

(See **page 11.6** for $F(a/b)$.)

Additional References to **page 11.6**: **Libatskii 1965, 1967; Yamera 1965**. For eccentrically located radial cracks, see **Tweed 1972a**.

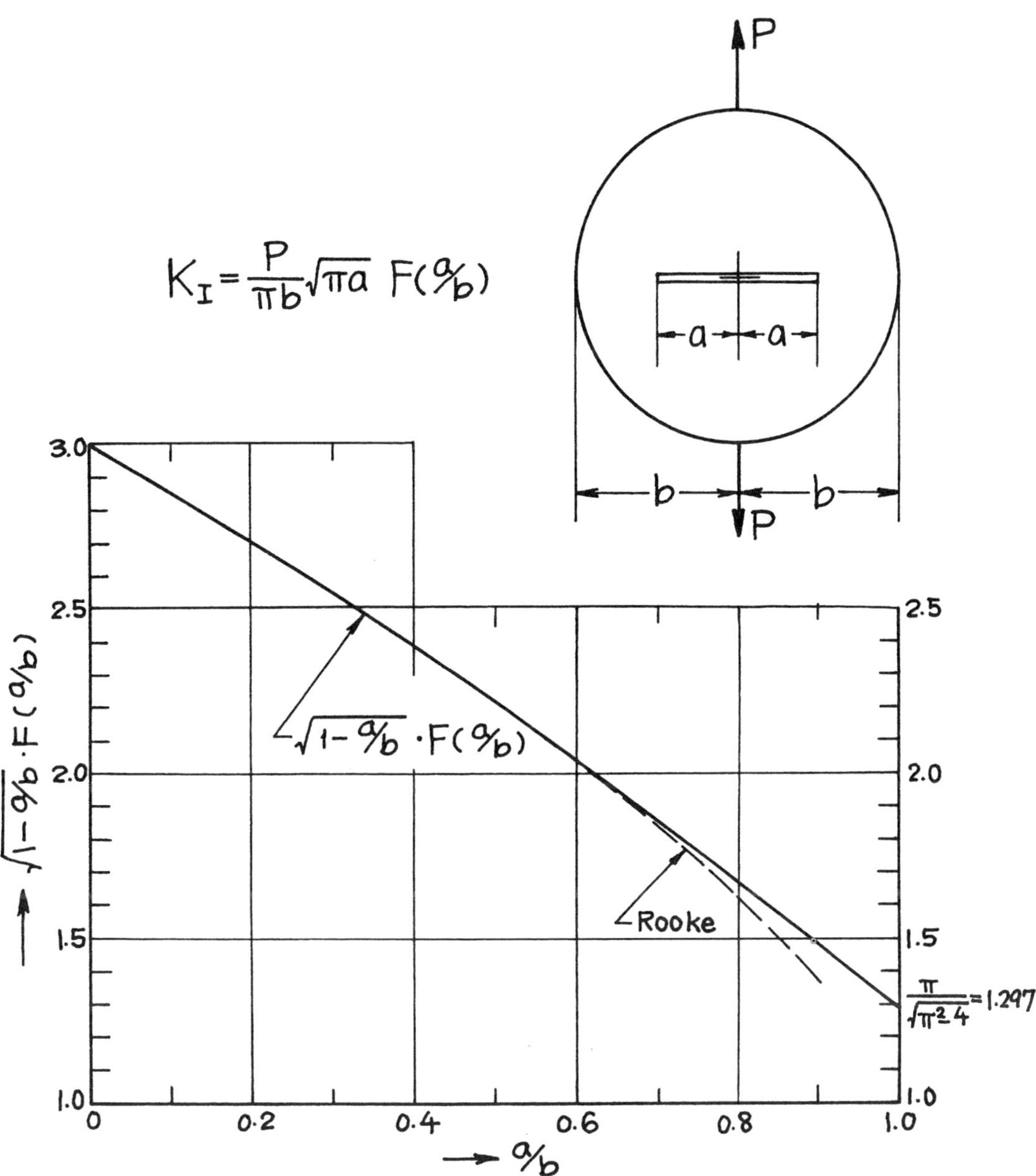

Methods: Mellin Transform and Integral Equation (Rooke, $a/b \leq 0.6$), Interpolated Asymptotically (Tada, $a/b > 0.6$)

Accuracy: Better than 1%

References: **Rooke 1973b; Tada 1973**

NOTE: For eccentrically located radial cracks, see **Rooke 1973b**.

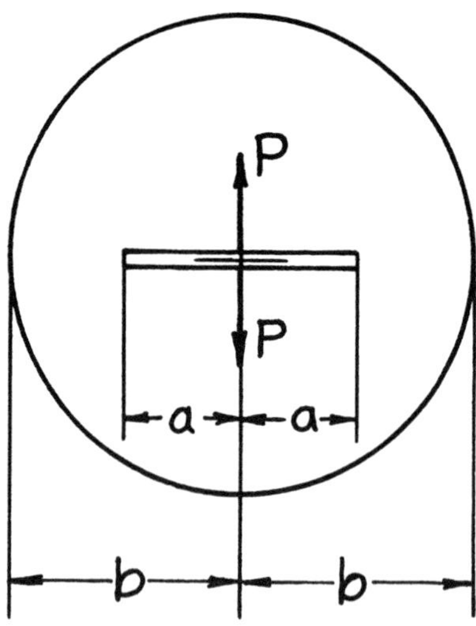

$$K_I = \frac{P}{\sqrt{\pi a}} \cdot F(a/b)$$

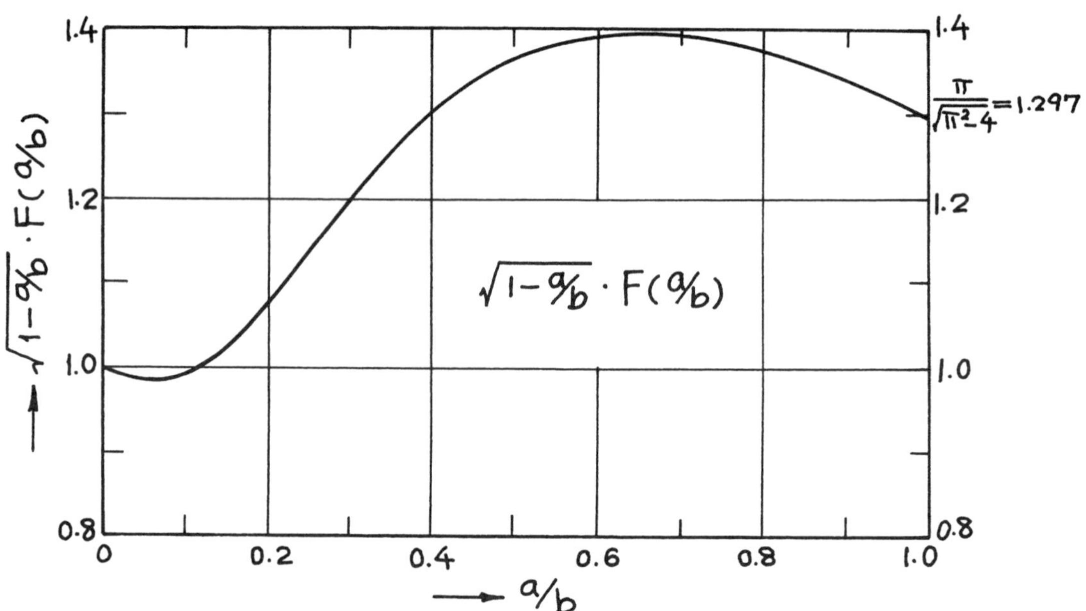

Method: Estimated Asymptotically
Accuracy: Expected to be better than 2%
Reference: **Tada 1973**

$$K_I = \frac{P}{\sqrt{\pi a}} \cdot F(a/b, y_0/b, \nu), \; 0 < y_0/b < 1$$

$$\left(K_I = \frac{P}{\sqrt{\pi a}} \cdot F(a/b), y_0/b = 0, 1\right)$$

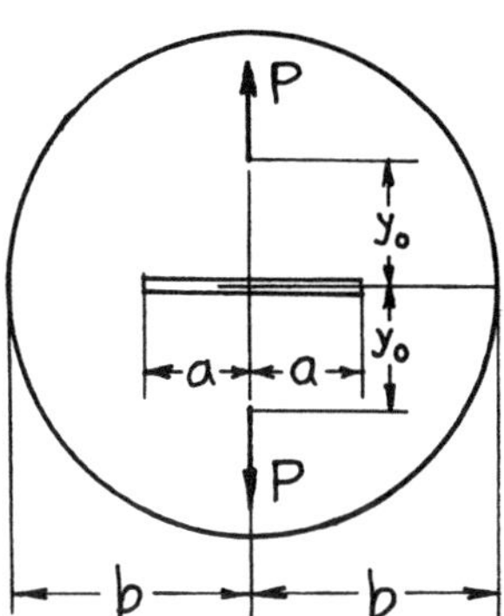

$$a/b \to 0: \quad F \to \frac{a}{b}\left\{(1+\alpha)\left(\frac{b}{y_0}+\frac{y_0}{b}\right) - (2\alpha - 1)\left(\frac{y_0}{b}\right)^3\right\}, \quad 0 < \frac{y_0}{b} \leqq 1$$

$$F \to 1, y_0/b = 0.$$

$$a/b \to 1: F \to \frac{\pi}{\sqrt{\pi^2 - 4}} \cdot \frac{1}{\sqrt{1 - a/b}}, 0 \leqq \frac{y_0}{b} \leqq 1$$

where

$$\alpha = \begin{cases} \frac{1}{2}(1+\nu) & \text{plane stress} \\ \frac{1}{2}\left(\frac{1}{1-\nu}\right) & \text{plane strain} \end{cases}$$

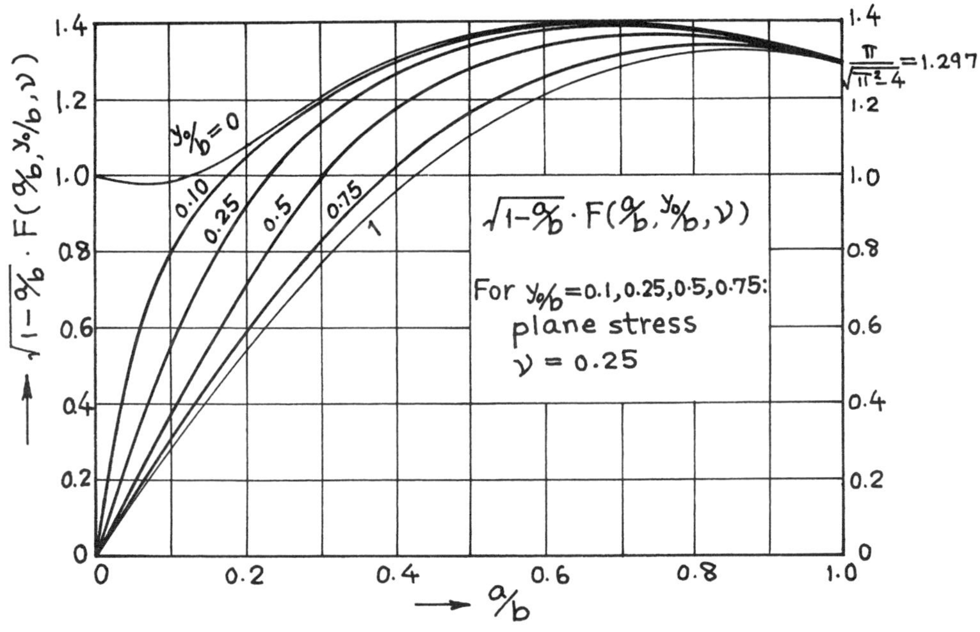

Methods: Integral Equation (**Likatskii,** $y_0/b = 0.25, 0.5, 0.75$, $a/b < 0.7$), Interpolated Asymptotically (Tada, $a/b \geq 0.7$ and $y_0/b = 0, 1$)

Accuracy: Better than 2% for any a/b and y_0/b

References: **Libatskii 1967; Tada 1973**

NOTE: For special cases $y_0/b = 0$ and 1, see **pages 11.9** and **page 11.8**, respectively.

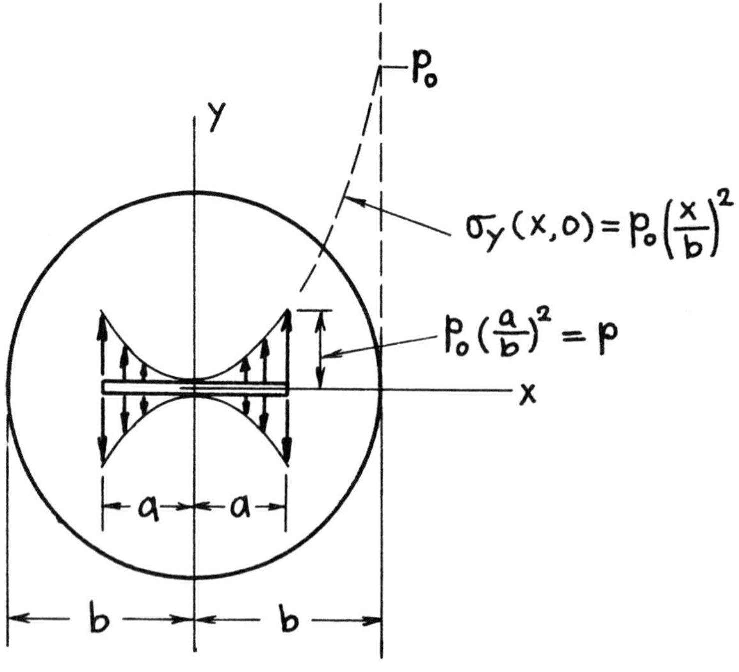

$$K_I = P\sqrt{\pi a} \cdot G\left(\frac{a}{b}\right)$$

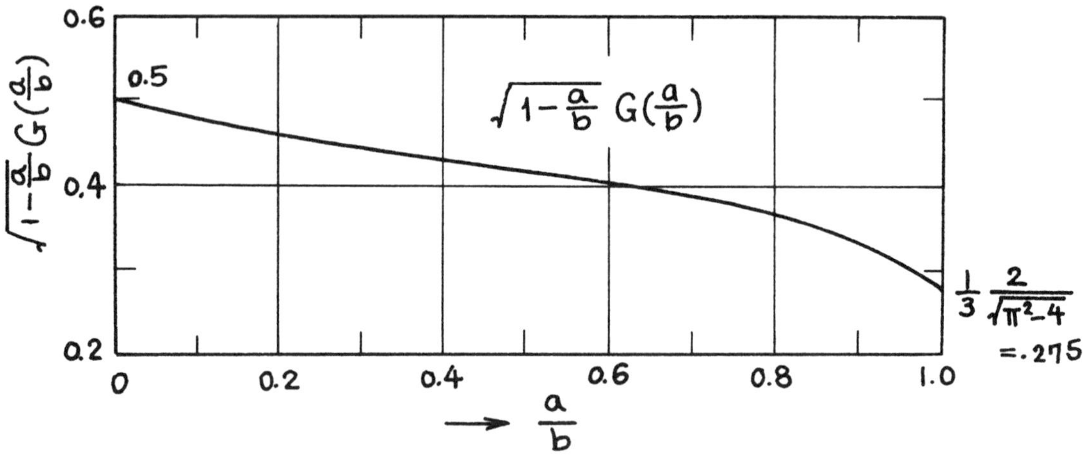

Method: Reduced from the results for **page 11.12** (Rooke: Integral Transform, Plane Strain, $\nu = 0.2,\ 0.3,\ 0.4$; Isida: Series Expansion, Plane Stress, $\nu = 0.3$)
Accuracy: 1%
References: **Rooke 1972; Isida 1974; Tada 1985**

NOTE: For an application, see **page 11.12**.

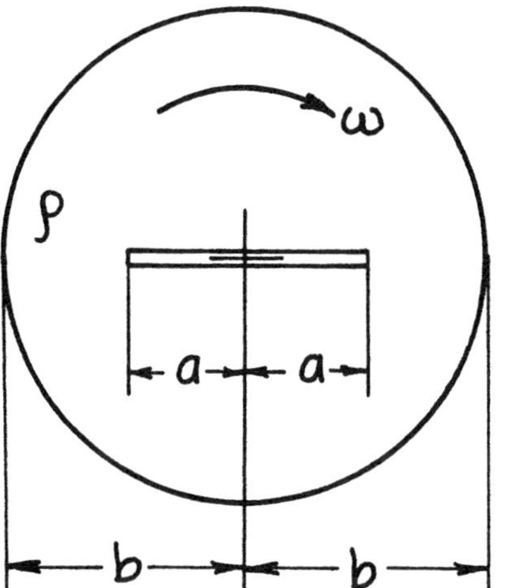

Rotating Disc or Shaft

$$K_I = \frac{\alpha+1}{4}\rho\omega^2 b^2\sqrt{\pi a}\cdot\left\{F(a/b) - \frac{3\alpha-1}{\alpha+1}\left(\frac{a}{b}\right)^2 G(a/b)\right\}$$

where

$$\omega = \text{angular velocity}$$

$$\rho = \text{density}$$

$$\alpha = \begin{cases} \frac{1}{2}(1+\nu) & \text{plane stress} \\ \frac{1}{2}\left(\frac{1}{1-\nu}\right) & \text{plane strain} \end{cases}$$

For numerical values of $F(a/b)$ and $G(a/b)$, and more information, see **pages 11.6 and 11.11**. For eccentrically located radial cracks, see **Rooke 1972**.

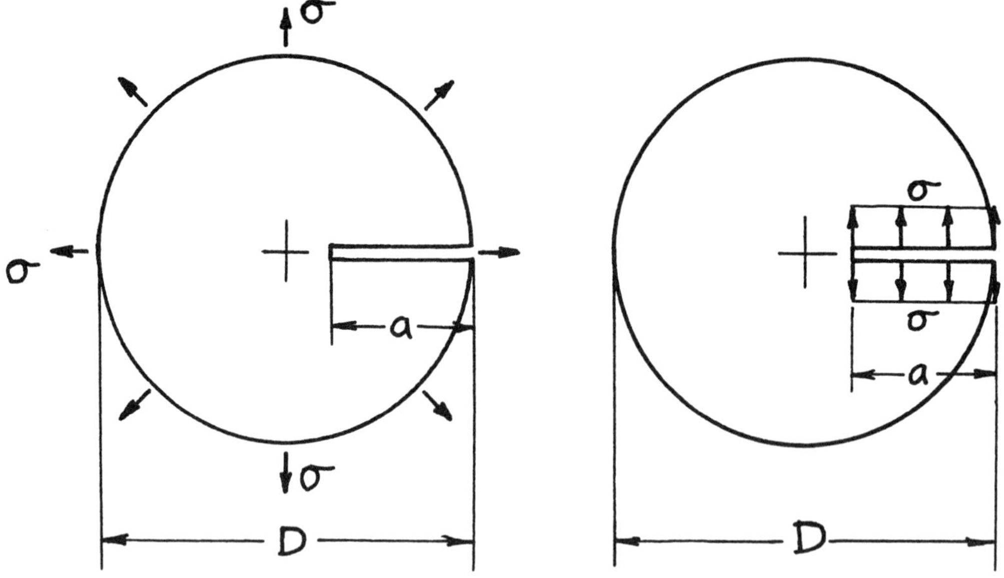

$$K_I = \sigma\sqrt{\pi a} \cdot F(a/D)$$

$$F(a/D) = \frac{1.122 + 0.140(a/D) - 0.545(a/D)^2 + 0.405(a/D)^3}{(1 - a/D)^{3/2}}$$

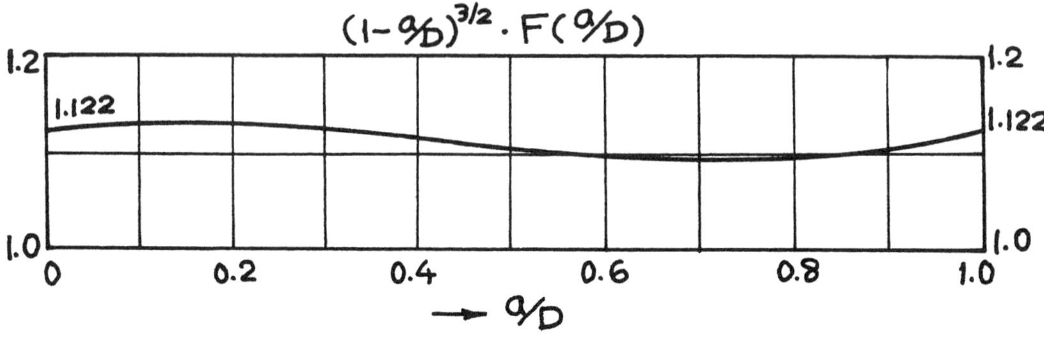

Methods: Mellin Transform and Integral Equation (Tweed, $a/D \leq 0.5$), Asymptotic Interpolation (**Tada**)
Accuracy: Better than 1%
References: **Tweed 1973; Tada 1973**

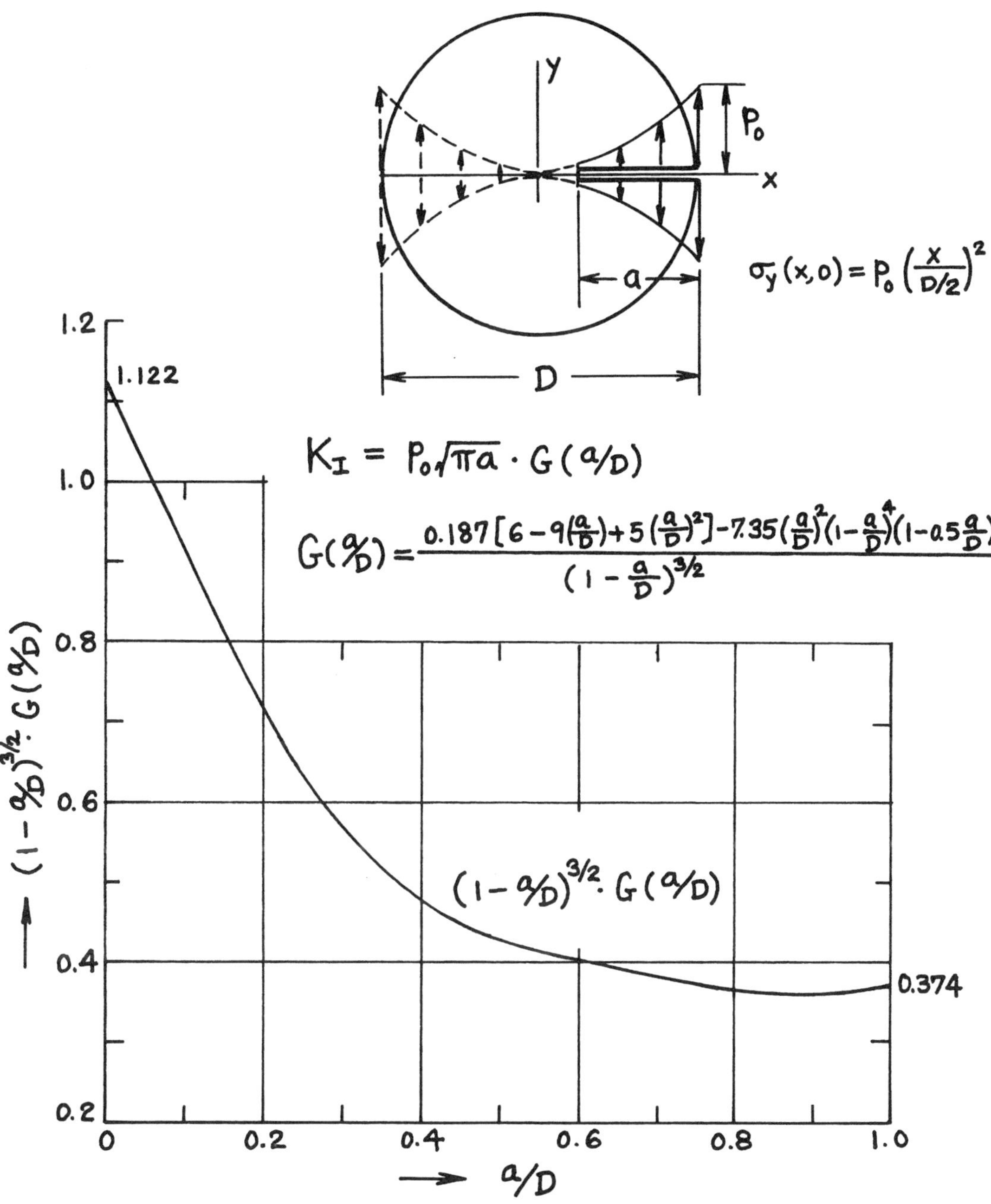

Methods: Extracted from the results of **page 11.15** (Rooke; Integral Transform; $a/D \leq 0.5$), Influence Function (Sire)

Accuracy: 1%

References: **Rooke 1973a; Sire 1989; Tada 2000**

NOTE: For an application, see **page 11.15.**

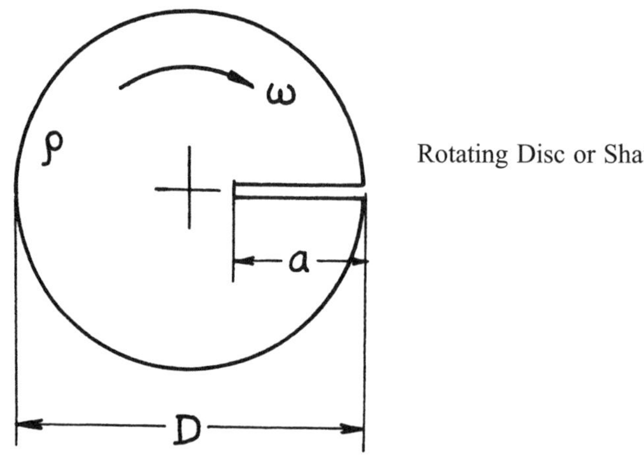

Rotating Disc or Shaft

$$K_I = \frac{1+\alpha}{16}\rho\omega^2 D^2\sqrt{\pi a}\left\{F(a/D) - \frac{3\alpha-1}{1+\alpha}G(a/D)\right\}$$

where

$$\omega = \text{angular velocity}$$

$$\rho = \text{density}$$

$$\alpha = \begin{cases} \frac{1}{2}(1+\nu) & \text{plane stress} \\ \frac{1}{2}\left(\frac{1}{1-\nu}\right) & \text{plane strain} \end{cases}$$

For numerical values of $F(a/D)$ and $G(a/D)$ and more information, see **pages 11.13 and 11.14**.

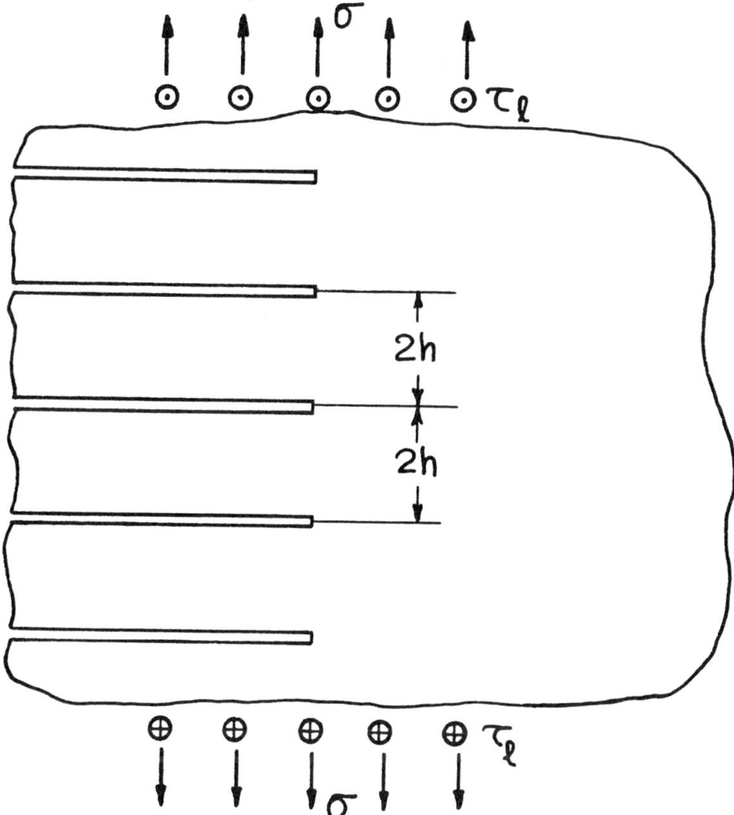

$$K_I = \sigma\sqrt{h}$$

$$K_{III} = \tau_\ell\sqrt{2h}$$

Methods: K_I Fourier Transform (Koiter, Benthem), Energy Consideration (Paris, Rice); K_{III} Westergaard Stress Function (Tada)
Accuracy: Exact
References: **Koiter 1956, 1961; Paris 1955, 1960; Rice 1967; Benthem 1972; Tada 1973**

NOTE: For complete Mode III stress functions, see **page 12.2**.

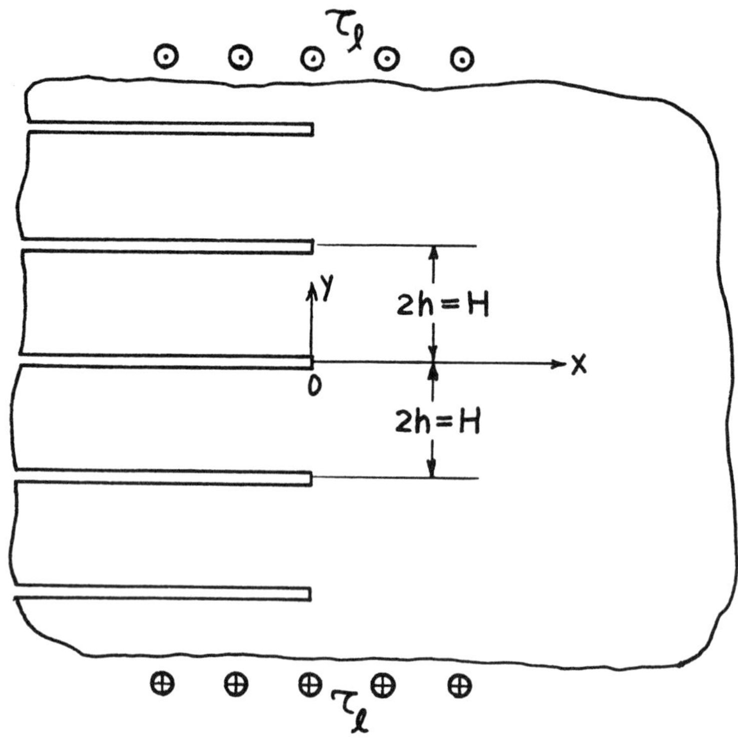

$$z = x + iy$$

$$Z_{III}(z) = \frac{\tau_\ell}{\sqrt{1 - e^{-\frac{2\pi z}{H}}}}$$

$$\overline{Z}_{III}(z) = \tau_\ell \cdot \frac{H}{\pi} \cosh^{-1}\left(e^{\frac{\pi z}{H}}\right)$$

$$K_{III} = \tau_\ell \sqrt{H}$$

$$\underset{x \leq 0}{w(x, 0)} = \frac{\tau_\ell}{G} \cdot \frac{H}{\pi} \cos^{-1}\left(e^{\frac{\pi x}{H}}\right)$$

Method: Westergaard Stress Function
Accuracy: Exact
Reference: **Tada 1973**

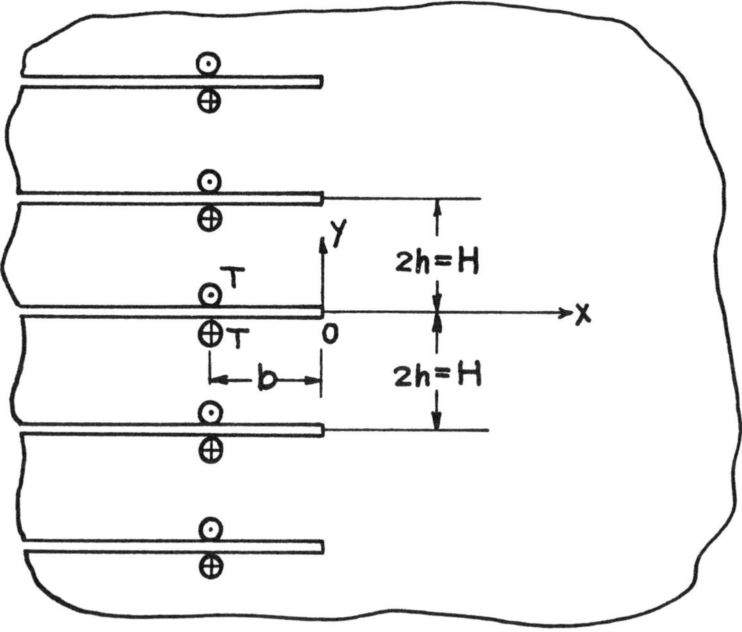

$$z = x + iy$$

$$Z_{III}(z) = \frac{2T}{H} \cdot \frac{e^{-\frac{\pi b}{H}}\sqrt{1 - e^{-\frac{2\pi b}{H}}}}{\left(e^{\frac{2\pi z}{H}} - e^{-\frac{2\pi b}{H}}\right)\sqrt{1 - e^{-\frac{2\pi z}{H}}}}$$

$$\overline{Z}_{III}(z) = \frac{2T}{\pi}\tan^{-1}\sqrt{\frac{e^{\frac{2\pi b}{H}} - 1}{1 - e^{-\frac{2\pi z}{H}}}}$$

$$K_{III} = \frac{2T}{\sqrt{H}}\frac{1}{\sqrt{e^{\frac{2\pi b}{H}} - 1}}$$

$$\underset{x \leq 0}{w(x, 0)} = \frac{2T}{\pi G}\begin{pmatrix}\tanh^{-1} \\ \coth^{-1}\end{pmatrix}\sqrt{\frac{1 - e^{\frac{2\pi b}{H}}}{1 - e^{-\frac{2\pi x}{H}}}} \begin{pmatrix} x < -b \\ -b < x < 0\end{pmatrix}$$

Method: Westergaard Stress Function
Accuracy: Exact
Reference: **Tada 1973**

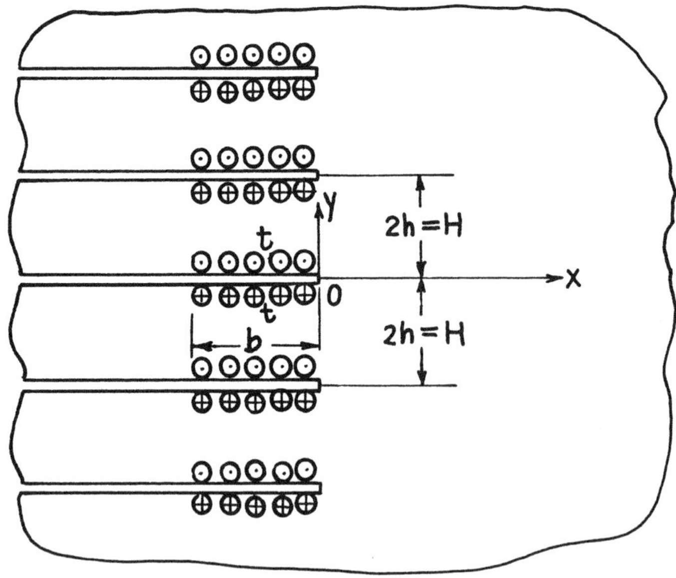

$$z = x + iy$$

$$Z_{III}(z) = \frac{2t}{\pi}\left\{\frac{\cos^{-1}\left(e^{-\frac{\pi b}{H}}\right)}{\sqrt{1-e^{-\frac{2\pi z}{H}}}} - \tan^{-1}\sqrt{\frac{e^{\frac{2\pi b}{H}}-1}{1-e^{-\frac{2\pi z}{H}}}}\right\}$$

$$\overline{Z}_{III}(z) = \frac{2t}{\pi}\int_0^b \tan^{-1}\sqrt{\frac{e^{\frac{2\pi x_0}{H}}-1}{1-e^{-\frac{2\pi z}{H}}}}\,dx_0$$

$$K_{III} = \frac{2t}{\pi}\sqrt{H}\cos^{-1}\left(e^{-\frac{\pi b}{H}}\right)$$

$$\underset{x\leq 0}{w(x,0)} = \frac{t}{\pi G}\int_0^b \ln\left|\frac{\sqrt{1-e^{\frac{2\pi x_0}{H}}}+\sqrt{1-e^{-\frac{2\pi x}{H}}}}{\sqrt{1-e^{\frac{2\pi x_0}{H}}}-\sqrt{1-e^{-\frac{2\pi x}{H}}}}\right|dx_0$$

Method: Westergaard Stress Function
Accuracy: Exact
Reference: **Tada 1973**

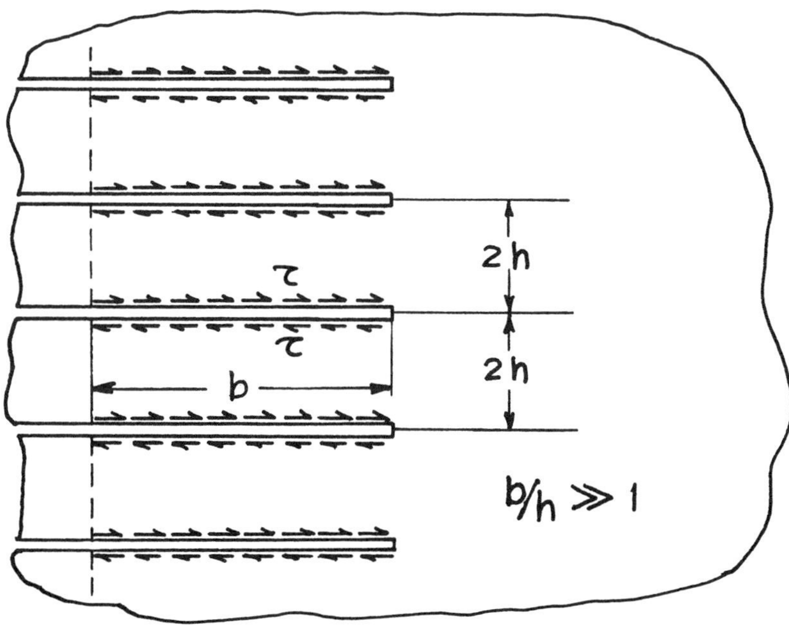

$$K_{II} = \tau\sqrt{3h}\left(\frac{b}{h} + 0.2865\right)$$

Method: Fourier Transform
Accuracy: Numerical value 0.2865 is believed to have accuracy within one unit of the last digit (obtained by numerical integration).
Reference: **Benthem 1972**

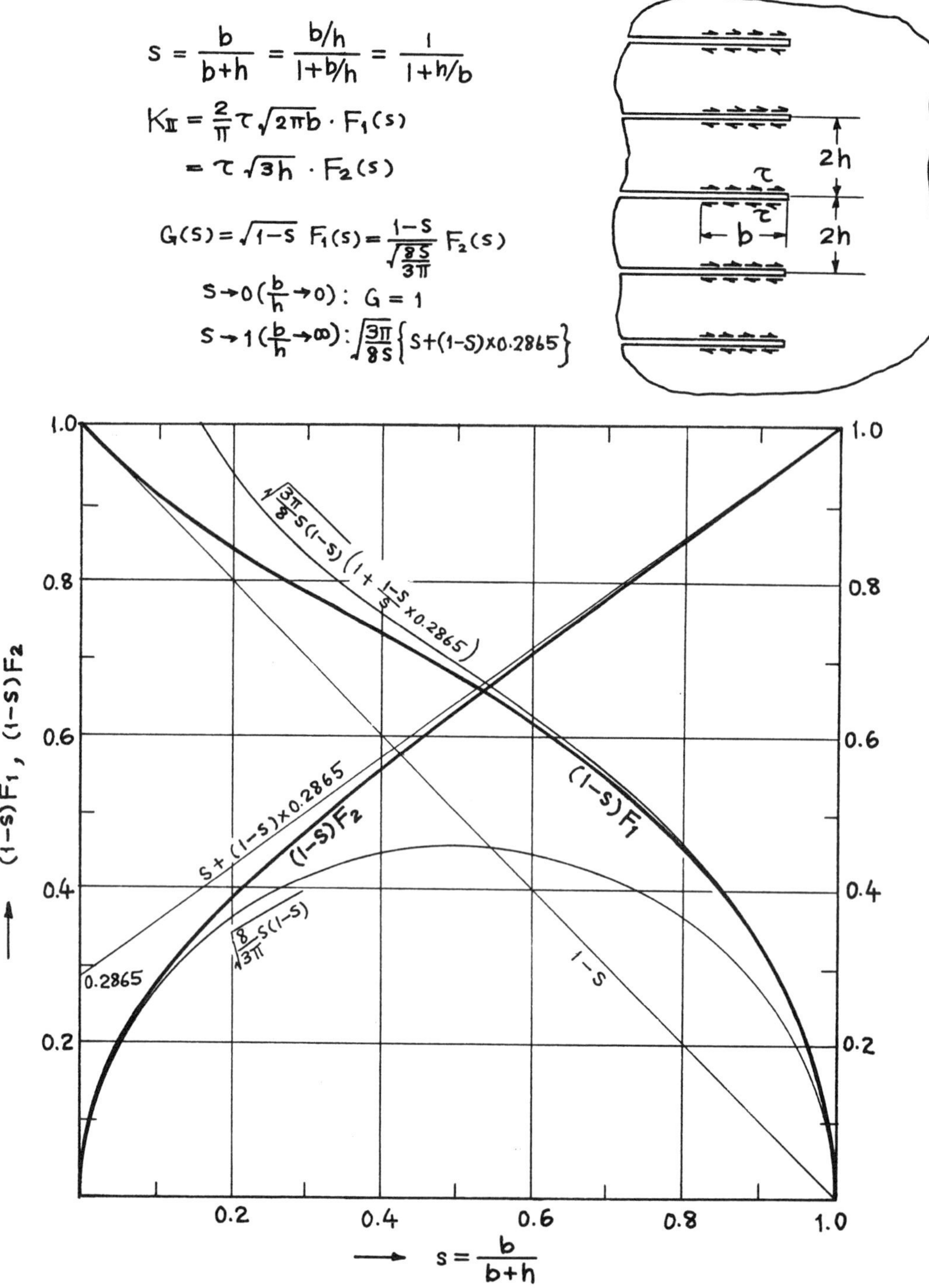

Method: Estimated Asymptotically
Accuracy: Expected to be better than 3%
Reference: **Tada 1973**

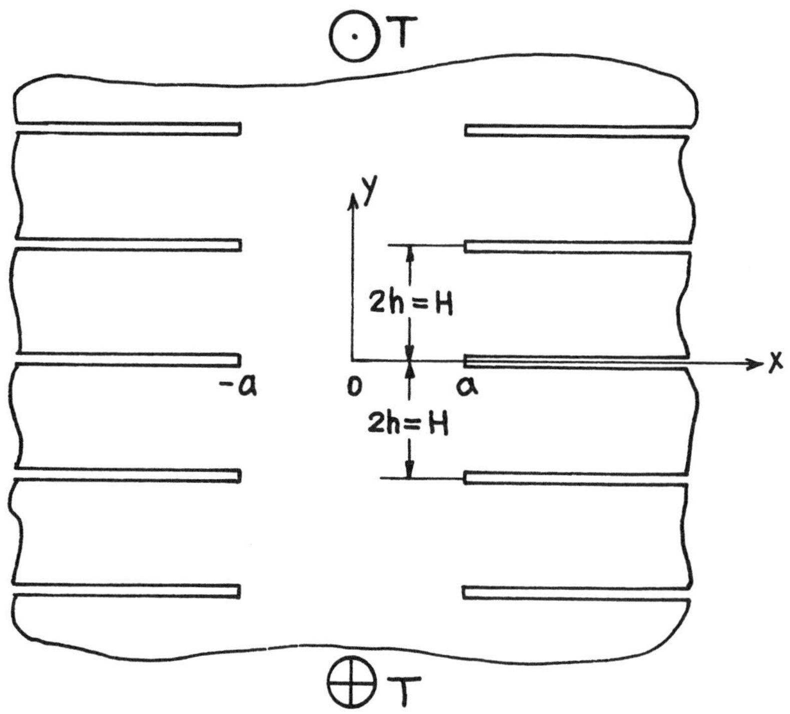

$$z = x + iy$$

$$Z_{III}(z) = \frac{2T}{H} \frac{1}{\sqrt{\left(\cosh \frac{\pi a}{H} \Big/ \cosh \frac{\pi z}{H}\right)^2 - 1}}$$

$$\overline{Z}_{III}(z) = \frac{2T}{\pi} \sin^{-1} \left(\frac{\sinh \frac{\pi z}{H}}{\sinh \frac{\pi a}{H}} \right)$$

$$K_{III} = \frac{2T}{H} \sqrt{H \coth \frac{\pi a}{H}}$$

$$\underset{|x| \geq a}{w(x, 0)} = \frac{2T}{\pi G} \sinh^{-1} \left(\frac{\sinh \frac{\pi x}{H}}{\sinh \frac{\pi a}{H}} \right)$$

Method: Westergaard Stress Function
Accuracy: Exact
Reference: **Tada 1973**

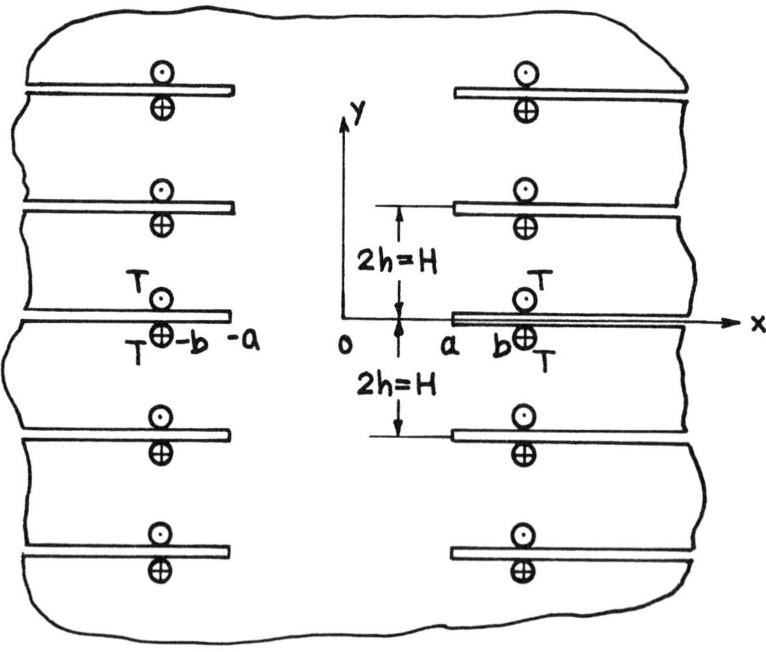

$$z = x + iy$$

$$Z_{III}(z) = \frac{2T}{H} \frac{\sinh\frac{\pi b}{H}\sqrt{\left(\cosh\frac{\pi b}{H}\right)^2 - \left(\cosh\frac{\pi a}{H}\right)^2}}{\left\{\left(\cosh\frac{\pi b}{H}\right)^2 - \left(\cosh\frac{\pi z}{H}\right)^2\right\}\sqrt{\left(\cosh\frac{\pi a}{H}\Big/\cosh\frac{\pi z}{H}\right)^2 - 1}}$$

$$\overline{Z}_{III}(z) = \frac{2T}{\pi}\tan^{-1}\sqrt{\frac{1-\left(\sinh\frac{\pi a}{H}\Big/\sinh\frac{\pi b}{H}\right)^2}{\left(\sinh\frac{\pi a}{H}\Big/\sinh\frac{\pi z}{H}\right)^2 - 1}}$$

$$K_{III} = \frac{2T}{H}\sqrt{H\coth\frac{\pi a}{H}} \frac{\sinh\frac{\pi b}{H}}{\sqrt{\left(\cosh\frac{\pi b}{H}\right)^2 - \left(\cosh\frac{\pi a}{H}\right)^2}}$$

$$\underset{|x|\geq a}{w(x,0)} = \frac{2T}{\pi G}\begin{pmatrix}\tanh^{-1}\\ \coth^{-1}\end{pmatrix}\sqrt{\frac{1-\left(\sinh\frac{\pi a}{H}\Big/\sinh\frac{\pi b}{H}\right)^2}{1-\left(\sinh\frac{\pi a}{H}\Big/\sinh\frac{\pi x}{H}\right)^2}}\begin{pmatrix}|x| > b\\ a \leq |x| < b\end{pmatrix}$$

Method: Westergaard Stress Function
Accuracy: Exact
Reference: **Tada 1973**

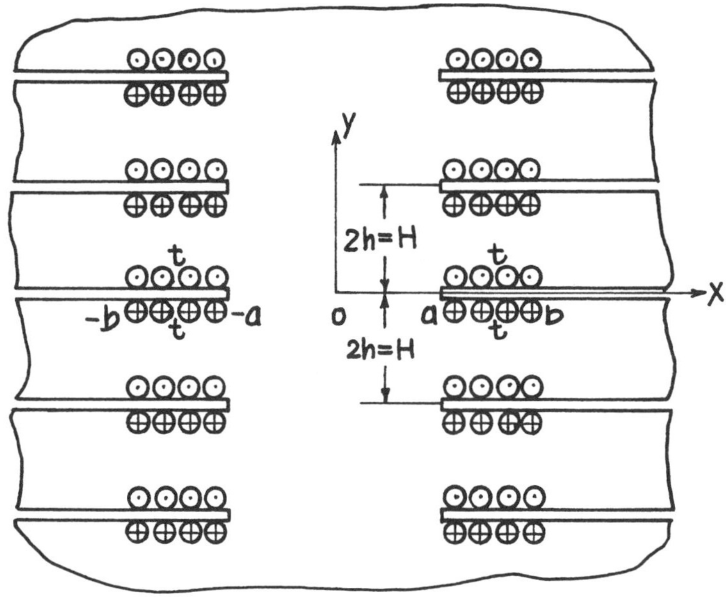

$$z = x + iy$$

$$Z_{III}(z) = \frac{2t}{\pi}\left\{\frac{\cosh^{-1}\left(\cosh\frac{\pi b}{H}\Big/\cosh\frac{\pi a}{H}\right)}{\sqrt{\left(\cosh\frac{\pi a}{H}\Big/\cosh\frac{\pi z}{H}\right)^2 - 1}} + \tan^{-1}\sqrt{\frac{\left(\cosh\frac{\pi a}{H}\Big/\cosh\frac{\pi b}{H}\right)^2 - 1}{1 - \left(\cosh\frac{\pi a}{H}\Big/\cosh\frac{\pi z}{H}\right)^2}}\right\}$$

$$\overline{Z}_{III}(z) = \frac{2t}{\pi}\int_a^b \tan^{-1}\sqrt{\frac{1 - \left(\sinh\frac{\pi a}{H}\Big/\sinh\frac{\pi x_0}{H}\right)^2}{\left(\sinh\frac{\pi a}{H}\Big/\sinh\frac{\pi z}{H}\right)^2 - 1}}\, dx_0$$

$$K_{III} = \frac{2t}{\pi}\sqrt{H\coth\frac{\pi a}{H}}\cosh^{-1}\left(\frac{\cosh\frac{\pi b}{H}}{\cosh\frac{\pi a}{H}}\right)$$

$$\underset{|x|\geq a}{w(x,0)} = \frac{t}{\pi G}\int_a^b \ln\left|\frac{\sqrt{1 - \left(\sinh\frac{\pi a}{H}\Big/\sinh\frac{\pi x_0}{H}\right)^2} + \sqrt{1 - \left(\sinh\frac{\pi a}{H}\Big/\sinh\frac{\pi x}{H}\right)^2}}{\sqrt{1 - \left(\sinh\frac{\pi a}{H}\Big/\sinh\frac{\pi x_0}{H}\right)^2} - \sqrt{1 - \left(\sinh\frac{\pi a}{H}\Big/\sinh\frac{\pi x}{H}\right)^2}}\right| dx_0$$

Method: Westergaard Stress Function
Accuracy: Exact
Reference: **Tada 1973**

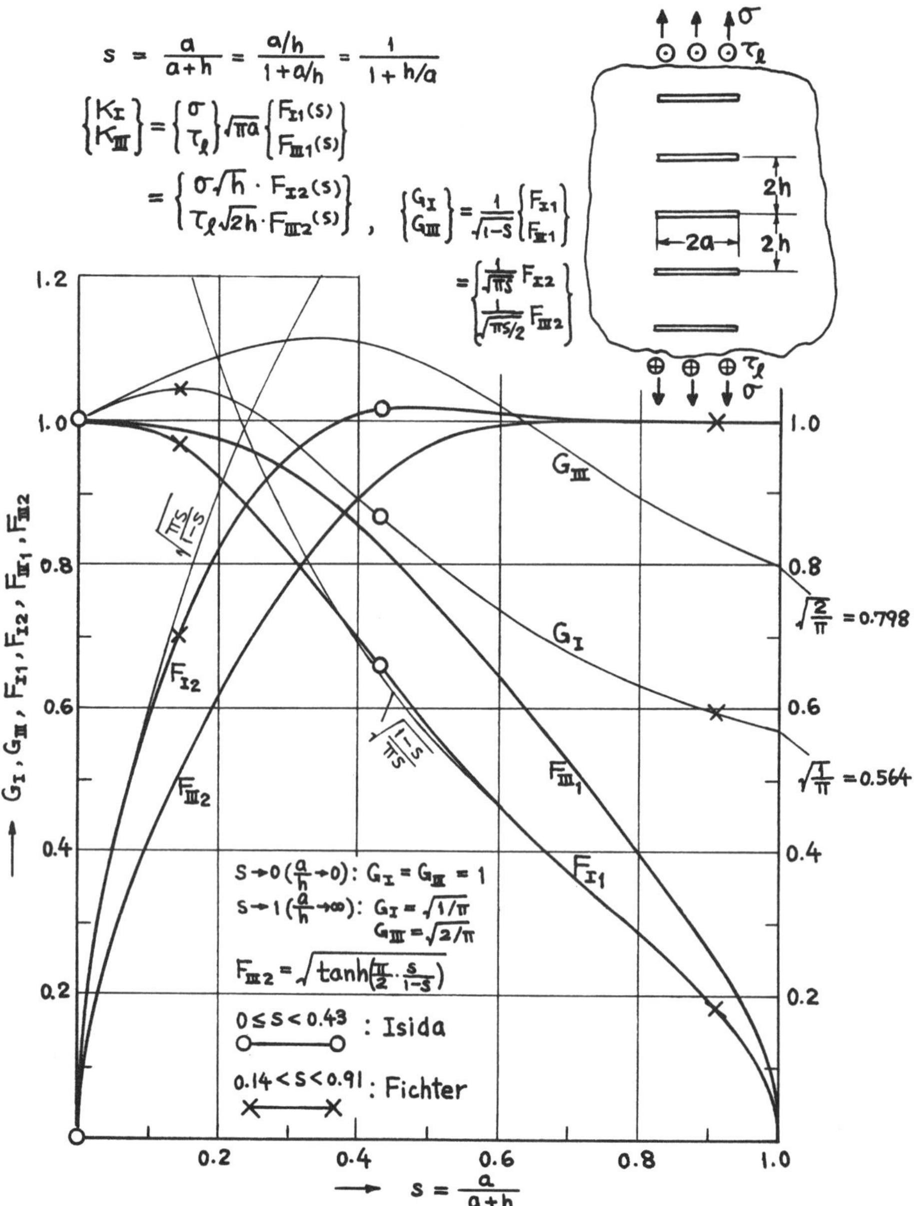

Method: K_I Fourier Transform (Fichter), Expansions of Complex Potentials (Isida), Asymptotic Approximation (polynomial of degree ten) (Benthem)

K_{III} Stress Concentration Factor (Neuber), Westergaard Stress Function (Tada)

Accuracy: K_I Estimated at 1%

K_{III} Exact

References: **Fichter 1967; Isida 1971a; Benthem 1972; Neuber 1937; Tada 1973**

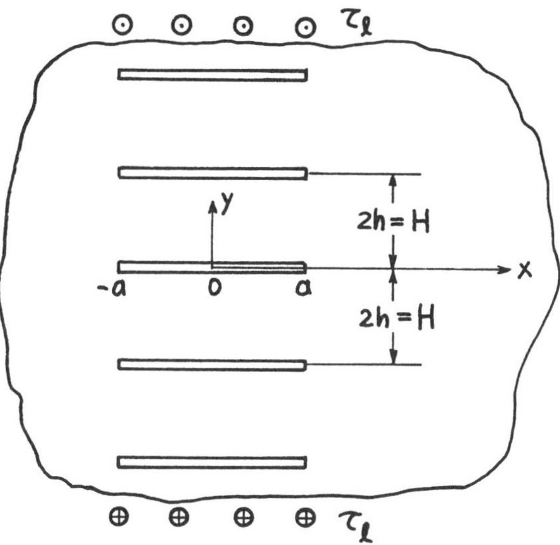

$$z = x + iy$$

$$Z_{III}(z) = \frac{\tau_\ell}{\sqrt{1 - \left(\sinh\frac{\pi a}{H} \Big/ \sinh\frac{\pi z}{H}\right)^2}}$$

$$\overline{Z}_{III}(z) = \frac{\tau_\ell H}{\pi} \cosh^{-1}\left(\frac{\cosh\frac{\pi z}{H}}{\cosh\frac{\pi a}{H}}\right)$$

$$K_{III} = \tau_\ell \sqrt{H \tanh\frac{\pi a}{H}}$$

$$\underset{|x| \le a}{w(x, 0)} = \frac{\tau_\ell}{G} \cdot \frac{H}{\pi} \cos^{-1}\left(\frac{\cosh\frac{\pi x}{H}}{\cosh\frac{\pi a}{H}}\right)$$

Method: Westergaard Stress Function
Accuracy: Exact
Reference: **Tada 1973**

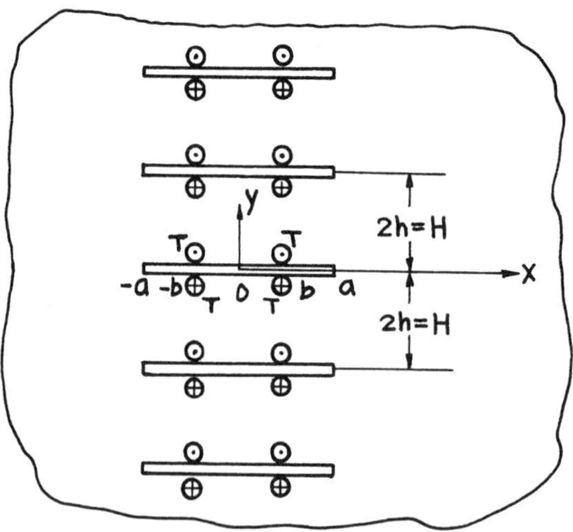

$$z = x + iy$$

$$Z_{III}(z) = \frac{2T}{H} \cdot \frac{\cosh\frac{\pi b}{H}\sqrt{\left(\sinh\frac{\pi a}{H}\right)^2 - \left(\sinh\frac{\pi b}{H}\right)^2}}{\left\{\left(\sinh\frac{\pi z}{H}\right)^2 - \left(\sinh\frac{\pi b}{H}\right)^2\right\}\sqrt{1 - \left(\sinh\frac{\pi a}{H}\Big/\sinh\frac{\pi z}{H}\right)^2}}$$

$$\overline{Z}_{III}(z) = \frac{2T}{\pi}\tan^{-1}\sqrt{\frac{1 - \left(\cosh\frac{\pi a}{H}\Big/\cosh\frac{\pi b}{H}\right)^2}{\left(\cosh\frac{\pi a}{H}\Big/\cosh\frac{\pi z}{H}\right)^2 - 1}}$$

$$K_{III} = \frac{2T}{H}\sqrt{H\tanh\frac{\pi a}{H}}\,\frac{\cosh\frac{\pi b}{H}}{\sqrt{\left(\sinh\frac{\pi a}{H}\right)^2 - \left(\sinh\frac{\pi b}{H}\right)^2}}$$

$$\underset{|x|\le a}{w(x,0)} = \frac{2T}{\pi G}\begin{pmatrix}\tanh^{-1}\\ \coth^{-1}\end{pmatrix}\sqrt{\frac{\left(\cosh\frac{\pi a}{H}\Big/\cosh\frac{\pi b}{H}\right)^2 - 1}{\left(\cosh\frac{\pi a}{H}\Big/\cosh\frac{\pi x}{H}\right)^2 - 1}} \qquad \begin{pmatrix}|x| < b\\ b < |x| \le a\end{pmatrix}$$

Method: Westergaard Stress Function
Accuracy: Exact
Reference: **Tada 1973**

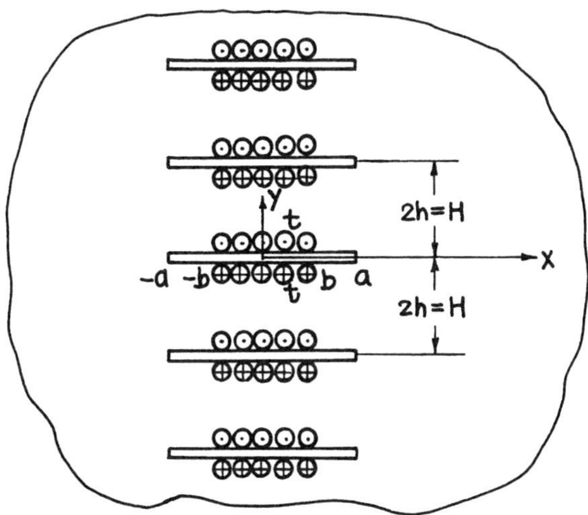

$$z = x + iy$$

$$Z_{III}(z) = \frac{2t}{\pi}\left\{\frac{\sin^{-1}\left(\sinh\frac{\pi b}{H}\Big/\sinh\frac{\pi a}{H}\right)}{\sqrt{1-\left(\sinh\frac{\pi a}{H}\Big/\sinh\frac{\pi z}{H}\right)^2}} + \tan^{-1}\sqrt{\frac{\left(\sinh\frac{\pi a}{H}\Big/\sinh\frac{\pi b}{H}\right)^2 - 1}{1-\left(\sinh\frac{\pi a}{H}\Big/\sinh\frac{\pi z}{H}\right)^2}}\right\}$$

$$\overline{Z}_{III}(z) = \frac{2t}{\pi}\int_0^b \tan^{-1}\sqrt{\frac{1-\left(\cosh\frac{\pi a}{H}\Big/\cosh\frac{\pi x_0}{H}\right)^2}{\left(\cosh\frac{\pi a}{H}\Big/\cosh\frac{\pi z}{H}\right)^2 - 1}}\, dx_0$$

$$K_{III} = \frac{2t}{\pi}\sqrt{H\tanh\frac{\pi a}{H}}\,\sin^{-1}\left(\frac{\sinh\frac{\pi b}{H}}{\sinh\frac{\pi a}{H}}\right)$$

$$\underset{|x|\le a}{w(x,0)} = \frac{t}{\pi G}\int_0^b \ln\left|\frac{\sqrt{\left(\cosh\frac{\pi a}{H}\Big/\cosh\frac{\pi x_0}{H}\right)^2 - 1} + \sqrt{\left(\cosh\frac{\pi a}{H}\Big/\cosh\frac{\pi x}{H}\right)^2 - 1}}{\sqrt{\left(\cosh\frac{\pi a}{H}\Big/\cosh\frac{\pi x_0}{H}\right)^2 - 1} - \sqrt{\left(\cosh\frac{\pi a}{H}\Big/\cosh\frac{\pi x}{H}\right)^2 - 1}}\right| dx_0$$

Method: Westergaard Stress Function
Accuracy: Exact
Reference: **Tada 1973**

$$s = \frac{a}{a+h} = \frac{a/h}{1+a/h} = \frac{1}{1+h/a}$$

$$K_I = \sigma_0\sqrt{\pi a}\cdot F_1(s)$$
$$= \sigma_0\sqrt{h}\cdot F_2(s)$$

$$s \to 0\left(\frac{a}{h}\to 0\right):\ F_1 = \frac{1}{2}$$
$$F_2 \to \frac{1}{2}\sqrt{\frac{\pi a}{h}}$$

$$s \to 1\left(\frac{a}{h}\to \infty\right):\ F_2 = 1$$
$$F_1 \to \sqrt{\frac{h}{\pi a}}$$

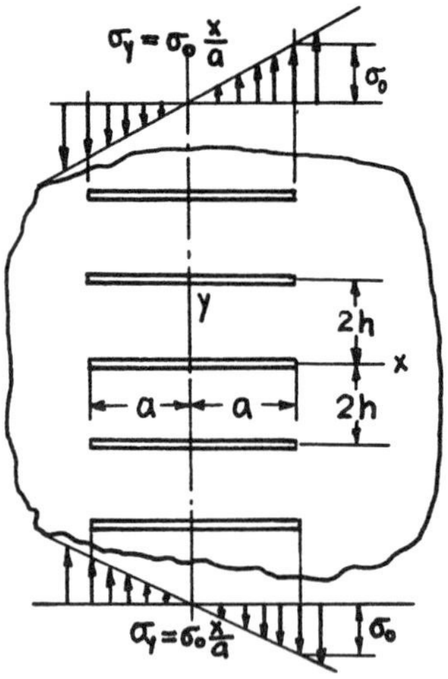

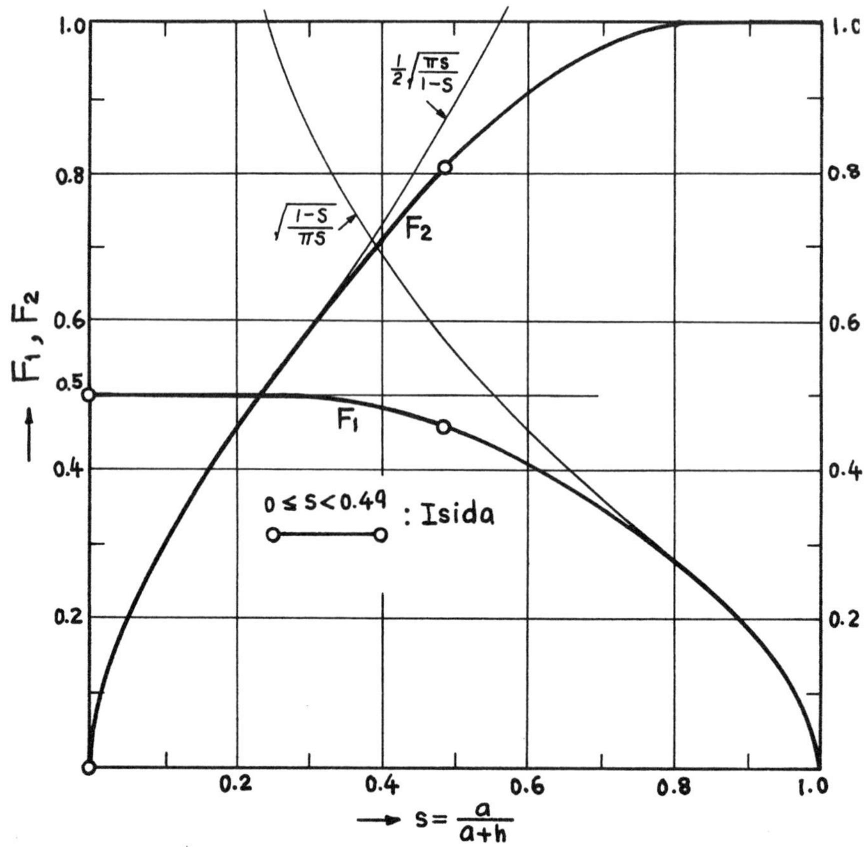

Methods: Expansions of Complex Potentials ($s \leq 0.49$, Isida), Interpolated Asymptotically ($s \geq 0.49$, Tada)

Accuracy: Better than 1% for $s \leq 0.49$, estimated at 3% for $s \geq 0.49$.

References: **Isida 1971a; Tada 1973**

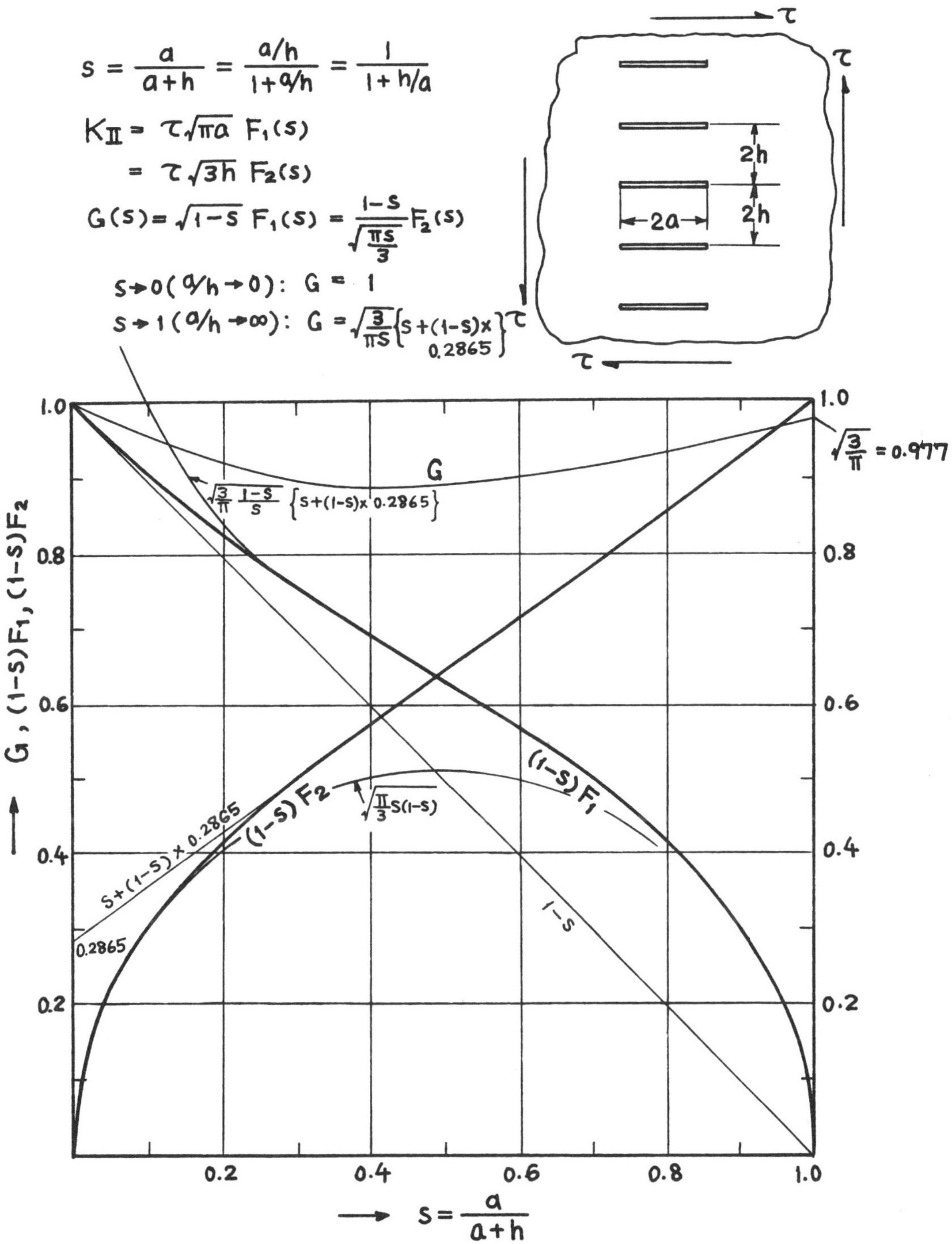

Methods: Asymptotic Approximation (Benthem), Dislocation Distribution (Kamei)
Accuracy: Better than 1%
References: **Benthem 1972; Kamei 1974.**

$$s = \frac{a}{a+h} = \frac{a/h}{1+a/h} = \frac{1}{1+h/a}$$

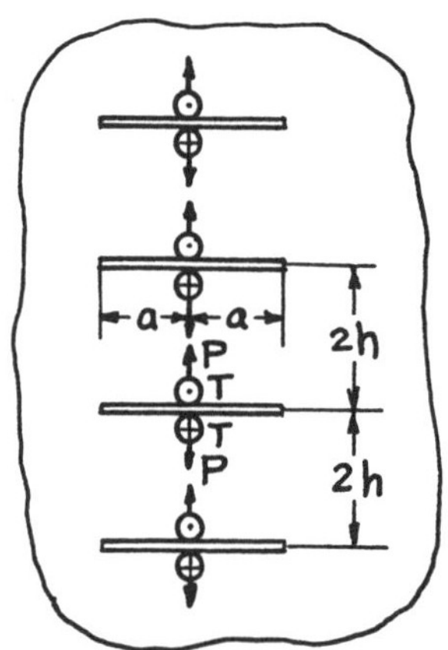

$$\left\{\begin{matrix} K_I \\ K_{III} \end{matrix}\right\} = \frac{1}{\sqrt{\pi a}}\left\{\begin{matrix} P \\ T \end{matrix}\right\}\left\{\begin{matrix} F_{I1}(s) \\ F_{III1}(s) \end{matrix}\right\}$$

$$= \frac{1}{\sqrt{2h}}\left\{\begin{matrix} P \\ T \end{matrix}\right\}\left\{\begin{matrix} F_{I2}(s) \\ F_{III2}(s) \end{matrix}\right\}$$

$$F_{III2} \equiv 1\Big/\sqrt{\sinh\frac{\pi a}{2h}}$$

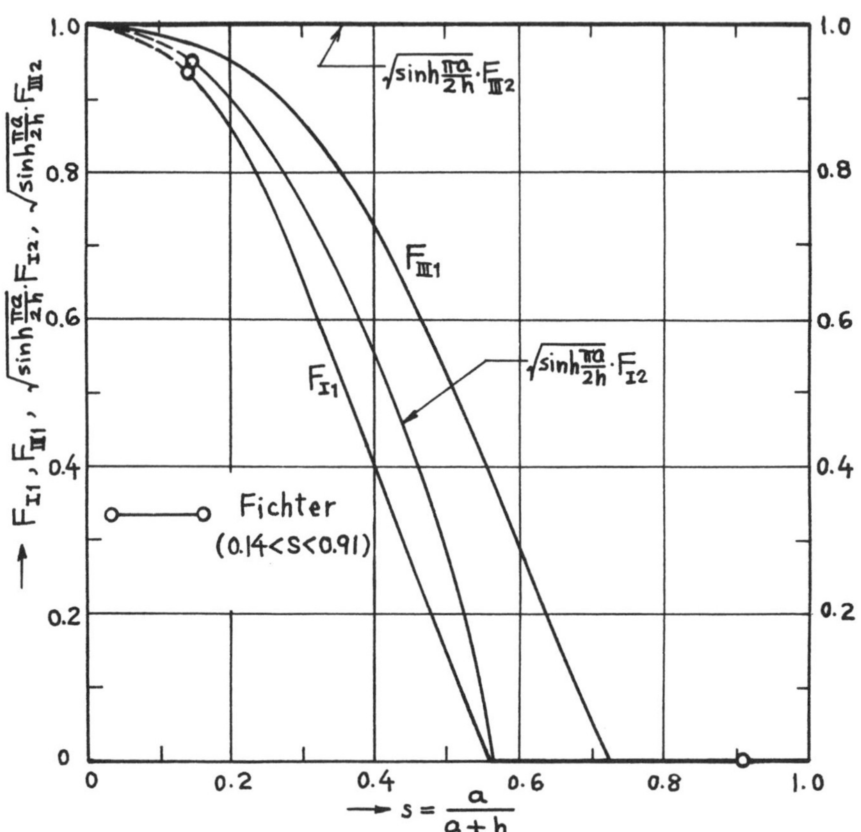

Methods: K_I Fourier Transform (Fichter)
K_{III} Westergaard Stress Function (Tada)
Accuracy: K_I 1% ,
K_{III} Exact
References: **Fichter 1967; Tada 1973**

$$K_I = \sigma\sqrt{\pi a} \cdot F_I(s)$$

$$K_{II} = \sigma\sqrt{\pi a} \cdot F_{II}(s)$$

$$K_{III} = \tau_\ell\sqrt{\pi a} \cdot F_{III}(s)$$

$$s = \frac{a}{a+h}$$

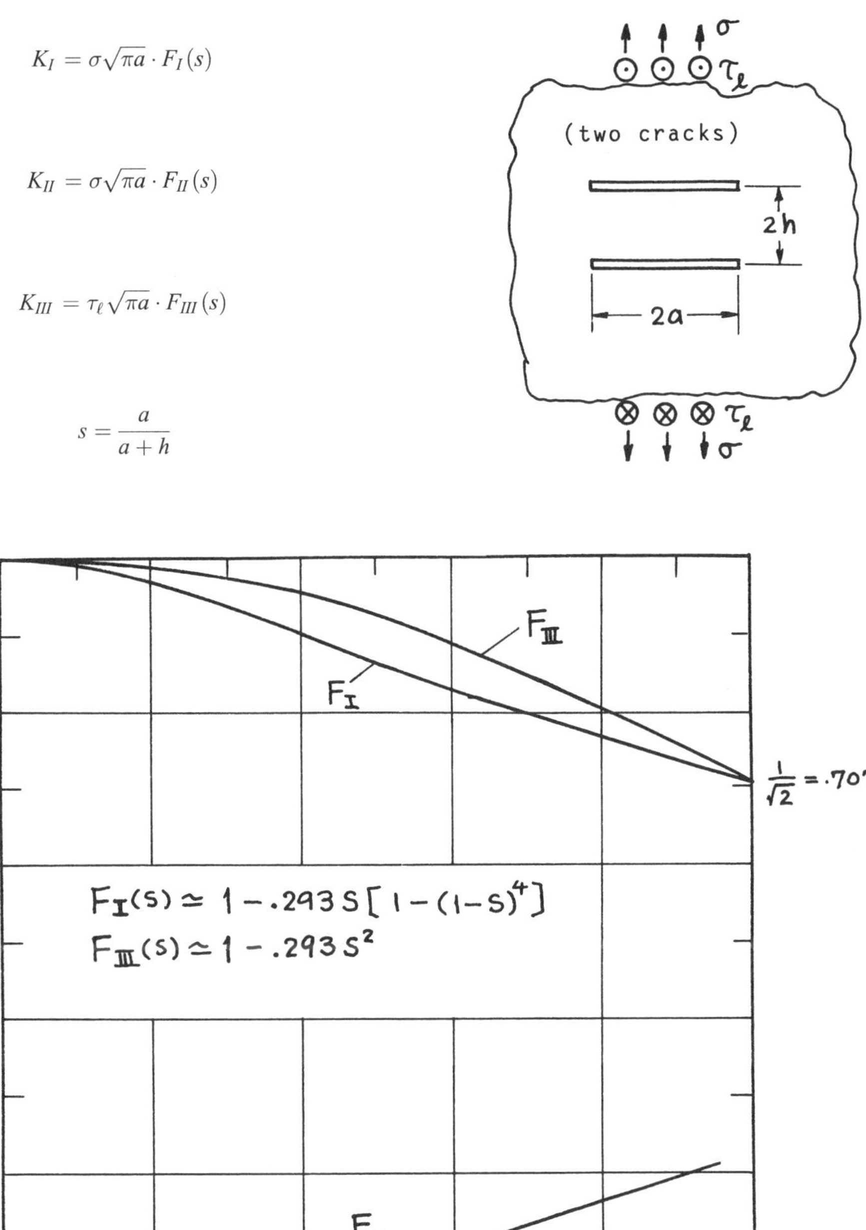

Methods: Series Expansion of Complex Potentials (Isida, $0 < s < 1/2$), Dislocation Distribution (Kamei)
Accuracy: F_I and F_{III} Better than 1%
F_{II} Expected to be better than 5%
References: **Isida 1972; Kamei 1974; Tada 1985**

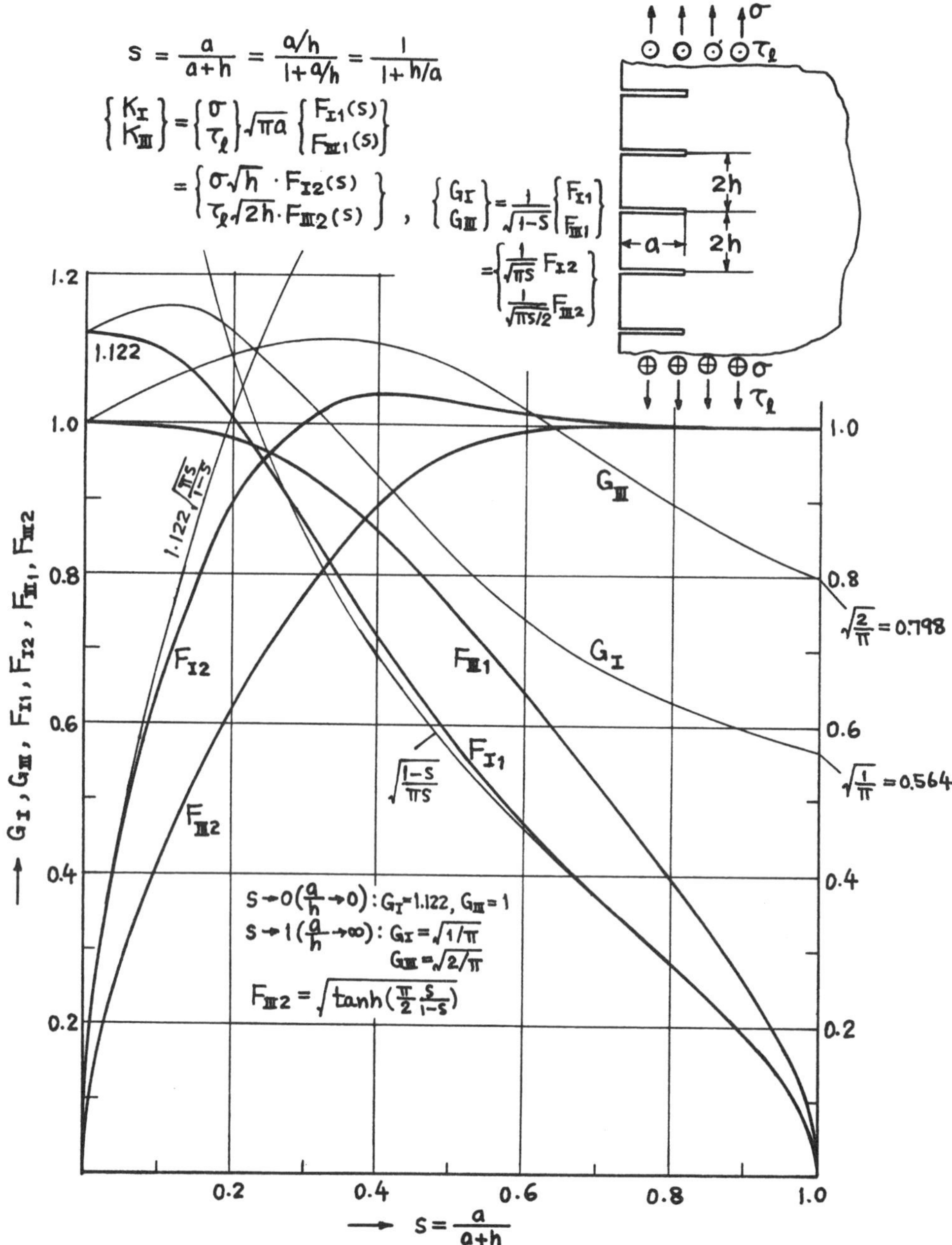

Methods: K_I Asymptotic Interpolation (Benthem), Mapping Function Method (Bowie), Body Force Method (Nishitani); K_{III} Stress Concentration Factor (Neuber), Westergaard Stress Function (Tada)

Accuracy: K_I 1% ,
K_{III} Exact

References: **Neuber 1937, 1958; Benthem 1972; Tada 1973; Bowie 1973; Nishitani 1976**

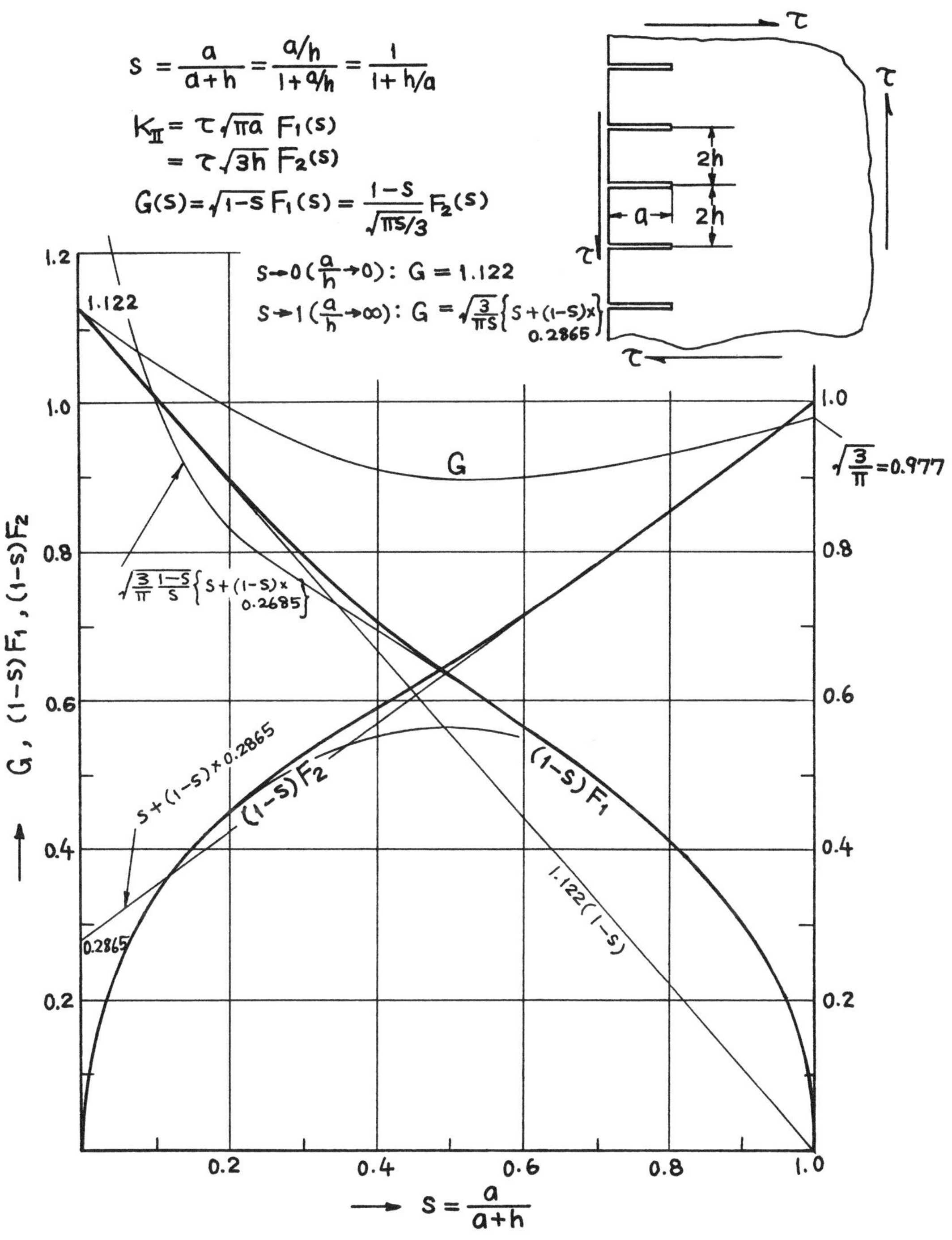

Method: Estimated Asymptotically
Accuracy: Estimated at better than 3%
Reference: **Tada 1973**

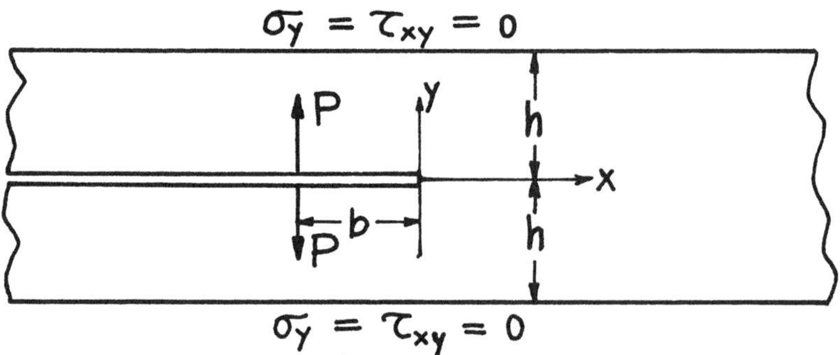

$$s = \frac{b}{b+h} = \frac{b/h}{1+b/h} = \frac{1}{1+h/b}$$

$$K_I = \sqrt{\frac{2}{\pi}}\frac{P}{\sqrt{b}} \cdot F_1(s)$$
$$= \frac{P}{\sqrt{h}}\left(\frac{b}{h}\right) F_2(s)$$

$$s \to 0\left(\frac{b}{h} \to 0\right) : \; F_1 = 1$$
$$F_2 \to \sqrt{\frac{2}{\pi}}\left(\frac{h}{b}\right)^{3/2} = \sqrt{\frac{2}{\pi}}\left(\frac{1-s}{s}\right)^{3/2}$$

$$s \to 1\left(\frac{b}{h} \to \infty\right) : F_2 = 2\sqrt{3}$$
$$F_1 \to \sqrt{6\pi}\left(\frac{b}{h}\right)^{3/2} = \sqrt{6\pi}\left(\frac{s}{1-s}\right)^{3/2}$$

Methods: $s \to 0$ Solution for a Semi-Infinite Crack (**page 3.6**), $s \to 1$ Theory of Beam Bending
Accuracy: Exact for both $s \to 0$ and $s \to 1$
References: $s \to 0$ **Irwin 1957, etc.**, $s \to 1$ **Gilman 1959; Barenblatt 1962**

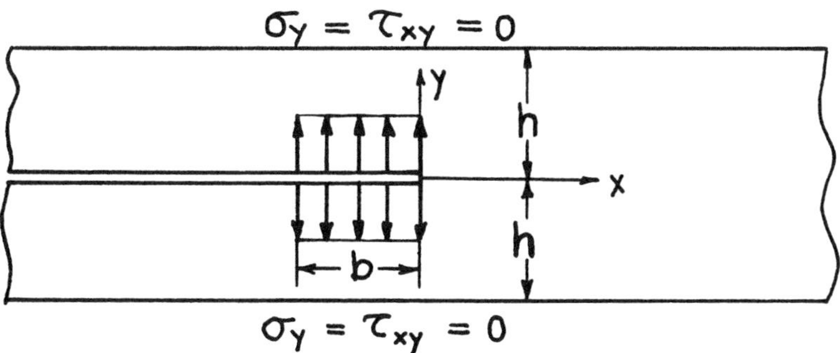

$$s = \frac{b}{b+h} = \frac{b/h}{1+b/h} = \frac{1}{1+h/b}$$

$$K_I = 2\sqrt{\frac{2}{\pi}}\,\sigma\sqrt{b}\cdot F_1(s)$$

$$= \sigma\sqrt{h}\left(\frac{b}{h}\right)^2 \cdot F_2(s)$$

$$s \to 0\left(\tfrac{b}{h} \to 0\right): \quad F_1 = 1$$

$$F_2 \to 2\sqrt{\tfrac{2}{\pi}}\left(\tfrac{h}{b}\right)^{3/2} = 2\sqrt{\tfrac{2}{\pi}}\left(\tfrac{1-s}{s}\right)^{3/2}$$

$$s \to 1\left(\tfrac{b}{h} \to \infty\right): \quad F_2 = \sqrt{\tfrac{3}{2}}$$

$$F_1 \to \tfrac{\sqrt{3\pi}}{4}\left(\tfrac{b}{h}\right)^{3/2} = \tfrac{\sqrt{3\pi}}{4}\left(\tfrac{s}{1-s}\right)^{3/2}$$

Methods: $s \to 0$ Solution for a Semi-Infinite Crack (**page 3.7**), $s \to 1$ Theory of Beam Bending
Accuracy: Exact for both $s \to 0$ and $s \to 1$
References: $s \to 0$ **Tada 1973**, $s \to 1$ **Gilman 1959; Barenblatt 1962**

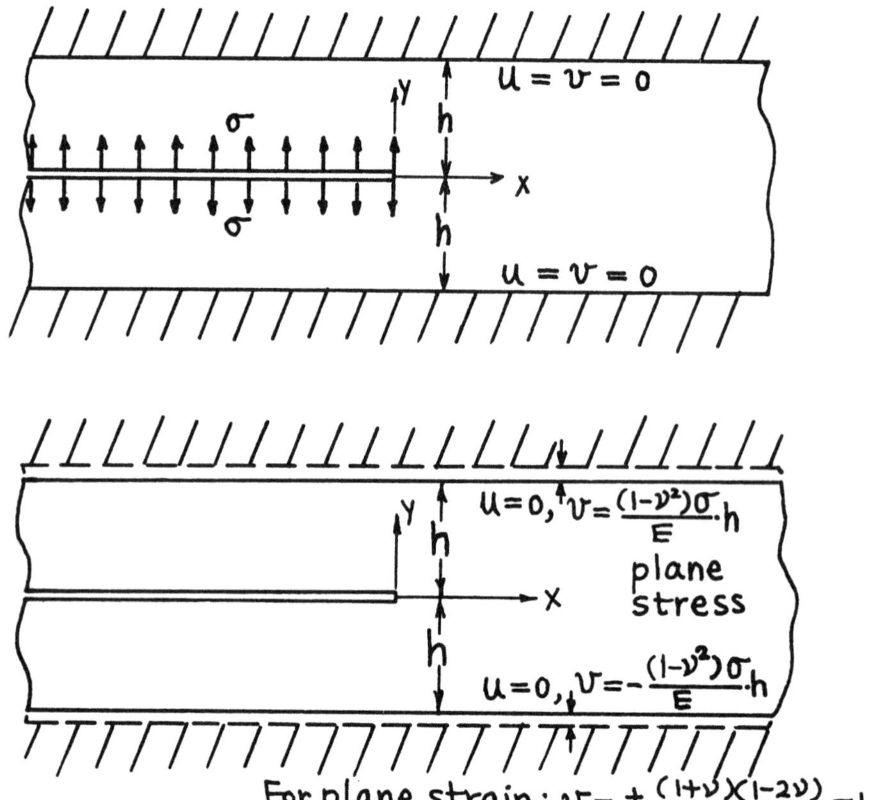

$$K_I = \sqrt{1 - \nu^2} \cdot \sigma\sqrt{h} \qquad \text{plane stress}$$

$$K_I = \frac{\sqrt{1 - 2\nu}}{1 - \nu} \cdot \sigma\sqrt{h} \qquad \text{plane strain}$$

Method: Energy Consideration
Accuracy: Exact
Reference: **Rice 1967**

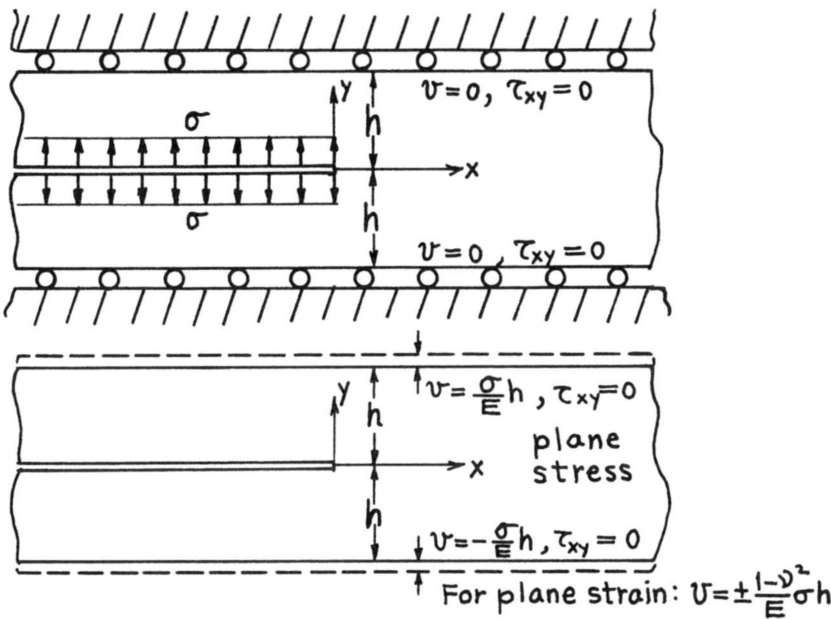

$$K_I = \sigma\sqrt{h}$$

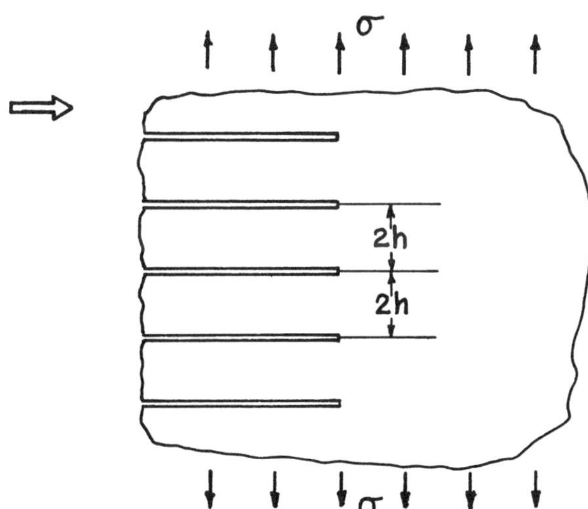

Methods: Energy Consideration (Paris, Rice), Fourier Transform (Benthem)
Accuracy: Exact
References: **Paris 1955, 1960; Rice 1967; Benthem 1972**

See **page 12.1**.

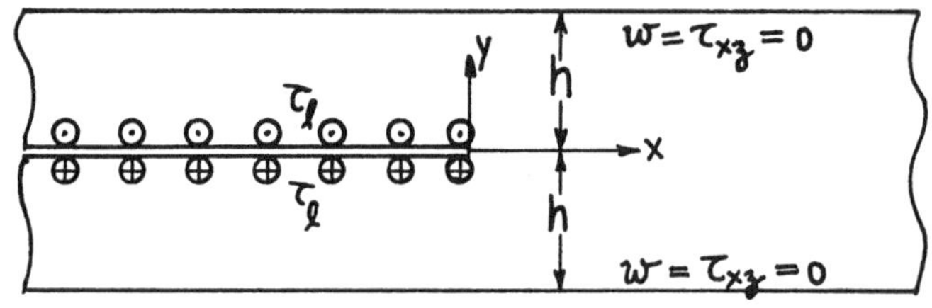

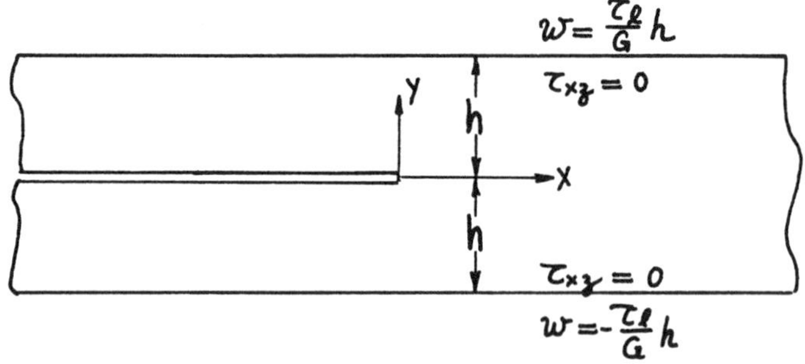

$$K_{III} = \tau_\ell \sqrt{2h}$$

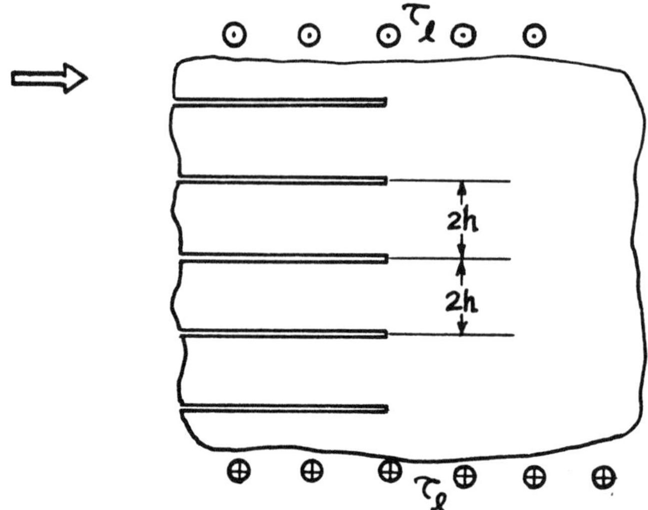

Method: Westergaard Stress Function
Accuracy: Exact
Reference: **Tada 1973**

See **page 12.2**.

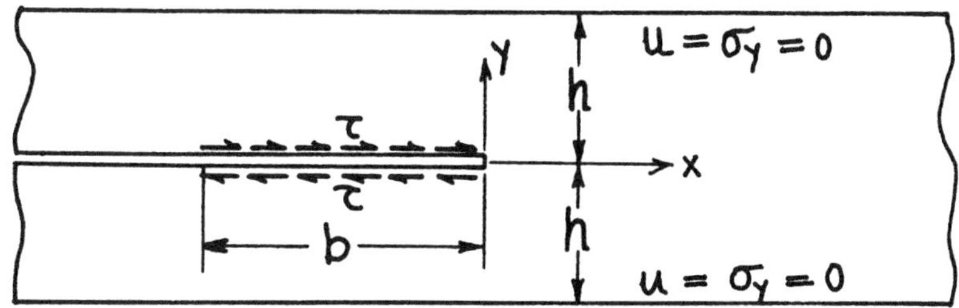

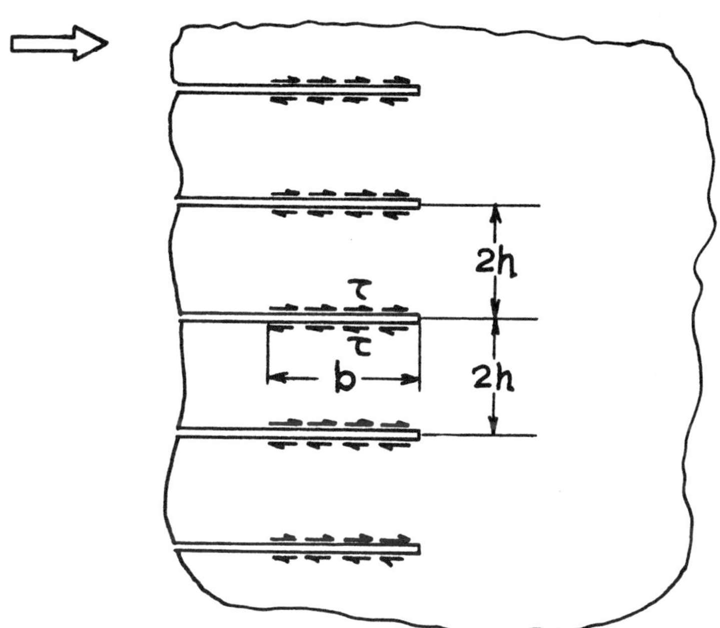

See **pages 12.5 and 12.6**.

$$s = \frac{a}{a+h} = \frac{a/h}{1+a/h} = \frac{1}{1+h/a}$$

$$K_I = \sigma\sqrt{\pi a}\cdot F_1(s)$$

$$= \sigma\sqrt{h}\left(\frac{a}{h}\right)^2 \cdot F_2(s)$$

$$s \to 0\left(a/h \to 0\right): \quad F_1 = 1$$

$$F_2 \to \sqrt{\pi}\left(\frac{h}{a}\right)^{3/2}$$

$$s \to 1\left(a/h \to \infty\right): \quad F_2 = 2/\sqrt{3}$$

$$F_1 \to \frac{2}{\sqrt{3\pi}}\left(\frac{a}{h}\right)^{3/2}$$

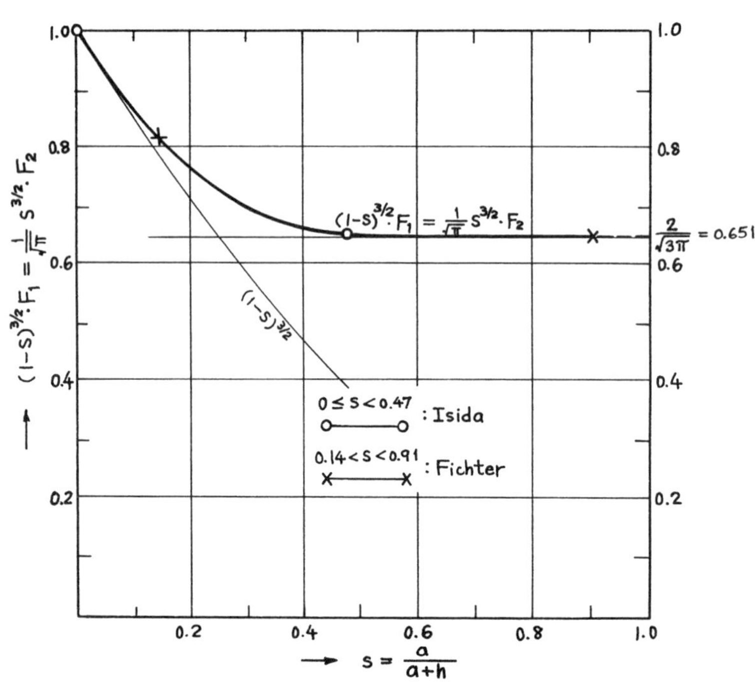

Methods: Expansions of Complex Stress Potentials (Isida), Fourier Transform (Fichter), Theory of Beam Bending ($s \to 1$)

Accuracy: Order of 1%

References: **Fichter 1967; Isida 1971a**

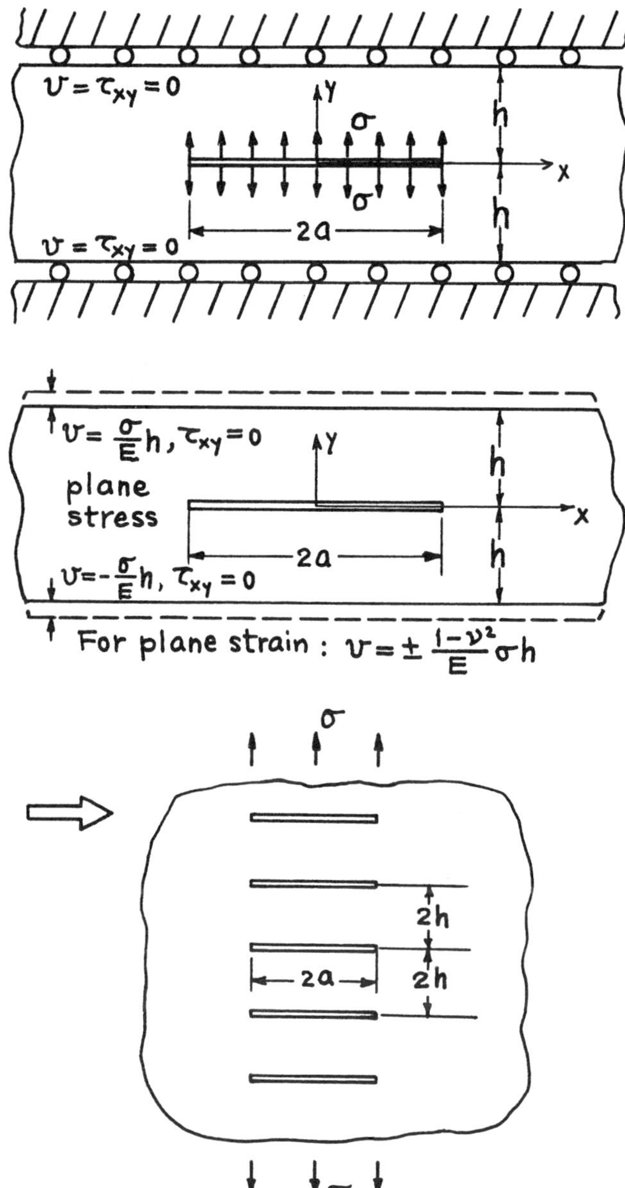

$v = \tau_{xy} = 0$
y
σ
h
x
σ
h
2a
$v = \tau_{xy} = 0$
$v = \frac{\sigma}{E}h, \tau_{xy} = 0$
y
h
plane
stress
x
2a
h
$v = -\frac{\sigma}{E}h, \tau_{xy} = 0$
For plane strain : $v = \pm \frac{1-\nu^2}{E}\sigma h$
σ
2h
2a
2h
σ

See **page 14.1**.

$$s = \frac{a}{a+h} = \frac{a/h}{1+a/h} = \frac{1}{1+h/a}$$

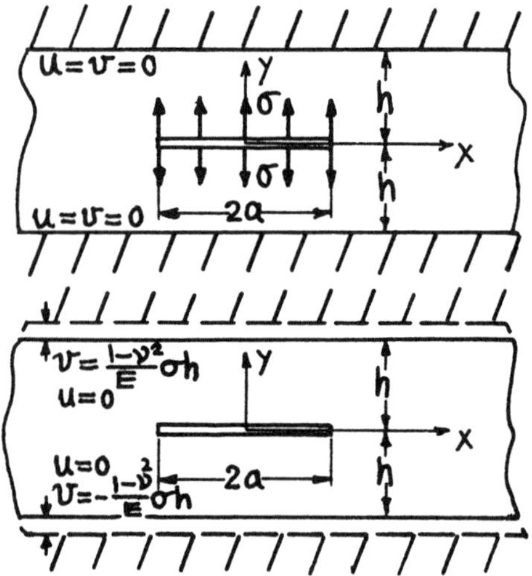

$$K_I = \sigma\sqrt{\pi a}\cdot F_1(s)$$

$$= \sigma\sqrt{h}\cdot F_2(s)$$

$$s \to 0(a/h \to 0): F_1 = 1, F_2 \to \sqrt{\frac{\pi a}{h}}$$

$$s \to 1\left(\frac{a}{h} \to \infty\right): \quad F_2 = \sqrt{1-\nu^2}$$

$$F_1 \to \sqrt{1-\nu^2}\sqrt{\frac{h}{\pi a}}$$

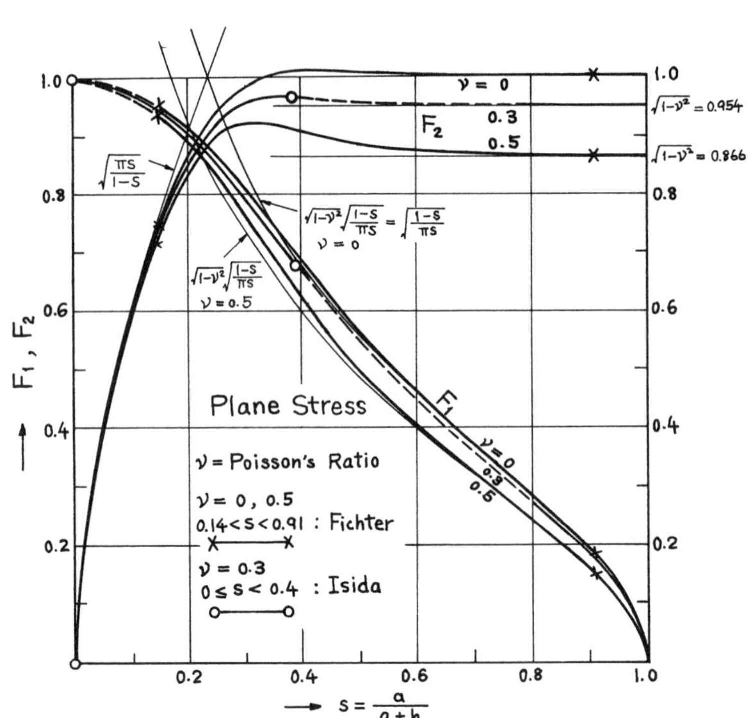

Methods: Expansions of Complex Stress Potentials (Isida), Fourier Transform (Fichter), $s \to 0$: Solution for Infinite Plate, $s \to 1$: Energy Balance (Rice)

Accuracy: Order of 1%

References: **Fichter 1967; Rice 1967; Isida 1971a**

NOTE: For plane strain $F_2(s \to 1) = \sqrt{1-2\nu}/(1-\nu)$, etc.

$$s = \frac{a}{a+h} = \frac{a/h}{1+a/h} = \frac{1}{1+h/a}$$

$$K_I = \frac{P}{\sqrt{\pi a}} \cdot F_1(s)$$
$$= \frac{P}{\sqrt{h}} \left(\frac{a}{h}\right) \cdot F_2(s)$$

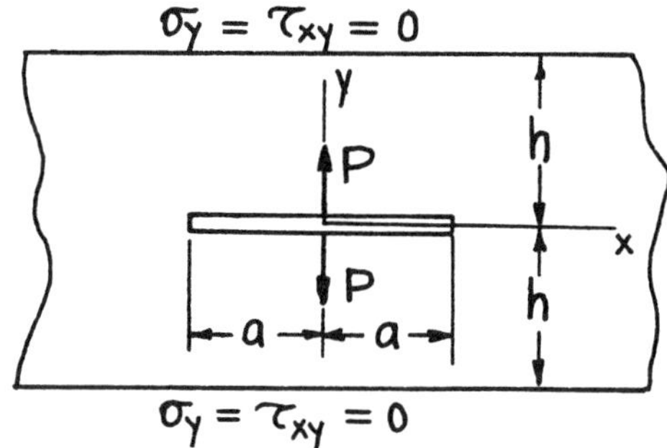

$$s \to 0 \left(\frac{a}{h} \to 0\right) : \ F_1 = 1, \ F_2 \to \frac{1}{\sqrt{\pi}} \left(\frac{h}{a}\right)^{3/2}$$

$$s \to 1 \left(\frac{a}{h} \to \infty\right) : F_2 = \frac{\sqrt{3}}{2}$$
$$F_1 \to \frac{\sqrt{3\pi}}{2} \left(\frac{a}{h}\right)^{3/2}$$

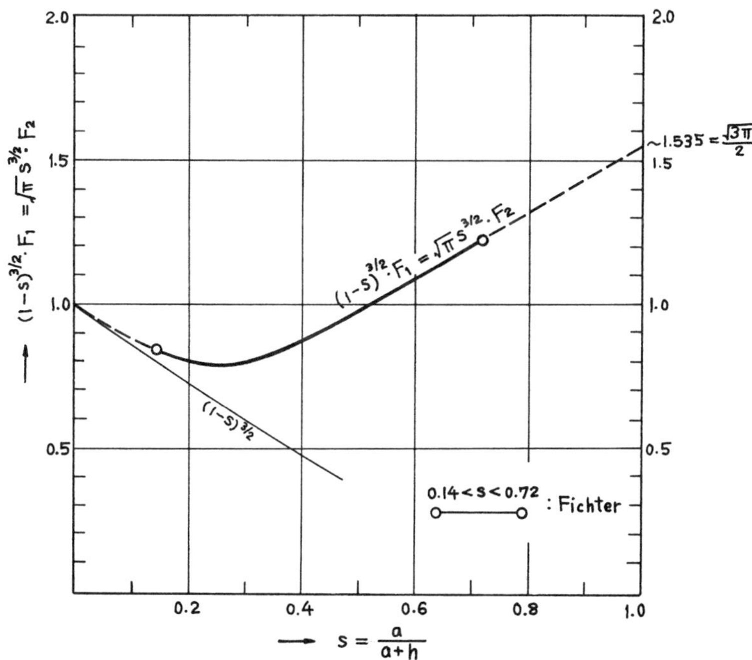

Methods: Fourier Transform (Fichter), $s \to 0$: Solution for infinite plate (**page 5.9**), $s \to 1$: Beam bending
Accuracy: Order of 1%
Reference: **Fichter 1967**

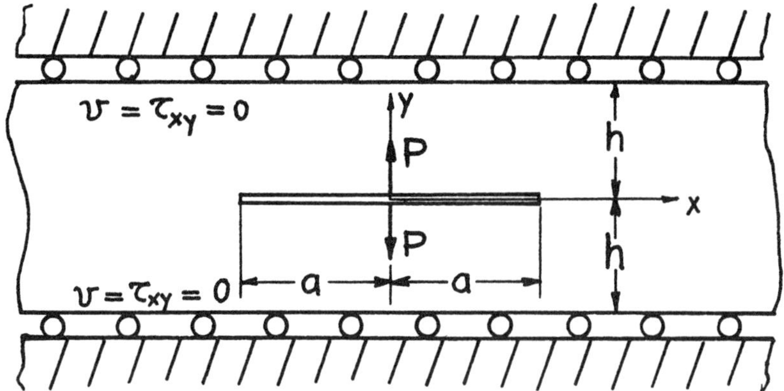

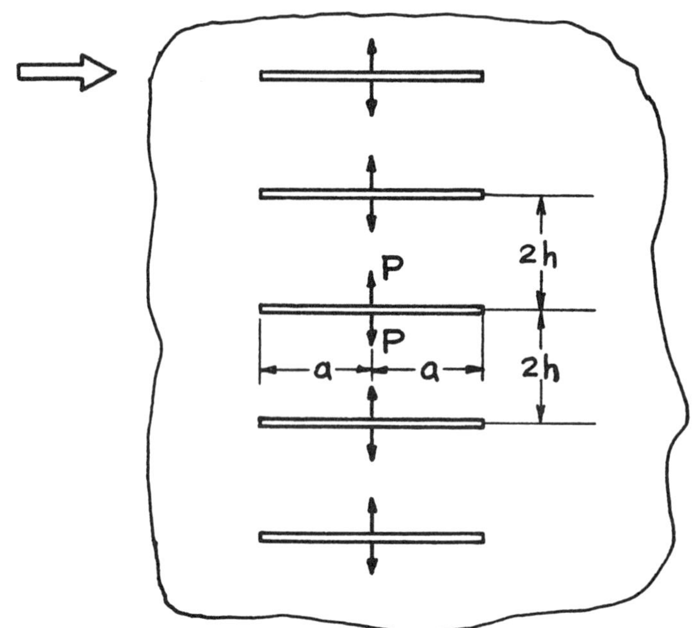

See **page 14.6**.

$$s = \frac{a}{a+h} = \frac{a/h}{1+a/h} = \frac{1}{1+h/a}$$

$$K_I = \frac{P}{\sqrt{\pi a}} \cdot F(s)$$

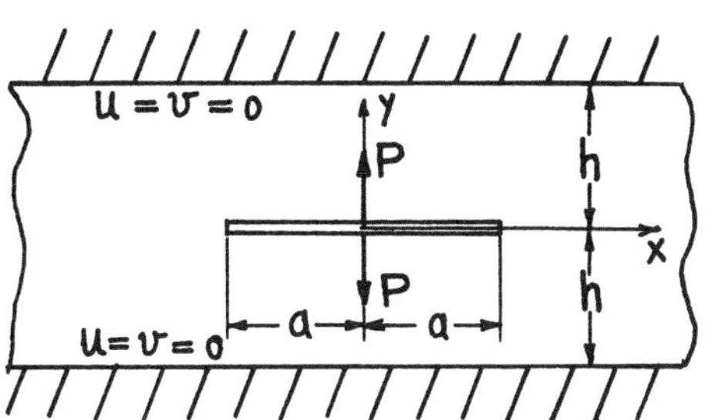

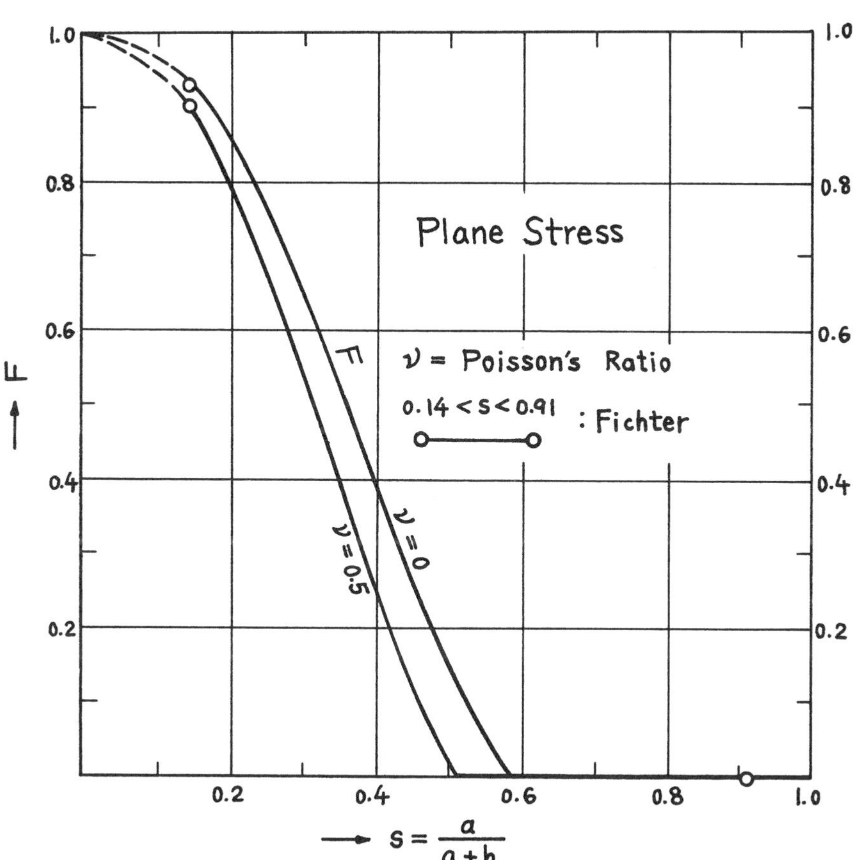

Methods: Fourier Transform (Fichter), $s \to 0$: Solution for an Infinite Plate (**page 5.9**)
Accuracy: Order of 1%
Reference: **Fichter 1967**

$$s = \frac{a}{a+h} = \frac{a/h}{1+a/h} = \frac{1}{1+h/a}$$

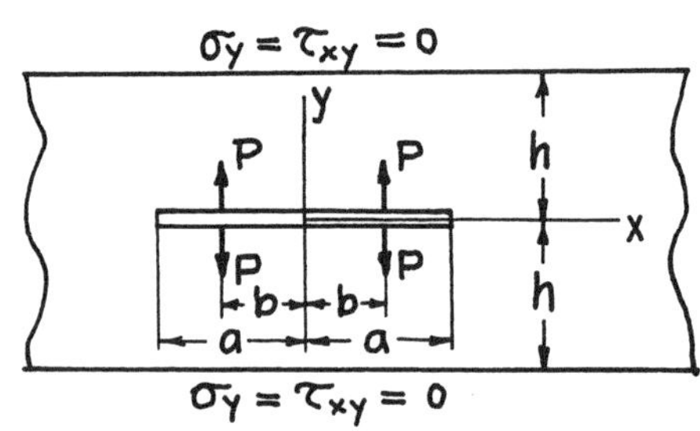

$$g\left(\frac{b}{a}\right) = \sqrt{1-\left(b/a\right)^2}$$

$$K_I = \frac{2P}{\sqrt{\pi a}} \cdot \frac{1}{g\left(b/a\right)} \cdot F_1\left(s, \frac{b}{a}\right)$$

$$= \frac{2P}{\sqrt{h}}\left(\frac{a}{h}\right)\left\{g\left(\frac{b}{a}\right)\right\}^2 F_2\left(s, \frac{b}{a}\right)$$

$$s \to 0\left(\frac{a}{h} \to 0\right) : \ F_1 = 1$$

$$F_2 \to \frac{1}{\sqrt{\pi}} \cdot \frac{1}{g^3}\left(\frac{h}{a}\right)^{3/2}$$

$$s \to 1\left(\frac{a}{h} \to \infty\right) : F_2 = \sqrt{3}/2$$

$$F_1 \to \frac{\sqrt{3\pi}}{2} \cdot g^3 \cdot \left(\frac{a}{h}\right)^{3/2}$$

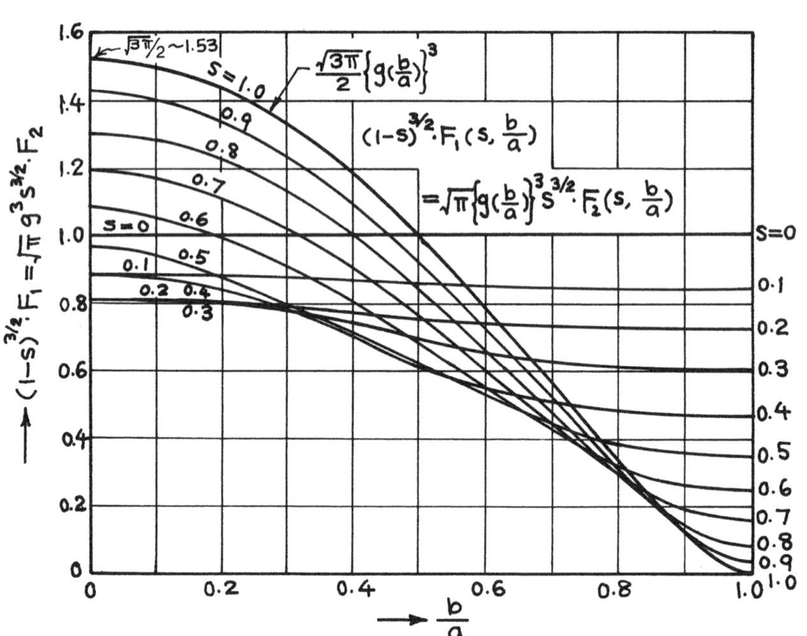

Method: Interpolated Asymptotically
Accuracy: Expected to be better than 5% for any s and b/a
Reference: **Tada 1973**

NOTE: $F(s \to 0, b/a)$ **(page 5.11)**, $F(s \to 1, b/a)$ (beam bending), $F(s, 0)$ **(page 17.4)**, $F(s, b/a \to 1)$ **(page 3.6)**.

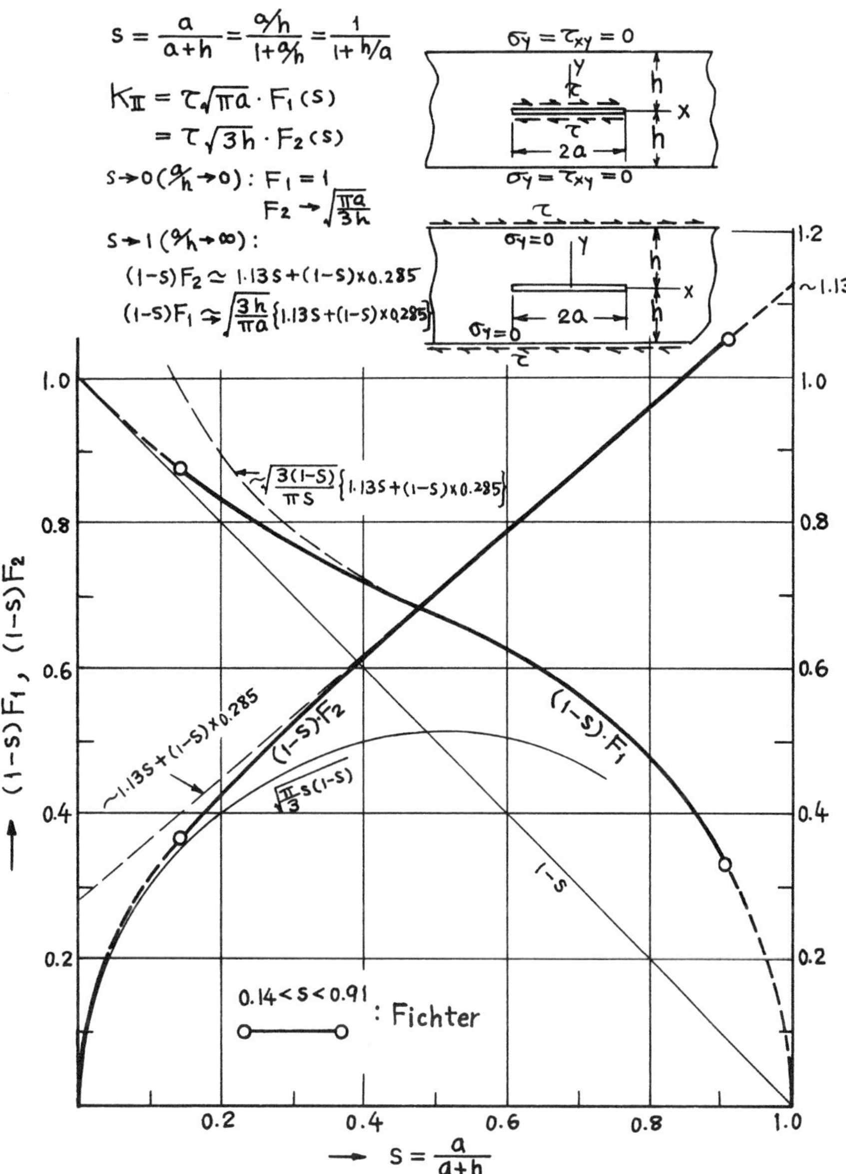

Method: Fourier Transform
Accuracy: Order of 1%
Reference: **Fichter 1967**

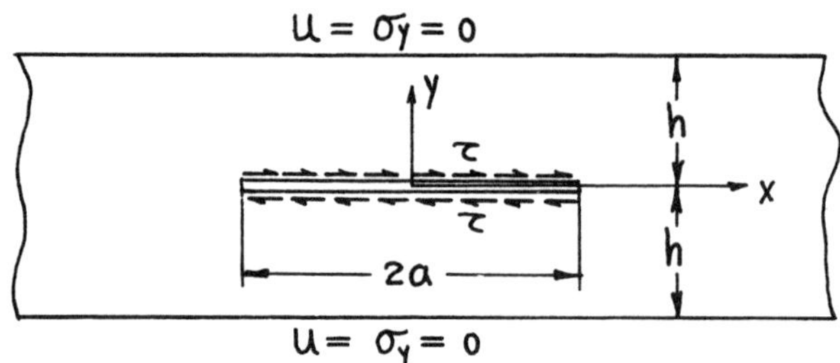

$u = \sigma_y = 0$
y
τ
x
τ
h
h
2a
$u = \sigma_y = 0$

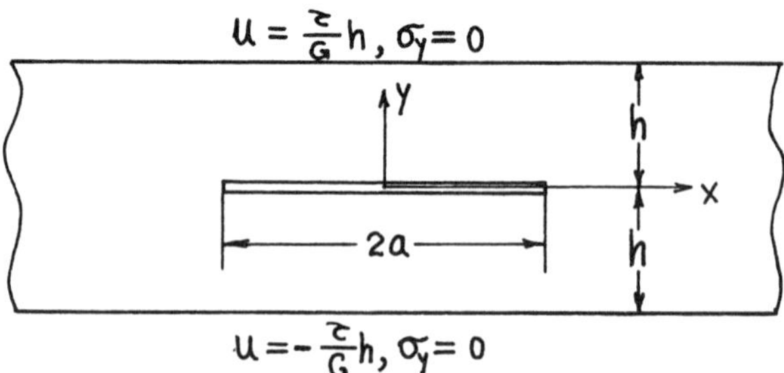

$u = \frac{\tau}{G}h, \sigma_y = 0$
y
h
x
h
2a
$u = -\frac{\tau}{G}h, \sigma_y = 0$

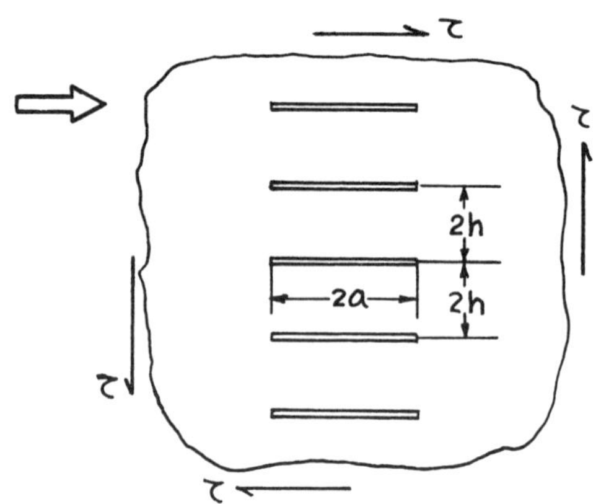

τ
τ
2h
2a
2h
τ
τ

See **page 14.5**.

$$s = \frac{a}{a+h} = \frac{a/h}{1+a/h} = \frac{1}{1+h/a}$$

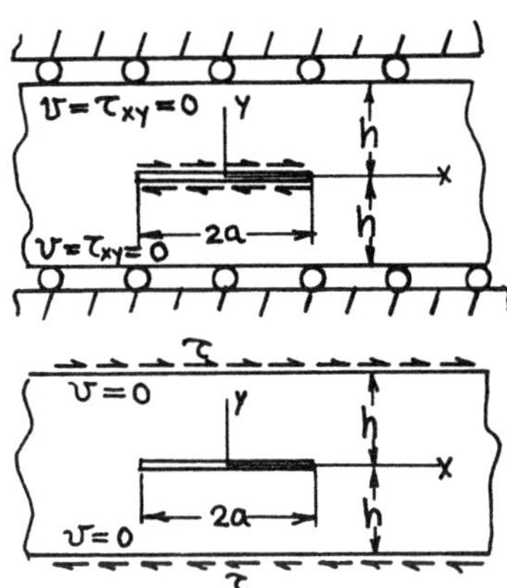

$$K_{II} = \tau\sqrt{\pi a}\cdot F_1(s)$$
$$= \tau\sqrt{3h}\cdot F_2(s)$$

$$s \to 0\left(\frac{a}{h} \to 0\right) : F_1 = 1,\ F_2 \to \sqrt{\frac{\pi a}{3h}}$$

$$s \to 1\left(\frac{a}{h} \to \infty\right) :$$
$$(1-s)F_2 \simeq 0.56s + (1-s)\times 0.43$$
$$(1-s)F_1 \simeq \sqrt{\frac{3h}{\pi a}}\{0.56s + (1-s)\times 0.43\}$$

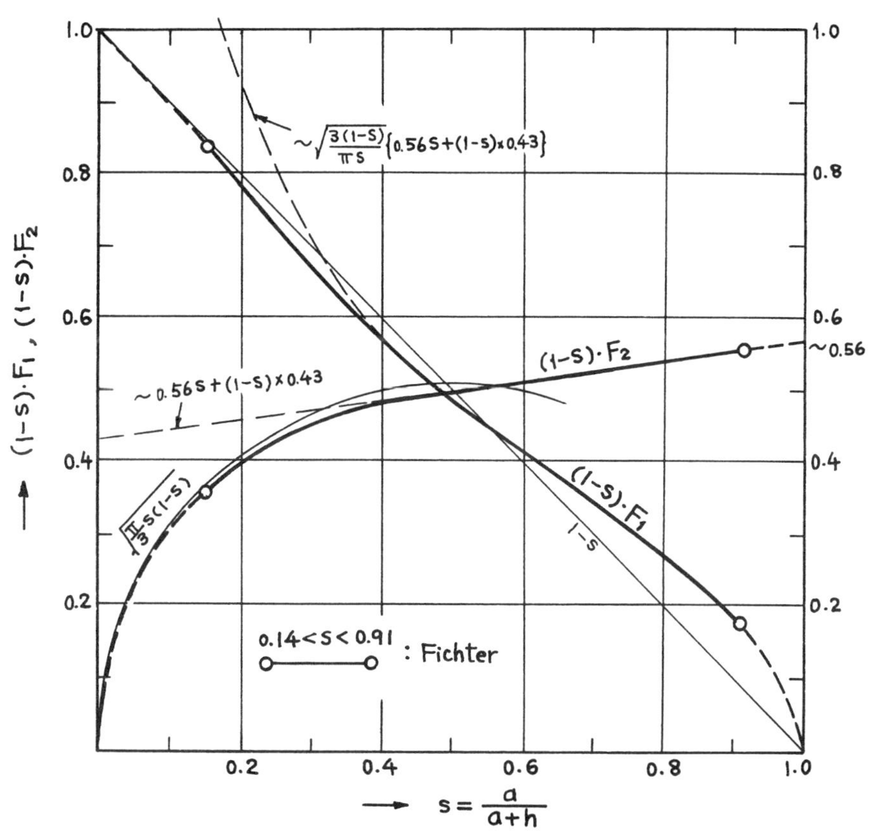

Method: Fourier Transform
Accuracy: Order of 1%
Reference: **Fichter 1967**

$$s = \frac{a}{a+h} = \frac{a/h}{1+a/h} = \frac{1}{1+h/a}$$

$$K_I = \sigma_0\sqrt{\pi a}\cdot F_1(s)$$
$$= \sigma_0\sqrt{h}\left(\frac{a}{h}\right)^2\cdot F_2(s)$$

$$s \to 0\left(\frac{a}{h}\to 0\right):\quad F_1 = \frac{1}{2}$$
$$F_2 \to \frac{\sqrt{\pi}}{2}\left(\frac{h}{a}\right)^{3/2}$$

$$s \to 1\left(\frac{a}{h}\to\infty\right):\quad F_2 \simeq \frac{2}{\sqrt{105}}$$
$$F_1 \simeq \frac{2}{\sqrt{105\pi}}\left(\frac{a}{h}\right)^{3/2}$$

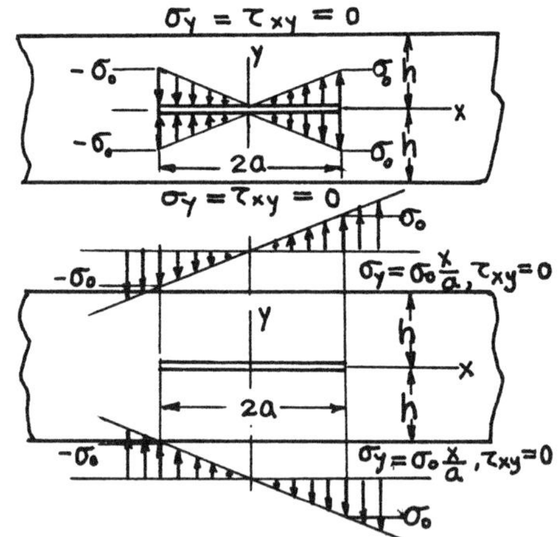

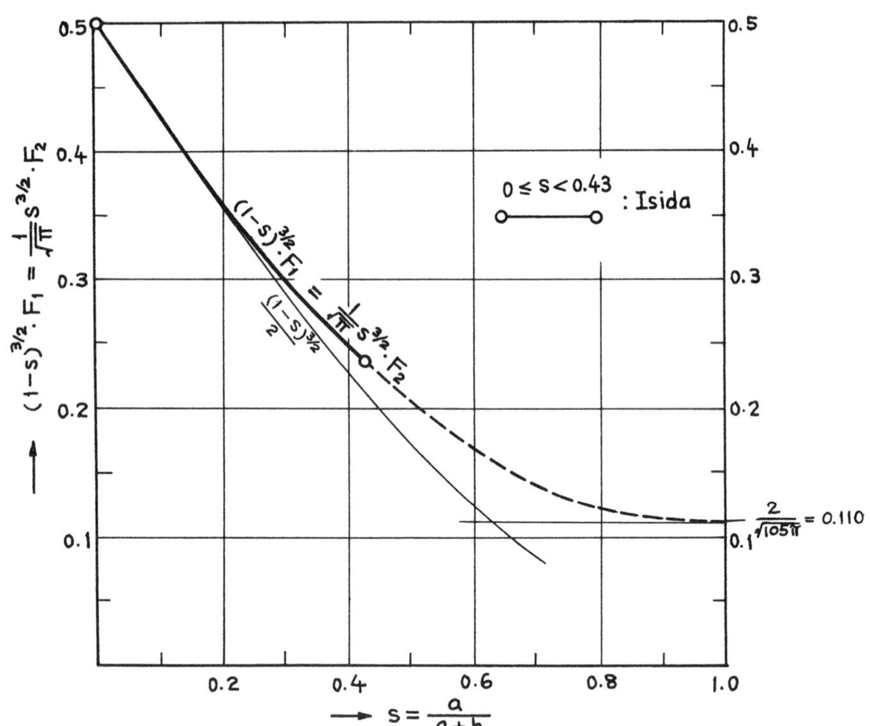

Methods: Expansions of Complex Stress Potentials (Isida, $0 < s < 0.43$, Interpolated Asymptotically (Tada $0.43 < s < 1$), $s \to 1$ Theory of Beam Bending

Accuracy: 1% for $0 < s < 0.43$, better than 5% for $0.43 < s < 1$ (exact for $s \to 0$ and $s \to 1$)

References: **Isida 1971a; Tada 1973**

$$s = \frac{a}{a+h} = \frac{a/h}{1+a/h} = \frac{1}{1+h/a}$$

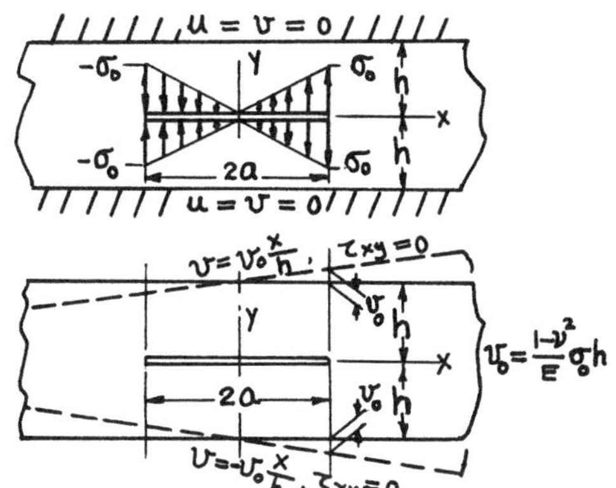

$$K_I = \sigma_0\sqrt{\pi a}\cdot F_1(s)$$
$$= \sigma_0\sqrt{h}\cdot F_2(s)$$

$s \to 0\left(\frac{a}{h} \to 0\right)$: $F_1 = \frac{1}{2}$

$F_2 \to \frac{1}{2}\sqrt{\frac{\pi a}{h}}$

$s \to 1\left(\frac{a}{h} \to \infty\right)$: $F_2 = \sqrt{1-\nu^2}$

$F_1 \to \sqrt{1-\nu^2}\sqrt{\frac{h}{\pi a}}$

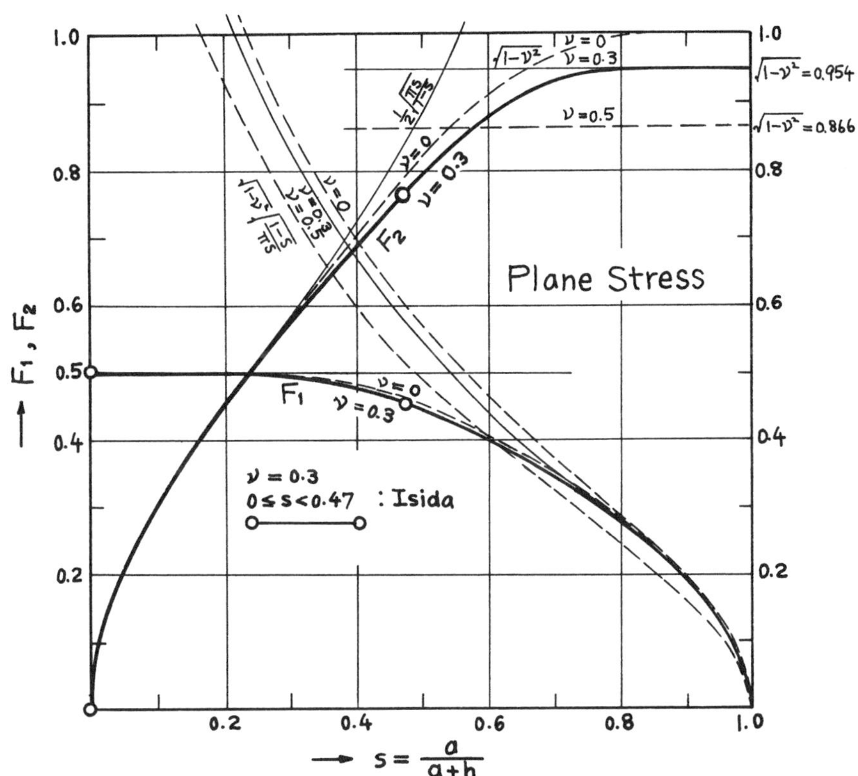

Methods: Expansions of Complex Stress Potentials (Isida, $s < 0.47$), Interpolated Asymptotically (Tada, $0.47\ s < 1$); $s \to 1$: Energy Balance (Rice)

Accuracy: 1% for $s < 0.47$, estimated at better than 3% for $0.47 < s < 1$ (exact for $s \to 0$ and $s \to 1$)

References: **Rice 1967; Isida 1971a; Tada 1973**

NOTE: For plane strain, $F_2(s \to 1) = \frac{\sqrt{1-2\nu}}{1-\nu}$, etc.; for $\nu = 0$ see **page 14.4**.

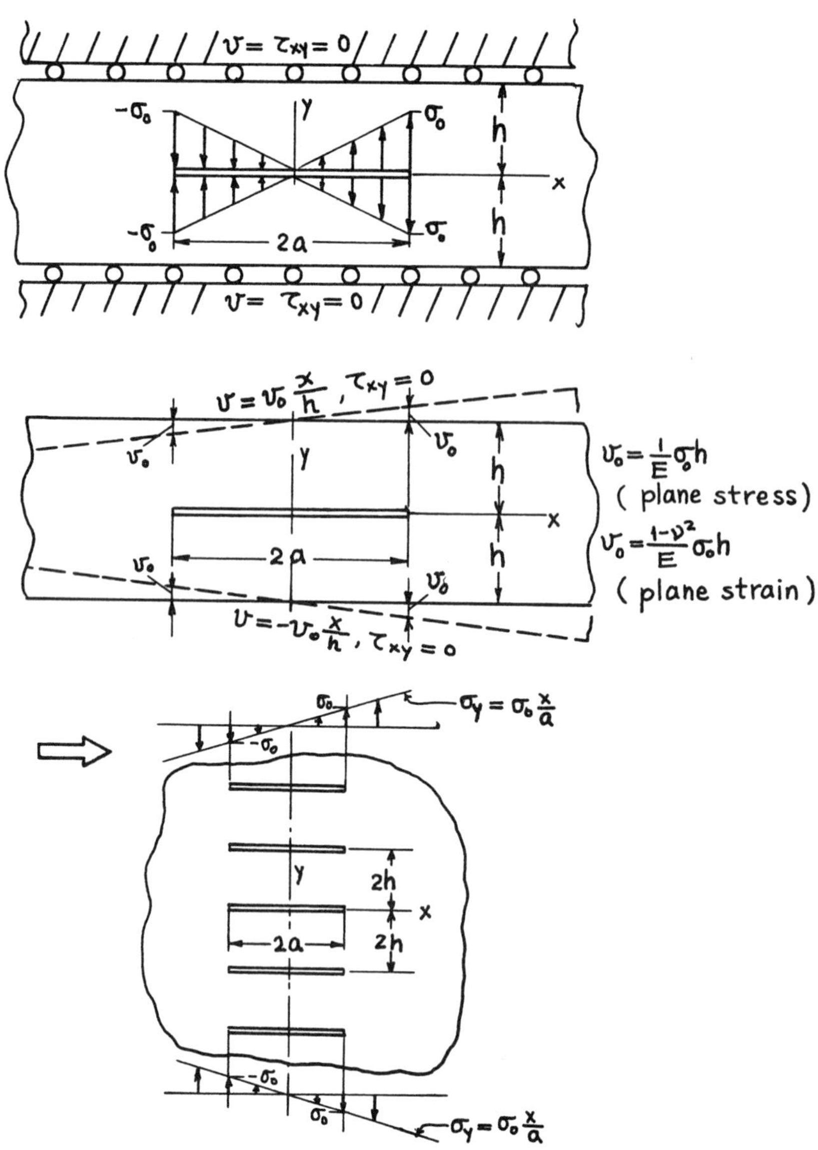
$v = \tau_{xy} = 0$
$-\sigma_0$
y
σ_0
h
x
$-\sigma_0$
2a
σ_0
h
$v = \tau_{xy} = 0$
$v = v_0 \frac{x}{h}$, $\tau_{xy} = 0$
v_0
y
v_0
h
$v_0 = \frac{1}{E}\sigma_0 h$
(plane stress)
x
$v_0 = \frac{1-\nu^2}{E}\sigma_0 h$
v_0
2a
h
v_0
(plane strain)
$v = -v_0 \frac{x}{h}$, $\tau_{xy} = 0$
σ_0
$\sigma_y = \sigma_0 \frac{x}{a}$
$-\sigma_0$
y
2h
x
2a
2h
$-\sigma_0$
σ_0
$\sigma_y = \sigma_0 \frac{x}{a}$

See **page 14.4**.

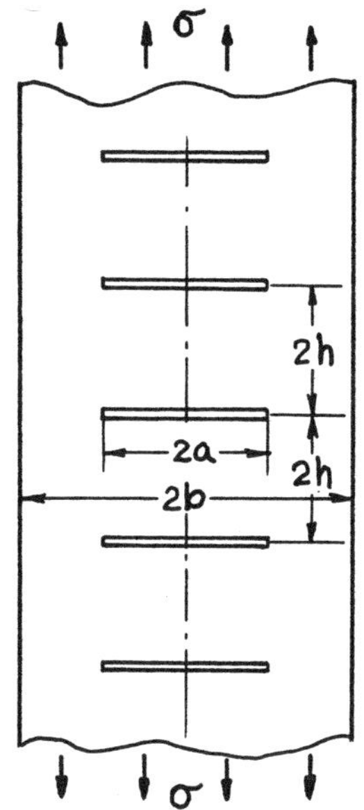
σ
2h
2h
2a
2b
σ

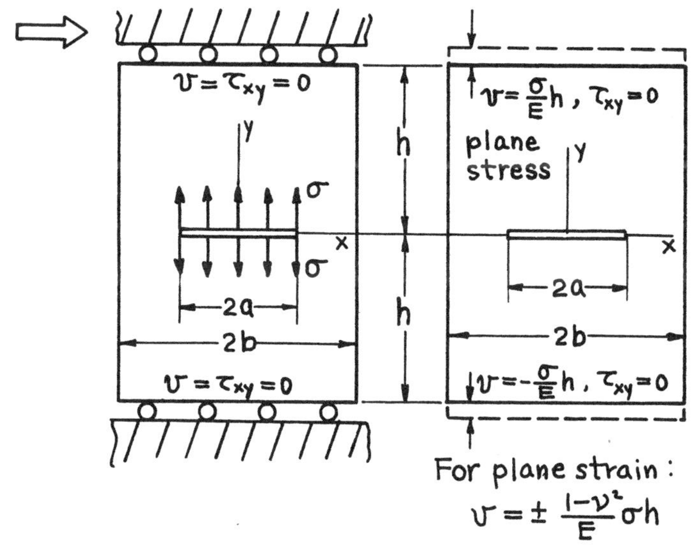
v = τxy = 0
y
σ
x
σ
2a
2b
v = τxy = 0
h
h
v = σ/E h , τxy = 0
plane
stress
y
x
2a
2b
v = -σ/E h , τxy = 0
For plane strain :
v = ± (1-ν²)/E σh

See **page 18.2**.

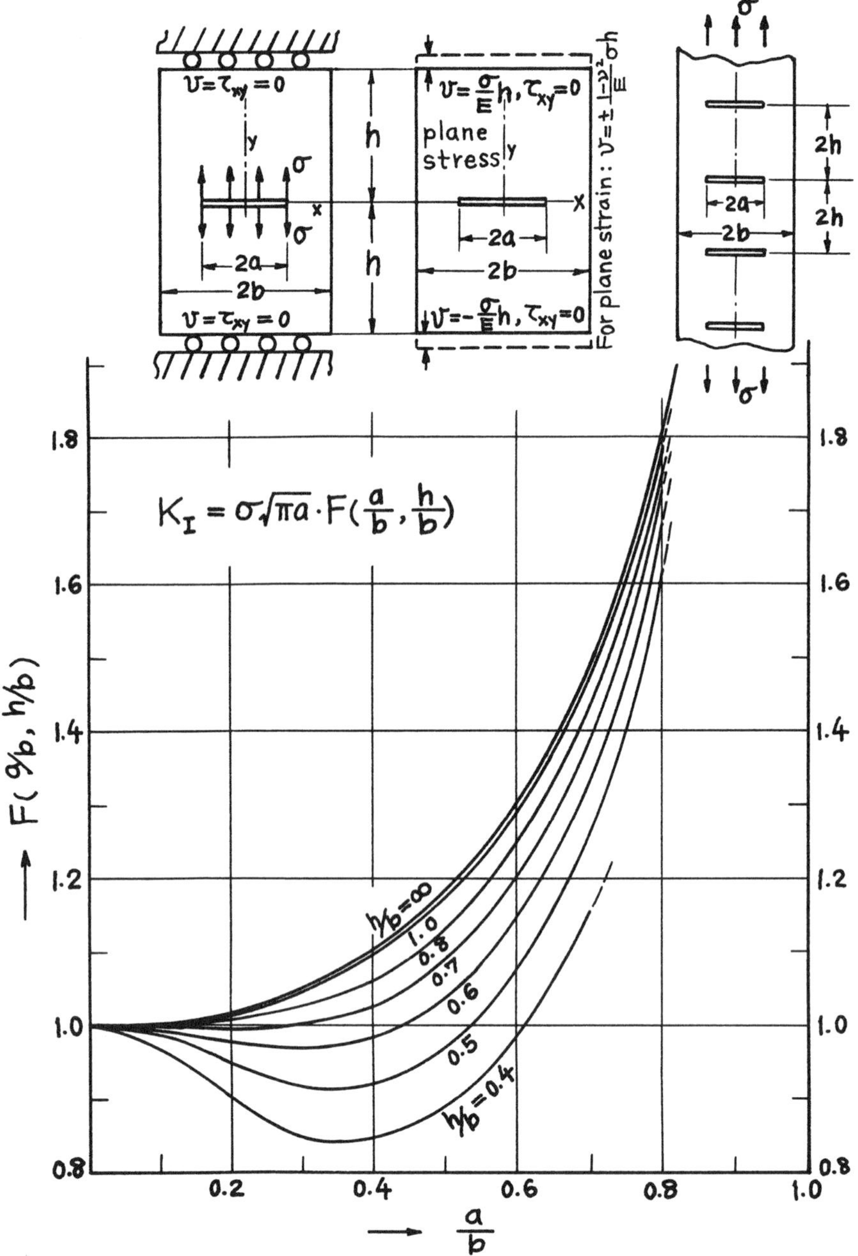

Method: Laurent's Expansion of Complex Stress Potentials
Accuracy: Better than 1%
Reference: **Isida 1971a,b**

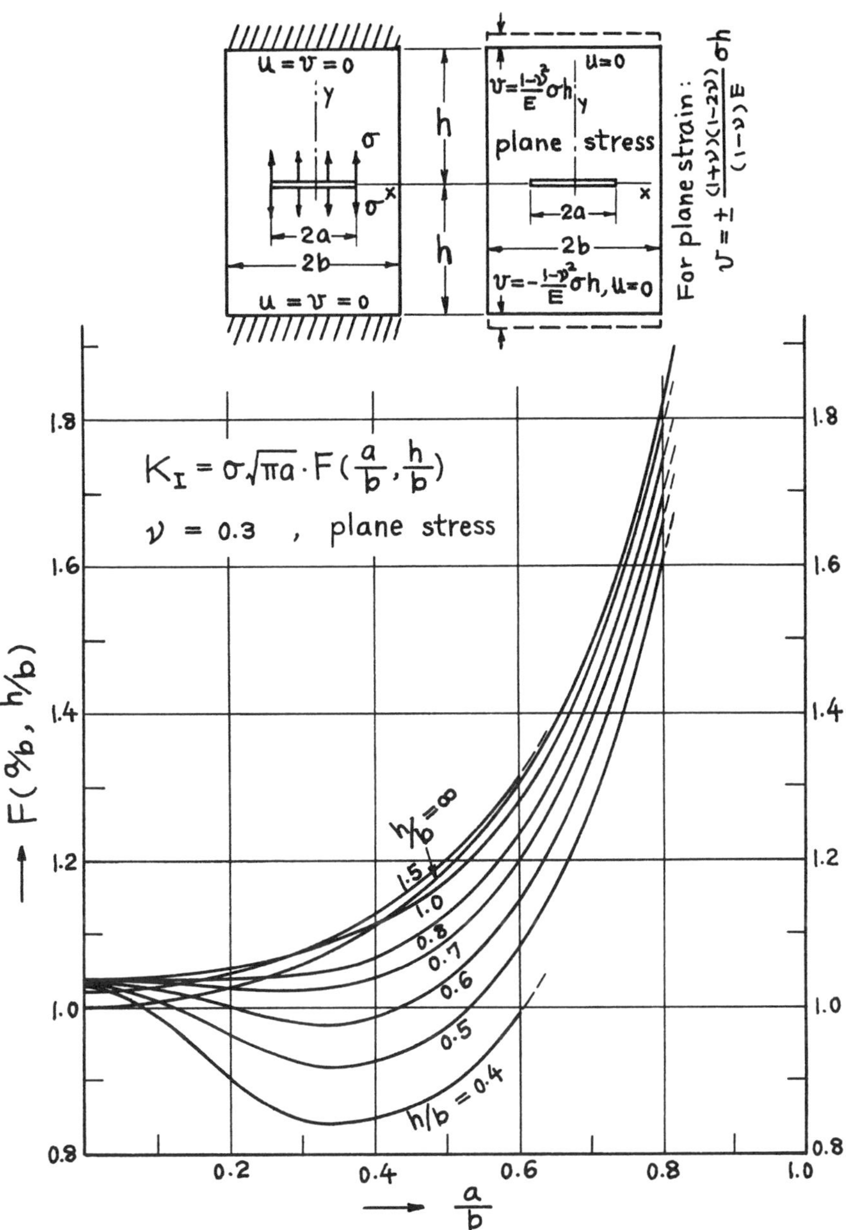

Method: Laurent's Expansion of Complex Stress Potentials Combined with a Boundary Collocation Method

Accuracy: Better than 1%

Reference: **Isida 1971a,b**

$$K_I = \sigma\sqrt{\pi a} \cdot F\left(\frac{a}{b}\right)$$

$$F\left(\frac{a}{b}\right) = 1 - .344\left(\frac{a}{b}\right)^2 - .156\left(\frac{a}{b}\right)^3 \left(\frac{a}{b} \leq .8\right)$$

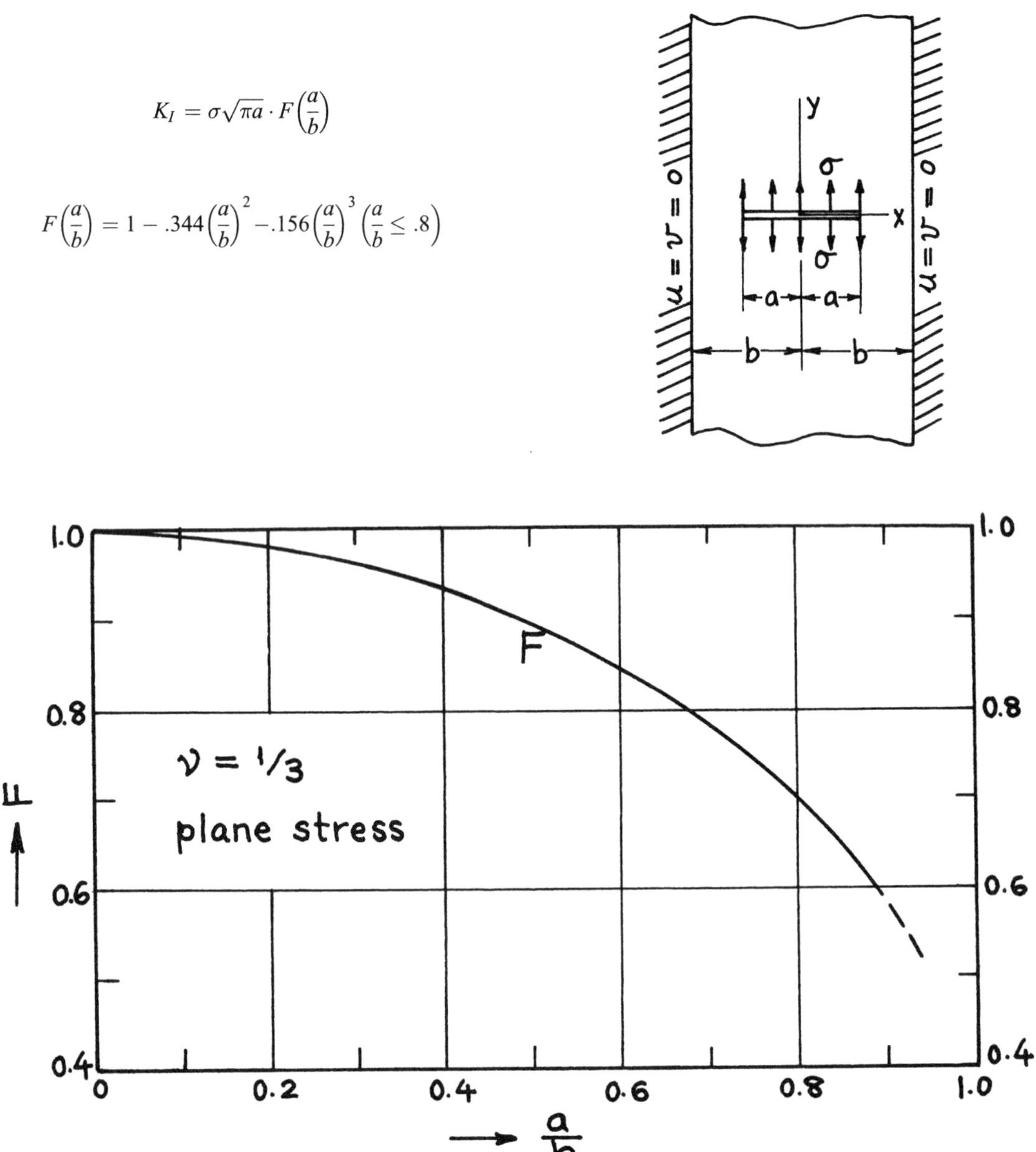

Method: Series Expansions of Complex Potentials
Accuracy: Curve is based on numerical values with 0.1% accuracy
Formula 1% for $a/b \leq 0.8$
References: **Isida 1970a; Tada 1985**

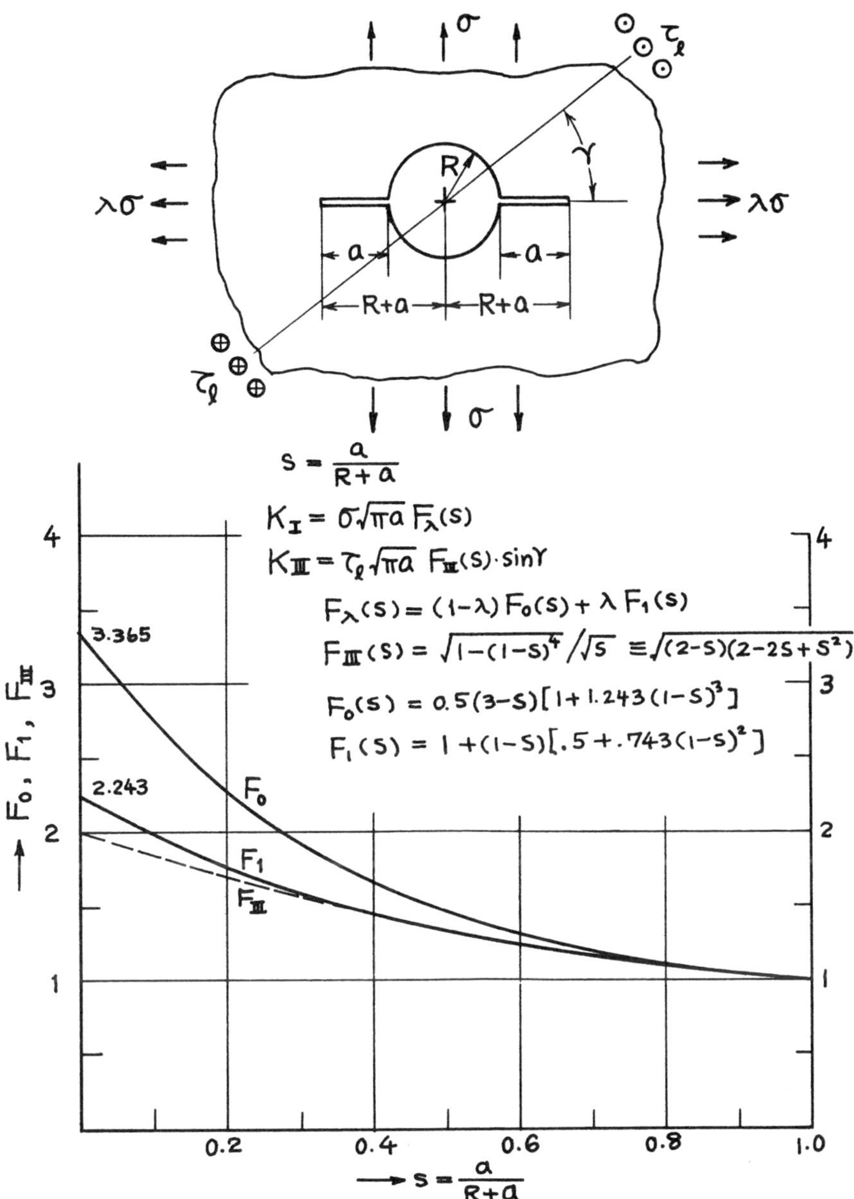

Methods: Mapping Function Methods (Bowie — Mode I; Sih — Mode III), Boundary Collocation Method (Newman)

Accuracy: F_0 and F_1 curves are based on numerical values with expected accuracy of 0.1%. Formulas F_0 and F_1 1%; F_{III} Exact

References: **Bowie 1956; Sih 1965a; Newman 1971; Tada 1985**

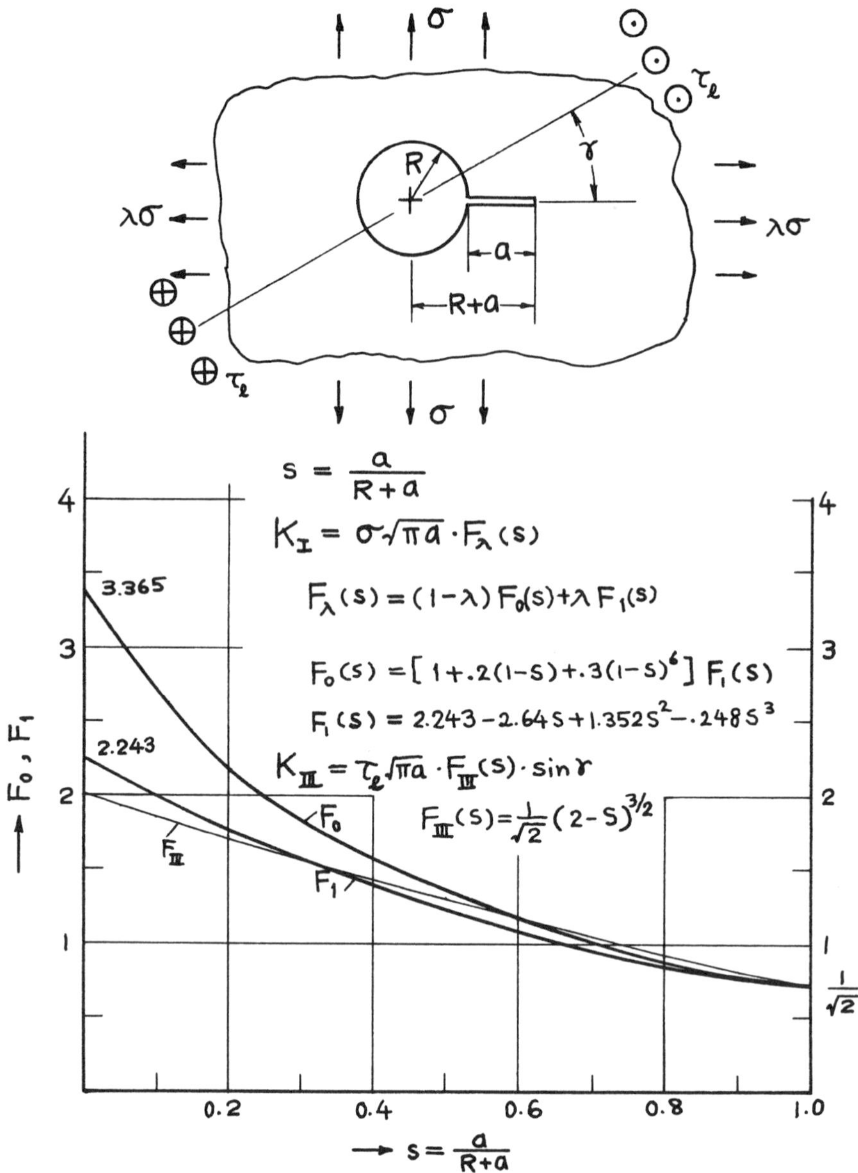

Method: Mapping Function Method
Accuracy: F_0 and F_I Better than 1%
F_{III} Exact
References: **Bowie 1956; Yokobori 1972 (or Kamei 1974); Tada 1985**

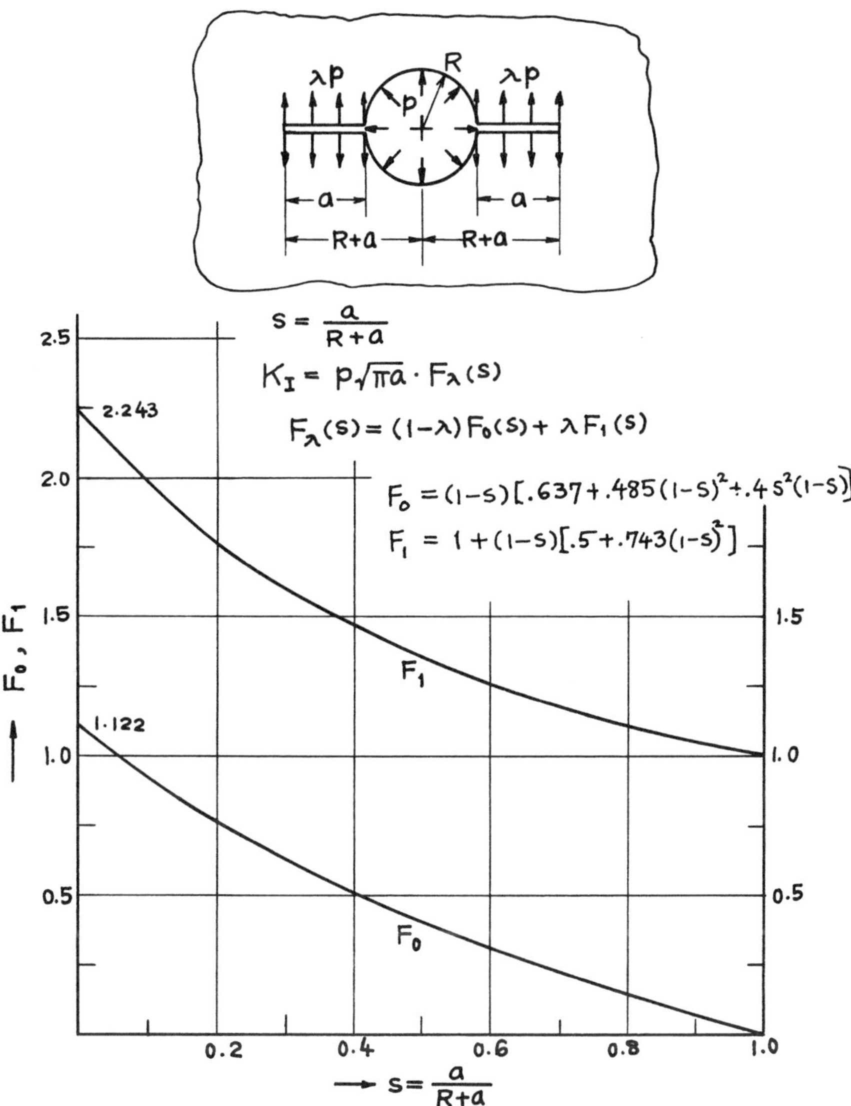

Method: Boundary Collocation Method

Accuracy: Curves are based on numerical values with 0.1% accuracy.
Formulas F_0 and F_1 1%

References: **Newman 1971; Tada 1985**

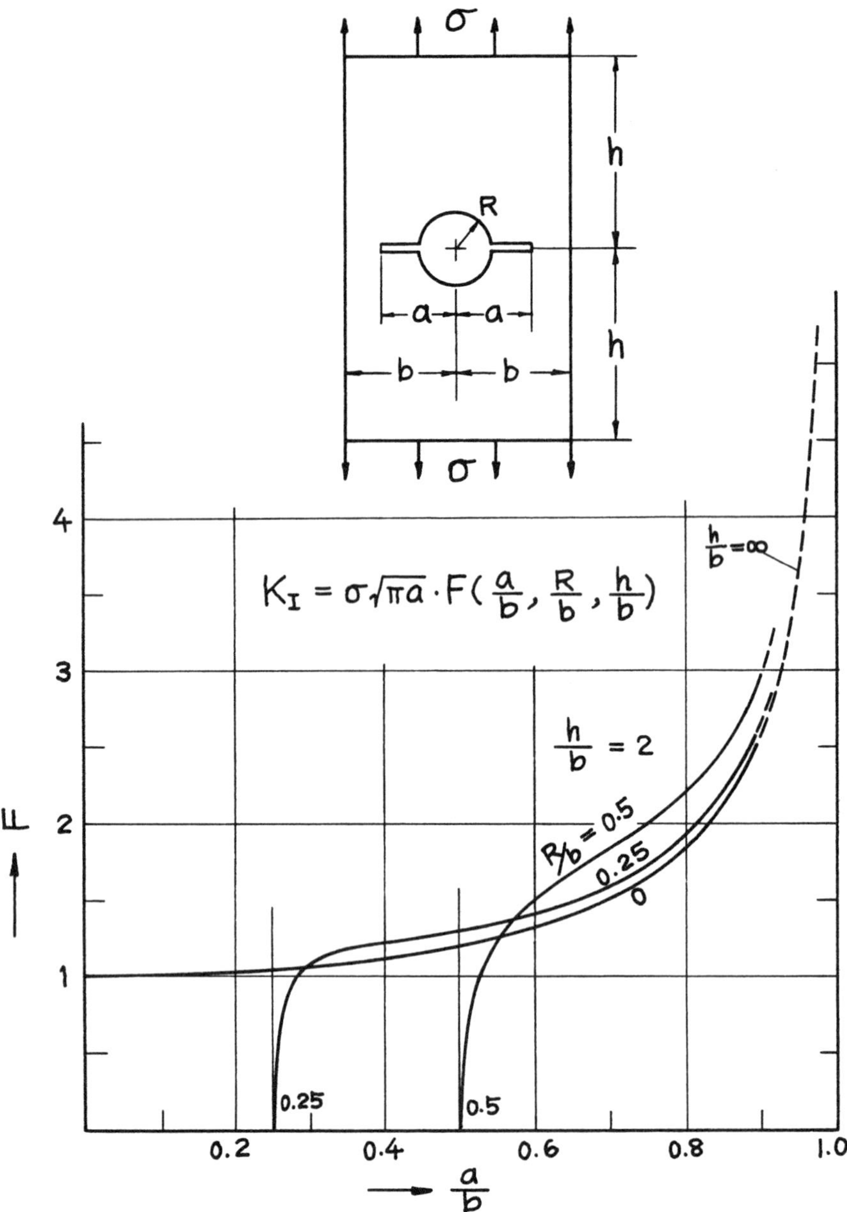

Method: Boundary Collocation Method.
Accuracy: Curves were drawn based on the results having better than 0.1% accuracy.
Reference: **Newman 1971**
See also **page 19.11**.

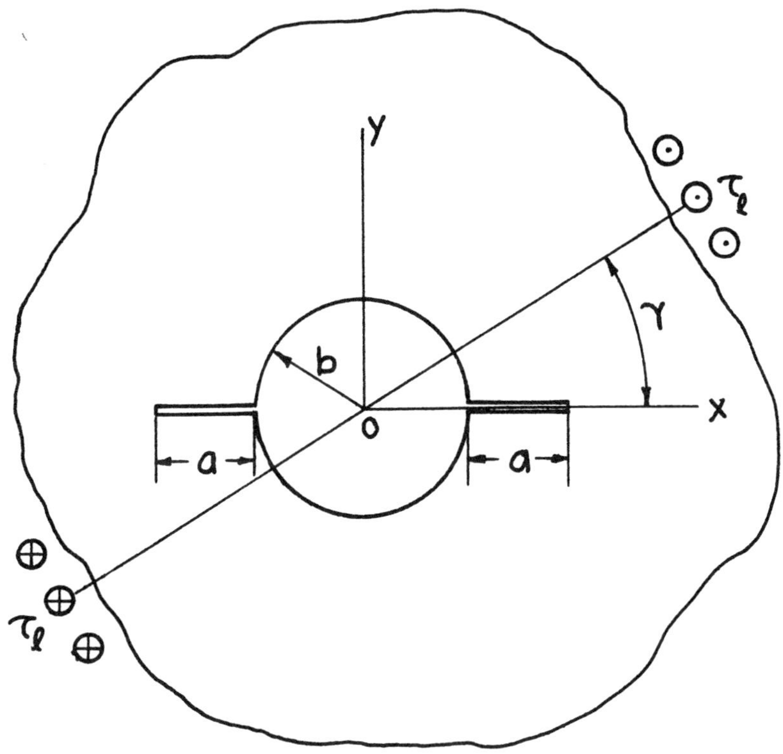

$$K_{III} = \tau_\ell \sqrt{\pi a} \frac{\sqrt{\left(1 + {}^{b}\!/_{a}\right)^4 - \left({}^{b}\!/_{a}\right)^4}}{\left(1 + {}^{b}\!/_{a}\right)^{3/2}} \sin \gamma$$

$$K_{III}\left({}^{b}\!/_{a} \rightarrow \infty\right) = 2\tau_\ell \sqrt{\pi a} \sin \gamma$$

$$K_{III}\left({}^{b}\!/_{a} \rightarrow 0\right) = \tau_\ell \sqrt{\pi a} \sin \gamma$$

Method: Conformal Mapping
Accuracy: Exact
Reference: **Sih 1965a**

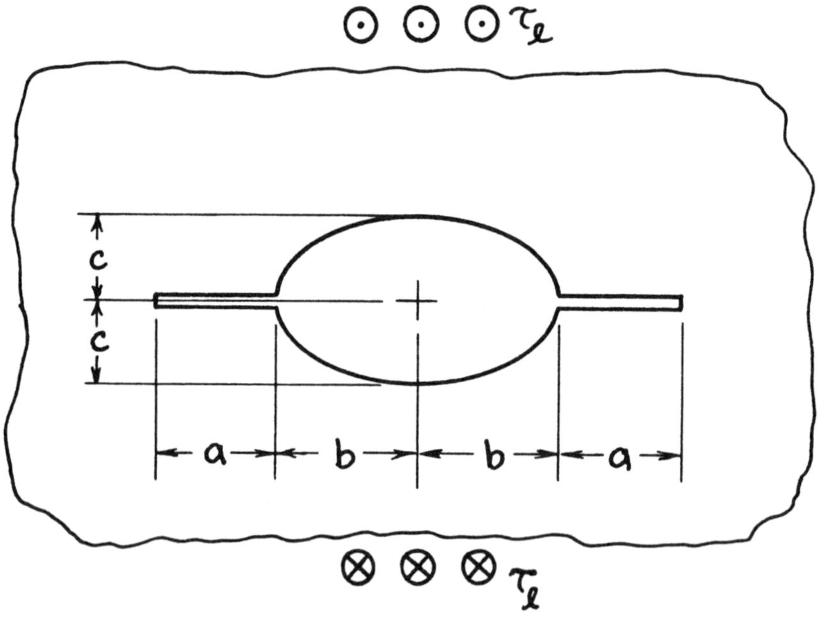

$$K_{III} = \tau_\ell \sqrt{\pi(b+c)} \cdot \sqrt{\frac{s^4 - 1}{2s\left(s^2 - \frac{b-c}{b+c}\right)}}$$

where

$$s = \frac{b + a + \sqrt{(b+a)^2 - \left(b^2 - c^2\right)}}{b+c}$$

For Circular Hole,

$$c = b = R : K_{III} = \tau_\ell \sqrt{\pi(R+a)} \sqrt{1 - \left(\frac{R}{R+a}\right)^4}$$

Method: Conformal Mapping
Accuracy: Exact
Reference: **Yokobori 1972 (or Kamei 1974)**

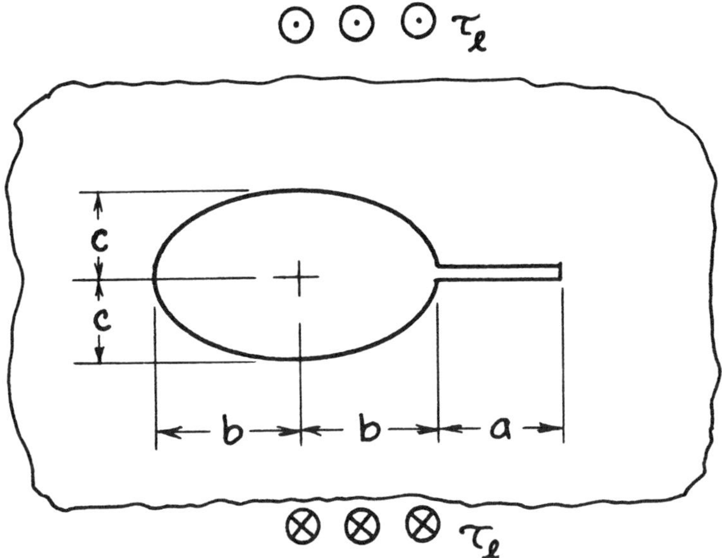

$$K_{III} = \tau_\ell \sqrt{\pi(b+c)}\,\frac{s+1}{2}\sqrt{\frac{s^2-1}{s\left(s^2-\frac{b-c}{b+c}\right)}}$$

where

$$s = \frac{b+a+\sqrt{(b+a)^2-\left(b^2-c^2\right)}}{b+c}$$

For Circular Hole,

$$c = b = R : K_{III} = \tau_\ell \sqrt{\pi a}\cdot\frac{1}{\sqrt{2}}\left(1+\frac{R}{R+a}\right)^{3/2}$$

Method: Conformal Mapping
Accuracy: Exact
Reference: **Yokobori 1972 (or Kamei 1974)**

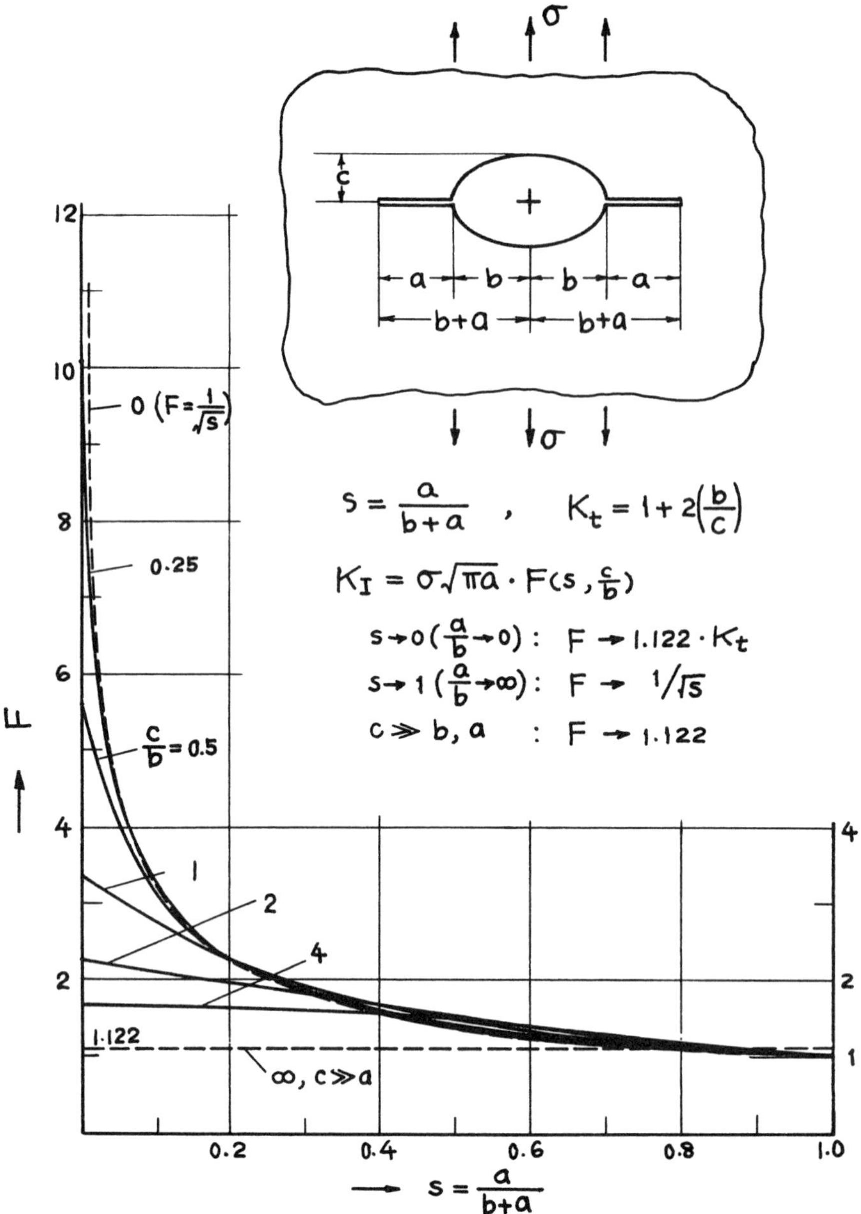

Methods: Boundary Collocation Method (Newman), Body Force Method (Nishitani)
Accuracy: Curves were drawn based on the results having better than 0.1% accuracy
References: **Nishitani 1969; Newman 1971**

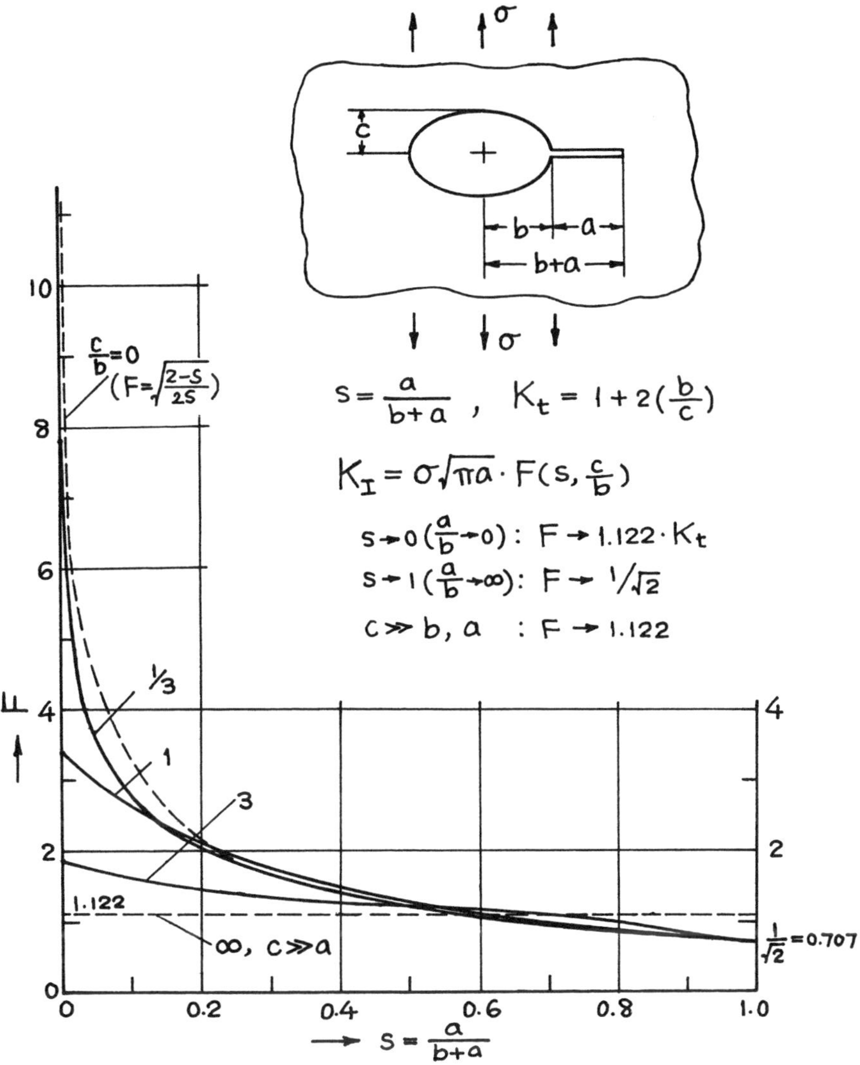

Method: Integral Equation
Accuracy: 3%
Reference: **Berezhnitskii 1966**

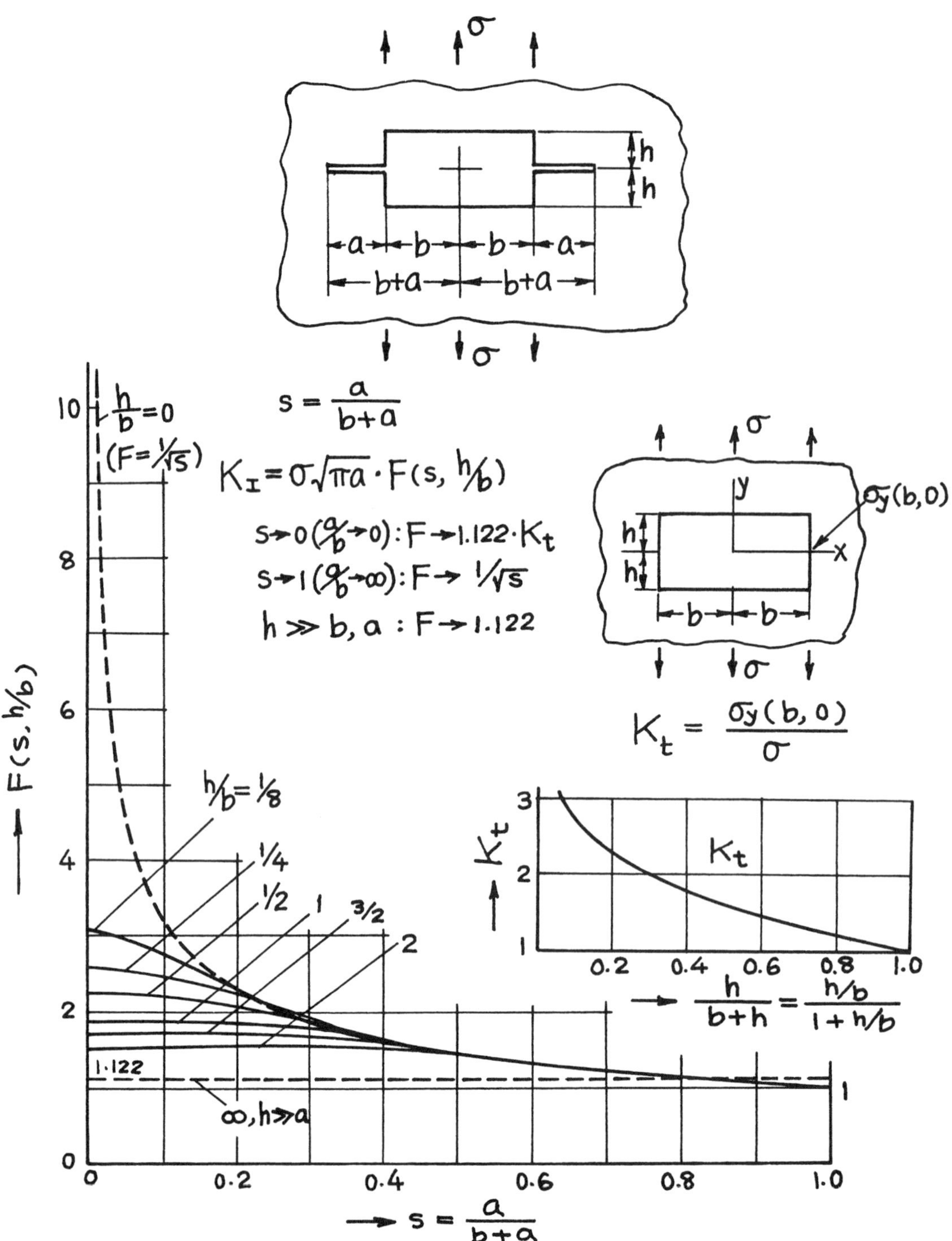

Methods: Mapping Collocation Method (Neal, $s \geq 0.25$), Estimated (Tada, $s < 0.25$)
Accuracy: Expected to be better than 5% for $s < 0.25$, better than 2% for $s \geq 0.25$
References: **Neal 1970; Tada 1973; Savin 1961, 1968** (K_t values)

NOTE: Neal's results for small cracks ($s < 0.25$) appear to be too large.

$\ell = c + a$

$$K_I = \sigma\sqrt{\pi a}\,F\left(\frac{c}{b},\frac{\ell}{b}\right)$$

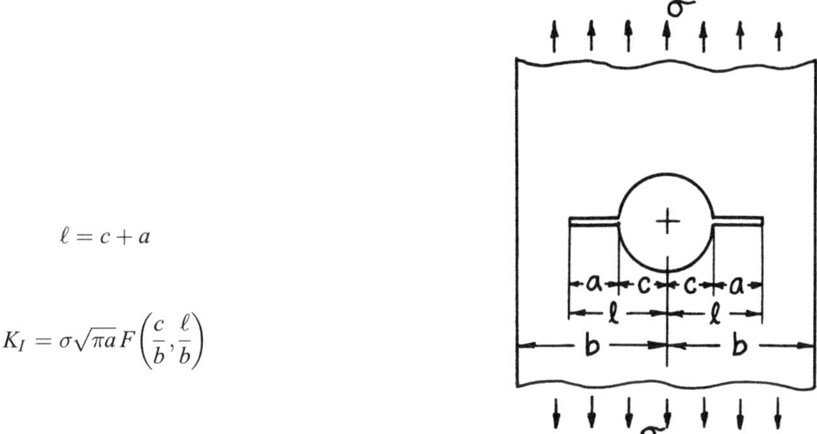

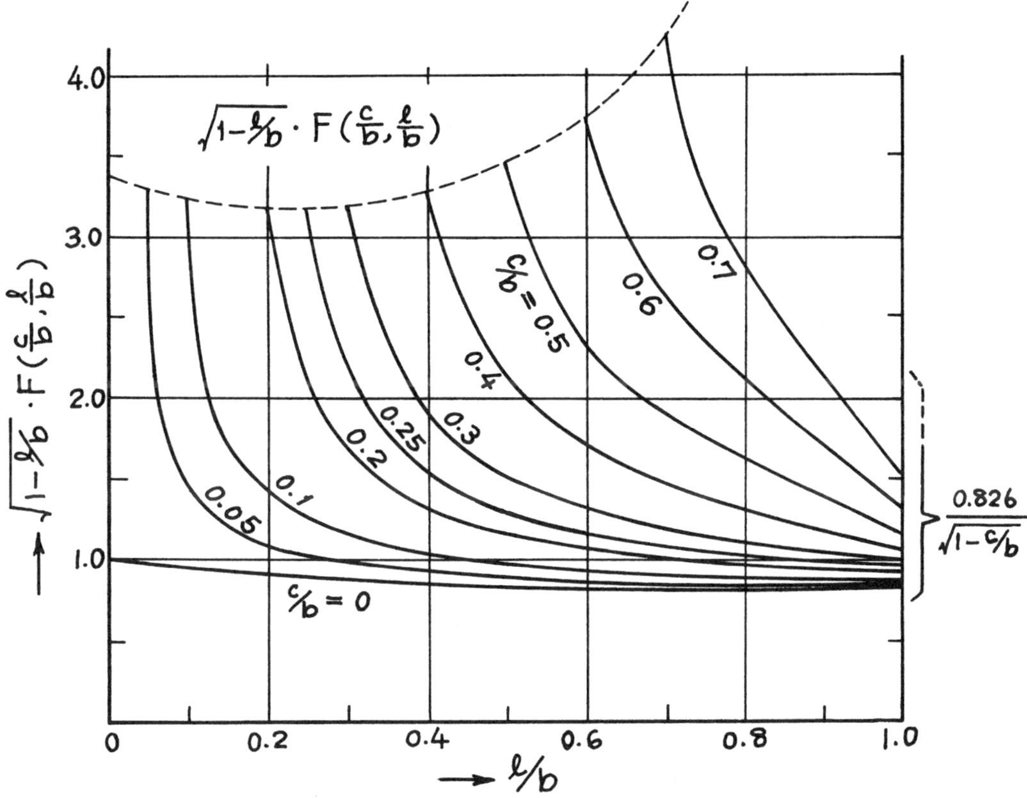

Methods: Boundary Cllocation ($c/b = 0.25,\ 0.5$; Newman), Estimated by Interpolation (c/b other than 0.25, 0.5; Tada)

Accuracy: Accurate for $c/b = 0.25,\ 0.5$; better than 5% for other values of c/b

References: **Newman 1971; Tada 1973**

NOTE: For $c/b << 1$, see **page 19.1 and 2.1**. See also **p. 19.4**.

$$\ell = c + a$$

$$K_I = \sigma\sqrt{\pi a} \cdot F\left(\frac{d}{c}, \frac{c}{b}, \frac{\ell}{b}\right)$$

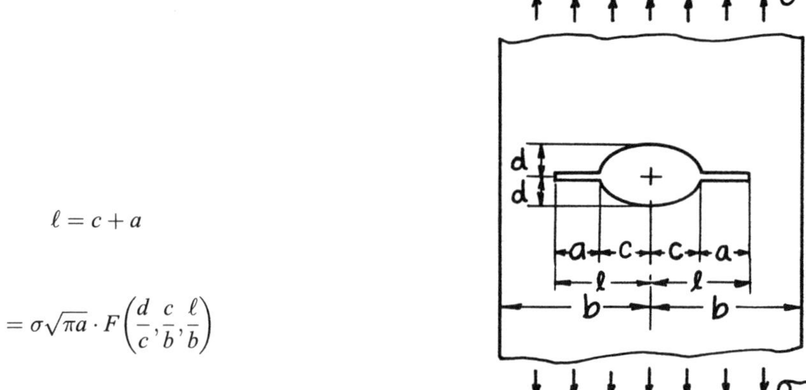

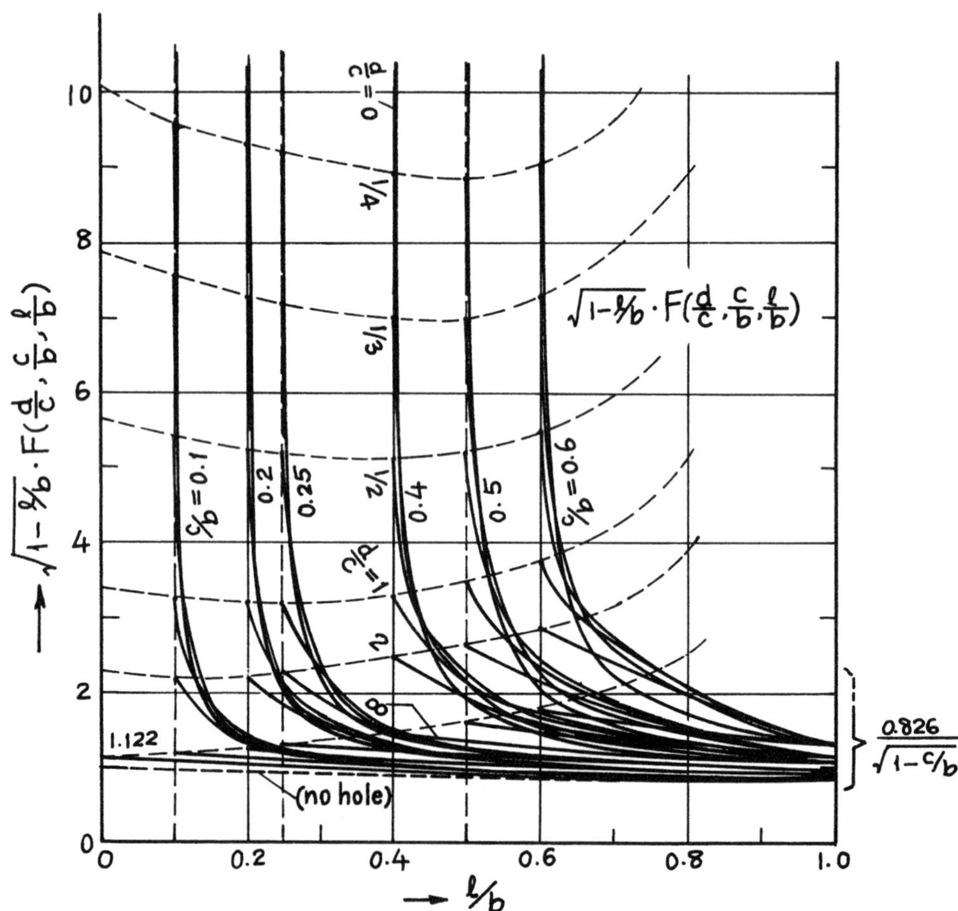

Method: Estimated by Interpolation
Accuracy: Better than 5 % for any d/c, c/b, and ℓ/b
Reference: **Tada 1973**

NOTE: For $c/b << 1$, see **page 19.8**.

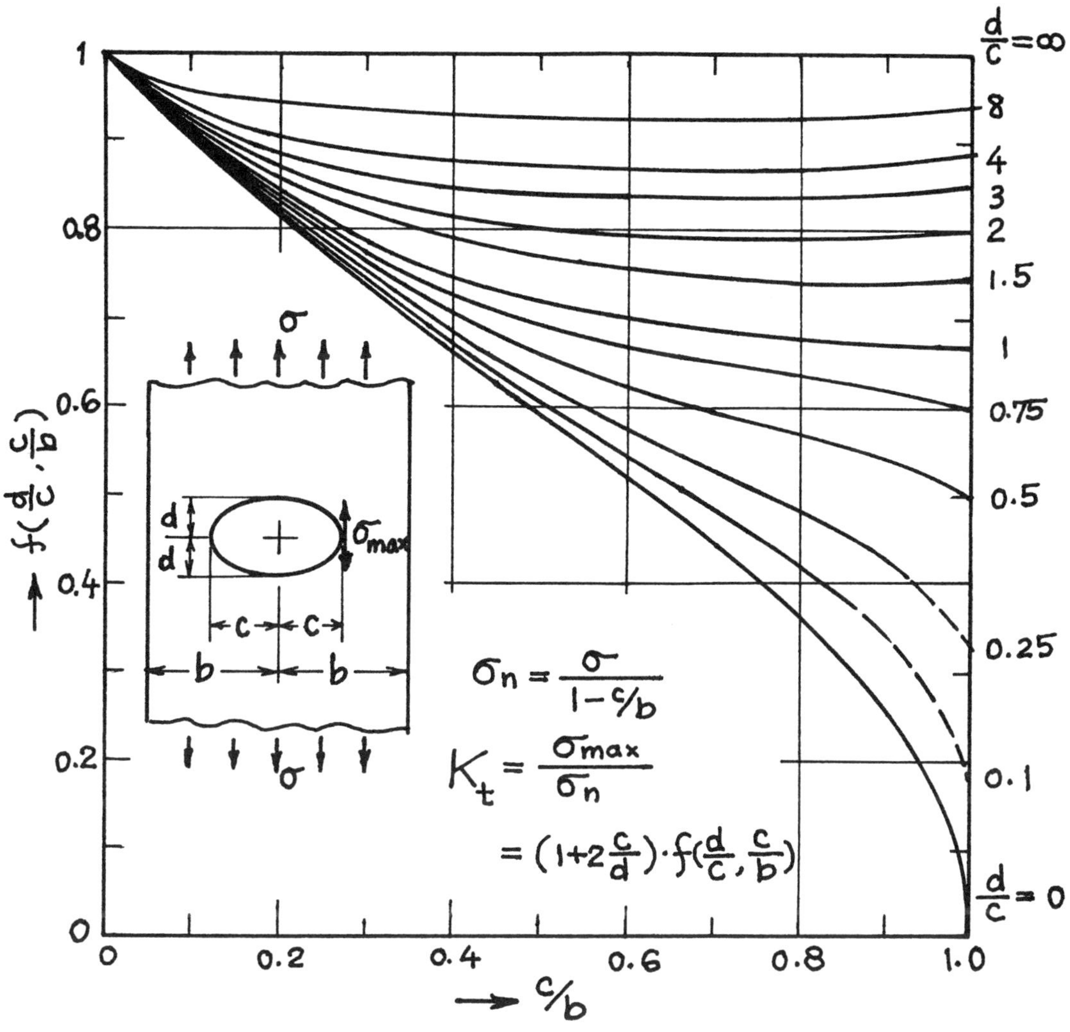

Method: Estimated by Interpolation

Accuracy: $f\left({}^{d}/{}_{c}=\infty,\ {}^{c}/{}_{b}\right)$ Exact

$f\left({}^{d}/{}_{c}=1,\ {}^{c}/{}_{b}\right), f\left({}^{d}/{}_{c}\rightarrow 0,\ {}^{c}/{}_{b}\right)$ Accurate

For all other values of ${}^{d}/{}_{c}$, accuracy is expected to be much better than 5 %.

References: **Tada 1974**

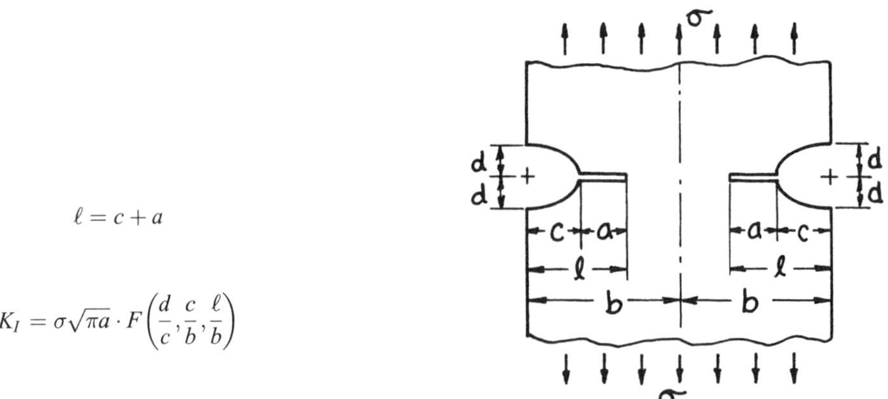

$$\ell = c + a$$

$$K_I = \sigma\sqrt{\pi a} \cdot F\left(\frac{d}{c}, \frac{c}{b}, \frac{\ell}{b}\right)$$

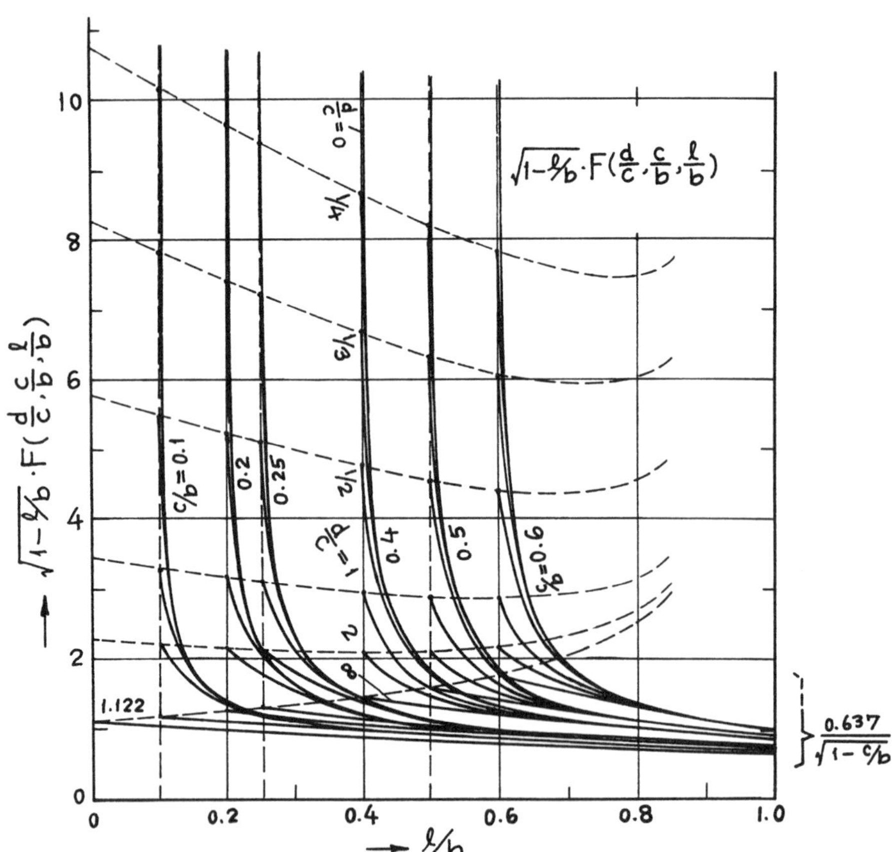

Method: Estimated by Interpolation
Accuracy: Better than 5 % for any d/c, c/b, and ℓ/b
Reference: **Tada 1973**

NOTE: For $c/b << 1$, see **pages 19.13 and 2.6**.

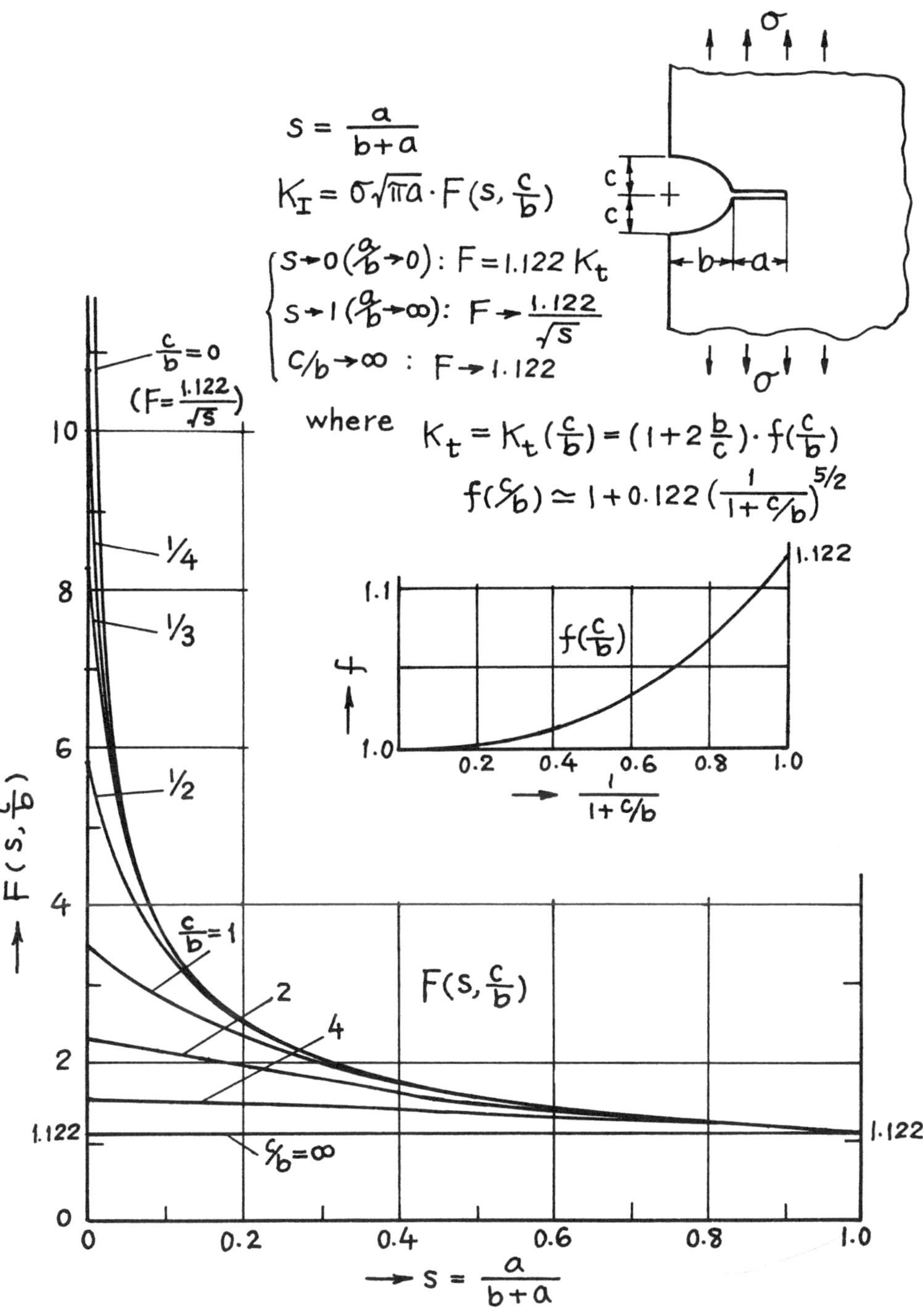

Methods: Stress Relaxation (Superposition) (Nishitani; $c/b = 1/2$, 1, 2 and $0.2 \leq a/b \leq 1$), Estimated by Interpolation (Tada)

Accuracy: Better than 2%

References: **Nishitani 1973; Tada 1973**

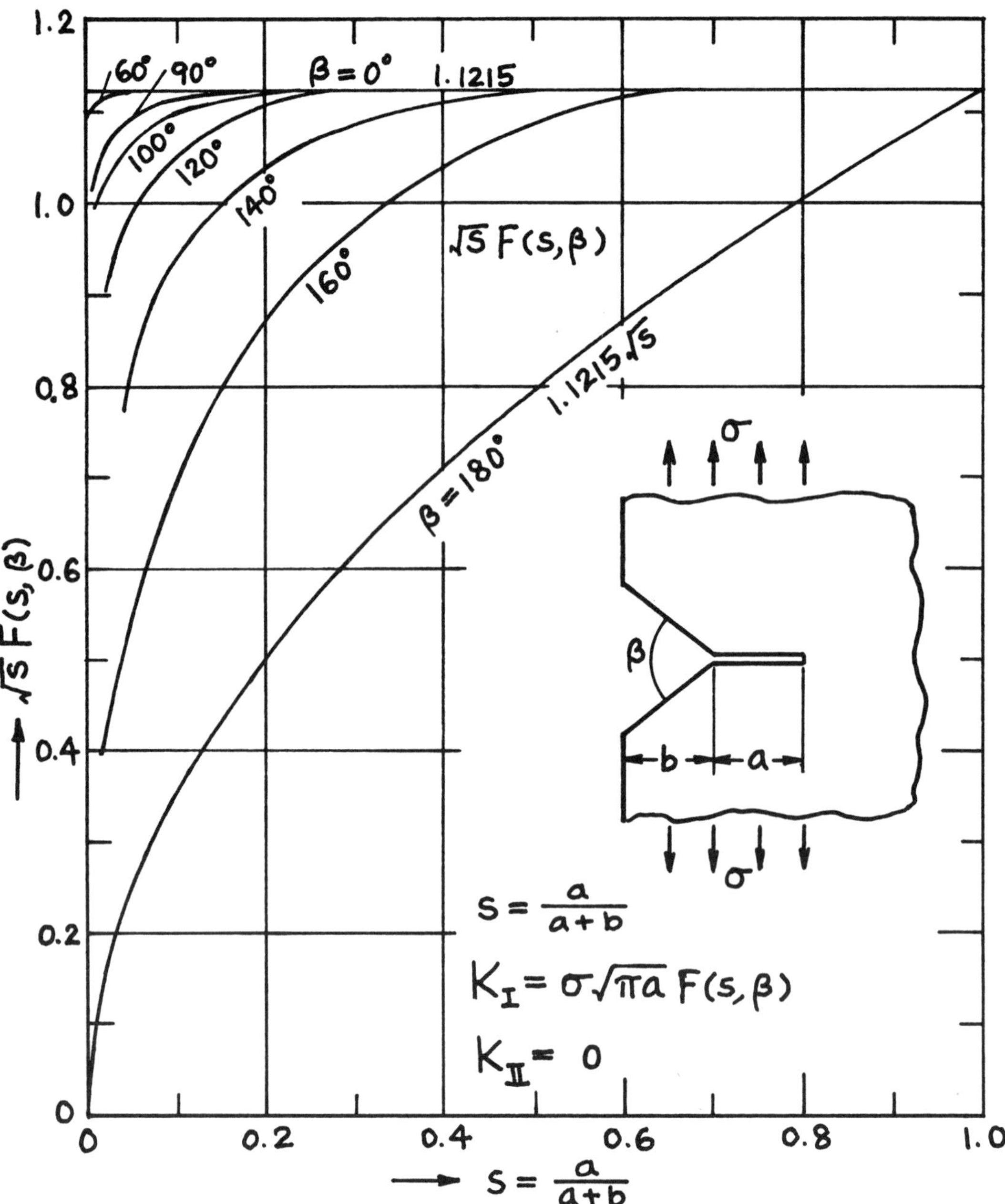

Method: Conformal Mapping

Accuracy: Curves are based on accurate (0.1%) numerical values (**Hasebe 1978**).

References: **Hasebe 1978; Tada 2000**

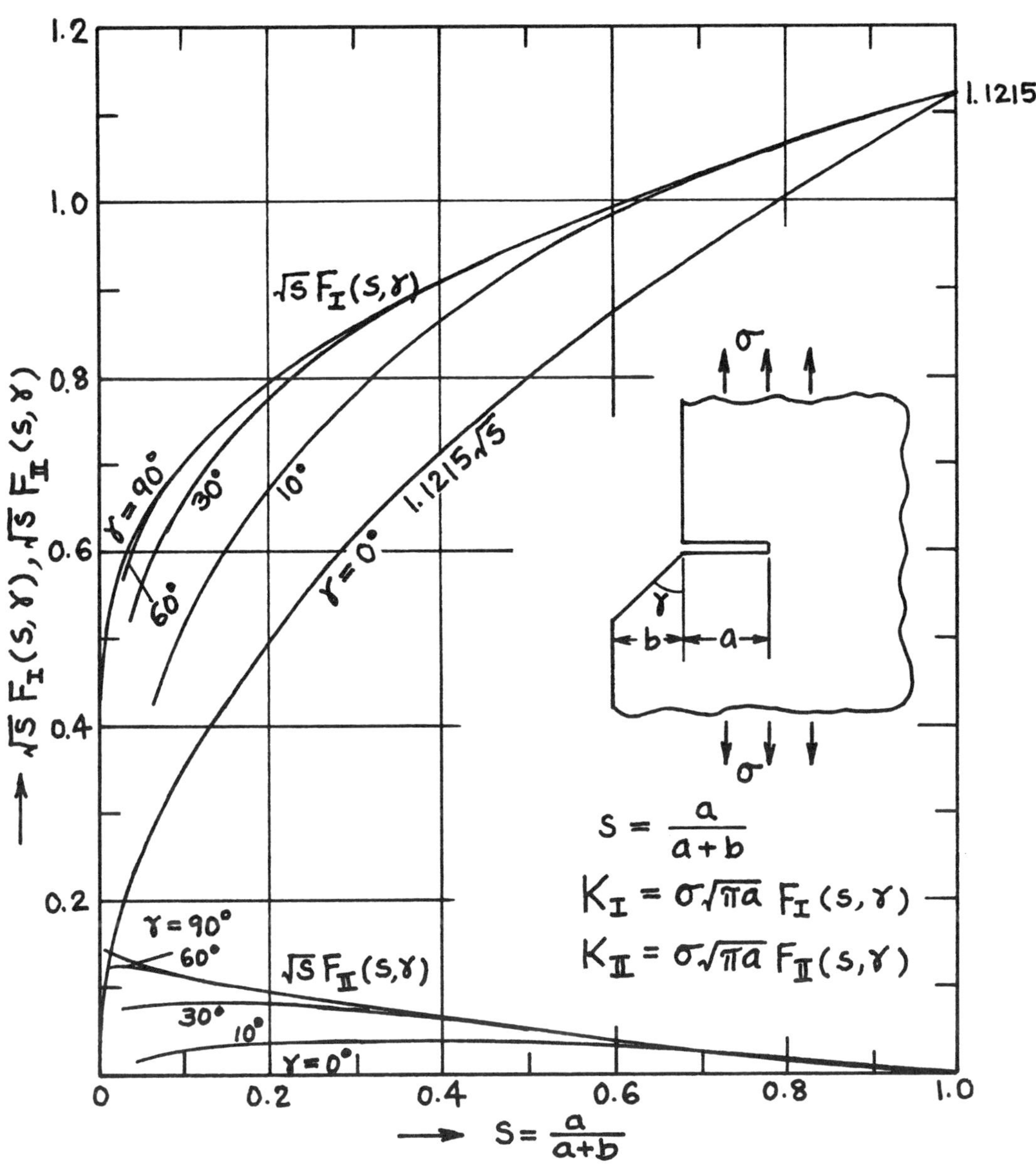

Method: Conformal Mapping

Accuracy: Curves are based on accurate (0.1%) numerical values (**Hasebe 1980**).

References: **Hasebe 1980; Tada 2000**

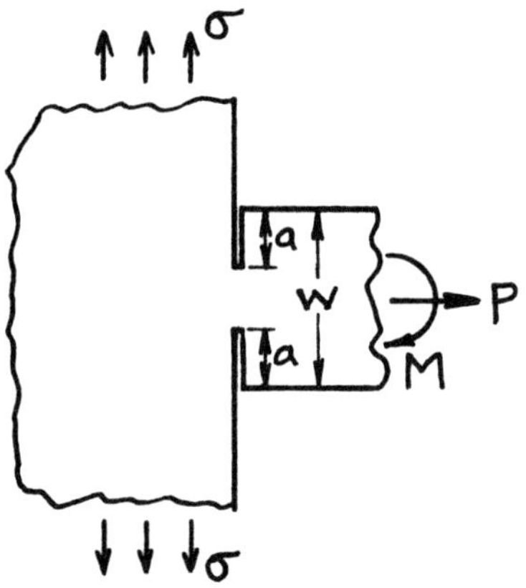

$$A = {}^{2a}/_{W}$$

$$K_I = \sqrt{\pi a}\left[\sigma F_{I\sigma}(A) + \frac{P}{W}F_{IP}(A) + \frac{M}{W^2}F_{IM}(A)\right]$$

$$K_{II} = \sqrt{\pi a}\left[\sigma F_{II\sigma}(A) + \frac{P}{W}F_{IIP}(A) + \frac{M}{W^2}F_{IIM}(A)\right]$$

$$F_{I\sigma}(A) = \sqrt{\frac{1-A}{A}} \cdot 0.15\left[(1-A)^5 - A(1-A)^8\right]$$

$$F_{II\sigma}(A) = \sqrt{\frac{1-A}{A}} \cdot \left[0.11 + 0.14\sqrt{1-(1-A)^2} - 0.016(1-A)^8\right]$$

$$F_{IP}(A) = \frac{1}{\sqrt{A(1-A)}} \cdot \left[0.637 - 0.224(1-A)^4 + 0.75\sqrt{A}(1-A)^3(0.3-A)\right]$$

$$F_{IIP}(A) = \frac{1}{\sqrt{A(1-A)}} \cdot 0.145(1-A)$$

$$F_{IM}(A) = \frac{1}{\sqrt{A}(1-A)^{3/2}} \cdot \left[0.424 - 0.15(1-A)^4 + 0.2\sqrt{A}(1-A)^{9/2}\right]$$

$$F_{IIM}(A) = \frac{1}{\sqrt{A}(1-A)^{3/2}} \cdot \left[0.85(1-A)^4 + 0.12A^2(1-A)^3\right]$$

Method: Conformal Mapping
Accuracy: F_{IP}, F_{IM} 0.5%; $F_{II\sigma}$, F_{IIM} 1%; $F_{I\sigma}$, F_{IIp} 2%
References: **Hasebe 1981; Tada 2000**

NOTE: All approximate formulas (**Tada 2000**) are based on accurate (0.1%) numerical values (**Hasebe 1981**).

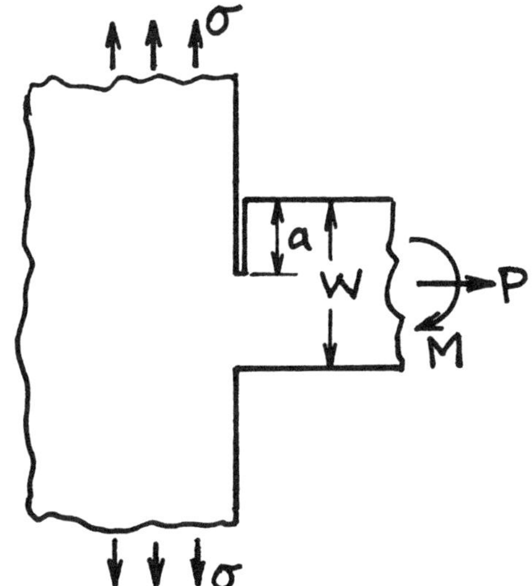

$$A = {}^{a}/_{W}$$

$$K_I = \sqrt{\pi a}\left[\sigma F_{I\sigma}(A) + \frac{P}{W}F_{IP}(A) + \frac{M}{W^2}F_{IM}(A)\right]$$

$$K_{II} = \sqrt{\pi a}\left[\sigma F_{II\sigma}(A) + \frac{P}{W}F_{IIP}(A) + \frac{M}{W^2}F_{IIM}(A)\right]$$

$$F_{I\sigma}(A) = \sqrt{\frac{1-A}{A}}\cdot\left[0.018 + 0.069\, e^{-12.5\left(\frac{A}{1-A}\right)}\right]$$

$$F_{II\sigma}(A) = \sqrt{\frac{1-A}{A}}\cdot\left[0.156 - 0.067\, e^{-8.9\left(\frac{A}{1-A}\right)}\right]$$

$$F_{IP}(A) = \frac{1}{\sqrt{A}(1-A)^{3/2}}\cdot\left[0.379 + 0.624A - 0.062\, e^{-12\left(\frac{A}{1-A}\right)}\right]$$

$$F_{IIP}(A) = \frac{1}{\sqrt{A}(1-A)^{3/2}}\cdot\left[0.126 - 0.24A - 0.023(1-A)^5\right]$$

$$F_{IM}(A) = \frac{1}{\sqrt{A}(1-A)^{3/2}}\cdot\left[2.005 - 0.72\, e^{-9\left(\frac{A}{1-A}\right)}\right]$$

$$F_{IIM}(A) = \frac{1}{\sqrt{A}(1-A)^{3/2}}\cdot\left[-0.228 + (1-A)^4\left(0.577 - 0.2A + 0.8A^2\right)\right]$$

Method: Conformal Mapping
Accuracy: F_{IP}, F_{IIP}, F_{IIM} 1%; $F_{I\sigma}$, $F_{II\sigma}$, F_{IM} 2%
References: **Hasebe 1987; Tada 2000**

NOTE: All approximate formulas (**Tada 2000**) are based on accurate (0.1%) numerical values (**Hasebe 1987**).

$$s = \frac{a}{R+a}$$

$$K_I = \sigma\sqrt{\pi a} \cdot F\left(s, \frac{d}{R}\right)$$

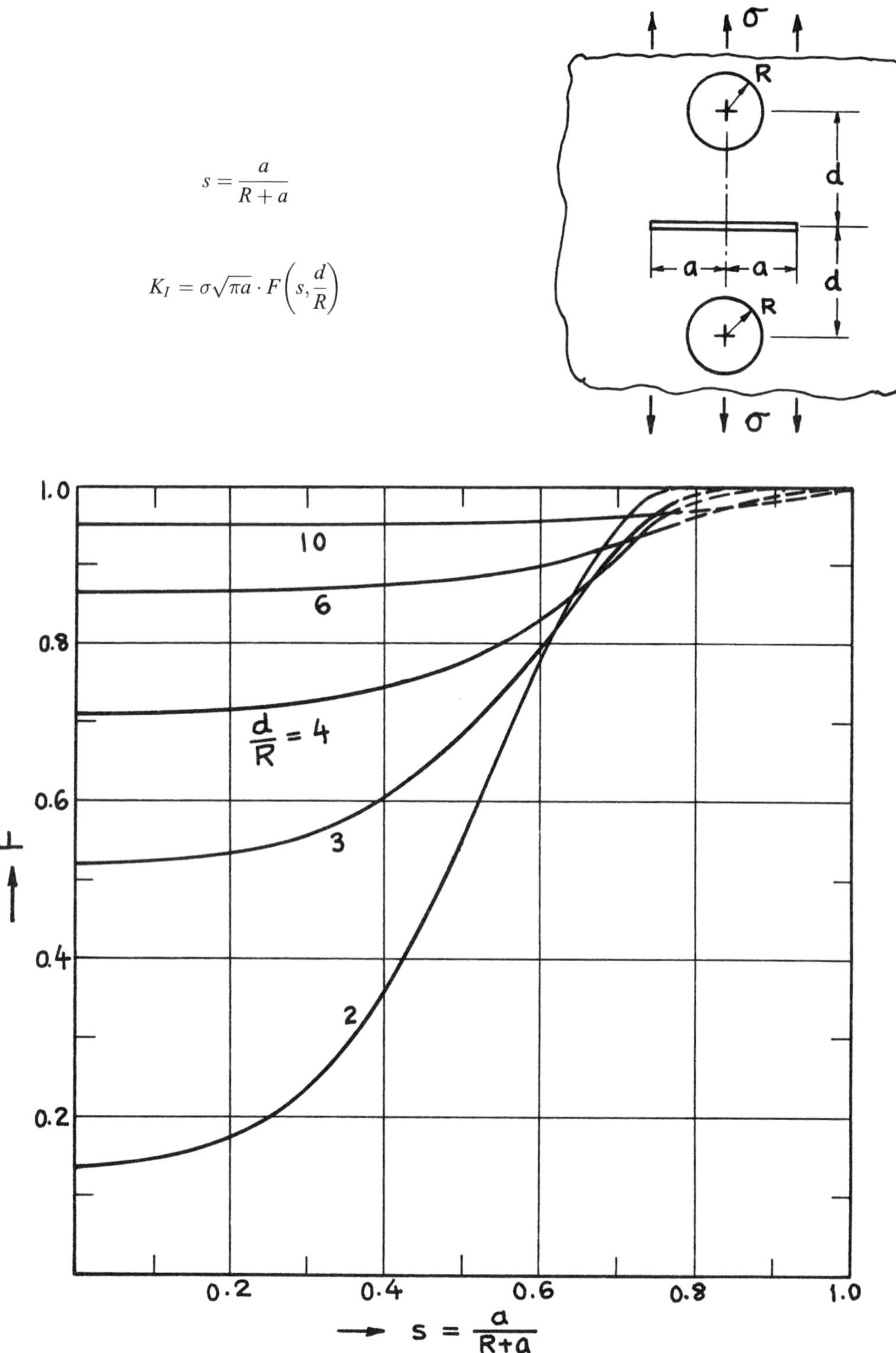

Method: Boundary Collocation Method
Accuracy: Curves (solid lines) were drawn based on the results having 0.1% accuracy.
Reference: **Newman 1971**

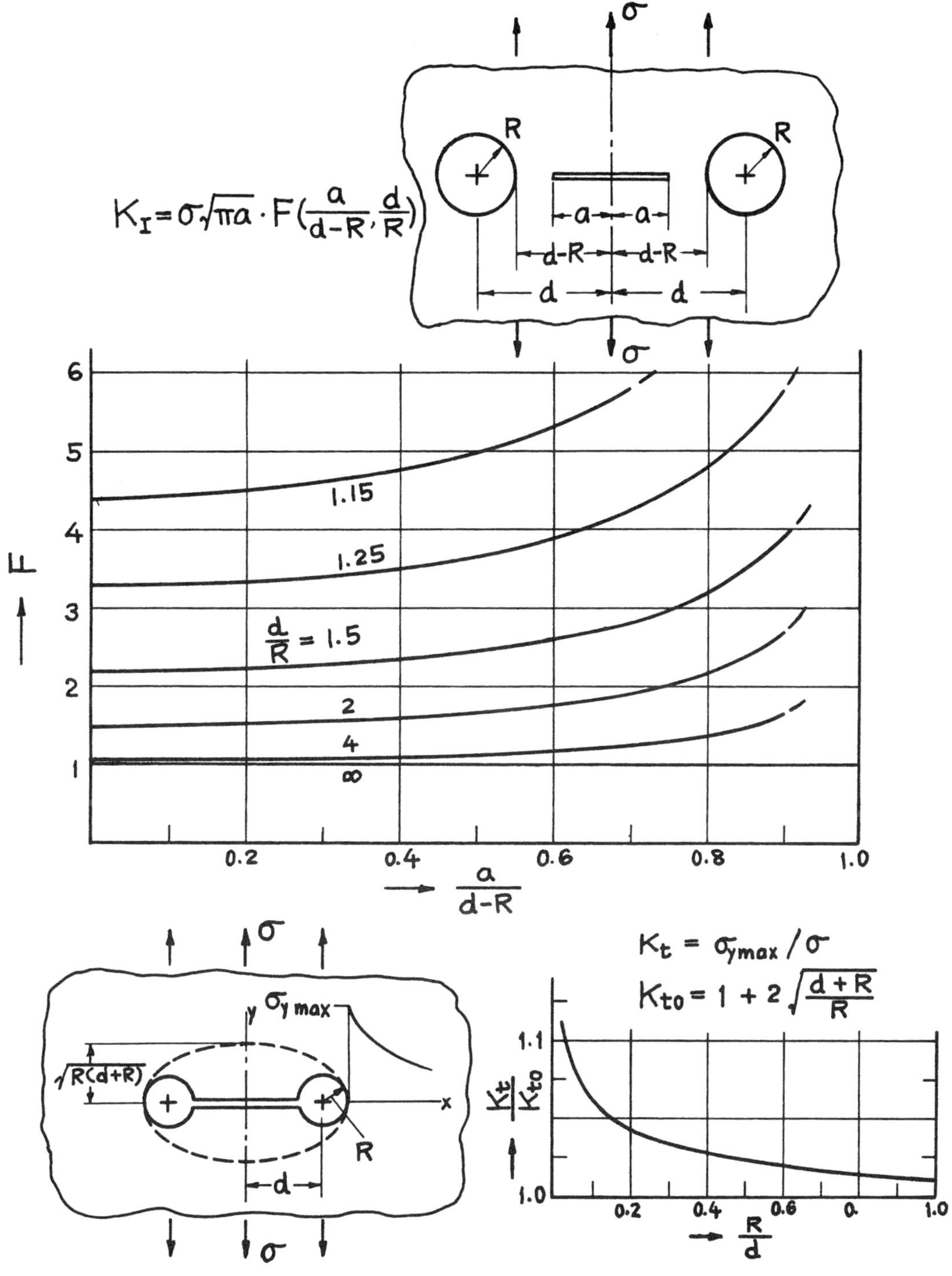

Methods: Boundary Collocation Method (Newman), Expansions of Complex Stress Potentials (Isida)
Accuracy: Curves were drawn based on the results having 0.1% accuracy.
References: **Newman 1971; Isida 1973**

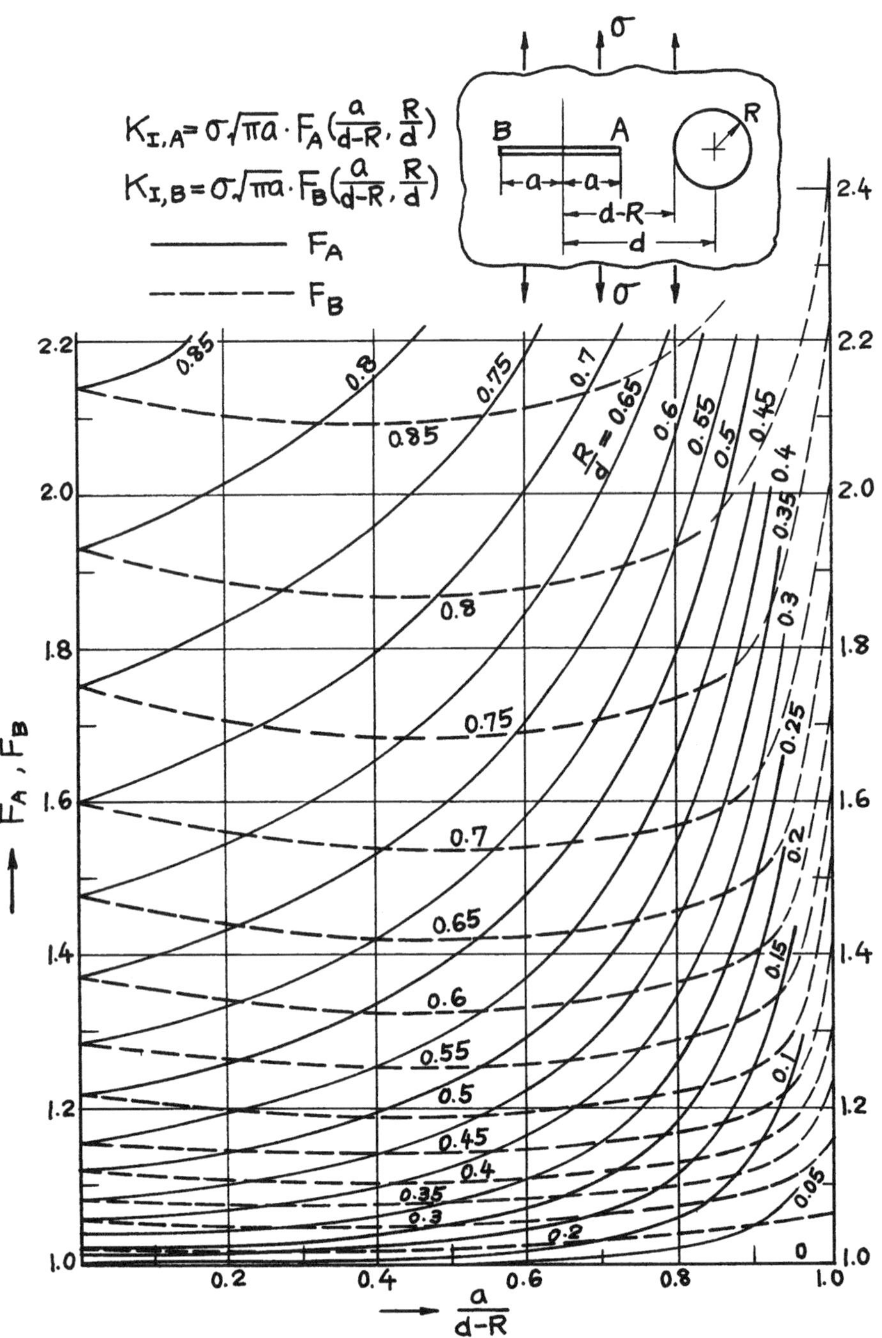

Method: Expansions of Complex Stress Potentials
Accuracy: Curves were drawn based on the results having 0.1% accuracy (thick solid and dashed lines).
Reference: **Isida 1970a**

NOTE: The values of F_B for $a/(d-R) \to 1$ were taken from **p. 19.2** where $\lambda = 0$.

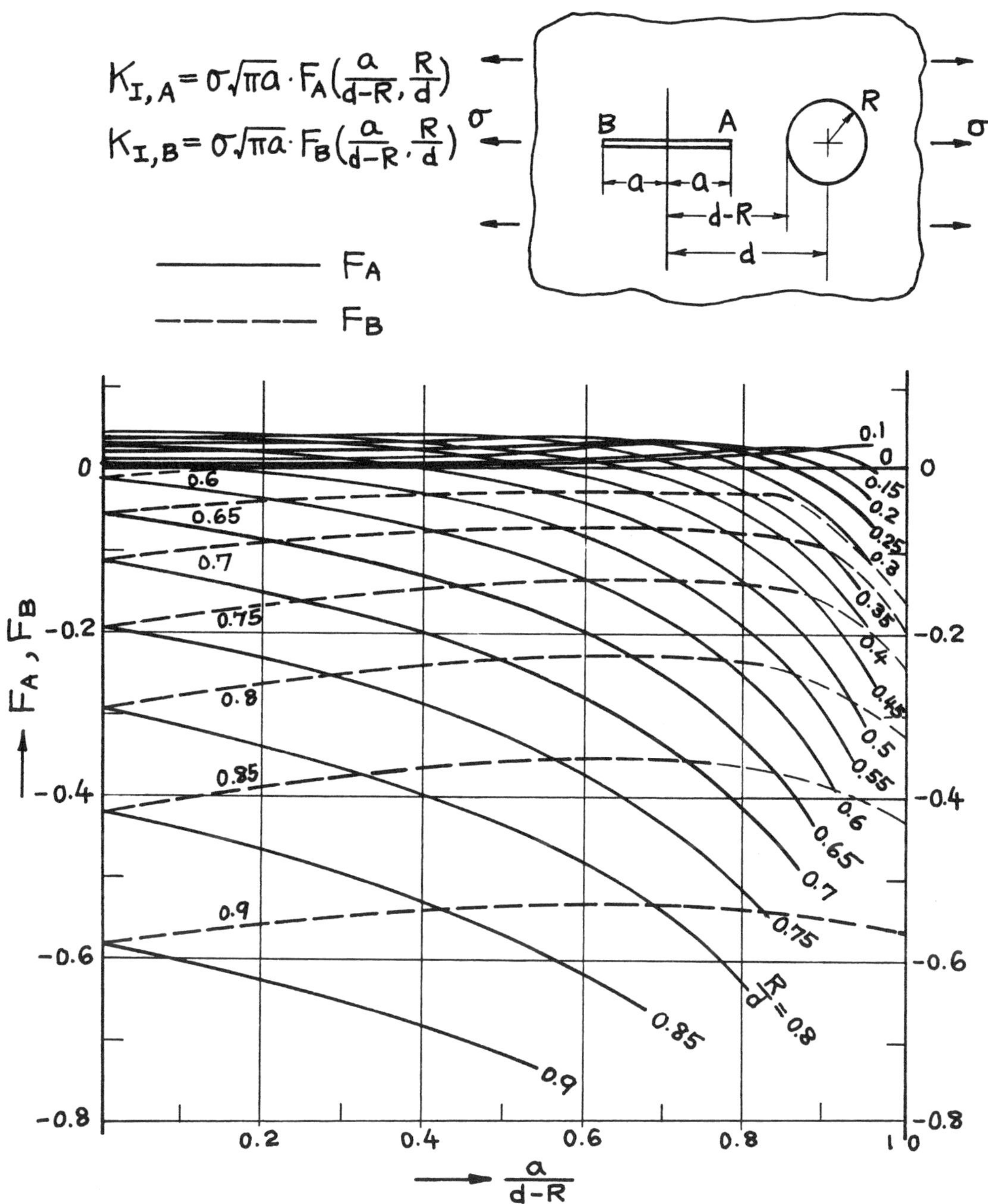

Method: Expansions of Complex Stress Potentials

Accuracy: Curves were drawn based on the results having 0.1% accuracy (thick solid and dashed lines).

Reference: **Isida 1970a**

NOTE: The values of F_B for $a/(d-R) \to 1$ were taken from **p. 19.2** where $\lambda = 1$.

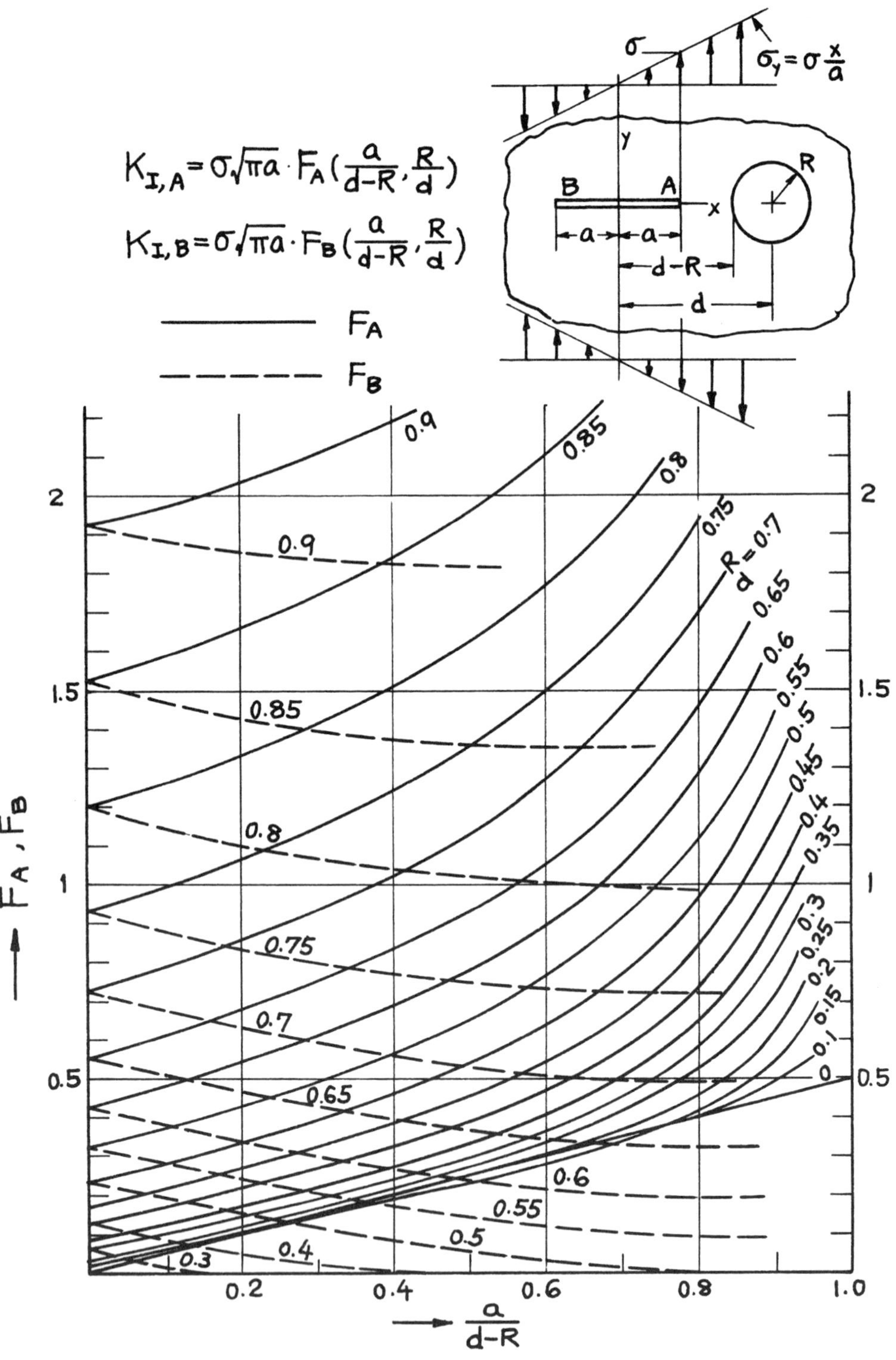

Method: Expansions of Complex Stress Potentials

Accuracy: Curves were drawn based on the results having 0.1% accuracy.

Reference: **Isida 1970a**

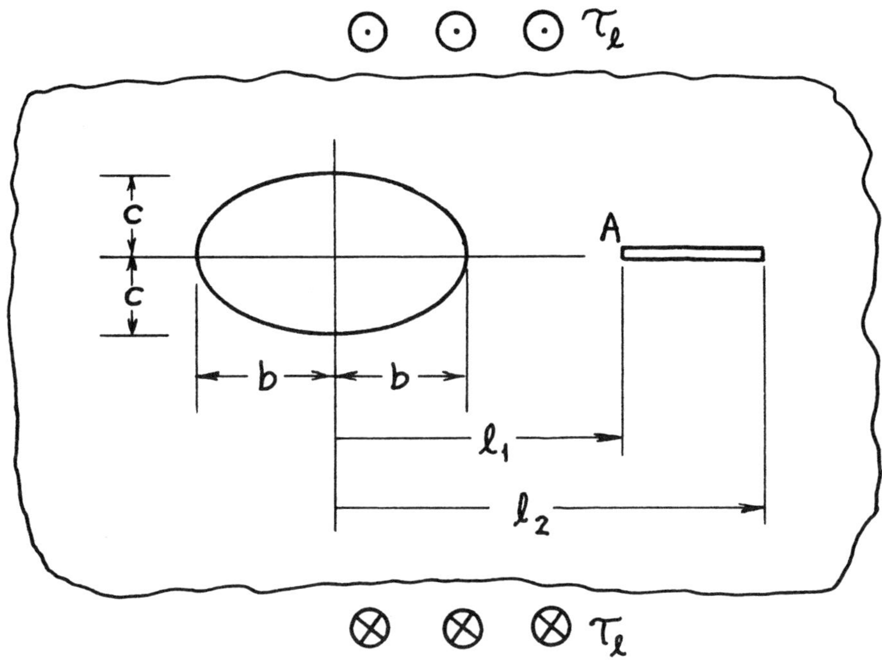

$$K_{III,A} = \tau_\ell \sqrt{\pi(b+c)} \frac{s_1 \left[-\frac{1}{2}\left(s_1 - \frac{1}{s_1}\right)\left(s_1 - \frac{1}{s_1} + s_2 - \frac{1}{s_2}\right) + \left(s_1 s_2 + \frac{1}{s_1 s_2} - 2\right) \frac{E(k)}{K(k)} \right]}{\sqrt{\left(s_1^2 - \frac{b-c}{b+c}\right)(s_2 - s_1)\left(s_1 - \frac{1}{s_1}\right)\left(s_1 - \frac{1}{s_2}\right)}}$$

where

$$s_i = \frac{1}{b+c}\left[\ell_i + \sqrt{\ell_i^2 - \left(b^2 - c^2\right)}\right], \quad i = 1, 2$$

$$k = \frac{s_2 - s_1}{s_1 s_2 - 1}$$

$$K(k) = \int_0^{\pi/2} \frac{d\varphi}{\sqrt{1 - k^2 \sin^2 \varphi}}$$

$$E(k) = \int_0^{\pi/2} \sqrt{1 - k^2 \sin^2 \varphi}\, d\varphi$$

For Circular Hole,

$$b = c = R: \quad s_1 = \frac{\ell_1}{R}, \quad s_2 = \frac{\ell_2}{R}$$

Method: Complex Potentials
Accuracy: Exact
Reference: **Yokobori 1972 (or Kamei 1974)**

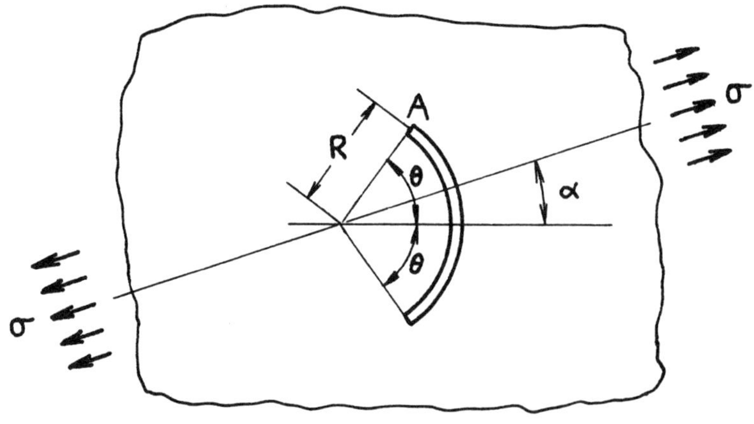

$$K_{IA} = \frac{\sigma\sqrt{\pi R \sin\theta}}{2}\left[\cos\frac{\theta}{2}\cdot\frac{1-\cos 2\alpha\left(\sin\frac{\theta}{2}\right)^2\left(\cos\frac{\theta}{2}\right)^2}{1+\left(\sin\frac{\theta}{2}\right)^2}+\sin 2\alpha\left(\sin\frac{\theta}{2}\right)^3+\cos\left(2\alpha-\frac{3\theta}{2}\right)\right]$$

$$K_{IIA} = \frac{\sigma\sqrt{\pi R \sin\theta}}{2}\left[\sin\frac{\theta}{2}\cdot\frac{1-\cos 2\alpha\left(\sin\frac{\theta}{2}\right)^2\left(\cos\frac{\theta}{2}\right)^2}{1+\left(\sin\frac{\theta}{2}\right)^2}-\sin 2\alpha\left(\sin\frac{\theta}{2}\right)^2\cos\frac{\theta}{2}-\sin\left(2\alpha-\frac{3\theta}{2}\right)\right]$$

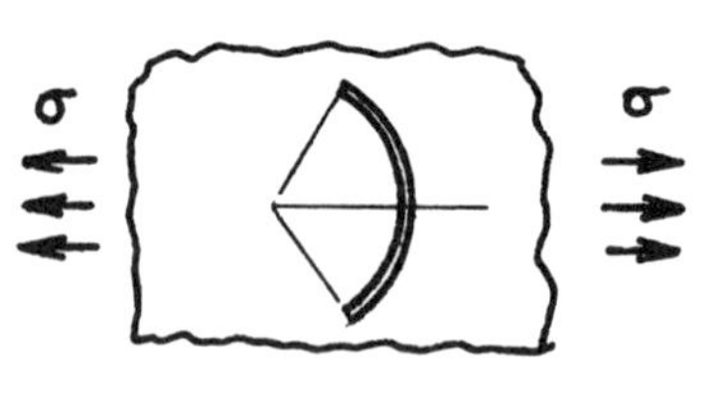

$$\begin{Bmatrix} K_I \\ K_{II} \end{Bmatrix} = \frac{\sigma\sqrt{\pi R \sin\theta}}{2\left\{1+\left(\sin\frac{\theta}{2}\right)^2\right\}} \cdot \begin{Bmatrix} \cos\frac{\theta}{2}\left[2-4\left(\sin\frac{\theta}{2}\right)^2-3\left(\sin\frac{\theta}{2}\right)^4\right] \\ \sin\frac{\theta}{2}\left[4-2\left(\sin\frac{\theta}{2}\right)^2-3\left(\sin\frac{\theta}{2}\right)^4\right] \end{Bmatrix}$$

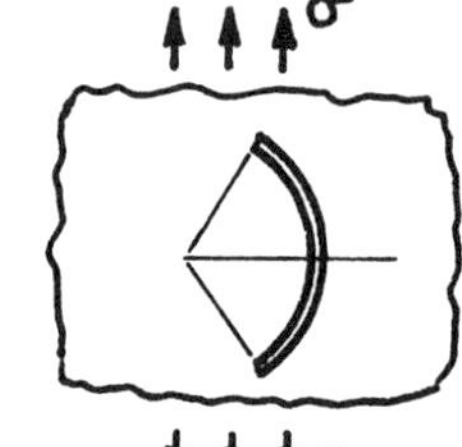

$$\begin{Bmatrix} K_I \\ K_{II} \end{Bmatrix} = \frac{\sigma\sqrt{\pi R \sin\theta}}{2\left\{1+\left(\sin\frac{\theta}{2}\right)^2\right\}} \cdot \begin{Bmatrix} \cos\frac{\theta}{2}\left(\sin\frac{\theta}{2}\right)^2\left[4+3\left(\sin\frac{\theta}{2}\right)^2\right] \\ \sin\frac{\theta}{2}\left[-2+2\left(\sin\frac{\theta}{2}\right)^2+3\left(\sin\frac{\theta}{2}\right)^4\right] \end{Bmatrix}$$

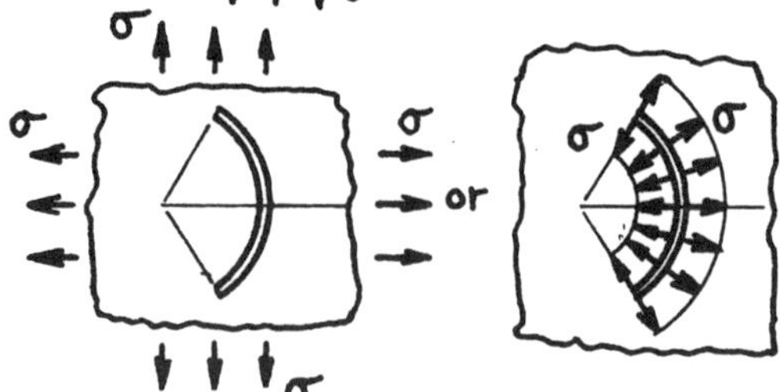

$$\begin{Bmatrix} K_I \\ K_{II} \end{Bmatrix} = \frac{\sigma\sqrt{\pi R \sin\theta}}{1+\left(\sin\frac{\theta}{2}\right)^2}\begin{Bmatrix} \cos\frac{\theta}{2} \\ \sin\frac{\theta}{2} \end{Bmatrix}$$

Method: Conformal Mapping (Muskhelishvili)
Accuracy: Exact
References: **Muskhelishvili 1933; Sih 1962b**

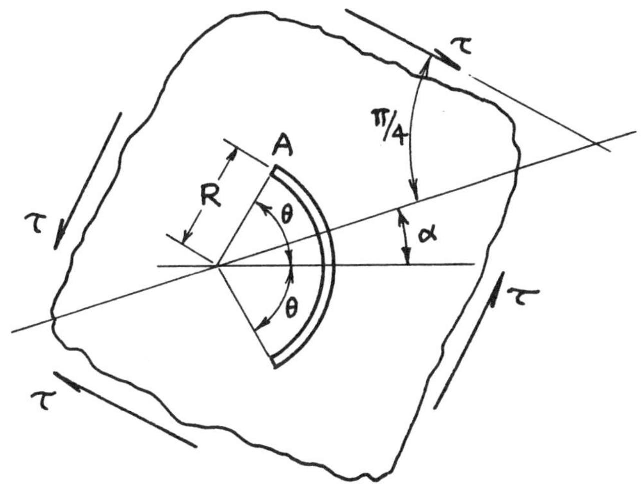

$$K_{IA} = \tau\sqrt{\pi R \sin\theta}\left[\frac{-\cos 2\alpha\left(\sin\frac{\theta}{2}\right)^{2}\left(\cos\frac{\theta}{2}\right)^{3}}{1+\left(\sin\frac{\theta}{2}\right)^{2}} + \sin 2\alpha\left(\sin\frac{\theta}{2}\right)^{3} + \cos\left(2\alpha - \frac{3\theta}{2}\right)\right]$$

$$K_{IIA} = \tau\sqrt{\pi R \sin\theta}\left[\frac{-\cos 2\alpha\left(\sin\frac{\theta}{2}\right)^{3}\left(\cos\frac{\theta}{2}\right)^{2}}{1+\left(\sin\frac{\theta}{2}\right)^{2}} - \sin 2\alpha\left(\sin\frac{\theta}{2}\right)^{2}\cos\frac{\theta}{2} - \sin\left(2\alpha - \frac{3\theta}{2}\right)\right]$$

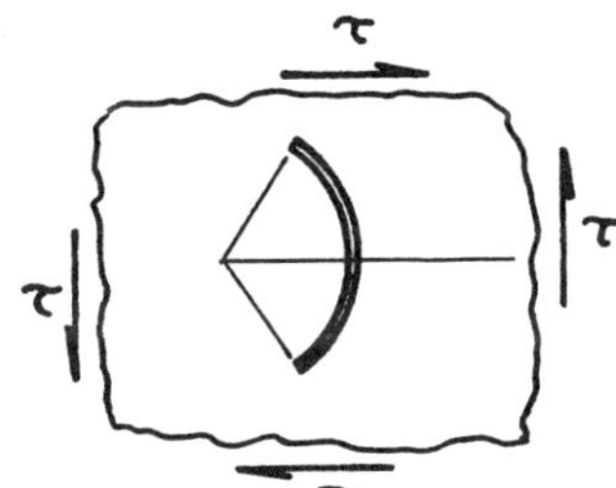

$$\begin{Bmatrix} K_I \\ K_{II} \end{Bmatrix} = \tau\sqrt{\pi R \sin\theta}\begin{Bmatrix} 3\sin\frac{\theta}{2}\left(\cos\frac{\theta}{2}\right)^{2} \\ \cos\frac{\theta}{2}\left[3\left(\sin\frac{\theta}{2}\right)^{2} - 1\right] \end{Bmatrix}$$

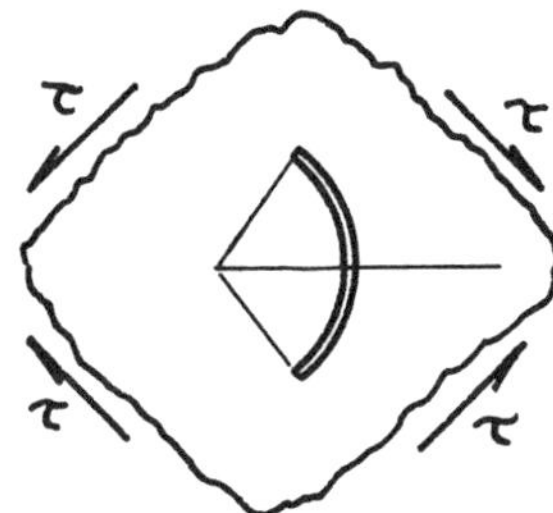

$$\begin{Bmatrix} K_I \\ K_{II} \end{Bmatrix} = \frac{\tau\sqrt{\pi R \sin\theta}}{1+\left(\sin\frac{\theta}{2}\right)^{2}}\begin{Bmatrix} \cos\frac{\theta}{2}\left[1 - 4\left(\sin\frac{\theta}{2}\right)^{2} - 3\left(\sin\frac{\theta}{2}\right)^{4}\right] \\ \sin\frac{\theta}{2}\left[3 - 2\left(\sin\frac{\theta}{2}\right)^{2} - 3\left(\sin\frac{\theta}{2}\right)^{4}\right] \end{Bmatrix}$$

Method: Superposition of **page 21.1**
Accuracy: Exact
Reference: **Tada 1985**

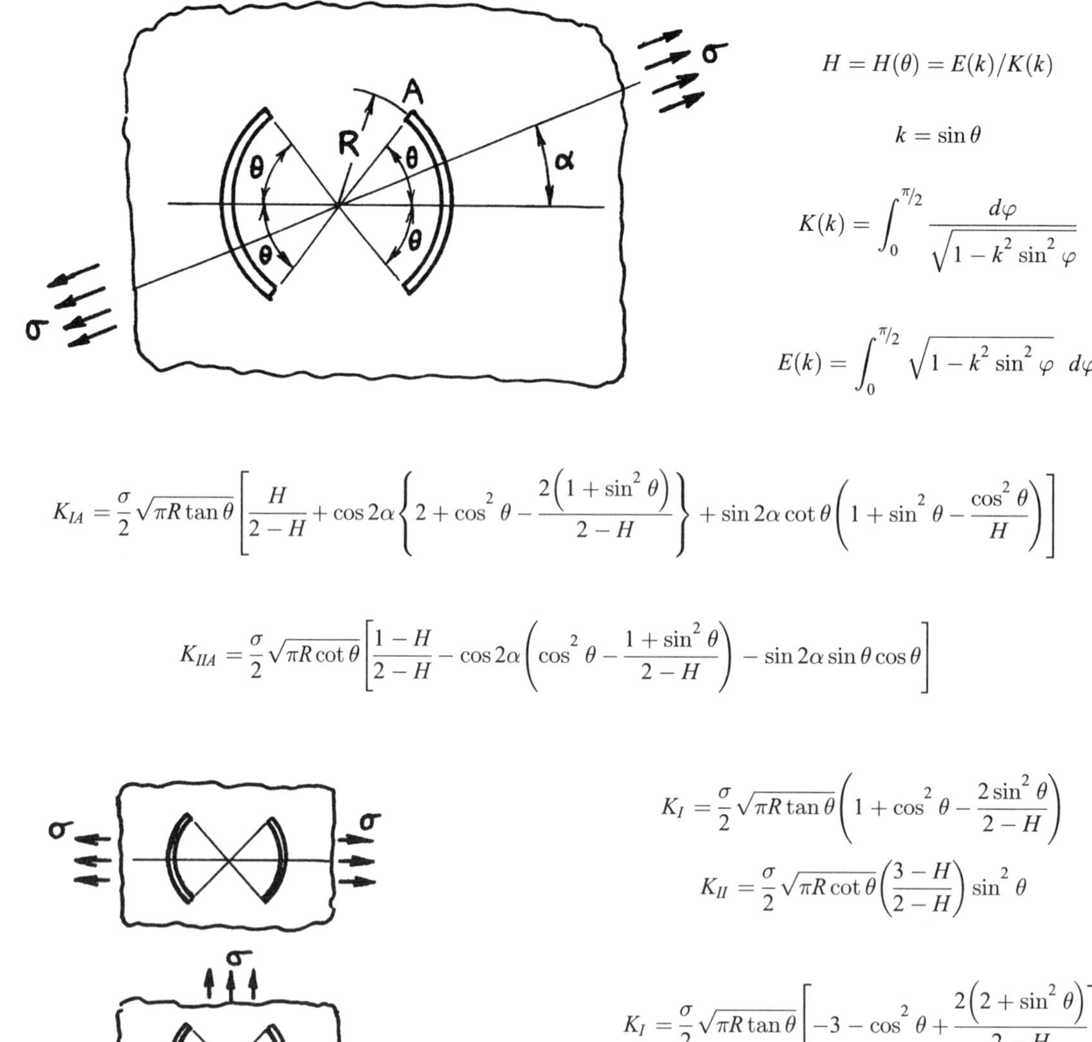

$$H = H(\theta) = E(k)/K(k)$$

$$k = \sin\theta$$

$$K(k) = \int_0^{\pi/2} \frac{d\varphi}{\sqrt{1 - k^2 \sin^2\varphi}}$$

$$E(k) = \int_0^{\pi/2} \sqrt{1 - k^2 \sin^2\varphi}\; d\varphi$$

$$K_{IA} = \frac{\sigma}{2}\sqrt{\pi R \tan\theta}\left[\frac{H}{2-H} + \cos 2\alpha\left\{2 + \cos^2\theta - \frac{2\left(1+\sin^2\theta\right)}{2-H}\right\} + \sin 2\alpha \cot\theta\left(1 + \sin^2\theta - \frac{\cos^2\theta}{H}\right)\right]$$

$$K_{IIA} = \frac{\sigma}{2}\sqrt{\pi R \cot\theta}\left[\frac{1-H}{2-H} - \cos 2\alpha\left(\cos^2\theta - \frac{1+\sin^2\theta}{2-H}\right) - \sin 2\alpha \sin\theta\cos\theta\right]$$

$$K_I = \frac{\sigma}{2}\sqrt{\pi R \tan\theta}\left(1 + \cos^2\theta - \frac{2\sin^2\theta}{2-H}\right)$$

$$K_{II} = \frac{\sigma}{2}\sqrt{\pi R \cot\theta}\left(\frac{3-H}{2-H}\right)\sin^2\theta$$

$$K_I = \frac{\sigma}{2}\sqrt{\pi R \tan\theta}\left[-3 - \cos^2\theta + \frac{2\left(2+\sin^2\theta\right)}{2-H}\right]$$

$$K_{II} = \frac{\sigma}{2}\sqrt{\pi R \cot\theta}\left[1 + \cos^2\theta - \frac{2+\sin^2\theta}{2-H}\right]$$

$$K_I = \sigma\sqrt{\pi R \tan\theta}\left(\frac{H}{2-H}\right)$$

$$K_{II} = \sigma\sqrt{\pi R \cot\theta}\left(\frac{1-H}{2-H}\right)$$

Method: Conformal Mapping (Muskhelishvili)
Accuracy: Exact
Reference: **Tada 1985**

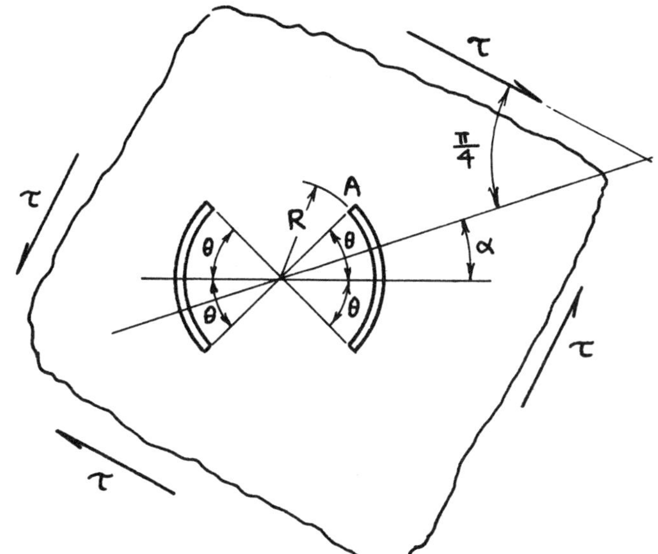

$$H = H(\theta) = E(k)/K(k)$$

$$k = \sin\theta$$

$$K(k) = \int_0^{\pi/2} \frac{d\varphi}{\sqrt{1 - k^2 \sin^2\varphi}}$$

$$E(k) = \int_0^{\pi/2} \sqrt{1 - k^2 \sin^2\varphi}\; d\varphi$$

$$K_{IA} = \tau\sqrt{\pi R \tan\theta}\left[\cos 2\alpha\left\{2 + \cos^2\theta - \frac{2\left(1 + \sin^2\theta\right)}{2 - H}\right\} + \sin 2\alpha \cot\theta\left(1 + \sin^2\theta - \frac{\cos^2\theta}{H}\right)\right]$$

$$K_{IIA} = \tau\sqrt{\pi R \cot\theta}\left[-\cos 2\alpha\left(\cos^2\theta - \frac{1 + \sin^2\theta}{2 - H}\right) - \sin 2\alpha \sin\theta\cos\theta\right]$$

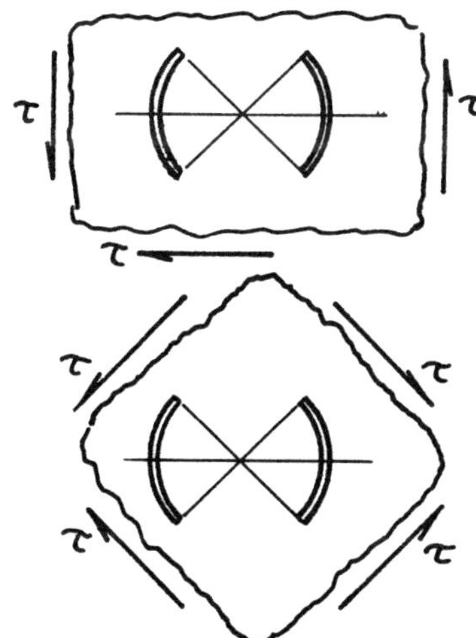

$$K_I = \tau\sqrt{\pi R \cot\theta}\left(1 + \sin^2\theta - \frac{\cos^2\theta}{H}\right)$$

$$K_{II} = \tau\sqrt{\pi R \cot\theta}(-\sin\theta\cos\theta)$$

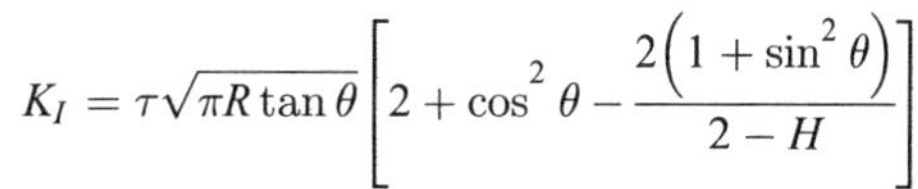

$$K_I = \tau\sqrt{\pi R \tan\theta}\left[2 + \cos^2\theta - \frac{2\left(1 + \sin^2\theta\right)}{2 - H}\right]$$

$$K_{II} = \tau\sqrt{\pi R \cot\theta}\left[-\cos^2\theta + \frac{1 + \sin^2\theta}{2 - H}\right]$$

Method: Superposition of **page 21.3**
Accuracy: Exact
Reference: **Tada 1985**

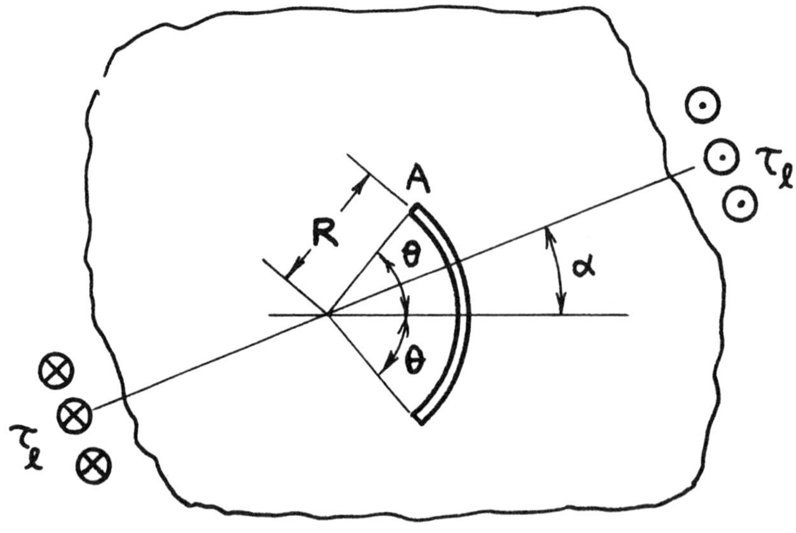

$$K_{IIIA} = \tau_\ell \sqrt{\pi R \sin\theta}\ \cos\left(\alpha - \frac{\theta}{2}\right)$$

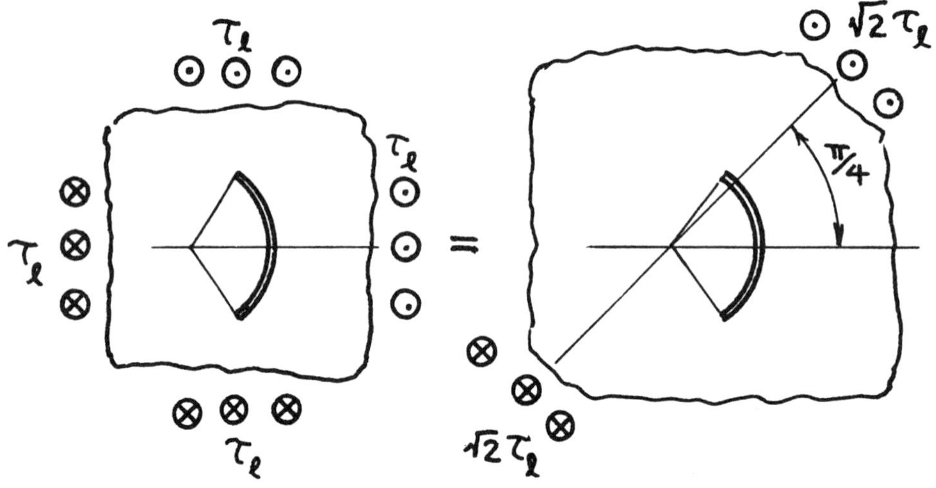

$$\begin{aligned} K_{III} &= \tau_\ell \sqrt{\pi R \sin\theta}\left(\cos\frac{\theta}{2} + \sin\frac{\theta}{2}\right) \\ &= \sqrt{2}\tau_\ell \sqrt{\pi R \sin\theta}\cos\left(\frac{\pi}{4} - \frac{\theta}{2}\right) \end{aligned}$$

Method: Conformal Mapping
Accuracy: Exact
References: **Sih 1965a**

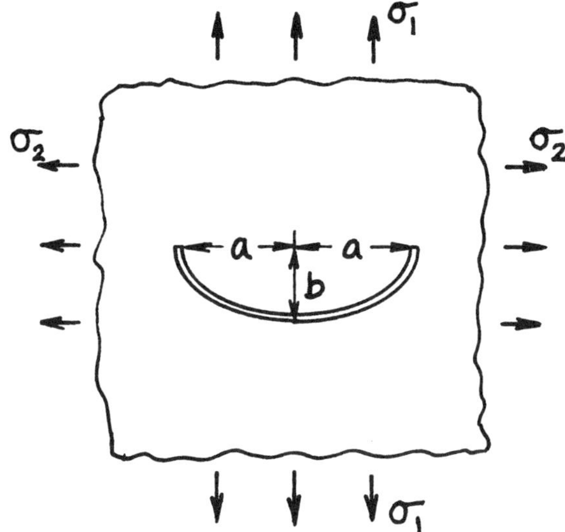

Semi-Elliptical Arc Crack

$$K_I = \sqrt{\pi a}\left[\sigma_1 F_{I1}\left(\frac{b}{a}\right) + \sigma_2 F_{I2}\left(\frac{b}{a}\right)\right]$$

$$K_{II} = \sqrt{\pi a}\left[\sigma_1 F_{II1}\left(\frac{b}{a}\right) + \sigma_2 F_{II2}\left(\frac{b}{a}\right)\right]$$

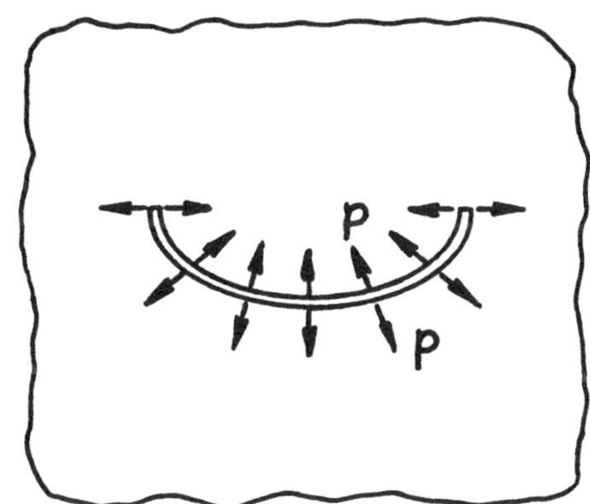

$$K_I = p\sqrt{\pi a}\left[F_{I1}\left(\frac{b}{a}\right) + F_{I2}\left(\frac{b}{a}\right)\right]$$

$$K_{II} = p\sqrt{\pi a}\left[F_{II1}\left(\frac{b}{a}\right) + F_{II2}\left(\frac{b}{a}\right)\right]$$

See **Narendran 1982**.

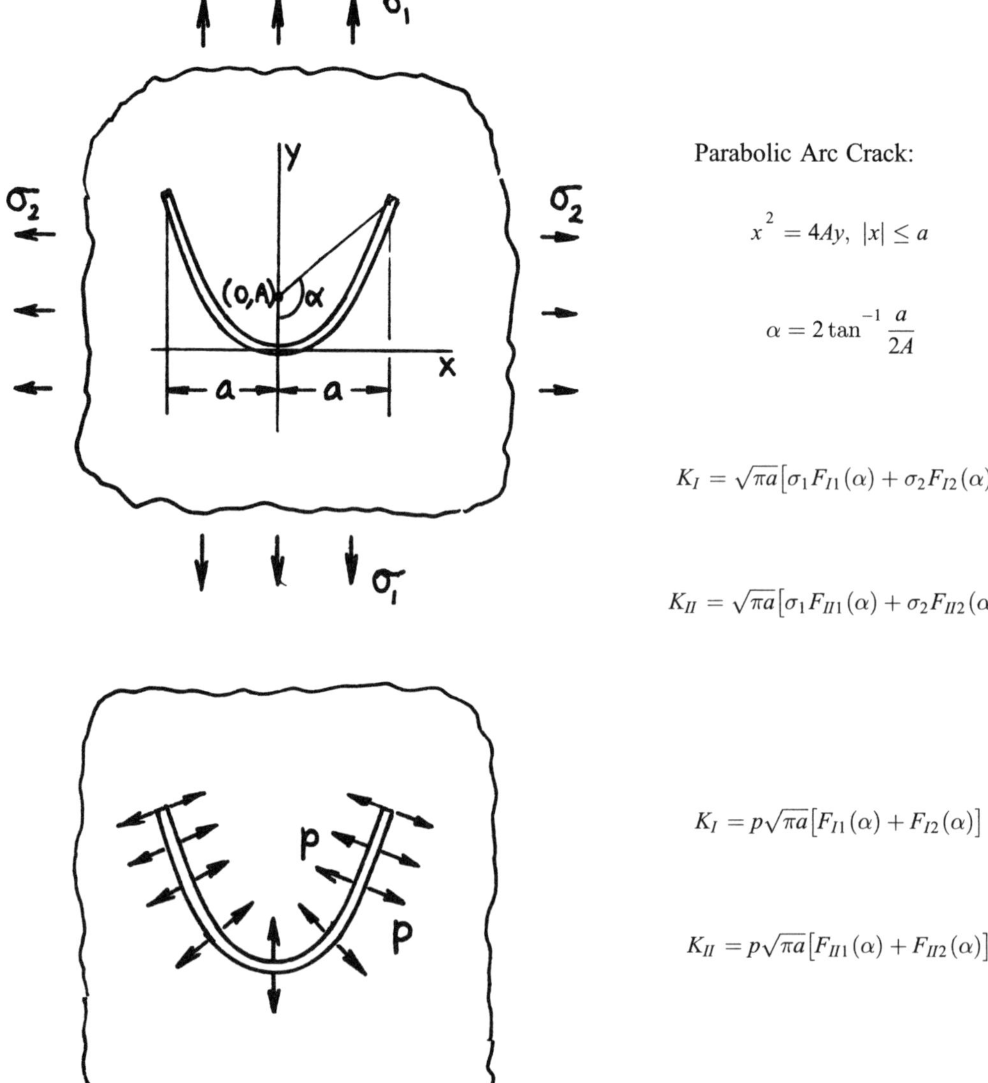

Parabolic Arc Crack:

$$x^2 = 4Ay, \ |x| \leq a$$

$$\alpha = 2\tan^{-1}\frac{a}{2A}$$

$$K_I = \sqrt{\pi a}\left[\sigma_1 F_{I1}(\alpha) + \sigma_2 F_{I2}(\alpha)\right]$$

$$K_{II} = \sqrt{\pi a}\left[\sigma_1 F_{II1}(\alpha) + \sigma_2 F_{II2}(\alpha)\right]$$

$$K_I = p\sqrt{\pi a}\left[F_{I1}(\alpha) + F_{I2}(\alpha)\right]$$

$$K_{II} = p\sqrt{\pi a}\left[F_{II1}(\alpha) + F_{II2}(\alpha)\right]$$

See **Narendran 1982**.

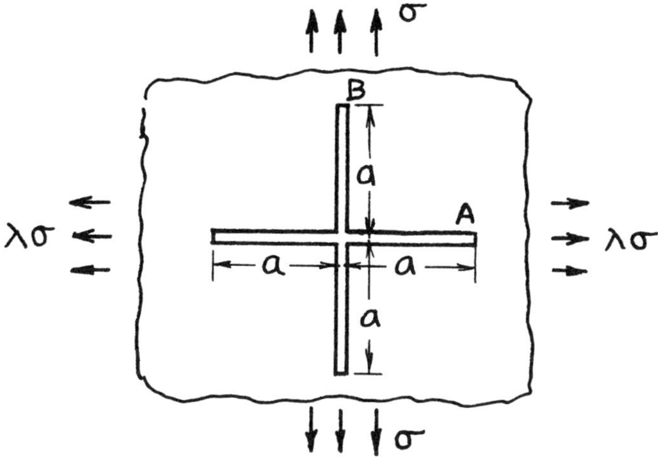

$$K_{IA} = \sigma\sqrt{\pi a}\,(1.0863 - .2227\lambda)$$

$$K_{IB} = \sigma\sqrt{\pi a}\,(-.2227 + 1.0863\lambda)$$

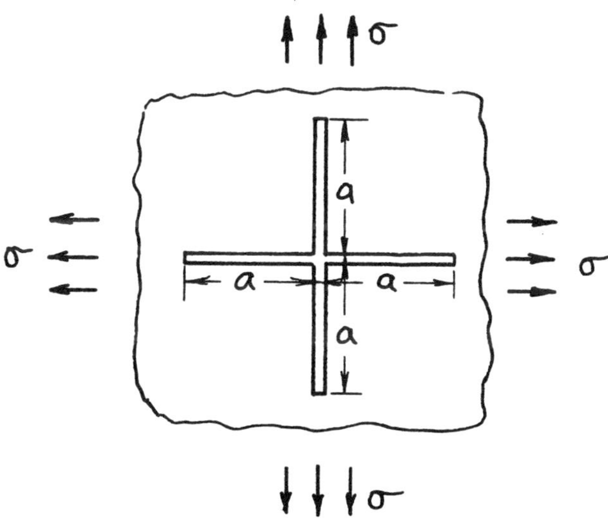

$$K_I = \sigma\sqrt{\pi a}\,(.8636)$$

Method: A Special Case of **page 21.10** and/or **page 21.9**
Accuracy: 0.1%
References: **Tada 1985**; also **Stallybrass 1969; Rooke 1969**

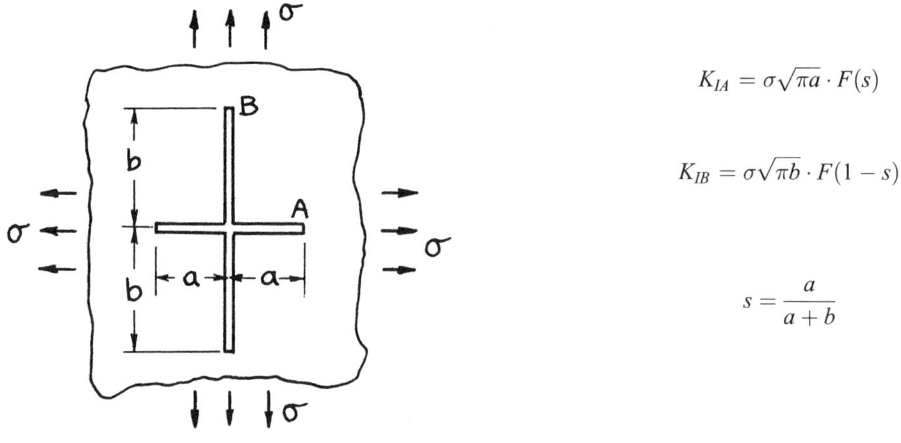

$$K_{IA} = \sigma\sqrt{\pi a}\cdot F(s)$$

$$K_{IB} = \sigma\sqrt{\pi b}\cdot F(1-s)$$

$$s = \frac{a}{a+b}$$

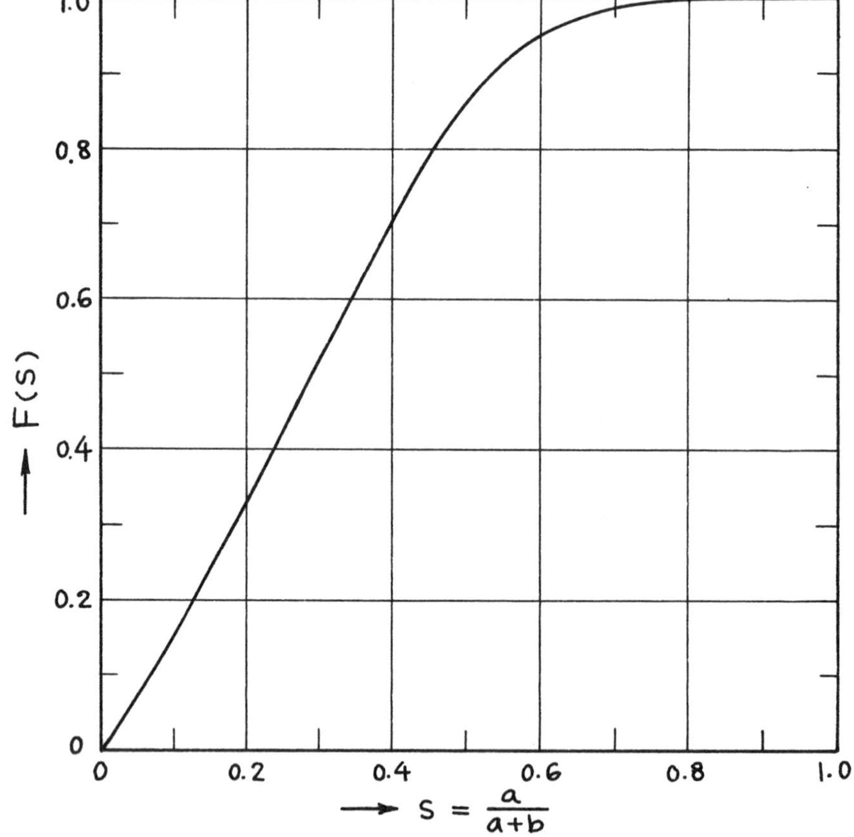

Method: Alternating Method (Special Case of **Page 21.10**; $\lambda = 1$)
Accuracy: Curve is based on numerical values with 0.1% accuracy.
Reference: **Tada 1985**

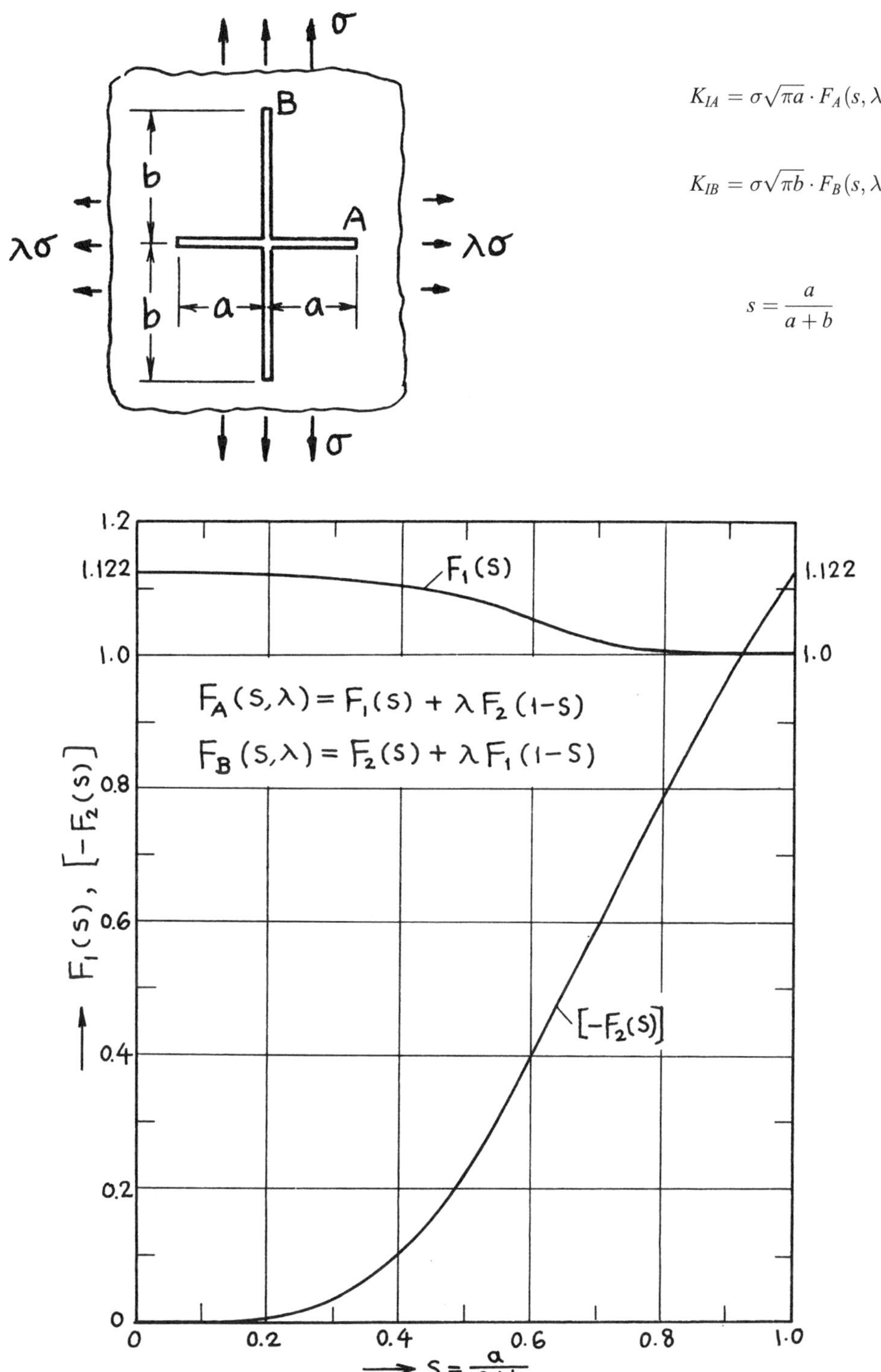

$$K_{IA} = \sigma\sqrt{\pi a} \cdot F_A(s, \lambda)$$

$$K_{IB} = \sigma\sqrt{\pi b} \cdot F_B(s, \lambda)$$

$$s = \frac{a}{a+b}$$

Method: Alternating Method (Simultaneous Integral Equations)
Accuracy: Curves are based on numerical values with 0.1% accuracy.
Reference: **Tada 1985**

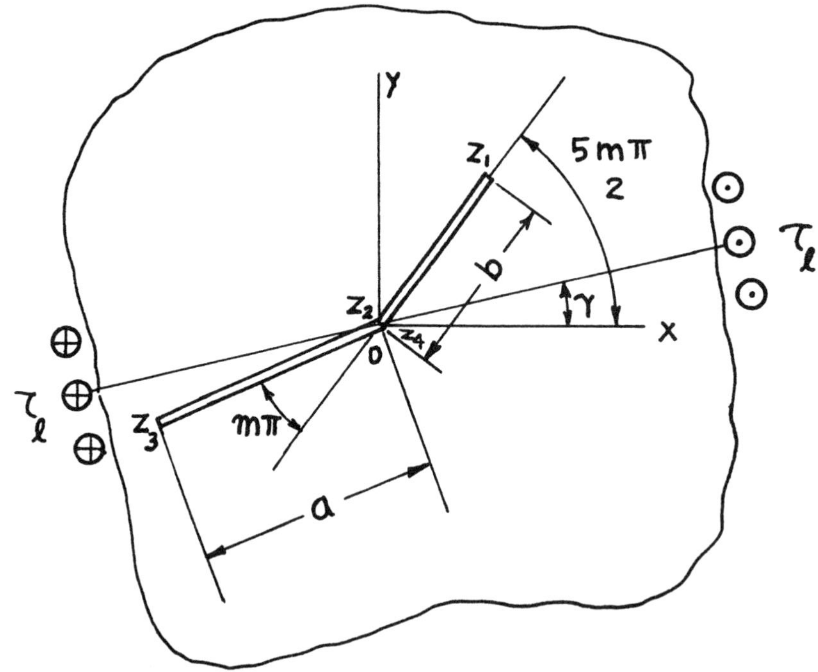

$$K_{IIIz_1} = \frac{\tau_\ell\sqrt{\pi b}\sin\left[(\gamma-\beta)-m\left(\frac{3\pi}{2}+\alpha\right)\right]}{\sqrt{\cos\beta(\cos\beta-\cos\alpha)}}\left\{\frac{\sin\left(\frac{\alpha-\beta}{2}\right)}{\sin\left(\frac{\alpha+\beta}{2}\right)}\right\}^m$$

$$K_{IIIz_3} = \frac{\tau_\ell\sqrt{\pi a}\sin\left[(\gamma+\beta)-m\left(\frac{3\pi}{2}+\alpha\right)\right]}{\sqrt{\cos\beta(\cos\beta+\cos\alpha)}}\left\{\frac{\sin\left(\frac{\alpha+\beta}{2}\right)}{\sin\left(\frac{\alpha-\beta}{2}\right)}\right\}^m$$

where:

$$\sin\beta = m\sin\alpha$$

$$\frac{b}{a} = \frac{\cos\beta-\cos\alpha}{\cos\beta+\cos\alpha}\left\{\frac{\sin(\alpha+\beta)}{\sin(\alpha-\beta)}\right\}^m \leq 1$$

$$0 \leq m \leq 1$$

Method: Conformal Mapping
Accuracy: Exact
Reference: **Sih 1965a**

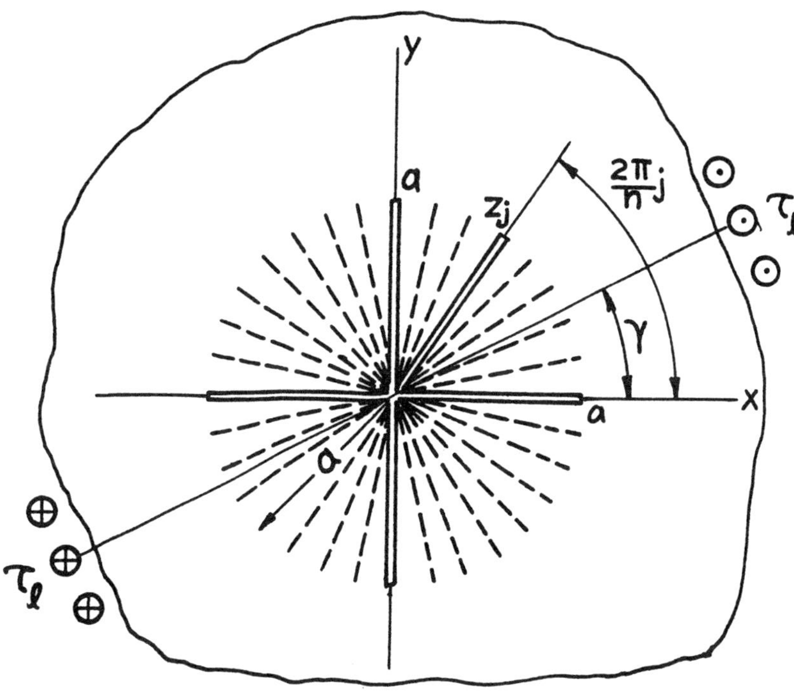

$$K_{IIIz_j} = (-1)^{1+j} 2\sqrt{2} \cdot 2^{-2/n} \tau_\ell \sqrt{\pi a} \frac{1}{\sqrt{n}} \sin\left(\gamma + \frac{2\pi}{n} j\right)$$

$$n = 1, 2, 3, \; \ldots\ldots$$

Method: Conformal Mapping
Accuracy: Exact
Reference: **Sih 1965a**

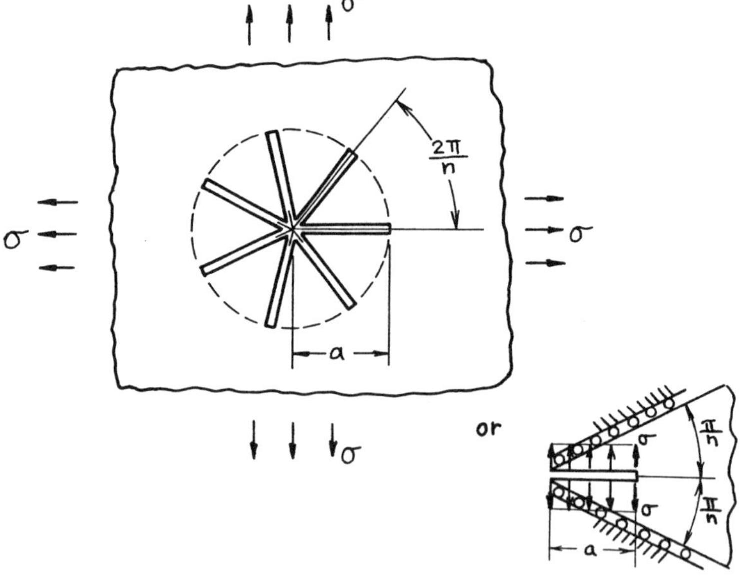

$$K_I = \sigma\sqrt{\pi a} \cdot F(n)$$

$$F(n) = \frac{1}{\sqrt{n}} G(n)$$

$$G(n) = 2\sqrt{1 - \frac{1}{n}}, \quad n \geq 2 \quad (\text{Ouchterlony})$$

or $$G(n) = 2 - 1.050\frac{1}{n} - .243\frac{1}{n^2}, \quad n \geq 1 \; (\text{Tada})$$

$$n \to \infty : \; F(n) \to \frac{2}{\sqrt{n}}$$

Total Area of Crack Opening:

$$A = \frac{\sigma \pi a^2}{E'} S(n)$$

$$S(n) = \{G(n)\}^2$$

$$n \to \infty : S(n) \to 4$$

or $$A_{n \to \infty} = 2A_{n=2}$$

Methods: Integral Transform and Integral Equation, Conformal mapping
Accuracy: Both formulas for K_I have better than 0.5% accuracy; A better than 1%
References: **Westmann 1964; Williams 1971; Kitagawa 1975; Ouchterlony 1975; Tada 1985**

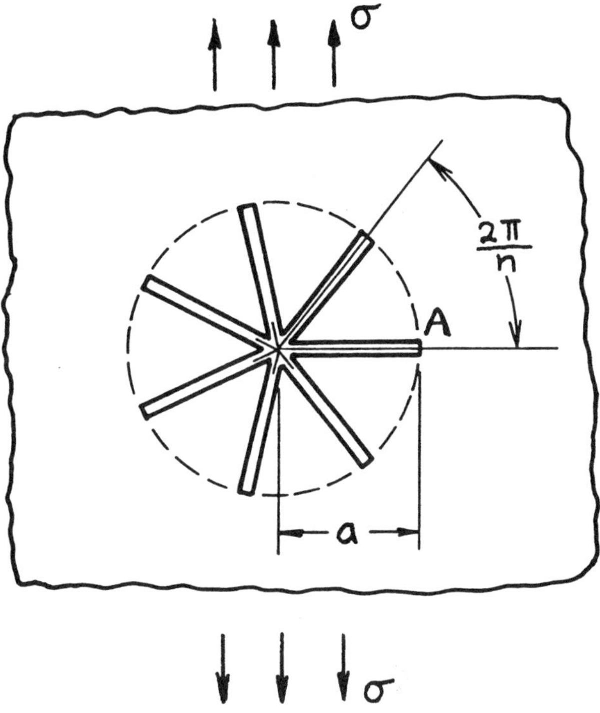

$$(K_I)_{max} = K_{IA} = \sigma\sqrt{\pi a} \cdot F_A(n)$$

$$F_A(n) = \frac{1}{\sqrt{n}}\left(3 - 3.25\frac{1}{n} - .64\frac{1}{n^2} + 1.6\frac{1}{n^3}\right) \quad n \geq 1$$

$$n \to \infty : \; F_A(n) = \frac{3}{\sqrt{n}}$$

Method: Estimated by Interpolation (based on results for $n = 1, 2, 4,$ and ∞)
Accuracy: Expected to be within 1%
Reference: **Tada 1985**

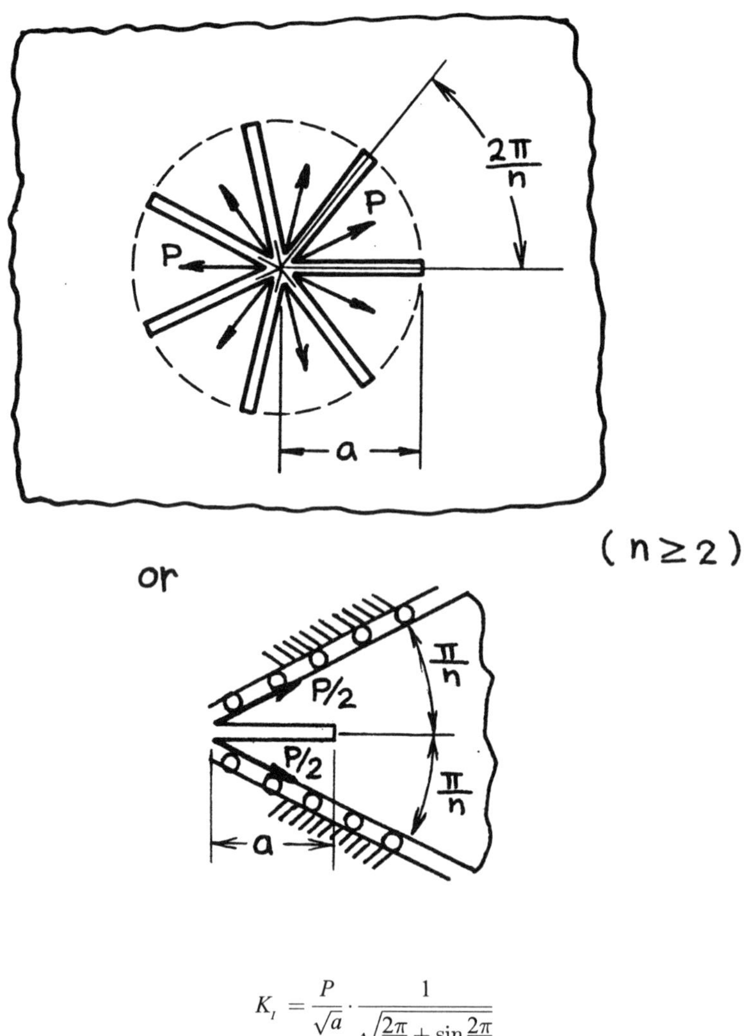

$$K_I = \frac{P}{\sqrt{a}} \cdot \frac{1}{\sqrt{\frac{2\pi}{n} + \sin\frac{2\pi}{n}}}$$

Method: Integral Transform
Accuracy: Exact
Reference: **Ouchterlony 1975, 1976**

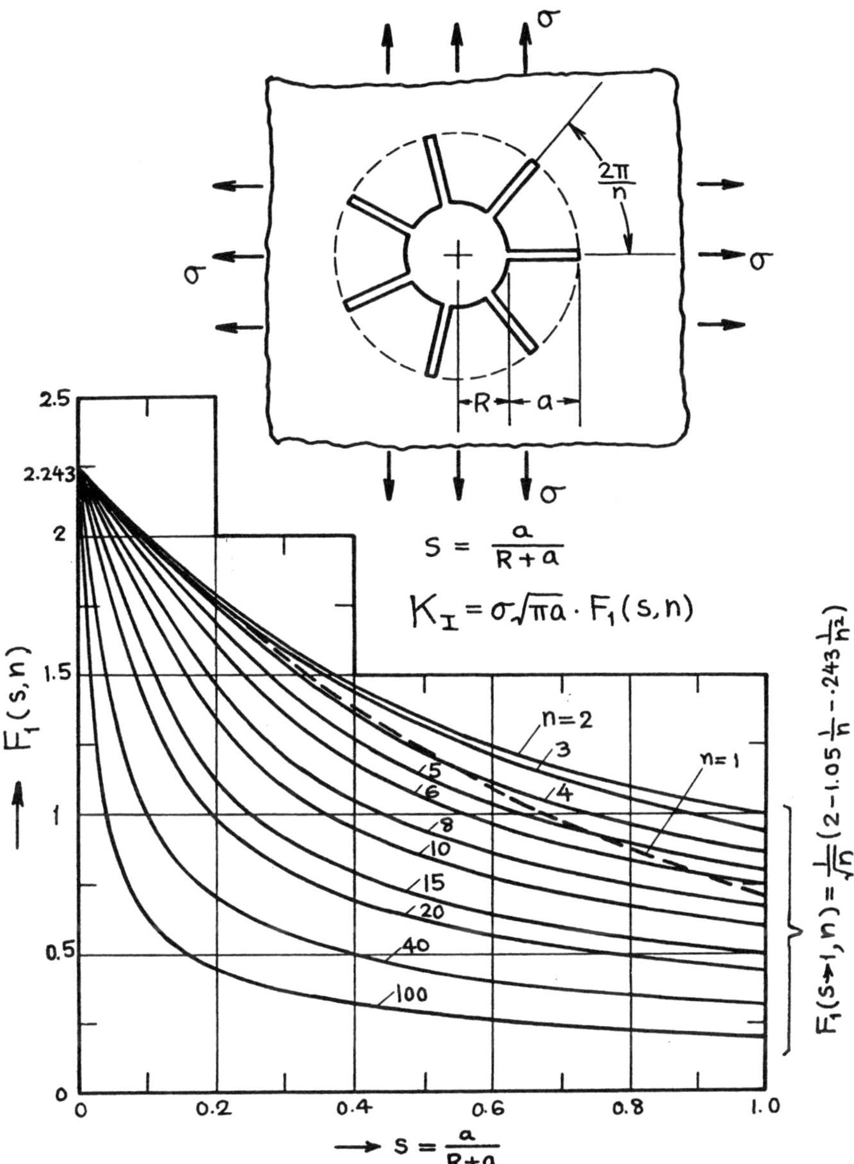

Methods: Conformal Mapping (Ouchterlony; $s \leq 0.6$, $n \leq 15$), Asymptotic Interpolation (Tada; $s > 0.6$, $n \leq 15$ and $0 < s < 1$, $n > 15$)

Accuracy: 1%

References: **Ouchterlony 1975; Tada 1985**

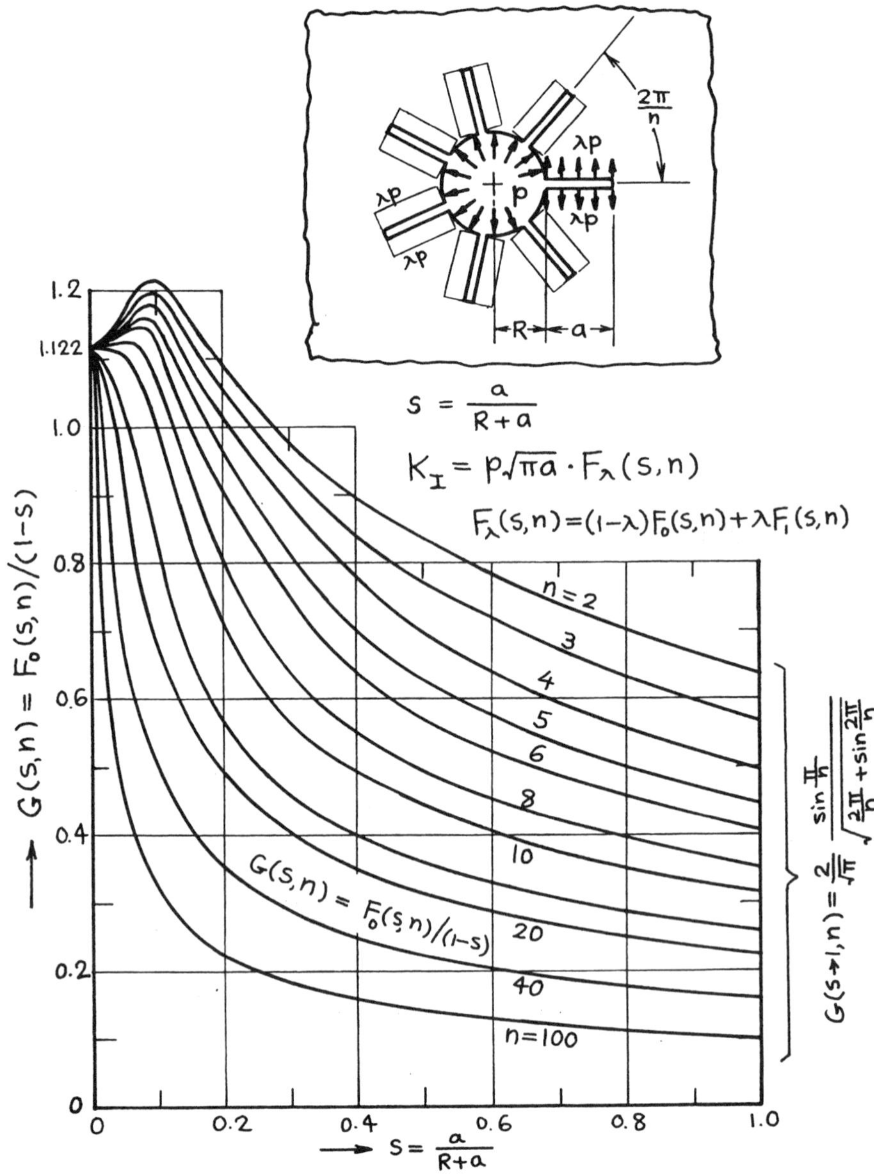

Methods: Conformal Mapping (Ouchterlony; $s \leq 0.6,\ n \leq 15$), Asymptotic Interpolation (Tada; $s > 0.6$, $n \leq 15$ and $0 < s < 1,\ n > 15$)

Accuracy: 1%

References: **Ouchterlony 1975; Tada 1985**

NOTE: For $F_1(s,n)$, see **page 21.16**.

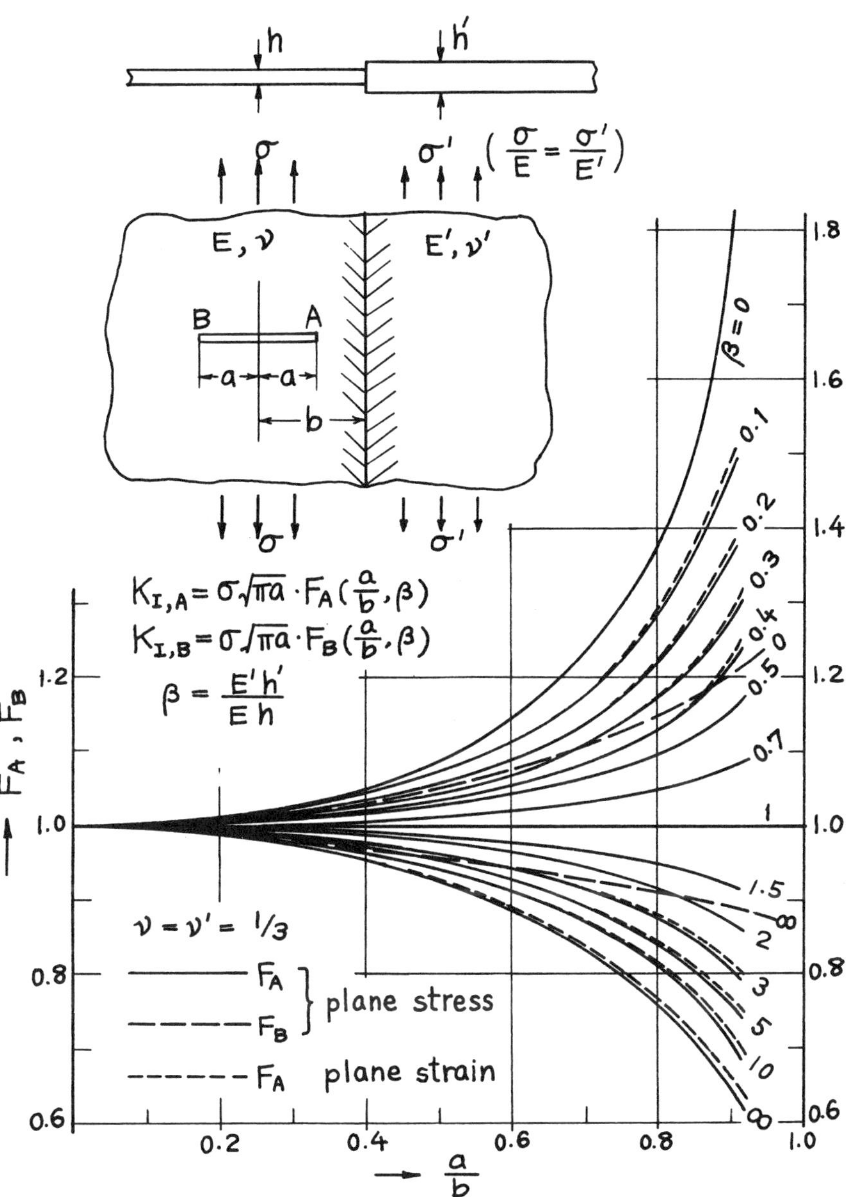

Method: Expansions of Complex Stress Potentials
Accuracy: Curves were drawn based on the results having 0.1% accuracy.
Reference: **Isida 1970a**

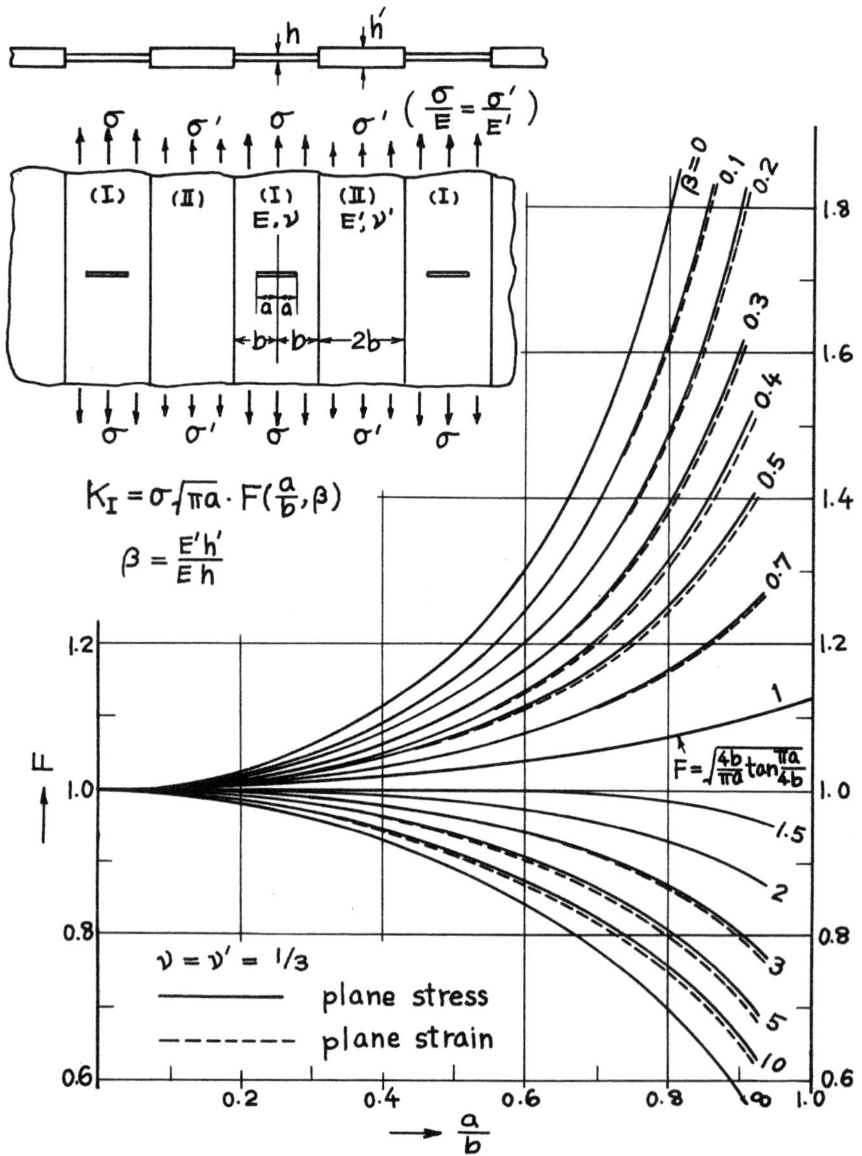

Method: Expansions of Complex Stress Potentials
Accuracy: Curves were drawn based on results with 0.1% accuracy.
Reference: **Isida 1970a**

PART

IV

THREE DIMENSIONAL CRACKED CONFIGURATIONS

- ❑ A Semi-Infinite Crack in an Infinite Body
- ❑ An Embedded Circular Crack in an Infinite Body
- ❑ An External Circular Crack (A Circular Net Section) or a Circular Ring (An Annular) Crack in an Infinite Body
- ❑ An Elliptical Crack or Net Section and a Parabolic Crack in an Infinite Body
- ❑ An External Circular Crack in a Round Bar
- ❑ An Internal Circular Crack in a Round Bar
- ❑ An Internal Circumferential Crack in a Thick-Walled Cylinder
- ❑ An External Circumferential Crack in a Thick-Walled Cylinder
- ❑ A Half-Circular Surface Crack in a Semi-Infinite Body
- ❑ A Quarter Circular Corner Crack in a Quarter-Infinite Body

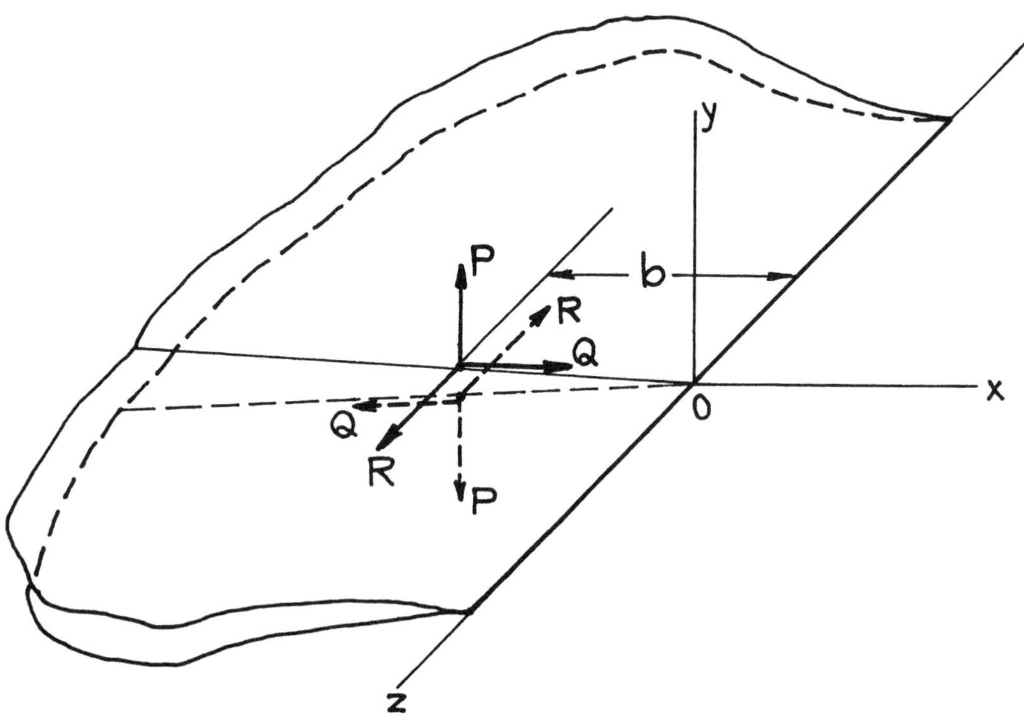

$$K_I = \frac{\sqrt{2}\,P}{(\pi b)^{3/2}} \cdot \frac{1}{1 + \left({}^{z}\!/_{b}\right)^2}$$

$$\begin{Bmatrix} K_{II} \\ K_{III} \end{Bmatrix} = \frac{\sqrt{2}}{(\pi b)^{3/2}} \cdot \begin{Bmatrix} Q \\ R \end{Bmatrix} \frac{1}{1 + \left({}^{z}\!/_{b}\right)^2} \left[1 \begin{Bmatrix} + \\ - \end{Bmatrix} \frac{2\nu}{2-\nu} \cdot \frac{1 - \left({}^{z}\!/_{b}\right)^2}{1 + \left({}^{z}\!/_{b}\right)^2} \right] + \frac{4\sqrt{2}}{(\pi b)^{3/2}} \begin{Bmatrix} R \\ Q \end{Bmatrix} \cdot \frac{\nu}{2-\nu} \cdot \frac{{}^{z}\!/_{b}}{\left[1 + \left({}^{z}\!/_{b}\right)^2\right]^2}$$

Method: Papkovich-Neuber Potentials
Accuracy: Exact
References: **Uflyand 1965; Sih 1968; Kassir 1973**

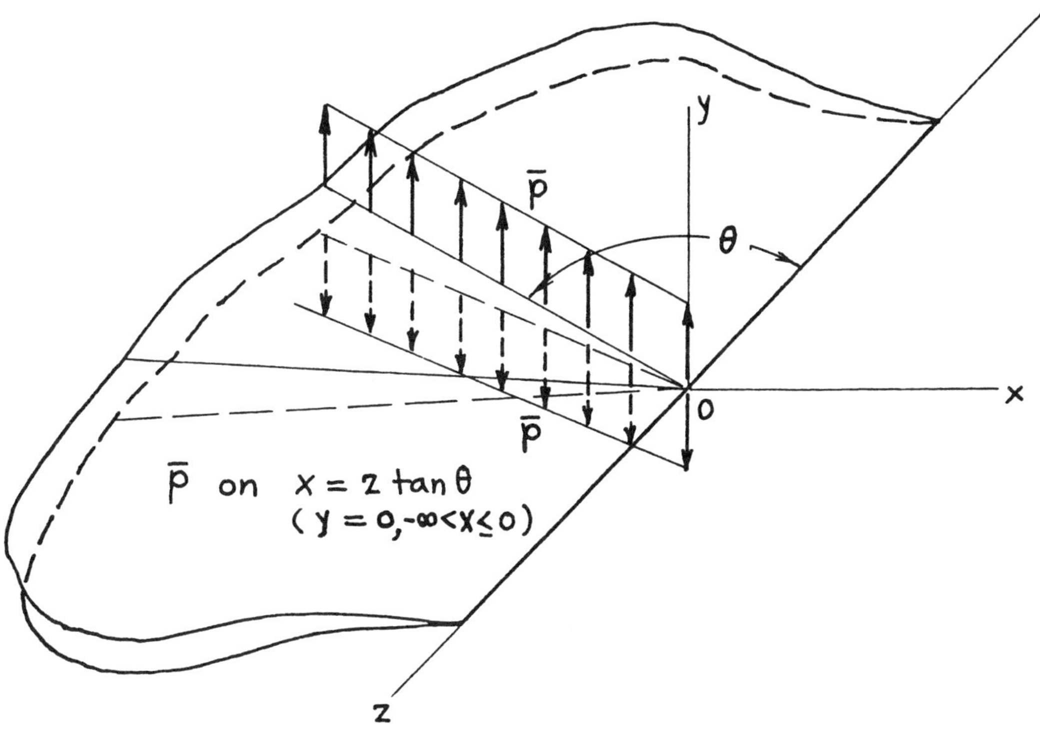

$$K_I(z;\ \theta) = \frac{\bar{p}}{\sqrt{\pi|z|}} \begin{cases} \sqrt{\tan\frac{\theta}{2}} & z > 0 \\ \sqrt{\cot\frac{\theta}{2}} & z < 0 \end{cases}$$

$$K_{II} = K_{III} = 0$$

Method: Integration of **page 23.1** or a Limiting Case of **page 24.11**
Accuracy: Exact
Reference: **Tada 1985**

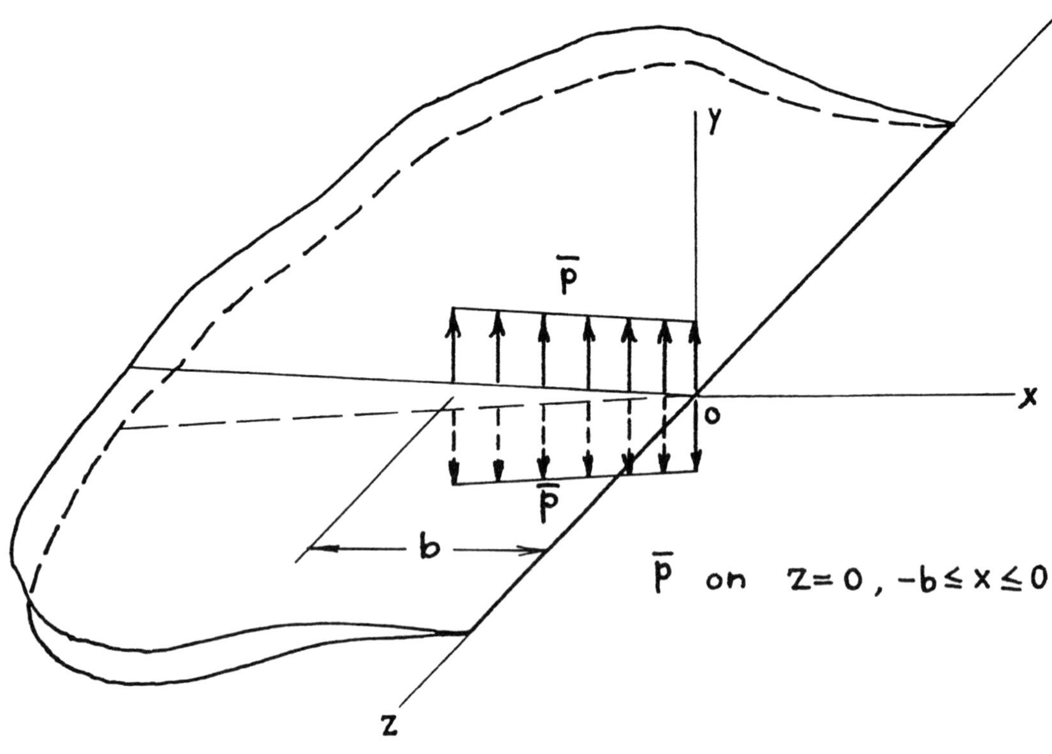

$$K_I(z\,;b) = \frac{\bar{p}}{\pi^{3/2}\sqrt{|z|}}\cdot\begin{cases} \pi - \sin^{-1}\sqrt{\dfrac{2\frac{|z|}{b}}{1+\left(\frac{z}{b}\right)^2}} - \sinh^{-1}\sqrt{\dfrac{2\frac{|z|}{b}}{1+\left(\frac{z}{b}\right)^2}} & \dfrac{|z|}{b}\leq 1 \\ \sin^{-1}\sqrt{\dfrac{2\frac{|z|}{b}}{1+\left(\frac{z}{b}\right)^2}} - \sinh^{-1}\sqrt{\dfrac{2\frac{|z|}{b}}{1+\left(\frac{z}{b}\right)^2}} & \dfrac{|z|}{b} > 1 \end{cases}$$

$$K_{II} = K_{III} = 0$$

Method: Integration of **page 23.1**
Accuracy: Exact
Reference: **Tada 1985**

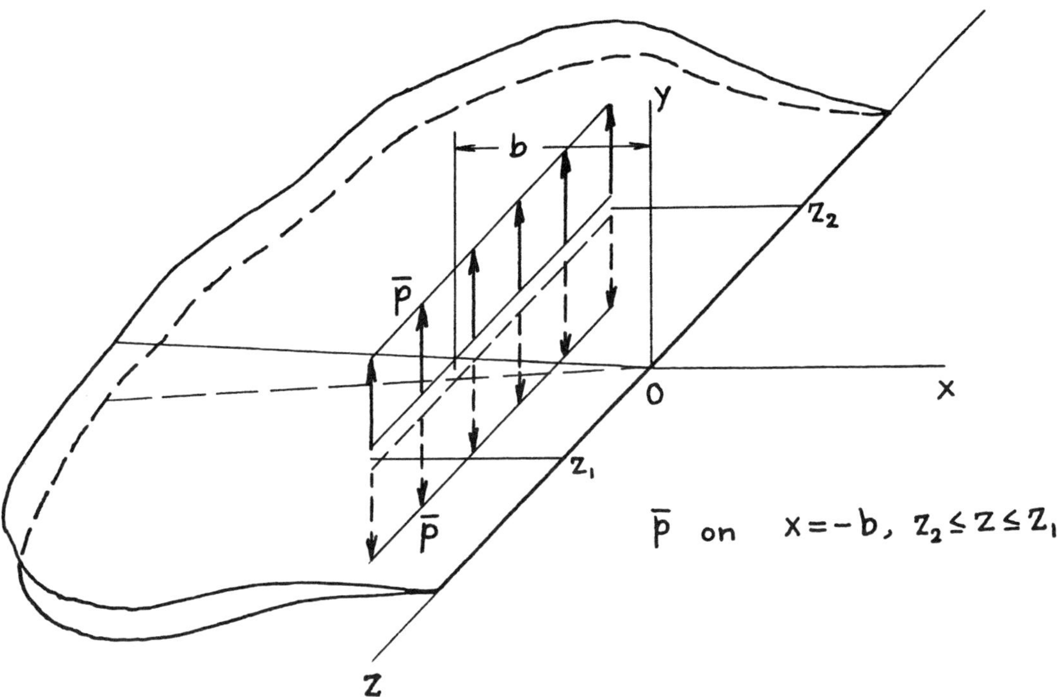

$$K_I(z) = \frac{\sqrt{2}\,\bar{p}}{\pi^{3/2}\sqrt{b}}\left[\tan^{-1}\frac{z-z_1}{b} - \tan^{-1}\frac{z-z_2}{b}\right]$$

$$K_{II} = K_{III} = 0$$

Method: Integration of **page 23.1** or a Limiting Case of **page 24.4**
Accuracy: Exact
Reference: **Tada 1985**

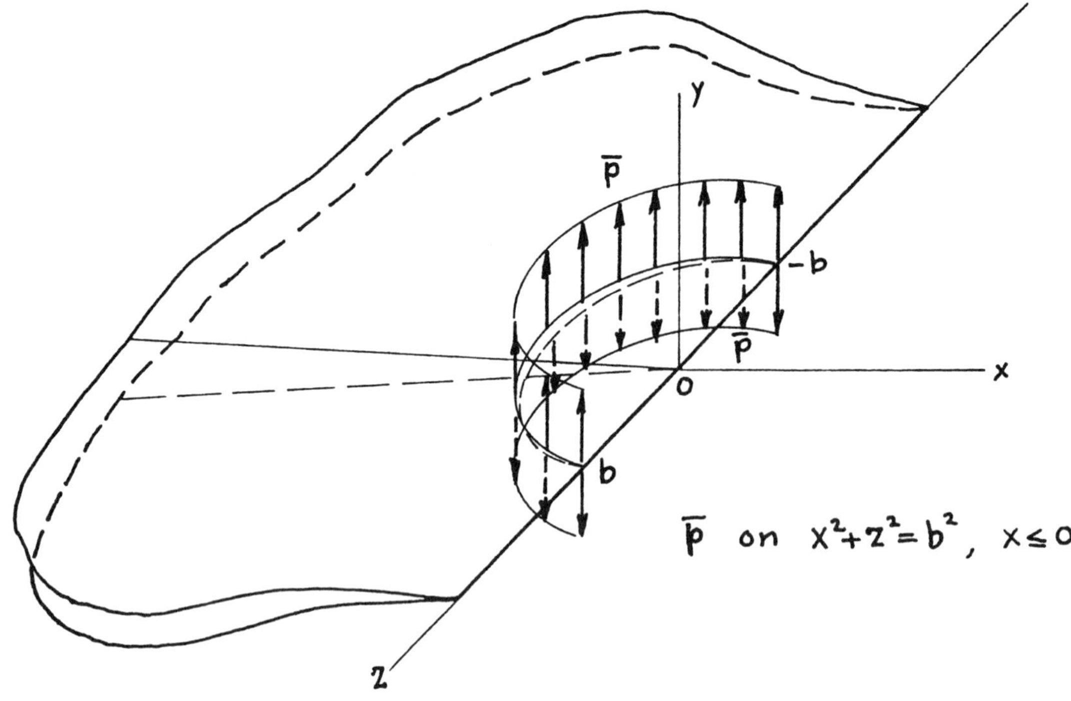

$$K_I(z) = \frac{\sqrt{2}\,\bar{p}}{\pi\sqrt{b}} \sum_{n=0}^{\infty} \frac{\Gamma\left(n+\frac{3}{4}\right)}{\Gamma\left(n+\frac{5}{4}\right)} \left(\frac{b}{z}\right)^{2n+2}$$

$$K_{II} = K_{III} = 0$$

Method: Integration of **page 23.1**
Accuracy: Exact
Reference: **Tada 1985**

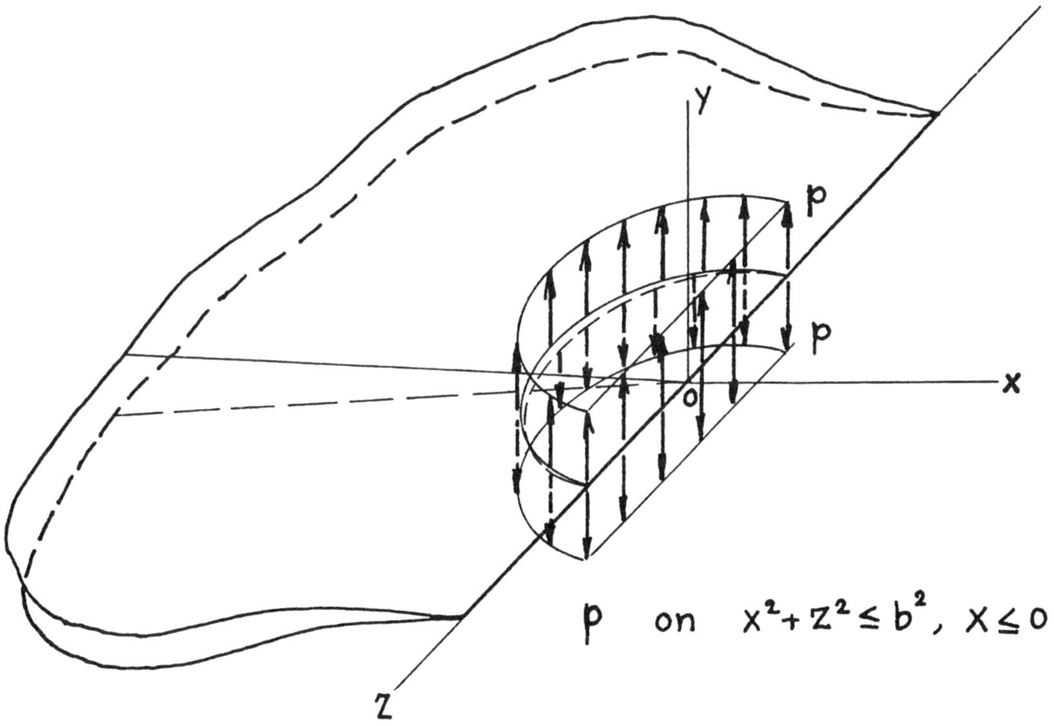

$$K_I(z) = \frac{P\sqrt{b}}{\sqrt{2}\pi} \sum_{n=0}^{\infty} \frac{\Gamma\left(n + \frac{3}{4}\right)}{\left(n + \frac{5}{4}\right)\Gamma\left(n + \frac{5}{4}\right)} \left(\frac{b}{z}\right)^{2n+2}$$

$$K_{II} = K_{III} = 0$$

Method: Integration of **page 23.1 (or 23.5)**
Accuracy: Exact
Reference: **Tada 1985**

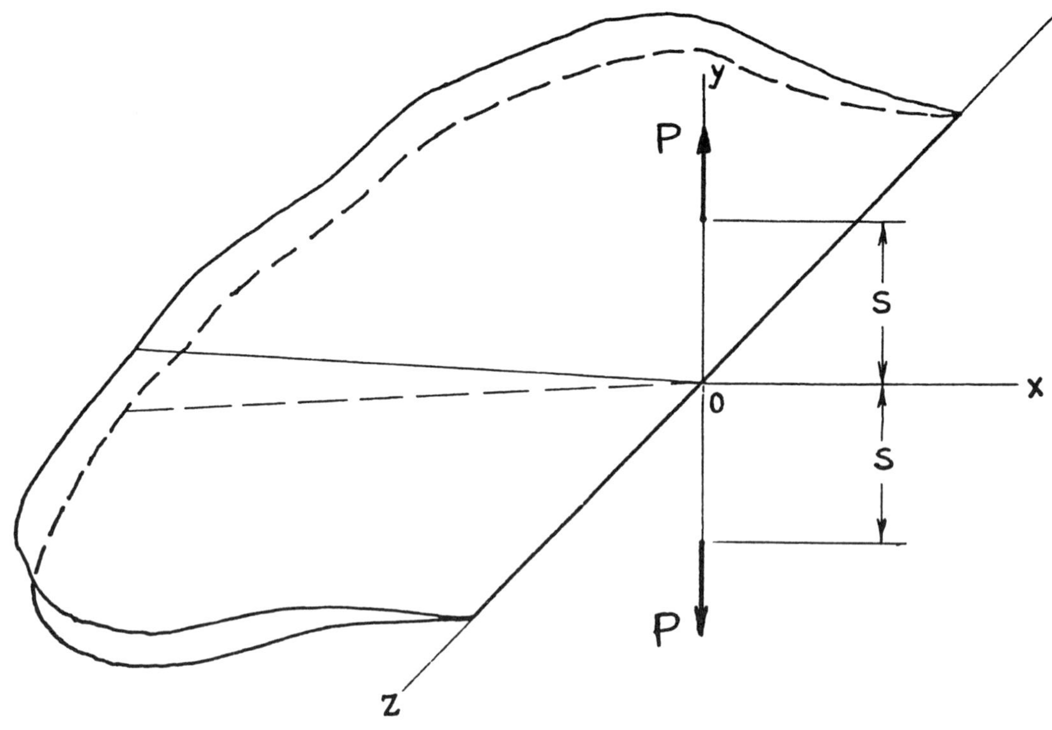

$$K_I(z) = \frac{P}{\pi^{3/2}}\left(1 - \alpha s \frac{\partial}{\partial s}\right)\frac{\sqrt{s}}{s^2 + z^2}$$

$$= \frac{P}{(\pi s)^{3/2}} \frac{1}{1 + \left(\frac{z}{s}\right)^2}\left[1 - \frac{\alpha}{2}\left\{1 - \frac{4}{1 + \left(\frac{z}{s}\right)^2}\right\}\right]$$

$$\text{where} \quad \alpha = \frac{1}{2(1 - \nu)}$$

$$K_{II} = K_{III} = 0$$

Method: Integration of **page 23.1 (or 23.5)**
Accuracy: Exact
References: **Tada 1985; also Kassir 1975**

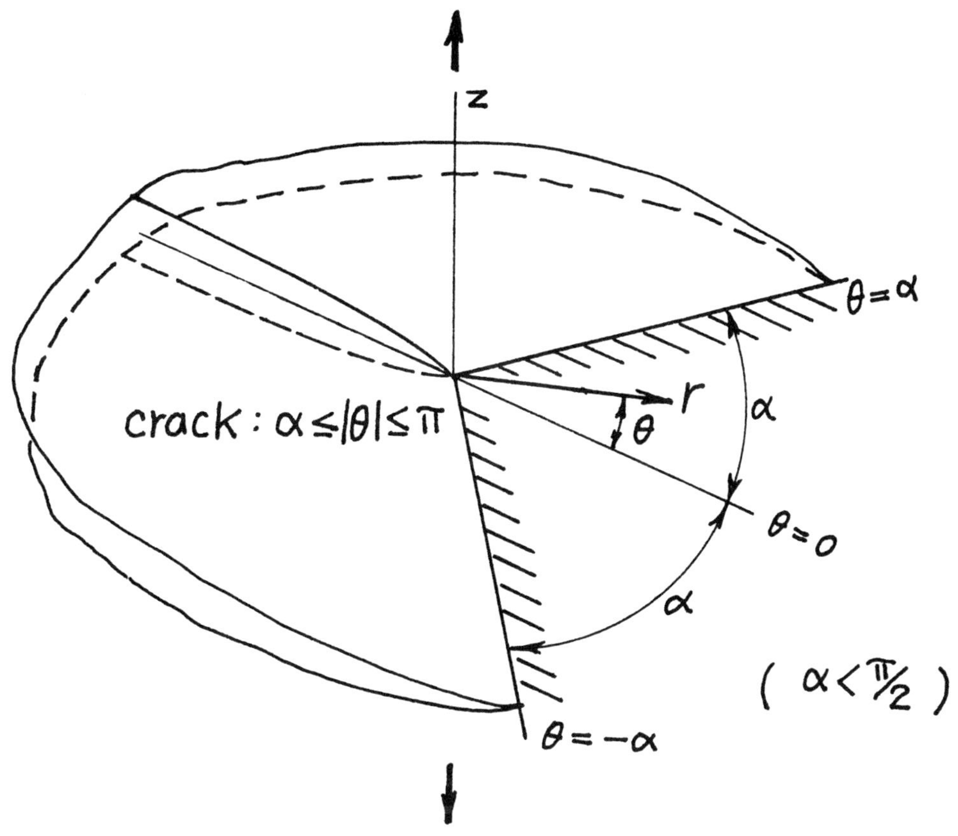

$$K_I = \frac{p}{\sqrt{\pi r} \sin \alpha}$$

$$K_{II} = K_{III} = 0$$

where $P =$ Total Line Load on Arc: $r =$ const., $-\alpha \leq \theta \leq \alpha$
= Constant for any r

Method: Neuber-Papkovich Potential
Accuracy: Exact
References: **Galin 1953; Tada 1974**

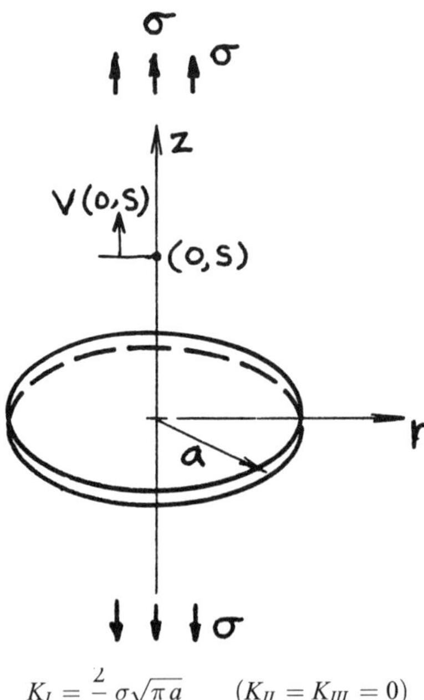

$$K_I = \frac{2}{\pi}\sigma\sqrt{\pi a} \qquad (K_{II} = K_{III} = 0)$$

Volume of Crack:

$$V = \frac{16\left(1-\nu^2\right)}{3E}\sigma a^3$$

Crack Opening Shape:

$$\underset{r \le a}{2v\,(r,0)} = \frac{8\left(1-\nu^2\right)}{\pi E}\sigma\sqrt{a^2 - r^2}$$

Opening at Center:

$$\delta_0 = 2v\,(0,0) = \frac{8\left(1-\nu^2\right)}{\pi E}\sigma a$$

Additional Displacement at $(0,s)$ due to Crack:

$$v\,(0,s) = \frac{4\left(1-\nu^2\right)}{\pi E}\sigma a\left\{1 - (1-\alpha)\frac{s}{a}\tan^{-1}\frac{a}{s} - \alpha\frac{s^2}{s^2+a^2}\right\}$$

where

$$\alpha = \frac{1}{2(1-\nu)}$$

Methods: Integral Transform, Integration of **page 24.5 or 24.11**, Paris' Equation (see **Appendix B**), Reciprocity (see **page 24.7**)
Accuracy: Exact
Reference: **Tada 1985**

NOTE: $V(0,s)$ Is the displacement at $(0,s)$ when uniform pressure σ is applied on crack surfaces.

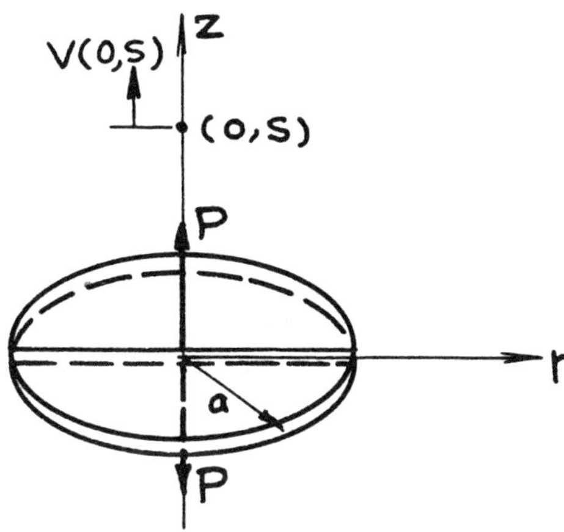

$$K_I = \frac{P}{(\pi a)^{3/2}} \qquad (K_{II} = K_{III} = 0)$$

Volume of Crack:

$$V = \frac{8\left(1-\nu^2\right)}{\pi E} P\, a$$

Crack Opening Shape:

$$2v \underset{r \le a}{(r, 0)} = \frac{4\left(1-\nu^2\right)}{\pi^2 E} P \frac{1}{r} \cos^{-1} \frac{r}{a}$$

Displacement at $(0, s)$:

$$v(0, s) = \frac{2\left(1-\nu^2\right)}{\pi^2 E} P \frac{1}{s} \left\{ (1+\alpha) \tan^{-1} \frac{a}{s} + \alpha \frac{as}{s^2 + a^2} \right\}$$

where

$$\alpha = \frac{1}{2(1-\nu)}$$

Methods: Special case of **page 24.3**, Reciprocity (see **page 24.3 or 24.7**)
Accuracy: Exact
References: **Galin 1953; Tada 1973, 1985**

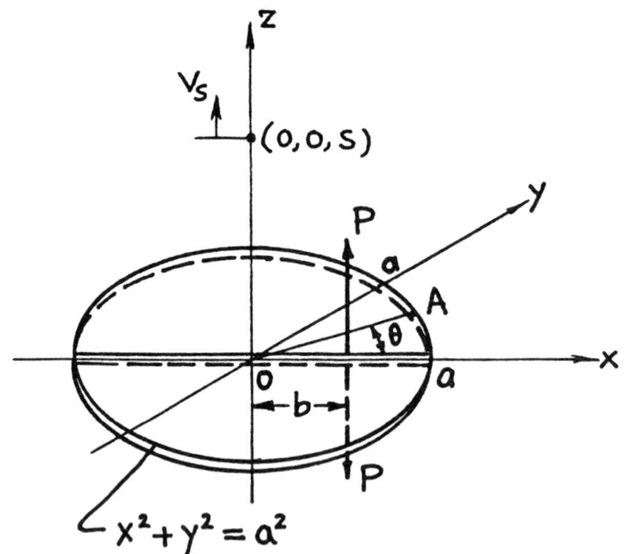

$$K_{IA} = K_I(\theta) = \frac{P}{\pi\sqrt{\pi a}} \cdot \frac{\sqrt{a^2 - b^2}}{a^2 + b^2 - 2ab\cos\theta}$$

$$(K_{II} = K_{III} = 0)$$

Volume of Crack:

$$V = \frac{8\left(1 - \nu^2\right)}{\pi E} P\sqrt{a^2 - b^2}$$

Crack Opening at Center:

$$\delta_0 = 2v(0,0,0) = \frac{4\left(1 - \nu^2\right)}{\pi^2 E} P \frac{1}{b} \cos^{-1}\frac{b}{a}$$

Vertical Displacement at $(0, 0, s)$:

$$v_s = v(0,0,s) = \frac{2\left(1 - \nu^2\right)}{\pi^2 E} P\left(1 - \alpha s \frac{\partial}{\partial s}\right) \frac{1}{\sqrt{s^2 + b^2}} \tan^{-1}\sqrt{\frac{a^2 - b^2}{s^2 + b^2}}$$

where

$$\alpha = \frac{1}{2(1 - \nu)}$$

Methods: Neuber-Papkovich Potentials, Reciprocity (see **pages 24.1, 24.2, 24.7**)
Accuracy: Exact
References: **Galin 1953; Tada 1985**

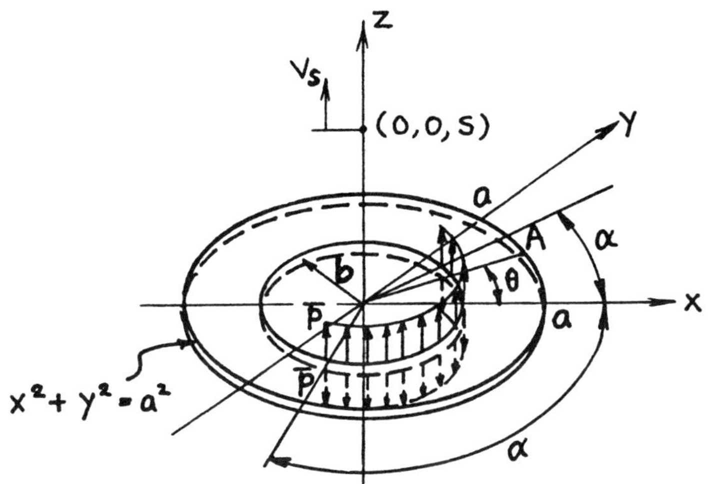

$$K_{IA} = K_I(\theta) = \frac{2P}{\pi\sqrt{\pi a}} \frac{b}{\sqrt{a^2 - b^2}} \left\{ \tan^{-1}\left(\frac{a+b}{a-b}\tan\frac{\theta+\alpha}{2}\right) - \tan^{-1}\left(\frac{a+b}{a-b}\tan\frac{\theta-\alpha}{2}\right) \right\}$$

$$(K_{II} = K_{III} = 0)$$

Volume of Crack:

$$V = \frac{16\left(1-\nu^2\right)\alpha}{\pi E} \bar{p} b \sqrt{a^2 - b^2}$$

Crack Opening at Center:

$$\delta_0 = 2v(0,0,0) = \frac{8\left(1-\nu^2\right)\alpha}{\pi^2 E} \bar{p} \cos^{-1}\frac{b}{a}$$

Vertical Displacement at $(0,0,s)$:

$$v_s = v(0,0,s) = \frac{4\left(1-\nu^2\right)\alpha}{\pi^2 E} \bar{p} \left(1 - \alpha' s \frac{\partial}{\partial s}\right) \frac{b}{\sqrt{s^2 + b^2}} \tan^{-1}\sqrt{\frac{a^2 - b^2}{s^2 + b^2}}$$

where

$$\alpha' = \frac{1}{2(1-\nu)}$$

Method: Integration of **page 24.3**
Accuracy: Exact
References: **Tada 1973, 1985**

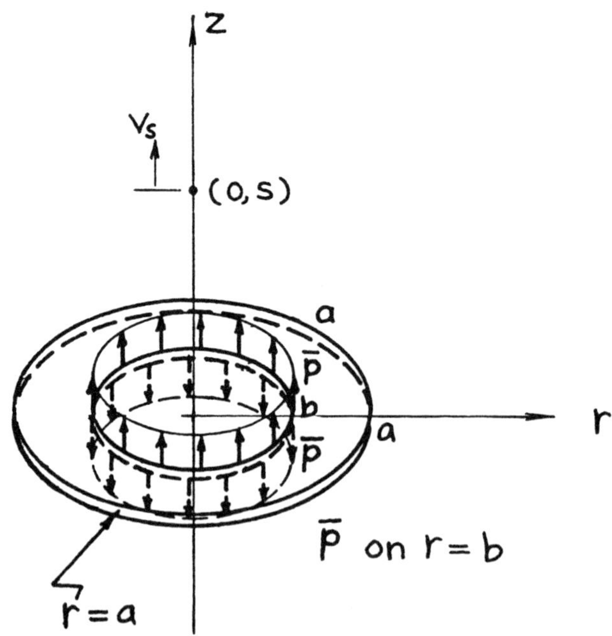

$$K_I = \frac{2\bar{p}}{\sqrt{\pi a}} \frac{b}{\sqrt{a^2 - b^2}} \qquad (K_{II} = K_{III} = 0)$$

Volume of Crack:

$$V = \frac{16\left(1 - \nu^2\right)}{E} \bar{p} b \sqrt{a^2 - b^2}$$

Crack Opening at Center:

$$\delta_0 = 2v(0,0) = \frac{8\left(1 - \nu^2\right)}{\pi E} \bar{p} \cos^{-1} \frac{b}{a}$$

Displacement at $(0, s)$:

$$v_s = v(0,s) = \frac{4\left(1 - \nu^2\right)}{\pi E} \bar{p} \left(1 - \alpha s \frac{\partial}{\partial s}\right) \frac{b}{\sqrt{s^2 + b^2}} \tan^{-1} \sqrt{\frac{a^2 - b^2}{s^2 + b^2}}$$

where

$$\alpha = \frac{1}{2(1 - \nu)}$$

Methods: Integral Transform, Weight Function Method, Special Case of **page 24.4**
Accuracy: Exact
References: **Sneddon 1946, 1951; Barenblatt 1962; Bueckner 1972; Tada 1985**

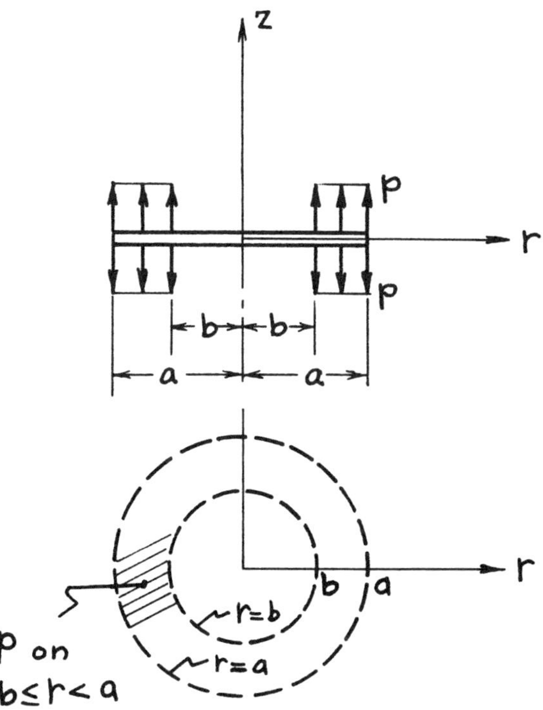

$$K_I = \frac{2p}{\sqrt{\pi a}}\sqrt{a^2 - b^2} \qquad (K_{II} = K_{III} = 0)$$

Volume of Crack:

$$V = \frac{16\left(1-\nu^2\right)}{3E}pa^3\left\{1-\left(\frac{b}{a}\right)^2\right\}^{3/2}$$

Crack Opening at Center:

$$\delta_0 = 2v(0,0) = \frac{8\left(1-\nu^2\right)}{\pi E}pa\left\{\sqrt{1-\left(\frac{b}{a}\right)^2} - \frac{b}{a}\cos^{-1}\frac{b}{a}\right\}$$

Crack Opening at $r = b$:

$$\delta_b = 2v(b,0) = \frac{8\left(1-\nu^2\right)}{\pi E}pa\left(1-\frac{b}{a}\right)$$

Methods: Integral Transform, Integration of **page 24.5**
Accuracy: Exact
References: **Sneddon 1951; Barenblatt 1962; Tada 1985**

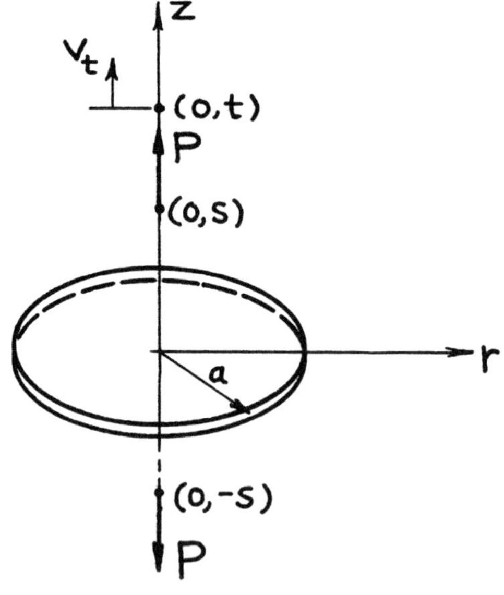

$$K_I = \frac{P}{\pi^2}\sqrt{\pi a}\left(1 - \alpha s\frac{\partial}{\partial s}\right)\frac{1}{a^2+s^2}$$

$$or \quad = \frac{P}{(\pi a)^{3/2}}\frac{a^2}{a^2+s^2}\left\{1 + \alpha\frac{2s^2}{a^2+s^2}\right\}$$

$$(K_{II} = K_{III} = 0)$$

Volume of Crack:

$$V = \frac{8\left(1-\nu^2\right)}{\pi E} Pa\left\{1 - (1-\alpha)\frac{s}{a}\tan^{-1}\frac{a}{s} - \alpha\frac{s^2}{a^2+s^2}\right\}$$

Crack Opening Shape:

$$\underset{r \le a}{2v(r,0)} = \frac{4\left(1-\nu^2\right)}{\pi^2 E} P\left(1 - \alpha s\frac{\partial}{\partial s}\right)\frac{1}{\sqrt{s^2+r^2}}\tan^{-1}\sqrt{\frac{a^2-r^2}{s^2+r^2}}$$

Opening at Center:

$$\delta_0 = 2v(0,0) = \frac{4\left(1-\nu^2\right)}{\pi^2 E} P\cdot\frac{1}{s}\left\{(1+\alpha)\tan^{-1}\frac{a}{s} + \alpha\frac{as}{a^2+s^2}\right\}$$

Displacement at $(0,t)$:

$$v_t = v(0,t) = \frac{4\left(1-\nu^2\right)}{\pi^2 E} P\left(1 - \alpha s\frac{\partial}{\partial s}\right)\left(1 - \alpha t\frac{\partial}{\partial t}\right)\left\{\frac{t}{t^2-s^2}\tan^{-1}\frac{a}{t} + \frac{s}{s^2-t^2}\tan^{-1}\frac{a}{s}\right\}$$

where

$$\alpha = \frac{1}{2(1-\nu)}$$

Methods: Integration of **page 24.5**, Paris' Equation (see **Appendix B**)
Accuracy: Exact
References: **Barenblatt 1962; Tada 1985**

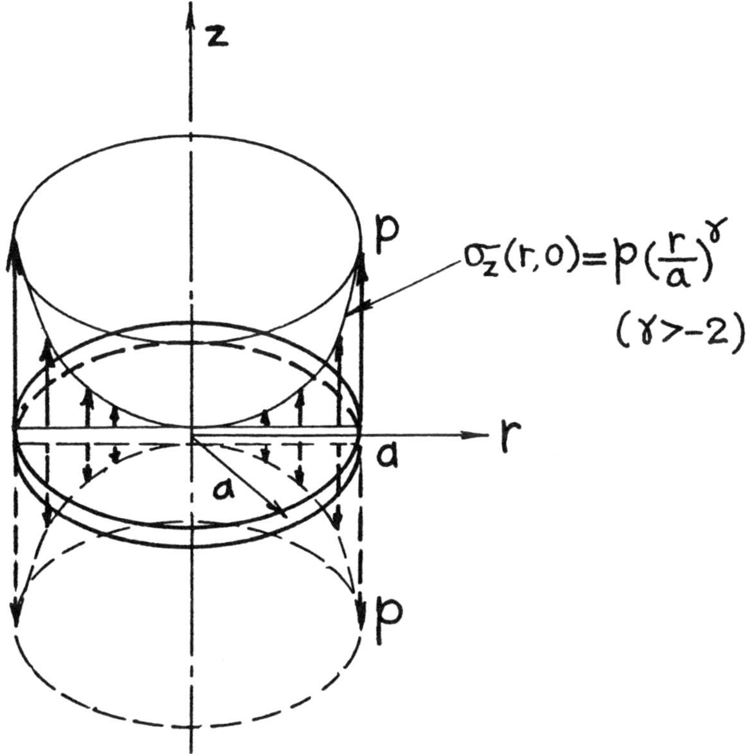

$$K_I = P\sqrt{a} \cdot \frac{\Gamma\left(\frac{\gamma}{2}+1\right)}{\Gamma\left(\frac{\gamma}{2}+\frac{3}{2}\right)}$$

$$(K_{II} = K_{III} = 0)$$

Volume of Crack:

$$V = \frac{4\left(1-\nu^2\right)}{E} pa^3 \sqrt{\pi} \cdot \frac{\Gamma\left(\frac{\gamma}{2}+1\right)}{\Gamma\left(\frac{\gamma}{2}+\frac{5}{2}\right)}$$

Crack Opening at Center:

$$\delta_0 = 2v(0,0) = \frac{4\left(1-\nu^2\right)}{E} pa \frac{1}{\sqrt{\pi}} \cdot \frac{\Gamma\left(\frac{\gamma}{2}+1\right)}{(\gamma+1)\Gamma\left(\frac{\gamma}{2}+\frac{3}{2}\right)}$$

where

$\Gamma(\gamma)$ = Gamma Function (see **Appendix M)**

Method: Integration of **page 24.5**, Paris' Equation (see **Appendix B**)
Accuracy: Exact
Reference: **Tada 1974, 1985**

NOTE: For special case of $\gamma = 0$, see **page 24.1**.

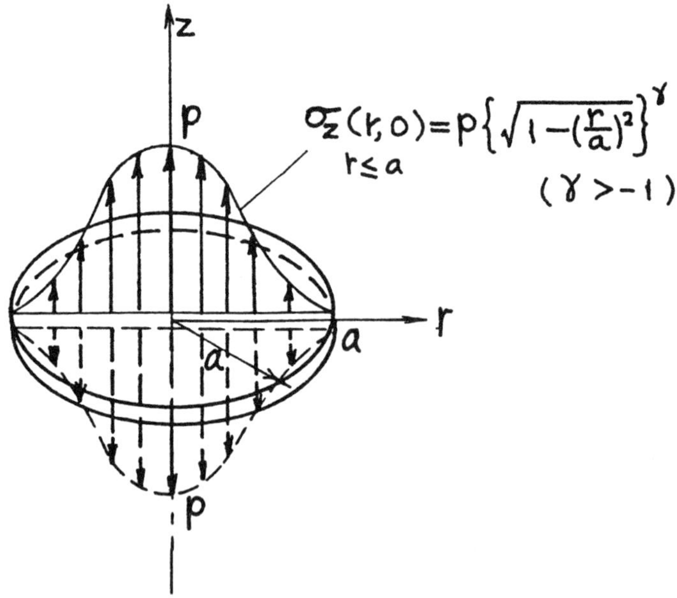

$$K_I = \frac{2}{\pi} P\sqrt{\pi a}\left(\frac{1}{\gamma+1}\right) \qquad (K_{II} = K_{III} = 0)$$

Volume of Crack:

$$V = \frac{16\left(1-\nu^2\right)}{3E} pa^3\left(\frac{3}{\gamma+3}\right)$$

Crack Opening at Center:

$$\delta_0 = 2v(0,0) = \frac{8\left(1-\nu^2\right)}{\pi E} pa \cdot D(\gamma)$$

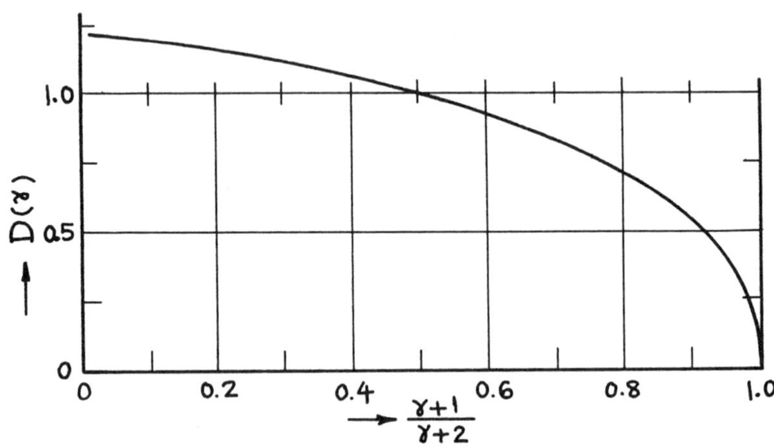

Method: Integration of **page 24.5**
Accuracy: K_I, V Exact; $D(\gamma)$ curve is based on accurate numerical values.
References: **Tada 1974, 1985**

NOTE: For special case of $\gamma = 0$, see **page 24.1**.

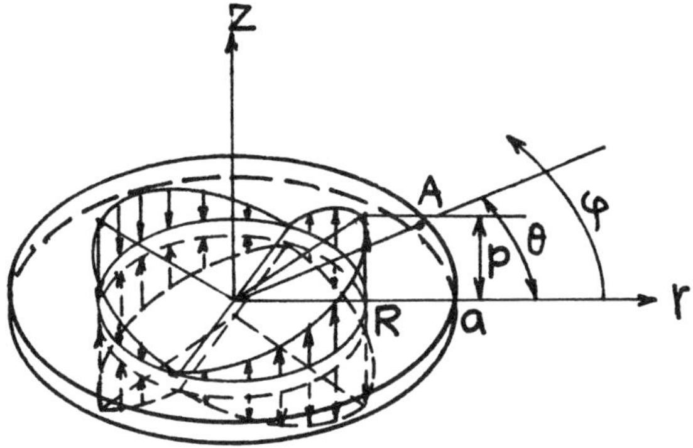

$$\sigma_z(r = R, \varphi) = p \cdot \cos\varphi$$

$$K_{IA} = \frac{2P}{a\sqrt{\pi a}} \frac{R^2}{\sqrt{a^2 - R^2}} \cos\theta$$

$$K_{IIA} = K_{IIIA} = 0$$

Method: Integration of **page 24.3**
Accuracy: Exact
Reference: **Tada 1973**

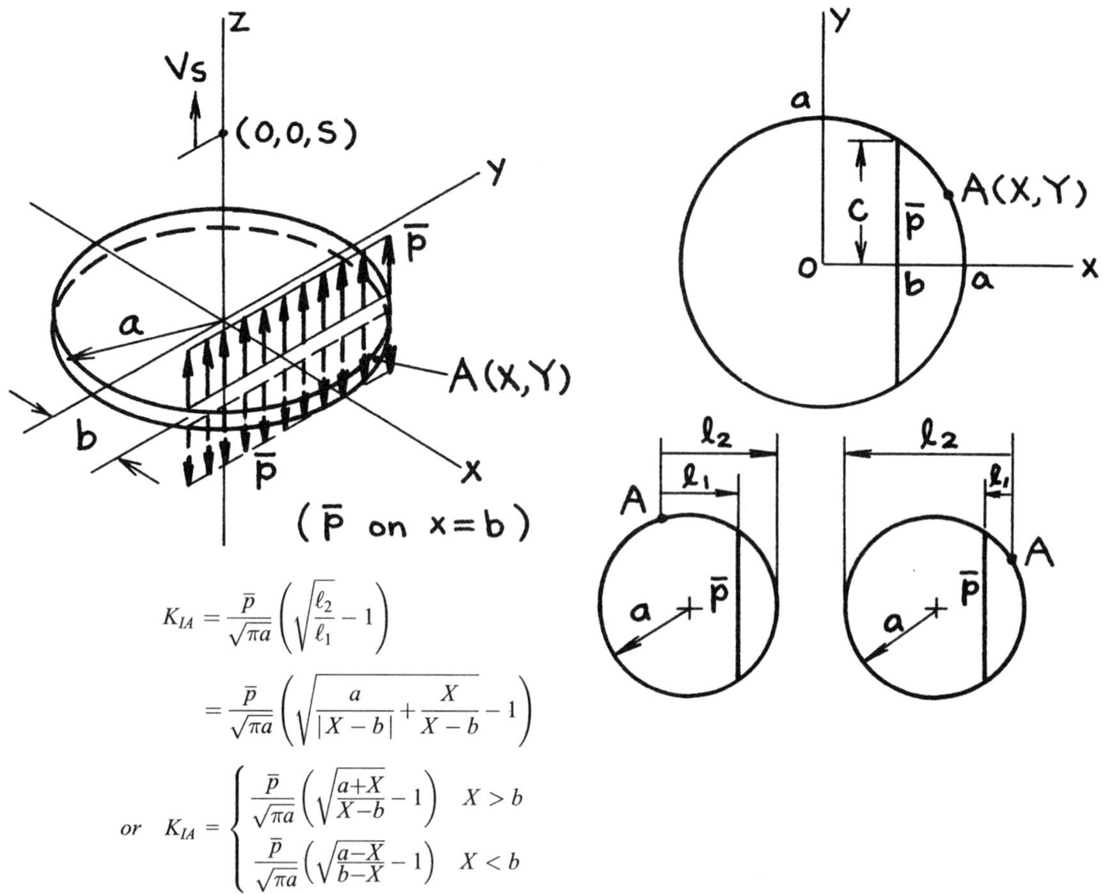

$$K_{IA} = \frac{\bar{p}}{\sqrt{\pi a}}\left(\sqrt{\frac{\ell_2}{\ell_1}} - 1\right)$$

$$= \frac{\bar{p}}{\sqrt{\pi a}}\left(\sqrt{\frac{a}{|X-b|} + \frac{X}{X-b}} - 1\right)$$

$$or \quad K_{IA} = \begin{cases} \dfrac{\bar{p}}{\sqrt{\pi a}}\left(\sqrt{\dfrac{a+X}{X-b}} - 1\right) & X > b \\ \dfrac{\bar{p}}{\sqrt{\pi a}}\left(\sqrt{\dfrac{a-X}{b-X}} - 1\right) & X < b \end{cases}$$

Volume of Crack:

$$V = \frac{4\left(1-\nu^2\right)}{E}\bar{p}\left(a^2 - b^2\right) = \frac{4\left(1-\nu^2\right)}{E}\bar{p}c^2$$

Crack Opening at Center:

$$\delta_0 = 2v(0,0,0) = \frac{4\left(1-\nu^2\right)}{\pi E}\bar{p}\,\ell n\frac{a}{|b|}$$

Vertical Displacement at $(0,0,s)$:

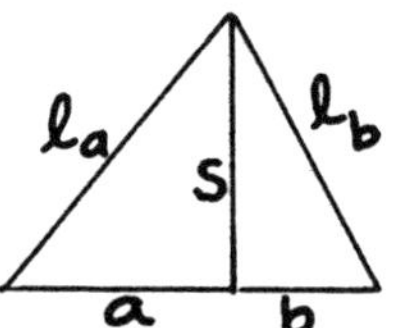

$$v_s = v(0,0,s) = \frac{2\left(1-\nu^2\right)}{\pi E}\bar{p}\left(1 - \alpha s\frac{\partial}{\partial s}\right)\ell n\frac{\ell_a}{\ell_b}$$

$$= \frac{2\left(1-\nu^2\right)}{\pi E}\bar{p}\left\{\ell n\sqrt{\frac{a^2+s^2}{b^2+s^2}} - \alpha\left(\frac{s^2}{a^2+s^2} - \frac{s^2}{b^2+s^2}\right)\right\}$$

$$\alpha = \frac{1}{2(1-\nu)}$$

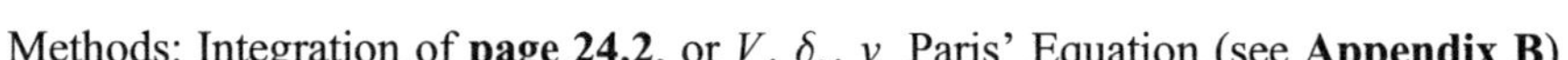

Methods: Integration of **page 24.2**, or V, δ_0, v_s Paris' Equation (see **Appendix B**)
Accuracy: Exact
References: **Tada 1975, 1985**

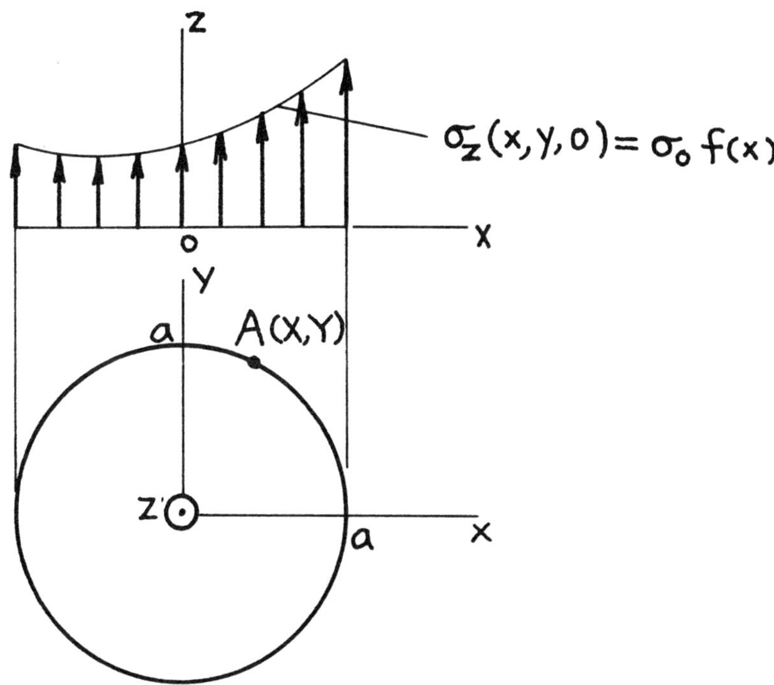

$$K_{IA} = \frac{\sigma_0}{\sqrt{\pi a}} \left\{ \sqrt{a+X} \int_{-a}^{X} \frac{f(x)}{\sqrt{X-x}}\, dx + \sqrt{a-X} \int_{X}^{a} \frac{f(x)}{\sqrt{x-X}}\, dx - \int_{-a}^{a} f(x) dx \right\}$$

Volume of Crack:

$$V = \frac{4\left(1-\nu^2\right)}{E} \sigma_0 \int_{-a}^{a} \left(a^2 - x^2\right) f(x)\, dx$$

Crack Opening at Center:

$$\delta_0 = 2v\,(0,0,0) = \frac{4\left(1-\nu^2\right)}{\pi E} \sigma_0 \int_{-a}^{a} f(x)\,\ell n \frac{a}{|x|}\, dx$$

Method: Integration of **page 24.11**
Accuracy: Exact
References: **Tada 1975, 1985**

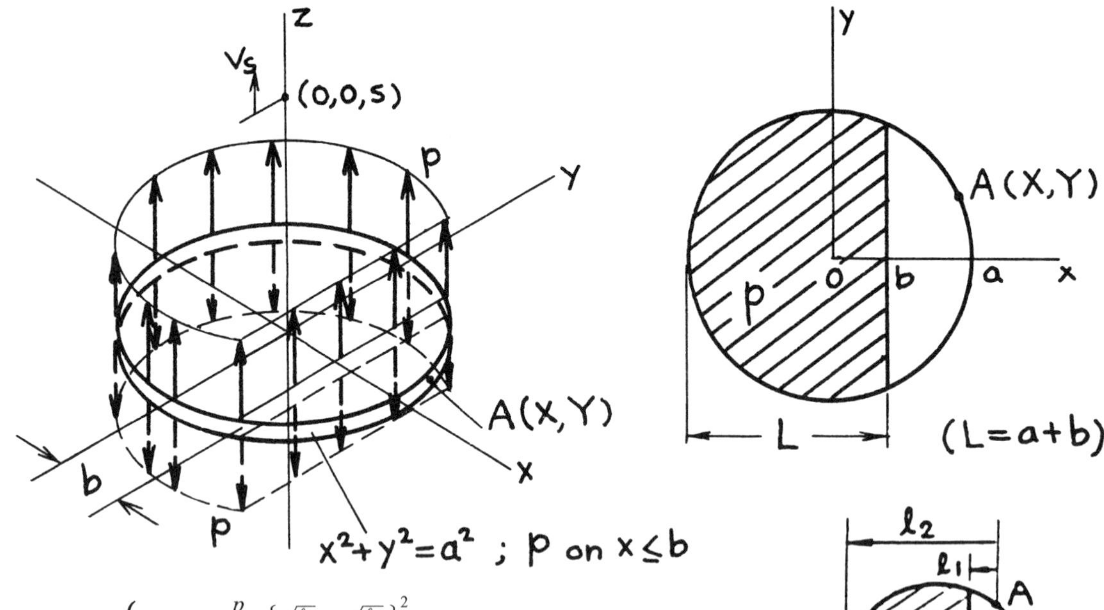

$$K_{IA} = \begin{cases} \frac{p}{\sqrt{\pi a}} \left\{ \sqrt{\ell_2} - \sqrt{\ell_1} \right\}^2 & \\ \left(= \frac{p}{\sqrt{\pi a}} \left\{ \sqrt{a+X} - \sqrt{X-b} \right\}^2 \right) & X > b \\ \frac{p}{\sqrt{\pi a}} \left\{ 2a - \left(\sqrt{\ell_2} - \sqrt{\ell_1} \right)^2 \right\} & \\ \left(= \frac{p}{\sqrt{\pi a}} \left\{ 2a - \left(\sqrt{a-X} - \sqrt{b-X} \right)^2 \right\} \right) & X < b \end{cases}$$

Volume of Crack:

$$V = \frac{4\left(1-\nu^2\right)}{3E} pL^2 (3a - L)$$

Crack Opening at Center:

$$\delta_0 = 2v(0,0,0) = \frac{4\left(1-\nu^2\right)}{\pi E} p \left\{ L - b \, \ell n \, \frac{a}{|b|} \right\}$$

Vertical Displacement at $(0,0,s)$:

$$v_0 = v(0,0,s) = \frac{2\left(1-\nu^2\right)}{\pi E} p \left(1 - \alpha s \frac{\partial}{\partial s} \right) \left\{ b \, \ell n \, \frac{\ell_a}{\ell_b} + L - L' \right\}$$

$$= \frac{2\left(1-\nu^2\right)}{\pi E} p \left(1 - \alpha s \frac{\partial}{\partial s} \right) \left\{ b \, \ell n \sqrt{\frac{a^2+s^2}{b^2+s^2}} + (a+b) - s\left(\tan^{-1} \frac{a}{s} + \tan^{-1} \frac{b}{s} \right) \right\}$$

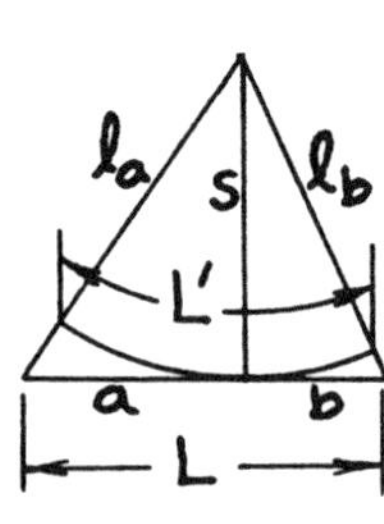

where

$$\alpha = \frac{1}{2(1-\nu)}$$

Method: Integration of **page 24.11**
Accuracy: Exact
Reference: **Tada 1985**

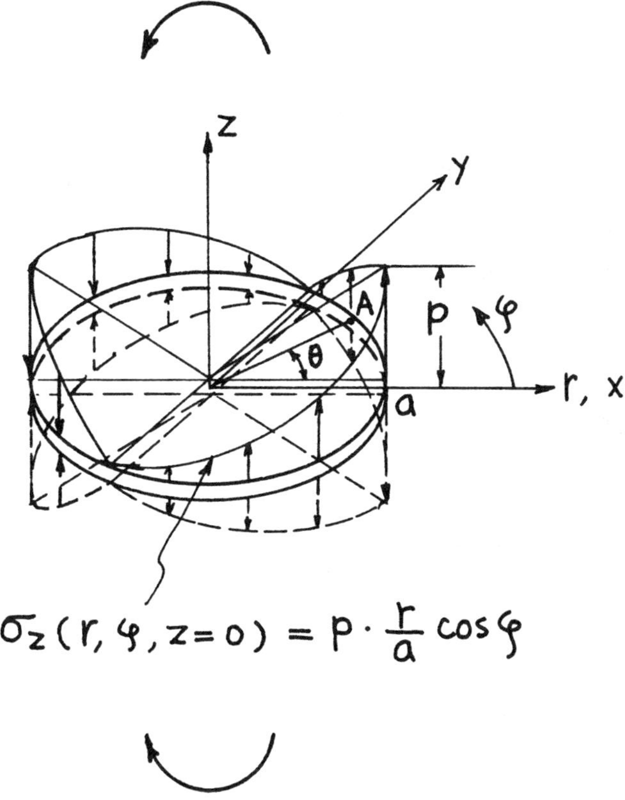

$$K_{IA} = \frac{4}{3\pi} p\sqrt{\pi a} \quad \cos\theta$$

$$K_{II} = K_{III} = 0$$

Method: Integral Transform (Hankel Transform) or Integration of **page 24.9 or 24.11**
Accuracy: Exact
Reference: **Benthem 1972**

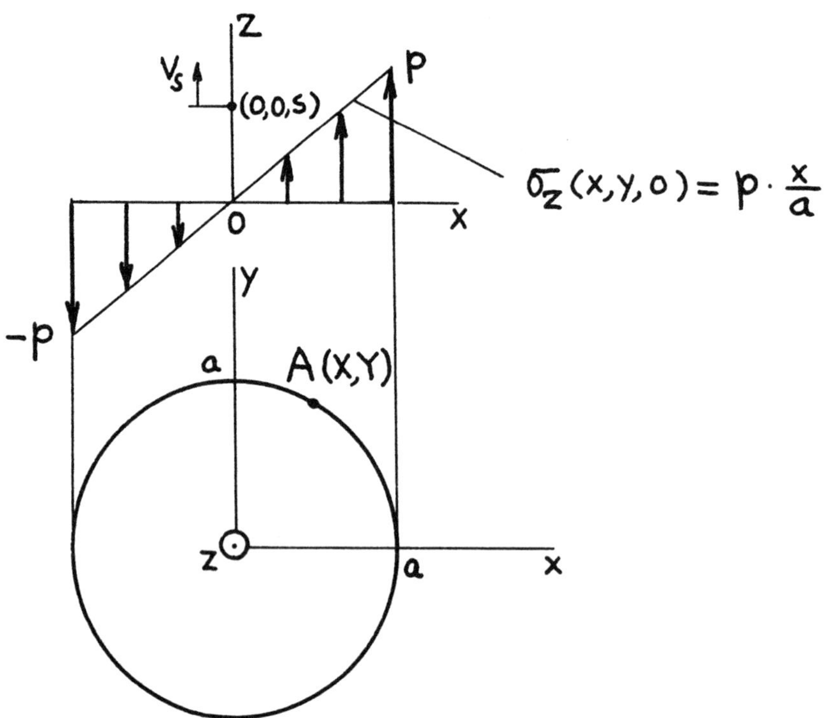

$$K_{IA} = \frac{4}{3\pi} p \sqrt{\pi a} \cdot \frac{X}{a}$$

Volume of Crack:

$$V = 0$$

Crack Opening along Diameter $x = 0$:

$$\delta = 2v(0, y, 0) = 0$$

Vertical Displacement at $(0, 0, s)$:

$$v_s = v(0, 0, s) = 0$$

Method: Integration of **page 24.11**
Accuracy: Exact
Reference: **Tada 1985**

NOTE: Crack surface interference ($x < 0$) was not considered.

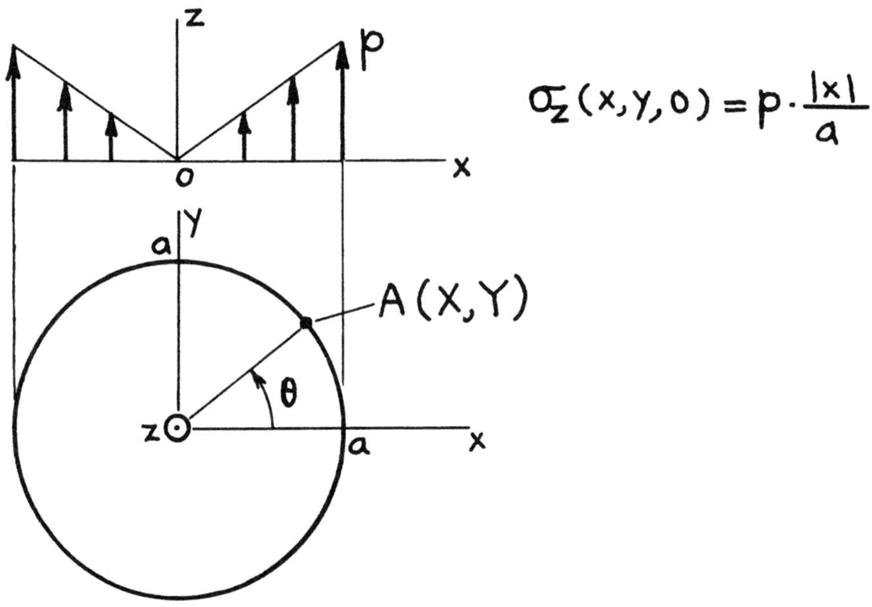

$$K_{IA} = \frac{2}{\pi} p\sqrt{\pi a}\left\{\frac{1}{6} + \frac{4}{3}\left(\frac{X}{a}\right)^{3/2}\left(\sqrt{1+\frac{X}{a}} - \sqrt{\frac{X}{a}}\right)\right\}$$

or

$$= \frac{2}{\pi} p\sqrt{\pi a}\left\{\frac{1}{6} + \frac{4}{3}(\cos\theta)^{3/2}\left(\sqrt{1+\cos\theta} - \sqrt{\cos\theta}\right)\right\}$$

Volume of Crack:

$$V = \frac{2\left(1-\nu^2\right)}{E} pa^3 \left[= \frac{16\left(1-\nu^2\right)}{3E} pa^3 \cdot \left(\frac{3}{8}\right)\right]$$

Crack Opening at Center:

$$\delta_0 = 2v(0,0,0) = \frac{2\left(1-\nu^2\right)}{\pi E} pa \left[= \frac{8\left(1-\nu^2\right)}{\pi E} pa \cdot \left(\frac{1}{4}\right)\right]$$

Displacement at $(0, 0, s)$:

$$v_s = v(0,0,s) = \frac{1-\nu^2}{\pi E} pa \left\{1 - (1-2\alpha)\frac{s^2}{a^2}\,\ell n\left(1+\frac{a^2}{s^2}\right) - 2\alpha\frac{s^2}{s^2+a^2}\right\}$$

where

$$\alpha = \frac{1}{2(1-\nu)}$$

Method: Integration of **page 24.11**
Accuracy: Exact
Reference: **Tada 1975, 1985**

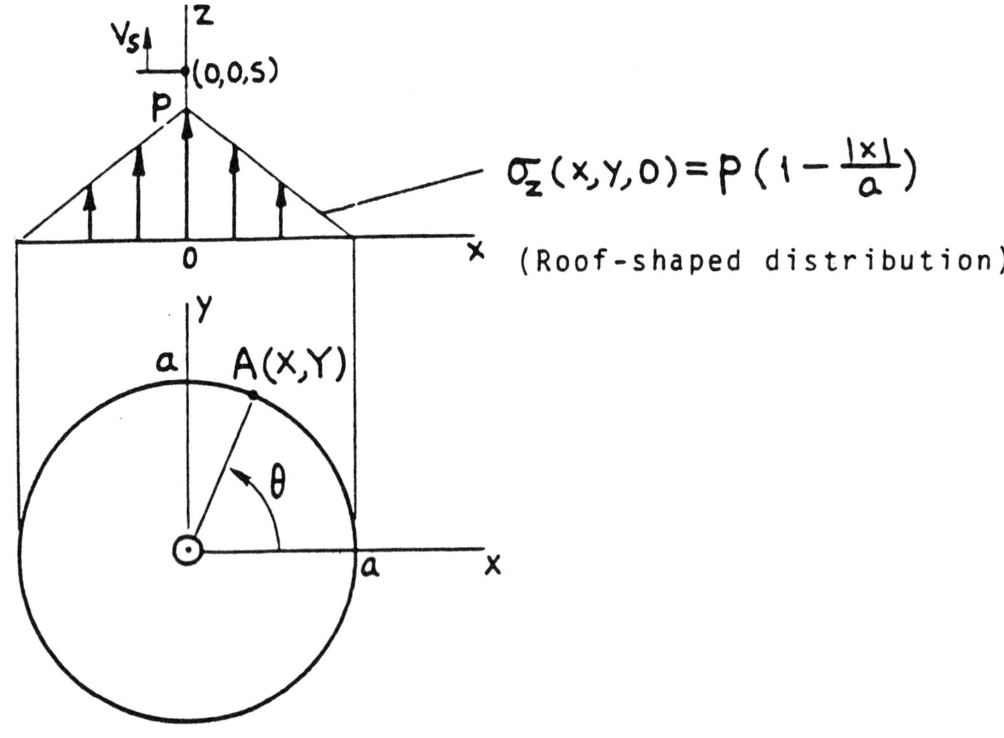

$$K_{IA} = \frac{2}{\pi} p\sqrt{\pi a} \left\{ \frac{5}{6} - \frac{4}{3} \left(\frac{X}{a} \right)^{3/2} \left(\sqrt{1 + \frac{X}{a}} - \sqrt{\frac{X}{a}} \right) \right\}$$

or

$$= \frac{2}{\pi} p\sqrt{\pi a} \left\{ \frac{5}{6} - \frac{4}{3} (\cos\theta)^{3/2} \left(\sqrt{1 + \cos\theta} - \sqrt{\cos\theta} \right) \right\}$$

Volume of Crack:

$$V = \frac{10\left(1 - \nu^2\right)}{3E} pa^3 \left[= \frac{16\left(1 - \nu^2\right)}{3E} pa^3 \cdot \left(\frac{5}{8}\right) \right]$$

Crack Opening at Center:

$$\delta_0 = 2v(0,0,0) = \frac{6\left(1 - \nu^2\right)}{\pi E} pa \left[= \frac{8\left(1 - \nu^2\right)}{\pi E} pa \cdot \left(\frac{3}{4}\right) \right]$$

Displacement at $(0, 0, s)$:

$$v_s = v(0,0,s) = \frac{1 - \nu^2}{\pi E} pa \left\{ 3 - 4(1 - \alpha) \frac{s}{a} \tan^{-1} \frac{a}{s} - (1 - 2\alpha) \frac{s^2}{a^2} \ell n \left(1 + \frac{a^2}{s^2} \right) - 2\alpha \frac{s^2}{s^2 + a^2} \right\}$$

where

$$\alpha = \frac{1}{2(1 - \nu)}$$

Method: Superposition of **pages 24.1 and 24.15**.
Accuracy: Exact
References: **Tada 1975, 1985**

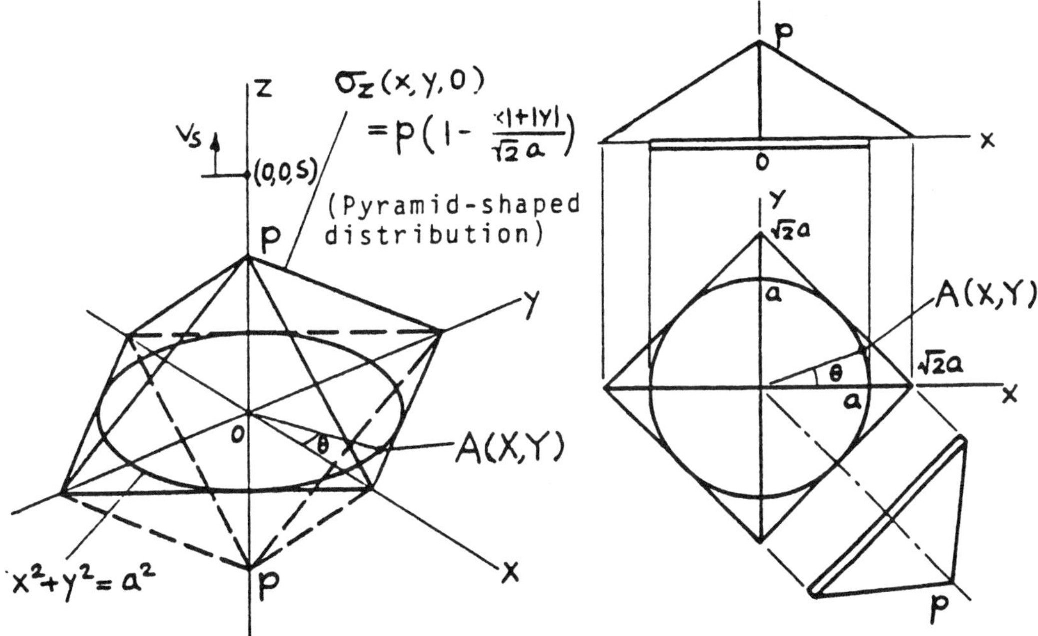

$$K_{IA}=\frac{2}{\pi}p\sqrt{\pi a}\left\{\left(1+\frac{1}{\sqrt{2}}\right)-\frac{2\sqrt{2}}{3}\left[\left(\frac{|X|}{a}\right)^{3/2}\sqrt{1+\frac{|X|}{a}}+\left(\frac{|Y|}{a}\right)^{3/2}\sqrt{1+\frac{|Y|}{a}}\right]\right\}$$

$$=\frac{2}{\pi}p\sqrt{\pi a}\left\{\left(1+\frac{1}{\sqrt{2}}\right)-\frac{2\sqrt{2}}{3}\left[(|\cos\theta|)^{3/2}\sqrt{1+|\cos\theta|}+(|\sin\theta|)^{3/2}\sqrt{1+|\sin\theta|}\right]\right\}$$

Volume of Crack:

$$V=\frac{16\left(1-\nu^2\right)}{3E}pa^3\cdot\left(1-\frac{3}{8}\sqrt{2}\right)\left[=\frac{16\left(1-\nu^2\right)}{3E}pa^3\cdot(.4697)\right]$$

Crack Opening at Center:

$$\delta_0=\frac{8\left(1-\nu^2\right)}{\pi E}pa\cdot\left(1-\frac{\sqrt{2}}{4}\right)\left[=\frac{8\left(1-\nu^2\right)}{\pi E}pa\cdot(.6464)\right]$$

Displacement at $(0,0,s)$:

$$v_s=v(0,0,s)$$

$$=\frac{1-\nu^2}{\pi E}pa\left\{\left(4-\sqrt{2}\right)-4(1-\alpha)\frac{s}{a}\tan^{-1}\frac{a}{s}+\sqrt{2}(1-2\alpha)\frac{s^2}{a^2}\ell n\left(1+\frac{a^2}{s^2}\right)-2\left(2-\sqrt{2}\right)\alpha\frac{s^2}{s^2+a^2}\right\}$$

where

$$\alpha=\frac{1}{2(1-\nu)}$$

Method: Superposition of **page 24.16**
Accuracy: Exact
Reference: **Tada 1985**

NOTE: Because of symmetry, it is only necessary to calculate K_{IA} for $0\le\theta\le\pi/4$.

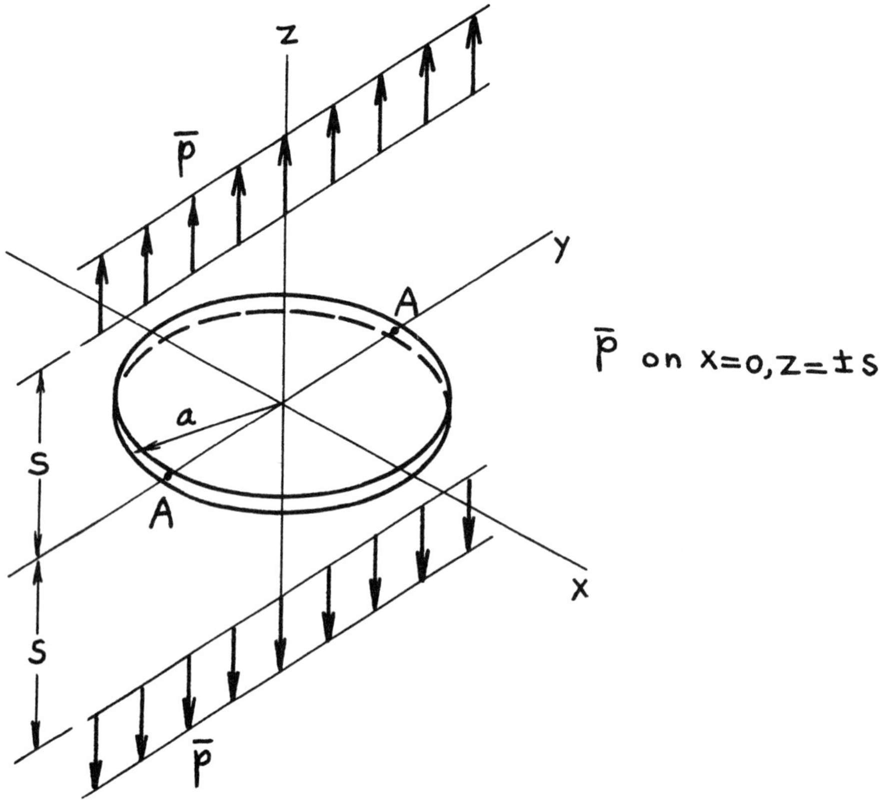

Maximum Value of K_I:

$$(K_I)_{\max} = K_{IA} = \frac{\sqrt{2}\,\bar{p}}{(\pi)^{3/2}}\left(1 - \alpha s\frac{\partial}{\partial s}\right)\left\{\frac{1}{\sqrt{s}}\left(\tanh^{-1}\frac{\sqrt{2as}}{s+a} + \tan^{-1}\frac{\sqrt{2as}}{s-a}\right) - \frac{\sqrt{2}}{\sqrt{a}}\tan^{-1}\frac{a}{s}\right\}$$

Volume of Crack:

$$V = \frac{8\left(1-\nu^2\right)}{\pi E}\bar{p}\,a^2\left\{\left[1 + (1-2\alpha)\frac{s^2}{a^2}\right]\tan^{-1}\frac{a}{s} - (1-2\alpha)\frac{s}{a}\right\}$$

where

$$\alpha = \frac{1}{2(1-\nu)}$$

Method: Integration of **page 24.11**
Accuracy: Exact
Reference: **Tada 1985**

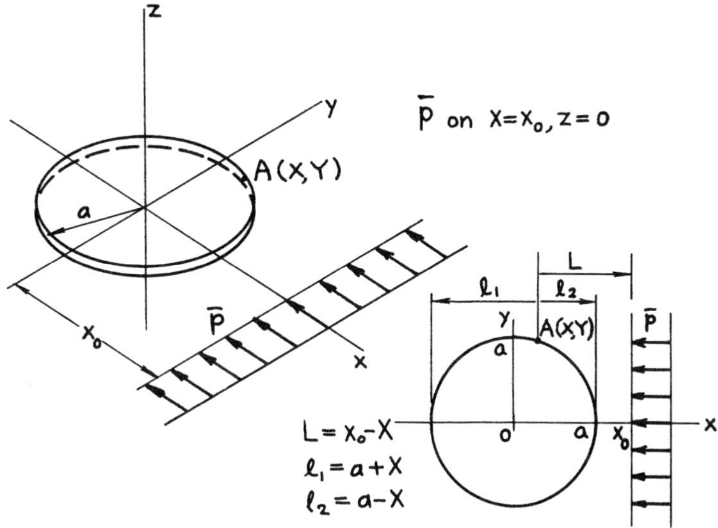

$$K_{IA} = \frac{\bar{p}(1-\alpha)}{\pi\sqrt{\pi a}}\left\{\sqrt{\frac{\ell_1}{L}}\tan^{-1}\sqrt{\frac{\ell_1}{L}} + \sqrt{\frac{\ell_2}{L}}\tanh^{-1}\sqrt{\frac{\ell_2}{L}} - \tanh^{-1}\frac{a}{x_0}\right\}$$

$$(K_{II} = K_{III} = 0)$$

Volume of Crack:

$$V = \frac{4\left(1-\nu^2\right)}{\pi E}\bar{p}(1-\alpha)a^2\left\{\frac{x_0}{a} - \left[\left(\frac{x_0}{a}\right)^2 - 1\right]\tanh^{-1}\frac{a}{x_0}\right\}$$

Crack Opening at Center:

$$\delta_0 = 2v(0,0,0) = \frac{8\left(1-\nu^2\right)}{\pi E}\left(\frac{\bar{p}(1-\alpha)}{2\pi}\cdot\frac{1}{x_0}\right)a\cdot D\left(\frac{x_0}{a}\right)$$

where

$$\alpha = \frac{1}{2(1-\nu)}$$

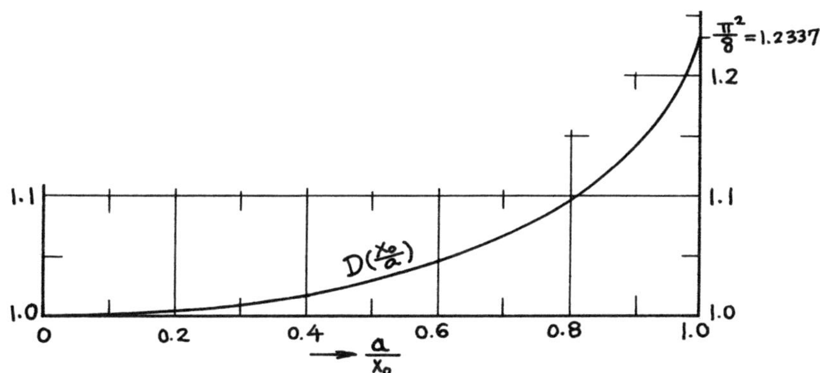

Method: Integration of **page 24.11**
Accuracy: K_{IA}, V Exact; $D(^{x_0}/_a)$ curve is based on accurate numerical values.
Reference: **Tada 1985**

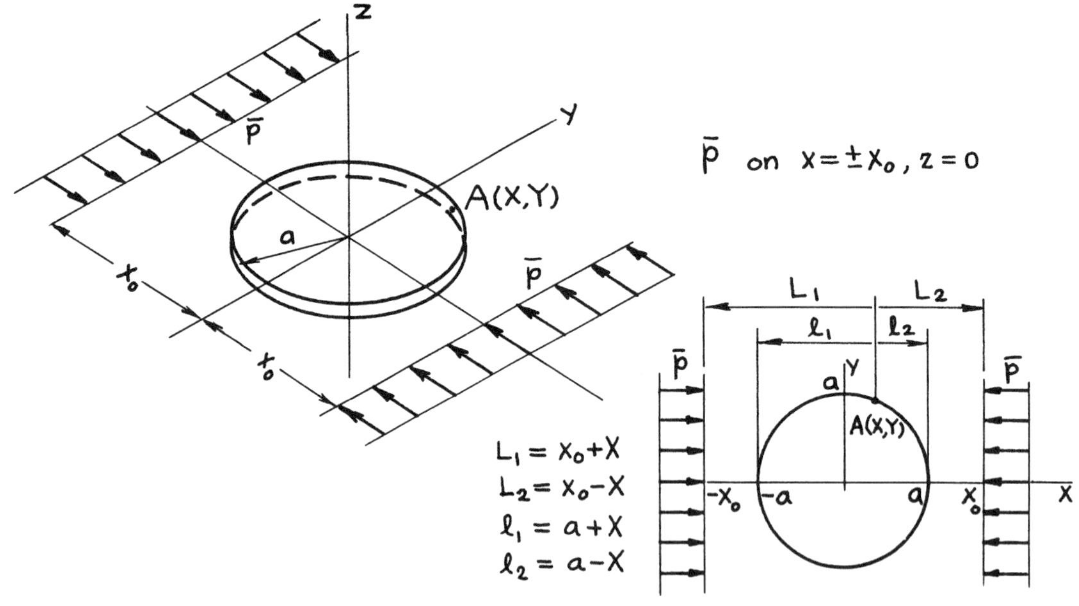

$$K_{IA} = \frac{\bar{p}(1-\alpha)}{\pi\sqrt{\pi a}}\left\{\left(\sqrt{\frac{\ell_1}{L_1}}\tanh^{-1}\sqrt{\frac{\ell_1}{L_1}} + \sqrt{\frac{\ell_2}{L_2}}\tanh^{-1}\sqrt{\frac{\ell_2}{L_2}}\right)\right.$$
$$\left. + \left(\sqrt{\frac{\ell_2}{L_1}}\tan^{-1}\sqrt{\frac{\ell_2}{L_1}} + \sqrt{\frac{\ell_1}{L_2}}\tan^{-1}\sqrt{\frac{\ell_1}{L_2}}\right) - 2\tanh^{-1}\frac{a}{x_0}\right\}$$

$$(K_{II} = K_{III} = 0)$$

Volume of Crack:

$$V = \frac{8\left(1-\nu^2\right)}{\pi E}\bar{p}(1-\alpha)a^2\left\{\frac{x_0}{a} - \left[\left(\frac{x_0}{a}\right)^2 - 1\right]\tanh^{-1}\frac{a}{x_0}\right\}$$

Crack Opening at Center:

$$\delta_0 = 2v(0,0,0) = \frac{8\left(1-\nu^2\right)}{\pi E}\left(\frac{\bar{p}(1-\alpha)}{\pi}\cdot\frac{1}{x_0}\right)a\cdot D\left(\frac{x_0}{a}\right)$$

where

$$\alpha = \frac{1}{2(1-\nu)}$$

Method: Superposition of **page 24.19**
Accuracy: K_{IA}, V Exact; $D(x_0/a)$ curve is based on accurate numerical values.
Reference: **Tada 1985**

NOTE: For numerical values of $D(x_0/a)$, see **page 24.19**.

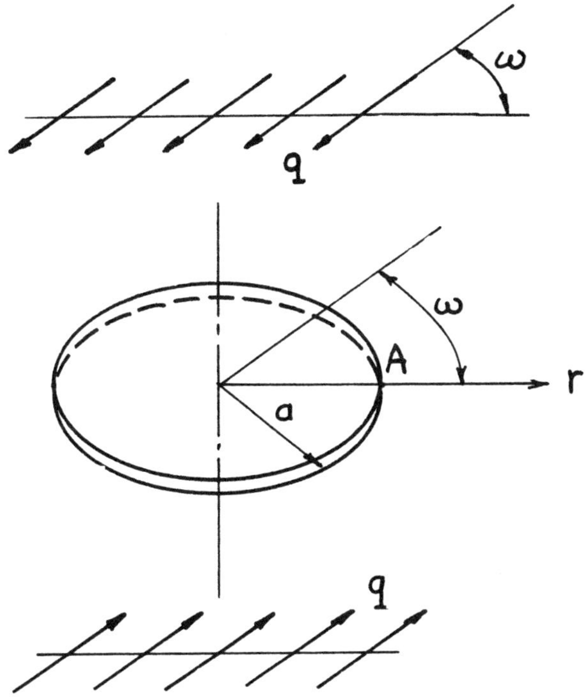

$$K_{IA} = 0$$

$$K_{IIA} = \frac{4}{\pi(2-\nu)}(q\cos\omega)\sqrt{\pi a}$$

$$K_{IIIA} = \frac{4(1-\nu)}{\pi(2-\nu)}(q\sin\omega)\sqrt{\pi a}$$

Method: Three-Dimensional Potential Functions or a Special Case of Elliptical Crack (**page 26.3**)
Accuracy: Exact
References: **Segedin 1950; Sih 1968**

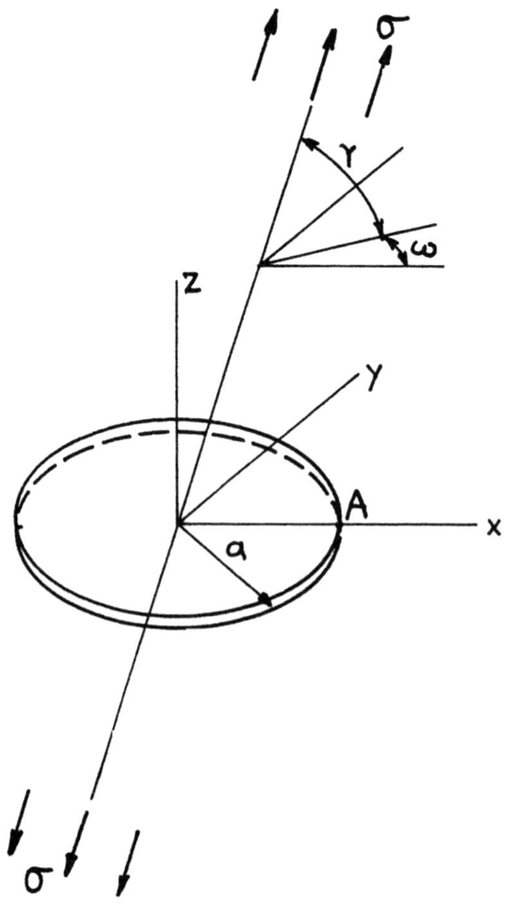

$$K_{IA} = \frac{2}{\pi}\left(\sigma \sin^2 \gamma\right)\sqrt{\pi a}$$

$$K_{IIA} = \frac{4}{\pi(2-\nu)}(\sigma \sin\gamma \cos\gamma) \cos\omega\sqrt{\pi a}$$

$$K_{IIIA} = \frac{4(1-\nu)}{\pi(2-\nu)}(\sigma \sin\gamma \cos\gamma) \sin\omega\sqrt{\pi a}$$

Method: Superposition of **pages 24.1 and 24.21** or Special Case of Elliptical Crack (**page 26.2**)
Accuracy: Exact
References (for Elliptical Crack): **Sadowski 1949; Green 1950; Irwin 1962b; Sih 1968**

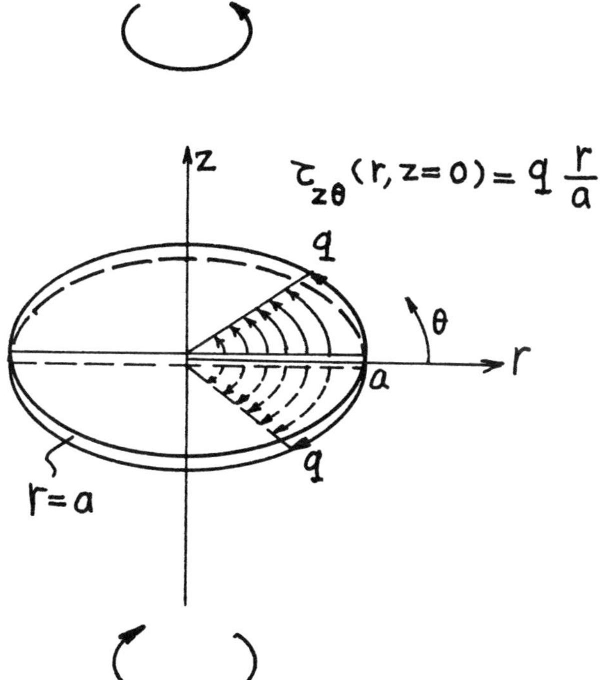

$$K_I = K_{II} = 0$$

$$K_{III} = \frac{4}{3\pi} q\sqrt{\pi a}$$

Method: Integral Transform (Hankel Transform) or from Stress Concentration Factor
Accuracy: Exact
References: **Neuber 1937; Weinstein 1952; Collins 1962; Benthem 1972**

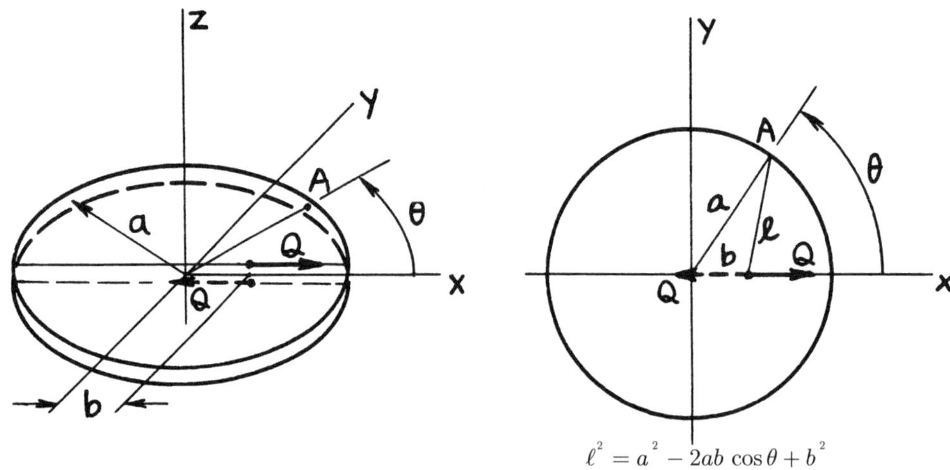

$$\ell^2 = a^2 - 2ab\cos\theta + b^2$$

$$B = {}^{b}\!/_{a}$$

$$L = {}^{\ell}\!/_{a}\left(L^2 = 1 - 2B\cos\theta + B^2\right)$$

$$K_I = 0$$

$$K_{IIA} = \frac{Q}{(\pi a)^{3/2}} \cdot \frac{1}{\sqrt{1-B^2}} \cdot \left\{ B + \frac{2}{2-\nu} \cdot \frac{1}{L^2} \left[\left\{1 + (1-\nu)B^2\right\}(\cos\theta - B) + \nu\left(1 - B^2\right) \frac{\left(1+B^2\right)\cos\theta - 2B}{L^2} \right] \right\}$$

$$K_{IIIA} = -\frac{2Q}{(\pi a)^{3/2}} \cdot \frac{\sqrt{1-B^2}}{2-\nu} \cdot \frac{\sin\theta}{L^2} \left\{ 1 - \nu - \nu \frac{1-B^2}{L^2} \right\}$$

<u>$b = 0$</u> $\quad K_{IIA} = \frac{2Q}{(\pi a)^{3/2}} \cdot \frac{1+\nu}{2-\nu} \cos\theta$

$$K_{IIIA} = -\frac{2Q}{(\pi a)^{3/2}} \cdot \frac{1-2\nu}{2-\nu} \sin\theta$$

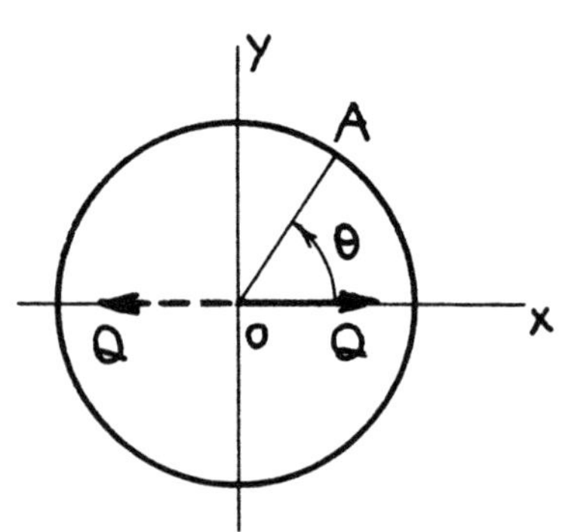

(See **page 24.25**)
Method: Fourier Series Expansions
Accuracy: Exact
References: **Kassir 1975; Tada 1985**

NOTE: A minor error in K_{III} (**Kassir 1975**) was corrected.

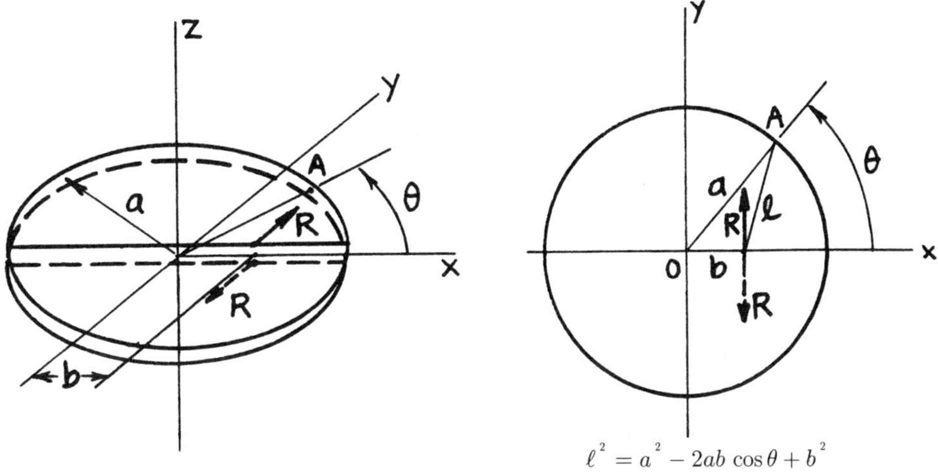

$$\ell^2 = a^2 - 2ab\cos\theta + b^2$$

$$B = b/a$$

$$L = \ell/a \left(L^2 = 1 - 2B\cos\theta + B^2\right)$$

$$K_I = 0$$

$$K_{IIA} = \frac{2R}{(\pi a)^{3/2}} \cdot \frac{1}{2-\nu} \cdot \frac{1+B^2}{\sqrt{1-B^2}} \cdot \frac{\sin\theta}{L^2}\left\{1 + \nu\frac{1-B^2}{L^2}\right\}$$

$$K_{IIIA} = \frac{2R}{(\pi a)^{3/2}} \cdot \frac{1}{2-\nu} \cdot \frac{1}{\sqrt{1-B^2}} \cdot \frac{1}{L^2}\left\{\left(1-\nu+B^2\right)\left(\cos\theta - B\right) - \nu\left(1-B^2\right)\frac{\left(1+B^2\right)\cos\theta - 2B}{L^2}\right\}$$

<u>$b = 0$</u> $\quad K_{IIA} = \frac{2R}{(\pi a)^{3/2}} \cdot \frac{1+\nu}{2-\nu}\sin\theta$

$$K_{IIIA} = \frac{2R}{(\pi a)^{3/2}} \cdot \frac{1-2\nu}{2-\nu}\cos\theta$$

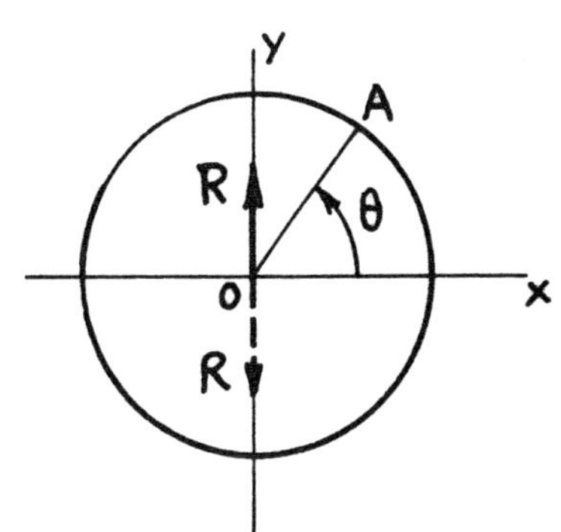

(See **page 24.24**)
Method: Fourier Series Expansions
Accuracy: Exact
References: **Kassir 1975; Tada 1985**

NOTE: The series form solutions in **Kassir (1975)** were converted into closed-form expressions.

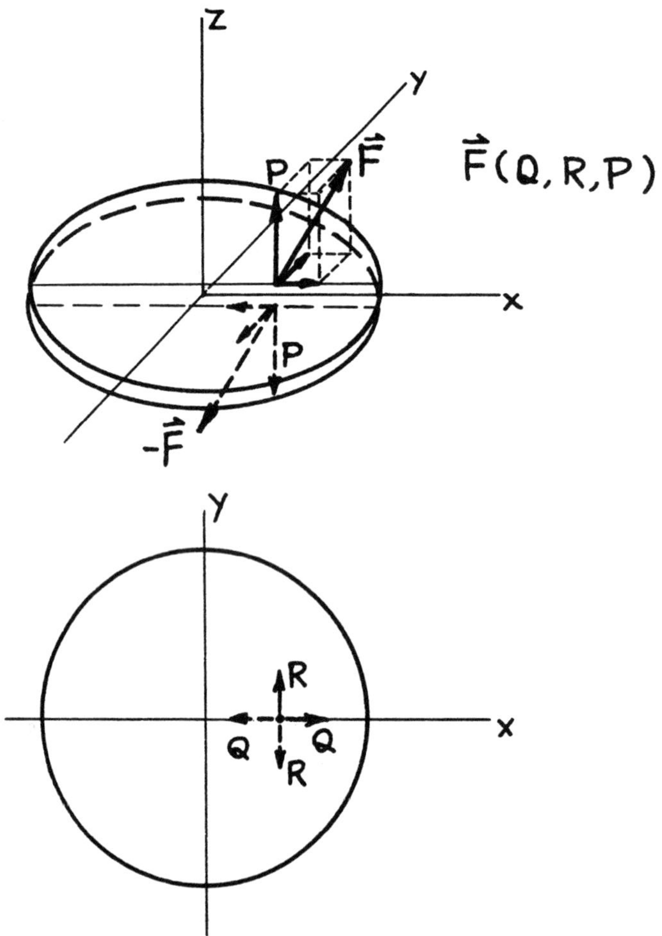

See **pages 24.3, 24.24, and 24.25**.

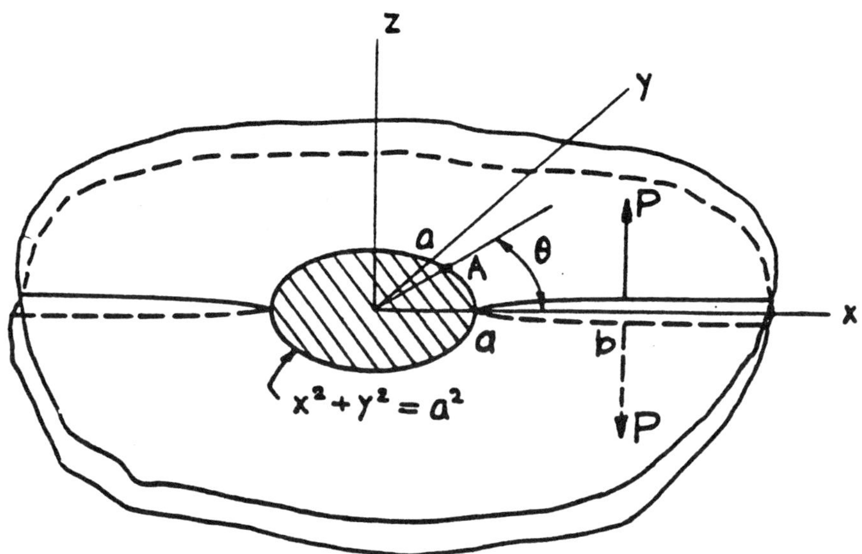

$$K_{IA} = \frac{P}{(\pi a)^{3/2}} \left\{ \cos^{-1}\frac{a}{b} + \frac{a\sqrt{b^2 - a^2}}{a^2 + b^2 - 2ab\cos\theta} \right\}$$

$$K_{IIA} = K_{IIIA} = 0$$

Method: Neuber-Papkovich potential
Accuracy: Exact
References: **Galin 1953; Tada 1973**

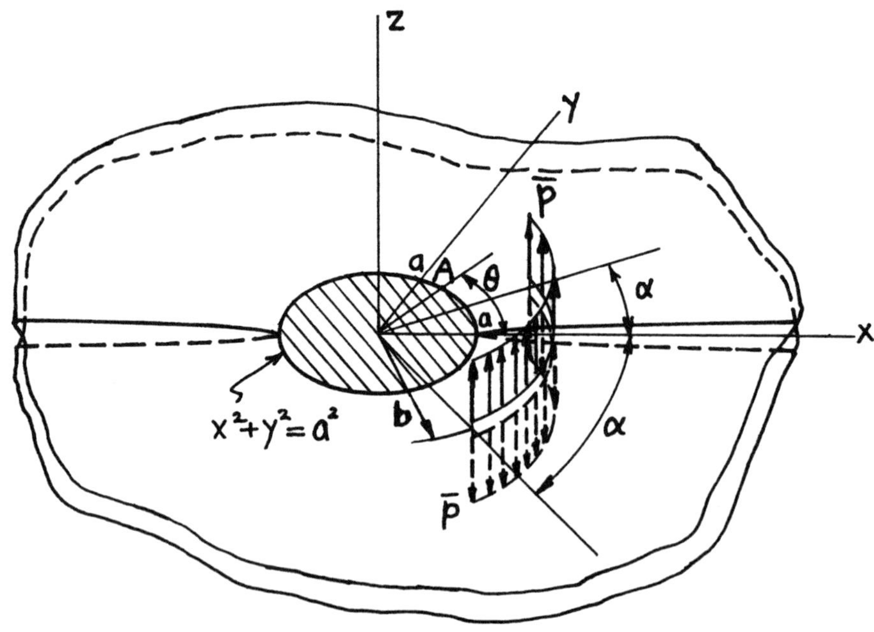

$$K_{IA} = \frac{2\bar{p}b}{(\pi a)^{3/2}}\left[\alpha \cos^{-1}\frac{a}{b} + \frac{a}{\sqrt{b^2 - a^2}}\left\{\tan^{-1}\left(\frac{b+a}{b-a}\tan\frac{\theta+\alpha}{2}\right) - \tan^{-1}\left(\frac{b+a}{b-a}\tan\frac{\theta-\alpha}{2}\right)\right\}\right]$$

$$K_{IIA} = K_{IIIA} = 0$$

Method: Integration of **page 25.1**
Accuracy: Exact
Reference: **Tada 1973**

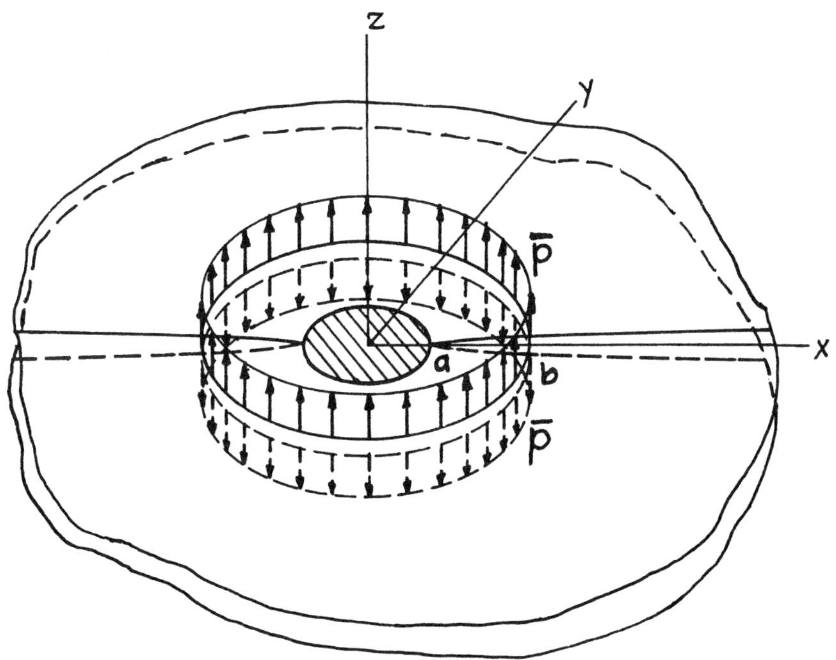

$$K_I = \frac{\bar{p}(2\pi b)}{(\pi a)^{3/2}} \left\{ \cos^{-1}\frac{a}{b} + \frac{a}{\sqrt{b^2 - a^2}} \right\}$$

$$K_{II} = K_{III} = 0$$

Methods: Boussinesq-Papkovich Potential or Special Case of **page 25.2**
Accuracy: Exact
References: **Bueckner 1972; Tada 1973**

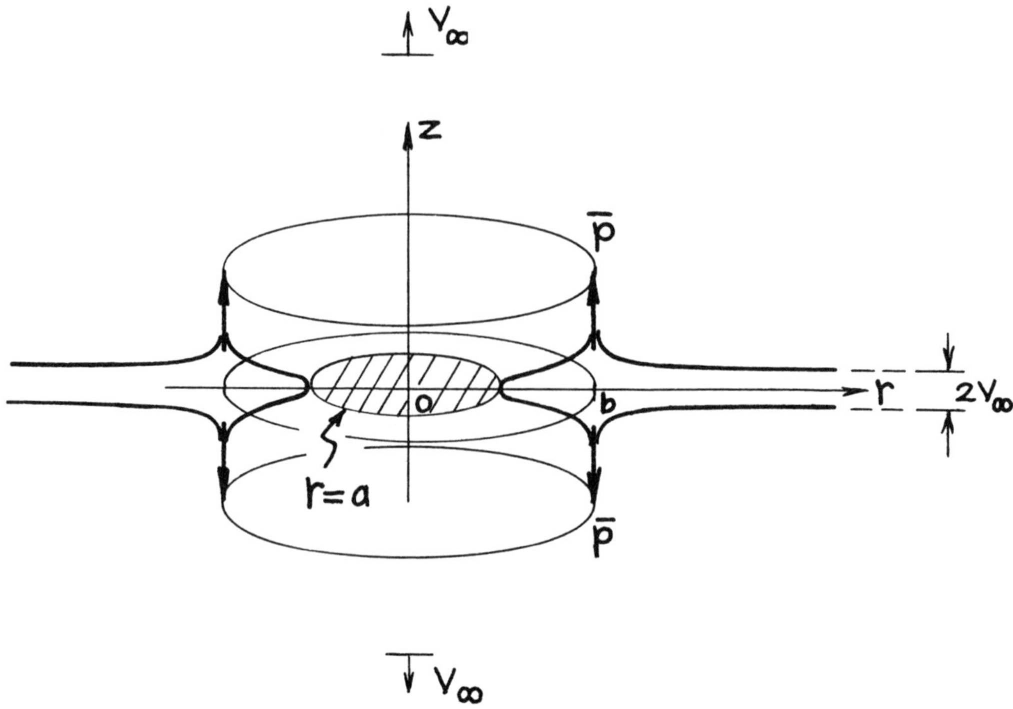

$$K_I = \frac{2\bar{p}}{\sqrt{\pi a}} \left\{ \frac{b}{a} \cos^{-1} \frac{a}{b} + \frac{b}{\sqrt{b^2 - a^2}} \right\}$$

Relative Displacement at Infinity:

$$2V_\infty = \frac{4\left(1 - \nu^2\right)}{E} \bar{p} \cdot \frac{b}{a} \cos^{-1} \frac{a}{b}$$

Method: Paris' Equation (see **Appendix B**)
Accuracy: Exact
Reference: **Tada 1985**

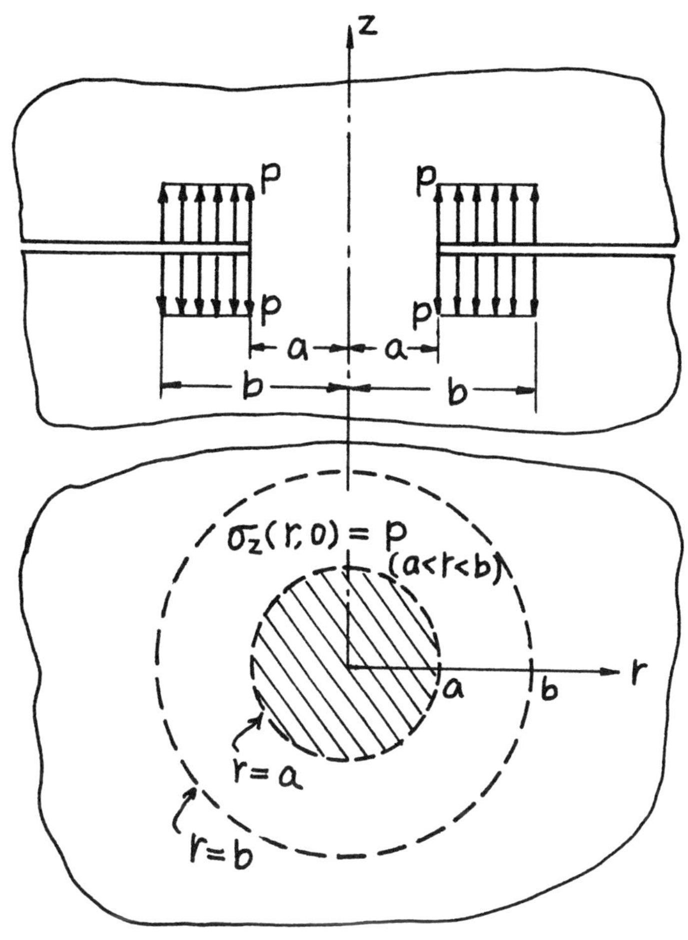

$$K_I = \frac{P\left(\pi b^2\right)}{(\pi a)^{3/2}}\left\{\cos^{-1}\frac{a}{b}+\frac{a}{b}\sqrt{1-\left(\frac{a}{b}\right)^2}\right\}$$

$$K_{II} = K_{III} = 0$$

Method: Integration of **page 25.3**
Accuracy: Exact
Reference: **Tada 1973**

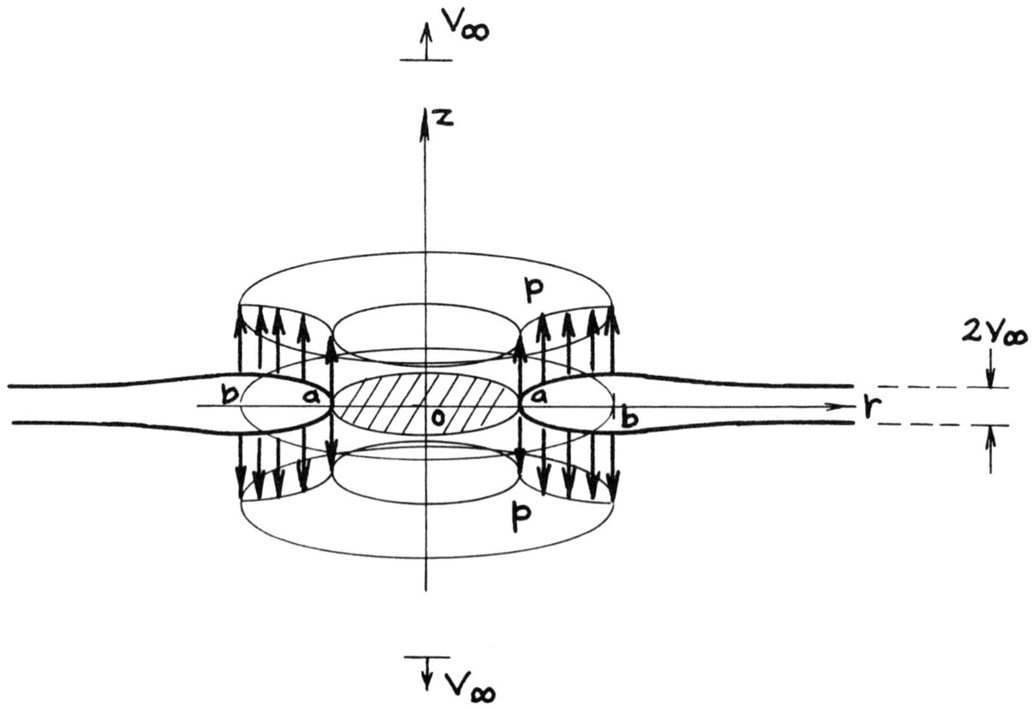

$$K_I = \frac{P}{\sqrt{\pi}}\sqrt{a}\left\{\left(\frac{b}{a}\right)^2 \cos^{-1}\frac{a}{b} + \sqrt{\left(\frac{b}{a}\right)^2 - 1}\right\}$$

Relative Displacement at Infinity:

$$2V_\infty = \frac{2(1-\nu^2)}{E}pa\left\{\left(\frac{b}{a}\right)^2 \cos^{-1}\frac{a}{b} - \sqrt{\left(\frac{b}{a}\right)^2 - 1}\right\}$$

Method: Integration of **page 25.3** or Paris' Equation (see **Appendix B**)
Accuracy: Exact
Reference: **Tada 1985**

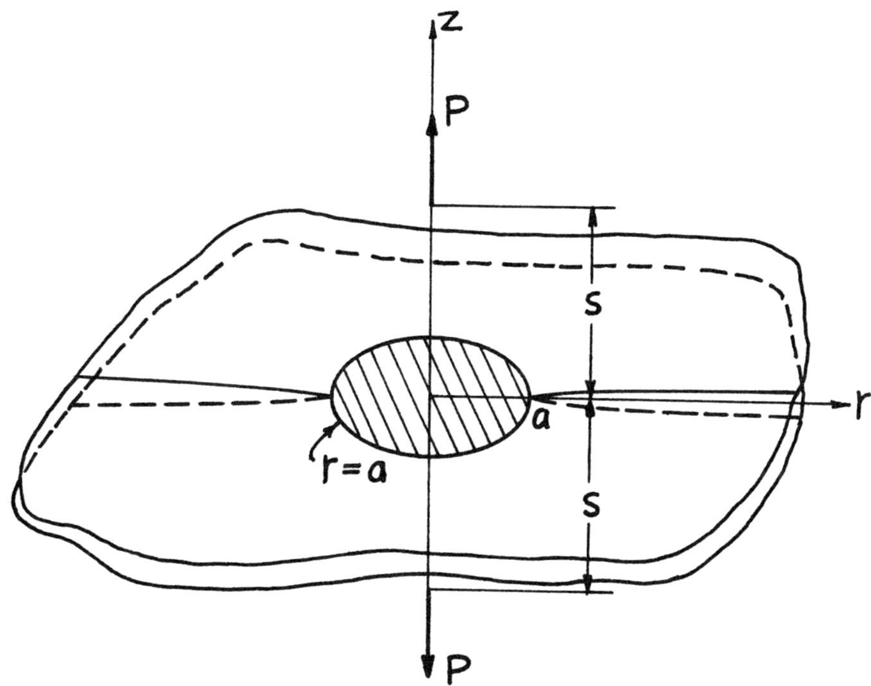

$$\underset{No\ Crack}{\sigma_z(r, z = 0)} = \frac{P}{2\pi}\left[1 - \alpha s \frac{\partial}{\partial s}\right] \frac{s}{\left(r^2 + s^2\right)^{3/2}}$$

$$K_I = \frac{P}{(\pi a)^{3/2}}\left[\tan^{-1}\frac{s}{a} + \frac{as}{a^2 + s^2}\left\{1 - \alpha\frac{2a^2}{a^2 + s^2}\right\}\right]$$

$$K_{II} = K_{III} = 0$$

where

$$\alpha = \frac{1}{2(1 - \nu)}$$

Method: Integration of **page 25.3**
Accuracy: Exact
Reference: **Tada 1973**

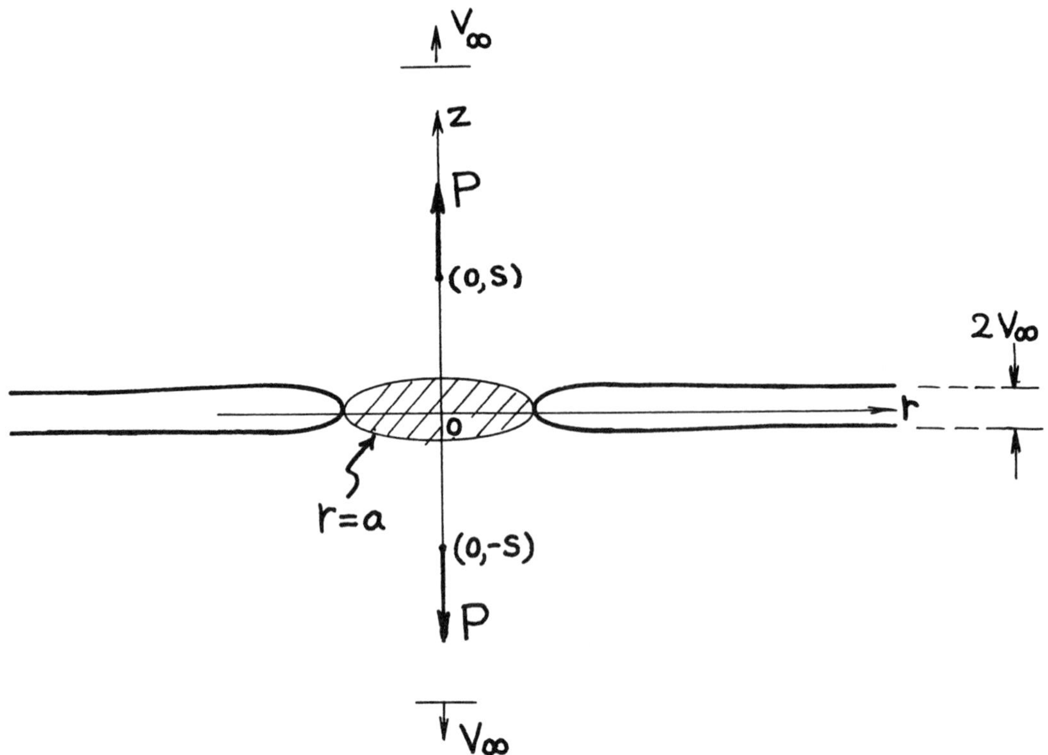

$$K_I = \frac{P}{(\pi a)^{3/2}}\left(1 - \alpha s\frac{\partial}{\partial s}\right)\left\{\tan^{-1}\frac{s}{a} + \frac{as}{a^2+s^2}\right\}$$

$$\text{or} \qquad = \frac{P}{(\pi a)^{3/2}}\left\{\tan^{-1}\frac{s}{a} + \frac{s/a}{1+(s/a)^2} - \alpha\frac{s/a}{\left[1+(s/a)^2\right]^2}\right\}$$

Relative Displacement at Infinity:

$$2V_\infty = \frac{2\left(1-\nu^2\right)}{\pi E}\frac{P}{a}\left\{\tan^{-1}\frac{s}{a} - \alpha\frac{s/a}{1+(s/a)^2}\right\}$$

where

$$\alpha = \frac{1}{2(1-\nu)}$$

Method: Reciprocity (see **page 25.6a**) or Paris' Equation (see **Appendix B**)
Accuracy: Exact
Reference: **Tada 1985**

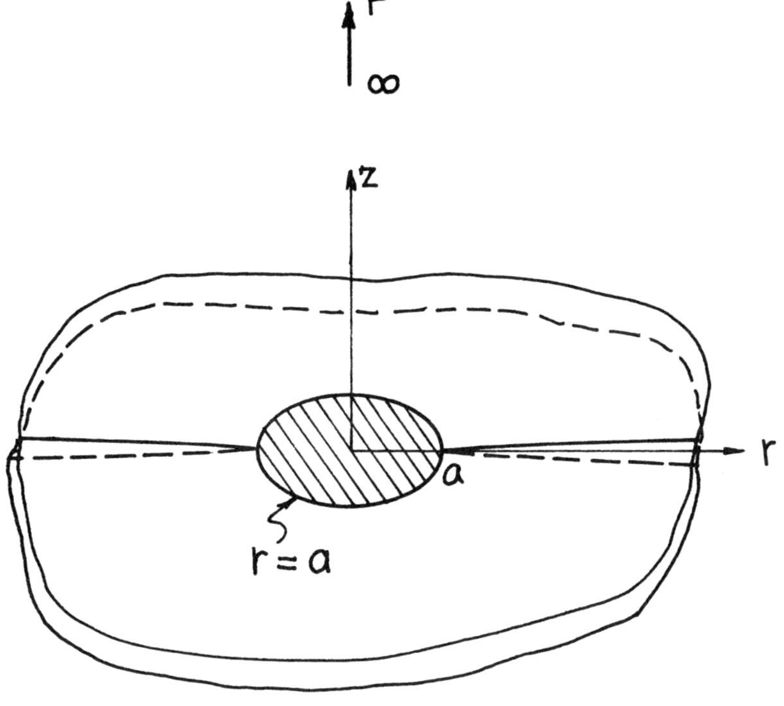

$$K_I = \frac{P}{2a\sqrt{\pi a}}$$

$$K_{II} = K_{III} = 0$$

Methods: Stress Concentration Factor (Neuber), Solution for a Stamp Problem (e.g., Sneddon), Special Case of Elliptical Net Ligament (**page 26.4**) or Special Case of **page 25.5**
Accuracy: Exact
References: **Neuber 1937; Sneddon 1951**

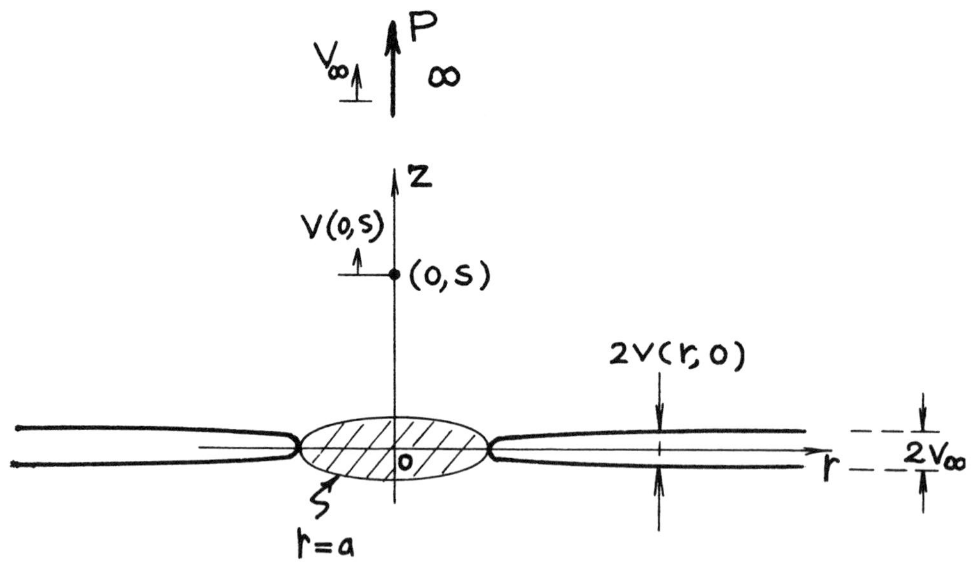

$$K_I = \frac{P}{2a\sqrt{\pi a}}$$

Crack Opening Profile:

$$\underset{r \geq a}{2v\,(r,0)} = \frac{2\left(1-\nu^2\right)}{\pi E}\frac{P}{a}\cos^{-1}\frac{a}{r}$$

Vertical Displacement at $(0,s)$:

$$v\,(0,s) = \frac{1-\nu^2}{\pi E}\frac{P}{a}\left\{\tan^{-1}\frac{s}{a} - \alpha\frac{s/a}{1+(s/a)^2}\right\}$$

Relative Displacement at Infinity:

$$2V_\infty\left(=\underset{r\to\infty}{\{2v(r,0)\}} = \underset{s\to\infty}{\{2v(0,s)\}}\right) = \frac{1-\nu^2}{E}\frac{P}{a}$$

where

$$\alpha = \frac{1}{2(1-\nu)}$$

Method: Paris' Equation (see **Appendix B**)
Accuracy: Exact
Reference: **Tada 1985**

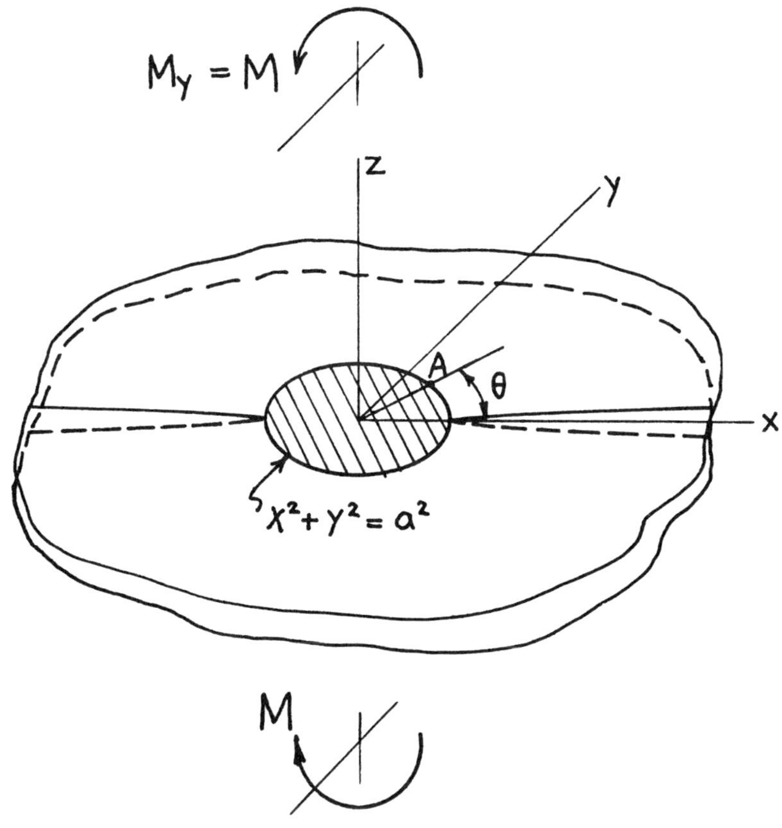

$$K_{IA} = \frac{3}{2}\frac{M}{a^2\sqrt{\pi a}}\cos\theta$$

$$= \frac{3}{8}\sigma_0\sqrt{\pi a}\cos\theta \qquad \left(\sigma_0 = \frac{4M}{\pi a^3}\right)$$

$$K_{II} = K_{III} = 0$$

Methods: Stress Concentration Factor (Neuber), Solution for a Stamp Problem (e.g., Sneddon)
Accuracy: Exact
References: **Neuber 1937; Sneddon 1951**

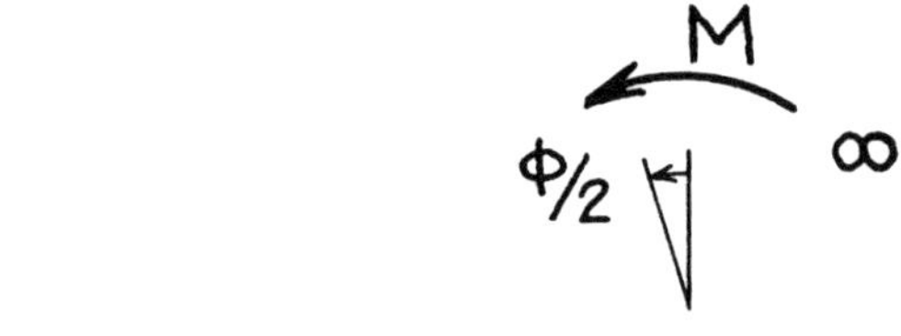

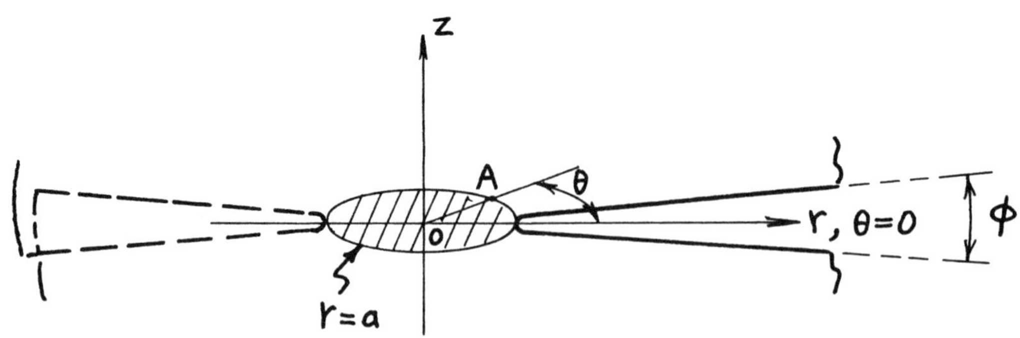

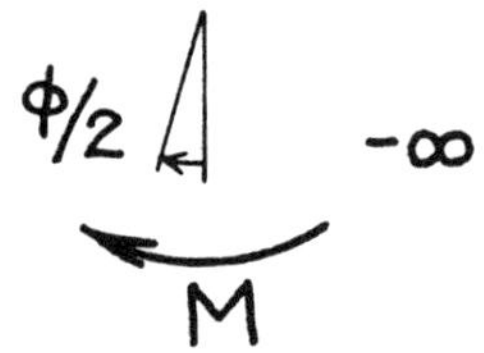

$$K_{IA} = \frac{3}{2} \frac{M}{a^2 \sqrt{\pi a}} \cos\theta$$

Relative Rotation at Infinity:

$$\phi = \frac{3(1-\nu^2)}{2E} \cdot \frac{M}{a^3}$$

K_{IA} in terms of ϕ:

$$K_{IA} = \frac{E}{\sqrt{\pi}\left(1-\nu^2\right)} \phi \sqrt{a} \cos\theta$$

Method: Paris' Equation (see **Appendix B**)
Accuracy: Exact
Reference: **Tada 1985**

NOTE: Crack surface interference ($\pi/2 < \theta < 3\pi/2$) was not considered.

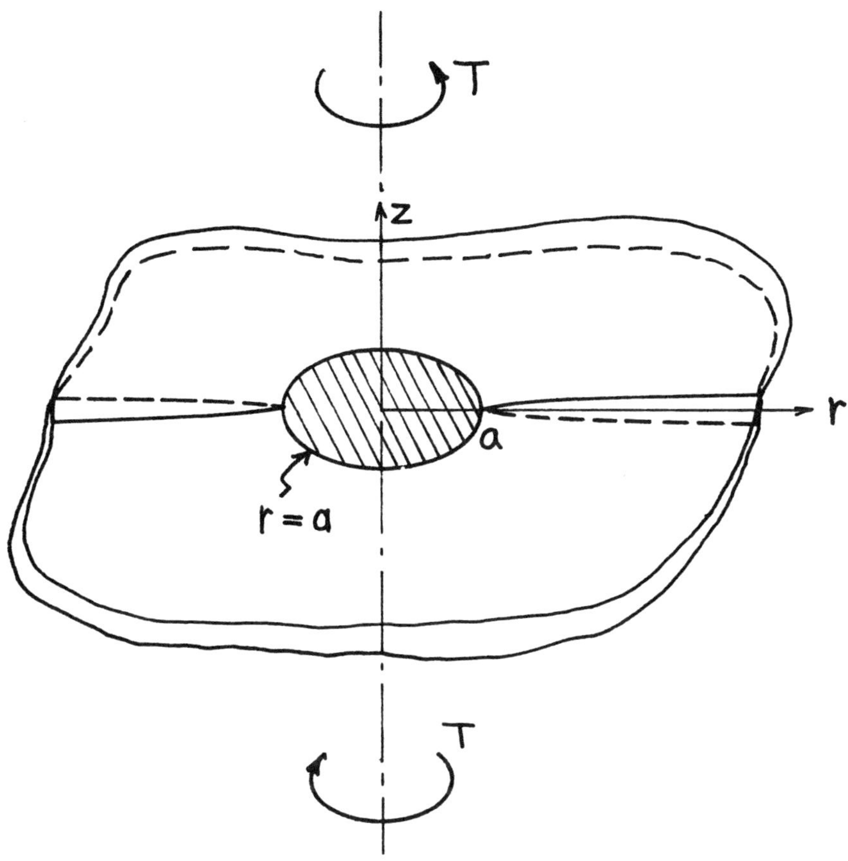

$$K_{III} = \frac{3}{4} \frac{T}{a^2 \sqrt{\pi a}}$$

$$= \frac{3}{8} \tau_0 \sqrt{\pi a} \qquad \left(\tau_0 = \frac{2T}{\pi a^3} \right)$$

$$K_I = K_{II} = 0$$

Methods: Stress Concentration Factor (Neuber), Solution for a Stamp Problem (e.g., Sneddon)
Accuracy: Exact
References: **Neuber 1937; Sneddon 1951**

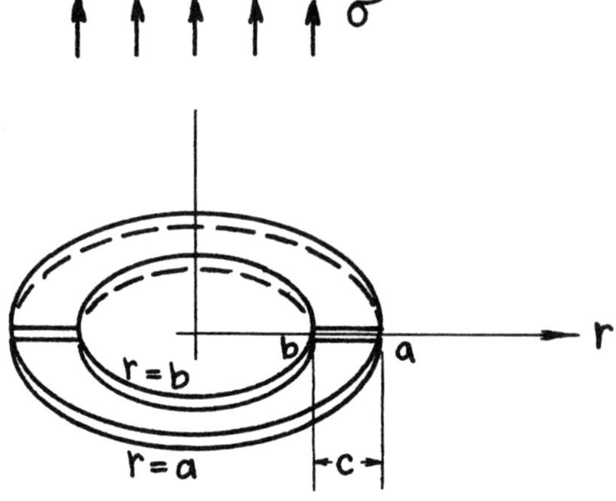

$$K_{I}\underset{r=a}{} = \sigma\sqrt{\pi\frac{c}{2}}\cdot F_a\left(\frac{c}{a}\right)$$

$$K_{I}\underset{r=b=a-c}{} = \sigma\sqrt{\pi\frac{c}{2}}\cdot F_b\left(\frac{c}{a}\right)$$

where

$$F_a\left(\frac{c}{a}\right) = 1 - .116\,\frac{c}{a} + .016\left(\frac{c}{a}\right)^2$$

$$F_b\left(\frac{c}{a}\right) = \frac{1 - .36\frac{c}{a} - .067\left(\frac{c}{a}\right)^2}{\sqrt{1-\frac{c}{a}}}$$

$$\left[F_a\left(\frac{c}{a}\right)\right]_{\frac{c}{a}\to 1} = \frac{2\sqrt{2}}{\pi} = .9002$$

$$\left[\sqrt{1-\frac{c}{a}}\,F_b\left(\frac{c}{a}\right)\right]_{\frac{c}{a}\to 1} = \frac{4\sqrt{2}}{\pi^2} = .573$$

Method: Singular Integral Equation
Accuracy: Better than 0.5% (formulas are based on Erdogan's numerical results)
References: **Erdogan 1982; Tada 1985**

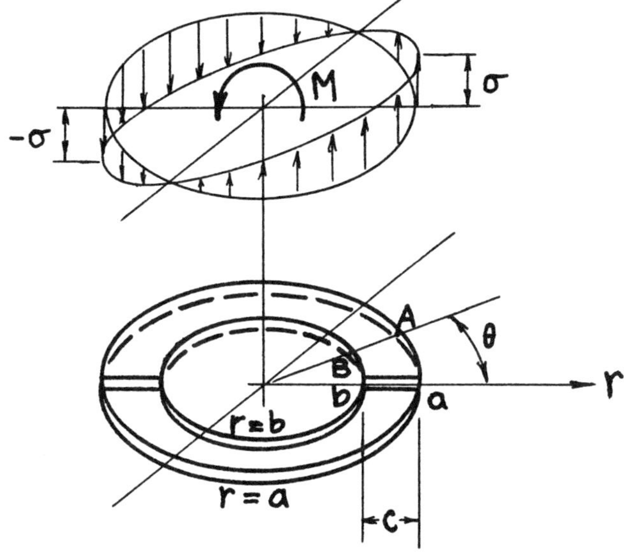

$$\underset{(r=a,0)}{K_{IA}} = \sigma\sqrt{\pi\frac{c}{2}}\cos\theta \cdot F_A\left(\frac{c}{a}\right)$$

$$\underset{(r=b=a-c,\theta)}{K_{IB}} = \sigma\sqrt{\pi\frac{c}{2}}\cos\theta \cdot F_B\left(\frac{c}{a}\right)$$

where

$$F_A\left(\frac{c}{a}\right) = 1 - .401\frac{c}{a} - .065\left(\frac{c}{a}\right)^2 + .066\left(\frac{c}{a}\right)^3$$

$$F_B\left(\frac{c}{a}\right) = \sqrt{1-\frac{c}{a}}\left[.573 + .427\left(1-\frac{c}{a}\right)^{1/4} - .26\left(\frac{c}{a}\right)^{5/2}\left(1-\frac{c}{a}\right)^{3/2}\right]$$

$$\left[F_A\left(\frac{c}{a}\right)\right]_{\frac{c}{a}\to 1} = \frac{4\sqrt{2}}{3\pi} = .6002$$

$$\left[\frac{F_B\left(\frac{c}{a}\right)}{\sqrt{1-\frac{c}{a}}}\right]_{\frac{c}{a}\to 1} = \frac{4\sqrt{2}}{\pi^2} = .573$$

Method: Singular Integral Equation
Accuracy: Better than 0.5% (formulas are based on Erdogan's numerical results)
References: **Erdogan 1982; Tada 1985**

Notes on Solutions for Elliptical Crack Problems

See **pages 26.2, 26.3, and 26.4** (Internal Cracks), and **page 26.5** (External Crack).

1. The angle θ is the parametric angle representing Point A on the crack front. That is, the coordinates of Point A are $[a\cos\theta,\ b\sin\theta,\ (0)]$, as shown in **Fig. 1**.
2. Note that

$$b\left(\sin^2\theta+\frac{b^2}{a^2}\cos^2\theta\right)^{1/2}=\ell$$

   is the length of the normal at Point A, as shown in **Fig. 2**.
3. A term $\left(\sin^2\theta+\frac{b^2}{a^2}\cos^2\theta\right)^{1/4}$ repeatedly appears in the solution. By replacing this term with $\sqrt{\ell/b}$, solutions may be expressed more concisely. For example,

$$K_{IA}=\frac{\sigma\sqrt{\pi\ell}}{E(k)}$$

   for **page 26.2** and

$$K_{IA}=\frac{P}{2a\sqrt{\pi\ell}}$$

   for **page 26.5**.
4. The numerical values of the complete elliptic integrals of the first kind, $K(k)$, and the second kind, $E(k)$, are tabulated in **Appendix L**. Accurate empirical formulas for the second complete elliptic integral, $E(k)$, are also given in **Appendix L**.

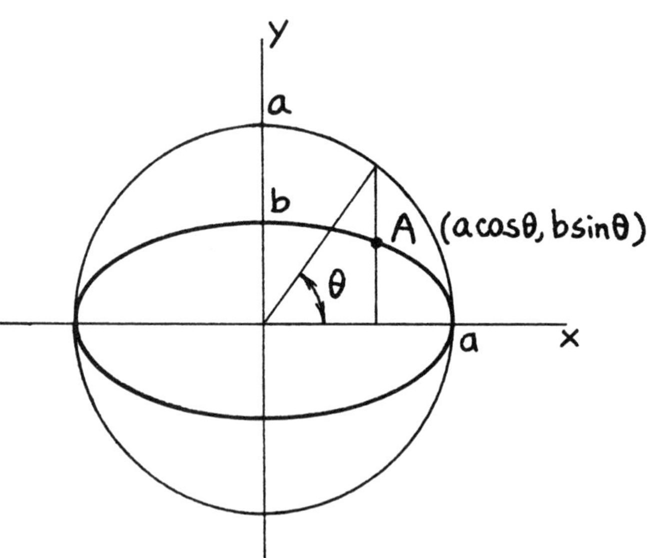

Fig. 1

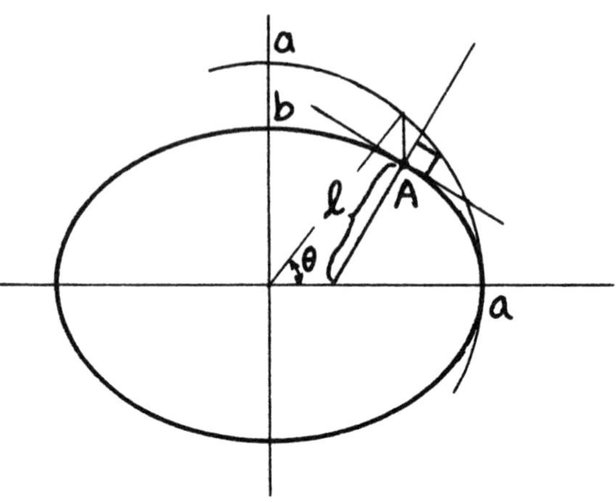

Fig. 2

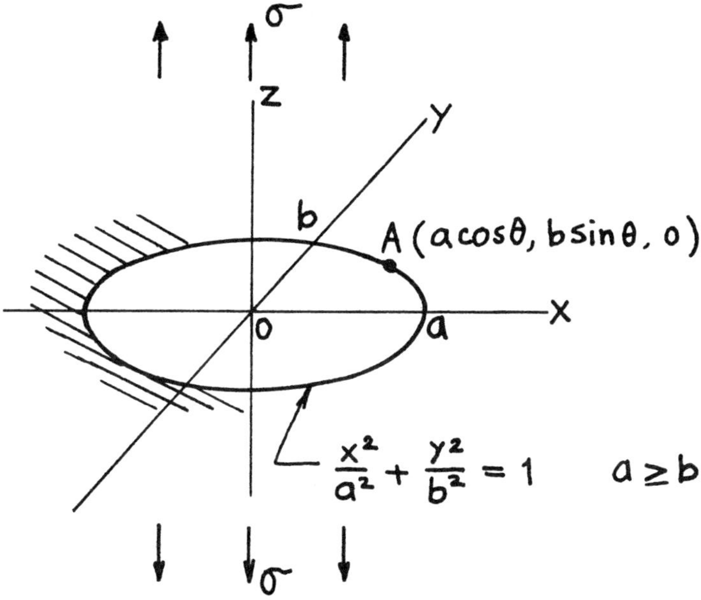

$$K_{IA} = \frac{\sigma\sqrt{\pi b}}{E(k)}\left\{\sin^2\theta + \frac{b^2}{a^2}\cos^2\theta\right\}^{1/4} = \frac{\sigma\sqrt{\pi \ell}}{E(k)}$$

$$E(k) = \int_0^{\pi/2} \sqrt{1 - k^2 \sin^2\varphi}\ \ d\varphi$$

$$k^2 = 1 - b^2/a^2$$

$$K_{I,\max} = K_I\left(\theta = \pm\pi/2\right) = \frac{\sigma\sqrt{\pi b}}{E(k)}$$

$$K_I(a = b) = \frac{2\sigma}{\pi}\sqrt{\pi a}$$

$$K_I(a \to \infty) = \sigma\sqrt{\pi b}$$

Method: Integral Transform (Three-Dimensional Potential Functions)
Accuracy: Exact
References: **Sadowsky 1949; Green 1950; Irwin 1962b**

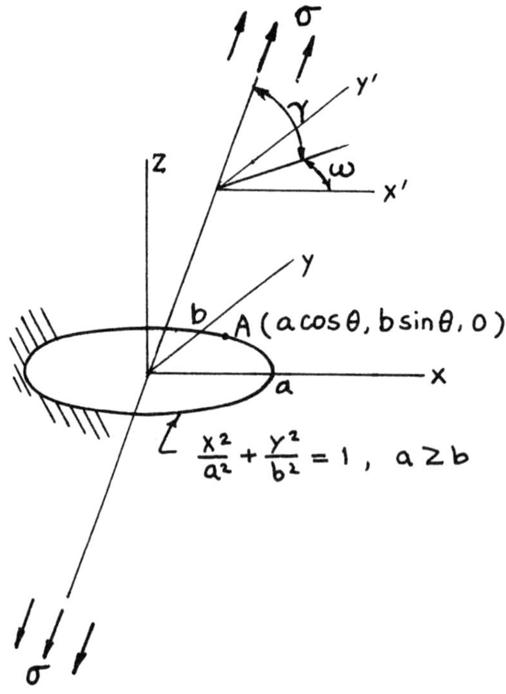

$$K_{IA} = \frac{\left(\sigma \sin^2 \gamma\right)\sqrt{\pi b}}{E(k)} \left\{ \sin^2 \theta + \left(\frac{b}{a}\right)^2 \cos^2 \theta \right\}^{1/4}$$

$$K_{IIA} = -\frac{(\sigma \sin \gamma \cos \gamma)\sqrt{\pi b}\, k^2}{\left\{ \sin^2 \theta + \left(\frac{b}{a}\right)^2 \cos^2 \theta \right\}^{1/4}} \left\{ \frac{k'}{B} \cos \omega \cos \theta + \frac{1}{C} \sin \omega \sin \theta \right\}$$

$$K_{IIIA} = \frac{(\sigma \sin \gamma \cos \gamma)\sqrt{\pi b}(1-\nu)\, k^2}{\left\{ \sin^2 \theta + \left(\frac{b}{a}\right)^2 \cos^2 \theta \right\}^{1/4}} \left\{ \frac{1}{B} \cos \omega \sin \theta - \frac{k'}{C} \sin \omega \cos \theta \right\}$$

$$B = \left(k^2 - \nu\right) E(k) + \nu k'^2 K(k)$$

$$C = \left(k^2 + \nu k'^2\right) E(k) - \nu k'^2 K(k)$$

$$k^2 = 1 - k'^2, \; k' = b/a$$

$$K(k) = \int_0^{\pi/2} \frac{d\varphi}{\sqrt{1 - k^2 \sin^2 \varphi}}, \quad E(k) = \int_0^{\pi/2} \sqrt{1 - k^2 \sin^2 \varphi} \; d\varphi$$

Methods: Three-Dimensional Potential Functions or Superposition of **pages 26.2 and 26.4**
Accuracy: Exact
References: **Kassir 1966; Sih 1968**

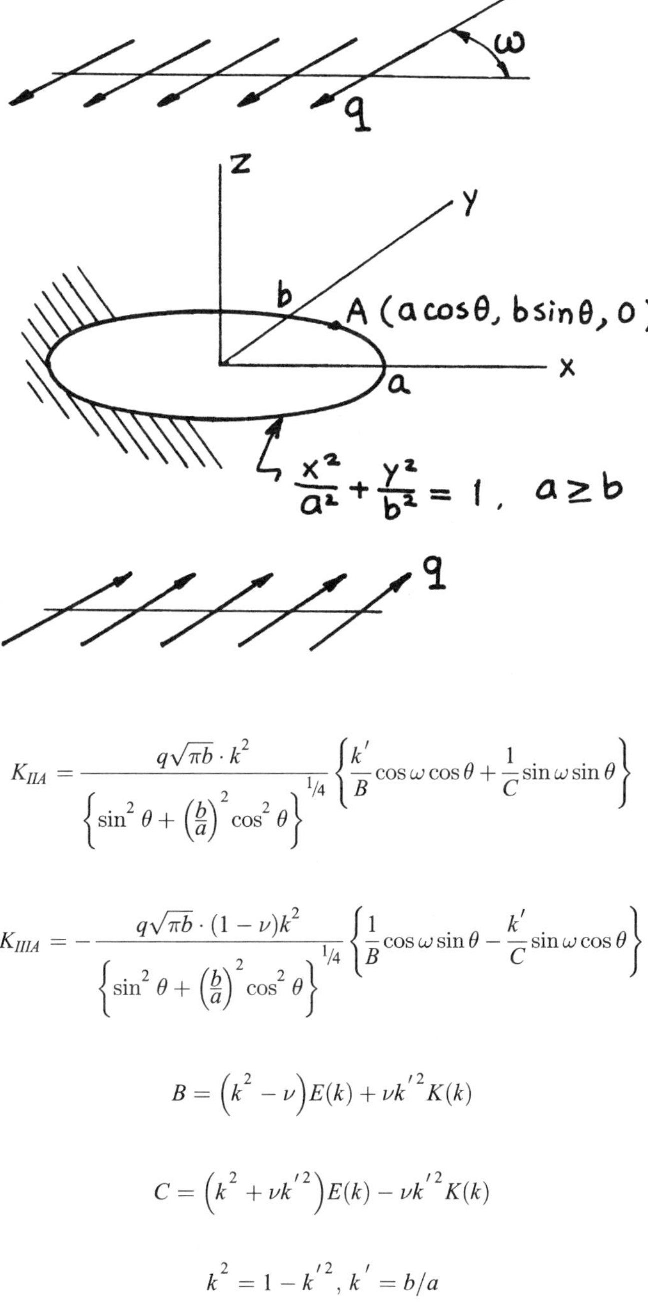

$$K_{IIA} = \frac{q\sqrt{\pi b}\cdot k^2}{\left\{\sin^2\theta + \left(\frac{b}{a}\right)^2\cos^2\theta\right\}^{1/4}}\left\{\frac{k'}{B}\cos\omega\cos\theta + \frac{1}{C}\sin\omega\sin\theta\right\}$$

$$K_{IIIA} = -\frac{q\sqrt{\pi b}\cdot(1-\nu)k^2}{\left\{\sin^2\theta + \left(\frac{b}{a}\right)^2\cos^2\theta\right\}^{1/4}}\left\{\frac{1}{B}\cos\omega\sin\theta - \frac{k'}{C}\sin\omega\cos\theta\right\}$$

$$B = \left(k^2 - \nu\right)E(k) + \nu k'^2 K(k)$$

$$C = \left(k^2 + \nu k'^2\right)E(k) - \nu k'^2 K(k)$$

$$k^2 = 1 - k'^2,\ k' = b/a$$

$$K(k) = \int_0^{\pi/2}\frac{d\varphi}{\sqrt{1-k^2\sin^2\varphi}},\quad E(k) = \int_0^{\pi/2}\sqrt{1-k^2\sin^2\varphi}\ \ d\varphi$$

Method: Three-Dimensional Potential Functions
Accuracy: Exact
References: **Kassir 1966; Sih 1968**

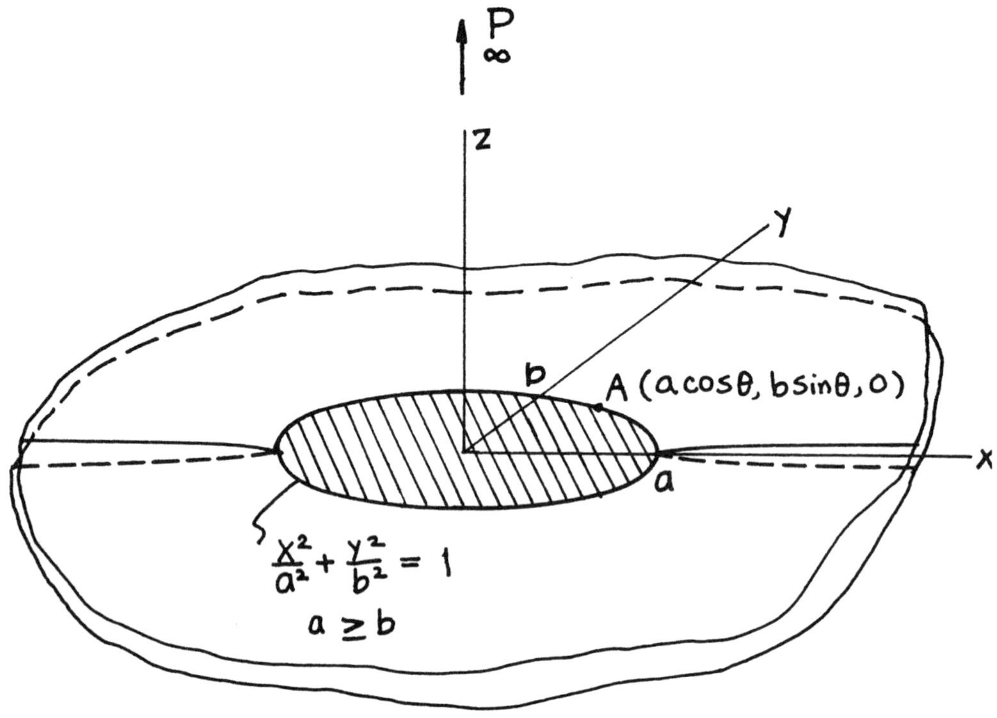

$$K_{IA} = \frac{P}{2a\sqrt{\pi b}} \frac{1}{\left\{\sin^2\theta + \left(\frac{b}{a}\right)^2 \cos^2\theta\right\}^{1/4}} = \frac{P}{2a\sqrt{\pi\ell}}$$

$$K_{IIA} = K_{IIIA} = 0$$

$$K_{I,\,\max} = K_I(\theta = 0) = \frac{P}{2b\sqrt{\pi a}}$$

$$K_I(a = b) = \frac{P}{2a\sqrt{\pi a}}$$

$$K_I(a \to \infty) = \left(\frac{P}{2a}\right)\frac{1}{\sqrt{\pi b}}$$

Method: Fourier Transform (Three-Dimensional Potential Functions)
Accuracy: Exact
References: **Green 1950; Westmann 1966**

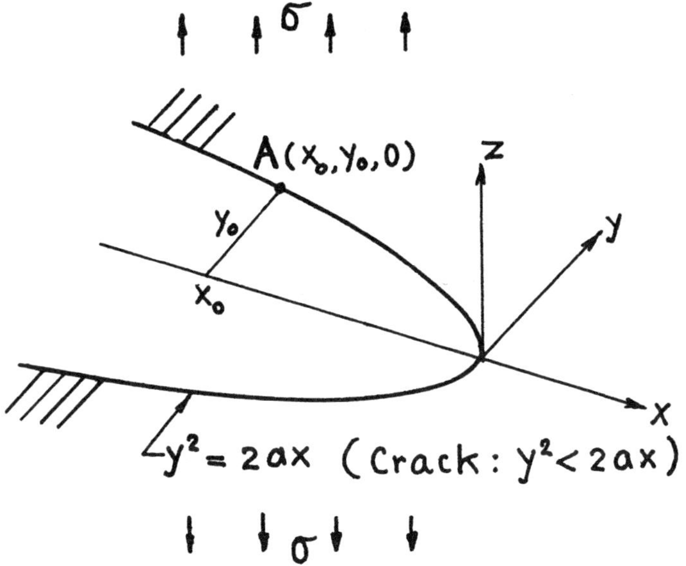

$$K_{IA} = \sigma\sqrt{\pi}\left(a^2 + y_0^2\right)^{1/4}$$

$$= \sigma\sqrt{\pi}\left(a^2 + 2ax_0\right)^{1/4}$$

$$K_{II} = K_{III} = 0$$

Method: Neuber-Papkovich Potential (or a Special Case of **page 26.2**)
Accuracy: Exact
Reference: **Shah 1968**

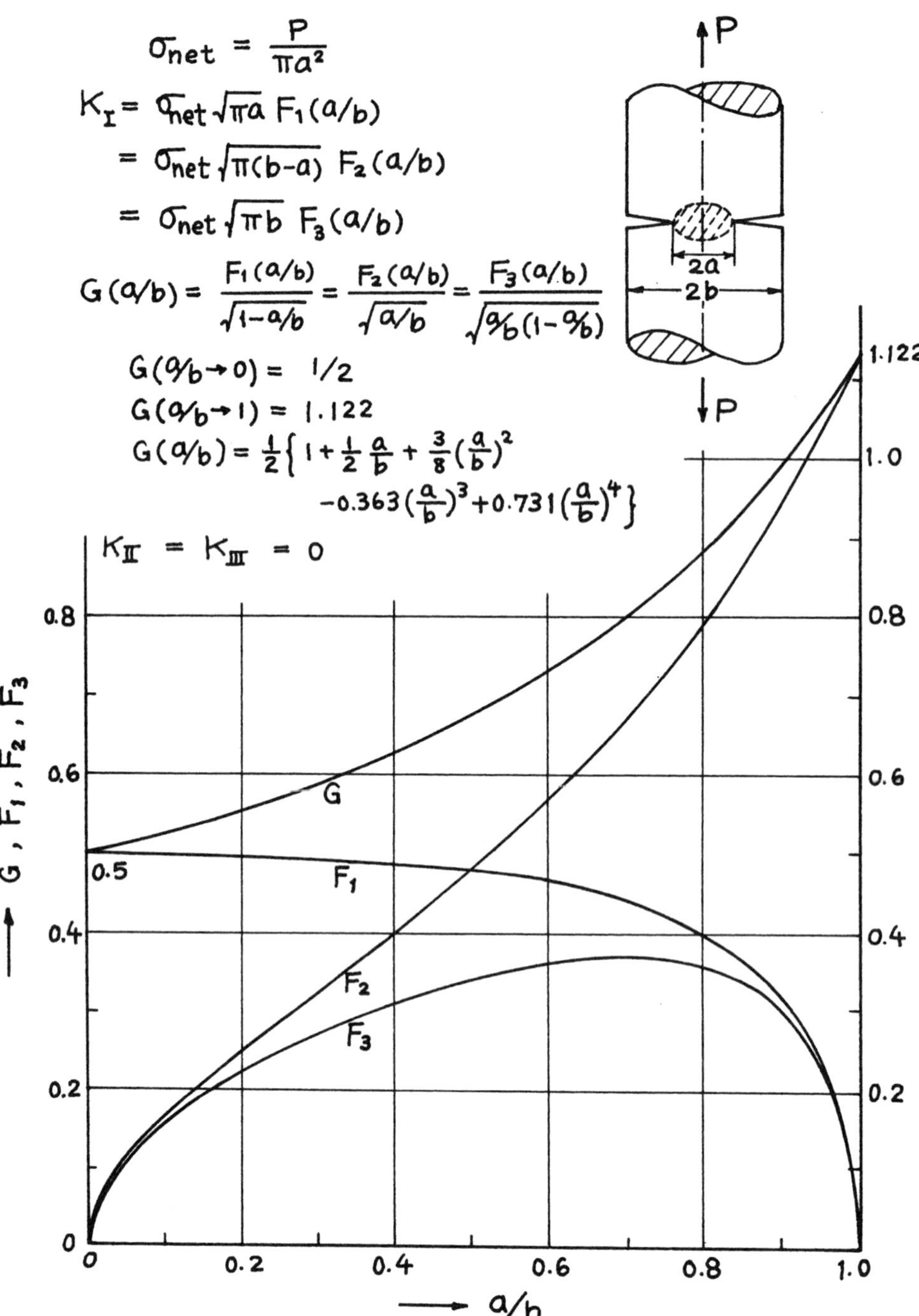

Method: Singular Integral Equation (Bueckner), Asymptotic Approximation (Benthem)
Accuracy: Better than 1%
Referencees: **Bueckner 1965, 1972; Benthem 1972**
Other References: **Lubahn 1959; Wundt 1959; Irwin 1961; Paris 1965; Zahn 1965; Harris 1967**

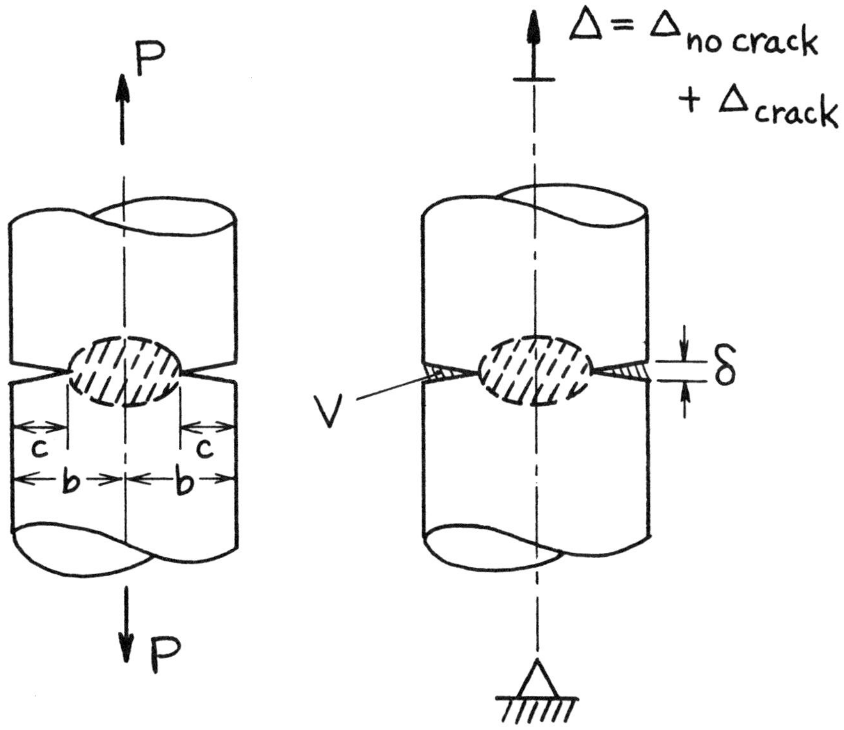

$$\sigma = \frac{P}{\pi b^2}$$

$$K_I = \sigma\sqrt{\pi c}\, F(c/b)$$

$$F(c/b) = \frac{1}{(1 - c/b)^{3/2}}\left\{1.122 - 1.302\,\frac{c}{b} + .988\left(\frac{c}{b}\right)^2 - .308\left(\frac{c}{b}\right)^3\right\}$$

Volume of Crack:

$$V = \frac{4(1-\nu^2)}{E}\sigma\pi^2 c^3 G(c/b)$$

Additional Displacement at Infinity due to Crack:

$$\Delta_{crack} = \frac{4(1-\nu^2)}{E}\sigma\pi c H(c/b)$$

Crack Opening at Edge:

$$\delta = \frac{4(1-\nu^2)}{E}\sigma c D(c/b)$$

where

$$G(c/b) = \frac{1}{3} \cdot \frac{1}{(1 - c/b)^2} \left\{ .375 + .383\left(1 - \frac{c}{b}\right) + .5\left(1 - \frac{c}{b}\right)^3 \right\}$$

$$H(c/b) = (c/b)^2 G(c/b)$$

$$D(c/b) = \frac{1}{(1 - c/b)^2} \left\{ 1.454 - 2.49\frac{c}{b} + 1.155\left(\frac{c}{b}\right)^2 \right\}$$

Methods: K_I, δ Integral Transform ($c/b \leq 0.6$), Interpolation ($c/b > 0.6$); V, Δ Paris' Equation (see **Appendix B**)
Accuracy: K_I, δ 1%; V, Δ 2%
References: **Erdogan 1982; Tada 1985**

NOTE: Δ_{crack} is the elongation at infinity when uniform pressure σ is applied on crack surfaces.

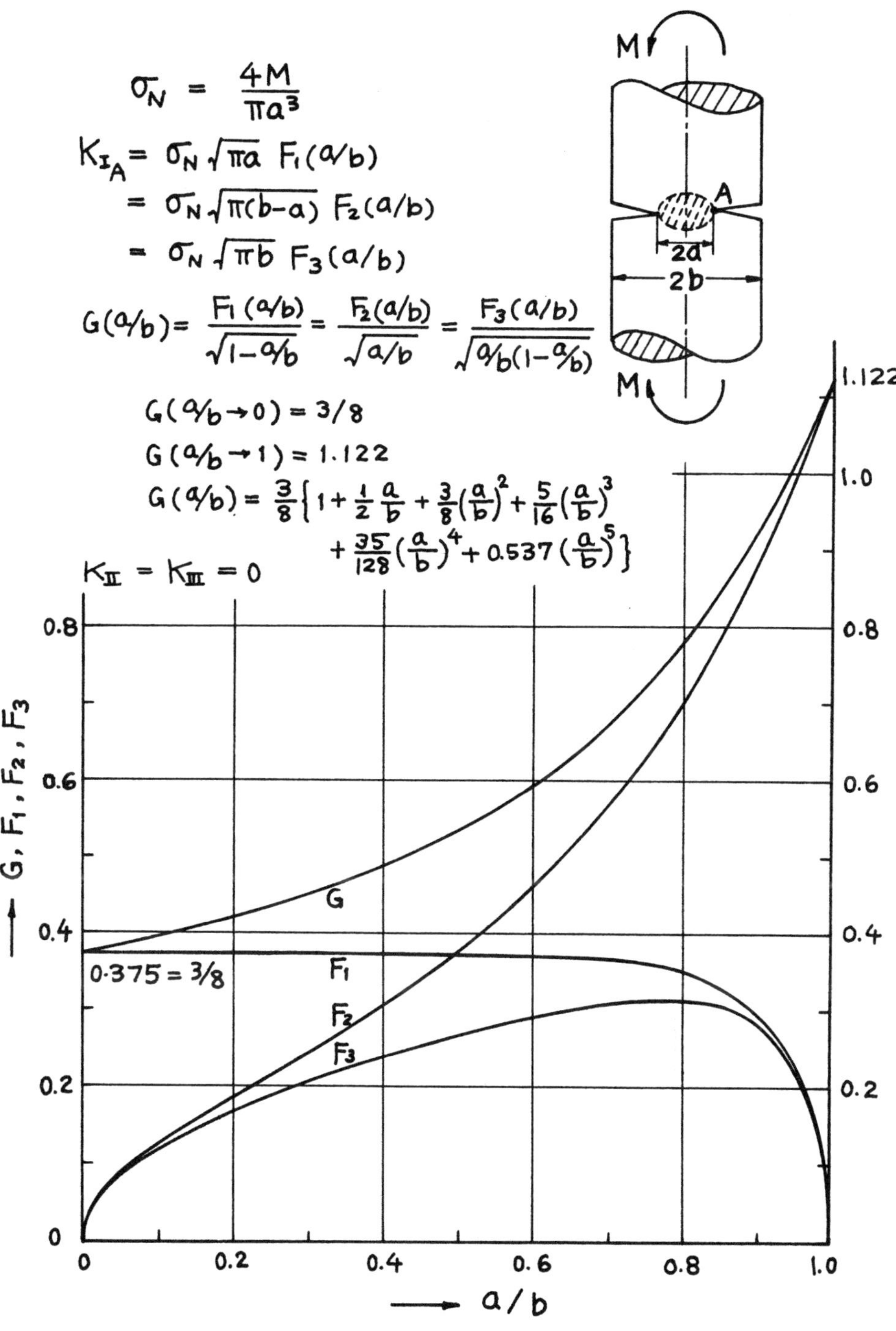

Method: Asymptotic Approximation
Accuracy: Better than 1%
Reference: **Benthem 1972**

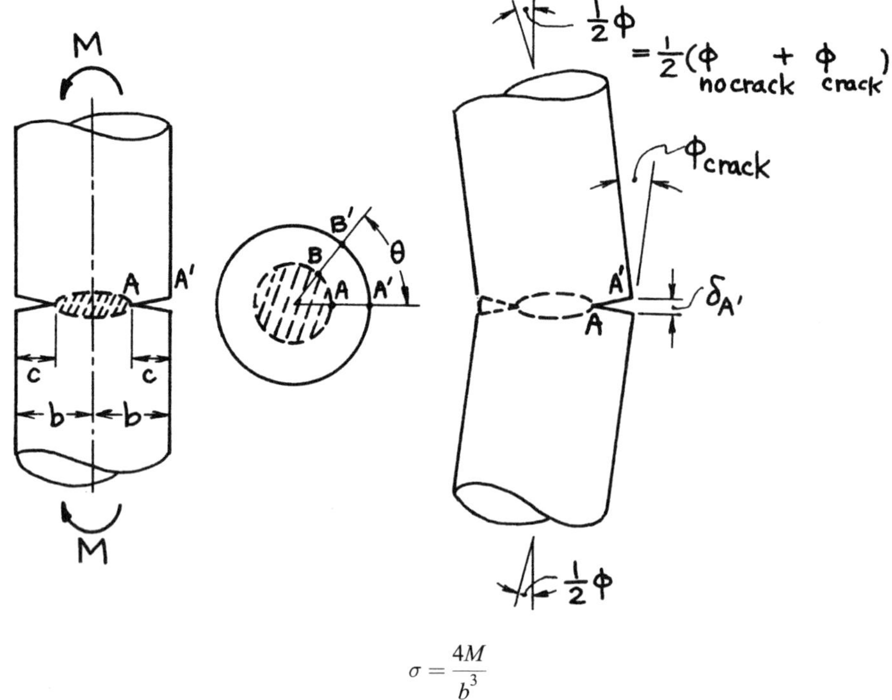

$$\sigma = \frac{4M}{b^3}$$

$$K_{IA} = \sigma\sqrt{\pi c}\ F(c/b); \quad K_{IB} = K_{IA}\ \cos\theta$$

$$F(c/b) = \frac{1}{\left(1 - c/b\right)^{5/2}}\left\{.563 - .188\frac{c}{b} + \left(1 - \frac{c}{b}\right)^2\left[.559 - 1.47\frac{c}{b} + 2.72\left(\frac{c}{b}\right)^2 - 2.40\left(\frac{c}{b}\right)^3\right]\right\}$$

Additional Rotation at Infinity or Kink at Cracked Section due to Crack:

$$\phi_{crack} = \frac{8\left(1 - \nu^2\right)}{E}\sigma\pi\left(\frac{c}{b}\right)^2\left\{1 - 1.244\frac{c}{b} + 2.11\left(\frac{c}{b}\right)^2 + \frac{.258 - .164\frac{c}{b}}{\left(1 - c/b\right)^3}\right\}$$

Crack Opening at Edge:

$$\delta_{A'} = \frac{4\left(1 - \nu^2\right)}{E}\sigma c\left\{\frac{.454 - .160\frac{c}{b}}{\left(1 - c/b\right)^3} + 1 - 2.6\frac{c}{b}\left(.31 - \frac{c}{b}\right)\left(1 - \frac{c}{b}\right)\left(1 + 3\frac{c}{b}\right)\right\}$$

$$\delta_{B'} = \delta_{A'}\cos\theta$$

Methods: K_I, δ Integral Transform ($c/b \leq 0.6$), Interpolation ($c/b > 0.6$); ϕ Paris' Equation (see **Appendix B**)
Accuracy: K_I, δ 1%; ϕ 2%
References: **Erdogan 1982; Tada 1985**

NOTE: Crack surface interference on compressive side is not considered.

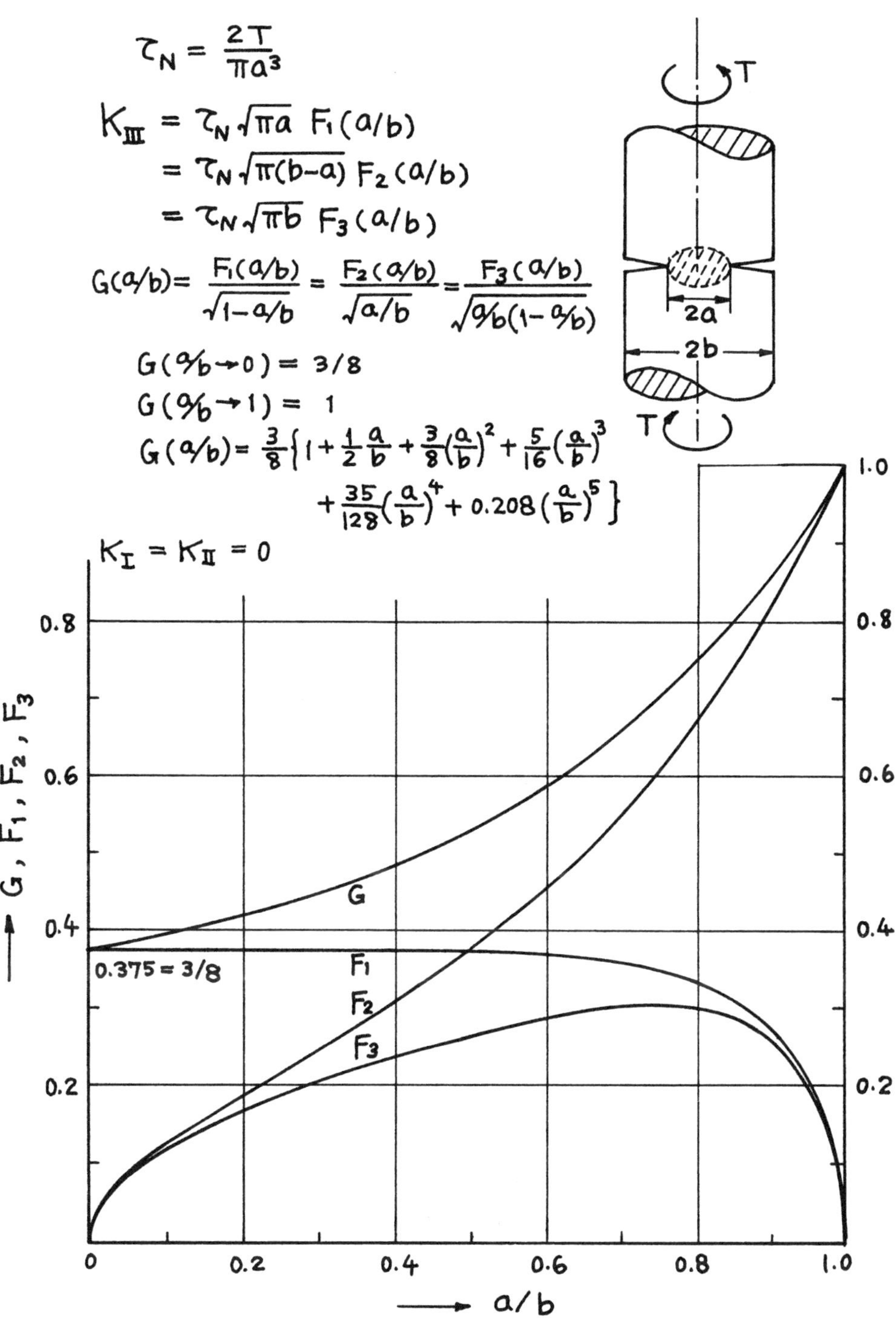

Method: Asymptotic Approximation
Accuracy: Better than 1%
Reference: **Benthem 1972**

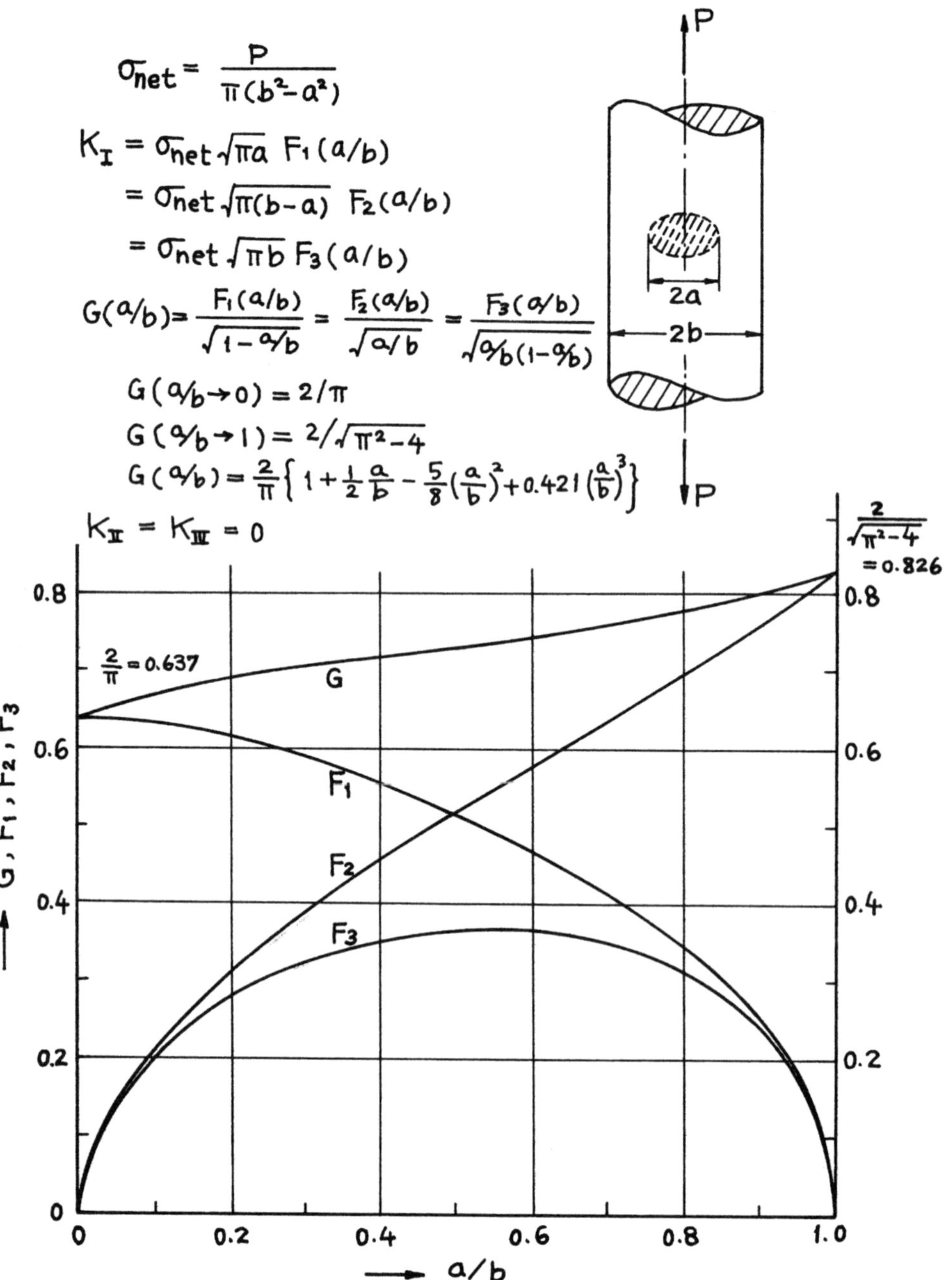

Method: Asymptotic Approximation
Accuracy: Better than 1%
Reference: **Benthem 1972**

NOTE: For an alternate solution and displacements, see **page 27.4a**.

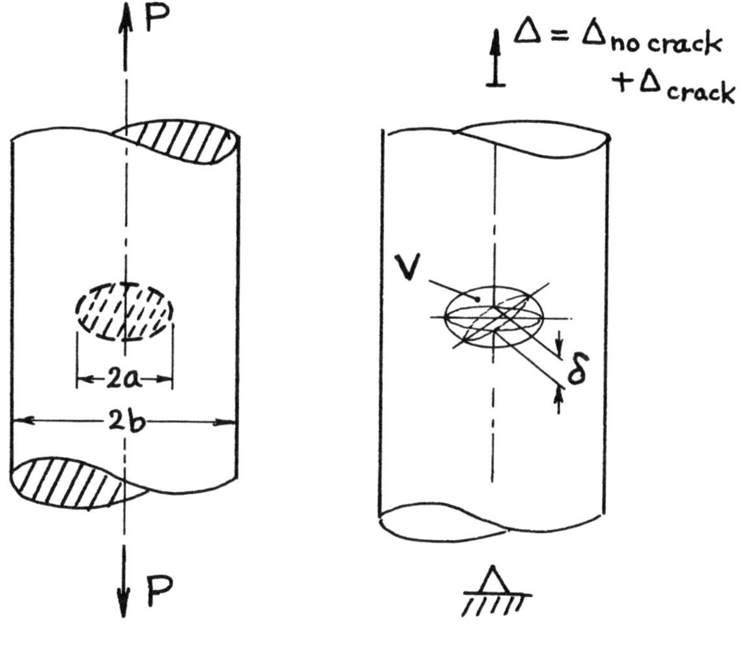

$$\sigma = \frac{P}{\pi b^2}$$

$$K_I = \frac{2}{\pi}\sigma\sqrt{\pi a}\cdot F\left(\frac{a}{b}\right)$$

$$F\left(\frac{a}{b}\right) = \frac{1 - \frac{1}{2}\frac{a}{b} + .148\left(\frac{a}{b}\right)^3}{\sqrt{1 - \frac{a}{b}}}$$

Volume of Crack:

$$V = \frac{16\left(1-\nu^2\right)}{3E}\sigma a^3 \cdot G\left(\frac{a}{b}\right)$$

$$G\left(\frac{a}{b}\right) = \frac{1}{\left(a/b\right)^3}\left[1.260\,\ell n\left(\frac{1}{1-\frac{a}{b}}\right) - 1.260\frac{a}{b} - .630\left(\frac{a}{b}\right)^2 + .580\left(\frac{a}{b}\right)^3 - .315\left(\frac{a}{b}\right)^4 - .102\left(\frac{a}{b}\right)^5 \right.$$
$$\left. + .063\left(\frac{a}{b}\right)^6 - .0093\left(\frac{a}{b}\right)^7 - .0081\left(\frac{a}{b}\right)^8\right]$$

Additional Elongation at Infinity due to Crack:

$$\Delta_{crack} = V / \left(\pi b^2\right)$$

Crack Opening at Center:

$$\delta = \frac{8\left(1 - \nu^2\right)}{\pi E} \sigma a \cdot H\left(\frac{a}{b}\right)$$

$$H\left(\frac{a}{b}\right) = \frac{1}{a/b} \ell n\left(\frac{1}{1 - a/b}\right)\left[1 - \frac{1}{2}\frac{a}{b} + .340\left(\frac{a}{b}\right)^{3.5}\right]$$

Method: K_I, δ Integral Transform; V, Δ Paris' Equation (see **Appendix B**)
Accuracy: K_I, δ 0.5%; V, Δ 1%
References: **Erdogan 1982; Tada 1985**

NOTE: Δ_{crack} is elongation at infinity when uniform pressure σ is applied on crack surfaces.

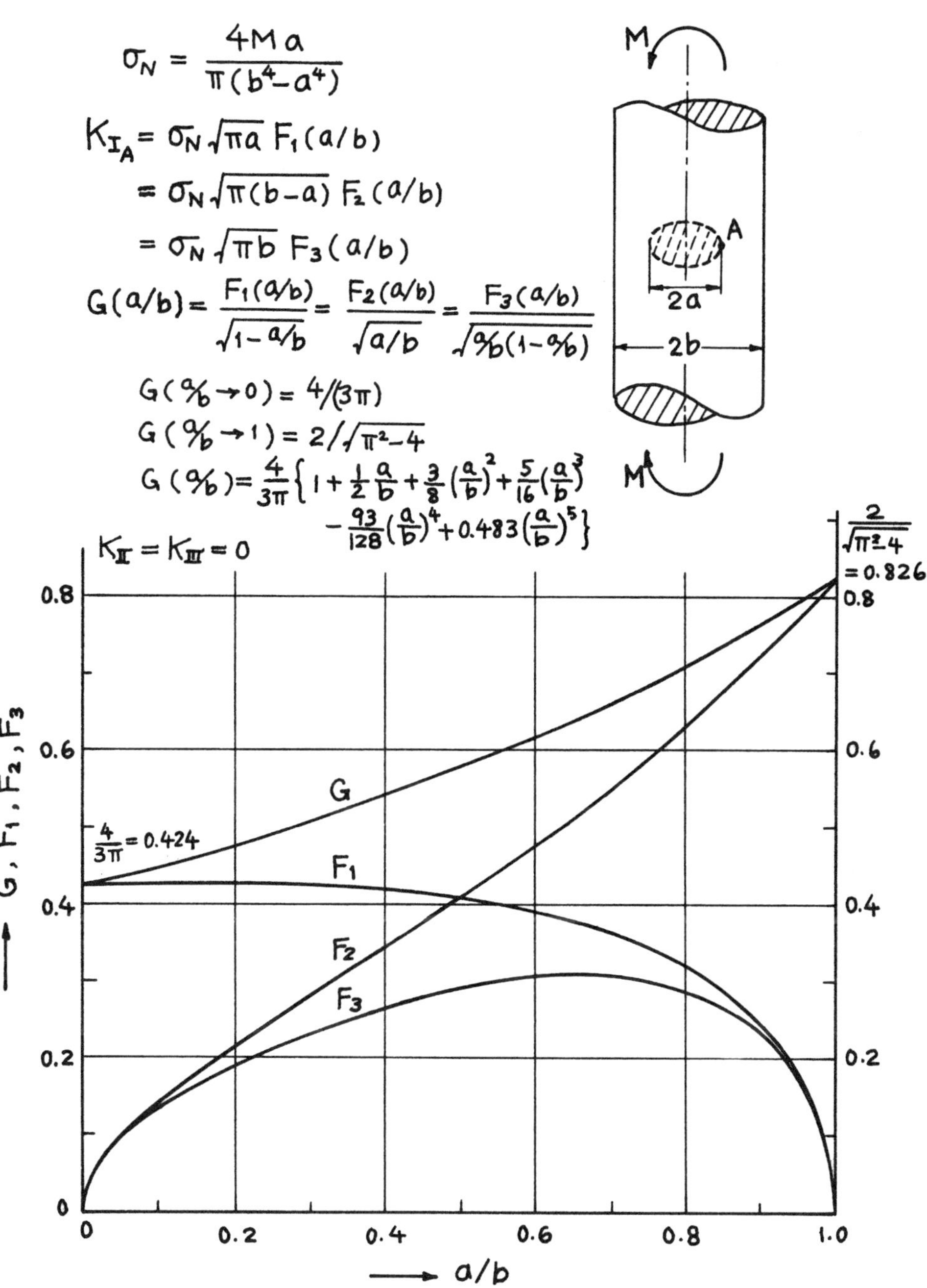

Method: Asymptotic Approximation
Accuracy: Better than 1%
Reference: **Benthem 1972**

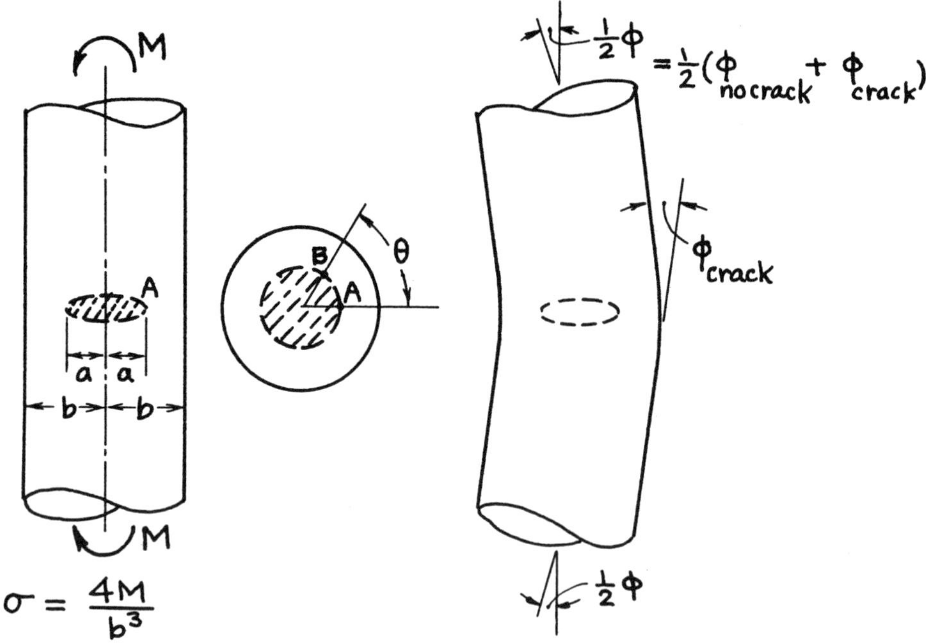

$$K_{IA} = \sigma\sqrt{\pi a}\,F(a/b); \quad K_{IB} = K_{IA}\cos\theta$$

$$F(a/b) = \frac{a}{b}\frac{\sqrt{1-a/b}}{\left\{1-(a/b)^4\right\}}\cdot\frac{4}{3\pi}\left\{1+\frac{1}{2}\frac{a}{b}+\frac{3}{8}\left(\frac{a}{b}\right)^2+\frac{5}{16}\left(\frac{a}{b}\right)^3-\frac{93}{128}\left(\frac{a}{b}\right)^4+.483\left(\frac{a}{b}\right)^5\right\}$$

Additional Rotation at Infinity or Kink at Cracked Section due to Crack:

$$\phi_{crack} = \frac{1280(1-\nu^2)}{9\pi E}\sigma(a/b)^4\,\ell n\,\frac{1}{1-a/b}\cdot\Phi(a/b)$$

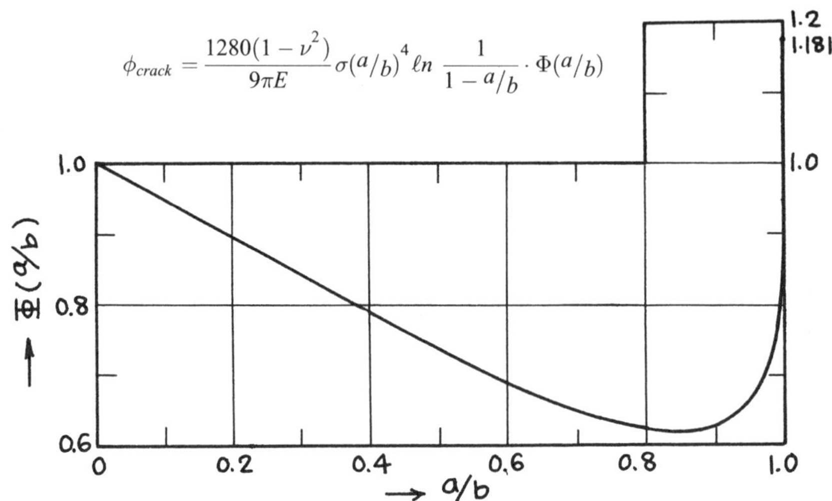

Methods: K_I Asymptotic Interpolation; ϕ Paris' Equation (see **Appendix B**)
Accuracy: K_I 1%; ϕ 2%
References: **Benthem 1972; Tada 1985**

NOTE: Crack surface interference on compressive side is not considered.

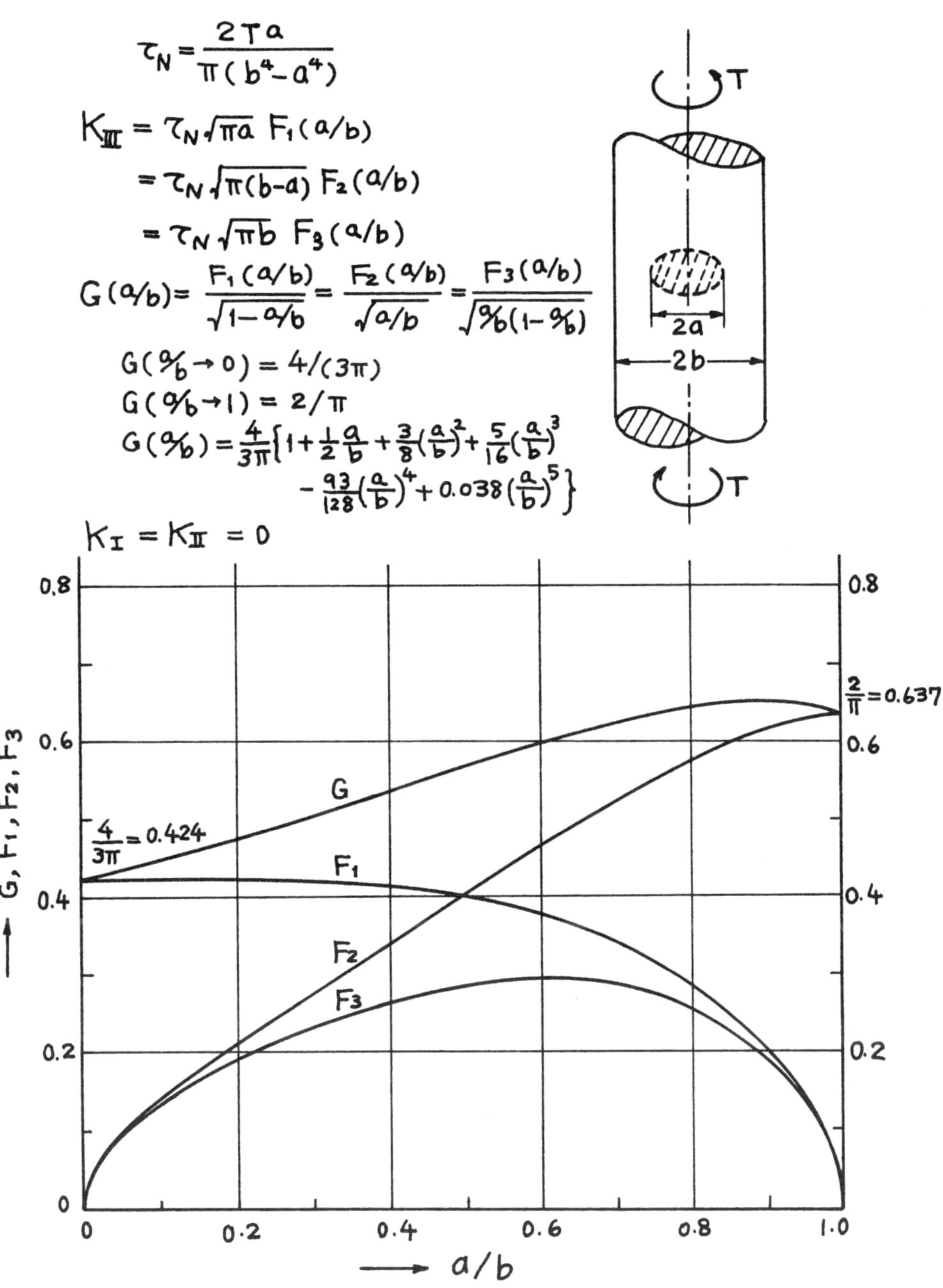

Method: Asymptotic Approximation
Accuracy: Better than 1%
Reference: **Benthem 1972**

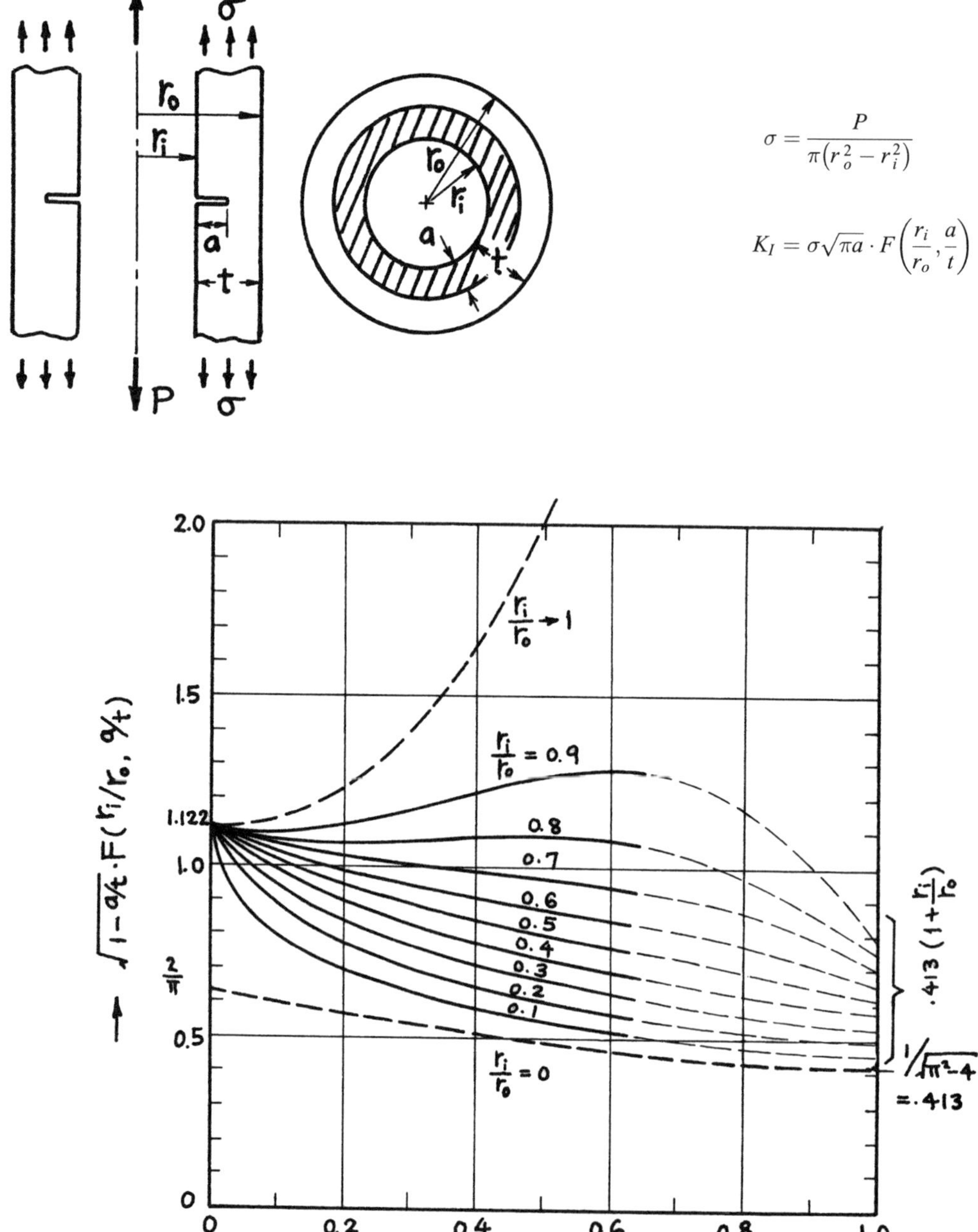

$$\sigma = \frac{P}{\pi\left(r_o^2 - r_i^2\right)}$$

$$K_I = \sigma\sqrt{\pi a}\cdot F\left(\frac{r_i}{r_o}, \frac{a}{t}\right)$$

Method: Integral Transform-Singular Integral Equation $(a/t \leq 0.6)$, Interpolation $(a/t > 0.6)$

Accuracy: Solid curves $\left(0.1 \leq r_i/r_o \leq 0.9, a/t \leq 0.6\right)$ are based on values with better than 1% accuracy; 2% for $a/t > 0.6$

References: **Erdogan 1982; Tada 1985**

NOTE: For $r_i/r_o = 0$ (solid cylinders), see **page 27.4**, and for $r_i/r_o \to 1$, see **page 2.10**.

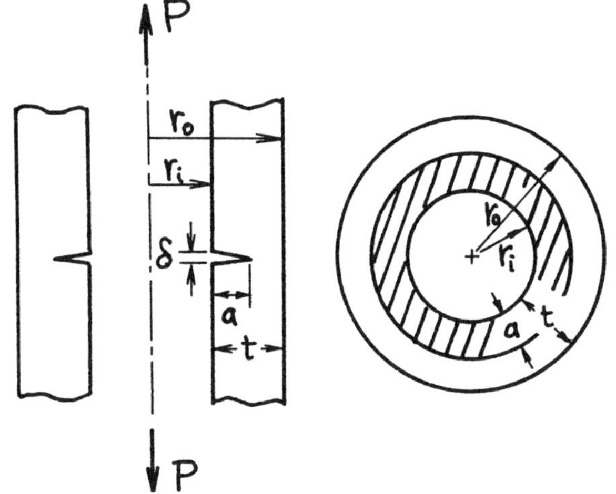

$$\sigma = \frac{P}{\pi\left(r_o^2 - r_i^2\right)}$$

Crack Opening at Edge:

$$\delta = \frac{4(1-\nu^2)}{E}\sigma a \cdot D\left(a/t, r_i/r_o\right)$$

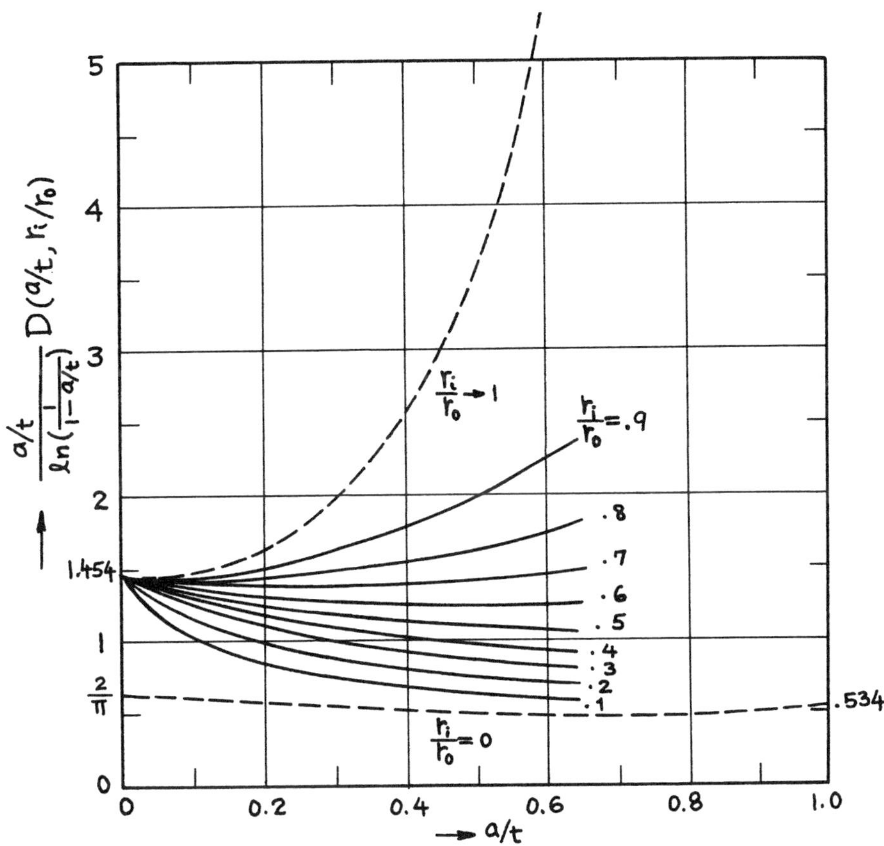

Method: Integral Transform-Integral Equations
Accuracy: Curves are based on values with better than 1% accuracy.
References: **Erdogan 1982; Tada 1985**

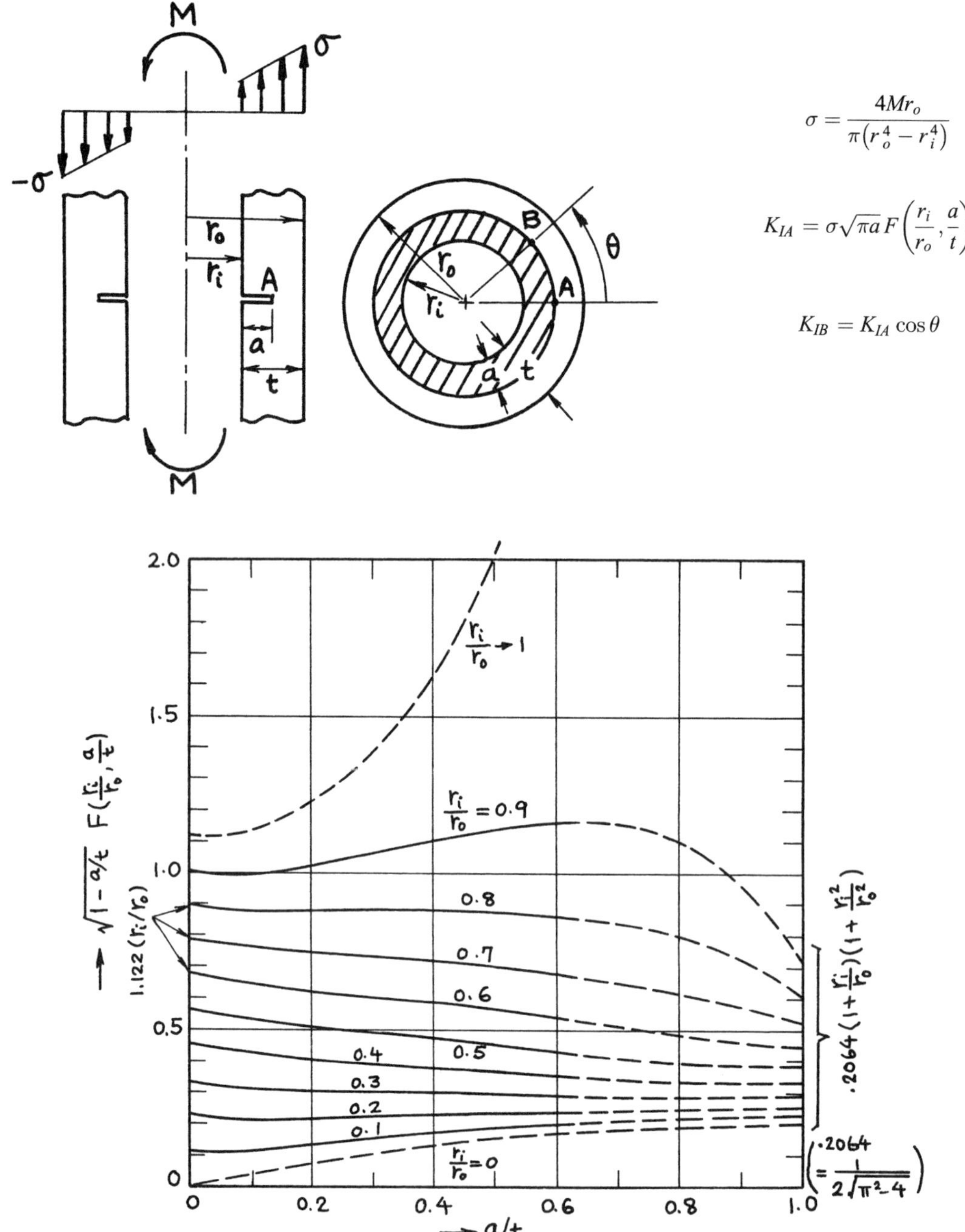

$$\sigma = \frac{4Mr_o}{\pi\left(r_o^4 - r_i^4\right)}$$

$$K_{IA} = \sigma\sqrt{\pi a}\, F\left(\frac{r_i}{r_o}, \frac{a}{t}\right)$$

$$K_{IB} = K_{IA}\cos\theta$$

Methods: Integral Transform-Integral Equations $(a/t \leq 0.6)$, Interpolation $(a/t > 0.6)$.
Accuracy: Solid curves $(0.1 \leq r_i/r_o < 0.9; a/t \leq 0.6)$ are based on values with better than 1% accuracy; 2% for $a/t > 0.6$.
References: **Erdogan 1982; Tada 1985**.

NOTE: Crack surface interference on compression side is not considered. For $r_i/r_o = 0$ (solid cylinder), see **page 27.5**; for $r_i/r_o \to 1$, see **page 2.10**.

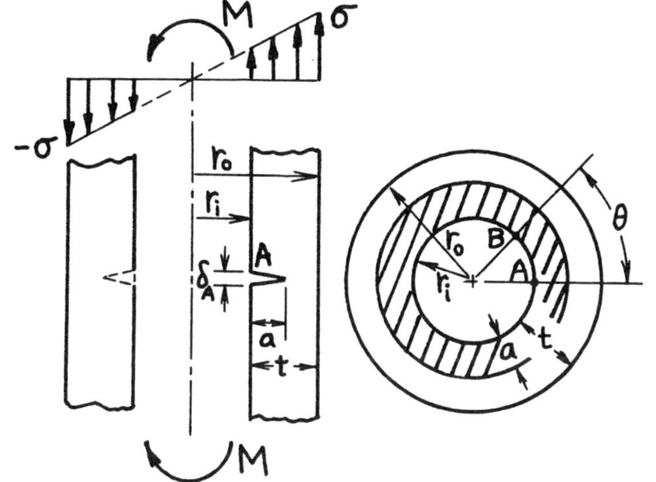

$$\sigma = \frac{4Mr_o}{\pi\left(r_o^4 - r_i^4\right)}$$

Crack Opening at Edge:

$$\delta_A = \frac{4\left(1-\nu^2\right)}{E}\sigma a \cdot D\left({}^{a}/_{t}, {}^{r_i}/_{r_o}\right)$$

$$\delta_B = \delta_A \cos\theta$$

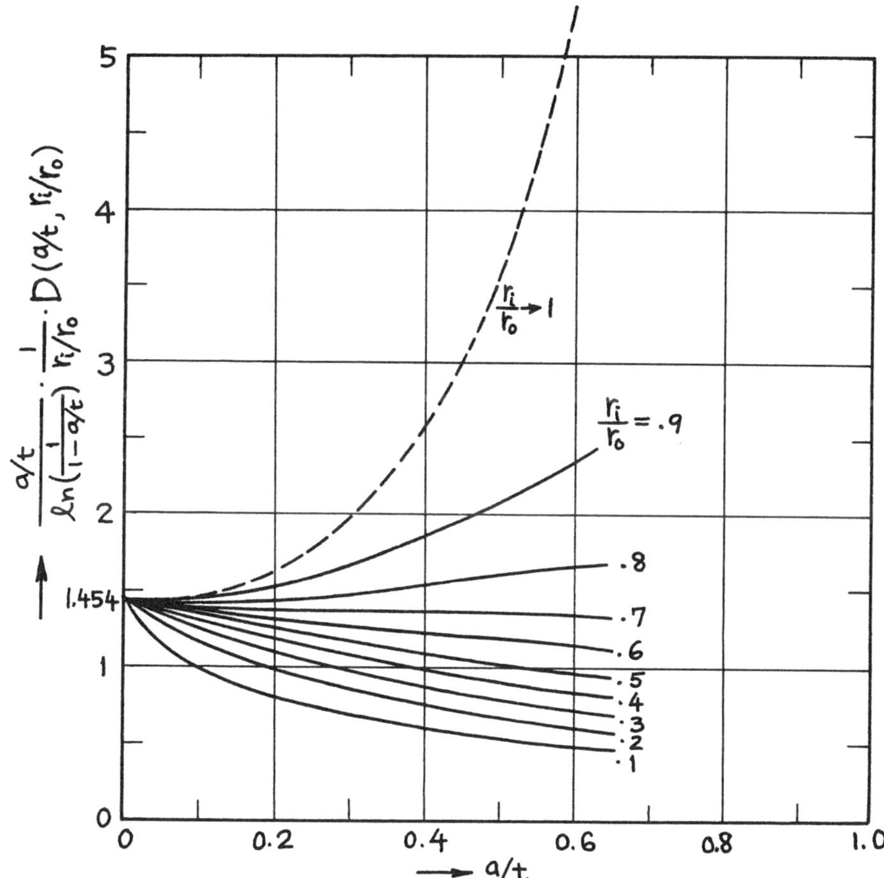

Method: Integral Transform - Integral Equations
Accuracy: Curves are based on values with better than 1% accuracy.
References: **Erdogan 1982; Tada 1985**

NOTE: Crack surface interference on compressive side is not considered.

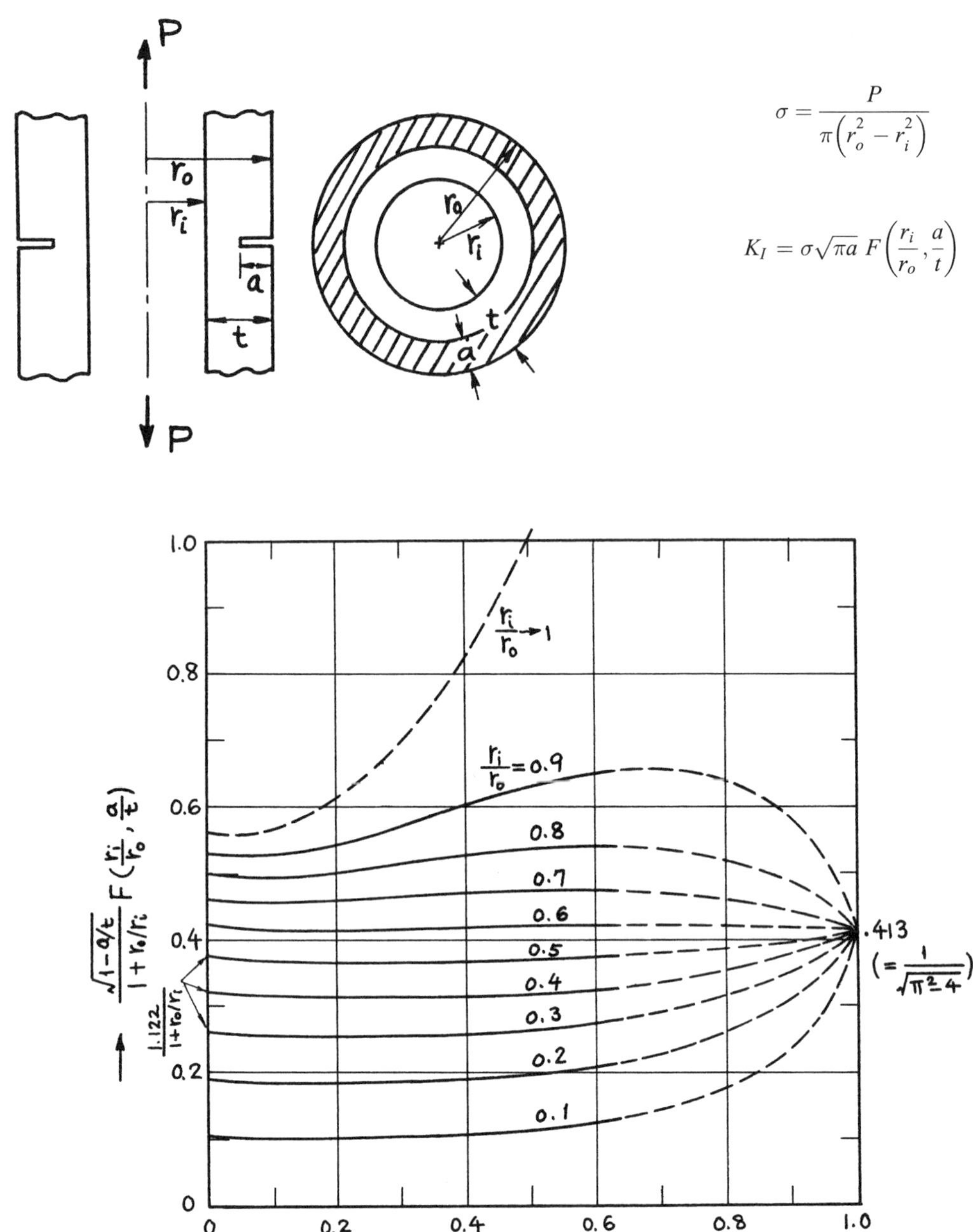

Methods: Integral Transform - Integral Equations $(a/t \leq 0.6)$, Interpolation $(a/t > 0.6)$

Accuracy: Solid curves $(0.1 \leq r_i/r_o \leq 0.9; a/t \leq 0.6)$ are based on values with better than 1% accuracy; 2% for $a/t > 0.6$.

References: **Erdogan 1982; Tada 1985**

NOTE: For $r_i/r_o = 0$ (solid cylinder), see **page 27.1**; for $r_i/r_o \to 1$, see **page 2.10**.

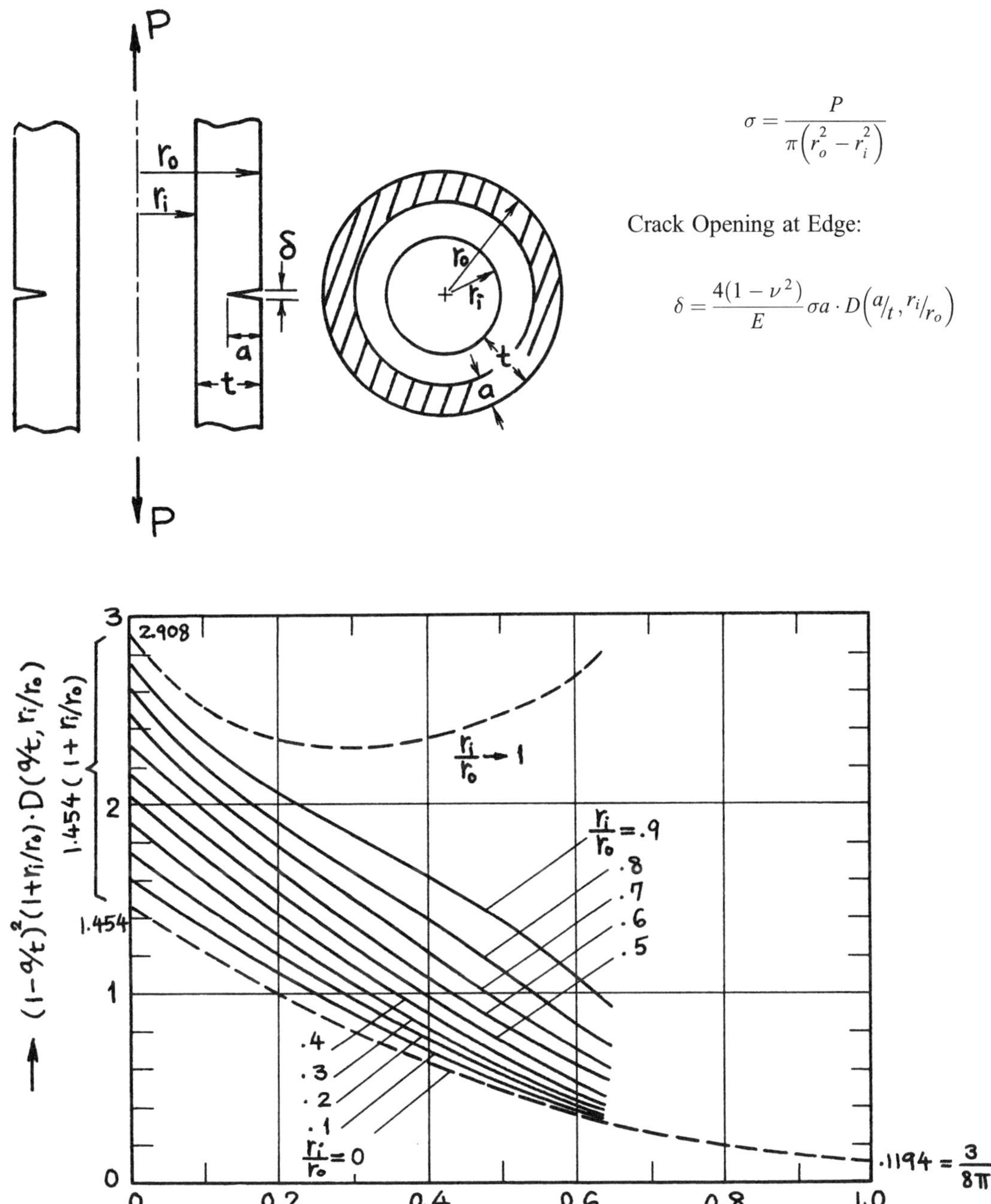

$$\sigma = \frac{P}{\pi\left(r_o^2 - r_i^2\right)}$$

Crack Opening at Edge:

$$\delta = \frac{4(1-\nu^2)}{E}\sigma a \cdot D\left(a/_t, r_i/_{r_o}\right)$$

Method: Integral Transform - Integral Equations
Accuracy: Curves are based on values with better than 1% accuracy.
References: **Erdogan 1982; Tada 1985**

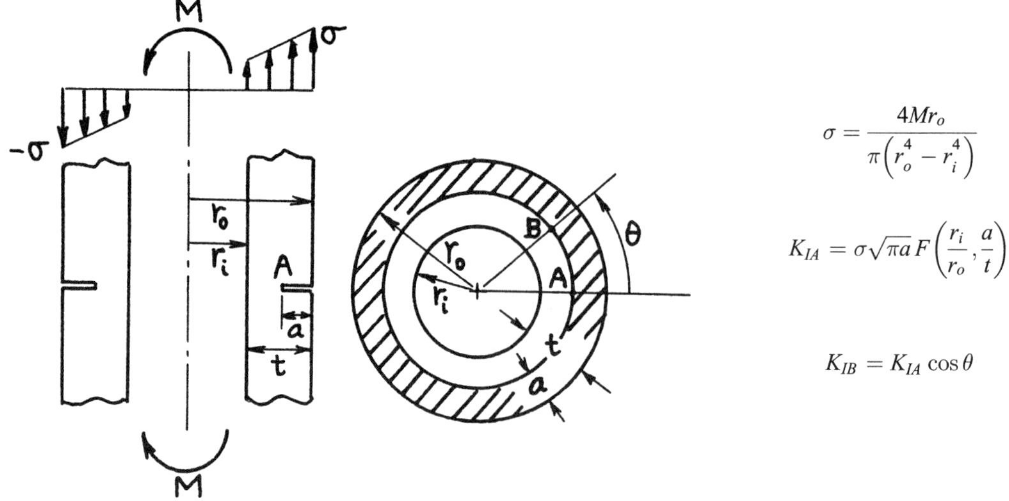

$$\sigma = \frac{4Mr_o}{\pi\left(r_o^4 - r_i^4\right)}$$

$$K_{IA} = \sigma\sqrt{\pi a}\,F\left(\frac{r_i}{r_o}, \frac{a}{t}\right)$$

$$K_{IB} = K_{IA}\cos\theta$$

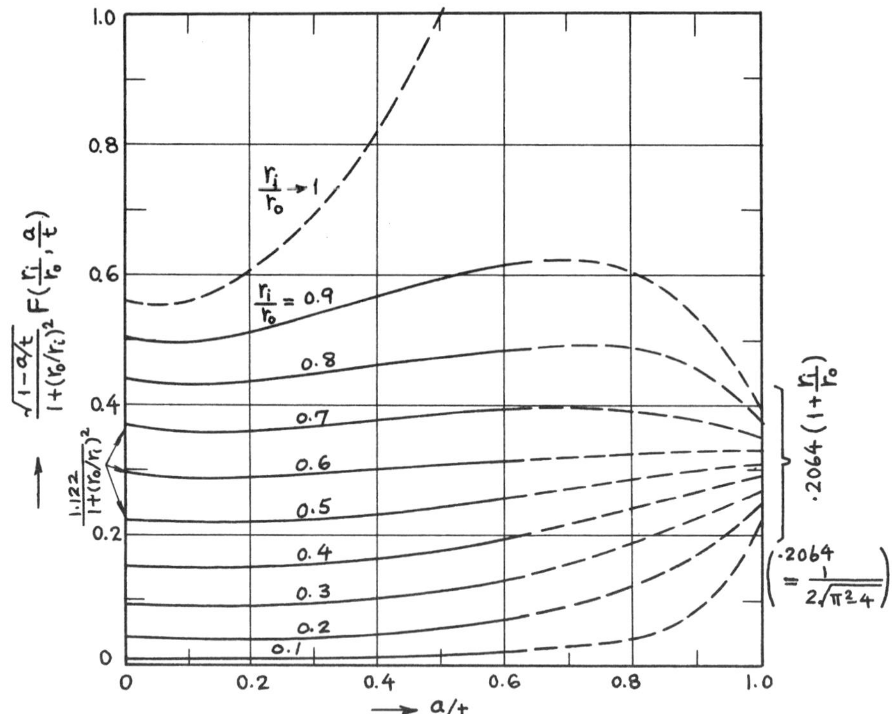

Methods: Integral Transform - Integral Equations $(a/t \leq 0.6)$, Interpolation $(a/t > 0.6)$

Accuracy: Solid curves $(0.1 \leq r_i/r_o \leq 0.9; a/t \leq 0.6)$ are based on values with better than 1% accuracy; 2% for $a/t > 0.6$.

References: **Erdogan 1982; Tada 1985**

NOTE: Crack surface interference on compression side is not considered. For $r_i/r_o = 0$ (solid cylinder), see **page 27.2**; for $r_i/r_o \to 1$, see **page 2.10**.

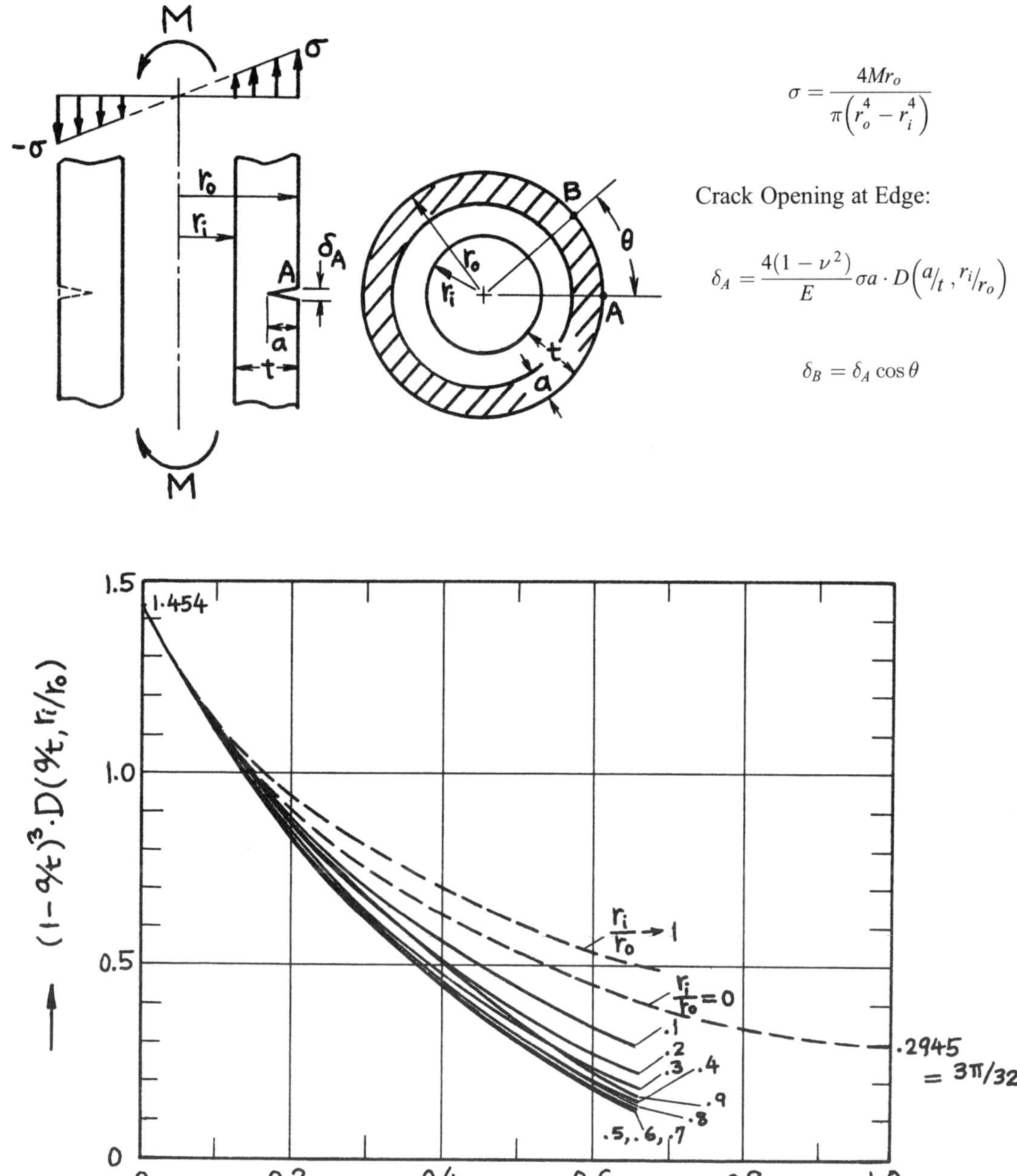

$$\sigma = \frac{4Mr_o}{\pi\left(r_o^4 - r_i^4\right)}$$

Crack Opening at Edge:

$$\delta_A = \frac{4(1-\nu^2)}{E}\sigma a \cdot D\left(a/t, r_i/r_o\right)$$

$$\delta_B = \delta_A \cos\theta$$

Method: Integral Transform - Integral Equations

Accuracy: Curves are based on values with better than 1% accuracy.

References: **Erdogan 1982; Tada 1985**

NOTE: Crack surface interference on compressive side is not considered.

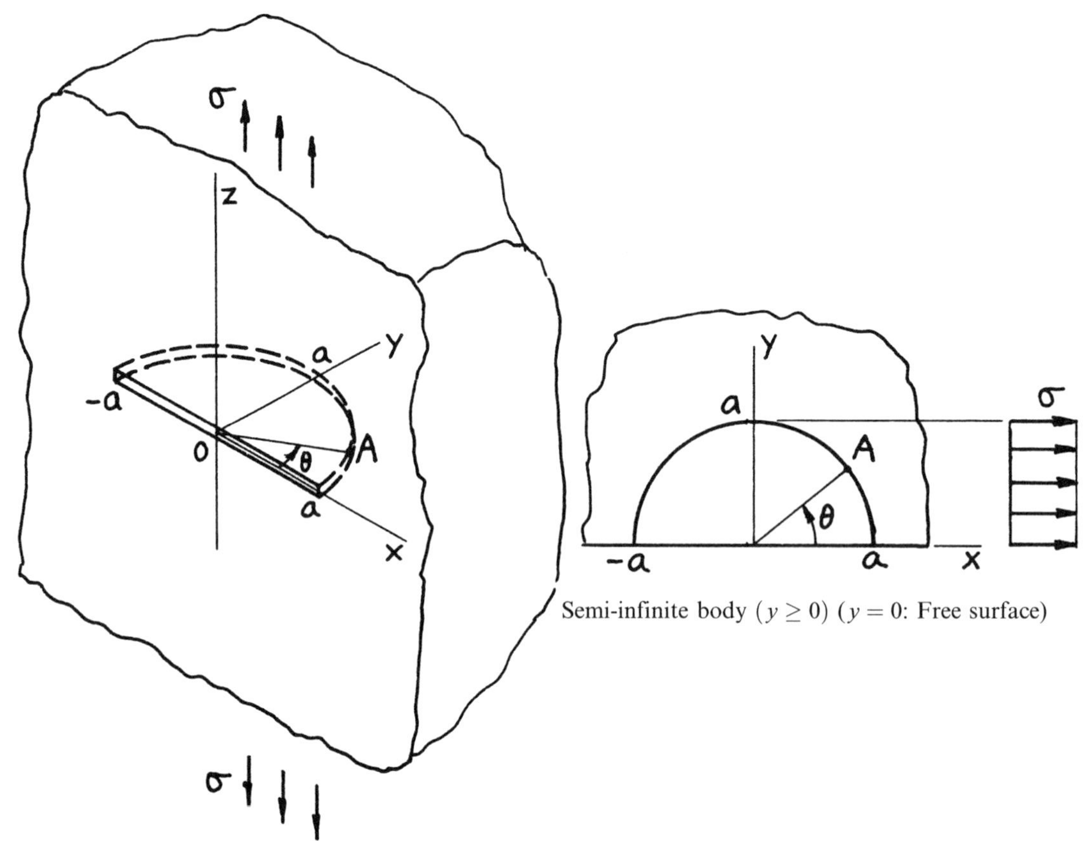

Semi-infinite body $(y \geq 0)$ $(y = 0$: Free surface)

$$K_{IA} = \frac{2}{\pi}\sigma\sqrt{\pi a}\,F(\theta)$$

$$F(\theta) = 1.211 - .186\sqrt{\sin\theta} \quad \left(10^{\circ} < \theta < 170^{\circ}\right)$$

Methods: Alternating Method (Smith, Hartranft), Finite Element Method (Tracey, Raju); $F\ (\theta)$ is based on Smith's result (Merkle)

Accuracy: 2%

References: **Smith 1967; Hartranft 1973; Tracey 1973; Merkle 1973; Raju 1979**

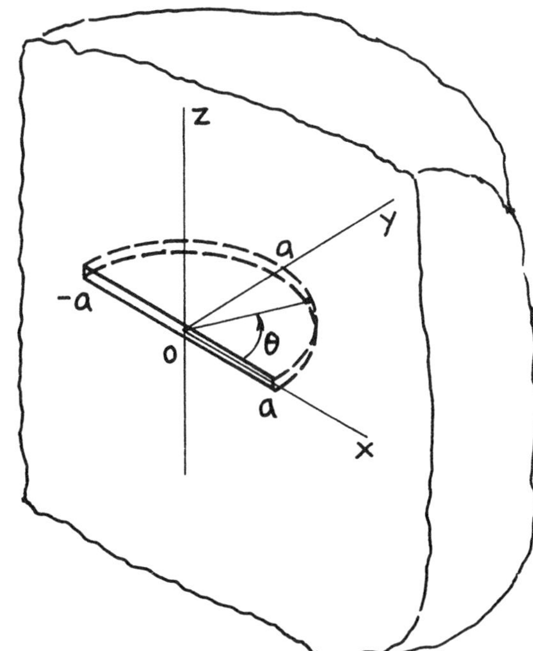

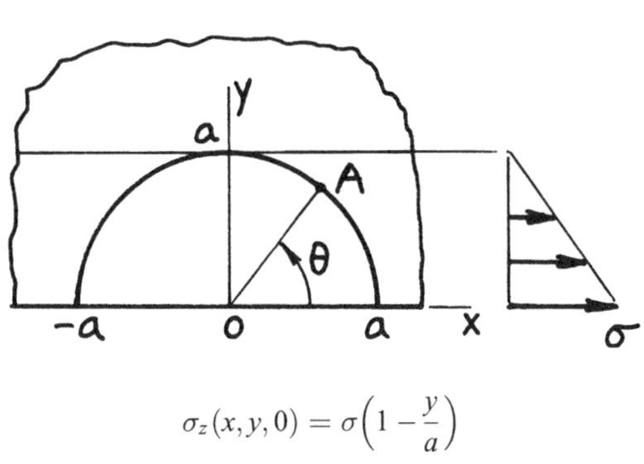

$$\sigma_z(x, y, 0) = \sigma\left(1 - \frac{y}{a}\right)$$

Semi-infinite body $(y \geq 0)$ $(y = 0$: Free surface)

$$K_{IA} = \frac{2}{\pi}\sigma\sqrt{\pi a} \cdot F(\theta)$$

$$or \quad = \frac{2}{\pi}\sigma\sqrt{\pi a}\left\{\frac{5}{6} - \frac{4}{3}(\sin\theta)^{3/2}\left(\sqrt{1+\sin\theta} - \sqrt{\sin\theta}\right)\right\} \cdot G(\theta)$$

$$F(\theta) = 1.031 - .186\sqrt{\sin\theta} - .54\sin\theta$$

$$(10^\circ < \theta < 170^\circ)$$

$$G(\theta) = 1.17 - .31\sin\theta + .23(\sin\theta)^2$$

(See **page 24.16**)
Methods: Approximations ($F(\theta)$, Merkle; $G(\theta)$, Tada) of Numerical Results by Alternating Method (Smith)
Accuracy: 3%
References: **Smith 1967; Merkle 1973; Tada 1975**

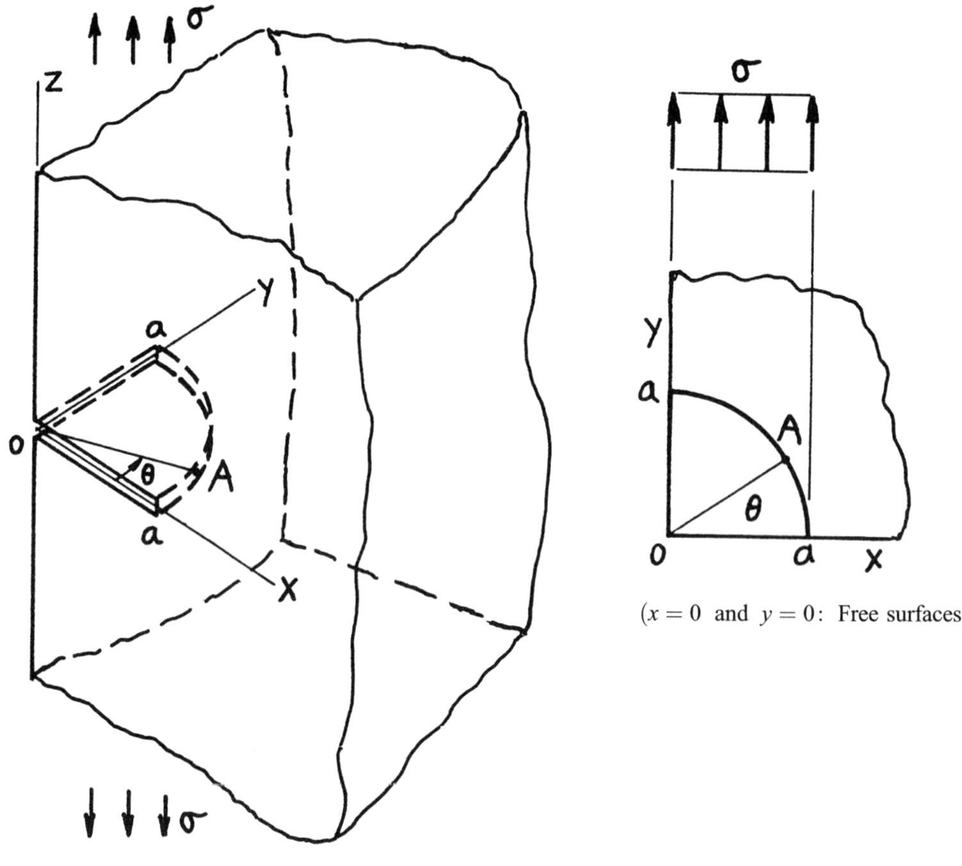

$(x = 0 \text{ and } y = 0: \text{ Free surfaces})$

$$K_{IA} = \frac{2}{\pi}\sigma\sqrt{\pi a}\, F_Q(\theta)$$

$$F_Q(\theta) = F(\theta) \cdot F\left(\frac{\pi}{2} - \theta\right)$$

$$F(\theta) = 1.211 - .186\sqrt{\sin\theta} \qquad \left(10^\circ < \theta < 80^\circ\right)$$

$$or \quad F_Q(\theta) = \left(1.211 - .186\sqrt{\sin\theta}\right)\left(1.211 - .186\sqrt{\cos\theta}\right)$$

Method: Alternating Method (Kobayashi), Finite Element Method (Tracey, Newman)
Accuracy: 3%
References: **Kobayashi 1976; Tracey 1973; Newman 1981a; Tada 1985**

NOTE: See **page 28.1**. $F(\theta)$ is the free surface correction for a half-circular surface crack. Total correction for a quarter-circular corner crack is approximately equal to product of surface corrections for a half-circular crack.

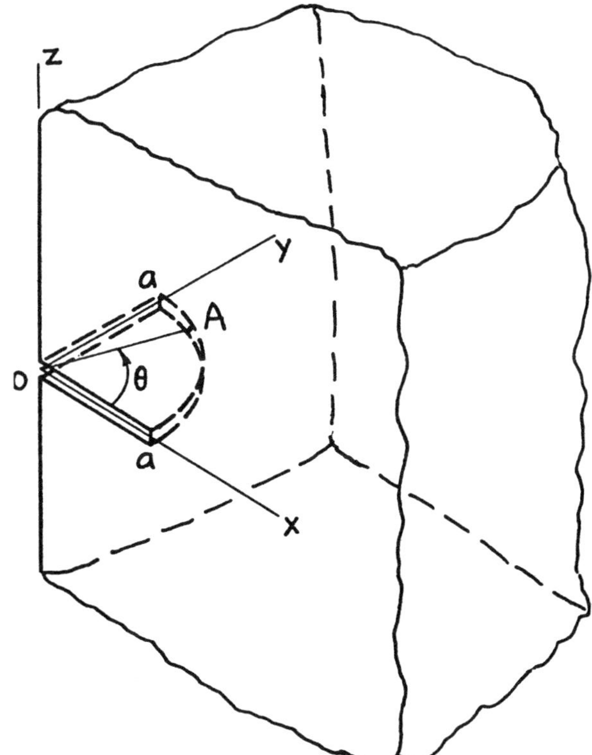

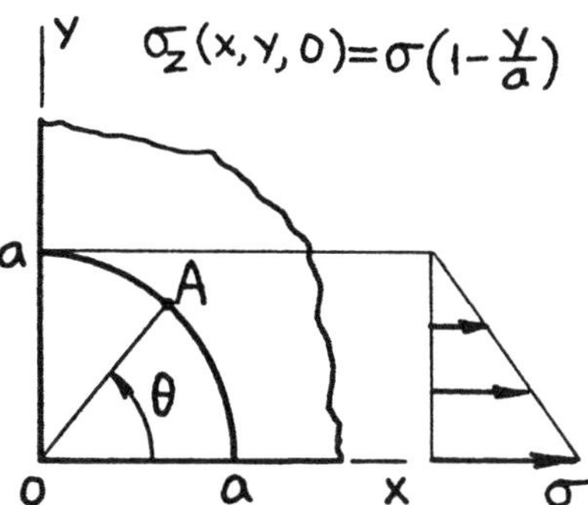

($x = 0$ and $y = 0$: Free surfaces)

$$K_{IA} = \frac{2}{\pi}\sigma\sqrt{\pi a} \cdot F(\theta)$$

$$or \quad = \frac{2}{\pi}\sigma\sqrt{\pi a}\left\{\frac{5}{6} - \frac{4}{3}(\sin\theta)^{3/2}\left(\sqrt{1+\sin\theta} - \sqrt{\sin\theta}\right)\right\} \cdot G(\theta)$$

$$F(\theta) = 1 - .72\sin\theta + .11(\sin\theta)^2$$

$$(10^\circ < \theta < 80^\circ)$$

$$G(\theta) = 1.22 - .56\sin\theta + .70(\sin\theta)^2$$

(See **page 24.16**)
Method: Approximation of Numerical Results by Alternating Method (Kobayashi)
Accuracy: 3%
References: **Kobayashi 1976; Tada 1975**

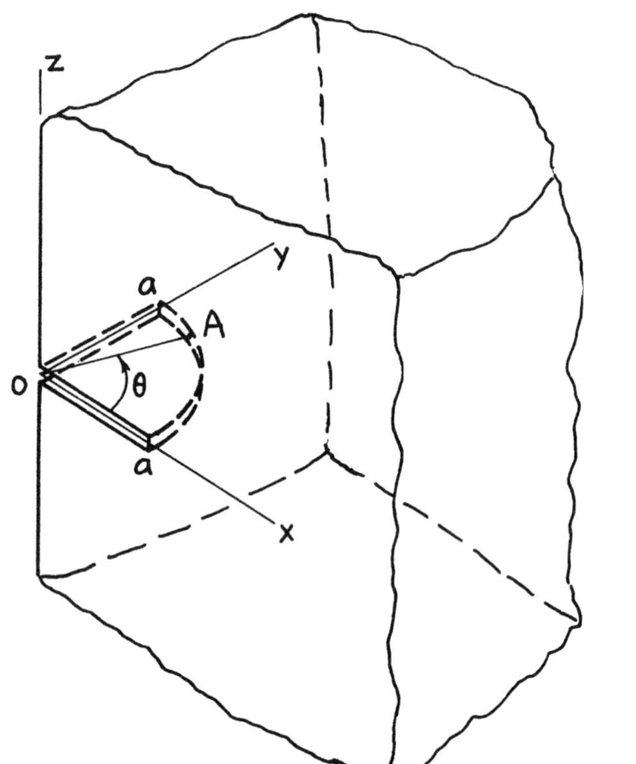

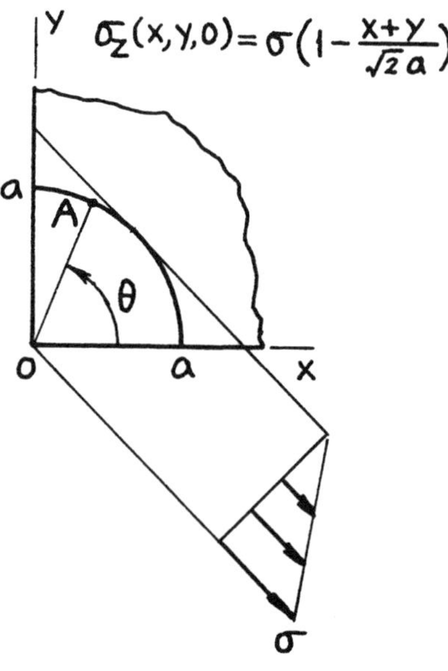

($x = 0$ and $y = 0$: Free surfaces)

$$K_{IA} = \frac{2}{\pi}\sigma\sqrt{\pi a} \cdot F(\theta)$$

or $$= \frac{2}{\pi}\sigma\sqrt{\pi a}\left\{\left(1+\frac{1}{\sqrt{2}}\right) - \frac{2\sqrt{2}}{3}\left[(\cos\theta)^{3/2}\sqrt{1+\cos\theta} + (\sin\theta)^{3/2}\sqrt{1+\sin\theta}\right]\right\} \cdot G(\theta)$$

$$F(\theta) = .311 + .154\frac{\left(1-\bar{\theta}^2\right)^2}{1+4\bar{\theta}^2}; \bar{\theta} = \frac{\theta}{\pi/4}$$

$$(10^\circ < \theta < 80^\circ)$$

$$G(\theta) = 1.245 + .04(\sin 2\theta)^2$$

(See **page 24.17**)
Method: Superposition of **page 28.4** (and also **page 28.3**)
Accuracy: $F(\theta)$ 4%; $G(\theta)$ 3%
Reference: **Tada 1985**

PART

V

CRACK(S) IN A ROD OR A PLATE BY ENERGY RATE ANALYSIS

❑ (Bending, Shearing, and Tension/Compression)

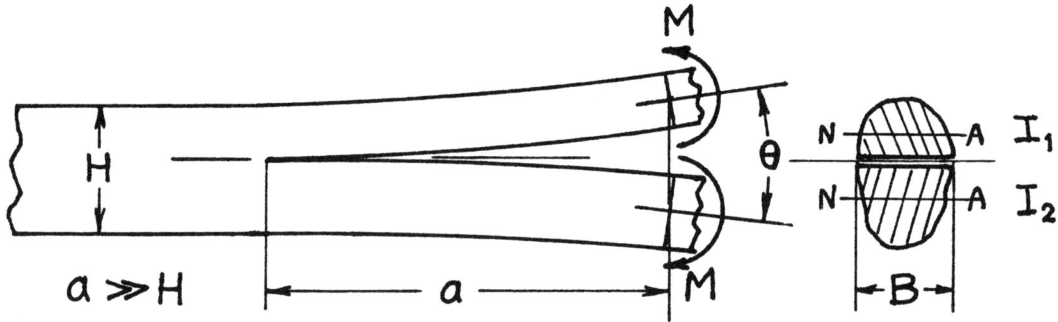

$$\mathcal{G} = \mathcal{G}_I + \mathcal{G}_{II}$$

$$= \frac{M^2}{2EIB}$$

or

$$= \frac{EI\theta^2}{2a^2B}$$

where $\frac{1}{I} = \frac{1}{I_1} + \frac{1}{I_2}$

$I_1, I_2 =$ moment of inertia of each cross section about its neutral axis

Method: Energy Balance
Accuracy: Approximation by Simple Beam Theory
References: **Tada 1974, 2000**

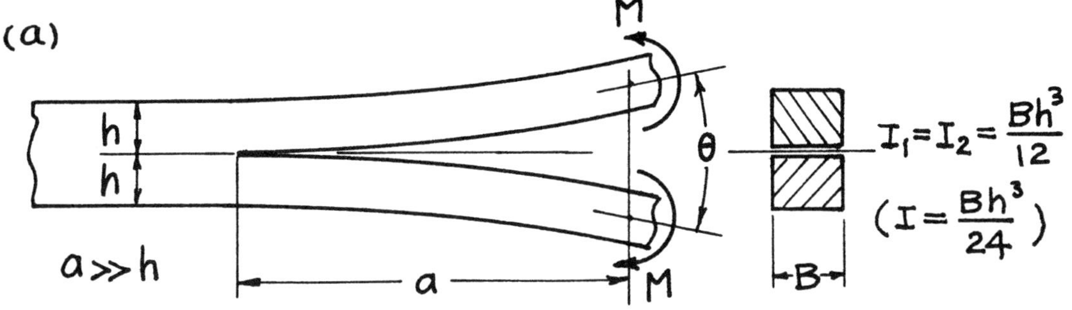

$$\mathcal{G} = \mathcal{G}_I = \frac{12\left(M/B\right)^2}{E'h^3} \overset{\text{or}}{=} \frac{E'h^3}{48\,a^2}\theta^2$$

$$K_I = \frac{2\sqrt{3}\left(M/B\right)}{h^{3/2}} \overset{\text{or}}{=} \frac{E'h^{3/2}}{4\sqrt{3}\,a}\theta$$

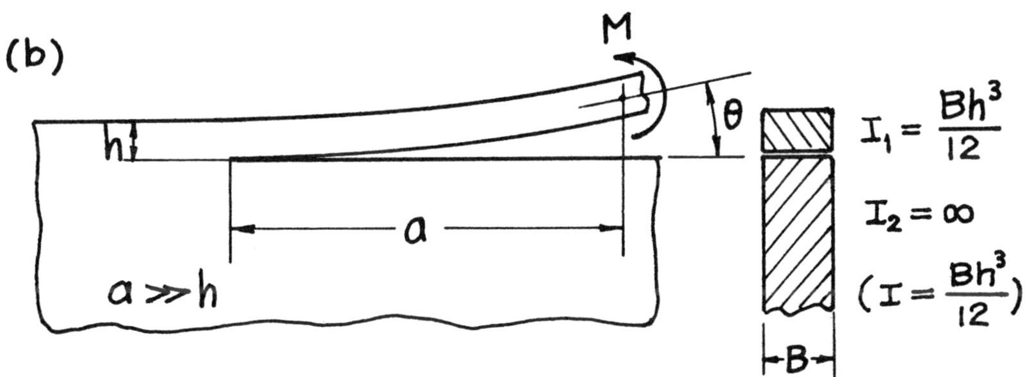

$$\mathcal{G} = \mathcal{G}_I + \mathcal{G}_{II} = \frac{6\left(M/B\right)^2}{E'h^3} \overset{\text{or}}{=} \frac{E'h^3}{24\,a^2}\theta^2$$

$$\mathcal{G}_I = 0.6222\mathcal{G}, \quad \mathcal{G}_{II} = 0.3778\mathcal{G}$$

$$K_I = 0.7888\frac{\sqrt{6}\left(M/B\right)}{h^{3/2}} \overset{\text{or}}{=} 0.7888\frac{E'h^{3/2}}{2\sqrt{6}\,a}\theta$$

$$K_{II} = -0.6147\frac{\sqrt{6}\left(M/B\right)}{h^{3/2}} \overset{\text{or}}{=} -0.6147\frac{E'h^{3/2}}{2\sqrt{6}\,a}\theta$$

(a) Special Case of **29.1**
(b) Special Case of **29.1** and **29.8**; see **Hutchinson 1992**

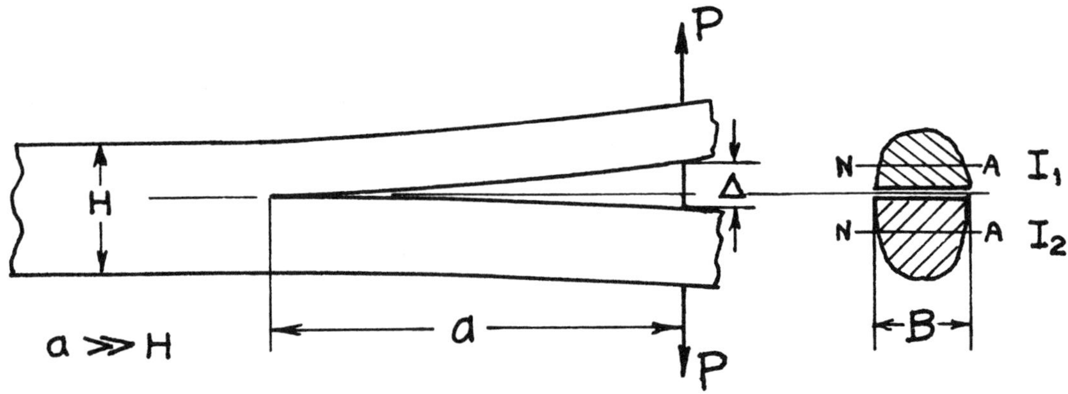

$$\mathcal{G} = \mathcal{G}_I + \mathcal{G}_{II}$$

$$= \frac{a^2}{2EIB} P^2$$

or

$$= \frac{9EI}{2a^4 B} \Delta^2$$

where $\frac{1}{I} = \frac{1}{I_1} + \frac{1}{I_2}$

$I_1, I_2 =$ moment of inertia of each cross section about its neutral axis

Method: Energy Balance
Accuracy: Approximation by Simple Beam Theory
References: **Tada 1974, 2000**

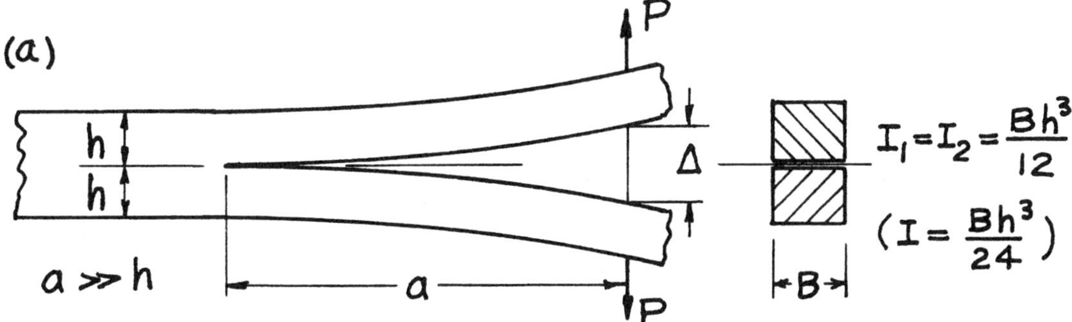

$$\mathcal{G} = \mathcal{G}_I = \frac{12\,a^2}{E'h^3}\left(\frac{P}{B}\right)^2 \overset{\text{or}}{=} \frac{3\,E'h^3}{16\,a^4}\Delta^2$$

$$K_I = \frac{2\sqrt{3}\,a}{h^{3/2}}\left(\frac{P}{B}\right) \overset{\text{or}}{=} \frac{\sqrt{3}\,E'h^{3/2}}{4\,a^2}\Delta$$

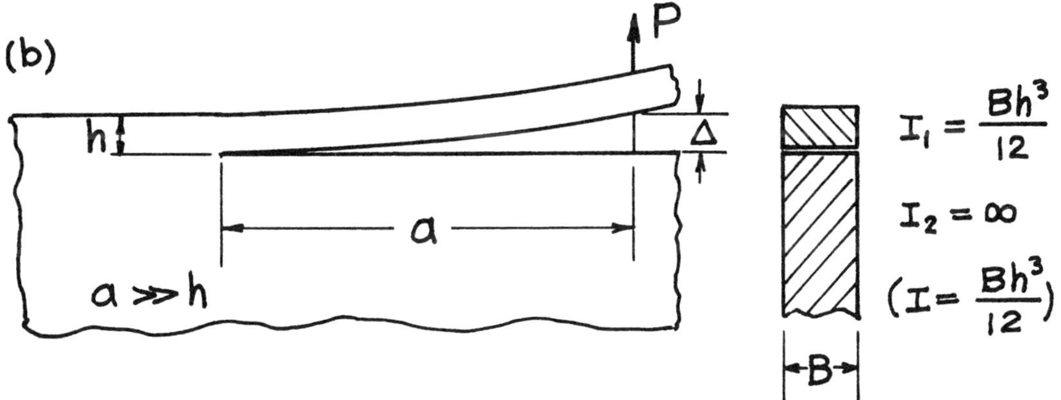

$$\mathcal{G} = \mathcal{G}_I + \mathcal{G}_{II}$$

$$= \frac{6\,a^2}{E'h^3}\left(\frac{P}{B}\right)^2 \overset{\text{or}}{=} \frac{3\,E'h^3}{8\,a^4}\Delta^2$$

Special cases of **29.3**

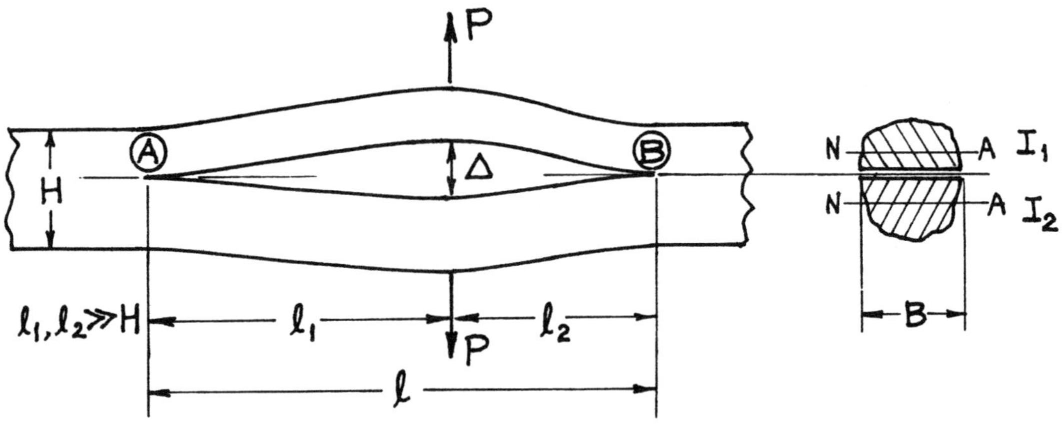

$$\mathcal{G} = \mathcal{G}_I + \mathcal{G}_{II}$$

$$\mathcal{G}_{Ⓐ} = \frac{1}{2EIB} \frac{\ell_1^2 \ell_2^4}{\ell^4} P^2 \overset{\text{or}}{=} \frac{9EI}{2B} \frac{\ell^2}{\ell_1^4 \ell_2^2} \Delta^2$$

$$\mathcal{G}_{Ⓑ} = \frac{1}{2EIB} \frac{\ell_1^4 \ell_2^2}{\ell^4} P^2 \overset{\text{or}}{=} \frac{9EI}{2B} \frac{\ell^2}{\ell_1^2 \ell_2^4} \Delta^2$$

where $\frac{1}{I} = \frac{1}{I_1} + \frac{1}{I_2}$

$I_1, I_2 =$ moment of inertia of each cross section about its neutral axis

Method: Energy Balance
Accuracy: Approximation by Simple Beam Theory
References: **Tada 1974, 2000**

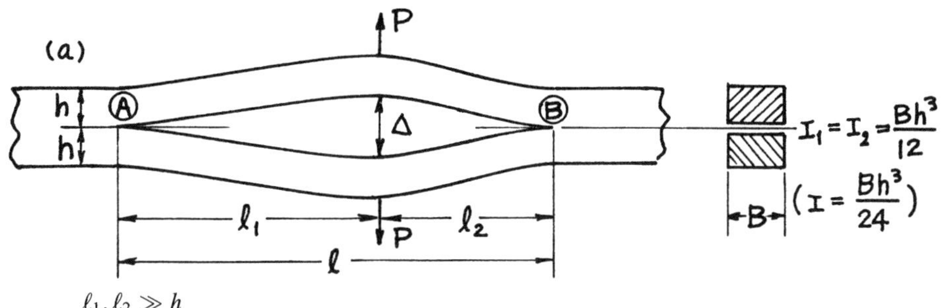

$\ell_1, \ell_2 \gg h$

$$\mathcal{G} = \mathcal{G}_I$$

$$\mathcal{G}_{I\text{Ⓐ}} = \frac{12}{E'h^3}\frac{\ell_1^2\,\ell_2^4}{\ell^4}\left(\frac{P}{B}\right)^2 \overset{\text{or}}{=} \frac{3\,E'h^3}{16}\frac{\ell^2}{\ell_1^4\,\ell_2^2}\Delta^2$$

$$\mathcal{G}_{I\text{Ⓑ}} = \frac{12}{E'h^3}\frac{\ell_1^4\,\ell_2^2}{\ell^4}\left(\frac{P}{B}\right)^2 \overset{\text{or}}{=} \frac{3\,E'h^3}{16}\frac{\ell^2}{\ell_1^2\,\ell_2^4}\Delta^2$$

$$K_{I\text{Ⓐ}} = \frac{2\sqrt{3}\,\ell_1\,\ell_2^2}{h^{3/2}\,\ell^2}\left(\frac{P}{B}\right) \overset{\text{or}}{=} \frac{\sqrt{3}\,E'h^{3/2}}{4}\frac{\ell}{\ell_1^2\,\ell_2}\Delta$$

$$K_{I\text{Ⓑ}} = \frac{2\sqrt{3}\,\ell_1^2\,\ell_2}{h^{3/2}\,\ell^2}\left(\frac{P}{B}\right) \overset{\text{or}}{=} \frac{\sqrt{3}\,E'h^{3/2}}{4}\frac{\ell}{\ell_1\,\ell_2^2}\Delta$$

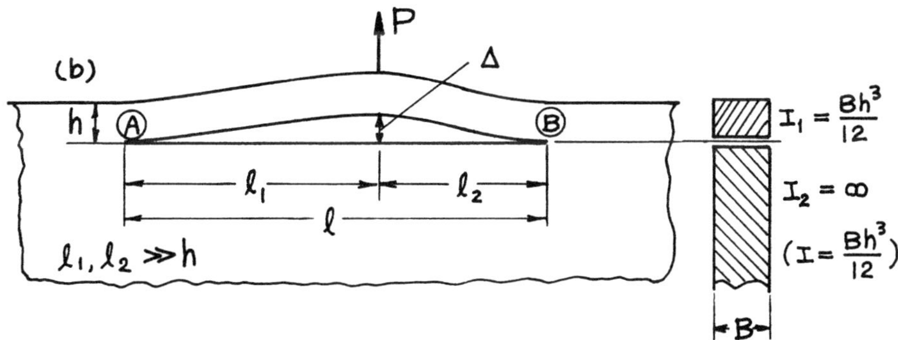

$$\mathcal{G} = \mathcal{G}_I + \mathcal{G}_{II}$$

$$\mathcal{G}_{\text{Ⓐ}} = \frac{6}{E'h^3}\frac{\ell_1^2\,\ell_2^4}{\ell^4}\left(\frac{P}{B}\right)^2 \overset{\text{or}}{=} \frac{3\,E'h^3}{8}\frac{\ell^2}{\ell_1^4\,\ell_2^2}\Delta^2$$

$$\mathcal{G}_{\text{Ⓑ}} = \frac{6}{E'h^3}\frac{\ell_1^4\,\ell_2^2}{\ell^4}\left(\frac{P}{B}\right)^2 \overset{\text{or}}{=} \frac{3\,E'h^3}{8}\frac{\ell^2}{\ell_1^2\,\ell_2^4}\Delta^2$$

Special cases of **29.5**

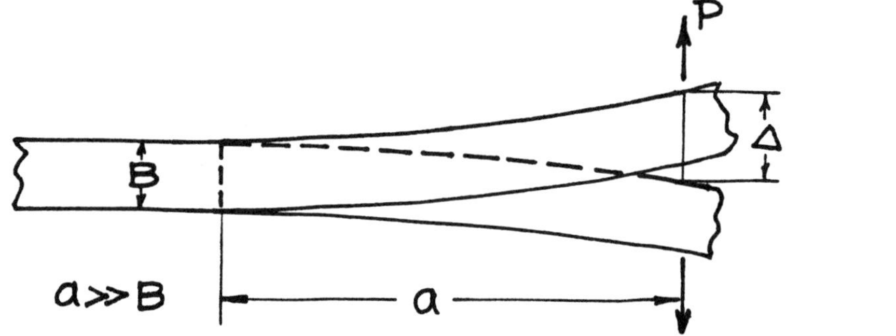

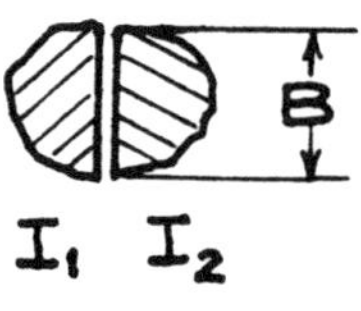

$$K_{III} = \frac{1}{\sqrt{1+\nu}} \cdot \frac{Pa}{\sqrt{2BI}}$$

or

$$K_{III} = \frac{1}{\sqrt{1+\nu}} \cdot \frac{3E\Delta}{a^2}\sqrt{\frac{I}{2B}}$$

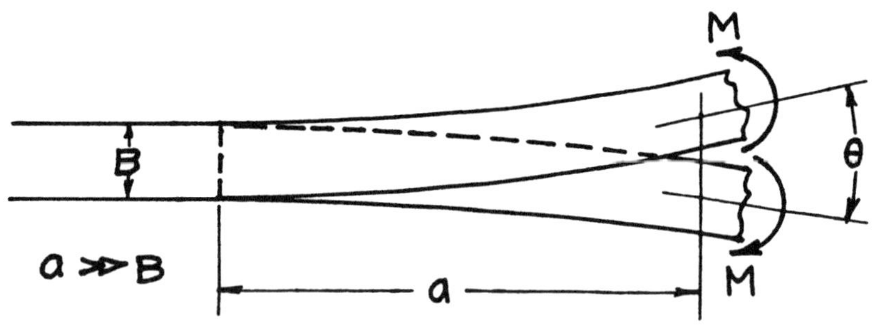

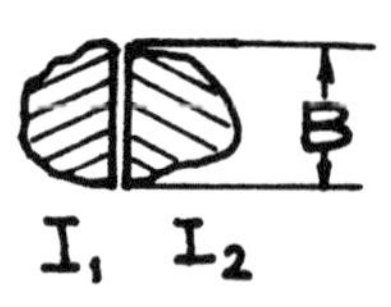

$$K_{III} = \frac{1}{\sqrt{1+\nu}} \cdot \frac{M}{\sqrt{2BI}}$$

or

$$K_{III} = \frac{1}{\sqrt{1+\nu}} \cdot \frac{E\theta}{a}\sqrt{\frac{I}{2B}}$$

where $$\frac{1}{I} = \frac{1}{I_1} + \frac{1}{I_2}$$

$I_1, I_2 =$ moment of inertia of each cross section about its neutral axis

Method: Energy Balance
Accuracy: Approximation by Simple Beam Theory
Reference: **Tada 1974**

NOTE: Loads are applied through the shear center of each beam so as to produce bending only with no torsion.

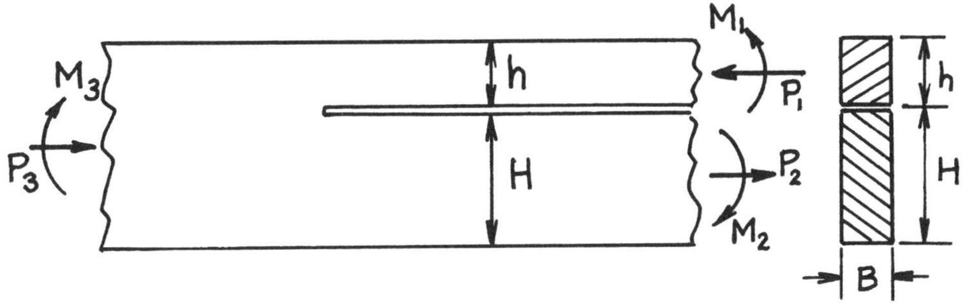

$$\mathcal{G} = \mathcal{G}_I + \mathcal{G}_{II}$$

$$= \frac{1}{E'B^2}\left[\frac{1}{2}\left(\frac{P_1^2}{h} + \frac{P_2^2}{H} + \frac{P_3^2}{h+H}\right) + 6\left(\frac{M_1^2}{h^3} + \frac{M_2^2}{H^3} + \frac{M_3^2}{(h+H)^3}\right)\right]$$

$$K_I = \frac{\left({}^P\!/_B\right)}{\sqrt{2h}} f(\eta)\cos\omega(\eta) + \frac{\left({}^M\!/_B\right)}{\sqrt{2h^3}} g(\eta)\sin[\omega(\eta) + \gamma(\eta)]$$

$$K_{II} = \frac{\left({}^P\!/_B\right)}{\sqrt{2h}} f(\eta)\sin\omega(\eta) - \frac{\left({}^M\!/_B\right)}{\sqrt{2h^3}} g(\eta)\cos[\omega(\eta) + \gamma(\eta)]$$

where $\eta = {}^h\!/_H, \quad 0 \le \eta \le 1$

$$P = P_1 - \frac{\eta}{1+\eta}P_3 - \frac{6\eta^2}{(1+\eta)^3}\frac{M_3}{h}, \quad M = M_1 - \left(\frac{\eta}{1+\eta}\right)^3 M_3$$

$$f(\eta) = \left(1 + 4\eta + 6\eta^2 + 3\eta^3\right)^{1/2}, \quad g(\eta) = 2\sqrt{3}\left(1 + \eta^3\right)^{1/2}$$

$$\omega(\eta) = 0.909 - 0.052\eta, \quad \gamma(\eta) = \sin^{-1}\left[\frac{6\eta^2(1+\eta)}{f(\eta)g(\eta)}\right]$$

Method: Energy Balance/Integral Equations
Accuracy: 1%
References: **Suo 1990a; Hutchinson 1992**

(a)

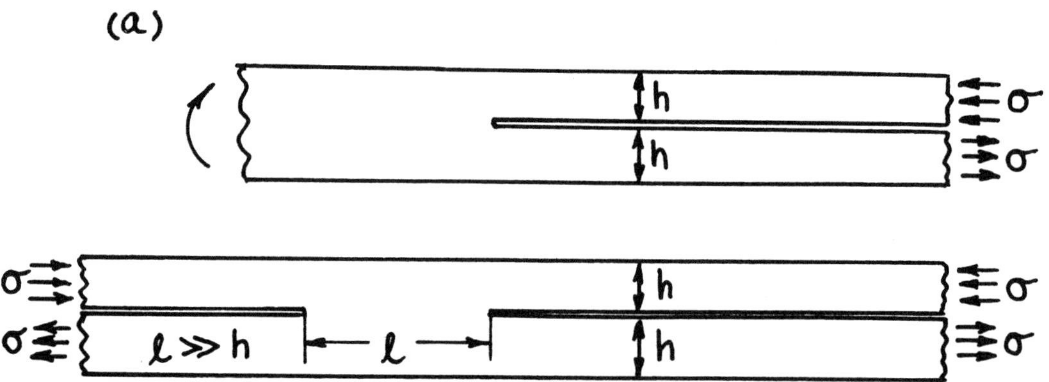

$$K_I = 0$$

$$K_{II} = \frac{1}{2}\sigma\sqrt{h}$$

(b)

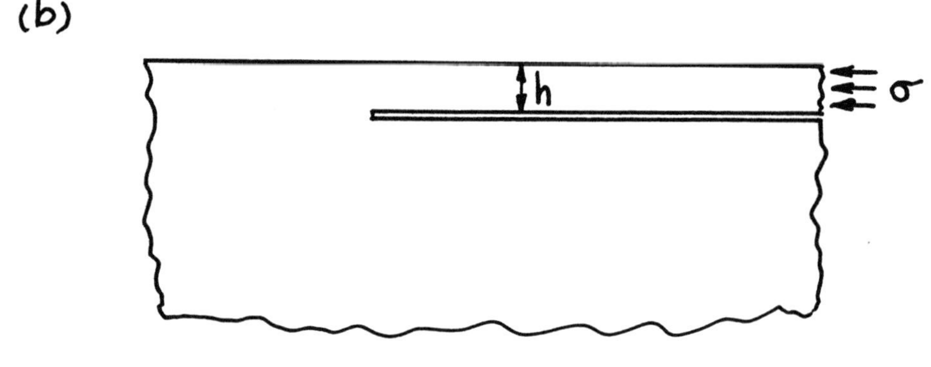

$$K_I = 0.4347 \ \sigma\sqrt{h}$$

$$K_{II} = 0.5578 \ \sigma\sqrt{h}$$

Special cases of **29.8**

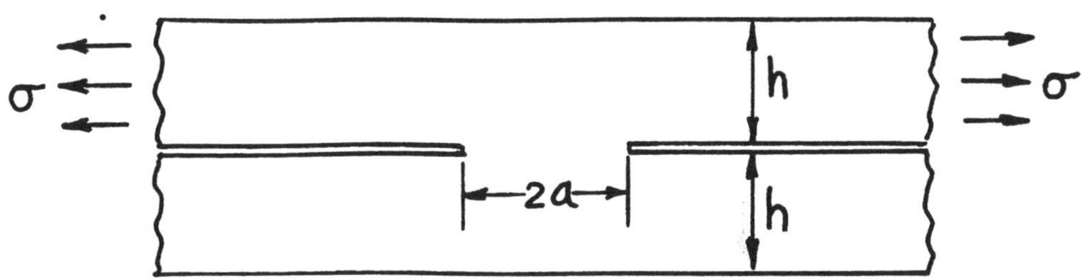

$$K_{II} = \frac{1}{4}\sigma\sqrt{h}\,F(s)$$

$$\text{where} \quad s = \frac{h}{a+h}$$

$$F(s) = \sqrt{1-s}\left(1 + 0.5s + 0.375s^2 - 1.081s^3 + 5.580s^4 - 4.601s^5\right)$$

Methods: Integral Transform Method (Keer; $s \leq 2/3$), Interpolation (Tada; $s > 2/3$)
Accuracy: 1%
References: **Keer 1974, 1989, 1990; Tada 2000**

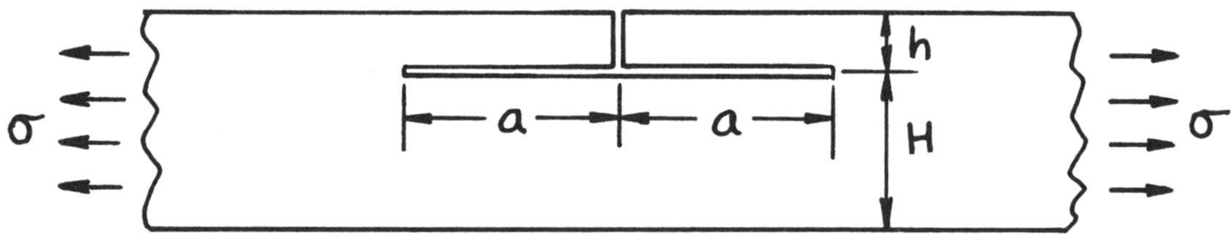

$$\mathcal{G} = \mathcal{G}_I + \mathcal{G}_{II} = \frac{\sigma^2 h}{2E'}\{f(\eta)\}^2$$

$$\mathcal{G}_I = \cos^2 \omega(\eta) \cdot \mathcal{G}$$

$$\mathcal{G}_{II} = \sin^2 \omega(\eta) \cdot \mathcal{G}$$

$$K_I = \frac{\sigma\sqrt{h}}{\sqrt{2}} f(\eta) \cos\omega(\eta)$$

$$K_{II} = \frac{\sigma\sqrt{h}}{\sqrt{2}} f(\eta) \sin\omega(\eta)$$

where $\eta = {}^h/_H$

$$f(\eta) = \left(1 + 4\eta + 6\eta^2 + 3\eta^3\right)^{1/2}$$

$$\omega(\eta) = 0.909 - 0.052\eta$$

Method: Energy Balance (Special Case of **29.8**)
Accuracy: Better than 1% for ${}^h/_a > 1.5$
Reference: **Suo 1990a**

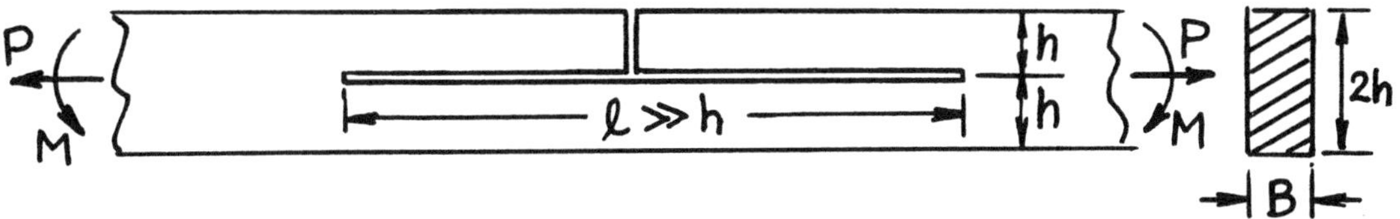

$$K_I = \frac{\sqrt{3}}{2}\frac{\left(P/B\right)}{\sqrt{h}} + \sqrt{3}\,\frac{\left(M/B\right)}{h^{3/2}}$$

$$K_{II} = \frac{\left(P/B\right)}{\sqrt{h}} + \frac{3}{2}\frac{\left(M/B\right)}{h^{3/2}}$$

Special cases of **29.8** and **29.11**

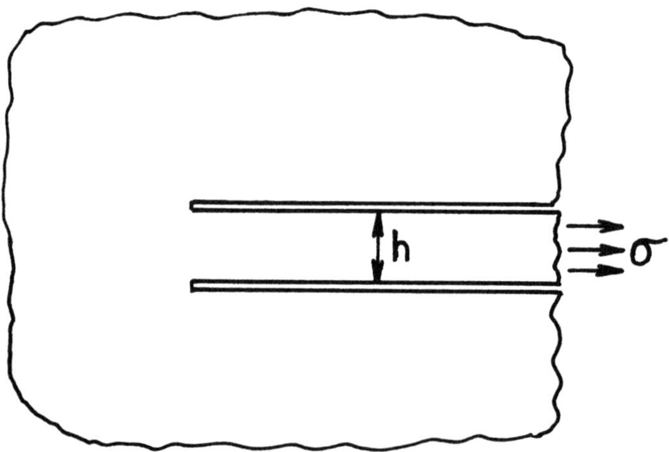

$$\mathcal{G} = \mathcal{G}_I + \mathcal{G}_{II} = \frac{\sigma^2 h}{4E'}$$

$$\mathcal{G}_I = 0.0904\mathcal{G}$$

$$\mathcal{G}_{II} = 0.9096\mathcal{G}$$

$$K_I = 0.3007 \cdot \frac{1}{2}\sigma\sqrt{h}$$

$$K_{II} = 0.9537 \cdot \frac{1}{2}\sigma\sqrt{h}$$

Method: Energy Balance/Integral Equations
Accuracy: Accurate numerical values
Reference: **Suo 1990b**

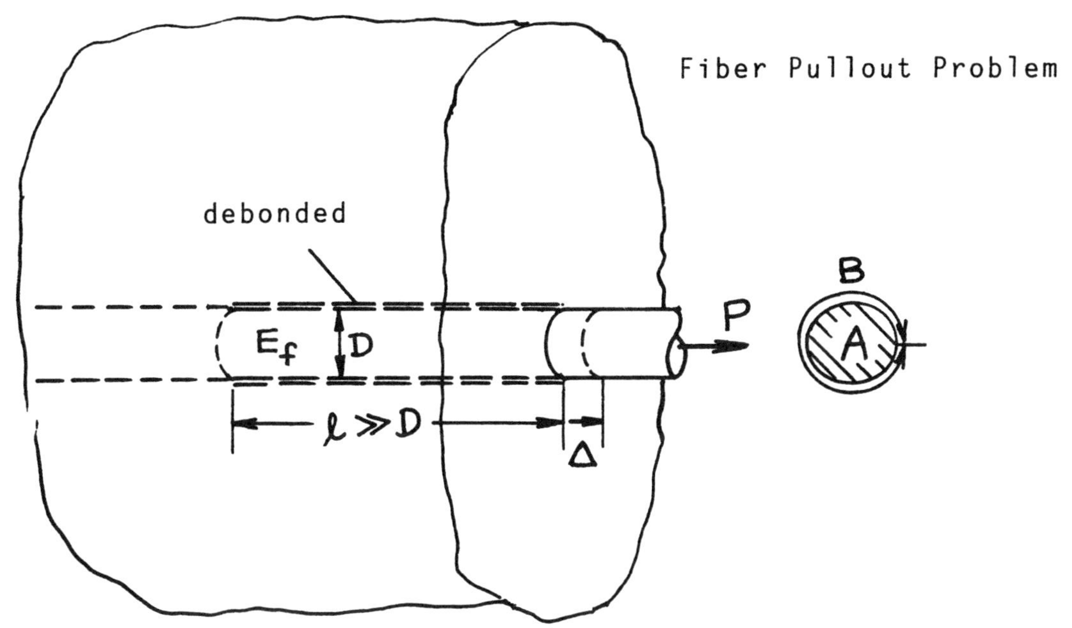

$$\mathcal{G}(\simeq \mathcal{G}_{II}) = \frac{P^2}{2E_f A} \cdot \frac{1}{B} = \frac{\sigma^2 A}{2E_f} \cdot \frac{1}{B}$$

$$= \frac{E_f A \Delta^2}{2\ell^2} \cdot \frac{1}{B}$$

where E_f = Young's modulus of fiber
A = cross sectional area of fiber
B = girth of fiber
ℓ = debonded length
$\sigma = P/A$

For circular fiber of diameter D:

$$\mathcal{G} = \frac{\sigma^2 D}{8E_f} = \frac{E_f D}{8\ell^2}\Delta^2$$

Method: Energy Balance
Accuracy: Approximation by simple tension for fiber
Reference: **Tada 2000**

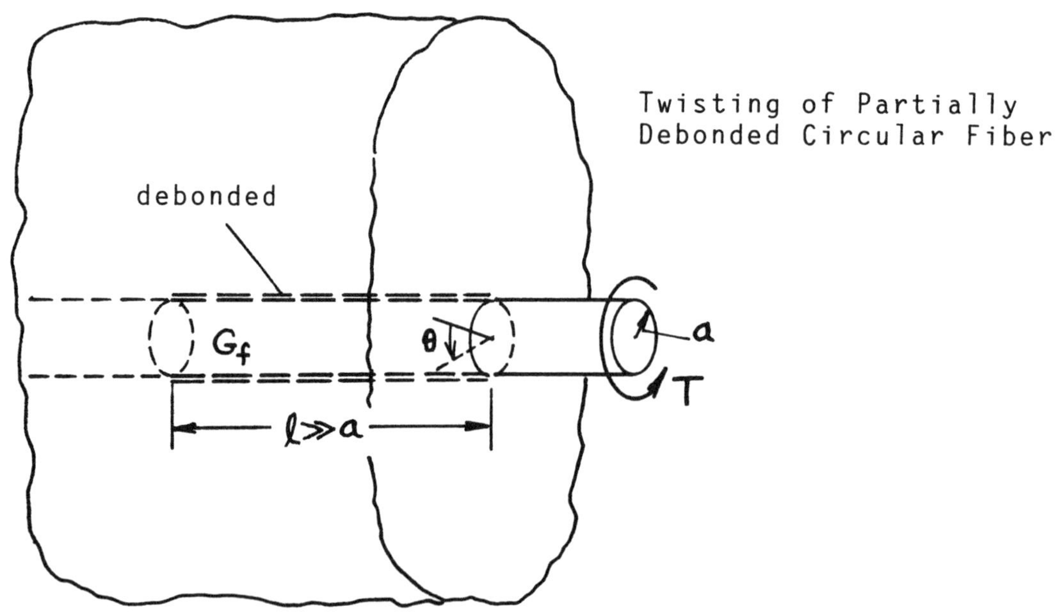

$$\mathcal{G}(\simeq \mathcal{G}_{III}) = \frac{T^2}{2G_f J}\cdot\frac{1}{B} = \frac{1}{2}G_f J\left(\frac{\theta}{\ell}\right)^2\cdot\frac{1}{B}$$

$$= \frac{\tau^2 a}{8G_f} = \frac{1}{8}G_f a^3\left(\frac{\theta}{\ell}\right)^2$$

where $G_f =$ shear modulus of fiber

$J =$ polar moment of inertia of fiber $= \dfrac{\pi a^4}{2}$

$B =$ circumference of fiber $= 2\pi a$

$\ell =$ debonded length

$\theta =$ angle of twist

$\tau = (T/J)a = 2T/(\pi a^3)$

Method: Energy Balance
Accuracy: Approximation by simple torsion of fiber
Reference: **Tada 2000**

PART

VI

STRIP YIELD MODEL SOLUTIONS

- Introduction to Strip Yield Model Analysis
- Additional Notes on Strip Yield Models
- Two-Dimensional Problems of Strip Yielding from Crack(s)
- Two-Dimensional Problems of Strip Yielding from a Hole with or without Crack(s)
- Three-Dimensional Strip Yielding Solutions

INTRODUCTION TO STRIP YIELD MODEL ANALYSIS

The following pages are devoted to strip yield model solutions (sometimes called the "Dugdale-Barrenblatt model"; e.g., **Barrenblatt 1962**). This model is useful in assessing the effects of finite configuration size and high net section stresses on the plastic zone size accompanying a crack tip. It does give plastic zone size estimates with some improvement over the small-scale yielding analysis represented by **Eq. (27) (p. 1.11)**, but falls short of a full plasticity solution. Consequently, the formulas on these solution pages should be regarded as improved estimates of plastic zone size, ℓ, as directly comparable to r_p of **Eq. (29) (p. 1.11)**.

The strip yield model replaces the actual crack and its plastic zone, **Fig. 30-1(a)**, with the superposition of two elastic crack solutions **Fig. 30-1 (b) and (c)**. The crack length in the model, (b) and (c), is taken to be the actual crack length plus the plastic zone size, ℓ. As in **Fig. 30-1(c)**, the plasticity effects are modeled by applying closing flow stresses, σ_Y, over the portion of the model crack surface where plasticity is occuring for the actual crack. At the new model crack tip the stresses must be finite; therefore the total crack tip stress intensity in the model must be zero, that is

$$K_{\text{total}} = K_{\text{applied}} + K_{\text{flow}} = 0$$

This relationship is used with the appropriate elastic stress solutions for the model cracks to determine ℓ and other features of the solution.

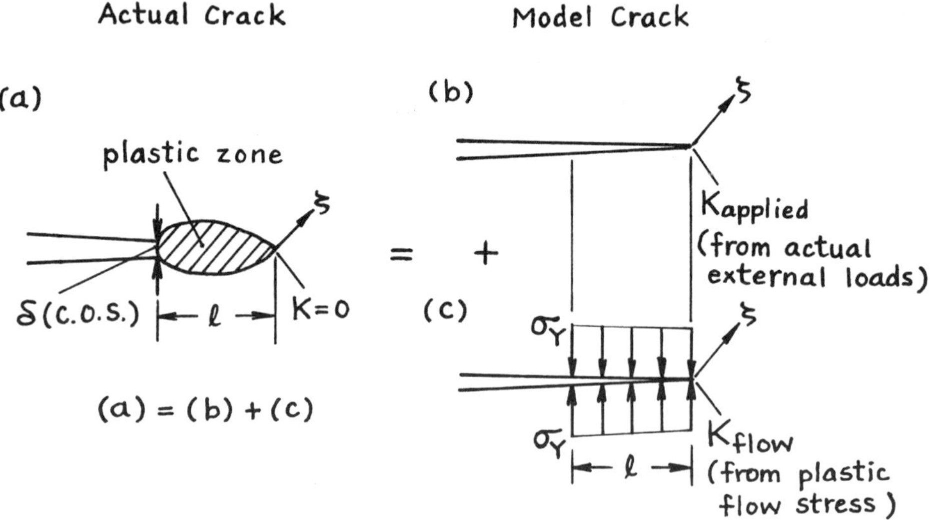

Fig. 30-1. Strip yield model.

As an example of the application of this model, consider the case of small-scale yielding, that is, infinite sheets with semi-infinite cracks where the actual external loads are applied remotely (at infinity). The solutions on **pages 3.1 and 3.7** are taken to correspond with **Fig. 30-1 (b) and (c)**, respectively (using Mode I only). Therefore

$$Z_{\text{applied}} = \frac{K_{\text{applied}}}{\sqrt{2\pi\zeta}}$$

$$K_{\text{applied}} = K_{\text{applied}}$$

and

$$Z_{\text{flow}} = -\frac{2}{\pi}\sigma_Y\left\{\sqrt{\frac{\ell}{\zeta}} - \tan^{-1}\sqrt{\frac{\ell}{\zeta}}\right\}$$

$$K_{\text{flow}} = -\frac{2}{\pi}\sigma_Y\sqrt{2\pi\ell}$$

Substituting the K formulas into the above expression for K_{total} gives

$$\ell = \frac{\pi}{8}\left(\frac{K_{\text{applied}}}{\sigma_Y}\right)^2$$

which corresponds well to the small-scale yield zone width r_p **(p. 1.11)**, considering that $\pi/8$ is nearly equal to the $1/\pi$ in the r_p formula letting σ_Y equal σ_{yp} for plane stress. (For plane strain, note that the flow stress, σ_Y, elevation due to constraint makes it about $\sqrt{3}\sigma_{yp}$, which should be inserted into the equation for ℓ.) The following solution pages give similar superpositions of pairs of stress solutions, corresponding to **Fig. 30-1 (b) and (c)**, but for the particular external configurations and loads indicated on the individual pages.

ADDITIONAL NOTES ON STRIP YIELD MODELS

From the preceding strip-yield analysis for small-scale yielding some additional notes are relevant. Since we are forming the solutions by superimposing two linear-elastic solutions, it is permissible to simply add stress functions for the model, that is

$$Z_{\text{total}} = Z_{\text{applied}} + Z_{\text{flow}}$$

For the particular case of small-scale yielding discussed above the results lead to

$$Z_{\text{total}} = \frac{2}{\pi}\sigma_Y\tan^{-1}\left(\frac{K_{\text{applied}}}{2\sigma_Y}\sqrt{\frac{\pi}{2\zeta}}\right)$$

This result represents the complete local stress distribution for the small-scale yielding strip yield model (for any external configuration and loading). Similar results for large-scale yielding may be derived from corresponding information in this handbook.

Note also that the above stress function, Z_{total}, for large ζ compared with ℓ, that is, away from the disturbance of the plastic zone (ℓ), reduces to

$$Z_{\text{total}\,(\zeta \gg \ell)} = \frac{K_{\text{applied}}}{\sqrt{2\pi\zeta}}$$

which corresponds to the original solution undisturbed by plasticity.

Moreover, the so-called "crack opening stretch" (C.O.S.) which corresponds physically to the opening displacement at the actual crack tip (see **Fig. 30-1**), can be found by integrating the stress function, Z_{total}, and evaluating the opening displacements, v, at the crack tip $\zeta = -\ell$, which for small-scale yielding results in

$$\delta = \text{C.O.S.} = 2v|_{\zeta=-\ell} = \frac{K^2_{\text{applied}}}{E'\sigma_Y} = \frac{\mathcal{G}}{\sigma_Y}$$

Indeed, other interesting and relevant calculations can be made from strip yield models. The technique has been shown here and the essentials are suggested on the pages to follow.

SOLUTION FOR SMALL SCALE YIELDING

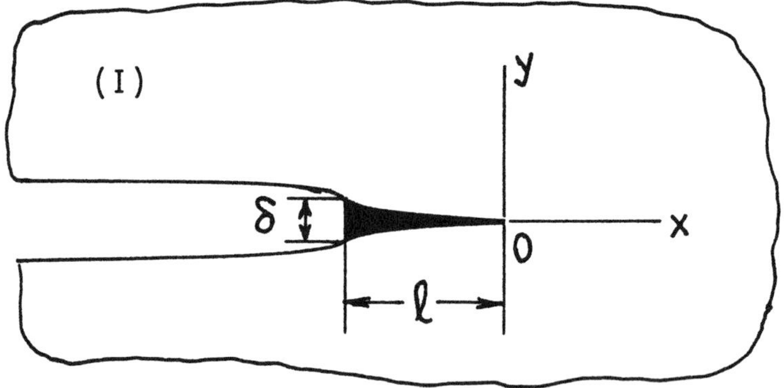

$$z = x + iy$$

$$Z(z) = \frac{2\sigma_Y}{\pi} \tan^{-1} \sqrt{\frac{\ell}{z}}$$

$$\overline{Z}(z) = \frac{2\sigma_Y \ell}{\pi} \left\{ \sqrt{\frac{z}{\ell}} + \left(1 + \frac{z}{\ell}\right) \tan^{-1} \sqrt{\frac{\ell}{z}} \right\}$$

$$\delta = 2v\Big|_{z=-\ell} = \frac{8\sigma_Y \ell}{\pi E'} = \frac{K_{I\,APPLIED}^2}{E'\sigma_Y} = \frac{\mathcal{G}}{\sigma_Y}$$

where

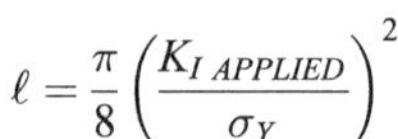

$$\ell = \frac{\pi}{8}\left(\frac{K_{I\,APPLIED}}{\sigma_Y}\right)^2$$

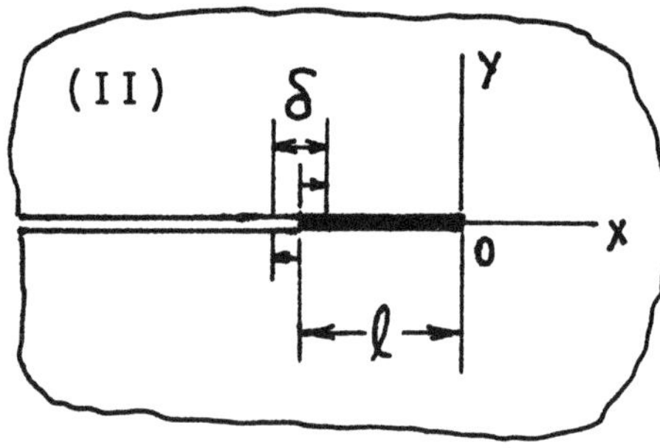

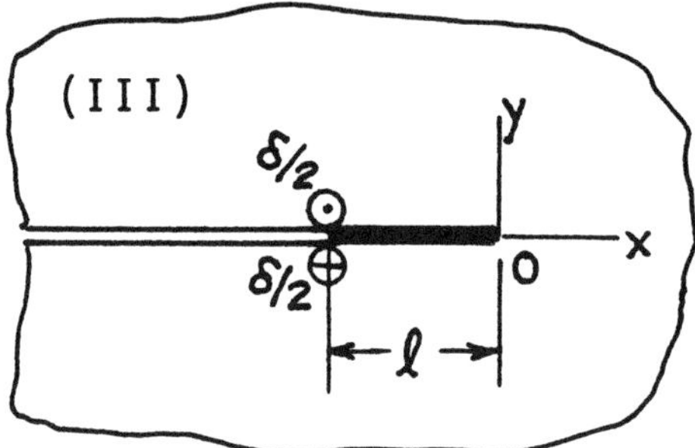

For (II) and (III), replace (K_I, σ_Y, v, E') by (K_{II}, τ_Y, u, E') and $(K_{III}, \tau_Y, w, 2G)$, respectively.
Method: Westergaard Stress Function (Superposition of **pages 3.1 and 3.7**)
Accuracy: Exact
References: **Barenblatt 1962; Irwin 1969; Tada 1974**

NOTE: See **pages 30.1 and 30.2**.

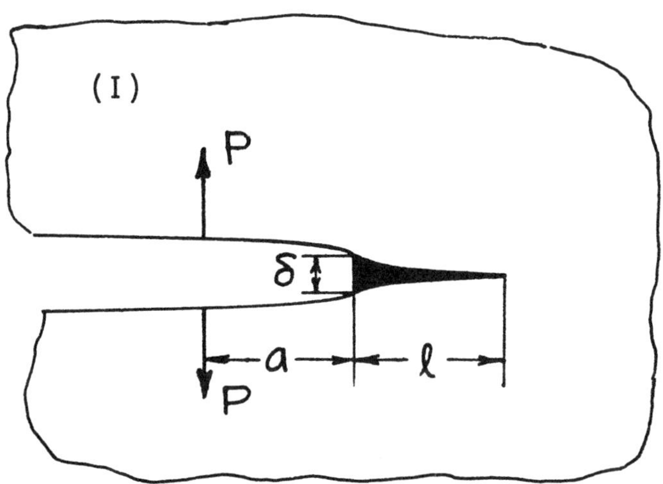

$$\frac{\ell}{a} = \frac{1}{2}\left(\sqrt{1 + \left(\frac{P}{\sigma_Y a}\right)^2} - 1\right)$$

$$\delta = \frac{8P}{\pi E'}\tanh^{-1}\sqrt{\frac{\ell}{a+\ell}} - \frac{8\sigma_Y \ell}{\pi E'}$$

$$\text{or} \qquad = \frac{8\sigma_Y \ell}{\pi E'}\left(2\sqrt{\frac{a+\ell}{\ell}}\tanh^{-1}\sqrt{\frac{\ell}{a+\ell}} - 1\right)$$

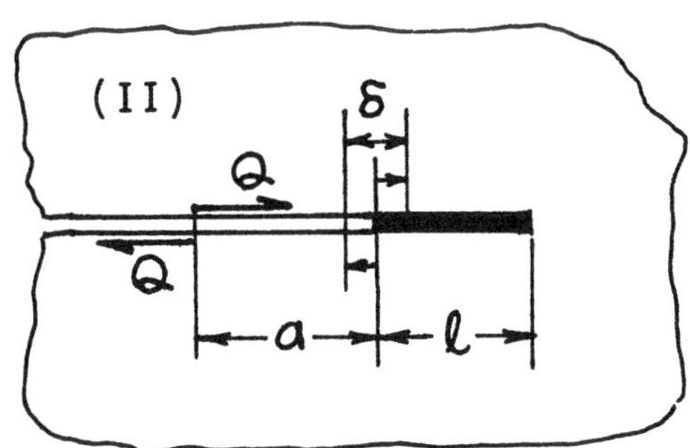

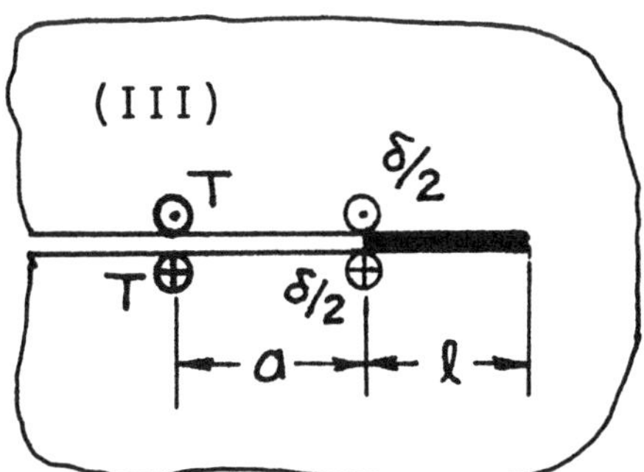

For (II) and (III), replace (P, σ_Y, E') by (Q, τ_Y, E') and $(T, \tau_Y, 2G)$, respectively.
Method: Superposition of **pages 3.6 and 3.7** (or a Limiting Case of **page 30.6**)
Accuracy: Exact
References: **Tada 1974**

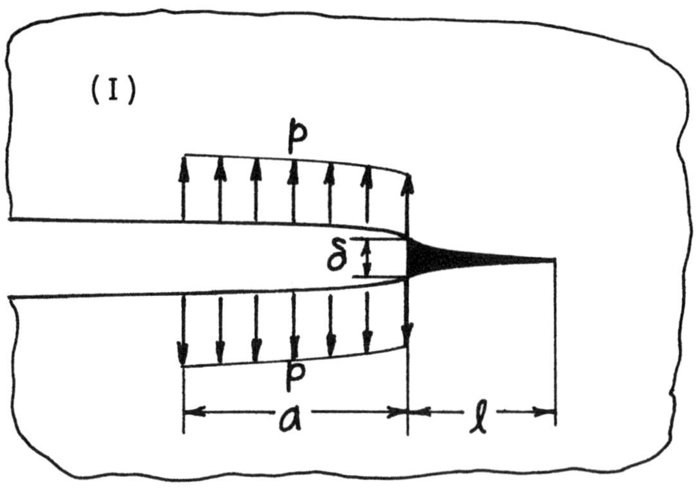

$$\frac{\ell}{a} = \frac{(p/\sigma_Y)^2}{2(p/\sigma_Y)+1}$$

$$\delta = \frac{8pa}{\pi E'} \tanh^{-1}\sqrt{\frac{\ell}{a+\ell}}$$

$$or \qquad = \frac{8pa}{\pi E'} \tanh^{-1}\left\{\frac{p/\sigma_Y}{1+(p/\sigma_Y)}\right\}$$

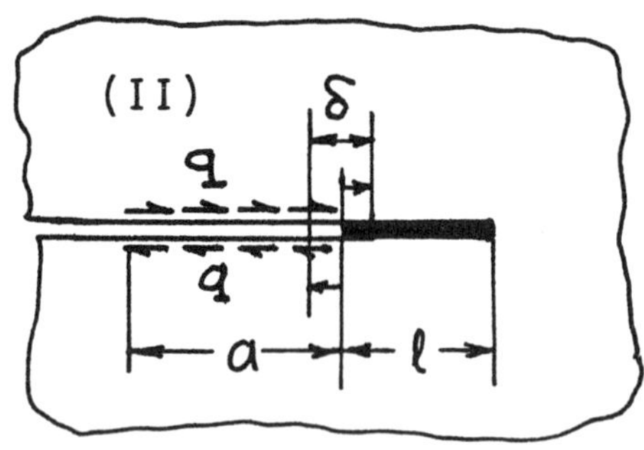

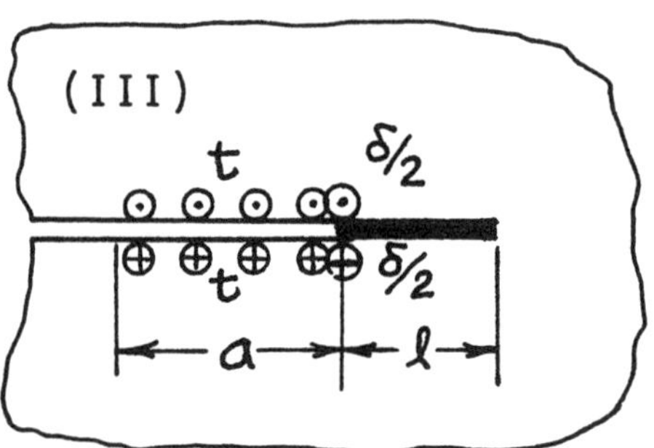

For (II) and (III), replace (p, σ_Y, E') by (q, τ_Y, E') and $(t, \tau_Y, 2G)$, respectively.
Method: Superposition of Two Cases of **page 3.7** (or a Special Case of **page 30.6**)
Accuracy: Exact
Reference: **Tada 1974**

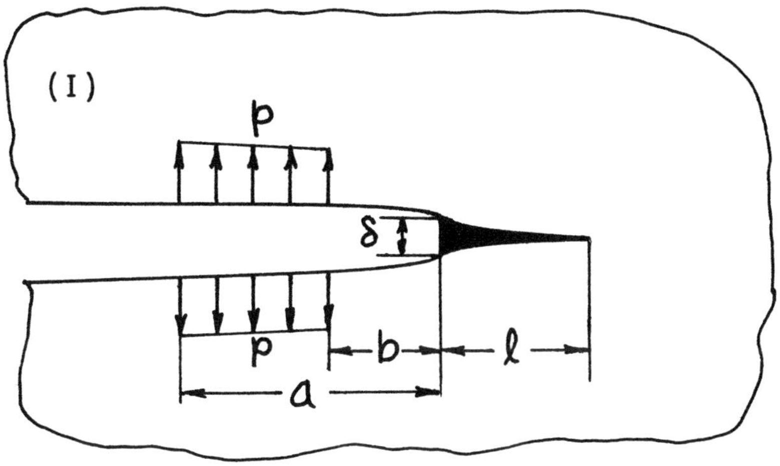

$$\frac{\ell}{a} = \frac{(p/\sigma_Y)^2}{1 - 4(p/\sigma_Y)^2}\left\{1 + \frac{b}{a} - 2\sqrt{\frac{b}{a} + \left(1 - \frac{b}{a}\right)^2 \left(\frac{p}{\sigma_Y}\right)^2}\right\}$$

$$\delta = \frac{8pa}{\pi E'}\left\{\tanh^{-1}\sqrt{\frac{\ell}{a+\ell}} - \frac{b}{a}\tanh^{-1}\sqrt{\frac{\ell}{b+\ell}}\right\}$$

When $\frac{p}{\sigma_Y} = \frac{1}{2}$: $\frac{\ell}{a} = \frac{1}{8}\frac{(1-b/a)^2}{1+b/a}$

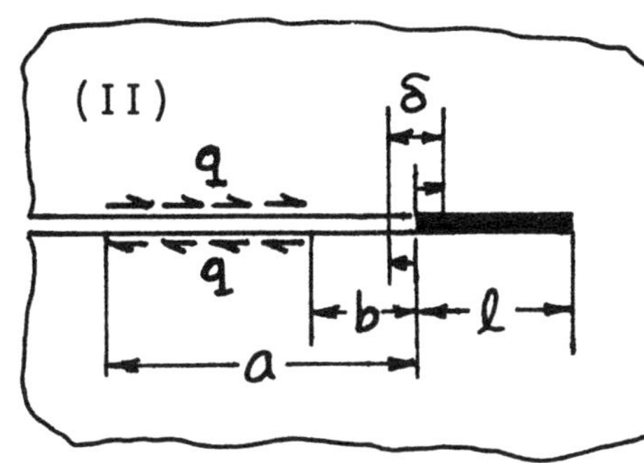

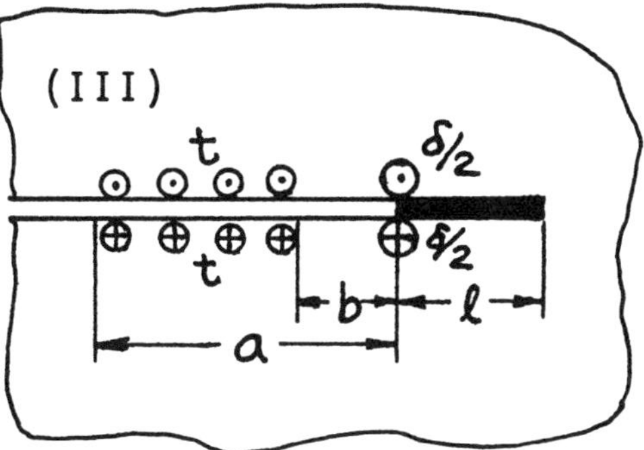

For (II) and (III), replace (p, σ_Y, E') by (q, τ_Y, E') and $(t, \tau_Y, 2G)$, respectively.
Method: Superposition of Three Cases of **page 3.7**
Accuracy: Exact
Reference: **Tada 1974**

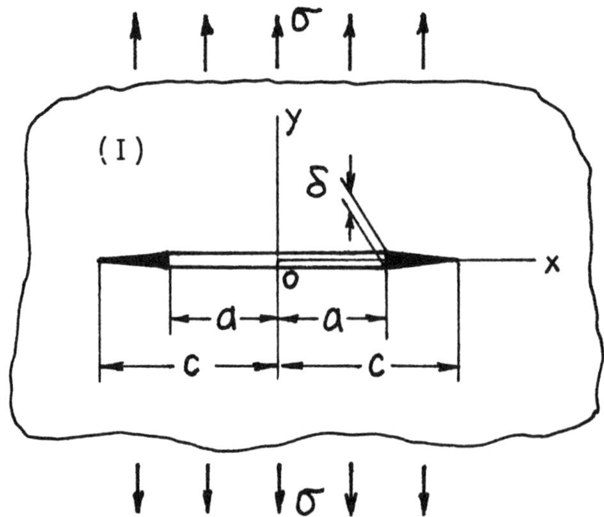

$$z = x + iy$$

$$Z(z) = \frac{2\sigma_Y}{\pi} \tan^{-1} \sqrt{\frac{(c/a)^2 - 1}{1 - (c/z)^2}}$$

$$\overline{Z}(z) = \frac{2\sigma_Y a}{\pi} \left\{ \tan^{-1} \sqrt{\frac{z^2 - c^2}{c^2 - a^2}} + \frac{z}{a} \tan^{-1} \sqrt{\frac{(c/a)^2 - 1}{1 - (c/z)^2}} \right\}$$

$$\delta = 2v\Big|_{z=\pm a} = \frac{8\sigma_Y a}{\pi E'} \ell n(c/a)$$

where

$$\frac{c}{a} = \sec\left(\frac{\pi\sigma}{2\sigma_Y}\right)$$

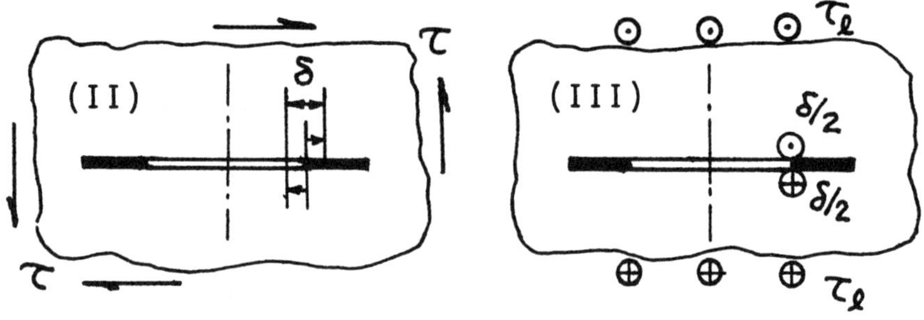

For (II) and (III), replace $(\sigma, \sigma_Y, v, E')$ by (τ, τ_Y, u, E') and $(\tau_\ell, \tau_Y, w, 2G)$, respectively.
Methods: Westergaard Stress Function (Superposition of **pages 5.1 and 5.12**) (Mode I: Dugdale; Mode III: Irwin), Dislocation Theory (Mode II, Bilby)
Accuracy: Exact
References: **Dugdale 1960; Bilby 1963; Irwin 1969**

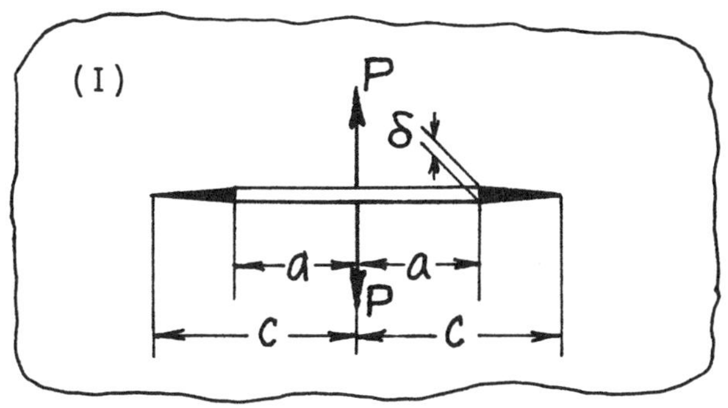

$$\delta = \frac{8\sigma_Y a}{\pi E'}\left\{\ell n\ \frac{c}{a} + \frac{P}{2a\sigma_Y}\left(\cosh^{-1}\frac{c}{a} - \sqrt{1-({}^{a}/_{c})^{2}}\right)\right\}$$

where
$$\frac{c}{a}\cos^{-1}\frac{a}{c} = \frac{P}{2a\sigma_Y}$$

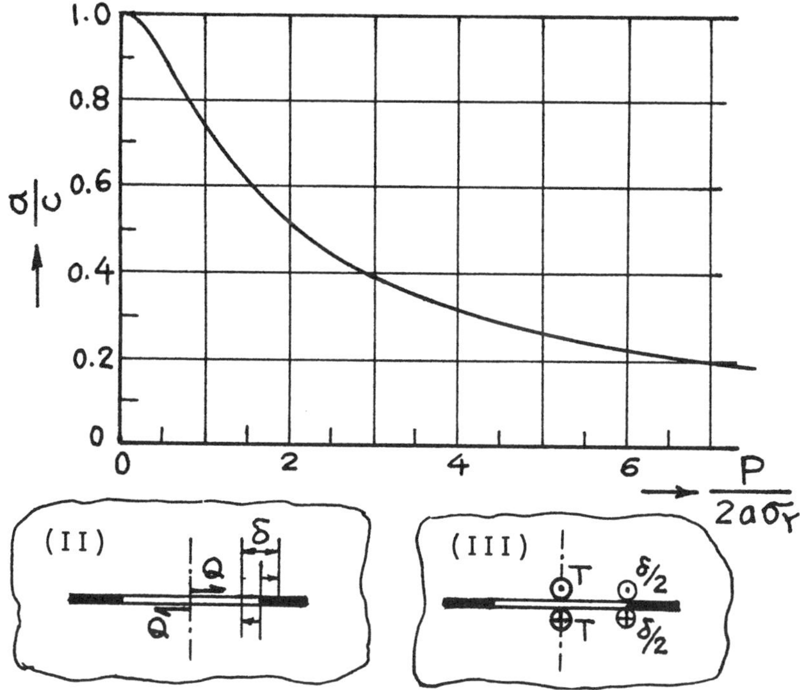

For (II) and (III), replace (P, σ_Y, E') by (Q, τ_Y, E') and $(T, \tau_Y, 2G)$, respectively.
Method: Superposition of **page 5.9 (or 5.10)** (b = 0) and **page 5.12** (or a Special Case of **page 30.9**)
Accuracy: Exact
Reference: **Tada 1974**

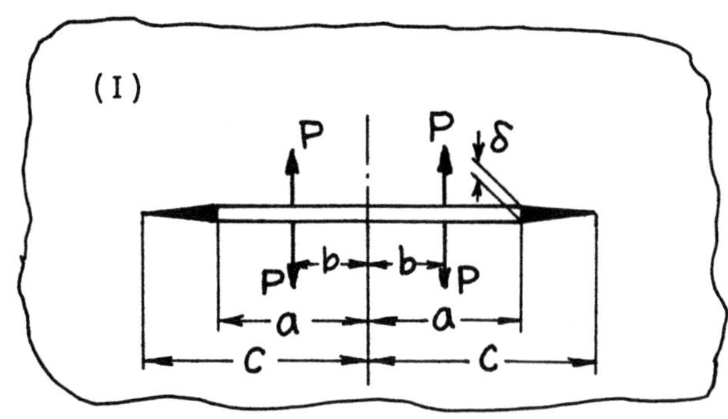

$$\delta = \frac{8\sigma_Y a}{\pi E'}\left\{\ell n\frac{c}{a} + \cos^{-1}\frac{a}{c}\left(\sqrt{\left(\frac{c}{a}\right)^2 - \left(\frac{b}{a}\right)^2}\tanh^{-1}\sqrt{\frac{(c/a)^2 - 1}{(c/a)^2 - (b/a)^2}} - \sqrt{1 - \left(\frac{a}{c}\right)^2}\right)\right\}$$

where $$\sqrt{\left(\frac{c}{a}\right)^2 - \left(\frac{b}{a}\right)^2}\cos^{-1}\frac{a}{c} = \frac{P}{a\sigma_Y}$$

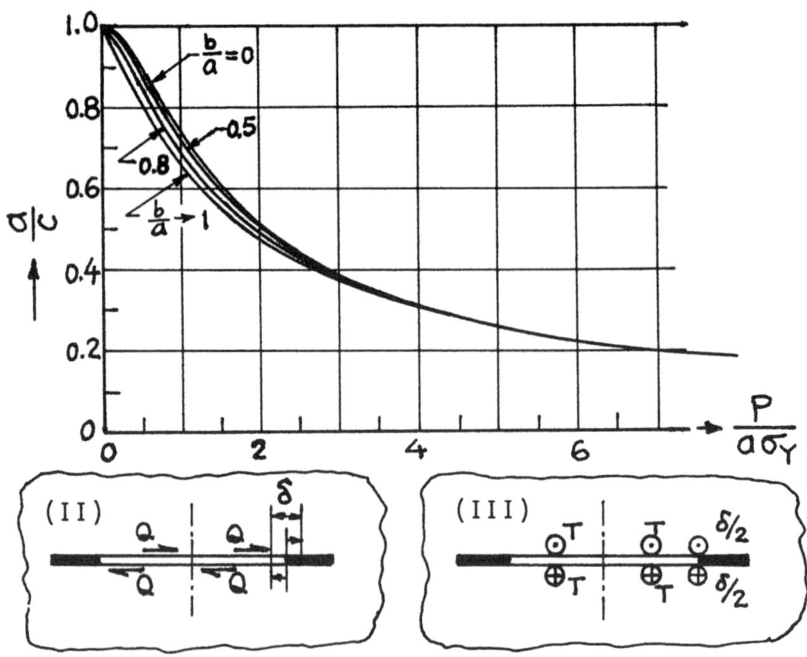

For (II) and (III), replace (P, σ_Y, E') by (Q, τ_Y, E') and $(T, \tau_Y, 2G)$, respectively.
Method: Superposition of **pages 5.10 and 5.12**
Accuracy: Exact
Reference: **Tada 1974**

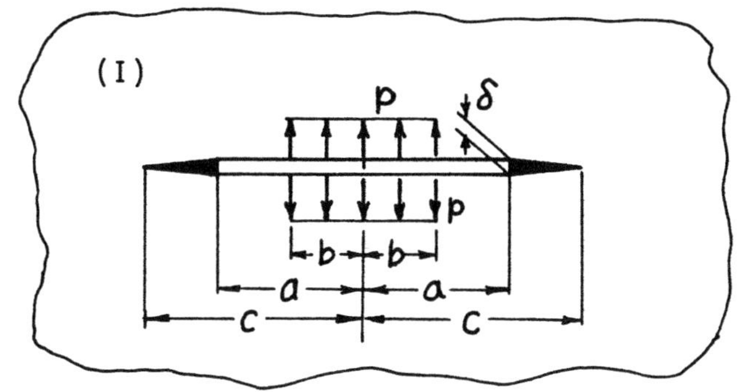

$$\delta = \frac{8\sigma_Y a}{\pi E'}\left\{ \ell n\frac{c}{a} - \frac{p}{\sigma_Y}\left(\tanh^{-1}\sqrt{\frac{(c/a)^2-1}{(c/b)^2-1}} - \frac{b}{a}\tanh^{-1}\sqrt{\frac{1-(a/c)^2}{1-(b/c)^2}} \right) \right\}$$

$$\text{where} \quad \frac{\cos^{-1}\frac{a}{c}}{\sin^{-1}\frac{b}{c}} = \frac{p}{\sigma_Y}$$

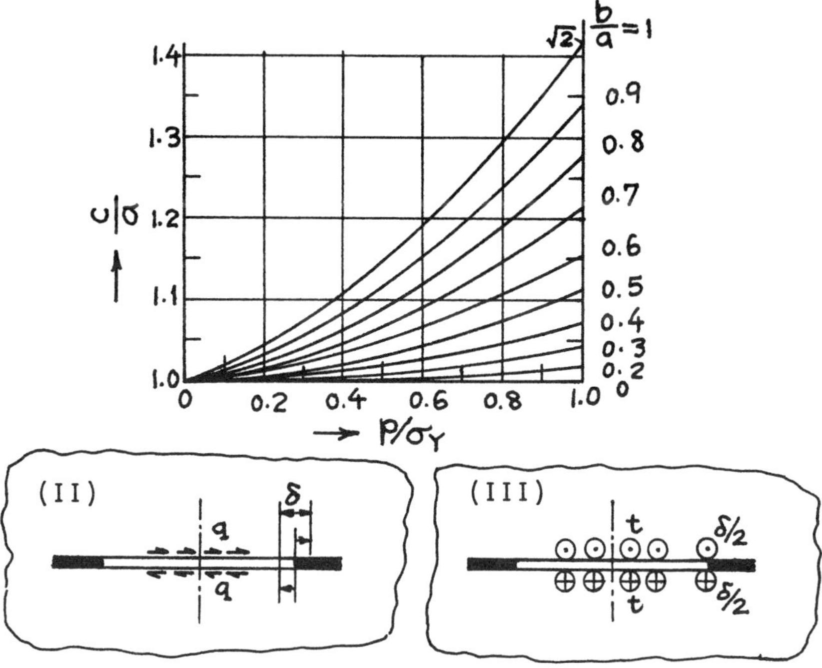

For (II) and (III), replace (p, σ_Y, E') by (q, τ_Y, E') and $(t, \tau_Y, 2G)$, respectively.
Method: Superposition of Two Cases of **page 5.12**
Accuracy: Exact
Reference: **Tada 1974**

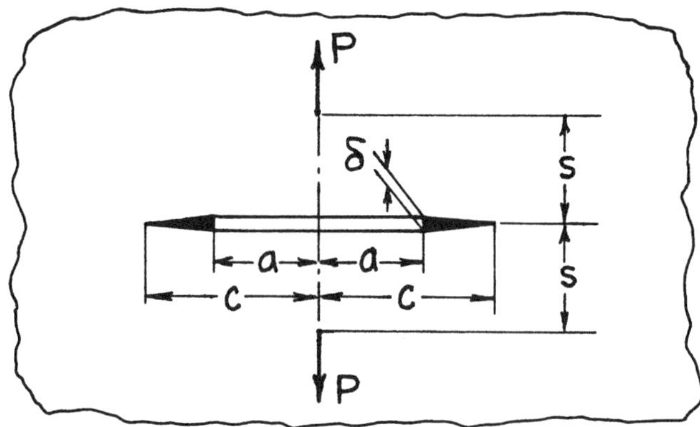

$$\delta = \frac{8\sigma_Y a}{\pi E'}\left\{\ell n\frac{c}{a} - \sqrt{\left(\frac{c}{a}\right)^2 - 1}\cos^{-1}\frac{a}{c} + \frac{P}{2a\sigma_Y}\left[\tanh^{-1}\sqrt{\frac{(c/a)^2 - 1}{(c/a)^2 + (s/a)^2}} + \alpha\frac{(s/a)^2}{1+(s/a)^2}\sqrt{\frac{(c/a)^2-1}{(c/a)^2+(s/a)^2}}\right]\right\}$$

$$\text{where} \quad \frac{\left\{\left(\frac{c}{a}\right)^2 + \left(\frac{s}{a}\right)^2\right\}^{3/2}}{\left(\frac{c}{a}\right)^2 + (1+\alpha)\left(\frac{s}{a}\right)^2}\cos^{-1}\frac{a}{c} = \frac{P}{2a\sigma_Y}$$

$$\alpha = \begin{cases} \frac{1}{2}(1+\nu) & \text{plane stress} \\ \frac{1}{2}\left(\frac{1}{1-\nu}\right) & \text{plane strain} \end{cases}$$

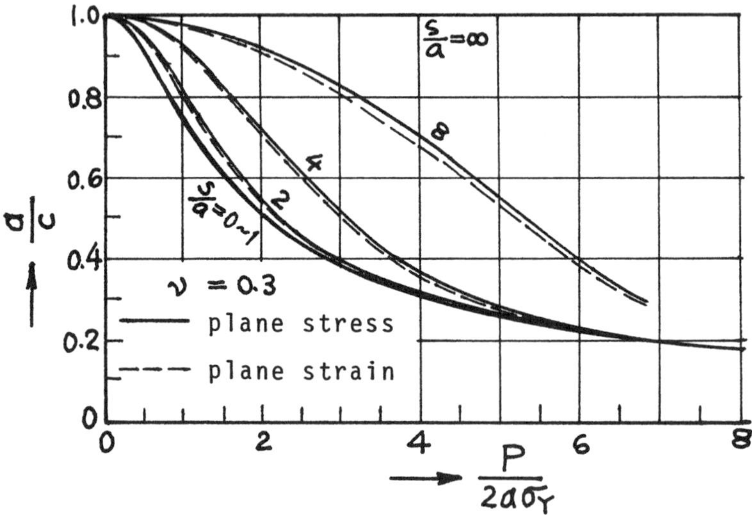

Method: Superposition of **pages 5.8 and 5.12**
Accuracy: Exact
Reference: **Tada 1974**

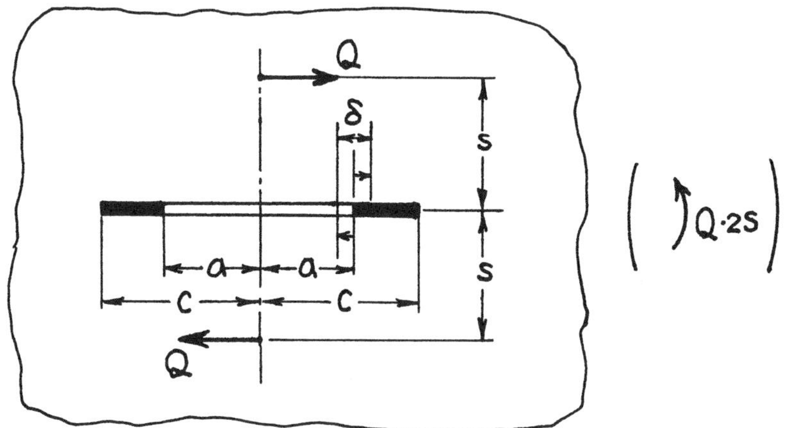

$$\delta = \frac{8\tau_Y a}{\pi E'}\left\{\ell n\frac{c}{a} - \sqrt{\left(\frac{c}{a}\right)^2 - 1}\cos^{-1}\frac{a}{c} + \frac{Q}{2a\tau_Y}\left[\tanh^{-1}\sqrt{\frac{(c/a)^2-1}{(c/a)^2+(s/a)^2}} - \alpha\frac{(s/a)^2}{1+(s/a)^2}\sqrt{\frac{(c/a)^2-1}{(c/a)^2+(c/a)^2}}\right]\right\}$$

where $\dfrac{\left\{\left(\frac{c}{a}\right)^2+\left(\frac{s}{a}\right)^2\right\}^{3/2}}{\left(\frac{c}{a}\right)^2+(1-\alpha)\left(\frac{s}{a}\right)^2}\cos^{-1}\frac{a}{c} = \frac{Q}{2a\tau_Y}$

$$\alpha = \begin{cases} \frac{1}{2}(1+\nu) & \text{plane stress} \\ \frac{1}{2}\left(\frac{1}{1-\nu}\right) & \text{plane strain} \end{cases}$$

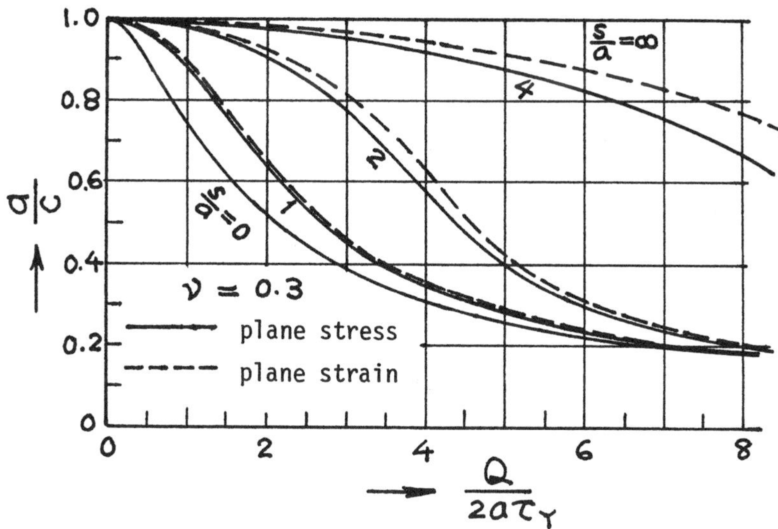

Method: Superposition of **pages 5.8 and 5.12**
Accuracy: Exact
Reference: **Tada 1974**

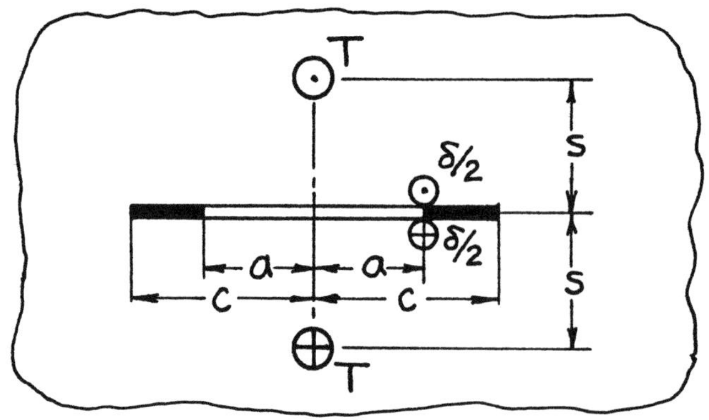

$$\delta = \frac{4\tau_Y a}{\pi G}\left\{\ell n\frac{c}{a} - \sqrt{\left(\frac{c}{a}\right)^2 - 1}\cos^{-1}\frac{a}{c} + \frac{T}{2a\tau_Y}\tanh^{-1}\sqrt{\frac{(c/a)^2 - 1}{(c/a)^2 + (s/a)^2}}\right\}$$

$$\text{where} \quad \sqrt{\left(\frac{c}{a}\right)^2 + \left(\frac{s}{a}\right)^2}\cos^{-1}\frac{a}{c} = \frac{T}{2a\tau_Y}$$

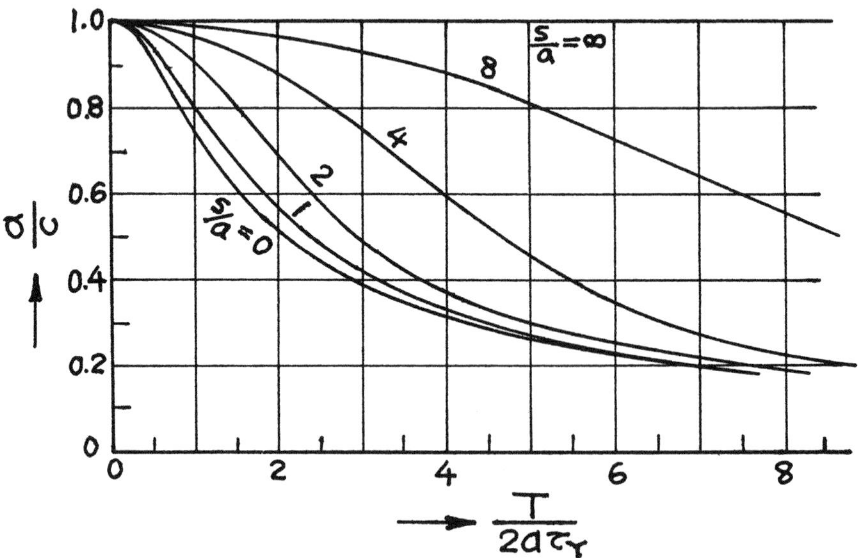

Method: Superposition of **page 5.12** and Mode III solution corresponding to **page 5.10** ($\alpha = 0$)
Accuracy: Exact
Reference: **Tada 1974**

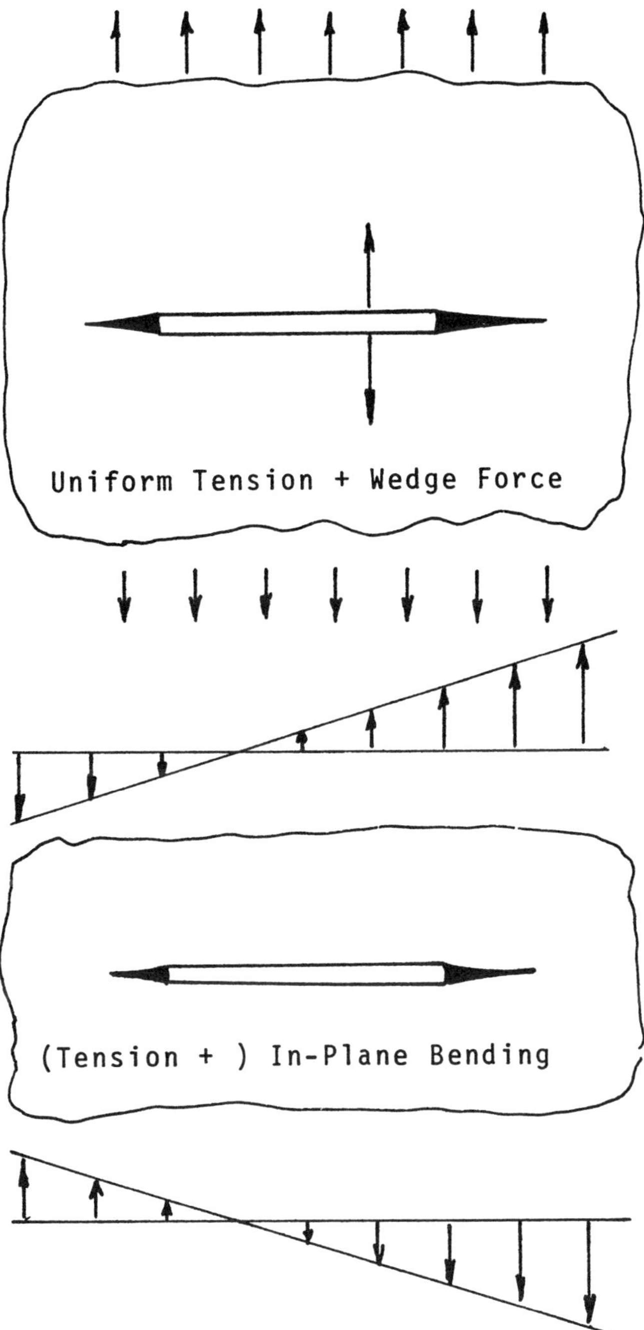

For some examples, see **Seeger 1973**.

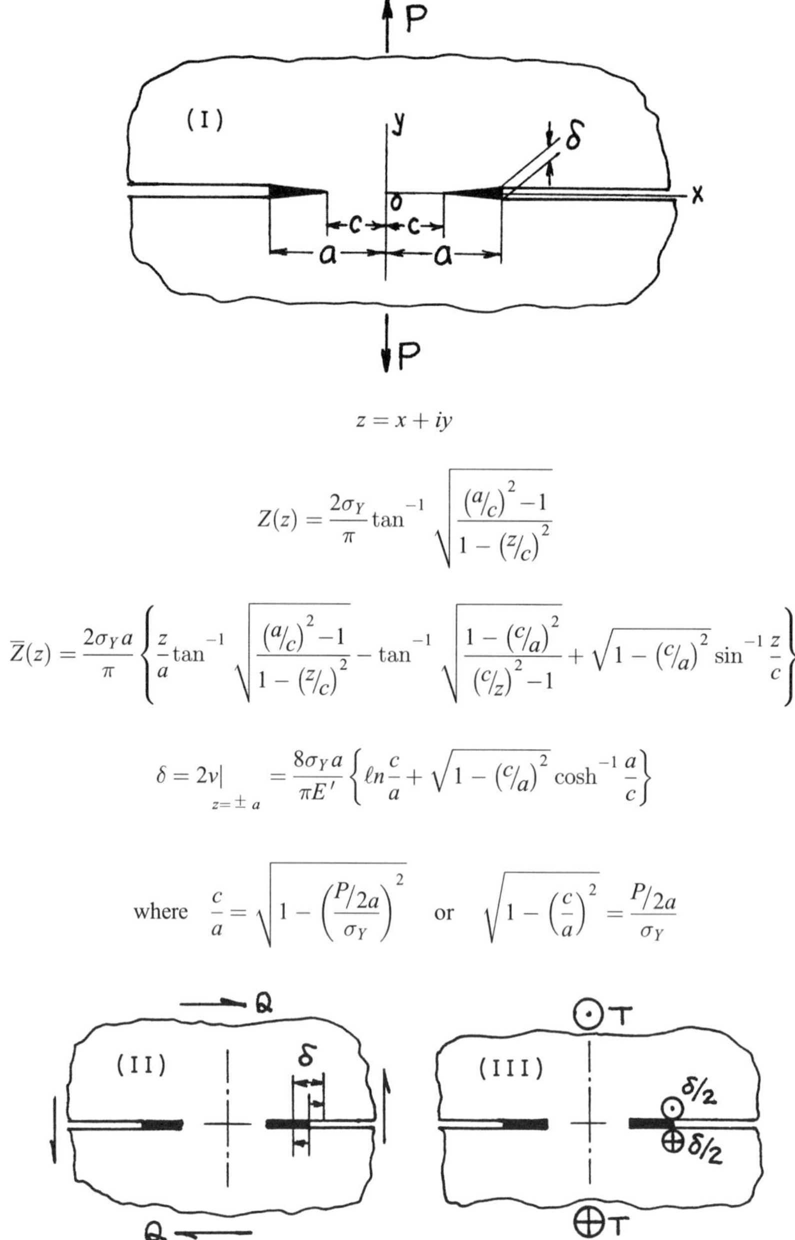

$$z = x + iy$$

$$Z(z) = \frac{2\sigma_Y}{\pi}\tan^{-1}\sqrt{\frac{\left({}^{a}\!/_{c}\right)^2 - 1}{1 - \left({}^{z}\!/_{c}\right)^2}}$$

$$\overline{Z}(z) = \frac{2\sigma_Y a}{\pi}\left\{\frac{z}{a}\tan^{-1}\sqrt{\frac{\left({}^{a}\!/_{c}\right)^2 - 1}{1 - \left({}^{z}\!/_{c}\right)^2}} - \tan^{-1}\sqrt{\frac{1 - \left({}^{c}\!/_{a}\right)^2}{\left({}^{c}\!/_{z}\right)^2 - 1}} + \sqrt{1 - \left({}^{c}\!/_{a}\right)^2}\,\sin^{-1}\frac{z}{c}\right\}$$

$$\delta = 2v\big|_{z=\pm a} = \frac{8\sigma_Y a}{\pi E'}\left\{\ell n\frac{c}{a} + \sqrt{1 - \left({}^{c}\!/_{a}\right)^2}\,\cosh^{-1}\frac{a}{c}\right\}$$

$$\text{where}\quad \frac{c}{a} = \sqrt{1 - \left(\frac{{}^{P}\!/_{2a}}{\sigma_Y}\right)^2} \quad\text{or}\quad \sqrt{1 - \left(\frac{c}{a}\right)^2} = \frac{{}^{P}\!/_{2a}}{\sigma_Y}$$

For (II) and (III), replace (P, σ_Y, v, E') by (Q, τ_Y, u, E') and $(T, \tau_Y, w, 2G)$, respectively.
Method: Westergaard Stress Function (Superposition of **pages 4.8 and 4.9**)
Accuracy: Exact
Reference: **Tada 1974**

NOTE: The Limiting Case of Net Section Yielding

$$\frac{{}^{P}\!/_{2a}}{\sigma_Y} = \frac{\sigma_{net}}{\sigma_Y} = 1\left(or\ \frac{c}{a} = 0\right) \rightarrow \delta = \frac{8\sigma_Y a}{\pi E'}\ell n 2$$

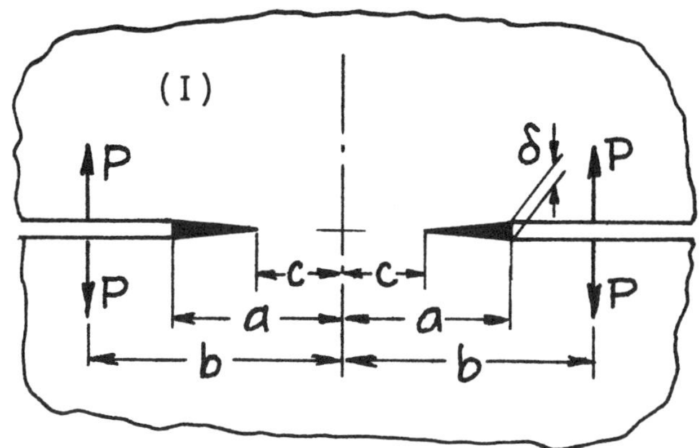

$$\frac{c}{a} = \frac{1}{\sqrt{2}}\left\{1 + \left(\frac{b}{a}\right)^2 - \sqrt{\left\{\left(\frac{b}{a}\right)^2 - 1\right\}^2 + 4\left(\frac{b}{a}\right)^2\left(\frac{P/a}{\sigma_Y}\right)^2}\right\}^{1/2}$$

$$\delta = \frac{8\sigma_Y a}{\pi E'}\left\{\ell n\frac{c}{a} + \frac{P/a}{\sigma_Y}\tanh^{-1}\sqrt{\frac{1-(c/a)^2}{1-(c/b)^2}}\right\}$$

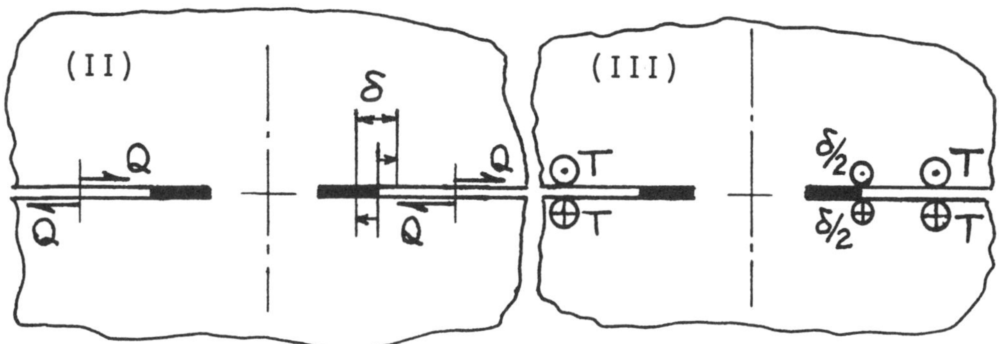

For (II) and (III), replace (P, σ_Y, E') by (Q, τ_Y, E') and $(T, \tau_Y, 2G)$, respectively.
Method: Superposition of **pages 4.6 and 4.8**
Accuracy: Exact
Reference: **Tada 1974**

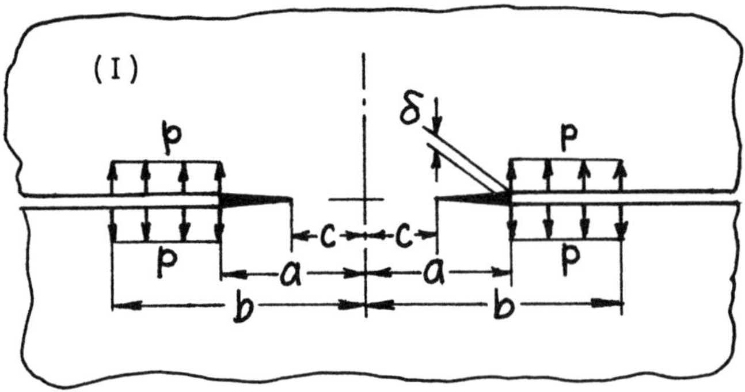

$$\frac{c}{a} = \sqrt{\frac{\left(1 + {}^{p}\!/_{\sigma_Y}\right)^2 - \left({}^{b}\!/_{a}\right)^2 \left({}^{p}\!/_{\sigma_Y}\right)^2}{1 + 2\,{}^{p}\!/_{\sigma_Y}}}$$

$$\delta = \frac{8\sigma_Y a}{\pi E'}\left\{\left(1 + \frac{p}{\sigma_Y}\right)\ell n\frac{c}{a} + \frac{p}{\sigma_Y}\left[\tanh^{-1}\sqrt{\frac{1 - \left({}^{c}\!/_{a}\right)^2}{\left({}^{b}\!/_{a}\right)^2 - \left({}^{c}\!/_{a}\right)^2}} + \frac{b}{a}\tanh^{-1}\sqrt{\frac{1 - \left({}^{c}\!/_{a}\right)^2}{1 - \left({}^{c}\!/_{b}\right)^2}}\right]\right\}$$

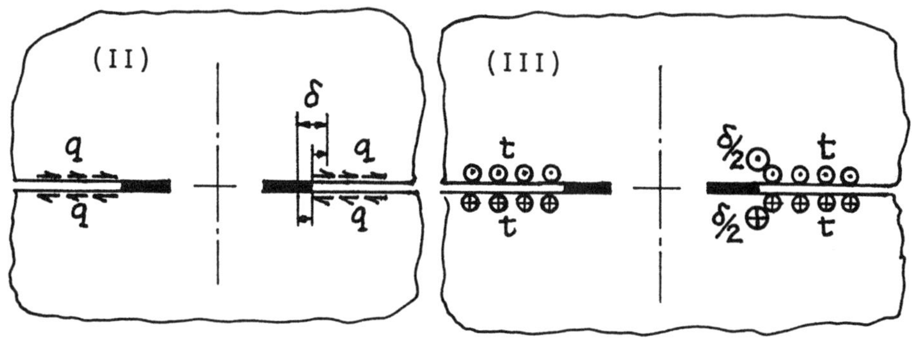

For (II) and (III), replace $(p,\ \sigma_Y,\ E')$ by $(q,\ \tau_Y,\ E')$ and $(t,\ \tau_Y,\ 2G)$, respectively.
Method: Superposition of Two Cases of **page 4.8**
Accuracy: Exact
Reference: **Tada 1974**

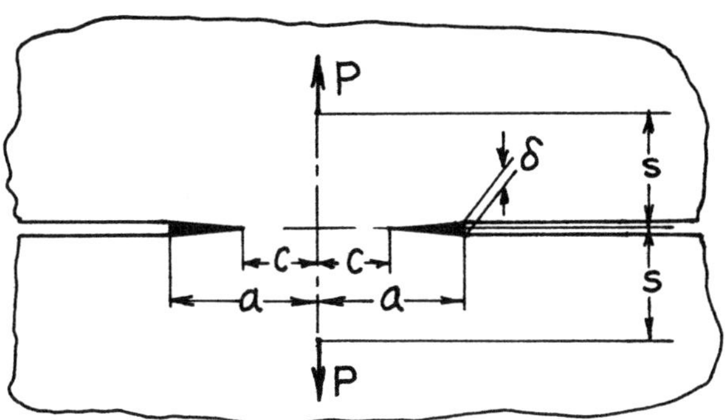

$$\delta = \frac{8\sigma_Y a}{\pi E'}\left\{\ell n\frac{c}{a} + \frac{P/2a}{\sigma_Y}\left[\tanh^{-1}\sqrt{\frac{1-(c/a)^2}{1+(c/s)^2}} - \alpha\frac{s/a}{1+(s/a)^2}\sqrt{\frac{1-(c/a)^2}{(c/a)^2+(s/a)^2}}\right]\right\}$$

$$\text{where}\quad \frac{\left\{\left(\frac{c}{a}\right)^2+\left(\frac{s}{a}\right)^2\right\}^{3/2}}{\frac{s}{a}\left\{(1-\alpha)\left(\frac{c}{a}\right)^2+\left(\frac{s}{a}\right)^2\right\}}\sqrt{1-\left(\frac{c}{a}\right)^2} = \frac{P/2a}{\sigma_Y}$$

$$\alpha = \begin{cases}\frac{1}{2}(1+\nu) & \text{plane stress}\\ \frac{1}{2}\left(\frac{1}{1-\nu}\right) & \text{plane strain}\end{cases}$$

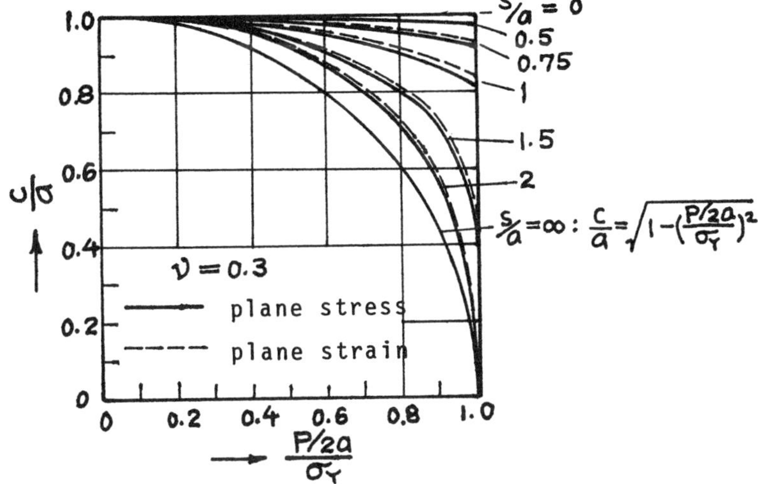

Method: Superposition of **pages 4.3 and 4.8**
Accuracy: Exact
Reference: **Tada 1974**

NOTE: For small values of s/a, c/a jumps to 0 when $\frac{P/2a}{\sigma_Y} = 1$.

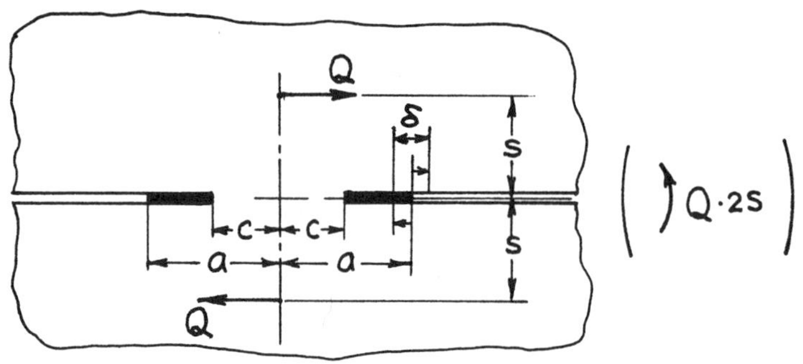

$$\delta = \frac{8\tau_Y a}{\pi E'}\left\{\ell n\frac{c}{a} + \frac{Q/2a}{\tau_Y}\left[\tanh^{-1}\sqrt{\frac{1-(c/a)^2}{1+(c/s)^2}} + \alpha\frac{s/a}{1+(s/a)^2}\sqrt{\frac{1-(c/a)^2}{(c/a)^2+(s/a)^2}}\right]\right\}$$

where
$$\frac{\left\{\left(\frac{c}{a}\right)^2+\left(\frac{s}{a}\right)^2\right\}^{3/2}}{\frac{s}{a}\left\{(1+\alpha)\left(\frac{c}{a}\right)^2+\left(\frac{s}{a}\right)^2\right\}}\sqrt{1-\left(\frac{c}{a}\right)^2} = \frac{Q/2a}{\tau_Y}$$

$$\alpha = \begin{cases} \frac{1}{2}(1+\nu) & \text{plane stress} \\ \frac{1}{2}\left(\frac{1}{1-\nu}\right) & \text{plane strain} \end{cases}$$

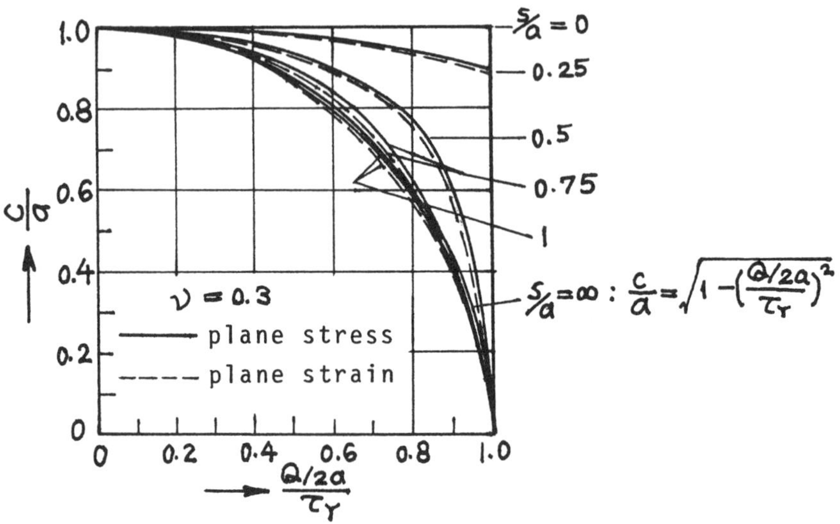

Method: Superposition of **pages 4.3 and 4.8**
Accuracy: Exact
Reference: **Tada 1974**

NOTE: For small values of s/a, c/a jumps to 0 when $\frac{Q/2a}{\tau_Y} = 1$.

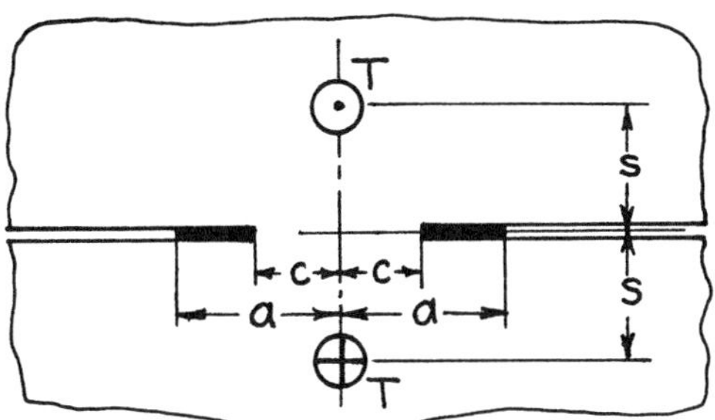

$$\delta = \frac{4\tau_Y a}{\pi G}\left\{\ell n\frac{c}{a} + \frac{T/2a}{\tau_Y}\tanh^{-1}\sqrt{\frac{1-(c/a)^2}{1+(c/s)^2}}\right\}$$

where:

$$\text{when}\quad \frac{s}{a} < 1 : \frac{c}{a} = \begin{cases} f\left(s/a, \dfrac{T/2a}{\tau_Y}\right) & 0 < \dfrac{T/2a}{\tau_Y} < 1 \\ 0 & \dfrac{T/2a}{\tau_Y} = 1 \end{cases}$$

$$\text{when}\quad \frac{s}{a} \geq 1 : \frac{c}{a} = f\left(s/a, \frac{T/2a}{\tau_Y}\right) \quad 0 < \frac{T/2a}{\tau_Y} \leq 1$$

$$f\left(\frac{s}{a}, \frac{T/2a}{\tau_Y}\right) = \frac{1}{\sqrt{2}}\left\{1 - \left(\frac{s}{a}\right)^2 + \sqrt{\left\{1 + \left(\frac{s}{a}\right)^2\right\}^2 - 4\left(\frac{s}{a}\right)^2\left(\frac{T/2a}{\tau_Y}\right)^2}\right\}^{1/2}$$

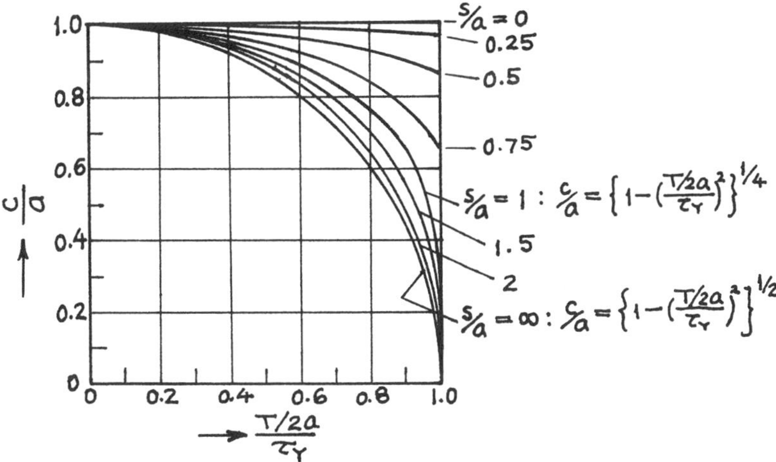

For $s/a < 1$, c/a jumps from $\sqrt{1-(s/a)^2}$ to 0 at $\frac{T/2a}{\tau_Y} = 1$.
Method: Superposition of **page 4.8** and Mode III solution corresponding to **page 4.3**$(\alpha = 0)$
Accuracy: Exact
Reference: **Tada 1974**

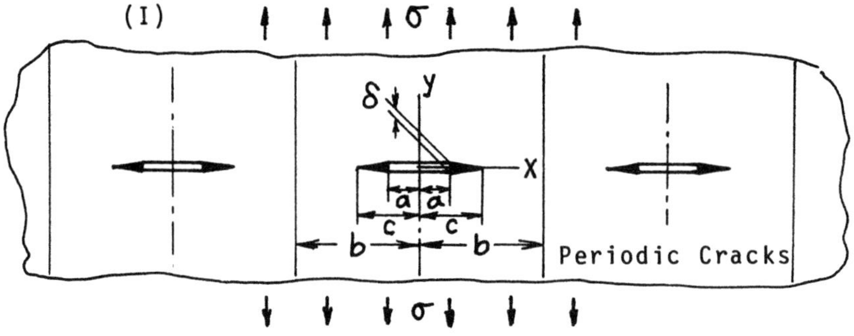

$$z = x + iy$$

$$Z(z) = \frac{2\sigma_Y}{\pi}\tan^{-1}\sqrt{\frac{\left(\sin\frac{\pi c}{2b}\Big/\sin\frac{\pi a}{2b}\right)^2 - 1}{1-\left(\sin\frac{\pi c}{2b}\Big/\sin\frac{\pi z}{2b}\right)^2}}$$

$$\delta = 2v\big|_{z=\pm a} = \frac{\sigma_Y(2b)}{E'}f\left(\frac{a}{b},\frac{c}{b}\right) \quad \text{or} \quad = \frac{\sigma_Y(2b)}{E'}g\left(\frac{a}{b},\frac{\sigma}{\sigma_Y}\right)$$

$$\text{where} \qquad \frac{\sin\frac{\pi a}{2b}}{\sin\frac{\pi c}{2b}} = \cos\frac{\pi\sigma}{2\sigma_Y}$$

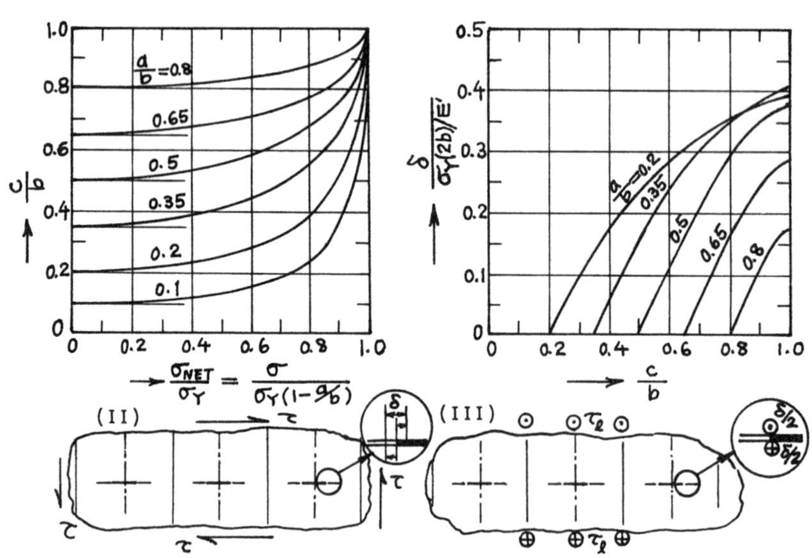

For (II) and (III), replace $(\sigma, \sigma_Y, v, E')$ by (τ, τ_Y, u, E') and $(\tau_\ell, \tau_Y, w, 2G)$, respectively.
Method: Westergaard Stress Function (Superposition of **pages 7.1 and 7.8**)
Accuracy: c/b Exact; δ 1%
References: **Irwin 1969; Tada 1974**

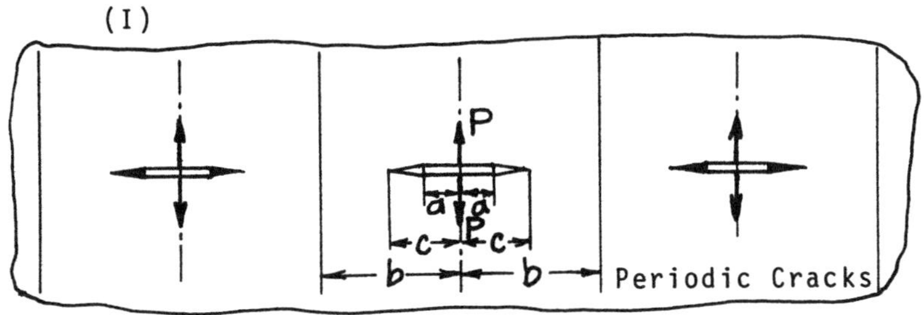

$$\frac{2}{\pi}\cos^{-1}\left(\frac{\sin\frac{\pi a}{2b}}{\sin\frac{\pi c}{2b}}\right)\cdot\sin\frac{\pi c}{2b}=\frac{\mathrm{P}}{\sigma_Y(2b)}$$

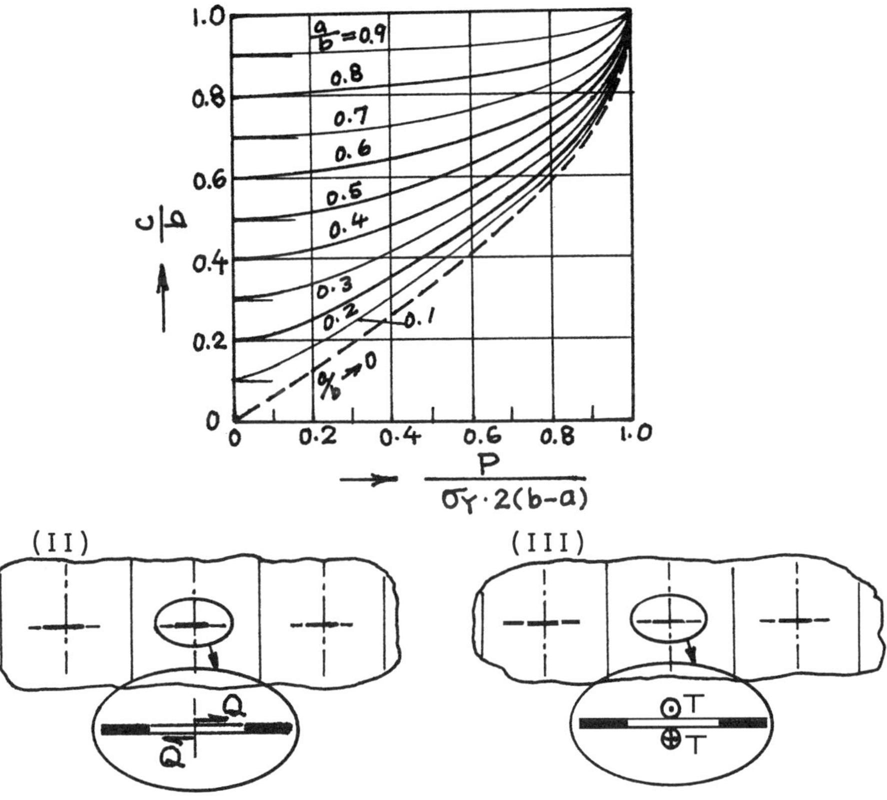

For (II) and (III), replace (P, σ_Y) by (Q, τ_Y) and (T, τ_Y), respectively.
Method: Superposition of **page 7.6**($b = 0$) and **page 7.8**
Accuracy: Exact
Reference: **Tada 1974**

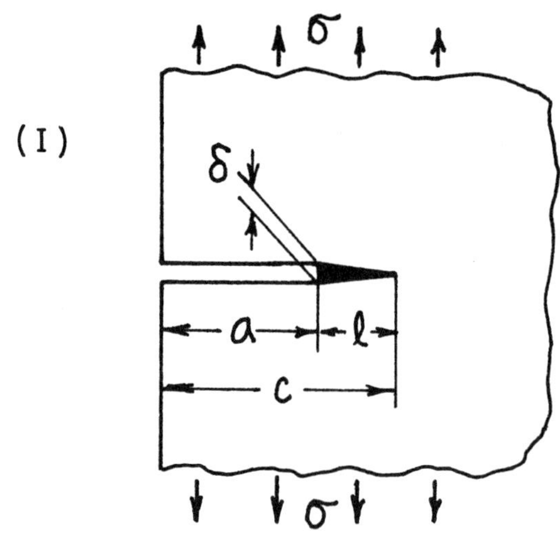

$$\frac{\ell}{c} = f\left(\sigma/\sigma_Y\right)$$

or $$\frac{\ell}{a} = \left(\sec\frac{\pi\sigma}{2\sigma_Y} - 1\right) g\left(\sigma/\sigma_Y\right)$$

$$\delta = \frac{\sigma_Y a}{E'} F\left(\ell/c\right)$$

or $$= \frac{8\sigma_Y a}{\pi E'} \ell n \frac{c}{a} \cdot G\left(\ell/c\right)$$

$$g\left(\sigma/\sigma_Y\right) \simeq 1.258 \quad \left(1.258 = 1.1215^2\right)$$

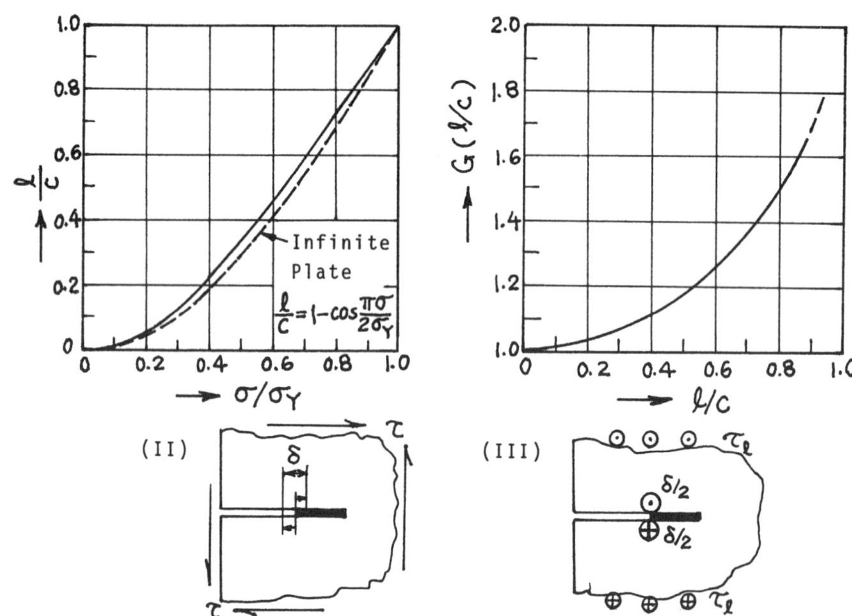

For (II), replace σ and σ_Y by τ and τ_Y, respectively.

For (III), $c/a = \sec\frac{\pi\tau_\ell}{2\tau_Y}$ and $\delta = \frac{4\tau_Y a}{\pi G}\ln(c/a)$ (Infinite plate solution, **page 30.7**)

Methods: Successive Stress Relaxation (I and II). c/a was also obtained by superposition of **pages 8.1 and 8.4**.

Accuracy: ℓ/c - curve is based on accurate values and $G(\ell/c)$-curve is based on the values having 1% accuracy.

References: **Nishitani 1971b; Tada 1972a, 1974**

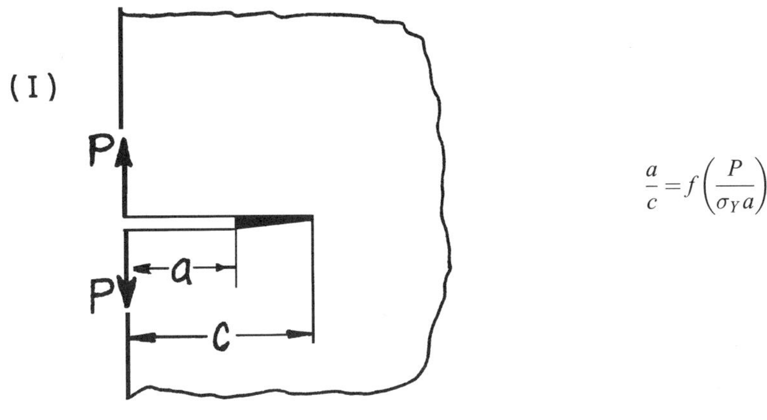

$$\frac{a}{c} = f\left(\frac{P}{\sigma_Y a}\right)$$

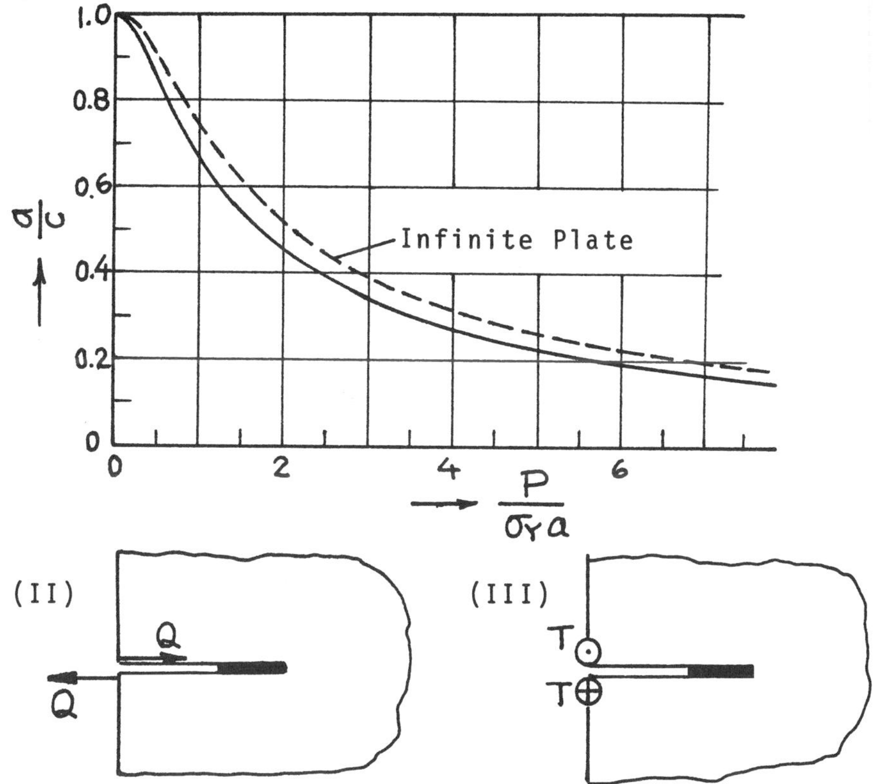

For (II), replace P and σ_Y by Q and τ_Y, respectively.
For (III), see the infinite plate solution (**page 30.8**).
Method: Superposition of **pages 8.2 and 8.4**
Accuracy: a/c - curve is based on the values having 1% accuracy.
Reference: **Tada 1974**

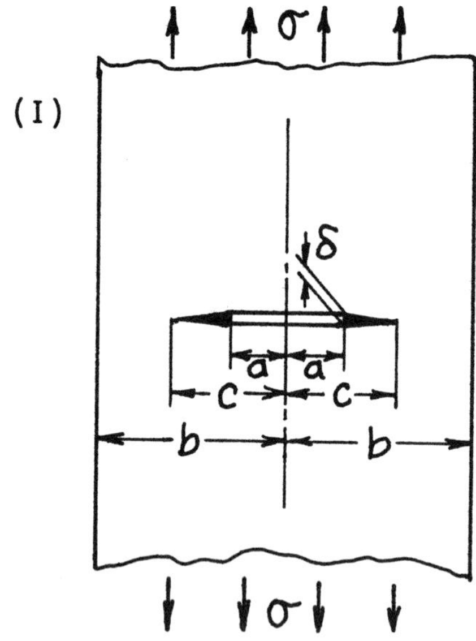

$$\frac{c}{b} = f\left(\frac{a}{b}, \frac{\sigma}{\sigma_Y}\right)$$

$$\delta = \frac{\sigma_Y(2b)}{E'} F\left(\frac{a}{b}, \frac{\sigma}{\sigma_Y}\right)$$

or

$$= \frac{\sigma_Y(2b)}{E'} G\left(\frac{a}{b}, \frac{c}{b}\right)$$

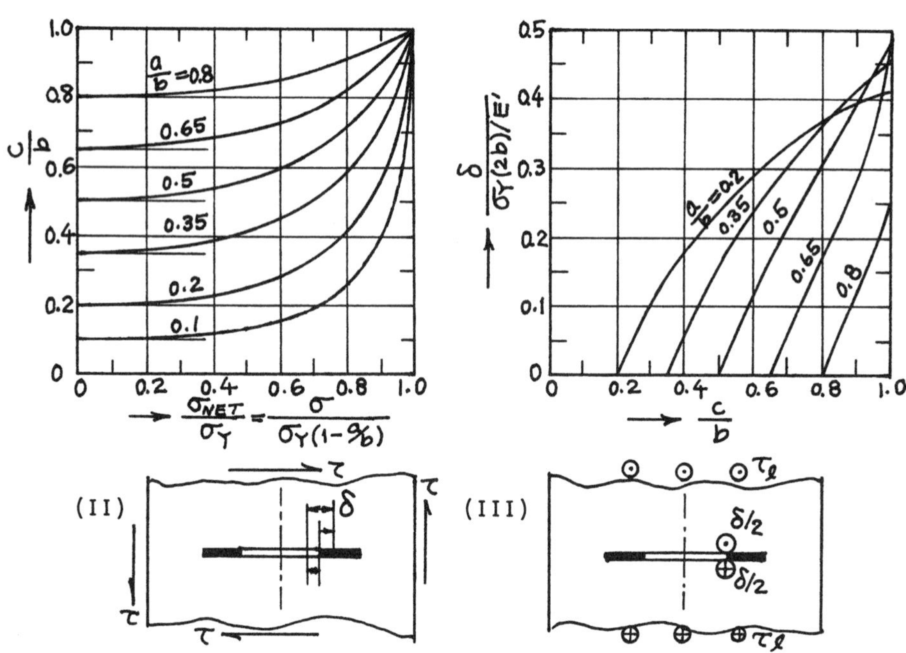

For (II), replace σ and σ_Y by τ and τ_Y, respectively.
For (III), see the solution for periodic cracks (**page 30.21**).
Methods: Expansion of Complex Stress Potentials (Isida), Successive Stress Adjustments (Tada)
Accuracy: Curves are based on values with better than 1% accuracy.
References: **Isida 1971b; Tada 1972a, 1974**

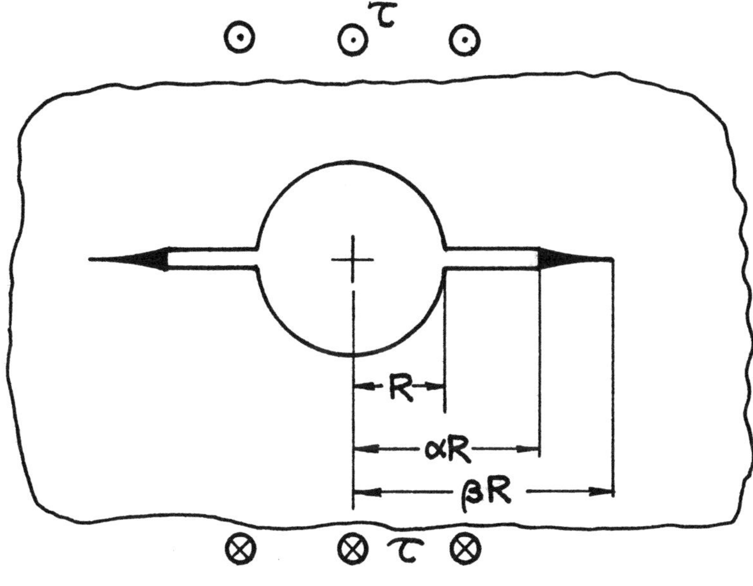

$$\sqrt{\frac{\alpha^2\beta^2 - 1}{\beta^4 - 1}} = \cos\left(\frac{\pi\tau}{2\tau_Y}\right)$$

or

$$\beta = \frac{1}{\sqrt{2}} \frac{\alpha}{\cos\frac{\pi\tau}{2\tau_Y}} \left\{ 1 + \sqrt{1 - \left(\frac{\sin\frac{\pi\tau}{\tau_Y}}{\alpha^2}\right)^2} \right\}^{1/2}$$

When $\alpha = 1$ (No Cracks):

$$\begin{cases} \beta = 1 & \textit{for } 0 \leq \frac{\tau}{\tau_Y} \leq 1/2 \\ \beta = \tan\frac{\pi\tau}{2\tau_Y} & \textit{for } 1/2 \leq \frac{\tau}{\tau_Y} \leq 1 \end{cases}$$

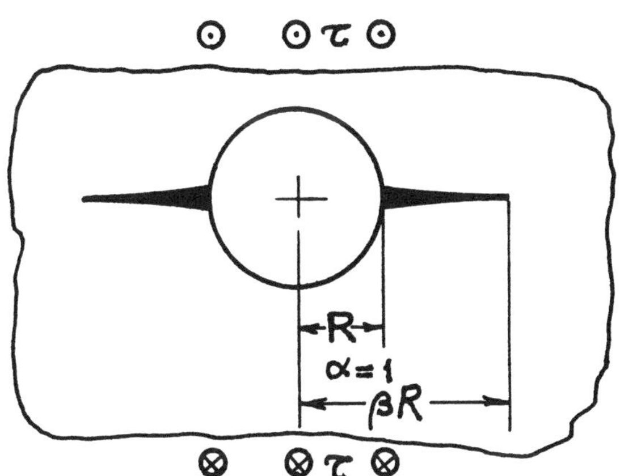

Method: Muskhelishvili's Method
Accuracy: Exact
References: **Erdogan 1966; Tada 1974**

NOTE: In **Erdogan 1966**, only an implicit solution was given in a more complicated form.

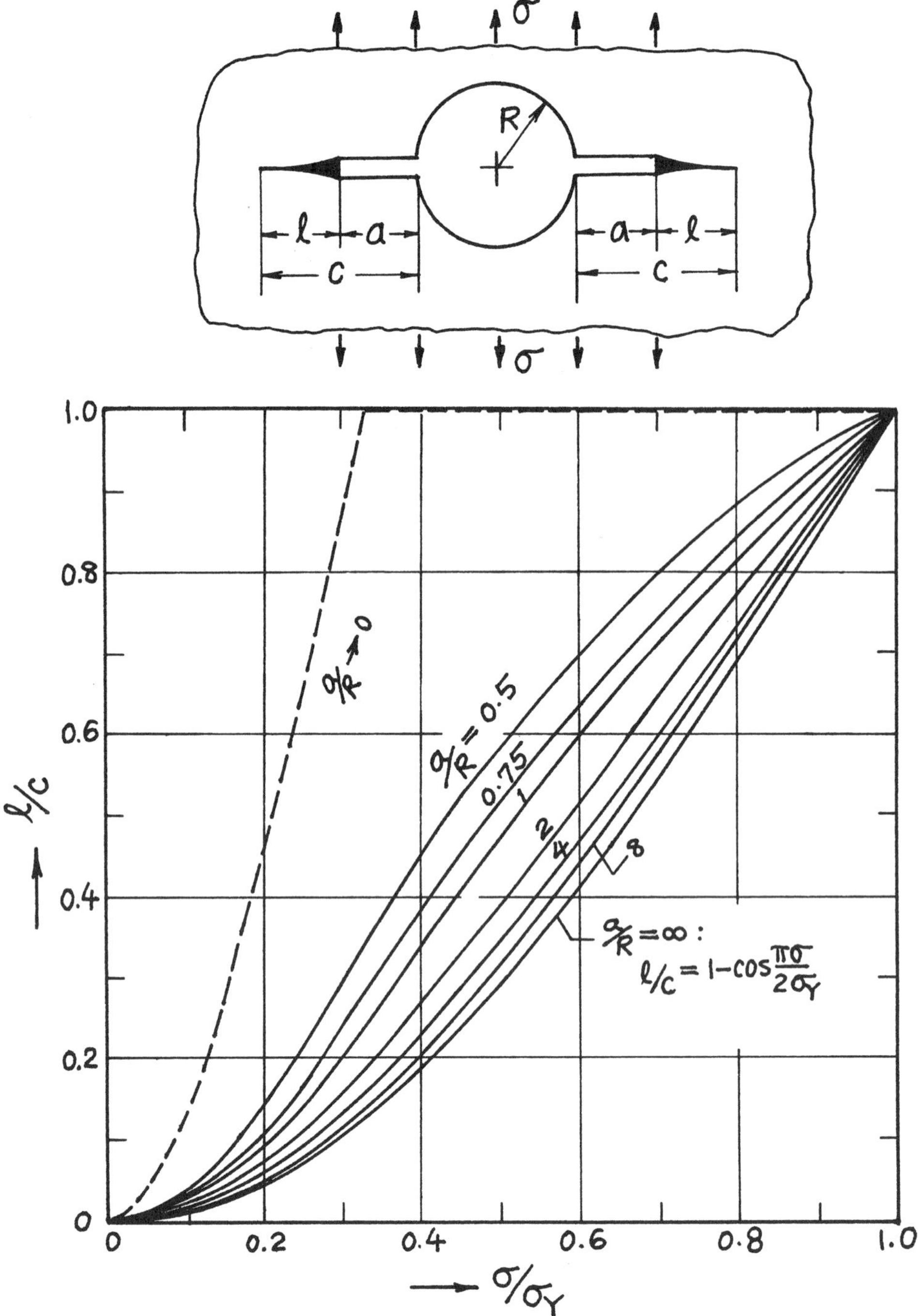

Method: Muskhelishvili's Method
Accuracy: Better than 1%
Reference: **Rich 1968**

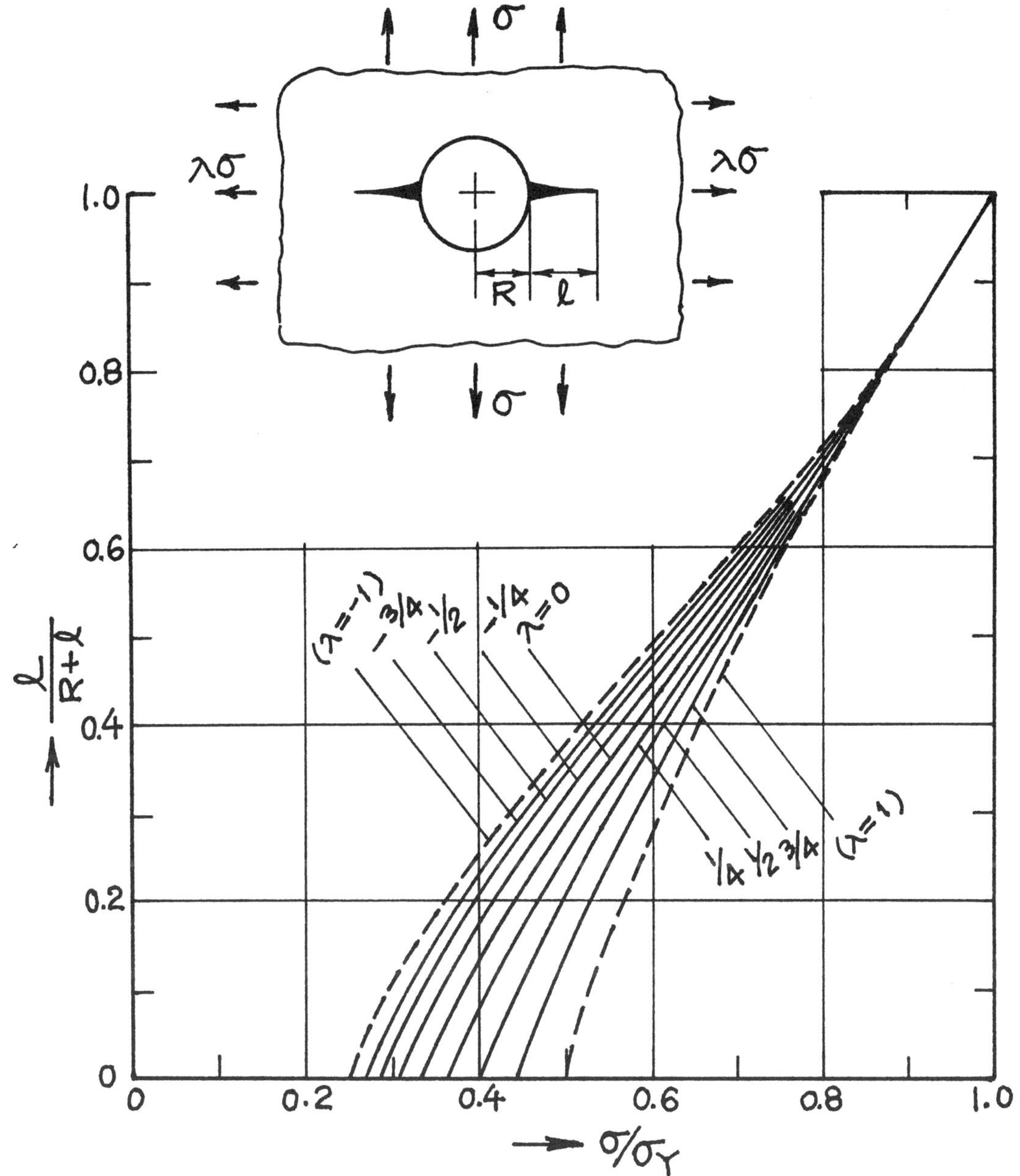

Method: Superposition of **page 19.1** ($\lambda = 1$) and **page 19.3** ($\lambda = 2,\ p = \sigma_Y$)
Accuracy: 1%
Reference: **Tada 1974**

NOTE: When $|\lambda| > 1$, use the curves above replacing ${}^{\sigma}/_{\sigma_Y}$ and λ by ${}^{\lambda\sigma}/_{\sigma_Y}$ and ${}^{1}/_{\lambda}$, respectively. Note that plastic zones develop at the top and bottom ends of the hole (when $\lambda > 1$, tensile plastic zones, and when $\lambda < -1$, compressive plastic zones).

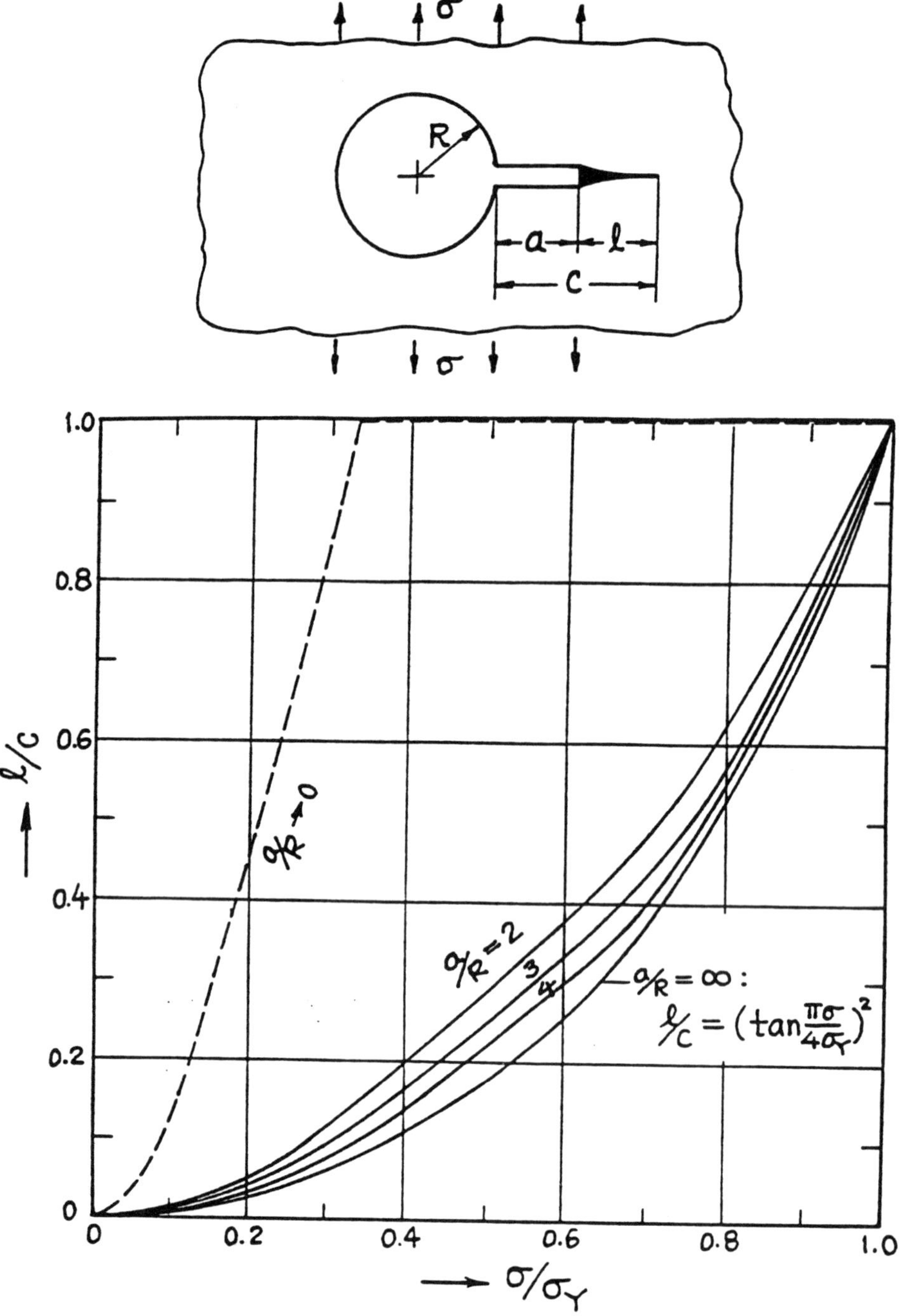

Method: Muskhelishvili's Method
Accuracy: Estimated at 2%
Reference: **Rich 1968**

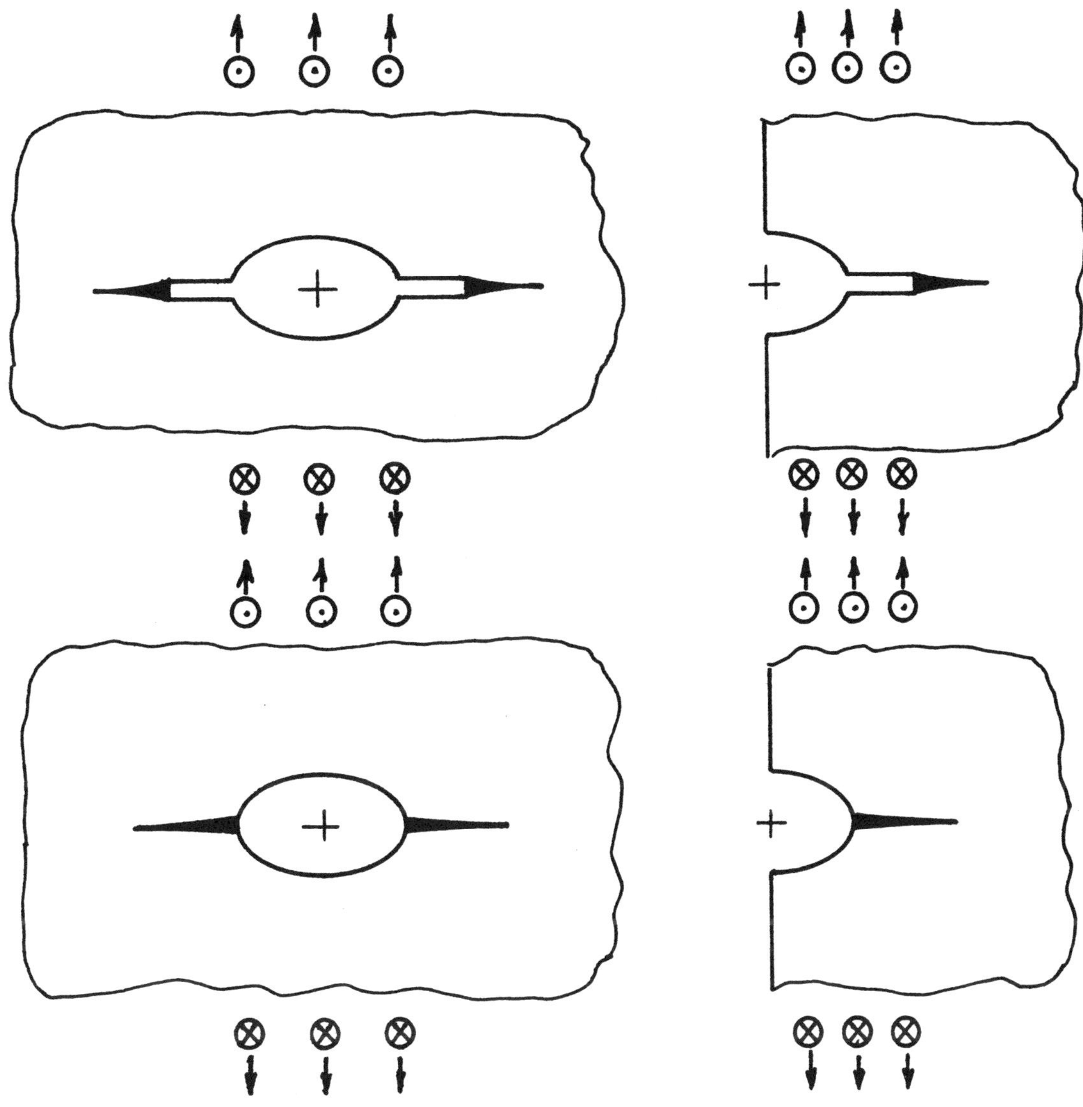

See **Nishitani 1973**.

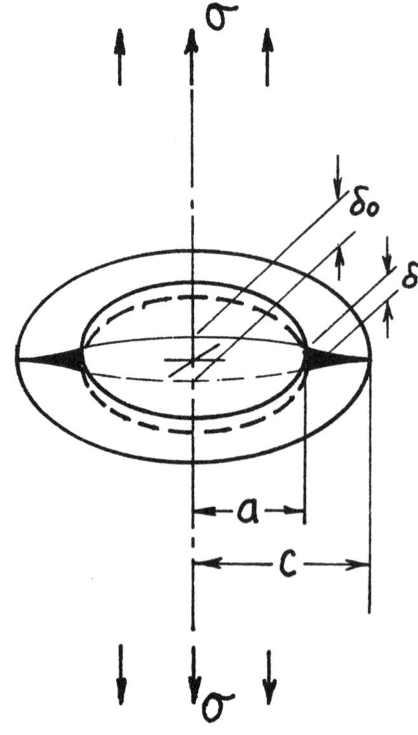

$$\frac{a}{c} = \sqrt{1 - \left(\sigma/\sigma_Y\right)^2}$$

Opening at Crack Tip:

$$\delta = \frac{8\left(1 - \nu^2\right)}{\pi E}\sigma_Y a\left(1 - \frac{a}{c}\right)$$

Opening at Center:

$$\delta_0 = \frac{8\left(1 - \nu^2\right)}{\pi E}\sigma_Y a \sin^{-1}\frac{\sigma}{\sigma_Y}; \left(\sin^{-1}\frac{\sigma}{\sigma_Y} = \cos^{-1}\frac{a}{c}\right)$$

Method: Superposition of **pages 24.1 and 24.6**
Accuracy: Exact
Reference: **Tada 1974, 1985**

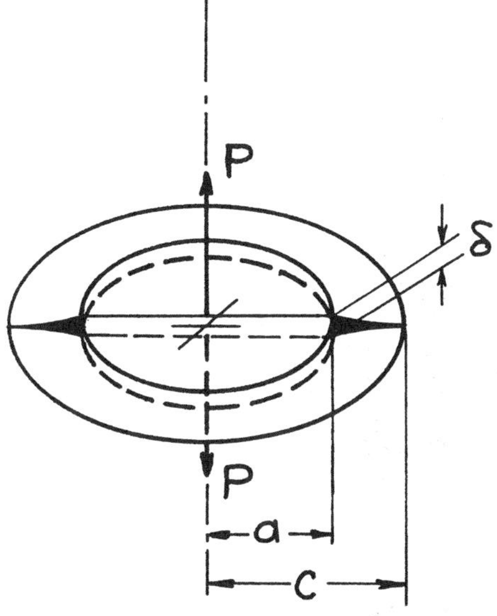

$$\frac{c}{a} = \frac{1}{\sqrt{2}}\left\{1 + \sqrt{1 + \left(\frac{P}{\sigma_Y \cdot \pi a^2}\right)^2}\right\}^{1/2}$$

Opening at Crack Tip:

$$\delta = \frac{8\left(1 - \nu^2\right)}{\pi E}\sigma_Y c\sqrt{\left(\frac{c}{a}\right)^2 - 1}\left\{\cos^{-1}\frac{a}{c} - \sqrt{\frac{c-a}{c+a}}\right\}$$

Method: Superposition of **page 24.2** or **page 24.7** ($s = 0$) and **page 24.6**
Accuracy: Exact
Reference: **Tada 1974**

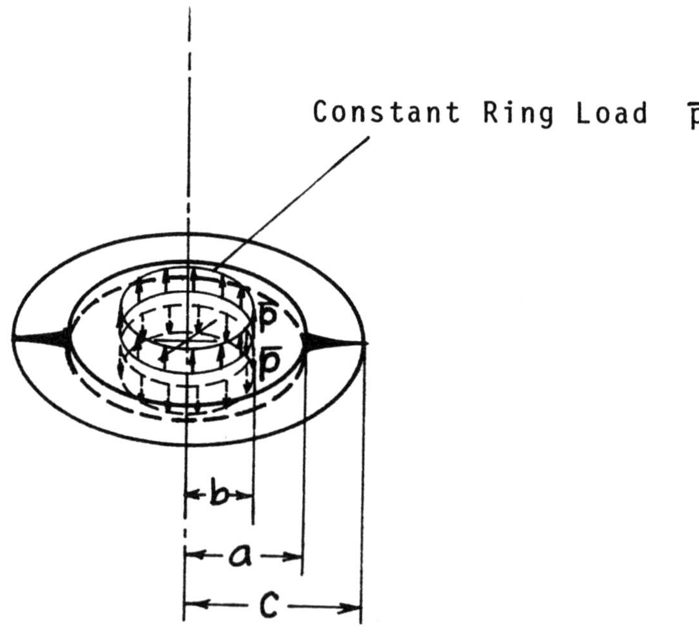

$$\frac{c}{a} = \frac{1}{\sqrt{2}} \left\{ 1 + \left(\frac{b}{a}\right)^2 + \sqrt{\left\{1 - \left(\frac{b}{a}\right)^2\right\}^2 + \left(\frac{\bar{p} \cdot 2\pi b}{\sigma_Y \cdot \pi a^2}\right)^2} \right\}^{1/2}$$

Method: Superposition of **pages 24.5 and 24.6**
Accuracy: Exact
Reference: **Tada 1974**

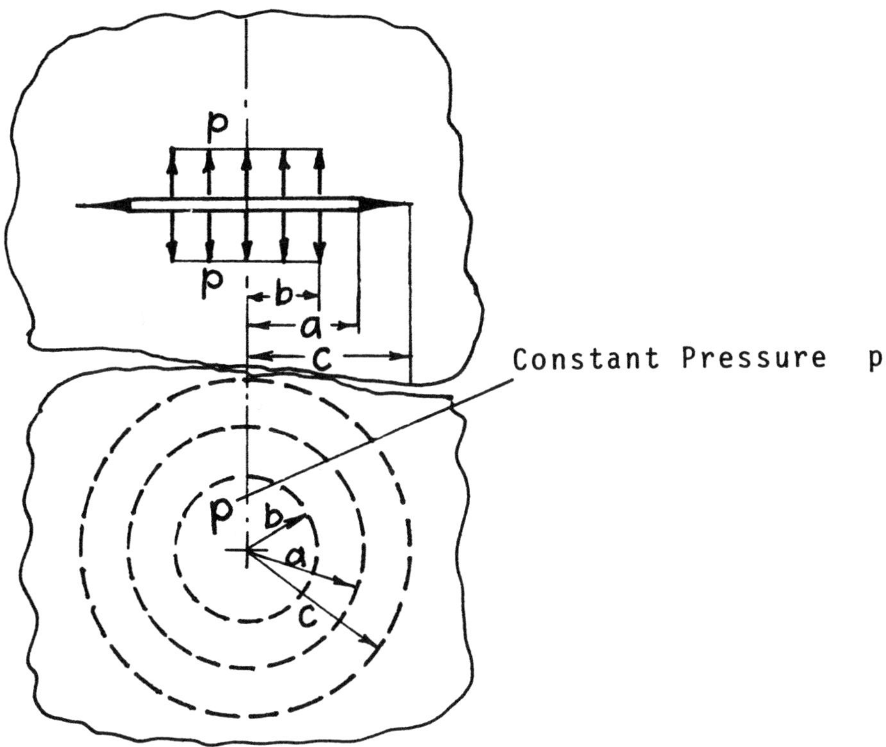

$$\frac{c}{a} = \left\{ \frac{1 - \left(2 + \frac{b^2}{a^2}\right)\left(\frac{P}{\sigma_Y}\right)^2 - 2\left(\frac{P}{\sigma_Y}\right)^2 \sqrt{1 - \frac{b^2}{a^2} + \left(\frac{Pb^2}{\sigma_Y a^2}\right)^2}}{1 - 4\left(P/\sigma_Y\right)^2} \right\}^{1/2}$$

$$\text{when} \quad P/\sigma_Y = 1/2: \quad c/a = \frac{1 - 1/4\left(b/a\right)^2}{\sqrt{1 - 1/2\left(b/a\right)^2}}$$

Method: Superposition of Two Cases of **page 24.6**
Accuracy: Exact
Reference: **Tada 1974**

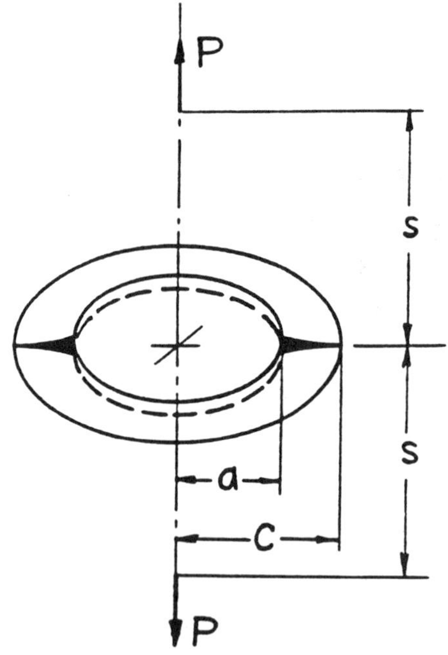

$$\sqrt{1-(a/c)^2}\,\frac{\left\{(c/a)^2+(s/a)^2\right\}^2}{(c/a)^2+\left(\frac{2-\nu}{1-\nu}\right)(s/a)^2}=\frac{1}{2}\left(\frac{P}{\sigma_Y\cdot\pi a^2}\right)$$

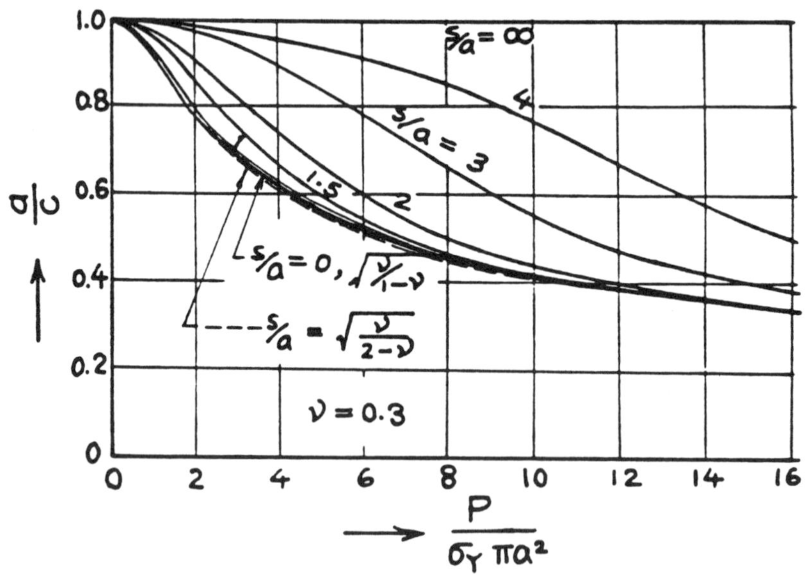

Method: Superposition of **pages 24.7 and 24.6**
Accuracy: Exact
Reference: **Tada 1974**

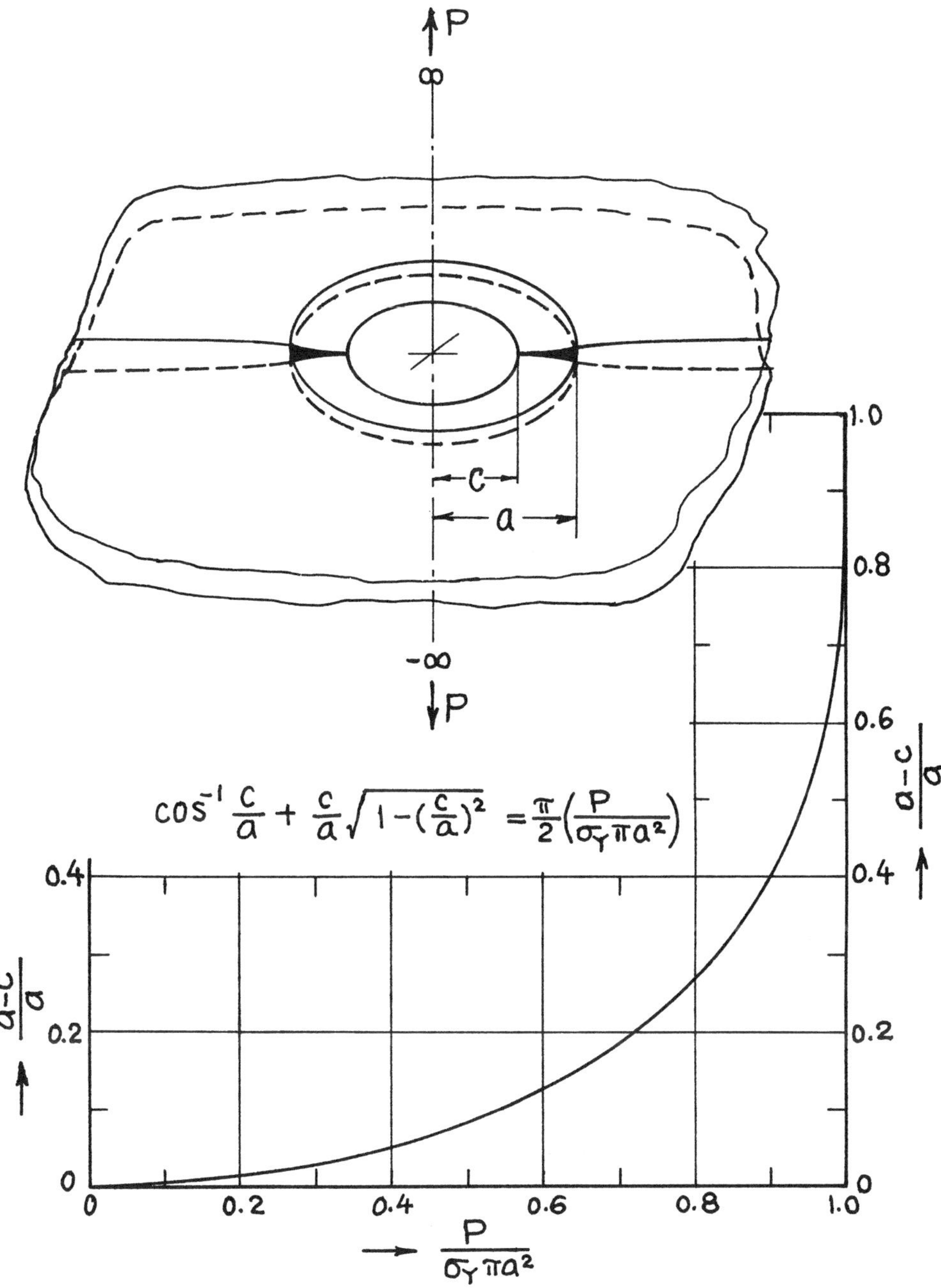

Method: Superposition of **pages 25.6 and 25.4**
Accuracy: Exact
Reference: **Tada 1974**

CRACK(S) IN A SHELL

- A Circumferential Crack in a Cylindrical Shell
- Multiple Circumferential Cracks in a Cylindrical Shell
- A Longitudinal Crack in a Cylindrical Shell
- A Crack in a Spherical Shell

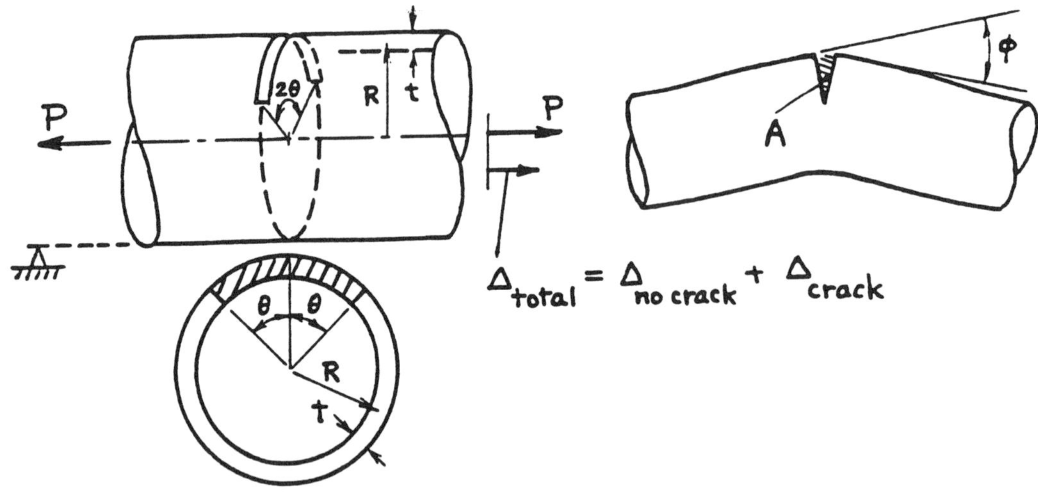

$R/t \simeq 10 \qquad (\theta < 110^\circ)$

$$\sigma = P/(2\pi Rt)$$

$$K_I = \sigma\sqrt{\pi(R\theta)} \cdot F(\theta)$$

$$F(\theta) = 1 + 7.5\left(\frac{\theta}{\pi}\right)^{3/2} - 15.0\left(\frac{\theta}{\pi}\right)^{5/2} + 33.0\left(\frac{\theta}{\pi}\right)^{7/2}$$

Crack Opening Area and Additional Extension due to Crack:

$$A = \frac{2\sigma\pi(R\theta)^2}{E'} \cdot I(\theta)$$

$$\Delta_{crack} = A/(2\pi R) = \frac{\sigma R\theta^2}{E'} \cdot I(\theta)$$

$$I(\theta) = 1 + \left(\frac{\theta}{\pi}\right)^{3/2}\left[8.6 - 13.3\left(\frac{\theta}{\pi}\right) + 24.0\left(\frac{\theta}{\pi}\right)^2\right] + \left(\frac{\theta}{\pi}\right)^3\left[22.5 - 75.0\left(\frac{\theta}{\pi}\right) + 205.7\left(\frac{\theta}{\pi}\right)^2 - 247.5\left(\frac{\theta}{\pi}\right)^3 + 242.0\left(\frac{\theta}{\pi}\right)^4\right]$$

Rotation (Kink) at Cracked Section due to Crack:

$$\phi = \frac{2\sigma\theta^2}{E'} \cdot \Phi(\theta)$$

$$\Phi(\theta) = 1 + \left(\frac{\theta}{\pi}\right)^{3/2} \left[8.2 - 12.7\left(\frac{\theta}{\pi}\right) + 19.3\left(\frac{\theta}{\pi}\right)^2\right]$$
$$+ \left(\frac{\theta}{\pi}\right)^3 \left[20.4 - 68.0\left(\frac{\theta}{\pi}\right) + 165.2\left(\frac{\theta}{\pi}\right)^2 - 187.2\left(\frac{\theta}{\pi}\right)^3 + 146.7\left(\frac{\theta}{\pi}\right)^4\right]$$

Method: Approximations of Sanders' Solution (for K_I) for $R/t \simeq 10$. A, Δ, ϕ Paris' Equation (see **Appendix B**)
Accuracy: K 1%; A, Δ, ϕ 2%
References: **Sanders 1982; Tada 1983a, 1985**

NOTE: Δ_{crack} is the extension at infinity when uniform pressure σ is applied on crack surfaces.

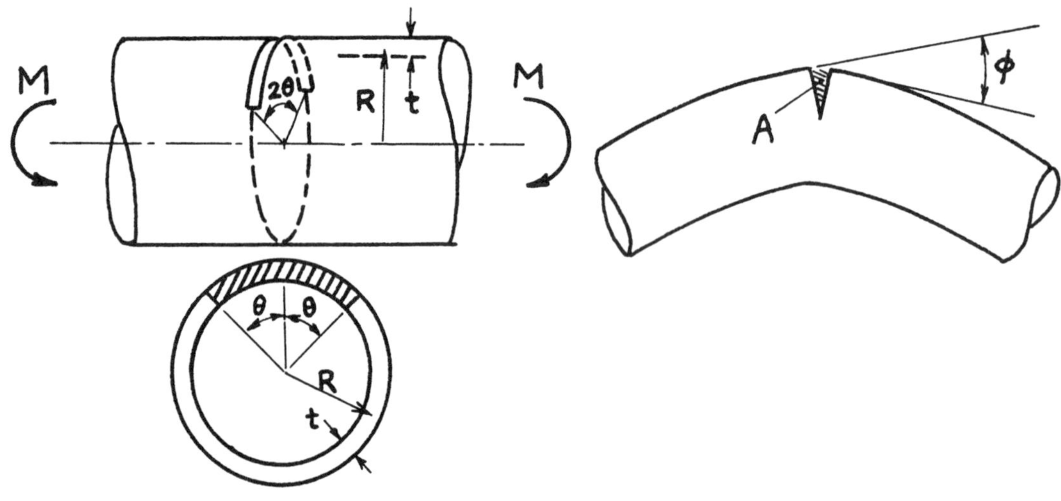

$$R/t \simeq 10 \qquad \left(\theta < 110^{\circ}\right)$$

$$\sigma = M/\left(\pi R^{2} t\right)$$

$$K_I = \sigma\sqrt{\pi(R\theta)} \cdot F(\theta)$$

$$F(\theta) = 1 + 6.8\left(\frac{\theta}{\pi}\right)^{3/2} - 13.6\left(\frac{\theta}{\pi}\right)^{5/2} + 20.0\left(\frac{\theta}{\pi}\right)^{7/2}$$

Crack Opening Area:

$$A = \frac{2\sigma\pi(R\theta)^{2}}{E'} \cdot I(\theta)$$

$$I(\theta) = 1 + \left(\frac{\theta}{\pi}\right)^{3/2}\left[8.2 - 12.7\left(\frac{\theta}{\pi}\right) + 19.3\left(\frac{\theta}{\pi}\right)^{2}\right] + \left(\frac{\theta}{\pi}\right)^{3}\left[20.4 - 68.0\left(\frac{\theta}{\pi}\right) + 165.2\left(\frac{\theta}{\pi}\right)^{2} - 187.2\left(\frac{\theta}{\pi}\right)^{3} + 146.7\left(\frac{\theta}{\pi}\right)^{4}\right]$$

Additional Rotation (Kink at Cracked Section) due to Crack:

$$\phi = \frac{2\sigma\theta^2}{E'} \cdot \Phi(\theta)$$

$$\Phi(\theta) = 1 + \left(\frac{\theta}{\pi}\right)^{3/2}\left[7.8 - 12.1\left(\frac{\theta}{\pi}\right) + 14.5\left(\frac{\theta}{\pi}\right)^2\right]$$
$$+\left(\frac{\theta}{\pi}\right)^3\left[18.5 - 61.7\left(\frac{\theta}{\pi}\right) + 130.5\left(\frac{\theta}{\pi}\right)^2 - 136.0\left(\frac{\theta}{\pi}\right)^3 + 88.9\left(\frac{\theta}{\pi}\right)^4\right]$$

Method: Approximation of Sanders' Solution (for K_I) for $R/t \simeq 10$. A, ϕ Paris' Equation (see **Appendix B**)
Accuracy: K_I 1%; A, ϕ 2%
References: **Sanders 1982, 1983; Tada 1983a, 1985**

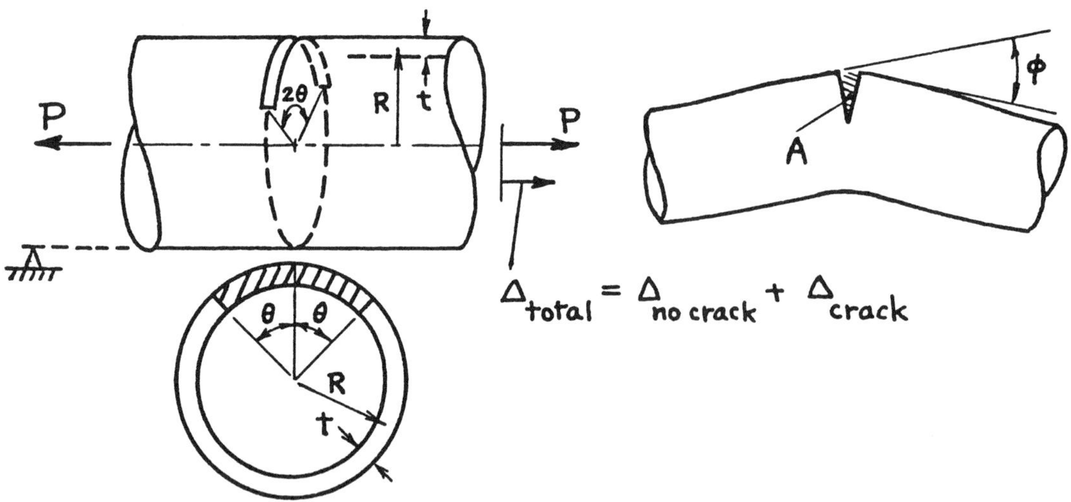

$$\sigma = P/(2\pi Rt)$$

$$\varepsilon^2 = \left({}^{t}\!/_{R}\right) \Big/ \sqrt{12\left(1-\nu^2\right)}$$

$$K_I = \sigma\sqrt{R}\left(\frac{\sqrt{2}}{\varepsilon}\right)^{1/2} \cdot F(\theta)$$

$$F(\theta) = \theta + \frac{1 - \theta\cot\theta}{2\cot\theta + \sqrt{2}\cot\left(\frac{\pi-\theta}{\sqrt{2}}\right)}$$

Crack Opening Area and Additional Axial Displacement at Infinity due to Crack:

$$A = \frac{4\sigma R^2}{E'}\left(\frac{\sqrt{2}}{\varepsilon}\right) \cdot I(\theta)$$

$$\Delta_{crack} = A/(2\pi R) = \frac{2\sigma R}{\pi E'}\left(\frac{\sqrt{2}}{\varepsilon}\right) \cdot I(\theta)$$

Bend (Kink) Angle at Cracked Section:

$$\phi = \frac{4\sigma}{\pi E'}\left(\frac{\sqrt{2}}{\varepsilon}\right) \cdot \Phi_1(\theta)$$

For numerical values of functions $F(\theta), I(\theta)$ and $\Phi_1(\theta)$, see **page 33.5**.

Methods: K Complete Shell Analysis (good for long cracks); A, Δ_{crack}, and ϕ Paris' Equation (see **Appendix B)**

Accuracy: 1% for $\lambda = \theta/\sqrt{t/R}(= a/\sqrt{Rt}) > 5$

References: **Sanders 1982, 1983; Tada 1985**

NOTE: Δ_{crack} is the axial displacement at infinity when uniform pressure σ is applied on crack surfaces.

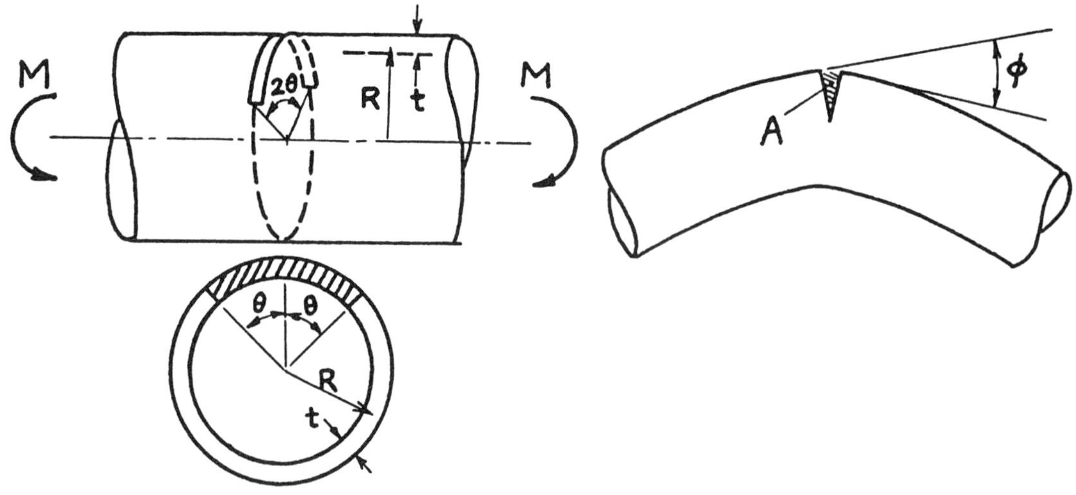

$$\sigma = M/\left(\pi R^2 t\right)$$

$$\varepsilon^2 = ({}^t/_R)\Big/\sqrt{12\left(1-\nu^2\right)}$$

$$K_I = \sigma\sqrt{R}\left(\frac{\sqrt{2}}{\varepsilon}\right)^{1/2} \cdot G(\theta)$$

$$G(\theta) = \sin\theta\left[1 + \frac{1}{2}\frac{\theta - \cot\theta(1-\theta\cot\theta)}{2\cot\theta + \sqrt{2}\cot\left(\frac{\pi-\theta}{\sqrt{2}}\right)}\right]$$

Crack Opening Area:

$$A = \frac{4\sigma R^2}{E'}\left(\frac{\sqrt{2}}{\varepsilon}\right) \cdot \Phi_1(\theta)$$

Bend (Kink) Angle at Cracked Section:

$$\phi = \frac{4\sigma}{\pi E'}\left(\frac{\sqrt{2}}{\varepsilon}\right) \cdot \Phi_2(\theta)$$

For numerical values of functions $G(\theta)$, $\Phi_1(\theta)$ and $\Phi_2(\theta)$, see **page 33.5**.
Methods: K Complete Shell Analysis (good for long cracks); A and ϕ Paris' Equation (see **Appendix B**)
Accuracy: 1% for $\lambda = \theta/\sqrt{{}^t/_R}\ (= a/\sqrt{Rt}) > 5$
References: **Sanders 1982, 1983; Tada 1985**

NOTE: ϕ is also the additional relative rotation at infinity due to the presence of crack ($\phi = \phi_{total} - \phi_{no\,crack}$).

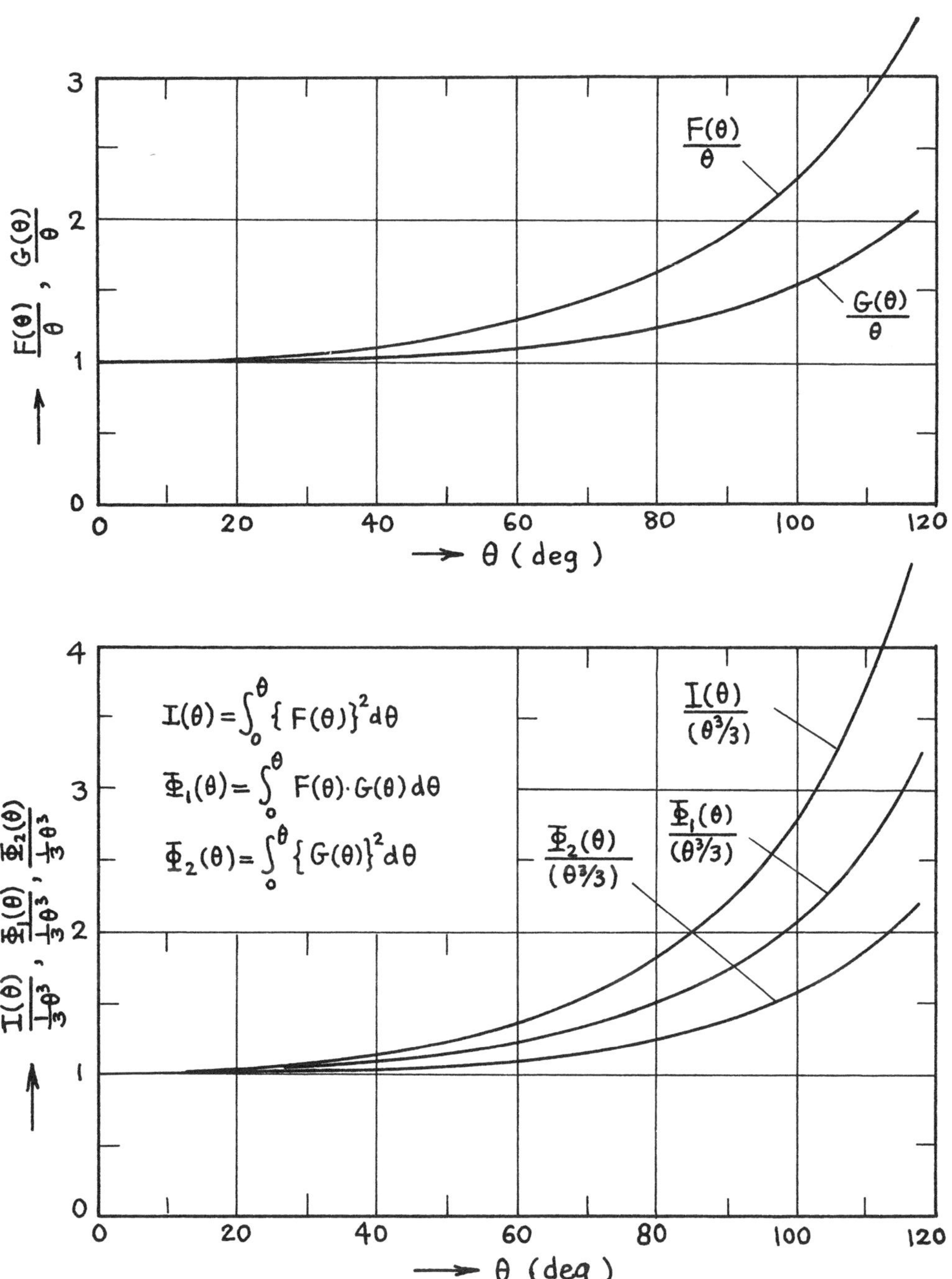

NOTE: Graphs are relevant for both **pages 33.3** and **33.4**.

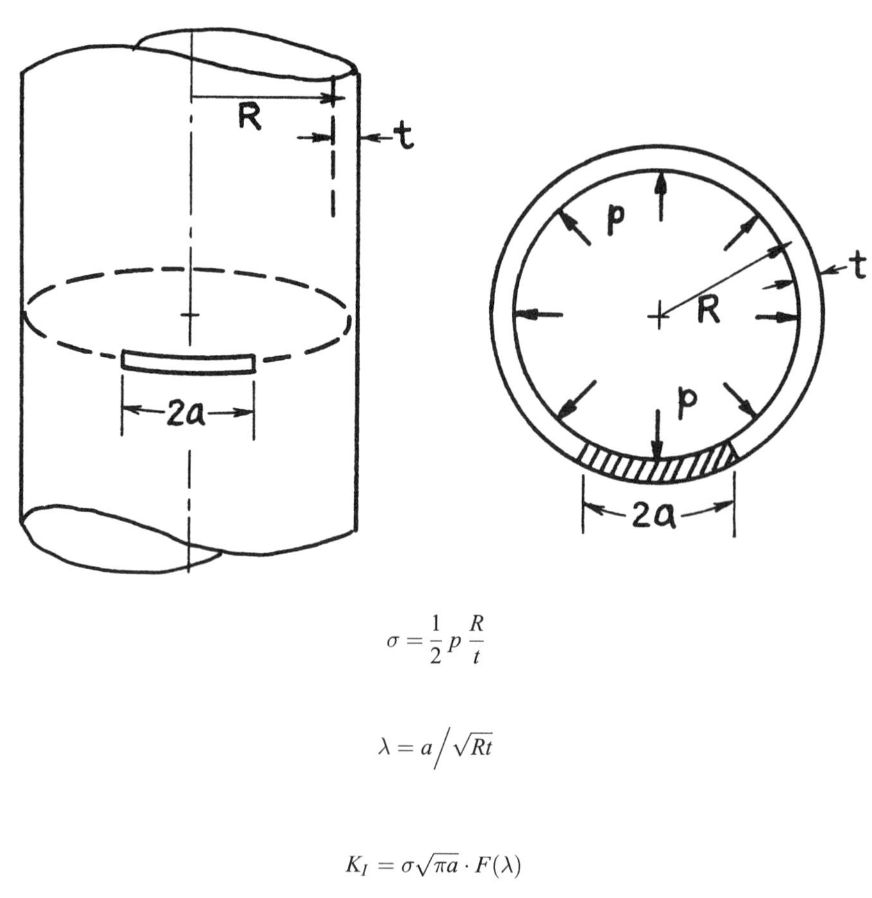

$$\sigma = \frac{1}{2} p \frac{R}{t}$$

$$\lambda = a \Big/ \sqrt{Rt}$$

$$K_I = \sigma \sqrt{\pi a} \cdot F(\lambda)$$

$$F(\lambda) = \left(1 + .3225\lambda^2\right)^{1/2} \qquad 0 < \lambda \leq 1$$

$$= 0.9 + 0.25\lambda \qquad 1 \leq \lambda \leq 5$$

Crack Opening Area:

$$A = \frac{\sigma}{E'} (2\pi Rt) \cdot G(\lambda)$$

$$G(\lambda) = \lambda^2 + 0.16\lambda^4 \qquad 0 < \lambda \leq 1$$

$$= .02 + .81\lambda^2 + .30\lambda^3 + .03\lambda^4 \qquad 1 \leq \lambda \leq 5$$

Methods: K_I Integral Equation; A Paris' Equation (see **Appendix B**)
Accuracy: K_I 1%; A 2%
References: **Folias 1967; Fama 1972; Tada 1983a**

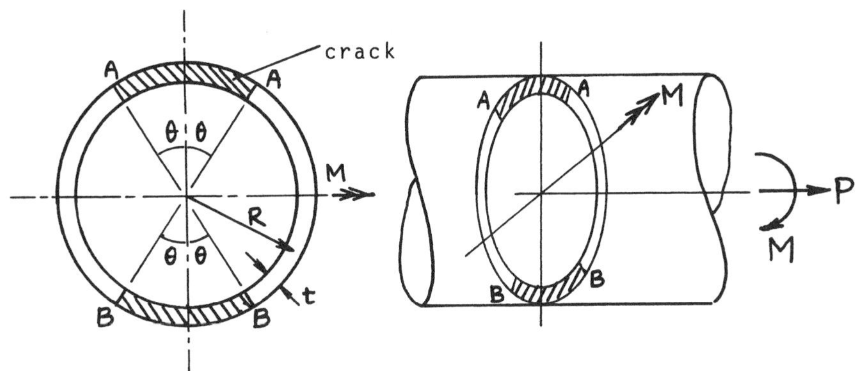

R = mean radius
t = wall thickness $\left(R/t > 5\right)$

$$\sigma_P = P/(2\pi Rt) \qquad \sigma_M = M/\left(\pi R^2 t\right)$$

$$K_{I\binom{A}{B}} = \sqrt{\pi R\theta}\{\sigma_P F_P(\theta)(\pm)\sigma_M F_M(\theta)\}$$

Crack Opening Area: $A = A_P + A_M$

$$A_{\binom{\widehat{AA}}{\widehat{BB}}} = \frac{4\pi R^2}{E'}\{\sigma_P I_P(\theta)(\pm)\sigma_M I_M(\theta)\}$$

where

$$F_P(\theta) = \sqrt{\tan\theta/\theta}$$

$$F_M(\theta) = \left(1 - 1.5\bar{\theta} + 0.905\bar{\theta}^2\right)/\left(1-\bar{\theta}\right)^{3/2}$$

$$I_P(\theta) = \ell n(\sec\theta)$$

$$I_M(\theta) = \frac{2}{\pi}\left\{\frac{1}{1-\bar{\theta}} + 1.411\ell n\left(1-\bar{\theta}\right) + \bar{\theta}\left(.411 + 1.643\bar{\theta} - .531\bar{\theta}^2 + .120\bar{\theta}^3\right)\right\}$$

$$\bar{\theta} = \theta/\left(\frac{\pi}{2}\right) = \frac{2}{\pi}\theta$$

Method: K Approximated with Periodic Cracks (for P), Estimated by Interpolation (for M); A Paris' Equation (see **Appendix B**)
Accuracy: K Expected to be better than 5%; A Expected to be better than 10%
Reference: **Tada 2000**

NOTE: No crack surface interference in compressive region is assumed.

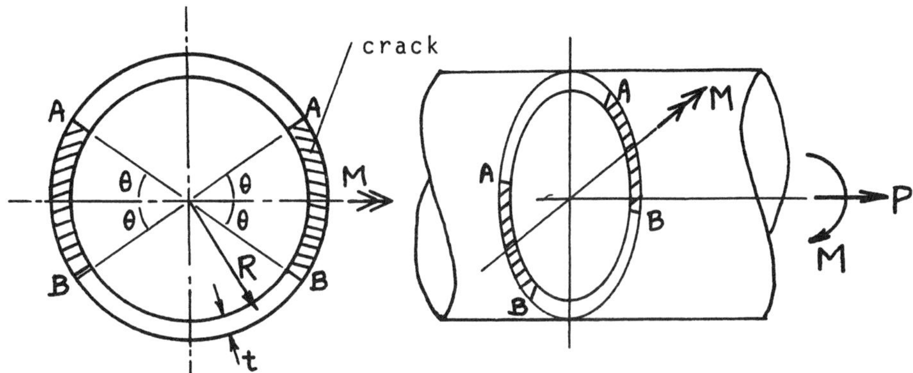

R = mean radius
t = wall thickness $\left(R/t > 5\right)$

$$\sigma_P = P/(2\pi Rt) \qquad \sigma_M = M\Big/\left(\pi R^2 t\right)$$

$$K_{I\binom{A}{B}} = \sqrt{\pi R\theta}\{\sigma_P F_P(\theta)(\pm)\sigma_M F_M(\theta)\}$$

Crack Opening Area: $A = A_P$

$$A_{\widehat{AB}} = \frac{4\sigma_P \pi R^2}{E'}\ell n(\sec\theta)$$

where

$$F_P(\theta) = \sqrt{\tan\theta/\theta}$$

$$F_M(\theta) = \frac{\theta}{2}\cdot\frac{1 - 0.5\bar{\theta} + 0.311\bar{\theta}^2}{\sqrt{1-\bar{\theta}}}$$

$$\bar{\theta} = \theta/\left(\frac{\pi}{2}\right) = \frac{2}{\pi}\theta$$

Method: Approximated with Periodic Cracks (for P), Estimated by Interpolation (for M)
Accuracy: K Expected to be better than 5%; A Expected to be better than 10%
Reference: **Tada 2000**

NOTE: No crack surface interference in compressive region is assumed.

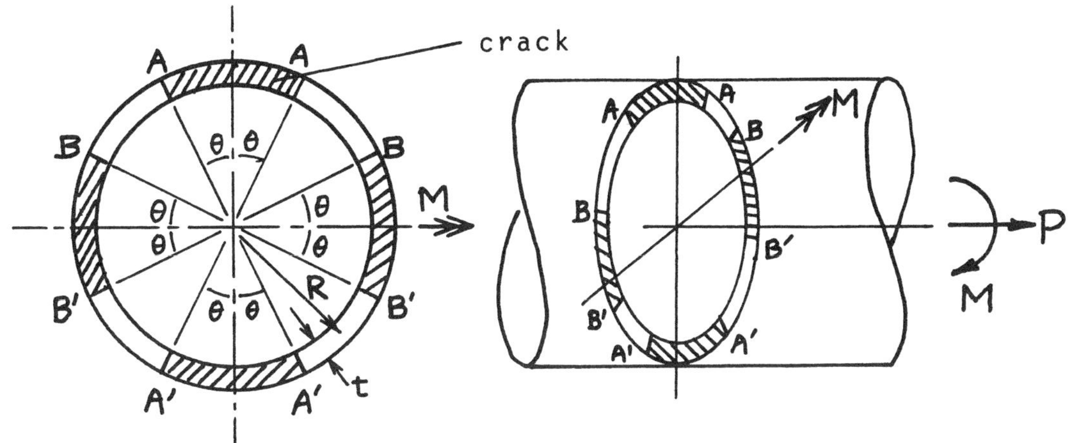

R = mean radius
t = wall thickness $\left(R/t > 5\right)$

$$\sigma_P = P/(2\pi Rt) \qquad \sigma_M = M/\left(\pi R^2 t\right)$$

$$K_{I\binom{A}{A'}} = \sqrt{\pi R\theta}\{\sigma_P F_P(\theta)(\pm)\sigma_M F_{MA}(\theta)\}$$

$$K_{II\binom{B}{B'}} = \sqrt{\pi R\theta}\{\sigma_P F_P(\theta)(\pm)\sigma_M F_{MB}(\theta)\}$$

Crack Opening area: $A = A_P + A_M$

$$A_{\binom{\widehat{AA}}{\widehat{A'A'}}} = A_P(\pm)A_{M\widehat{AB}} \qquad \left(A_{\widehat{BB'}} = A_P\right)$$

$$= \frac{\pi R^2}{E'}\{\sigma_P I_P(\theta)(\pm)\sigma_M I_M(\theta)\}$$

where

$$F_P(\theta) = \sqrt{\tan 2\theta / 2\theta}$$

$$F_{MA}(\theta) = \left(1 - 0.5\bar{\theta} - 0.05\bar{\theta}^2\right) \Big/ \sqrt{1 - \bar{\theta}}$$

$$F_{MB}(\theta) = \theta\left(0.5 - 0.25\bar{\theta} + 0.323\bar{\theta}^2\right) \Big/ \sqrt{1 - \bar{\theta}}$$

$$I_P(\theta) = \ell n(\sec 2\theta)$$

$$I_M(\theta) = \frac{1}{\sqrt{2}}\left\{\ell n \frac{1}{1 - \bar{\theta}} - \bar{\theta}\left(1 - 1.245\bar{\theta} + .333\bar{\theta}^2 - .044\bar{\theta}^3 - .005\bar{\theta}^4\right)\right\}$$

$$\bar{\theta} = \theta / \left(\frac{\pi}{4}\right) = \frac{4}{\pi}\theta$$

Method: K Approximated with Periodic Cracks (for P), Estimated by Interpolation (for M); A Paris' Equation (see **Appendix B**)
Accuracy: K Expected to be better than 5%; A Expected to be better than 10%
Reference: **Tada 2000**

NOTE: No crack surface interference in compressive region is assumed.

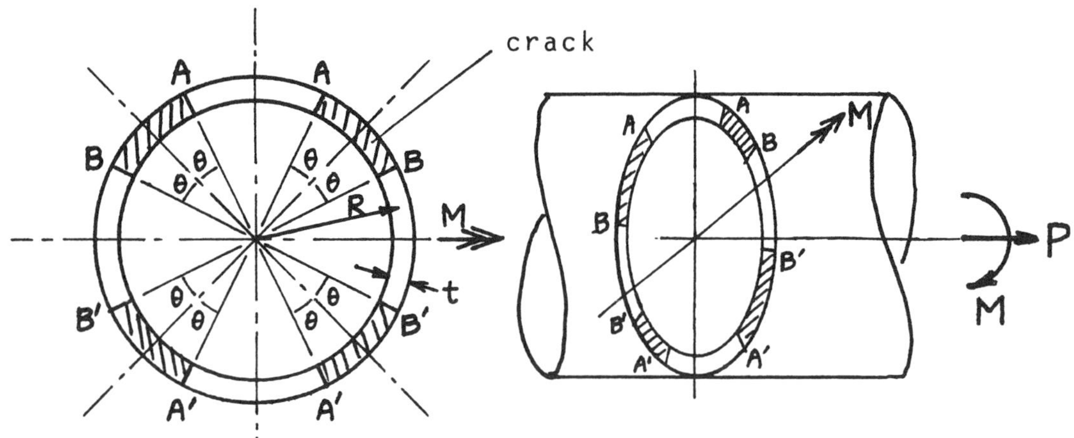

R = mean radius
t = wall thickness $\quad \left(R/t > 5\right)$

$$\sigma_P = P/(2\pi R t) \qquad \sigma_M = M/\left(\pi R^2 t\right)$$

$$K_{I\binom{A}{A'}} = \sqrt{\pi R \theta}\{\sigma_P F_P(\theta)(\pm)\sigma_M F_{MA}(\theta)\}$$

$$K_{I\binom{B}{B'}} = \sqrt{\pi R \theta}\{\sigma_P F_P(\theta)(\pm)\sigma_M F_{MB}(\theta)\}$$

Crack Opening Area: $\quad A = A_P + A_M$

$$A_{\binom{\widehat{AB}}{\widehat{A'B'}}} = A_P(\pm)A_{M\widehat{AB}}$$

$$= \frac{\pi R^2}{E'}\{\sigma_P I_P(\theta)(\pm)\sigma_M I_M(\theta)\}$$

where

$$F_P(\theta) = \sqrt{\tan 2\theta / 2\theta}$$

$$F_{MA}(\theta) = \left(0.707 - 0.070\bar{\theta}\right) \Big/ \sqrt{1 - \bar{\theta}}$$

$$F_{MB}(\theta) = \left(0.707 + 0.076\bar{\theta} - 0.450\bar{\theta}^2\right) \cdot \sqrt{1 - \bar{\theta}}$$

$$I_P(\theta) = \ell n(\sec 2\theta)$$

$$I_M(\theta) = \frac{1}{2}\left\{\ell n \frac{1}{1-\bar{\theta}} - \bar{\theta}\left(1 - 1.244\bar{\theta} + .329\bar{\theta}^2 + .235\bar{\theta}^3 - .116\bar{\theta}^4 + .025\bar{\theta}^5\right)\right\}$$

$$\bar{\theta} = \theta / \left(\frac{\pi}{4}\right) = \frac{4}{\pi}\theta$$

Method: K Approximated with Periodic Cracks (for P), Estimated by Interpolation (for M); A Paris' Equation (see **Appendix B**)
Accuracy: K Expected to be better than 5%; A Expected to be better than 10%
Reference: **Tada 2000**

NOTE: No crack surface interference in compressive region is assumed.

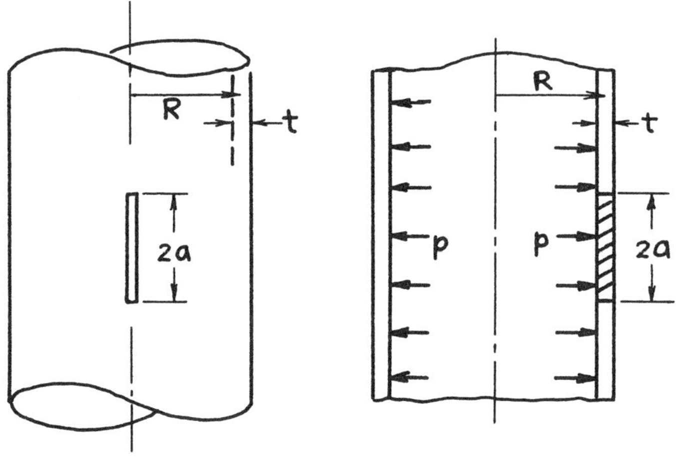

$$\sigma = p\frac{R}{t}$$

$$\lambda = a \big/ \sqrt{Rt}$$

$$K_I = \sigma\sqrt{\pi a} \cdot F(\lambda)$$

$$F(\lambda) = (1 + 1.25\lambda^2)^{1/2} \qquad 0 < \lambda \leq 1$$

$$= 0.6 + 0.9\,\lambda \qquad 1 \leq \lambda \leq 5$$

Crack Opening Area:

$$A = \frac{\sigma}{E'}(2\pi Rt) \cdot G(\lambda)$$

$$G(\lambda) = \lambda^2 + .625\lambda^4 \qquad 0 < \lambda \leq 1$$

$$= .14 + .36\lambda^2 + .72\lambda^3 + .405\lambda^4 \qquad 1 \leq \lambda \leq 5$$

Methods: K Integral Transform; A Paris' Equation (see **Appendix B**)
Accuracy: K_I 1%; A 2%
References: **Folias 1965; Erdogan 1969; Tada 1983a**

NOTE: As $a \to \infty$, $K_I \to \sigma\sqrt{R}\left(\frac{\sqrt{27\pi}}{2} \cdot \frac{R}{t}\right)$ **(Harris 1997).**

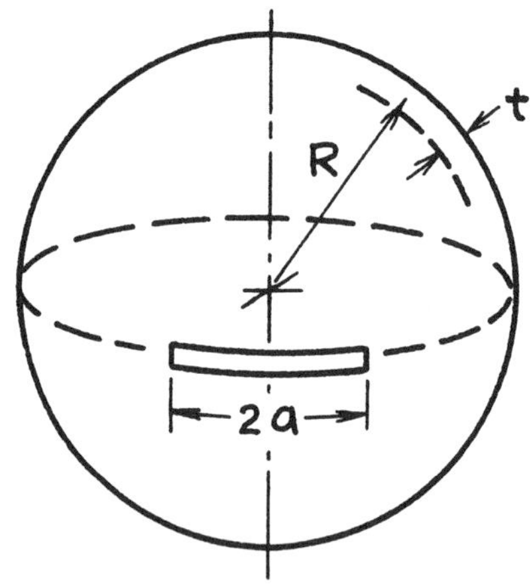

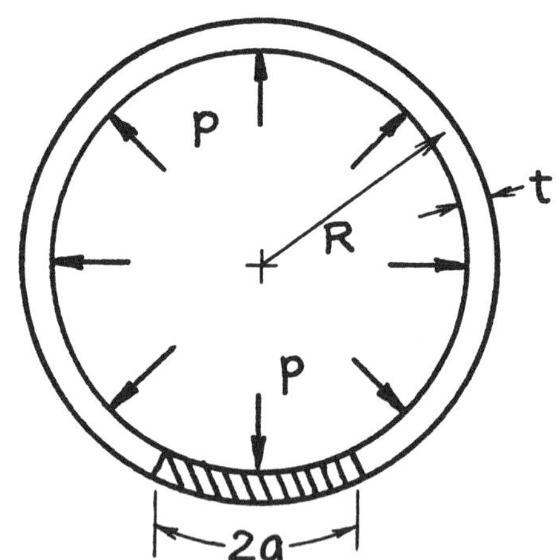

$$\sigma = \frac{1}{2} p \frac{R}{t}$$

$$\lambda = a \Big/ \sqrt{Rt}$$

$$K_I = \sigma\sqrt{\pi a} \cdot F(\lambda)$$

$$F(\lambda) = \left(1 + 1.41\lambda^2 + 0.04\lambda^3\right)^{1/2} \qquad 0 < \lambda \leq 3$$

Crack Opening Area:

$$A = \frac{\sigma}{E'}(2\pi Rt) \cdot G(\lambda)$$

$$G(\lambda) = \lambda^2 + 0.705\lambda^4 + 0.016\lambda^5 \qquad 0 < \lambda \leq 3$$

Methods: K_I Integral Equations; A Paris' Equation (see **Appendix B**)
Accuracy: K_I 1%; A 2%
References: **Erdogan 1969; Tada 1985**

APPENDIX A

COMPLIANCE CALIBRATION METHODS

A. DETERMINATIONS OF $\mathcal{G}$ AND K^2

As defined earlier and in **Irwin (1954)**, the crack extension force, $\mathcal{G}$, can be given by

$$\mathcal{G} = -\frac{dU_T}{dA}\,(\text{system isolated}) \tag{A1}$$

where U_T is the total strain energy of elastic deformation in the solid containing a crack, and dA is an infinitesimal (virtual) increment of new severed area. The system isolated requirement can be removed (**Paris 1965**) by subtracting from U_T the increment of energy, $\sum P_i d\Delta_i$, which enters the body during the movement, dA, of the crack. P_i is one of the applied loads and Δ_i is the corresponding load displacement (parallel to the load). Thus, $\mathcal{G}$ is given by

$$\mathcal{G} = \sum P_i \frac{d\Delta_i}{dA} - \frac{dU_T}{dA} \tag{A2}$$

Because the generalization from a single pair of loading points with oppositely directed loads, P, to multiple loading points is obvious, consider **Eq. (A2)** in the simpler form

$$\mathcal{G}\, dA = P d\Delta - dU_T \tag{A3}$$

Assuming linear-elastic behavior, the load point displacement, Δ (total between the pair of loading points) and the strain energy, U_T, are given by

$$\Delta = CP \tag{A4}$$

where C is the compliance, and

$$U_T = \frac{1}{2}\Delta P \tag{A5}$$

Substitution of (A4) and (A5) into (A3) provides

$$\mathcal{G}\,dA = \frac{1}{2}P^2\,dC \tag{A6}$$

Note that **Eq. (A5)** implies that the plate specimen does not contain a system of self-balanced (residual) stresses. In the case of a plate-type specimen of thickness B, containing only one straight leading edge of the crack, and taking the crack size to be a, dA equals Bda.
Thus **Eq. (A6)** becomes

$$\mathcal{G} = \frac{1}{2}P^2\frac{dC}{Bda} \tag{A7}$$

Further, if the plate-type specimen has a characteristic width W, and a Young's Modulus, E, **Eq. (A7)** can then be written in the form [see **Eq. (20)**]

$$K^2 = E\mathcal{G} = \frac{1}{2}\left(\frac{P}{BW}\right)^2 W\frac{d(EBC)}{d(^a/_W)} \tag{A8}$$

where the derivative term in the equation is dimensionless.

Eq. (A7) suggests an experimental method for determining the proportionality of $\mathcal{G}$ to P^2. Measurements of the ratio $\Delta/P = C$ for a series of crack lengths (a-values) can be made over a range of interest. The required derivative dC/da, as a function of a, can than be obtained from these results. Early trials of this method for notched round bars and edge notched bars in bending are found in **Lubahn (1959)**. In later efforts (**Srawley 1964**), a long, single-edge-notched plate in tension was used and the results were directly compared to numerically computed values. In terms of the information necessary for the computation of K values, it appeared that an accuracy of better than 3% could be achieved by the experimental method in central portions of the series of crack lengths used in the calibration. Potential improvement in accuracy of the compliance calibration method is to be expected with increases of dC/da relative to the value of C. A list of fracture test specimens, in order of decreasing expected potential compliance calibration accuracy, is:

1. Double-cantilever specimens
2. Single-edge-notched bend bars
3. Compact tension specimens
4. Long edge-notched plates in tension
5. Long centrally notched plates in tension

With regard to accuracy, since the end result required is generally a K calibration, and K^2 does not depend E, a low modulus plate material such as a high-strength aluminum alloy can be used to increase the observed displacements. However, the value of E for the calibration material must be adequately known or measured.* Comments on the accuracy of various displacement gages are given in **Srawley (1965)**. Contributions of plastic deformation at the loading points should be eliminated and influences of any plastic strain at the tip of the crack-simulating notch should be minimized. After establishing calibration, values of K can be obtained from various pairs of observations such as P and a, P and C, and Δ and a.

* **It is not sufficient to use a literature value.**

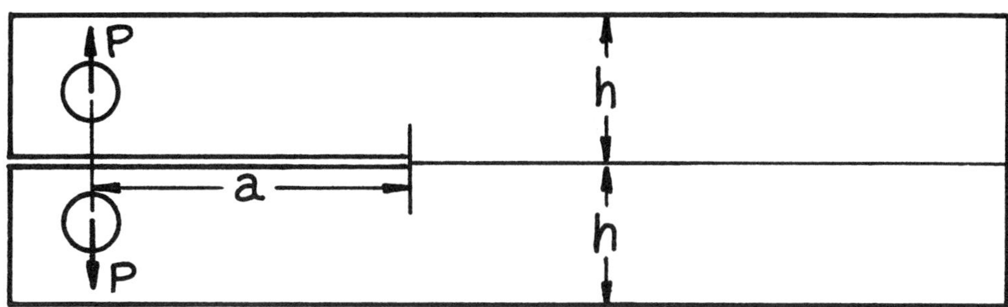

Fig. (A1). DCB specimen for measuring the tensile fracture toughness of an adhesive bond.

Eq. (A7) and (A8), coupled with compliance measurements, have been frequently used for double-cantilever (DCB) specimens. The conditions pertaining to the use of DCB specimens for measurements of tensile fracture strength of adhesive joints and, with face-grooving, for homogeneous solids are reviewed in **Irwin (1971)**. Consider two adherent bars joined by a thin layer of adhesive as shown in **Fig. (A1)**.

If an estimate of C is made using simple beam theory with a correction for deflection due to shear, the following expression for $\mathcal{G}$ can be derived.

$$\mathcal{G} = \frac{4P^2}{EhB^2}\left\{3\left(\frac{a+a_0}{h}\right)^2 + 1\right\} \tag{A9}$$

Eq. (A9) assumes each beam is "built-in" at a distance $(a + a_0)$ from the loading line. The second term in the bracketed expression results from the shear correction and takes the simple form shown if one assumes Poisson's ratio is 1/3. Experimental results suggest a value of about *h/3* for a_0. These assumptions appear to be adequate for estimates of $\mathcal{G}$ so long as a is larger than *2h* and the unbroken ligament is larger than *3h*. However, a calibration of improved accuracy can be readily developed through careful compliance measurements (**Ripling 1964**).

Ripling (1964) discusses a modified (tapered) DCB specimen shape such that dC/da does not depend crack size across a certain range of a values. In the case of all of the DCB specimens discussed above, the experimentors have relied primarily compliance calibrations rather than results derived from numerical analysis of the stress fields.

Compliance calibration methods are applicable to plate-type specimens with through-the-thickness cracks for which thickness average values of $\mathcal{G}$ are sufficient for experimental purposes. Compliance calibration can be used, with limited accuracy, for notched round bars in tension because $\mathcal{G}$ is constant around the leading edge of the crack-simulating notch. In the case of a crack for which the variation of $\mathcal{G}$ along various regions of the leading edge is appreciable and important, it must be remembered that measurements of the change of compliance with enlargement of the crack can provide only $\mathcal{G}$ value averages. With regard to the relationship between the thickness average $\mathcal{G}$ and K^2, the assumption, $K^2 = E\,\mathcal{G}$, implied in the K_{IC} test method, ASTM E-399, can be regarded as adequate for practical applications.

When a plate specimen is used as a model of a service component that contains a system of self-balanced (residual) stresses, attention must be given to the possibility that the residual stresses may hold the crack partially closed during initial portions of the externally applied loading. In the absence of this complexity, a plate specimen can be used to determine that portion of K that is due to the external loading. The remaining portion of K, due to the residual stresses, can be found from knowledge of the residual stresses using the analytical or numerical methods discussed in **Part 1**.

B. RELATIONSHIP OF $\mathcal{G}$ TO J

In this section, J is given a restricted interpretation as merely an expression for $\mathcal{G}$ applicable to a non-linear-elastic as well as to a linear-elastic solid containing a crack. Thus **Eqs. (A1) and (A2)** may be regarded as definitions of J (**Rice 1968a**). For simplicity this discussion is restricted to two-dimensional crack stress fields. **Eq. (A2)** suggests a plan for representing J as a path-independent contour integral enclosing the crack tip, as follows. Because the solid is elastic, energy disappearance can occur only at the crack tip singularity during its forward increment of (virtual) displacement. Thus the net flux of energy inward across the boundary of any area enclosing the crack tip must give the same value, J. In the construction of a J contour integral, the summation of input work, $\sum P_i d\Delta_i$, can be replaced by its equivalent in terms of the integral of each boundary stress times the parallel increment of displacement for the individual segments of the contour. In addition the change of stress field energy that occurs inside of the contour must be subtracted.

Assuming an increment, da, of forward (x-direction) displacement of the crack tip, and assuming the coordinate normal to the crack is y, one finds

$$Jda = -da\iint dxdy\left(\frac{\partial U}{\partial a}\right) + da\oint\left[dy\left(\sigma_x\frac{\partial u}{\partial a} + \tau_{xy}\frac{\partial v}{\partial a}\right) - dx\left(\tau_{xy}\frac{\partial u}{\partial a} + \sigma_y\frac{\partial v}{\partial a}\right)\right] \tag{A10}$$

where U is the strain energy density.

Obviously da can be eliminated from each side of **Eq. (A10)**. Next, consider the fact that the location of the contour relative to the crack tip, after the crack tip has been displaced an infinitesimal amount da in the x-direction, is equivalent to rigid displacement of the contour by an equal infinitesimal amount, dx, in the opposite direction. Thus the calculations indicated in **Eq. (A10)** are unchanged if $\frac{\partial}{\partial a}$ is replaced by $-\frac{\partial}{\partial x}$. With regard to the preceding term of **Eq. (A10)**, note that the substitution of $-\frac{\partial}{\partial x}$ for $\frac{\partial}{\partial a}$ permits completion of the x-direction portion of the integral or

$$\iint dxdy\left(\frac{\partial U}{\partial x}\right) = \oint U\,dy \tag{A11}$$

The final result is

$$J = \oint\left[dy\left\{U - \sigma_x\frac{\partial u}{\partial x} - \tau_{xy}\frac{\partial v}{\partial x}\right\} + dx\left\{\tau_{xy}\frac{\partial u}{\partial x} + \sigma_y\frac{\partial v}{\partial x}\right\}\right] \tag{A12}$$

Eq. (A12) can be expressed in more compact form. However, the form given above is convenient for purposes of numerical computation. As a check on the path independence of **Eq. (A12)**, one can equate $\frac{\partial}{\partial x}$ of the coefficient of dy to $\frac{\partial}{\partial y}$ of the coefficient of dx. After use of stress compatibility relations, the result is

$$\frac{\partial U}{\partial x} = \sigma_x\frac{\partial \varepsilon_x}{\partial x} + \sigma_y\frac{\partial \varepsilon_y}{\partial x} + \tau_{xy}\frac{\partial \gamma_{xy}}{\partial x} \tag{A13}$$

Given the existence of a set of stress–strain relations, linear or nonlinear, which are strain-path independent, U can be represented as

$$U = \int_0^{\varepsilon_x}\sigma_x d\varepsilon_x' + \int_0^{\varepsilon_y}\sigma_y d\varepsilon_y' + \int_0^{\gamma_{xy}}\tau_{xy} d\gamma'_{xy} \tag{A14}$$

From the assumptions, the integrand of each integral can be regarded as a function only of the corresponding integration variable, ε_x', ε_y' or γ_{xy}'. The derivative of **Eq. (A14)** then results in **Eq. (A13)**.

In later applications, the path independence of **Eq. (A12)** will be of special interest because the path of integration can be taken around the fracture process zone, well within the region of pronounced nonlinearity, yet outside the region containing advance separations and localized slip displacements special to the progressive fracturing process. Thus J can serve as a characterization of the stress–strain environment enclosing the fracture process zone.

Anticipating a later discussion of J with plasticity present, without discussing applicability of J inside a plastic zone, it is clear that the J-Integral is path independent for any contour on or outside the elastic–plastic boundary. When the plastic zone of an opening mode crack is modeled as a Dugdale or strip-yield zone extending directly forward from the crack tip, it is easily shown that $J = \sigma_Y \delta$, where δ is nominally the crack opening stretch and σ_Y is a flow stress property assumed to be constant along the strip plastic zone. **Tada (1972a)** presents calculations that tend to support the likelihood of close equivalence between J and values of $\mathcal{G}$ computed using linear-elasticity with an adjustment of the crack size. Given a crack-tip plastic zone which does not contact a free surface of the specimen, other than the crack surfaces, and drawing the J-Integral contour as a circle enclosing the crack-tip plastic zone, calculation of $\mathcal{G}$ using linear-elastic assumptions but with the crack-tip moved into a central position within the plastic zone may give a result similar to the J-value.

Returning to compliance calibration methods, it is evident that **Eq. (A3)**, rather than its equivalent, in the form of **Eq. (A12)**, is of most interest as a basis for compliance calibration. Assuming only two oppositely directed loads, P, as before, a series of P versus Δ lines for a series of crack sizes, a, will not be straight lines, as with linear behavior. However, values of J times da (where da is the difference of the crack lengths for adjacent curves) can be determined as a function of Δ by careful planimeter measurements, by curve fitting and integration, and so on. In the case of nonlinear behaviors such that slope $dP/d\Delta$ is small in the region of measurement, it is most convenient to plot J as a function of Δ for the series of crack lengths of interest. The J determinations can then, if desired, be based observations of a and Δ.

C. CRACK SIZE DETERMINATIONS

Part A of this section discusses determinations of C as a function of crack size. In the case of DCB specimens, the sensitivity of Δ at fixed load to crack size is adequate for the purpose of using the Δ/P ratio to determine the crack size. In other specimens, determinations of crack size from displacement–load ratios are usually more accurate when a different choice is made of the displacement measurement position. For example, with a centrally notched tensile specimen, the opening at the notch center is a convenient and appropriate choice. In the case of long single-notched tensile specimens, the opening at the free edge intersection of the crack is often used. In order to use a displacement–load ratio to estimate crack size, an accurate calibration based measurements of this ratio in the range of linear behavior for a series of crack sizes can be used.

Moreover, many analytical expressions for these displacements are found in this handbook. In the case of thin-sheet testing to determine cracking resistance, restraints against out-of-plane deformation (buckling) may be advisable. Even when such restraints are used, the displacement-measuring device should be designed so as to indicate only center-plane displacement or so as to compensate for small amounts of out-of-plane bending.

Testing or application situations where the plastic zone size is appreciable relative to the crack size are not uncommon. In such cases it is a common practice to compute K on the basis of an effective crack size. The addition of r_Y to the crack size at each crack-tip involves uncertainties with regard to choice of α and σ_Y in the equation

$$r_Y = \alpha\left(\frac{K}{\sigma_Y}\right)^2 \tag{A15}$$

As an alternative, one can determine the effective crack size from the "effective compliance," where the effective compliance is simply the ratio of deflection to load at the load for which a calculation of K is desired. Trials with analytical elastic–plastic models have shown that different choices of the displacement measurement position will cause differences in the effective crack size estimate. However, these differences are relatively small. Any disadvantage of this kind appears to be outweighed by the advantage of avoiding an arbitrary choice of α and σ_Y in the r_Y equation.

APPENDIX B

A METHOD FOR COMPUTING CERTAIN DISPLACEMENTS RELEVANT TO CRACK PROBLEMS

In earlier sections, $\mathcal{G}$ was shown (see **page 1.6**) to be equivalent to the rate of increase of the total strain energy, U_T, with increase in crack area, dA, and loading forces constant. That is to say

$$\mathcal{G} = \left.\frac{\partial U_T}{\partial A}\right|_{\text{Forces Constant}} \tag{B1}$$

Consider a body loaded by forces, P, which in addition has virtual forces, F, applied as in **Fig. (B1)**. (No generality is lost in considering a single force, P, and its support reaction and a single pair of forces, F, in equilibrium any distance, d, apart in the undeformed position.) The total energy rate, $\mathcal{G}$, can be, summed for the three modes, that is,

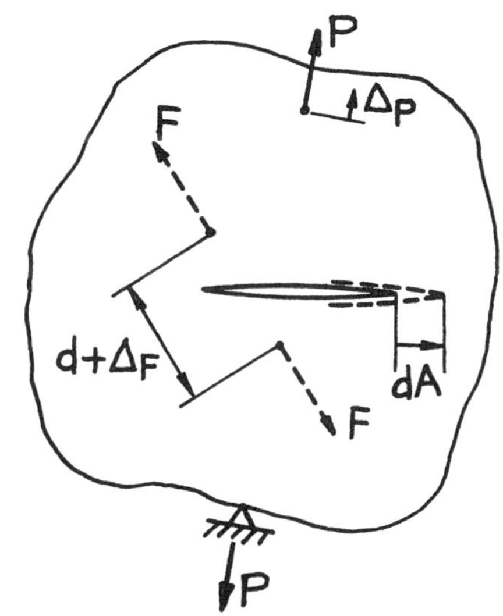

Fig. B1

$$\mathcal{G} = \mathcal{G}_I + \mathcal{G}_{II} + \mathcal{G}_{III} \tag{B2}$$

as was illustrated earlier (see **page 1.7**). Moreover, the resulting K_i values for each mode due to loading forces and virtual forces may be added, or

$$\left.\begin{aligned} K_I &= K_{IP} + K_{IF} \\ K_{II} &= K_{IIP} + K_{IIF} \\ K_{III} &= K_{IIIP} + K_{IIIF} \end{aligned}\right\} \tag{B3}$$

using the relationships between K_i and energy rates $\mathcal{G}_i$ (see **page 1.7**), that is,

$$E'\mathcal{G}_I = K_I^2, \quad E'\mathcal{G}_{II} = K_{II}^2, \quad E'\mathcal{G}_{III} = \alpha K_{III}^2 \tag{B4}$$

where E' and α are elastic constants depending on stress conditions, plane stress or plane strain. Combining **Eqs. (B2), (B3), and (B4)** gives

$$E'\mathcal{G} = (K_{IP} + K_{IF})^2 + (K_{IIP} + K_{IIF})^2 + (K_{IIIP} + K_{IIIF})^2 \cdot \alpha \tag{B5}$$

With the above information and Castigliano's theorem, it is possible to compute certain displacements in the manner suggested by **Paris (1957)**. Castigliano's theorem states that displacement of any load Q (in its own direction) may be computed by

$$\Delta_Q = \frac{\partial U_T}{\partial Q} \tag{B6}$$

The total strain energy may be regarded as that due to applying loading forces with no crack present plus that due to introducing the crack while holding forces constant, or

$$U_T = U_{No\,Crack} + \int_0^A \frac{\partial U_T}{\partial A} dA \tag{B7}$$

Introducing **Eqs. (B1) and (B6)**, displacement can be computed from

$$\Delta_Q = \frac{\partial U_T}{\partial Q} = \frac{\partial U_{No\,Crack}}{\partial Q} + \frac{\partial}{\partial Q}\int_0^A \mathcal{G} dA \tag{B8}$$

where it may be interpreted that

$$\frac{\partial U_{No\,Crack}}{\partial Q} = \Delta_{Q\,No\,Crack} \tag{B9}$$

Thus substituting **Eq. (B5) into (B8)**, and allowing the virtual forces to approach zero $(F \to 0)$, gives

$$\Delta_P = \Delta_{P\,No\,Crack} + \frac{2}{E'}\int_0^A \left(K_{IP}\frac{\partial K_{IP}}{\partial P} + K_{IIP}\frac{\partial K_{IIP}}{\partial P} + \alpha K_{IIIP}\frac{\partial K_{IIIP}}{\partial P}\right) dA$$

or (B10)

$$\Delta_F = \Delta_{F\,No\,Crack} + \frac{2}{E'}\int_0^A \left(K_{IP}\frac{\partial K_{IF}}{\partial F} + K_{IIP}\frac{\partial K_{IIF}}{\partial F} + \alpha K_{IIIP}\frac{\partial K_{IIIF}}{\partial F}\right) dA$$

Therefore displacements Δ_P or Δ_F may be computed from displacements with no crack present and a knowledge of K-formulas by using **Eq. (B10)**. This result adds interest in concentrated force K-formula solutions for forces such as F, as their displacement may be computed directly.

Special Case of Opening Displacements of Crack Surfaces

When computing relative displacements of crack surfaces, the displacements with no crack present are zero, that is,

$$\Delta_{No\,Crack} = 0 \tag{B11}$$

Moreover, if opening mode (Mode I) displacements are desired, as indicated in **Fig. (B2)**, then normally

$$\frac{\partial K_{IIF}}{\partial F} = \frac{\partial K_{IIIF}}{\partial F} = 0 \tag{B12}$$

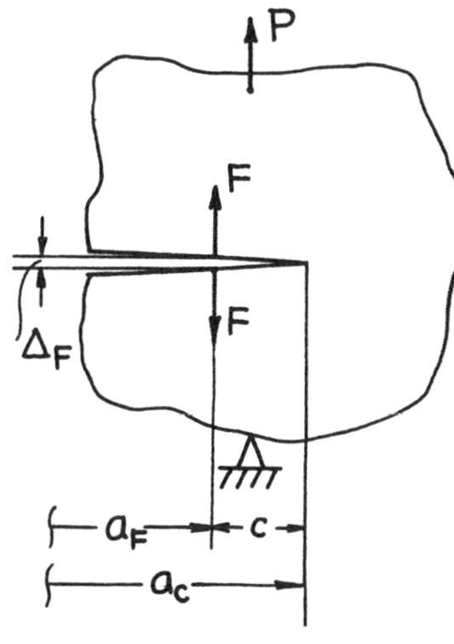

Fig. B2

Furthermore, until the crack, a, grows beyond the forces, F, it is observed that

$$\frac{\partial K_{IF}}{\partial F} = 0 \quad (a \leq a_F) \tag{B13}$$

Using **Eqs. (B11), (B12), and (B13)**, in conjunction with **Eq. (B10)**, gives

$$\Delta_F = \frac{2}{E'} \int_{a_F}^{a_c} K_{IP} \frac{\partial K_{IF}}{\partial F} da \tag{B14}$$

Now both factors in the integrand are functions of crack size a and load P (but not F). By inserting formulas for K_{IP} and K_{IF} in **Eq. (B14)**, it can be directly integrated for many cases. Integration can be assisted by series or polynomial expansions in difficult cases.

Opening Displacements Near a Crack Tip

For displacements near a crack tip, that is, $c << (a_F$ and other planar dimensions), then the form for K_{IF} is (see **page 3.6**)

$$K_{IF} = \frac{\sqrt{2}F}{\sqrt{\pi(a - a_F)}} \tag{B15}$$

The expression for K_{IP} may be replaced by a power series about a_F (or in the variable $a - a_F$)

$$K_{IP}(P, a) = K_{IP}(P, a_F) + \frac{\partial K_{IP}(P, a_F)}{\partial a} \cdot (a - a_F) + \frac{\partial^2 K_{IP}(P, a_F)}{\partial a^2} \cdot \frac{(a - a_F)^2}{2!} + \dots \tag{B16}$$

Substituting **Eqs. (B15) and (B16)** into **Eq. (B14)** with the variable $x = a - a_F$ results in

$$\Delta_F = \frac{2\sqrt{2}}{\sqrt{\pi}E'} \int_0^c \left[K_{IP}(P, a_F) + \frac{\partial K_{IP}(P, a_F)}{\partial a} \cdot x + \frac{\partial^2 K_{IP}(P, a_F)}{\partial a^2} \cdot \frac{x^2}{2!} + \dots \right] \frac{dx}{\sqrt{x}} \tag{B17}$$

Performing the integration

$$\Delta_F = \frac{4\sqrt{2}}{\sqrt{\pi}E'} \left[K_{IP}(P, a_F)\sqrt{c} + \frac{1}{3} \frac{\partial K_{IP}(P, a_F)}{\partial a} \cdot c^{3/2} + \dots \right] \tag{B18}$$

If c is very small compared with a_F and other planar dimensions, then $K_{IP}(P, a_F) \cong K_{IP}(P, a_c) = K_{IP}$, and neglecting second-order terms, **Eq. (B18)** becomes

$$\Delta_F = \frac{4\sqrt{2}}{\sqrt{\pi}E'} K_{IP}\sqrt{c} \tag{B19}$$

This result, **Eq. (B19)**, is an example of the use of **Eq. (B10)** to obtain a particular result, the elastic opening of a crack near its tip. Other examples are easy to add, such as those opening displacements of cracks used for experimental purposes when measurements are made with "clip gages" (e.g., see **pages 2.4, 2.8, 2.11, 2.17, 2.20**).

Example I

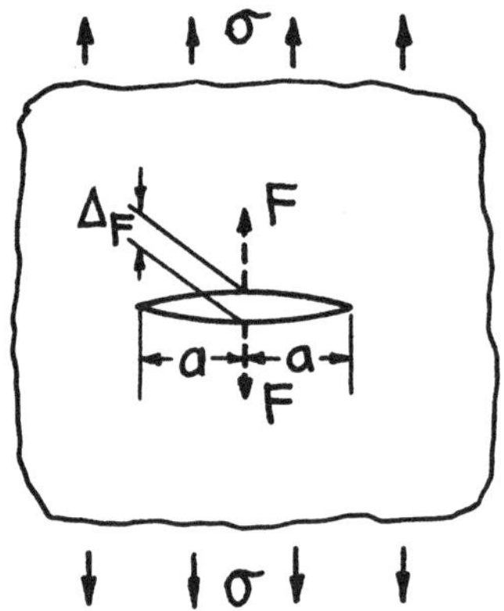

For the figure shown, the K-formulas (see **pages 5.1 and 5.9**) are

$$K_{I\sigma} = \sigma\sqrt{\pi a}$$
$$K_{IF} = \frac{F}{\sqrt{\pi a}}$$
$$K_{II} = K_{III} = 0$$

Note again that $\Delta_{no\,crack}$ is zero. Then **Eq. (B10)** reduces to

$$\Delta_F = \left\{\frac{2\sigma}{E'}\int_0^a da\right\} \times 2 = \frac{4\sigma a}{E'}$$

where the additional factor of 2 occurs because two crack tips contribute to the displacement computation.

Example II

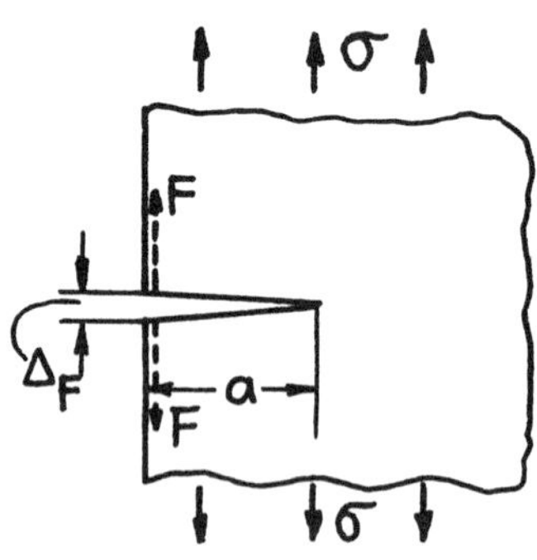

Similar to Example I (**pages 8.1 and 8.2**):

$$K_{I\sigma} = 1.1215\,\sigma\sqrt{\pi a}$$
$$K_{IF} = 1.30\frac{2F}{\sqrt{\pi a}}$$

Proceeding as before

$$\Delta_F = 5.83\frac{\sigma a}{E'}$$

Example III

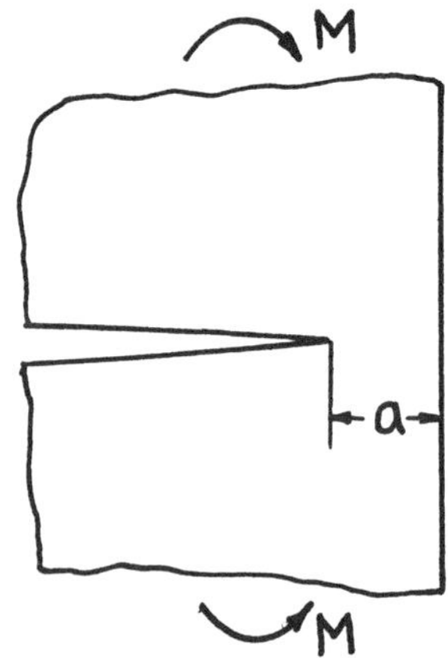

For a semi-infinite plate with a semi-infinite crack leaving a neck of width a, subjected to pure moment M, applied at infinity as shown (see **page 9.1**):

$$K_{IM} = \frac{3.975\,M}{a^{3/2}}$$
$$K_{II} = K_{III} = 0$$

Applying the first form of **Eq. (B10)** to obtain the relative rotation, θ_M, of applied moments at infinity, note that $\theta_{No\,Crack}$ is zero and then

$$\theta_M = \frac{2}{E'}\int_a^{\infty}\frac{(3.975)^2 M}{a^3}(-da) = \frac{15.8\,M}{E'\,a^2}$$

APPENDIX C

THE WEIGHT FUNCTION METHOD FOR DETERMINING STRESS INTENSITY FACTORS[1]

Bueckner (1970, 1971, 1972) devised a method of determining stress intensity factor solutions,[2] which has also been discussed by **Rice (1972)**. The method depends mainly on the reciprocal theorem and other energy-method-like considerations. The method may also be extended to computations of displacements in a manner almost identical to that in **Appendix B**. For an elegant presentation of the method in all generality, the reader is refered to **Bueckner (1970, 1971, 1972) and Rice (1972)**. Here, the most important results are presented in a simplified manner.

The following analysis shows that if the complete solution (for its stress intensity factor and displacements) to a crack problem for one loading system is known, then the solution (for K) for the same cracked configuration with any other loading may be obtained directly from the known solution. To show this, consider a cracked body with loads $P_1, P_2 \ldots, P_N$ as the independently applied loads. From previous results and definitions (see **pp. 1.6 – 1.8** and **Appendixes A and B**), the Griffith energy rate is

$$\mathcal{G} = \left.\frac{\partial U}{\partial a}\right|_P = \frac{1}{2}\sum_{i=1}^{N}\sum_{j=1}^{N}\frac{\partial C_{ij}(a)}{\partial a}P_iP_j = \frac{1}{2}\sum_{i=1}^{N}P_i\frac{\partial u_i}{\partial a} \tag{C1}$$

where the displacements of loading points, u_i, can be written in terms of elastic compliance coefficients, $C_{ij}(a)$, as functions of crack length, a

$$u_i = \sum_{j=1}^{N}\overset{j}{u}_i = \sum_{j=1}^{N}C_{ij}(a)P_j \tag{C2}$$

Because of the reciprocal theorem $C_{ij} = C_{ji}$, which was used in writing the above equations.

On the other hand, the Griffith energy rate may be written in terms of stress-intensity factors (see **p. 1.8**) as

$$\mathcal{G} = \frac{K^2}{E'} = \frac{1}{E'}\sum_{i=1}^{N}\sum_{j=1}^{N}k_i(a)k_j(a)P_iP_j \tag{C3}$$

[1] **Paris 1976.**

[2] **For earlier "Green's Function" from concentrated force solutions with results similar to weight functions, see Paris (1957, 1960), Barenblatt (1962), and Sih (1962b).**

since the stress intensity factor, K, is linearly dependent on the loads, P_i, or

$$K = \sum_{i=1}^{N} K_i = \sum_{i=1}^{N} k_i(a)P_i \tag{C4}$$

Equating the double sums in both results for $\mathcal{G}$ above,[3] that is, **(C1) and (C4)**, and noting that since this must be true for any values of the loads, $P_1, P_2, \ldots, P_N$, the coefficients must be identical, term by term, then

$$\frac{k_i(a)k_j(a)}{E'} = \frac{1}{2}\frac{\partial C_{ij}(a)}{\partial a} \tag{C5}$$

Let a full solution be known for just one of the loads, say P_m. Then, rearranging the latest result

$$k_i(a) = \frac{E'}{2}\frac{\partial C_{im}(a)}{\partial a}\frac{1}{k_m(a)}$$

or from **Eq. (C4)**

$$k_i(a) = \frac{E'}{2}\frac{\partial C_{im}(a)}{\partial a}\frac{P_m}{K_m} \tag{C6}$$

By saying the solution is known for a load, P_m, it means that K_m is known and C_{im} is known, since the displacements, u_i^m, for the load at P_m are presumed to be known and from **Eq. (C2)**

$$C_{im} = \frac{u_i^m}{P_m}$$

Combining **(C4) and (C6)**

$$K = \sum_{i=1}^{N} k_i(a)P_i = \frac{E'}{2}\frac{P_m}{K_m}\sum_{i=1}^{N}\frac{\partial C_{im}(a)}{\partial a}P_i \tag{C7}$$

Thus K can be found for loads, P_i, from results obtained from just one load, P_m. This is the desired result.

For arbitrary distributed tractions, $T(s)$, over a surface, s, instead of discrete forces, P_i, the form of the result **(C7)** becomes

$$K = \int f_m(s,a)T(s)ds \tag{C8}$$

where $f_m(s,a)$ is the "weight function" as determined entirely from the solution for a load (or loading system) characterized by "m." For this result, note that

$$f_m(s,a) = \frac{E'}{2K_m^{(a)}} \cdot \frac{\partial u^m(s,a)}{\partial a} \tag{C9}$$

where $u^m(s,a)$ are displacements at s in the direction of the tractions $T(s)$ but caused only by the loading system characterized by "m."

[3] As suggested by J. R. Rice, private communication, 1974.

A Special Method of Determining the Weight Function — Mode I

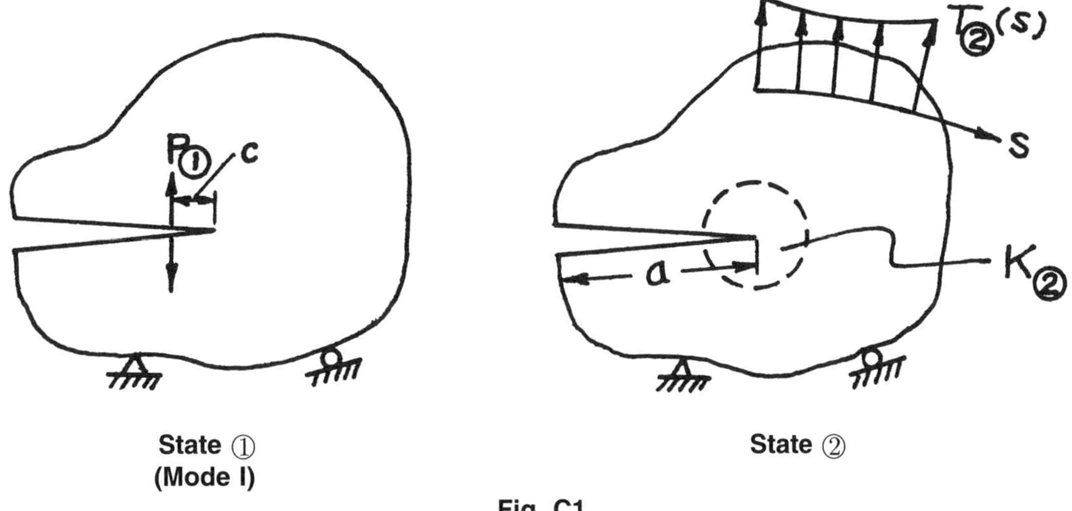

Fig. C1

In a two-dimensional problem, let loading state ① be the known loading state (corresponding to "m" above) where concentrated forces, $P_{\textcircled{1}}$, are applied on the crack surface at a distance, c, from the crack tip. Let loading state ②(for the same configuration) be one of arbitrary tractions for which it is desired to determine $K_{\textcircled{2}}$. By reciprocal theorem:

$$P_{\textcircled{1}}\, u_{\textcircled{2}} = \int T_{\textcircled{2}} u_{\textcircled{1}}\, ds \tag{C10}$$

where the displacements, u, form reciprocal work products (e.g., $u_{\textcircled{2}}$ is the displacement at the location of $P_{\textcircled{1}}$ in the direction of $P_{\textcircled{1}}$ but due to loading state ②).

Now, presume that the distance, c, from the crack tip to the loading forces in state ①is very small (i.e., approaching zero compared with other dimensions). Then the displacement $u_{\textcircled{2}}$ will be within the crack-tip stress field for state ②or (see **p. 1.2**, **Eq. (1)**, etc.)

$$u_{\textcircled{2}} = 2v\Big|_{\substack{r=c \\ \theta=\pi}} = \frac{4\sqrt{2}}{\sqrt{\pi}E'} K_{I\textcircled{2}}\sqrt{c} \tag{C11}$$

Note that, due to the symmetrical (with respect to the crack) force system selected for state ①, only Mode I fields contribute work-producing displacements, $u_{\textcircled{2}}$, thus the K being computed here is only the Mode I component. Substituting **(C11)** and rearranging **(C10)** gives

$$K_{I\textcircled{2}} = \frac{\sqrt{\pi}E'}{4\sqrt{2}} \frac{1}{P_{\textcircled{1}}\sqrt{c}} \int T_{\textcircled{2}} u_{\textcircled{1}}\, ds \tag{C12}$$

As c diminishes to zero, let $P_{①}\sqrt{c}$ remain a finite constant, choosing for later convenience

$$\frac{P_{①}\sqrt{c}}{\pi} = B_I \tag{C13}$$

Considering the results on page 3.6, this leads to a local situation in state ① of

$$Z_{I①}(z) = B_I \Big/ z^{3/2} \tag{C14}$$

As c approaches zero, the boundaries by comparison become infinitely far away. Again, substituting **(C13)** into **(C12)** gives

$$K_{I②} = \frac{E'}{4\sqrt{2\pi}B_I} \int T_{②} u^I_{①} ds \tag{15}$$

where $T_{②}$ are the applied tractions and $u^I_{①}$ are now the corresponding displacements without applied loads but with insertion of a local Mode I singularity of $z^{-3/2}$ type [4] of strength "B_I" as in **(C14)**, at the crack tip where $K_{②}$ is desired. The weight function is then

$$f_{\mathrm{Im}}(a,s) = \frac{E' u^I_{①}(a,s)}{4\sqrt{2\pi}B_I} \tag{C16}$$

Bueckner (1970, 1971, 1972) obtained a similar result by a less direct approach.

It is easy to visualize inserting the local singularity described by **(C14)** into a finite element scheme to determine resulting displacements, $u^I_{①}(a,s)$, for all mesh points with no other loads present. This has distinct advantages over other methods, as the solution generated applies to <u>all</u> possible loadings. Other results are possible using collocation or direct solutions using this method.

This derivation considers Mode I only, and the resulting stress intensity factor in **(C15)** is of a Mode I type. However, it is possible to replace state ① with its Mode II or Mode III counterparts (see **p. 3.6**) and rederive results for K_{II} and K_{III}, as well as K_I, as follows.

[4] **Let the $z^{-3/2}$ singularity be known as the "Bueckner Type," with its strength appropriately denoted as "B."**

Mode II and Mode III Weight Functions

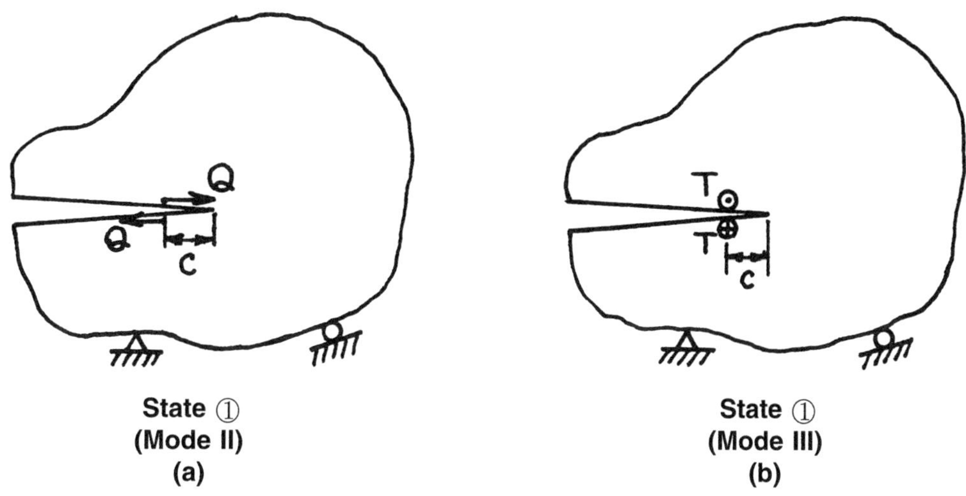

Fig. C2

By repeating the derivation of the preceding section, that is, **(C10)** through **(C16)**, but replacing state ①in **Fig. C1** with its Mode II or Mode III counterpart, as shown in **Fig. C2 (a) and (b)**, then weight functions may be developed directly for Mode II and Mode III stress intensity factors. The results are as follows.

Mode II

$$K_{II②} = \frac{E'}{4\sqrt{2\pi}B_{II}} \int T_{②} u^{II}_{①} ds \tag{C17}$$

where $u^{II}_{①}$ is now caused by inserting a local field of the Bueckner type at the crack tip of interest, that is,

$$Z_{II①}(z) = B_{II} \Big/ z^{3/2} \tag{C18}$$

Mode III

$$K_{III②} = \frac{G}{2\sqrt{2\pi}B_{III}} \int T_{②} u^{III}_{①} ds \tag{C19}$$

where $u^{III}_{①}$ is caused by inserting the Bueckner singularity

$$Z_{III①}(z) = B_{III} \Big/ z^{3/2} \tag{C20}$$

Therefore it is equally easy to insert Mode II and Mode III singularities in problems to obtain the Mode II and Mode III weight functions

$$f_{IIm}(a,s) = \frac{E' u^{II}_{①}(a,s)}{4\sqrt{2\pi}B_{II}} \tag{C21}$$

and

$$f_{IIIm}(a,s) = \frac{G u_{①}^{III}(a,s)}{2\sqrt{2\pi}B_{III}} \tag{C22}$$

where now **Eq. (C8)** may be applied individually for each of the three modes of stress intensity factors.

Near Tip Bueckner Displacement Fields

The displacement fields may be computed for Bueckner-type singularities, that is [see **Eqs. (C14), (C18) and (C20)**],

$$Z(z) = B \Big/ z^{3/2}$$

for each mode, making use of the usual r, θ coordinates (see **Fig. 2, p. 1.2**) and **Eqs. (39), (56), or (59)** as is appropriate for each mode. The results (for plane strain for Modes I and II) are as follows.

Mode I

$$\begin{aligned} u &= \frac{B_I}{G\sqrt{r}}\cos\frac{\theta}{2}\left[2\nu - 1 + \sin\frac{\theta}{2}\sin\frac{3\theta}{2}\right] \\ v &= \frac{B_I}{G\sqrt{r}}\sin\frac{\theta}{2}\left[2 - 2\nu - \cos\frac{\theta}{2}\cos\frac{3\theta}{2}\right] \\ w &= 0 \end{aligned} \tag{C23}$$

Mode II

$$\begin{aligned} u &= \frac{B_{II}}{G\sqrt{r}}\sin\frac{\theta}{2}\left[2 - 2\nu + \cos\frac{\theta}{2}\cos\frac{3\theta}{2}\right] \\ v &= \frac{B_{II}}{G\sqrt{r}}\cos\frac{\theta}{2}\left[1 - 2\nu + \sin\frac{\theta}{2}\sin\frac{3\theta}{2}\right] \\ w &= 0 \end{aligned} \tag{C24}$$

Mode III

$$\begin{aligned} u &= 0 \\ v &= 0 \\ w &= \frac{B_{III}}{G\sqrt{r}} 2\sin\frac{\theta}{2} \end{aligned} \tag{C25}$$

For plane stress for Modes I and II, replace ν by $\nu/(1+\nu)$ and also note that $w \neq 0$.

For generating weight functions, these fields should be inserted locally at the crack tip where K is desired. They are then the actual weight function displacements that should be used for loads near the crack tip.

For a semi-infinite crack in an infinite body, these fields are the weight function displacements for the whole problem (two-dimensional problems). Some examples follow, in which these results are used for very simple problems so as to illustrate the method. The power of the method, however, is only fully appreciated with more complicated problems, when combined with numerical, finite element, and other procedures.

Closed-Form Weight Functions

Closed forms for Westergaard stress functions, Z and $\overline{Z}$, can be written to form weight function displacement fields throughout a body. The technique of finding such stress functions is much the same as for normal crack stress analysis problems, except that the crack tip for which the weight function is desired will have a $z^{-3/2}$ (in Z) of strength, B_I. Some examples are as follows (each applies to all three modes, I, II, III):

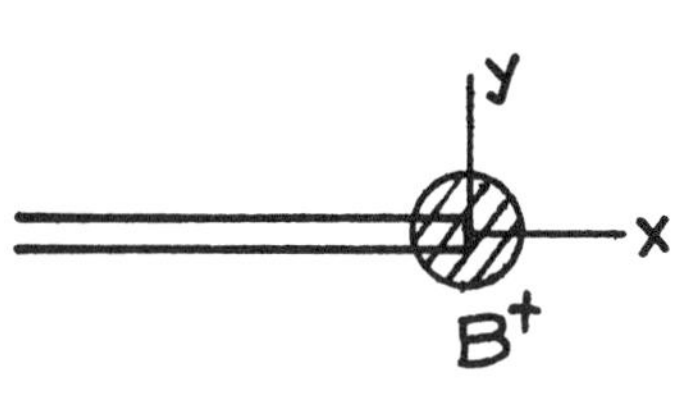

$$\begin{Bmatrix} Z_I(z) \\ Z_{II}(z) \\ Z_{III}(z) \end{Bmatrix} = \begin{Bmatrix} B_I \\ B_{II} \\ B_{III} \end{Bmatrix} \frac{1}{z^{3/2}}$$

$$\begin{Bmatrix} \overline{Z}_I(z) \\ \overline{Z}_{II}(z) \\ \overline{Z}_{III}(z) \end{Bmatrix} = -\begin{Bmatrix} B_I \\ B_{II} \\ B_{III} \end{Bmatrix} \frac{2}{z^{1/2}}$$

NOTE: All Z and $\overline{Z}$ expressions below apply to all three modes as indicated here.

$$Z(z) = \frac{iB}{z^{3/2}}$$

$$\overline{Z}(z) = -\frac{2iB}{z^{1/2}}$$

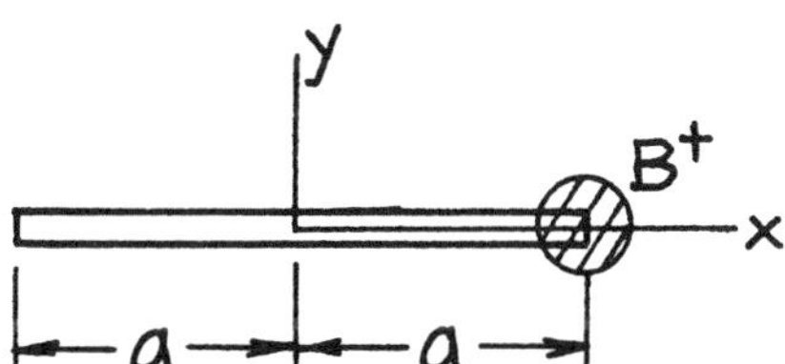

$$Z(z) = \frac{B\sqrt{2a}}{(z-a)^{3/2}(z+a)^{1/2}}$$

$$\overline{Z}(z) = -B\sqrt{\frac{2}{a}}\left(\frac{z+a}{z-a}\right)^{1/2}$$

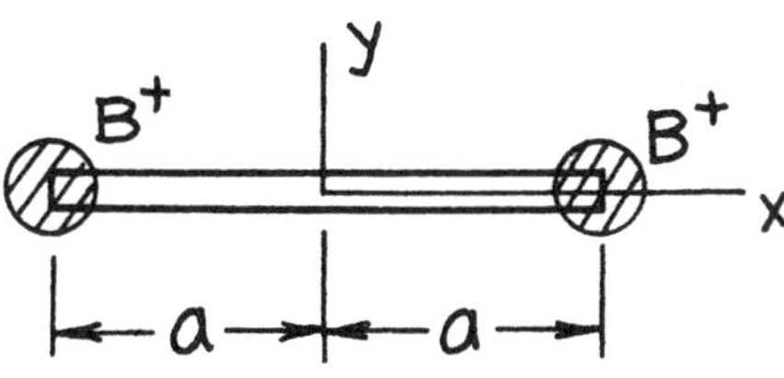

symmetric about y-axis

$$Z(z) = \frac{2B\sqrt{2a}\,z}{\left(z^2-a^2\right)^{3/2}}$$

$$\overline{Z}(z) = \frac{-2B\sqrt{2a}}{\left(z^2-a^2\right)^{1/2}}$$

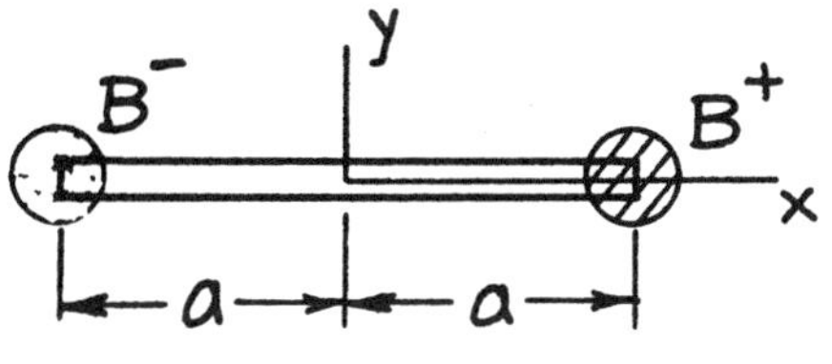

skew-symmetric about y-axis

$$Z(z) = \frac{B(2a)^{3/2}}{\left(z^2-a^2\right)^{3/2}}$$

$$\overline{Z}(z) = \frac{-2B\sqrt{2a}}{\left(z^2-a^2\right)^{1/2}} \cdot \frac{z}{a}$$

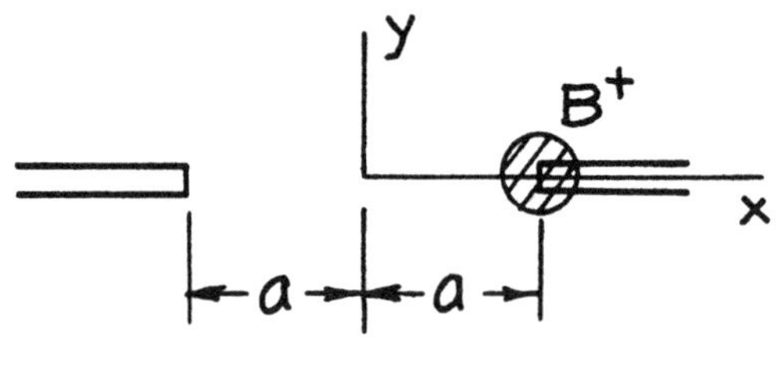

$$Z(z) = \frac{B\sqrt{2a}}{(a-z)^{3/2}(a+z)^{1/2}}$$

$$\overline{Z}(z) = B\sqrt{\frac{2}{a}}\left(\frac{a+z}{a-z}\right)^{1/2}$$

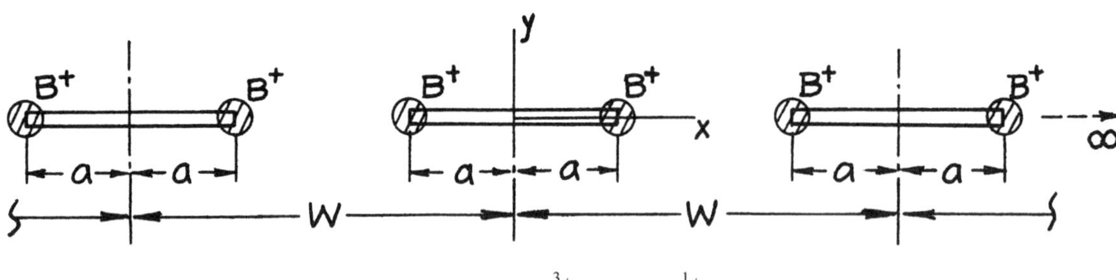

$$Z(z) = \frac{2B\left(\frac{\pi}{W}\cos\frac{\pi a}{W}\right)^{3/2}\left(2\sin\frac{\pi a}{W}\right)^{1/2}\sin\frac{\pi z}{W}}{\left(\sin\frac{\pi z}{W}-\sin\frac{\pi a}{W}\right)^{3/2}\left(\sin\frac{\pi z}{W}+\sin\frac{\pi a}{W}\right)^{3/2}}$$

$$\overline{Z}(z) = \frac{-2B\left(\frac{2\pi}{W}\tan\frac{\pi a}{W}\right)^{1/2}\cos\frac{\pi z}{W}}{\left[\left(\sin\frac{\pi z}{W}\right)^2-\left(\sin\frac{\pi a}{W}\right)^2\right]^{1/2}}$$

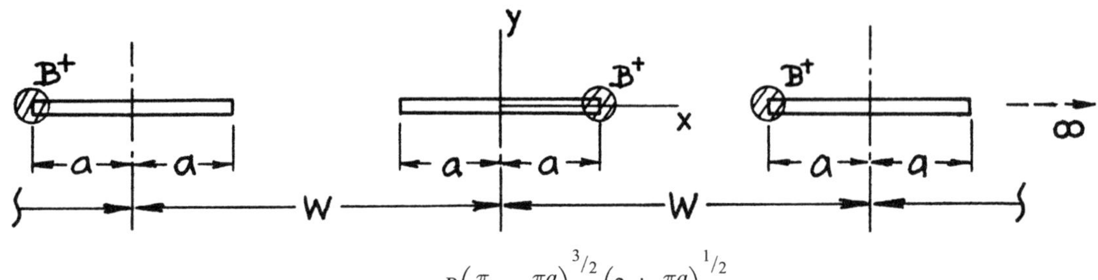

$$Z(z) = \frac{B\left(\frac{\pi}{W}\cos\frac{\pi a}{W}\right)^{3/2}\left(2\sin\frac{\pi a}{W}\right)^{1/2}}{\left(\sin\frac{\pi z}{W}-\sin\frac{\pi a}{W}\right)^{3/2}\left(\sin\frac{\pi z}{W}+\sin\frac{\pi a}{W}\right)^{1/2}}$$

$$\overline{Z}(z) = B\left\{\frac{-\left(\frac{2\pi}{W}\tan\frac{\pi a}{W}\right)^{1/2}\cos\frac{\pi z}{W}}{\left[\left(\sin\frac{\pi z}{W}\right)^2-\left(\sin\frac{\pi a}{W}\right)^2\right]^{1/2}} + i\ \cot\frac{\pi a}{W}\left(\frac{\pi}{W}\sin\frac{2\pi a}{W}\right)^{1/2}\Pi_c\left[\sin^{-1}\left(\frac{\sin\frac{\pi z}{W}}{\sin\frac{\pi a}{W}}\right), 1, \sin\frac{\pi a}{W}\right]\right\}$$

$$\Pi_c[\varphi, n, k] = \int\frac{d\varphi}{\left(1-n\ \sin^2\varphi\right)\left(1-k^2\ \sin^2\varphi\right)^{1/2}}$$

An Example of Finite Element Results

As mentioned earlier, weight function displacements can be obtained using finite element analysis by putting a small hole at the crack tip and inserting the Bueckner-type field as boundary conditions on the hole, that is, by using **Eqs. (C23) [or (C24) or (C25)]** as boundary conditions on the hole, with all other surfaces stress free, in order to determine the state ① displacements, $u_{①}$.

As an example, consider the single edge cracked strip for which a variety of loadings are avaliable with previously tabulated results (see **Paris 1975a** and **pages 2.10, 2.13, 2.16, 2.27, 2.29**, etc.). The particular strip selected here (see the figure on **page 2.10**) is $a = b/2$, $2h = 6b$, with a circular hole at the crack of radius $r = .006944b$ and $\nu = 0.3$. A mesh of 352 elements and 398 nodes was used, as shown in Fig. C3 (with the inner circles of elements removed and enlarged for clarity).

Table C1 gives results obtained by simply inserting the displacement field, **Eqs. (C23)**, at 25 points on the upper half of the hole, and constraining points on the crack plane ahead of the crack to remain on that plane. As implied, only one-half of the problem requires treatment. The points on the table are numbered as indicated on the mesh in Fig. C3 and located by coordinates x/b and y/b. The corresponding components of the displacements, $u_{①}$, for use in the integral to obtain $K_{I②}$, are

$$\bar{u}/b = \frac{E'\left(u_{①}\right)_x}{2\sqrt{2\pi}B_I b}$$

$$\bar{v}/b = \frac{E'\left(u_{①}\right)_y}{2\sqrt{2\pi}B_I b}$$

Thus $K_{I②}$ may be obtained from components of tractions for one-half of a problem (with respect to crack-plane symmetry) from

$$K_{I②} = \int (T_x\bar{u} + T_y\bar{v})ds$$

Using this formula and the table, the following accuracies are observed in comparative results for various loadings:

a. Uniform tension applied at the ends — 2.9% (see **p. 2.10**)
b. Uniform pressure on the crack surface — <4.9%
c. Pure bending (moment applied anywhere beyond $y/b = 1.5$) — 3.2% (see **p. 2.13**)
d. Four-point bending (with loads at $y/b = 1$ and 3) — 3.2%
e. Three-point bending (with a span of $4b$) — 3.6% (see **p. 2.16**)

This is only an example and better accuracies may be produced by finer finite element meshes. However, the table allows the reader to calculate K_I results of comparable accuracies for any loading for this configuration. Its advantages and generality are apparent.

The same method applies to Modes II and III simply by using the proper fields, **Eqs. (C24) and (C25)**. Moreover, it is also applicable to residual stress, thermal stress, and body force problems via simple superposition or direct application. The method is also applicable to three-dimensional finite element analysis.

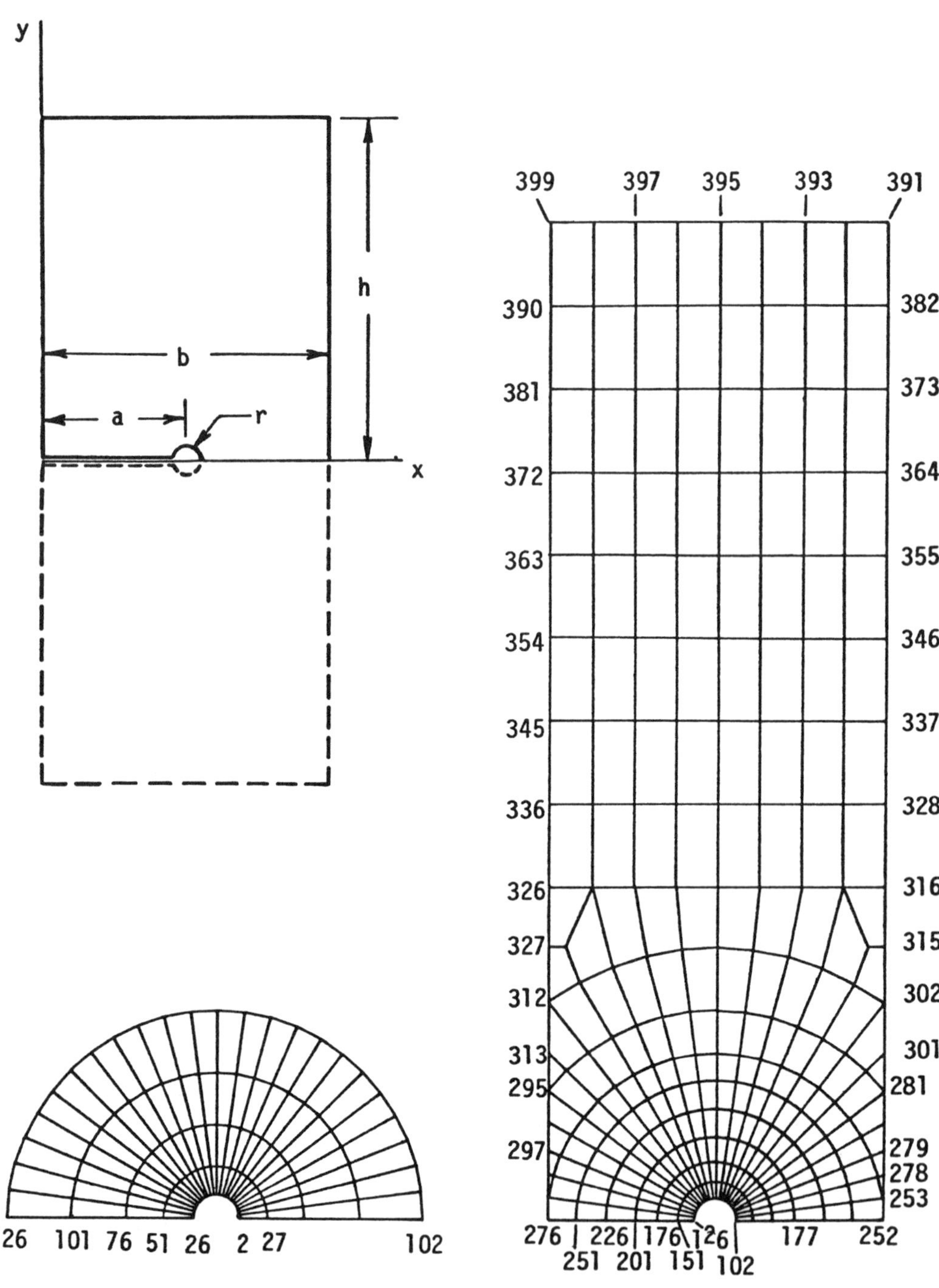
y
h
b
a
r
x
399
397
395
393
391
390
382
381
373
372
364
363
355
354
346
345
337
336
328
326
316
327
315
312
302
313
301
295
281
297
279
278
253
126
101
76
51
26
2
27
102
276
226
176
126
177
252
251
201
151
102

Fig. C3

Weight Function Displacements for an Edge Cracked Strip

$$\left(\frac{a}{b} = 0.5,\ \frac{h}{b} = 3,\ \nu = 0.3,\ \frac{r}{b} = .006944\right)$$

Point	x/b	y/b	$\bar{u}/b$	$\bar{v}/b$
2	0.5069	0	- 2.736	0
27	0.5156	0	- 1.767	0
102	0.5625	0	- 0.595	0
177	0.75	0	0.285	0
26	0.4931	0	- 0.000003	9.575
51	0.4844	0	- 0.232	7.066
76	0.4722	0	- 0.269	5.930
101	0.4566	0	- 0.297	5.395
126	0.4375	0	- 0.319	5.187
151	0.3888	0	- 0.319	5.229
176	0.3264	0	- 0.346	5.670
201	0.25	0	- 0.368	6.374
226	0.1666	0	- 0.385	7.214
251	0.0833	0	- 0.393	8.084
276	0	0	- 0.396	8.960
297	0	0.207	1.795	8.971
295	0	0.3837	3.719	8.956
313	0	0.5	4.988	8.933
312	0	0.6516	6.634	8.908
327	0	0.820	8.453	8.890
326	0	1.00	10.401	8.887
336	0	1.25	13.116	8.887
345	0	1.50	15.838	8.889
354	0	1.75	18.563	8.891
363	0	2.00	21.289	8.892
372	0	2.25	24.015	8.892
381	0	2.50	26.740	8.892
390	0	2.75	29.466	8.892
399	0	3.00	32.191	8.892
397	0.25	3.00	32.191	6.166
395	0.50	3.00	32.191	3.441
393	0.75	3.00	32.191	0.715
391	1.0	3.00	32.191	-2.010
382	1.0	2.75	29.466	-2.010
373	1.0	2.50	26.740	-2.010
364	1.0	2.25	24.015	-2.010
355	1.0	2.00	21.289	-2.010
346	1.0	1.75	18.564	-2.011
337	1.0	1.50	15.838	-2.012
328	1.0	1.25	13.109	-2.015
316	1.0	1.0	10.371	-2.017
315	1.0	0.820	8.386	-2.015
302	1.0	0.6516	6.515	-1.980
301	1.0	0.50	4.819	-1.912
281	1.0	0.3837	3.525	-1.744
279	1.0	0.207	1.774	-1.249
278	1.0	0.134	1.228	-0.865
253	1.0	0.0653	0.904	-0.437
252	1.0	0	0.805	0

Closed-Form Examples

Example (1)

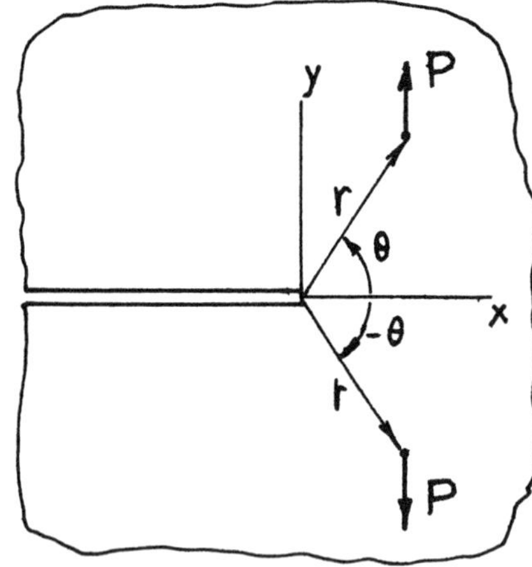

Consider an infinite sheet with a semi-infinite crack subjected to a pair of equal and opposite forces (per unit thickness), P, as shown. The stress solution with a Bueckner-type Mode I singularity (and no loading forces) is

$$Z_{I①}(z) = \frac{B_I}{z^{3/2}} \text{(Any } z\text{)}$$

The stress intensity factor for the loads shown may be computed from

$$K_I = \frac{E'}{4\sqrt{2\pi}B_I} \int T_{②} u^I_{①} ds$$

Now because of x-axis symmetry

$$\int T_{②} u^I_{①} ds = 2P \cdot v_{①}(r, \theta)$$

From **Eqs. (39)**, for plane strain conditions,

$$2Gv_{①} = 2(1-\nu)\ \text{Im}\bar{Z}_{I①} - y\ \text{Re}Z_{I①}$$

where for plane stress, ν can be replaced by $\nu/(1+\nu)$. Substituting $Z_{I①}$ and taking $z = re^{i\theta}$, etc. [see also **Eqs. (C23)**]

$$v_{①} = \frac{B_I}{G\sqrt{r}} \sin\frac{\theta}{2}\left[2(1-\nu) - \cos\frac{\theta}{2}\cos\frac{3\theta}{2}\right]$$

Continuing the substitutions

$$K_I = \frac{P}{(1-\nu)\sqrt{2\pi r}} \sin\frac{\theta}{2}\left[2(1-\nu) - \cos\frac{\theta}{2}\cos\frac{3\theta}{2}\right]$$

This result is a somewhat simpler form of the solution of **page 3.4**.

For the special case $r = b$ and $\theta = \pi$, the result is the same as **page 3.6**, or

$$K_I = \frac{2P}{\sqrt{2\pi b}}$$

and for the special case $r = y_0$ and $\theta = \frac{\pi}{2}$, the result is the same as **page 3.5**, or

$$K_I = \frac{P}{\sqrt{\pi y_0}}\left[\frac{5-4\nu}{4-4\nu}\right]$$

Therefore in this case getting the K_I through the weight function method gives direct algebraic access to a simpler form for some uses.

Note also from (C24) and (C25) that no Mode II or Mode III work is done by the forces, P, with these displacements. Thus $K_{II} = K_{III} = 0$.

Example (2)

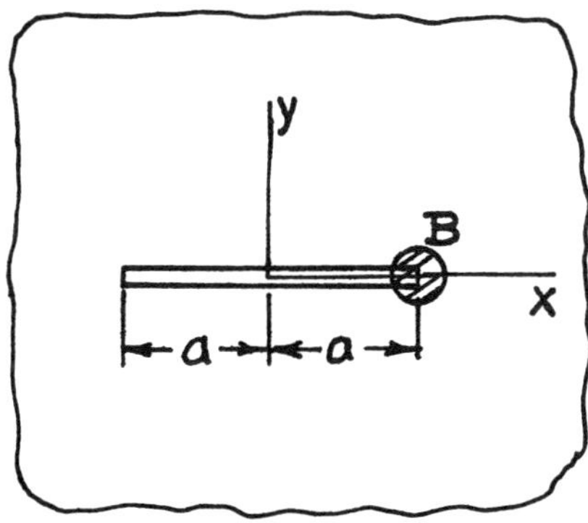

Consider a finite crack of length 2a, with a Bueckner-type singularity at one end. The stress function is [as verified from **page 5.9** and **Eq. (C13)**]

$$Z_{I①}(z) = \frac{B_I\sqrt{2a}}{(z-a)^{3/2}(z+a)^{1/2}}$$

Consider now the problem of uniform stress, σ, parallel to the y-axis at infinity. Then

$$\int T_{②} u^I_{①}\, ds = 2\int_{-\infty}^{\infty} \sigma v \,\Big|_{y\gg a}\, dx$$

At great distances from the crack ($y>>a$)

$$Z_{I①\infty}(z) \to \frac{B_I\sqrt{2a}}{z^2}$$

and integrating

$$\bar{Z}_{I①\infty}(z) \to -\frac{B_I\sqrt{2a}}{z}$$

Substituting these expressions for Z and $\bar{Z}$ into the expression for ν [**Eqs. (39)**, **page 1.13**] gives

$$\int T_{②} u^I_{①}\, ds = \frac{4\pi\sigma B_I\sqrt{2a}}{E'}$$

Thus from **Eq. (C15)**

$$K_I = \sigma\sqrt{\pi a}$$

Example (3)

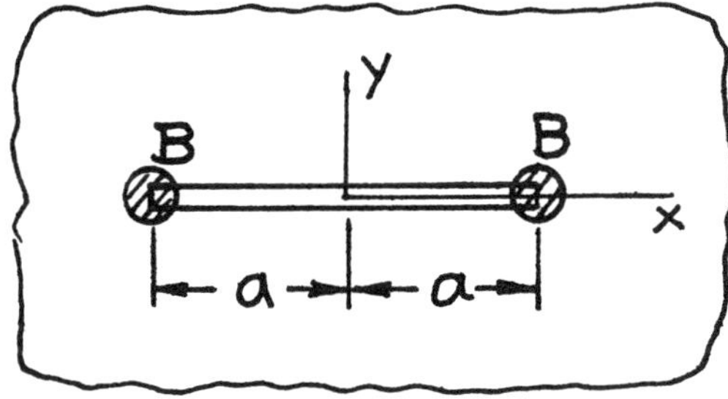

For problems like Example (2), with y-axis symmetry, it is permitted to insert Bueckner singularities at both crack tips or

$$Z_{I①}(z) = \frac{B_I\sqrt{2a}\,2z}{\left(z^2-a^2\right)^{3/2}}$$

Note that for $z>>a$, the stress function becomes twice the result in the previous example, that is,

$$Z_{I①\infty}(z) \to \frac{2B_I\sqrt{2a}}{z^2}(z\gg a)$$

where the analysis would follow in the same manner for $2K_I$ instead of K_I, as two crack tips are simultaneously considered.

However, here it is very easy to integrate $Z_{I①}$ (for any z) to obtain

$$\overline{Z}_{I①}(z) = -\frac{2B_I\sqrt{2a}}{\left(z^2 - a^2\right)^{1/2}} \qquad (\text{Any } z)$$

Hence it is possible to compute the weight function displacements easily for any point.

As a special case consider the displacement only on the crack surface, that is,

$$2Gv = 2(1-\nu)\,\mathrm{Im}\overline{Z}_I + 0 \quad (y = 0, |x| < a)$$

$$2Gu = 0$$

or, substituting for $\overline{Z}$

$$v = \frac{2(1-\nu)B_I\sqrt{2a}}{G}\left(a^2 - x^2\right)^{-1/2}$$

For a crack surface loaded by compressive stresses symmetric with respect to both axes, $\sigma(x) = \sigma(-x)$, then

$$2K_I = \frac{E'}{4\sqrt{2\pi}B_I}\int_0^a 2\sigma(x)\cdot 2v(x)dx$$

which upon substitution gives

$$K_I = \frac{2\sqrt{a}}{\sqrt{\pi}}\int_0^a \frac{\sigma(x)}{\left(a^2 - x^2\right)^{1/2}}dx$$

and

$$K_{II} = 0$$

which are well known results.

As another application consider equal and opposite forces P in the y direction at the origin, or

$$2K_I = \frac{E'}{4\sqrt{2\pi}B_I}P\cdot 2v\Big|_{\substack{x=0\\y=0}}$$

which gives

$$K_I = \frac{P}{\sqrt{\pi a}}$$

A result that is easily checked by **page 5.9**

Example (4)

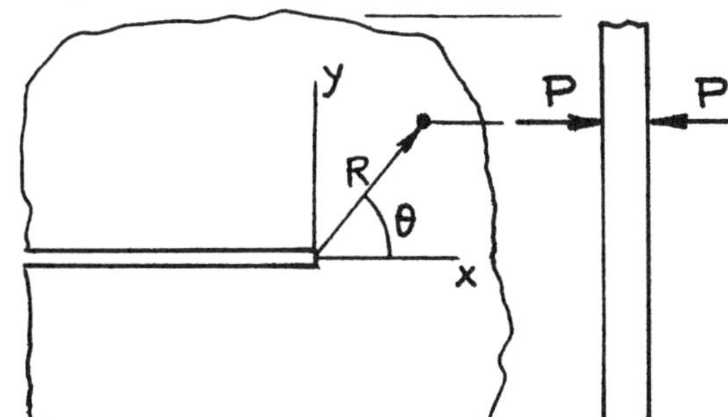

Consider a crack tip in a sheet that is subject to pinching forces P as shown. The displacement per unit thickness is

$$u_{①} = \varepsilon_z = \frac{-\nu}{E}(\sigma_x + \sigma_y)$$

which from **Eqs. (38) and (55)** becomes

Mode I

$$u_{①} = -\frac{2\nu}{E}\ \mathrm{Re}Z_I$$

Mode II

$$u_{①} = -\frac{2\nu}{E}\ \mathrm{Im}Z_{II}$$

and for both modes

$$Z(z) = \frac{B}{z^{3/2}} = \frac{B}{r^{3/2}}\left(\cos\frac{3\theta}{2} - i\ \sin\frac{3\theta}{2}\right)$$

where substituting these results into (C15) and (C17) gives

$$K_I = \frac{\nu P}{2\sqrt{2\pi}R^{3/2}}\cos\frac{3\theta}{2}$$

$$K_{II} = \frac{-\nu P}{2\sqrt{2\pi}R^{3/2}}\sin\frac{3\theta}{2}$$

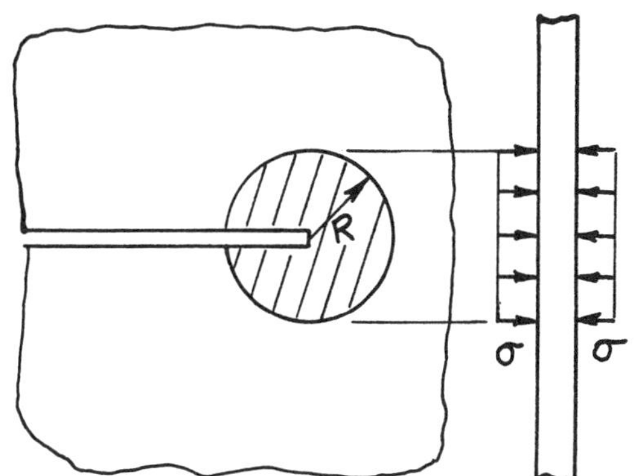

These expressions can be used to obtain results for a crack tip centered in a zone of radius R and pinching stress σ to give

$$K_I = \frac{-4\nu}{3\sqrt{2\pi}}\sigma\sqrt{R}$$

$$K_{II} = 0$$

Many other solutions for pinching load problems are compiled in **Appendix E**.

Example (5)

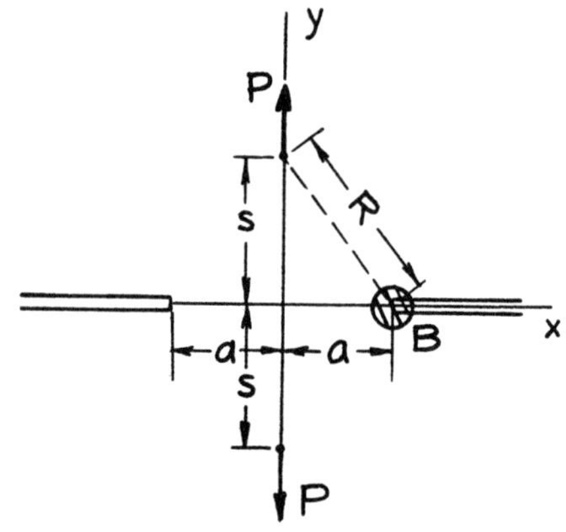

Consider two opposing semi-infinite cracks in a plate leaving a neck of width $2a$, loaded by opposing concentrated forces P, each centered on the neck at a distance s from the plane of the neck. The stress function solution with a Bueckner-type-singularity at the right-hand crack tip is

$$Z(z) = \frac{B\sqrt{2a}}{(a-z)^{3/2}(a+z)^{1/2}}$$

$$\overline{Z}(z) = \int Z(z)dz = B\sqrt{\frac{2}{a}}\left(\frac{a+z}{a-z}\right)^{1/2}$$

Now from **Eq. (C15)**

$$K = \frac{E'}{4\sqrt{2\pi}B}(2Pv)$$

where v is the displacement at the load point to be computed using the above Z and $\overline{Z}$ in **Eq. (39)** or

$$2\,Gv = 2(1-\nu)\ \mathrm{Im}\overline{Z} - y\ \mathrm{Re}Z$$

The calculation could be carried out for any position of the loading forces (not necessarily on the y-axis), but the algebra becomes messy. Thus for the location of loads as shown in the figure

$$\mathrm{Im}\overline{Z} = B\sqrt{\frac{2}{a}}\frac{s}{\left(a^2+s^2\right)^{1/2}}$$

and

$$\mathrm{Re}Z = \frac{B\sqrt{2a}\,a}{\left(a^2+s^2\right)^{3/2}}$$

Defining the distance between a load point and a crack tip as R or $R^2 = a^2 + s^2$, and combining expressions

$$K = \frac{P}{\sqrt{\pi a}}\left[\frac{s}{R} + \frac{(1+\nu)}{2}\frac{a^2 s}{R^3}\right]$$

This result is noted as a simpler way to obtain K compared with the solution on **page 4.3** (and **pages 4.9, 4.11**, etc.).

APPENDIX D

ANISOTROPIC LINEAR-ELASTIC CRACK-TIP STRESS FIELDS

Hooke's law for a homogeneous (rectilinear) anisotropic material is

$$\left.\begin{aligned}
\varepsilon_x &= a_{11}\sigma_x + a_{12}\sigma_y + a_{13}\sigma_z + a_{14}\tau_{yz} + a_{15}\tau_{zx} + a_{16}\tau_{xy} \\
\varepsilon_y &= a_{21}\sigma_x + \cdots \\
\varepsilon_z &= a_{31}\sigma_x + \cdots \\
\gamma_{yz} &= a_{41}\sigma_x + \cdots \\
\gamma_{zx} &= a_{51}\sigma_x + \cdots \\
\gamma_{xy} &= a_{61}\sigma_x + a_{62}\sigma_y + a_{63}\sigma_z + a_{64}\tau_{yz} + a_{65}\tau_{zx} + a_{66}\tau_{xy}
\end{aligned}\right\} \qquad \text{(D1)}$$

where, from reciprocity, $a_{ij} = a_{ji}$.

The generalized Hooke's law may be reduced by imposing conditions of plane strain (or stress) and pure shear to lead to tractable problems of determining crack-tip stress fields. Proceeding as in **Sih (1965b) or Paris (1965)**, the crack-tip stress field equations can be written as follows.

For plane strain

$$\left.\begin{aligned}
\sigma_x &= \frac{K_{Ia}}{\sqrt{2\pi r}}\,\mathrm{Re}\left\{\frac{\mu_1\mu_2}{\mu_1-\mu_2}\left(\frac{\mu_2}{\sqrt{\cos\theta+\mu_2\sin\theta}}-\frac{\mu_1}{\sqrt{\cos\theta+\mu_1\sin\theta}}\right)\right\} \\
&\quad+\frac{K_{IIa}}{\sqrt{2\pi r}}\,\mathrm{Re}\left\{\frac{1}{\mu_1-\mu_2}\left(\frac{\mu_2^2}{\sqrt{\cos\theta+\mu_2\sin\theta}}-\frac{\mu_1^2}{\sqrt{\cos\theta+\mu_1\sin\theta}}\right)\right\} \\
\sigma_y &= \frac{K_{Ia}}{\sqrt{2\pi r}}\,\mathrm{Re}\left\{\frac{1}{\mu_1-\mu_2}\left(\frac{\mu_1}{\sqrt{\cos\theta+\mu_2\sin\theta}}-\frac{\mu_2}{\sqrt{\cos\theta+\mu_1\sin\theta}}\right)\right\} \\
&\quad+\frac{K_{IIa}}{\sqrt{2\pi r}}\,\mathrm{Re}\left\{\frac{1}{\mu_1-\mu_2}\left(\frac{1}{\sqrt{\cos\theta+\mu_2\sin\theta}}-\frac{1}{\sqrt{\cos\theta+\mu_1\sin\theta}}\right)\right\} \\
\tau_{xy} &= \frac{K_{Ia}}{\sqrt{2\pi r}}\,\mathrm{Re}\left\{\frac{\mu_1\mu_2}{\mu_1-\mu_2}\left(\frac{1}{\sqrt{\cos\theta+\mu_1\sin\theta}}-\frac{1}{\sqrt{\cos\theta+\mu_2\sin\theta}}\right)\right\} \\
&\quad+\frac{K_{IIa}}{\sqrt{2\pi r}}\,\mathrm{Re}\left\{\frac{1}{\mu_1-\mu_2}\left(\frac{\mu_1}{\sqrt{\cos\theta+\mu_1\sin\theta}}-\frac{\mu_2}{\sqrt{\cos\theta+\mu_2\sin\theta}}\right)\right\}
\end{aligned}\right\} \qquad \text{(D2)}$$

where (other corresponding components of stress and displacement omitted) μ_1 and μ_2 are from each of the conjugate pairs of roots of

$$A_{11}\mu^4 - 2A_{16}\mu^3 + (2A_{12}+A_{66})\mu^2 - 2A_{26}\mu + A_{22} = 0 \qquad \text{(D3)}$$

and where A_{ij} are obtained by using the restrictions on strain to eliminate the appearance of z-components of stress to give

$$\left.\begin{aligned}\varepsilon_x &= A_{11}\sigma_x + A_{12}\sigma_y + A_{16}\tau_{xy}\\ \varepsilon_y &= A_{21}\sigma_x + A_{22}\sigma_y + A_{26}\tau_{xy}\\ \gamma_{xy} &= A_{61}\sigma_x + A_{62}\sigma_y + A_{66}\tau_{xy}\end{aligned}\right\} \quad \text{(D4)}$$

where again, $A_{ij} = A_{ji}$.

For pure shear

$$\left.\begin{aligned}\tau_{yz} &= \frac{K_{IIIa}}{\sqrt{2\pi r}}\,\mathrm{Re}\left\{\frac{1}{\sqrt{\cos\theta+\mu_5\ \sin\theta}}\right\}\\ \tau_{xz} &= \frac{-K_{IIIa}}{\sqrt{2\pi r}}\,\mathrm{Re}\left\{\frac{\mu_5}{\sqrt{\cos\theta+\mu_5\ \sin\theta}}\right\}\end{aligned}\right\} \quad \text{(D5)}$$

where μ_5 is one of the conjugate roots of

$$A_{55}\mu^2 - 2A_{45}\mu + A_{44} = 0 \quad \text{(D6)}$$

and where A_{ij} are obtained by using the restrictions for pure shear to give

$$\left.\begin{aligned}\gamma_{yz} &= A_{44}\tau_{yz} + A_{45}\tau_{xz}\\ \gamma_{xz} &= A_{45}\tau_{yz} + A_{55}\tau_{xz}\end{aligned}\right\} \quad \text{(D7)}$$

The resulting stress intensity factors, K_{Ia}, K_{IIa}, and K_{IIIa}, as defined above, are represented by K-formulas for the corresponding isotropic boundary value problems, except for cases of non-self-equilibrating loadings. Refer to **Sih (1965b) or Paris (1965)** for complete details.

Finally, for the Hooke's laws defined above, the elastic coefficients, C, relating energy rates to stress intensity factors are

$$\mathcal{G}_i = CK_i^2 \quad \text{(D8)}$$

Values of C (for the Case of Plane Stress[1])

Mode i	Isotropic	Orthotropic $A_{16} = A_{26} = A_{45} = 0$	Anisotropic[2]
I	$1/E$	$\sqrt{\frac{A_{11}A_{22}}{2}}\left[\sqrt{\frac{A_{22}}{A_{11}}} + \frac{2A_{12}+A_{66}}{2A_{11}}\right]^{1/2}$	$-\frac{A_{11}}{2}\,\mathrm{Im}\left(\frac{\mu_1+\mu_2}{\mu_1\mu_2}\right)$
II	$1/E$	$\frac{A_{11}}{\sqrt{2}}\left[\sqrt{\frac{A_{22}}{A_{11}}} + \frac{2A_{12}+A_{66}}{2A_{11}}\right]^{1/2}$	$\frac{A_{11}}{2}\,\mathrm{Im}(\mu_1+\mu_2)$
III	$(1+\nu)/E$	$\frac{1}{2}\sqrt{A_{44}A_{55}}$	$\frac{1}{2}\frac{\left(A_{44}A_{55}-A_{45}^2\right)^{3/2}}{A_{44}A_{55}}$

The resulting values for C have been obtained in the same manner as outlined earlier, **Eq. (16)** (see **page 1.7**) but using **Eqs. (D1), (D2), and (D3)** of this appendix.

[1] For plane strain (Mode I and Mode II), replace E by $E/\left(1-\nu^2\right)$ for an isotropic material, and replace A_{ij} by $A_{ij} - A_{i3}A_{j3}/A_{33}\,(i,j = 1,2,6)$ for an anisotropic material.

[2] For general isotropy, one mode is not decoupled from the other two modes. The expressions given in the table are the coefficients for each mode in the absence of the other two modes.

APPENDIX E

STRESS INTENSITY FACTORS FOR CRACKS IN A PLATE SUBJECTED TO PINCHING LOADS

Analysis of cracks in an elastic plate near transverse loads (pinching loads) is of practical interest in, for example, accounting for the effect of spot weldings or contriving crack arresters.

The fundamental solution for a pinching force applied near the tip of a crack is presented in **Appendix C**. Namely, the stress intensity factors K_I and K_{II} are given for a semi-infinite crack in a plate subjected to a concentrated pinching force P in plane stress conditions [**Example (4), Appendix C**]. The derivation of this solution is one illustration of the usefulness of the weight function method discussed in **Appendix C**. The solution is collectively given by

$$K_I + i\,K_{II} = \frac{\nu P}{2\sqrt{2\pi}} \cdot \frac{1}{r^{3/2}} e^{-i\frac{3\theta}{2}} \tag{E1}$$

where ν is the Poisson's ratio and the coordinates (r, θ) are shown in Solution (I-1).

This solution serves as the Green's function for a semi-infinite crack subjected to an arbitrarily distributed pinching load. That is, in the (r, θ) coordinate system, if the pinching load distributed on an area S is expressed as $p = p(r, \theta)$, then the stress intensity factors are determined by

$$K_I + i\,K_{II} = \frac{\nu}{2\sqrt{2\pi}} \int_s \frac{p(r,\theta)}{r^{1/2}} e^{-i\frac{3\theta}{2}} drd\theta \tag{E2}$$

Most solutions for a semi-infinite crack in this appendix were obtained by using **Eq. (E2)**.

The solutions for concentrated pinching forces for other crack geometries (II-1, II-2, III-1, III-2) were derived using the weight functions in **Appendix C**, and then several additional solutions were obtained by superposition or integration (II-3, II-4; III-3, III-4, etc.).

When the pinching load is a uniform compression over circular areas centered on the line of cracks, the customary method of integrating off the stress distributed in the absence of the cracks seems more convenient than the weight function method. Solutions (I-10), (II-5, 6, 7), and (III-5) were obtained by this method. **Harris (1972)** obtained the complete K_I solution by this method for the entire process for a semi-infinite crack to extend through a circular region under uniform pinching pressure initially located ahead of the crack (Solution I-11). Harris also analyzed the effect of plate thickness (i.e., of the departure from the plane stress conditions).

All solutions included in this appendix are exact in the plane stress conditions and represent the upper bounds of the effect of pinching loads in actual plates.

The following are a few remarks on the crack-arresting effect of pinching loads. Only Mode I is addressed.

1. As observed from Solution (I-1), pinching forces near the crack tip can have the crack-arresting effect $(K_I < 0)$ only when they are applied behind the rays $|\theta| = \pi/3$ (i.e., in the regions $\frac{\pi}{3} \leq |\theta| \leq \pi$). For example, when a circular area ahead of the crack is uniformly pinched, Solution (I-10), the pinching load contributes positive K_I until the crack extends some distance into the pinched zone. See the sequence of Solutions (I-10), (I-9), and (I-8); and also **Harris (1972)**.
2. The maximum crack-arresting effect is gained when pinching forces are applied on the rays $|\theta| = 2\pi/3$. The dependency on the distance from the crack tip is common for all directions.
3. As readily observed from the comparison of Solutions (I-3) and (I-3a), when the pinching load distribution on a circular area $r \leq R$ is given by a pressure $p(r)$ or a line load $\overline{p}(r)$, for example, Solutions (I-3a), (I-4a), (I-5), and (I-6), the crack-arresting effect becomes 1.5 times as large if the load on the sector $|\theta| \leq \pi/3$ is removed. Or, if the same total pinching load can be distributed on sectors $\pi/3 \leq |\theta| \leq \pi$ only, the crack-arresting effect becomes 2.25 times as large.

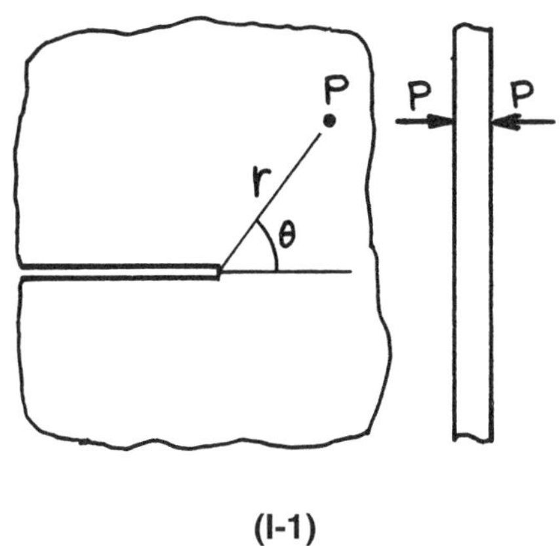

(I-1)

Pinching force P at (r, θ)

$$K_I = \frac{\nu P}{2\sqrt{2\pi}\, r^{3/2}} \cos\frac{3\theta}{2}$$

$$K_{II} = -\frac{\nu P}{2\sqrt{2\pi}\, r^{3/2}} \sin\frac{3\theta}{2}$$

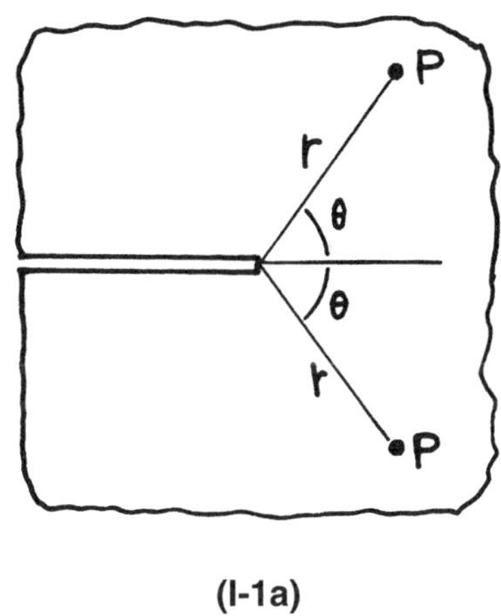

(I-1a)

Pinching forces P at $(r, \pm\theta)$

$$K_I = \frac{\nu P}{\sqrt{2\pi} r^{3/2}} \cos\frac{3\theta}{2}$$

$$K_{II} = 0$$

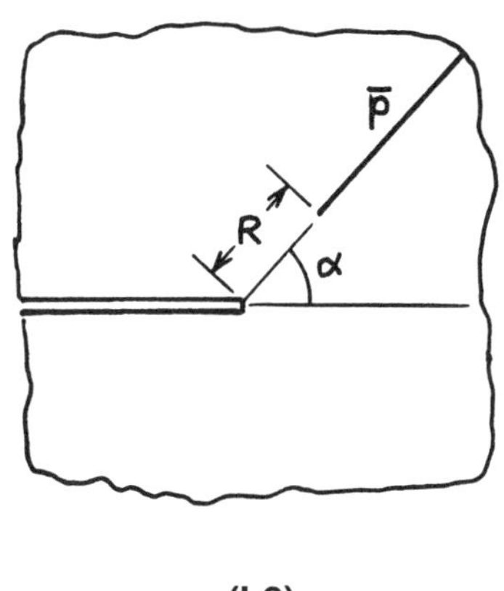

(I-2)

Uniform pinching line force $\bar{p}$ on $r \geq R,\ \theta = \alpha$

$$K_I = \frac{\nu\bar{p}}{\sqrt{2\pi R}} \cos\frac{3\alpha}{2}$$

$$K_{II} = -\frac{\nu\bar{p}}{\sqrt{2\pi R}} \sin\frac{3\alpha}{2}$$

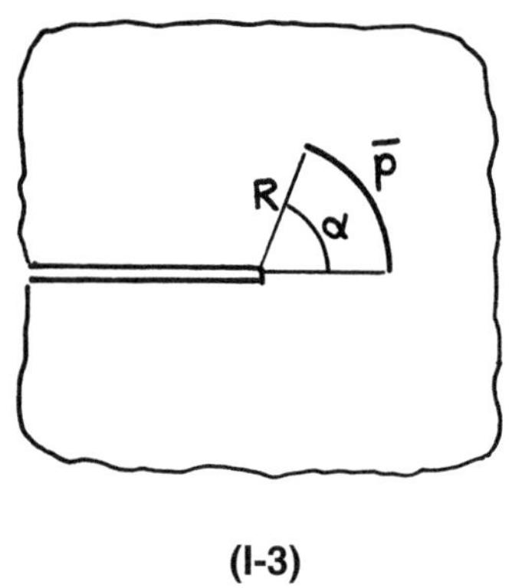

(I-3)

Uniform pinching line force $\overline{p}$ on circular arc $r = R,\ 0 \leq \theta \leq \alpha$

$$K_I = \frac{\nu\overline{p}}{3\sqrt{2\pi R}} \sin\frac{3\alpha}{2}$$

$$K_{II} = -\frac{\nu\overline{p}}{3\sqrt{2\pi R}}\left(1 - \cos\frac{3\alpha}{2}\right)$$

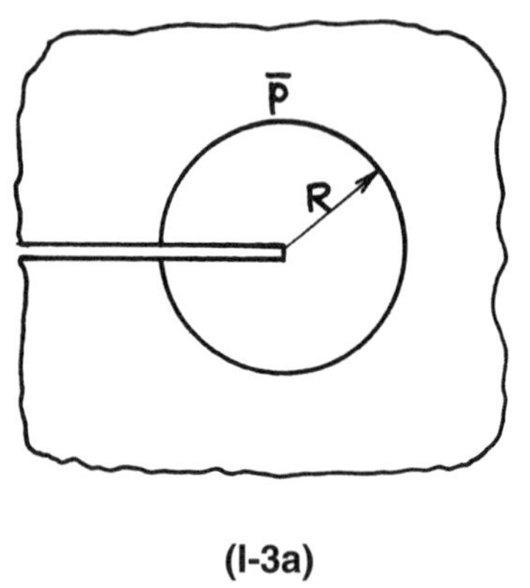

(I-3a)

Uniform pinching line force $\overline{p}$ on circle $r = R$

$$K_I = -\frac{2\nu\overline{p}}{3\sqrt{2\pi R}}$$

$$K_{II} = 0$$

(I-4)

Uniform pinching pressure p on sector $r \leq R,\ 0 \leq \theta \leq \alpha$

$$K_I = \frac{2\nu P}{3\sqrt{2\pi}} \sqrt{R} \sin\frac{3\alpha}{2}$$

$$K_{II} = -\frac{2\nu p}{3\sqrt{2\pi}} \sqrt{R}\left(1 - \cos\frac{3\alpha}{2}\right)$$

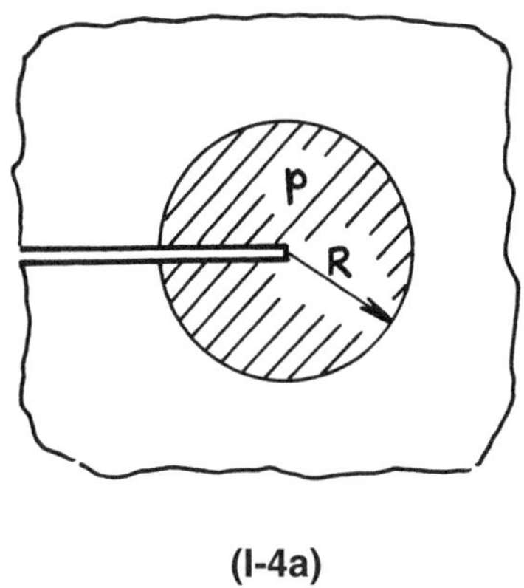

(I-4a)

Uniform pinching pressure p on circular area $r \leq R$

$$K_I = -\frac{4\nu P}{3\sqrt{2\pi}}\sqrt{R}$$

$$K_{II} = 0$$

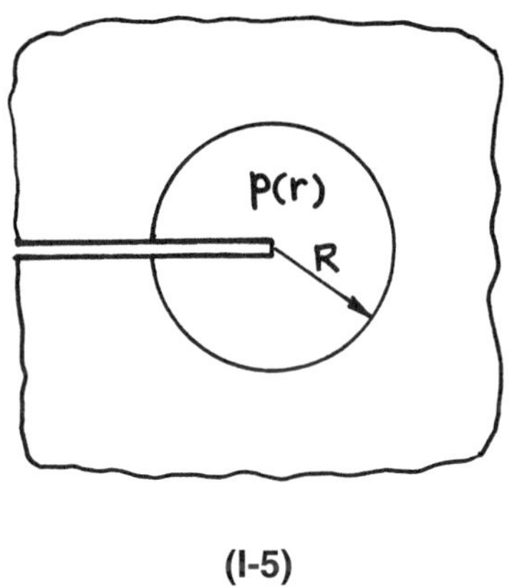

(I-5)

Pinching force P distributed on circular area $r \leq R$

$$p(r) = \frac{P}{2\pi R^2} \cdot \frac{1}{\sqrt{1 - (r/R)^2}}$$

$$K_I = -\frac{\nu P}{6\sqrt{2\pi} R^{3/2}} \cdot \frac{\Gamma\left(1/4\right)}{\Gamma\left(3/4\right)} = -.1110 \frac{\nu P}{R^{3/2}}$$

$$K_{II} = 0$$

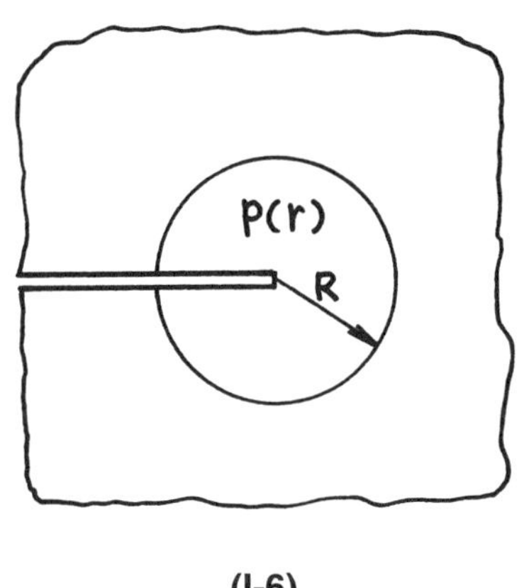

(I-6)

Pinching force P distributed on circular area $r \leq R$

$$p(r) = \frac{3P}{2\pi R^2}\sqrt{1 - (r/R)^2}$$

$$K_I = -\frac{\nu P}{\sqrt{2\pi} R^{3/2}} \cdot \frac{\Gamma(5/4)}{\Gamma(7/4)} = - \cdot 2220 \frac{\nu P}{R^{3/2}}$$

$$K_{II} = 0$$

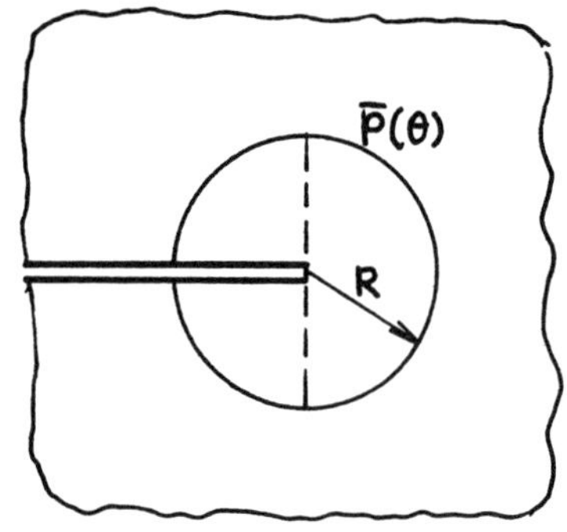

(I-7)

Pinching line force on circle $r = R$

$$\bar{p}(R, \theta) = \bar{p}_0 \cos \theta$$

$$K_I = \frac{6\nu\bar{p}_0}{5\sqrt{2\pi R}}$$

$$K_{II} = 0$$

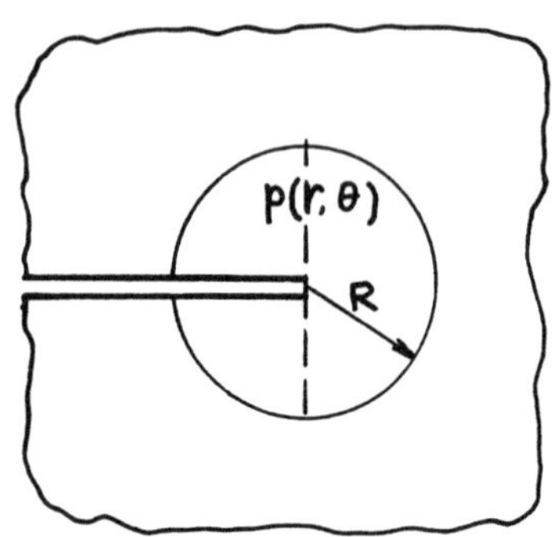

(I-8)

Pinching pressure distributed on circular area $r \leq R$

$$p(r, \theta) = p_0 \frac{r}{R} \cos \theta$$

$$K_I = \frac{4\nu p_0}{5\sqrt{2\pi}} \sqrt{R}$$

$$K_{II} = 0$$

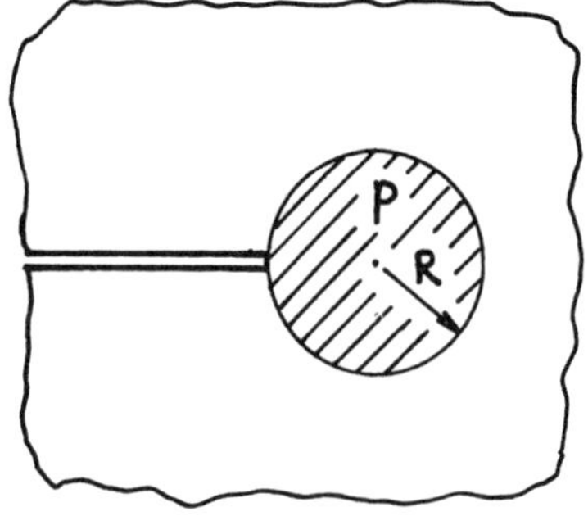

(I-9)

Uniform pinching pressure p on circular area $r \leq 2R \cos \theta$

$$K_I = \frac{\nu p}{2\sqrt{2}} \sqrt{\pi R}$$

$$K_{II} = 0$$

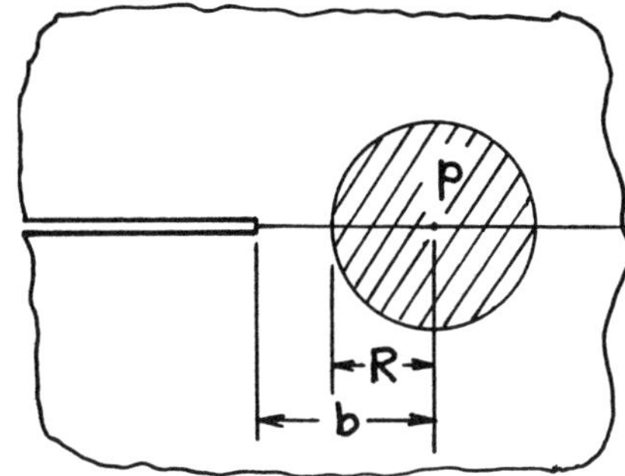

(I-10)

Uniform pinching pressure p on circular area ahead of crack

$$K_I = \frac{\nu p}{2\sqrt{2}}\sqrt{\pi R}\left(\frac{R}{b}\right)^{3/2}$$

$$K_{II} = 0$$

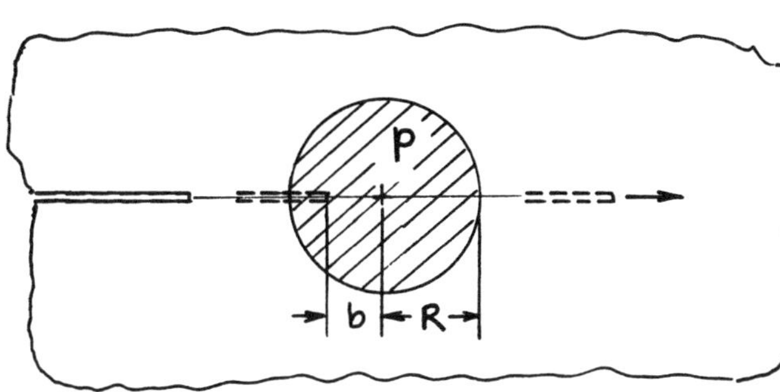

(I-11)

Crack in (I-10) extending through circular zone under uniform pinching pressure p

$$K_I = \nu p\sqrt{R}\, f\left(\frac{b}{R}\right)$$

$$K_{II} = 0$$

Complete solution $f(^{b}/_{R})$ is given in **Harris (1972)**. (I-4a), (I-9), and (I-10) are special cases. Effect of plate thickness is also accounted for in **Harris (1972)**.

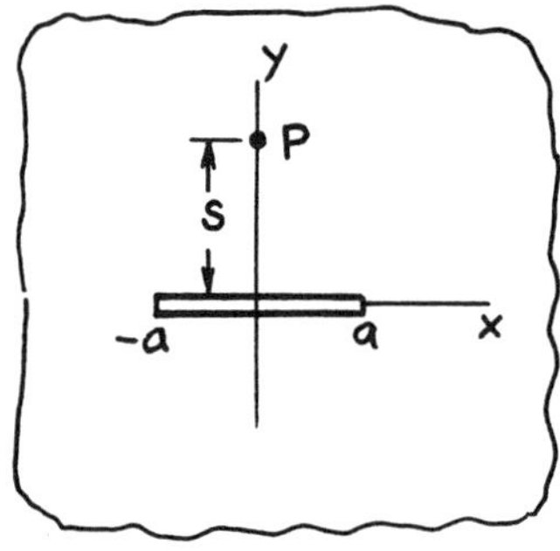

(II-1)

Pinching force P at $(0, s)$

$$K_I = -\frac{\nu P}{2}\sqrt{\frac{a}{\pi}}\frac{s}{\left(a^2+s^2\right)^{3/2}}$$

$$K_{II\pm a} = \pm\frac{\nu P}{2}\sqrt{\frac{a}{\pi}}\frac{a}{\left(a^2+s^2\right)^{3/2}}$$

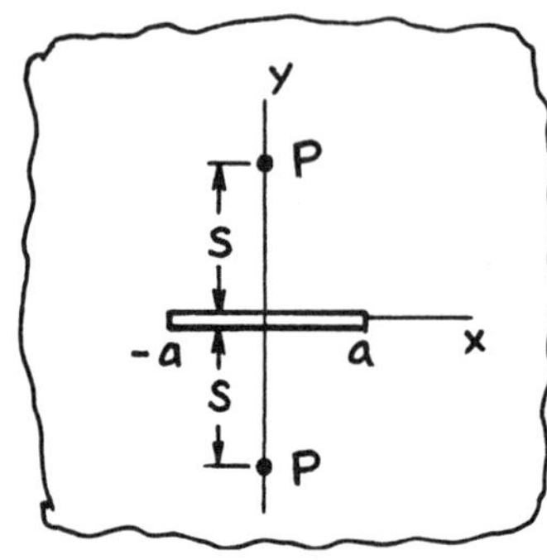

(II-1a)

Pinching forces P at $(0, \pm s)$

$$K_I = -\nu P\sqrt{\frac{a}{\pi}}\frac{s}{\left(a^2+s^2\right)^{3/2}}$$

$$K_{II} = 0$$

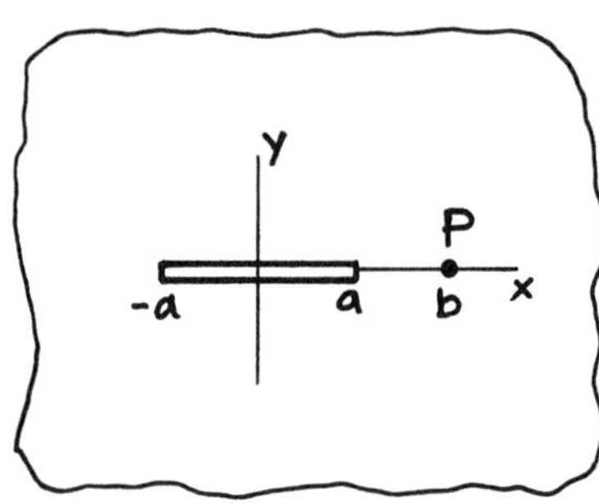

(II-2)

Pinching force P at $(b, 0)$

$$K_{I\pm a} = \pm\frac{\nu P}{2}\sqrt{\frac{a}{\pi}}\frac{a}{(b\mp a)\sqrt{b^2-a^2}}$$

$$K_{II} = 0$$

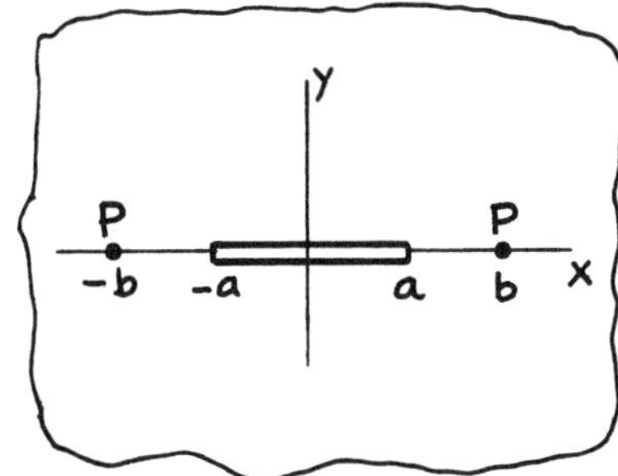

(II-2a)

Pinching forces P at $(\pm b, 0)$

$$K_I = \nu P \sqrt{\frac{a}{\pi}} \frac{b}{\left(b^2 - a^2\right)^{3/2}}$$

$$K_{II} = 0$$

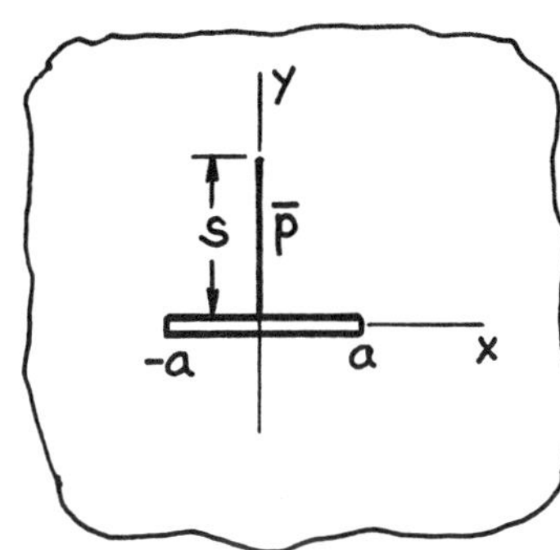

(II-3)

Uniform pinching line force $\bar{p}$ on $x = 0$, $0 \leq y \leq s$

$$K_I = -\frac{\nu \bar{p}}{2\sqrt{\pi a}} \left(1 - \frac{a}{\sqrt{a^2 + s^2}}\right)$$

$$K_{II \pm a} = \pm \frac{\nu \bar{p}}{2\sqrt{\pi a}} \frac{s}{\sqrt{a^2 + s^2}}$$

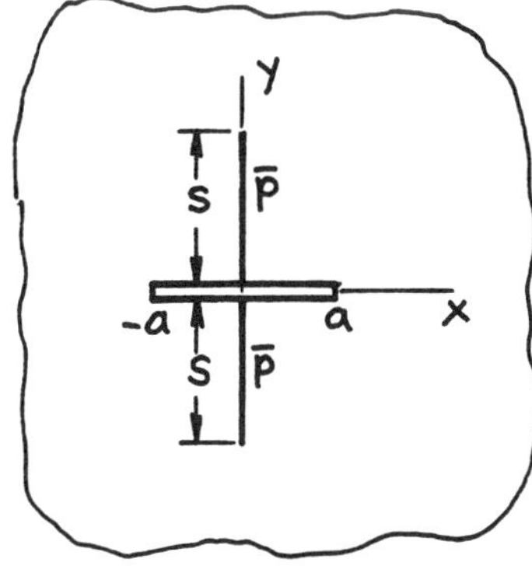

(II-3a)

Uniform pinching line force $\bar{p}$ on $x = 0$, $|y| \leq s$

$$K_I = -\frac{\nu \bar{p}}{\sqrt{\pi a}} \left(1 - \frac{a}{\sqrt{a^2 + s^2}}\right)$$

$$K_{II} = 0$$

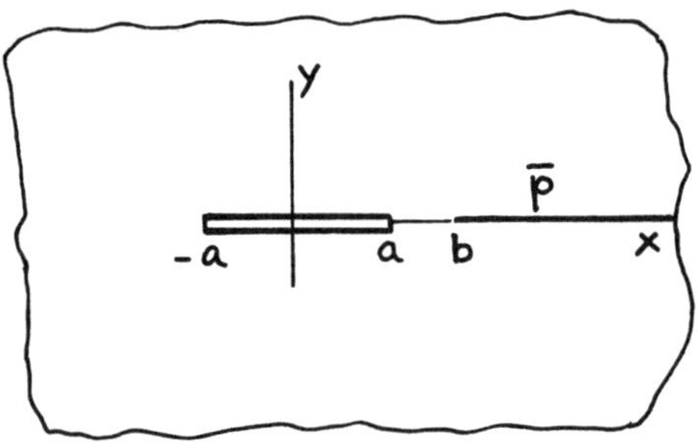

(II-4)

Uniform pinching line force $\bar{p}$ on $x \geq b$, $y = 0$

$$K_{I\pm a} = \pm \frac{\nu\bar{p}}{2\sqrt{\pi a}}\left(\sqrt{\frac{b \pm a}{b \mp a}} - 1\right)$$

$$K_{II} = 0$$

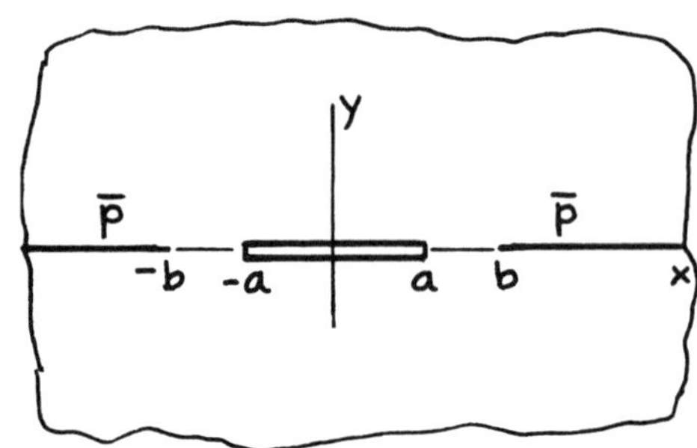

(II-4a)

Uniform pinching line force $\bar{p}$ on $|x| \geq b$, $y = 0$

$$K_I = \frac{\nu\bar{p}}{\sqrt{\pi a}} \frac{a}{\sqrt{b^2 - a^2}}$$

$$K_{II} = 0$$

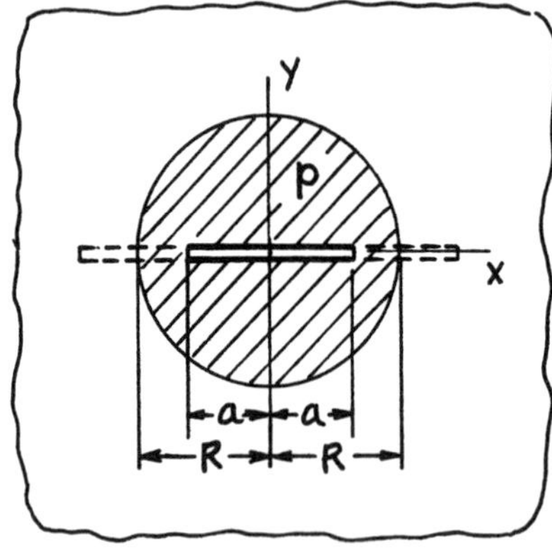

(II-5)

Uniform pinching pressure p on circle $r \leq R$

$$a \leq R: \; K_I = -\frac{\nu P}{2}\sqrt{\pi a}$$

$$K_{II} = 0$$

$$a \geq R: \; K_I = -\frac{\nu P}{2}\sqrt{\pi a} \cdot \frac{2}{\pi}\left[\sin^{-1}\frac{R}{a} - \frac{R}{a}\sqrt{1 - \left(\frac{R}{a}\right)^2}\right]$$

$$K_{II} = 0$$

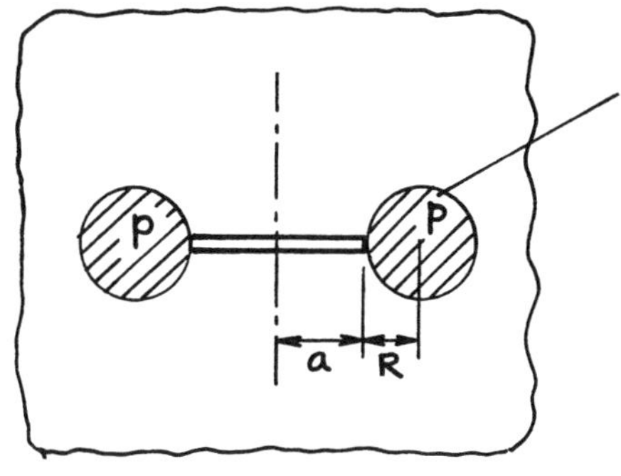

Uniform pinching pressure p

$$K_I = \nu p\sqrt{\pi a}\ \frac{\left(\frac{R}{a+R}\right)^2}{\left[1-\left(\frac{a}{a+R}\right)^2\right]^{3/2}}$$

$$K_{II} = 0$$

(II-6)

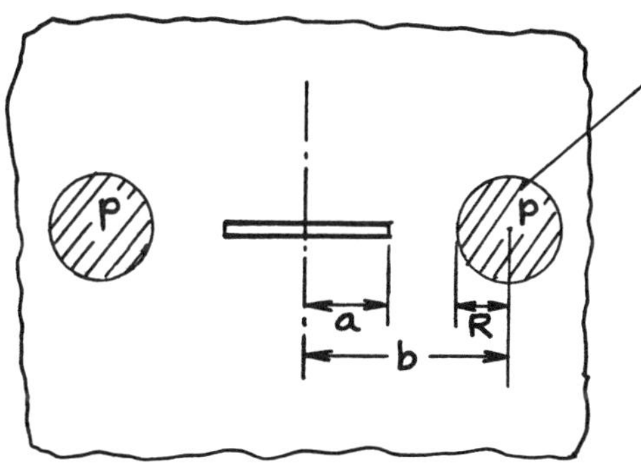

Uniform pinching pressure p

$$K_I = \nu p\sqrt{\pi a}\ \frac{bR^2}{\left(b^2-a^2\right)^{3/2}}$$

$$K_{II} = 0$$

(II-6) is a special case with $b = a + R$.

(II-7)

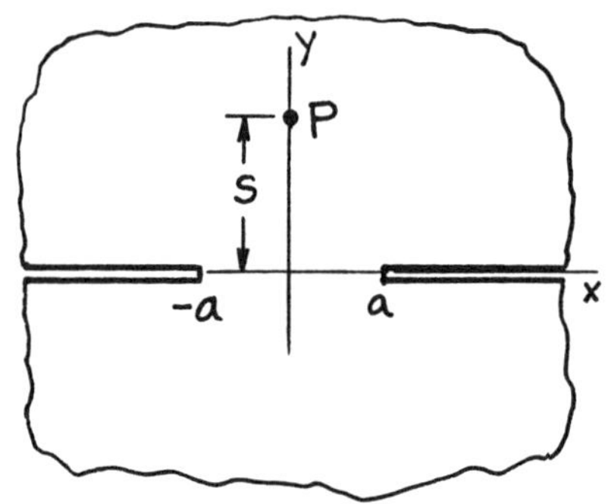

(III-1)

Pinching force P at $(0, s)$

$$K_I = \frac{\nu P}{2}\sqrt{\frac{a}{\pi}}\frac{a}{\left(a^2 + s^2\right)^{3/2}}$$

$$K_{II} = \frac{\nu P}{2}\sqrt{\frac{a}{\pi}}\frac{s}{\left(a^2 + s^2\right)^{3/2}}$$

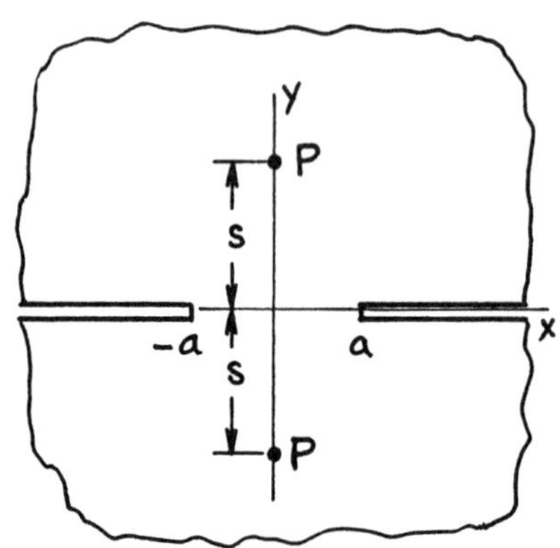

(III-1a)

Pinching forces P at $(0, \pm s)$

$$K_I = \nu P\sqrt{\frac{a}{\pi}}\frac{a}{\left(a^2 + s^2\right)^{3/2}}$$

$$K_{II} = 0$$

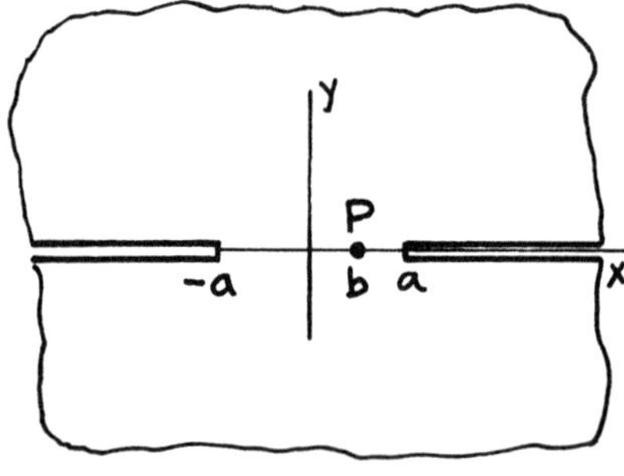

(III-2)

Pinching force P at $(b, 0)$

$$K_{I\pm a} = \frac{\nu P}{2}\sqrt{\frac{a}{\pi}}\frac{1}{(a \mp b)\sqrt{a^2 - b^2}}$$

$$K_{II} = 0$$

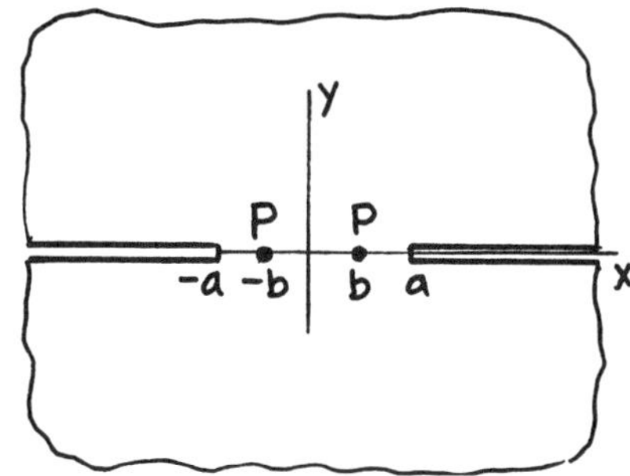

(III-2a)

Pinching forces P at $(\pm b, 0)$

$$K_I = \nu P \sqrt{\frac{a}{\pi}} \frac{a}{\left(a^2 - b^2\right)^{3/2}}$$

$$K_{II} = 0$$

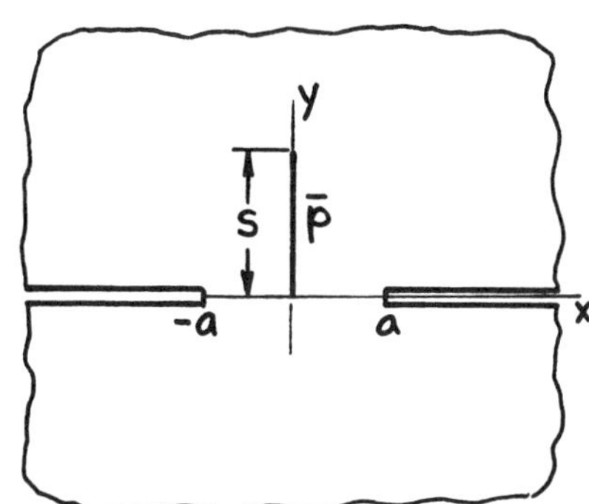

(III-3)

Uniform pinching line force $\bar{p}$ on $x = 0$, $0 \leq y \leq s$

$$K_I = \frac{\nu \bar{p}}{2\sqrt{\pi a}} \frac{s}{\sqrt{a^2 + s^2}}$$

$$K_{II \pm a} = \pm \frac{\nu \bar{p}}{2\sqrt{\pi a}} \left(1 - \frac{a}{\sqrt{a^2 + s^2}}\right)$$

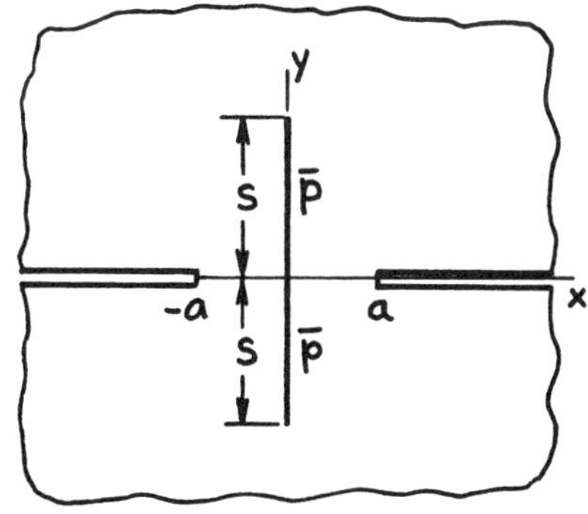

(III-3a)

Uniform pinching line force $\bar{p}$ on $x = 0$, $|y| \leq s$

$$K_I = \frac{\nu \bar{p}}{\sqrt{\pi a}} \frac{s}{\sqrt{a^2 + s^2}}$$

$$K_{II} = 0$$

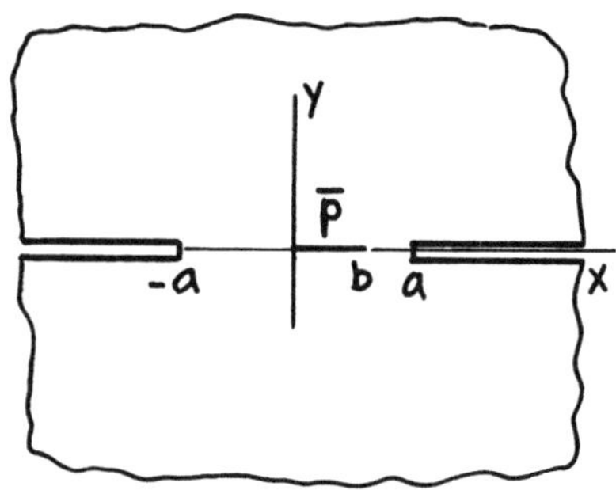

(III-4)

Uniform pinching line force $\bar{p}$ on $0 \le x \le b$, $y = 0$

$$K_{I\pm a} = \pm \frac{\nu \bar{p}}{2\sqrt{\pi a}} \left(\sqrt{\frac{a \pm b}{a \mp b}} - 1 \right)$$

$$K_{II} = 0$$

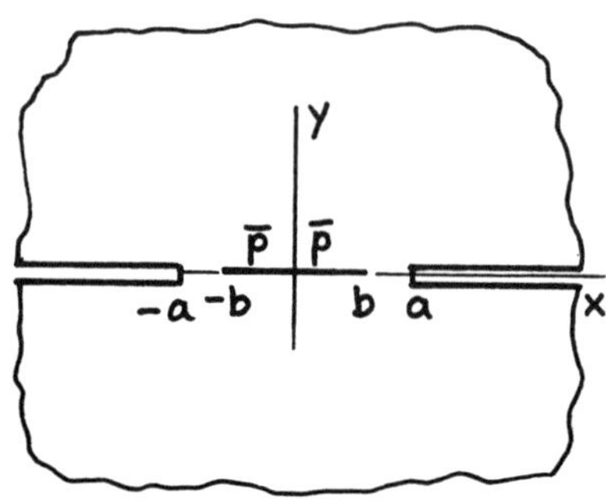

(III-4a)

Uniform pinching line force $\bar{p}$ on $|x| \le b$, $y = 0$

$$K_I = \frac{\nu \bar{p}}{\sqrt{\pi a}} \frac{b}{\sqrt{a^2 - b^2}}$$

$$K_{II} = 0$$

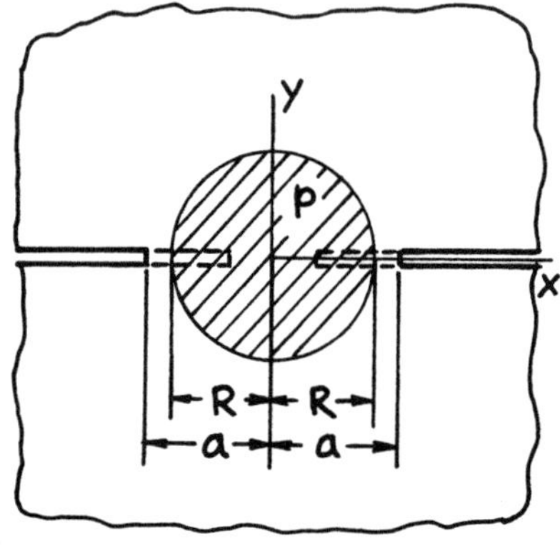

(III-5)

Uniform pinching pressure p on circular area $r \le R$

$$a \ge R : K_I = \frac{\nu P}{2} \sqrt{\pi a} \left(\frac{R}{a} \right)^2$$

$$K_{II} = 0$$

$$a \le R : \; K_I = \frac{\nu P}{2} \sqrt{\pi a} \left(\frac{R}{a} \right)^2 \cdot \frac{2}{\pi} \left[\sin^{-1} \frac{a}{R} - \frac{a}{R} \sqrt{1 - \left(\frac{a}{R} \right)^2} \right]$$

$$K_{II} = 0$$

NOTE: K_I is always positive.

APPENDIX F

CRACKS IN RESIDUAL STRESS FIELDS

Analyses of cracks developed in residual stress fields (including thermal stress fields, etc.) are of practical importance in welded structures and in many other applications. The residual stress field is, in the absence of external loads, a self-balanced, built-in field. Determination of the intensity of the crack-tip elastic field for cracks introduced into the residual stress field requires no special treatment. The method is based on the superposition principle.

The following two stress states should be superimposed to satisfy the free surface condition on the crack:

a. The residual stress field in the absence of the crack
b. The opening pressure applied on the crack surfaces whose distribution corresponds to a.

That is, the effect of the presence of the crack on the resultant stress field is determined solely from stress state b. For example, the determination of stress intensity factor K_I is reduced to no more than the evaluation of an integral if the Green's function for K_I(K_I solution for a pair of concentrated splitting forces on crack surfaces) is known for the crack geometry of interest. For the examples on **pages F.2 through F.5** in this appendix, the K_I solution on **page 5.11** is most conveniently used as the Green's function K_{IG}.

$$K_{IG}(a, b; P) = \frac{2P}{\sqrt{\pi a}} \cdot \frac{1}{\sqrt{1 - \left(b/a\right)^2}} \qquad \textbf{(F1)}$$

Then the stress intensity factor for the crack absent stress distribution $\sigma_y(x)$ is readily determined by integration.

$$K_I = \int_0^a K_{IG}(a, x; \sigma_y(x)dx)$$

$$= \frac{2}{\sqrt{\pi a}} \int_0^a \frac{\sigma_y(x)\, dx}{\sqrt{1 - \left(x/a\right)^2}}$$

Similarly, for displacement solutions (crack opening area, crack opening displacement, etc.), the solutions on **page 5.11a** may be used as Green's functions. Use of Paris' equation (see Appendix B), however, is often more convenient (i.e., integrations are simpler) for this purpose.

The following pages contain solutions for various "crack-absent" residual stress distributions and crack geometries. An actual, practical situation may be approximately represented by one of these solutions. When additional external loads are also present, the total solution is obtained by superposition.

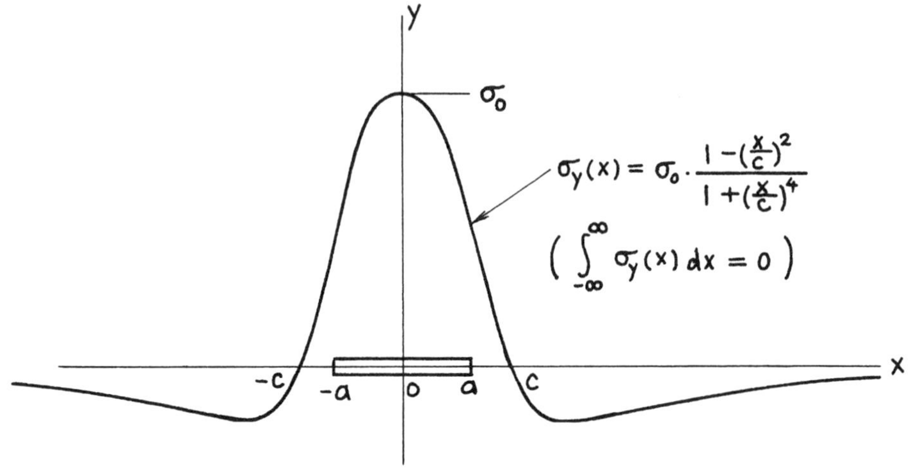

$$\alpha = \frac{a}{c}$$

$$K_I = \sigma_0\sqrt{\pi a}\left\{\frac{\sqrt{1+\alpha^4}-\alpha^2}{1+\alpha^4}\right\}^{1/2}$$

Crack Opening Area:

$$A = \frac{4\sigma_0 \pi c^2}{E'}\left\{1-\left(\sqrt{1+\alpha^4}-\alpha^2\right)^{1/2}\right\}$$

Opening at Center: $\delta = 2v(0,0)$

$$\delta = \frac{4\sigma_0\, c}{E'} \cdot \frac{1}{\sqrt{2}}\cos^{-1}\left(\sqrt{1+\alpha^4}-\alpha^2\right)$$

Method: K_I Integration of **page 5.11**

A and δ Paris' Equation (see **Appendix B**)

Accuracy: Exact

References: **Tada 1983b, 1985**

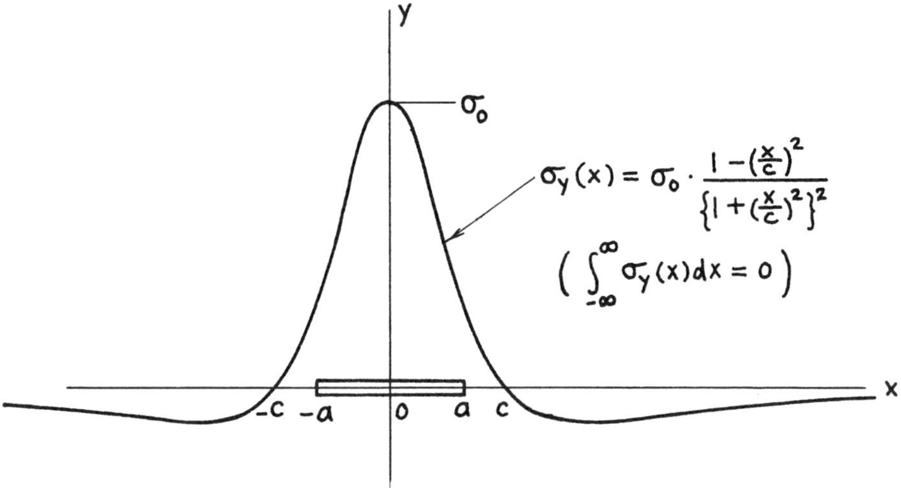

$$\alpha = \frac{a}{c}$$

$$K_I = \sigma_0 \sqrt{\pi a} \frac{1}{\left(1+\alpha^2\right)^{3/2}}$$

Crack Opening Area:

$$A = \frac{4\sigma_0 \pi c^2}{E'} \left\{ 1 - \frac{1}{\sqrt{1+\alpha^2}} \right\}$$

Opening at Center: $\delta = 2v(0, 0)$

$$\delta = \frac{4\sigma_0 a}{E'} \cdot \frac{1}{\sqrt{1+\alpha^2}}$$

Method: K_I Integration of **page 5.11**

A and δ Paris' Equation (see **Appendix B**)

Accuracy: Exact

Reference: **Tada 1985**

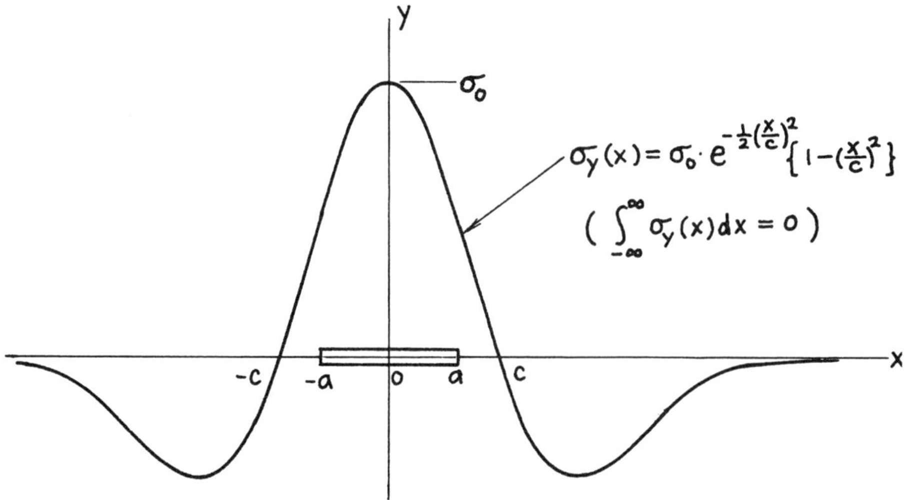

$$\alpha = \frac{a}{c}$$

$$K_I = \sigma_0 \sqrt{\pi a}\, e^{-.42\alpha^2}\left(1 - \frac{1}{\pi}\alpha^2\right)$$

Crack Opening Area:

$$A = \frac{2\sigma_0 \pi a^2}{E'} e^{-.36\alpha^2}$$

Opening at Center: $\delta = 2v(0, 0)$

$$\delta = \frac{4\sigma_0 a}{E'} \cdot \frac{1}{1 + \frac{1}{4}\alpha^2}$$

Method: K_I Integration of **page 5.11**
A and δ Paris' Equation (see **Appendix B**)
Accuracy: K_I 0.5 %
A 1 % for $a/c \leq \sqrt{\pi}$
δ 2 % for $a/c \leq \sqrt{\pi}$
References: **Terada 1976, Tada 1983b, 1985**

NOTE: $K_I < 0$ for $a/c \gtrsim \sqrt{\pi}$

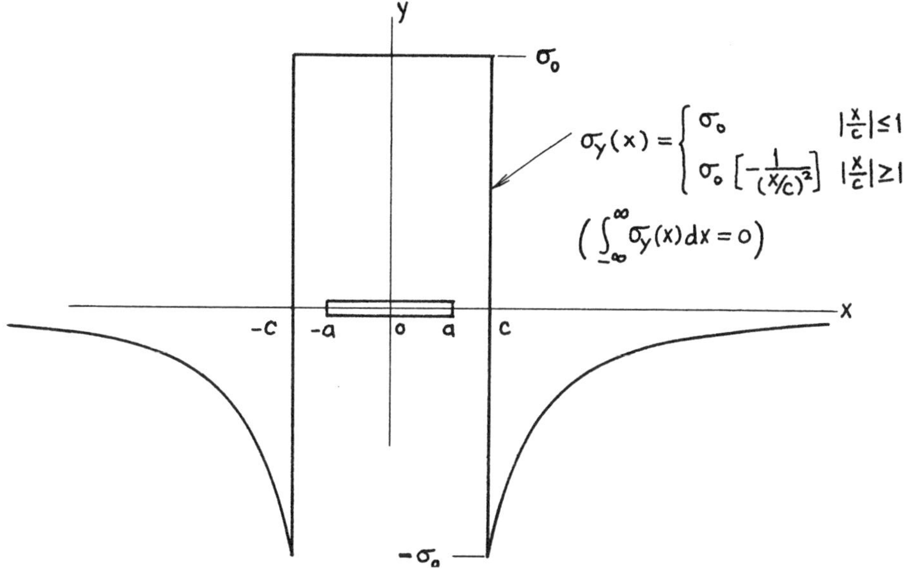

$$\alpha = \frac{a}{c}$$

$$K_I = \sigma_0\sqrt{\pi a}\begin{cases} 1 & \alpha \le 1 \\ \frac{2}{\pi}\left(\sin^{-1}\frac{1}{\alpha} - \frac{\sqrt{\alpha^2-1}}{\alpha^2}\right) & \alpha \ge 1 \end{cases}$$

Crack Opening Area:

$$A = \frac{2\,\sigma_0\,\pi a^2}{E'}\begin{cases} 1 & \alpha \le 1 \\ \frac{2}{\pi}\left(\frac{2}{\alpha^2}\cos^{-1}\frac{1}{\alpha} + \sin^{-1}\frac{1}{\alpha} - \frac{\sqrt{\alpha^2-1}}{\alpha^2}\right) & \alpha \ge 1 \end{cases}$$

Opening at Center: $\delta = 2v(0,0)$

$$\delta = \frac{4\,\sigma_0\,a}{E'}\begin{cases} 1 & \alpha \le 1 \\ \frac{2}{\pi}\left(\sin^{-1}\frac{1}{\alpha} + \frac{\sqrt{\alpha^2-1}}{\alpha^2}\right) & \alpha \ge 1 \end{cases}$$

Method: K_I Integration of **page 5.11**
A and δ Paris' Equation (see **Appendix B**)
Accuracy: Exact
Reference: **Tada 1985**

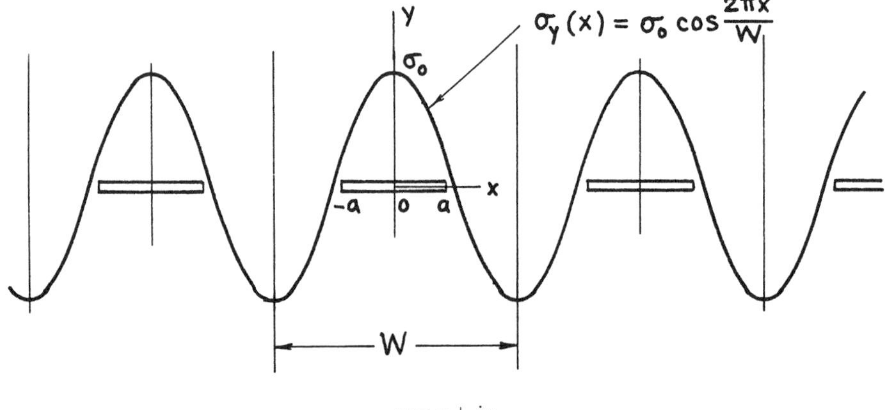

$$z = x + iy$$

$$Z_I(z) = \sigma_0 \frac{\sin\dfrac{\pi z}{W}}{\sqrt{\left(\sin\dfrac{\pi z}{W}\right)^2 - \left(\sin\dfrac{\pi a}{W}\right)^2}} \left[2\left(\cos\frac{\pi z}{W}\right)^2 - \left(\cos\frac{\pi a}{W}\right)^2\right]$$

$$\overline{Z}_I(z) = \sigma_0 \frac{W}{\pi} \cos\frac{\pi z}{W} \sqrt{\left(\sin\frac{\pi z}{W}\right)^2 - \left(\sin\frac{\pi a}{W}\right)^2}$$

$$K_I = \sigma_0 \sqrt{\pi a} \sqrt{\frac{W}{\pi a} \tan\frac{\pi a}{W}} \left(\cos\frac{\pi a}{W}\right)^2$$

$$\underset{|x| \leq a}{\mathrm{Im}\left\{\overline{Z}(x)\right\}} = \frac{\sigma_0 W}{\pi} \cos\frac{\pi x}{W} \sqrt{\left(\sin\frac{\pi a}{W}\right)^2 - \left(\sin\frac{\pi x}{W}\right)^2}$$

Crack Opening Area:

$$A = \frac{2\sigma_0 W^2}{\pi E'} \left(\sin\frac{\pi a}{W}\right)^2$$

Opening at Center: $\delta = 2v(0, 0)$

$$\delta = \frac{4\sigma_0 W}{\pi E'} \sin\frac{\pi a}{W}$$

Relative Vertical Displacement at Infinity due to Crack:

$$\Delta\,(= \Delta_{crack}) = A/W \left(= \frac{1}{2}\delta \sin\frac{\pi a}{W}\right)$$

Methods: Westergaard Stress Function (Integration of **page 7.7**)
A, δ, Δ also Paris' Equation (see **Appendix B**)
Accuracy: Exact
Reference: **Tada 1985**

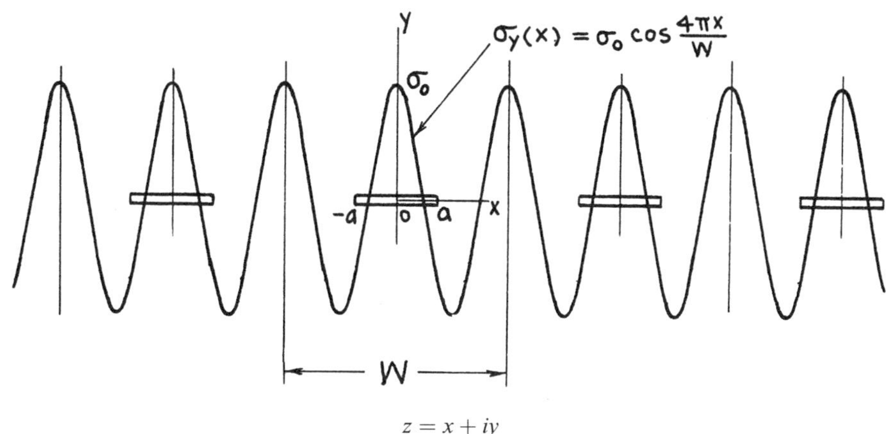

$z = x + iy$

$$Z_I(z) = \sigma_0 \frac{\sin \frac{\pi z}{W}}{\sqrt{\left(\sin \frac{\pi z}{W}\right)^2 - \left(\sin \frac{\pi a}{W}\right)^2}} \left[8\left(\cos\frac{\pi z}{W}\right)^4 - 4\left(\cos\frac{\pi z}{W}\right)^2 \left\{1 + \left(\cos\frac{\pi a}{W}\right)^2\right\} + 2\left(\cos\frac{\pi a}{W}\right)^2 - \left(\cos\frac{\pi a}{W}\right)^4\right]$$

$$\overline{Z}_I(z) = \sigma_0 \frac{W}{\pi} \cos\frac{\pi z}{W} \sqrt{\left(\sin\frac{\pi z}{W}\right)^2 - \left(\sin\frac{\pi a}{W}\right)^2} \left[\left(\cos\frac{\pi a}{W}\right)^2 - 2\left(\sin\frac{\pi z}{W}\right)^2\right]$$

$$K_I = \sigma_0 \sqrt{\pi a} \sqrt{\frac{W}{\pi a} \tan\frac{\pi a}{W}} \left(\cos\frac{\pi a}{W}\right)^2 \left[3\left(\cos\frac{\pi a}{W}\right)^2 - 2\right]$$

$$\operatorname*{Im}_{|x| \le a} \left\{\overline{Z}_I(x)\right\} = \frac{\sigma_0 W}{\pi} \cos\frac{\pi x}{W} \sqrt{\left(\sin\frac{\pi a}{W}\right)^2 - \left(\sin\frac{\pi x}{W}\right)^2} \left[\left(\cos\frac{\pi a}{W}\right)^2 - 2\left(\sin\frac{\pi x}{W}\right)^2\right]$$

Crack Opening Area:

$$A = \frac{\sigma_0 W^2}{\pi E'} \left(\sin\frac{\pi a}{W}\right)^2 \left[3\left(\cos\frac{\pi a}{W}\right)^2 - 1\right]$$

Opening at Center: $\delta = 2v(0, 0)$

$$\delta = \frac{4\sigma_0 W}{\pi E'} \sin\frac{\pi a}{W} \left(\cos\frac{\pi a}{W}\right)^2$$

Relative Vertical Displacement at Infinity due to Crack:

$$\Delta (= \Delta_{crack}) = A/W$$

Method: Westergaard Stress Function (Integration of **page 7.7**)
A, δ, Δ also Paris' Equation (see **Appendix B**)
Accuracy: Exact
Reference: **Tada 1985**

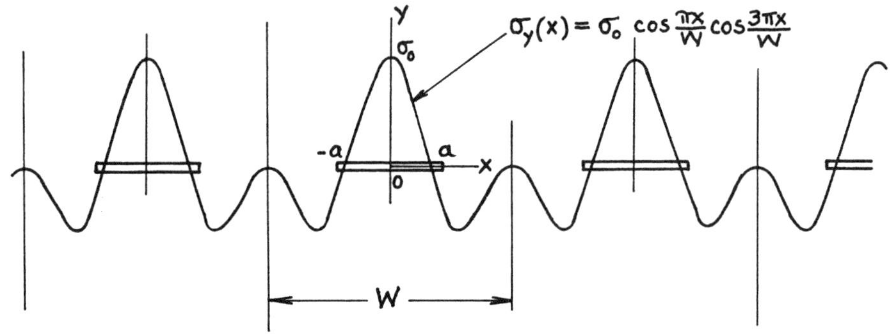

$$z = x + iy$$

$$Z_I(z) = \sigma_0 \frac{\sin\frac{\pi z}{W}}{\sqrt{\left(\sin\frac{\pi z}{W}\right)^2 - \left(\sin\frac{\pi a}{W}\right)^2}} \left[4\left(\cos\frac{\pi z}{W}\right)^4 - \left(\cos\frac{\pi z}{W}\right)^2 \left\{1 + 2\left(\cos\frac{\pi a}{W}\right)^2\right\} + \frac{1}{2}\left(\cos\frac{\pi a}{W}\right)^2 \left(\sin\frac{\pi a}{W}\right)^2\right]$$

$$\overline{Z}_I(z) = \sigma_0 \frac{W}{\pi} \cos\frac{\pi z}{W} \sqrt{\left(\sin\frac{\pi z}{W}\right)^2 - \left(\sin\frac{\pi a}{W}\right)^2} \left\{\left(\cos\frac{\pi z}{W}\right)^2 - \frac{1}{2}\left(\sin\frac{\pi a}{W}\right)^2\right\}$$

$$K_I = \sigma_0 \sqrt{\pi a} \sqrt{\frac{W}{\pi a} \tan\frac{\pi a}{W}} \cdot \frac{1}{2}\left(\cos\frac{\pi a}{W}\right)^2 \left\{3\left(\cos\frac{\pi a}{W}\right)^2 - 1\right\}$$

$$\operatorname*{Im}_{|x|\le a} \left\{\overline{Z}_I(x)\right\} = \sigma_0 \frac{W}{\pi} \cos\frac{\pi x}{W} \sqrt{\left(\sin\frac{\pi a}{W}\right)^2 - \left(\sin\frac{\pi x}{W}\right)^2} \left\{\left(\cos\frac{\pi x}{W}\right)^2 - \frac{1}{2}\left(\sin\frac{\pi a}{W}\right)^2\right\}$$

Crack Opening Area:

$$A = \frac{\sigma_0 W^2}{\pi E'} \cdot \frac{1}{2}\left(\sin\frac{\pi a}{W}\right)^2 \left[3\left(\cos\frac{\pi a}{W}\right)^2 + 1\right]$$

Opening at Center: $= 2v(0,0)$

$$\delta = \frac{2\sigma_0 W}{\pi E'} \cdot \sin\frac{\pi a}{W} \left[\left(\cos\frac{\pi a}{W}\right)^2 + 1\right]$$

Relative Vertical Displacement at Infinity due to Crack:

$$\Delta (= \Delta_{crack}) = A/W$$

Method: Superposition of **pages F.6 and F.7**: $\frac{1}{2}$ **(F.6 + F.7)** (or Integration of **page 7.7**)
A, δ, Δ also Paris' Equation (see **Appendix B**)
Accuracy: Exact
Reference: **Tada 2000**

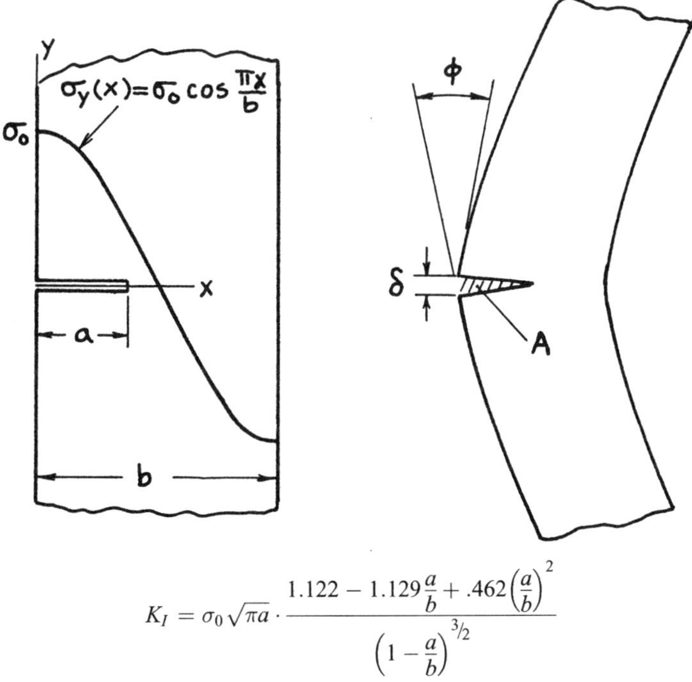

$$K_I = \sigma_0\sqrt{\pi a}\cdot\frac{1.122 - 1.129\frac{a}{b} + .462\left(\frac{a}{b}\right)^2}{\left(1-\frac{a}{b}\right)^{3/2}}$$

Crack Opening Area:

$$A = \frac{\sigma_0\,\pi a^2}{E'}\cdot\frac{1.258 - 1.62\frac{a}{b} + 1.49\left(\frac{a}{b}\right)^2 - .62\left(\frac{a}{b}\right)^3}{\left(1-\frac{a}{b}\right)^2}$$

Opening at Edge:

$$\delta = \frac{4\sigma_0\,a}{E'}\cdot\frac{1.164 - .36\frac{a}{b} + .29\left(1-\frac{a}{b}\right)^6}{\left(1-\frac{a}{b}\right)^2}$$

Rotation (Kink) at Cracked Section due to Crack:

$$\phi = \frac{\sigma_0}{E'}\cdot\left(\frac{\frac{a}{b}}{1-\frac{a}{b}}\right)^2\left[23.8 - 40.15\frac{a}{b} + 19.55\left(\frac{a}{b}\right)^2 - \left(15.73 - 11.24\frac{a}{b}\right)\frac{a}{b}\left(1-2\frac{a}{b}\right)\left(1-\frac{a}{b}\right)\right]$$

Methods: K_I Integration of **page 2.27**
A, δ, ϕ Paris' Equation (see **Appendix B**)
Accuracy: K_I 1%; A, δ, ϕ 2%
Reference: **Tada 1985**

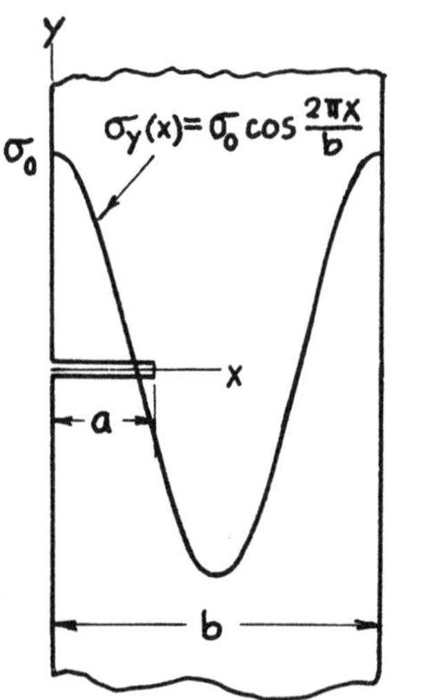

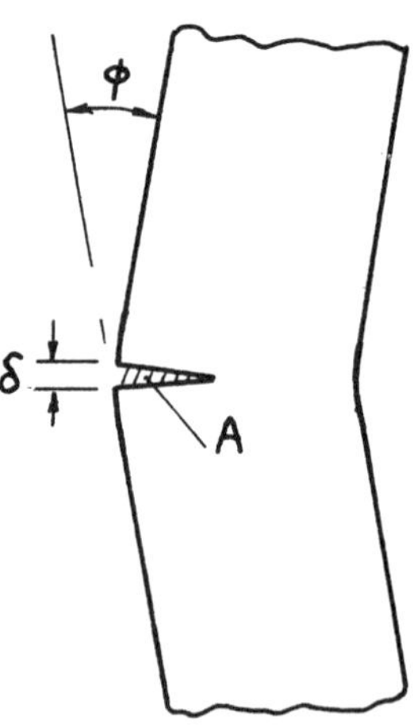

$$K_I = \sigma_0\sqrt{\pi a}\left[\frac{1.122 - 4\left(\frac{a}{b}\right)^3}{\left\{1+2\left(\frac{a}{b}\right)^2\right\}^2} + 1.5\frac{a}{b}\left(.375 - \frac{a}{b}\right)\left(1 - \frac{a}{b}\right)^3\right]$$

Crack Opening Area:

$$A = \frac{\sigma_0\,\pi a^2}{E'}\cdot\frac{1.258 - .057\frac{a}{b} - 1.76\left(\frac{a}{b}\right)^2}{\sqrt{1-\frac{a}{b}}}$$

Opening at Edge:

$$\delta = \frac{4\sigma_0\,a}{E'}\cdot\frac{1.454 + .13\frac{a}{b} - 1.72\left(\frac{a}{b}\right)^3 - .76\left(\frac{a}{b}\right)^4}{\sqrt{1-\frac{a}{b}}}$$

Angle of Rotation (Kink at Cracked Section) due to Crack:

$$\phi = \frac{\sigma_0}{E'}\cdot\frac{\left(\frac{a}{b}\right)^2\left(23.8 - 28\frac{a}{b}\right)}{\sqrt{1-\frac{a}{b}}}$$

Methods: K_I Integration of **page 2.27**
A, δ, ϕ Paris' Equation (see **Appendix B**)
Accuracy: K_I 1 %; A, δ 2% for $a/b \lesssim 0.65$; ϕ 3% for $a/b \lesssim 0.65$
Reference: **Tada 1985**

NOTE: $K_I < 0$ for $a/b \gtrsim 0.65$

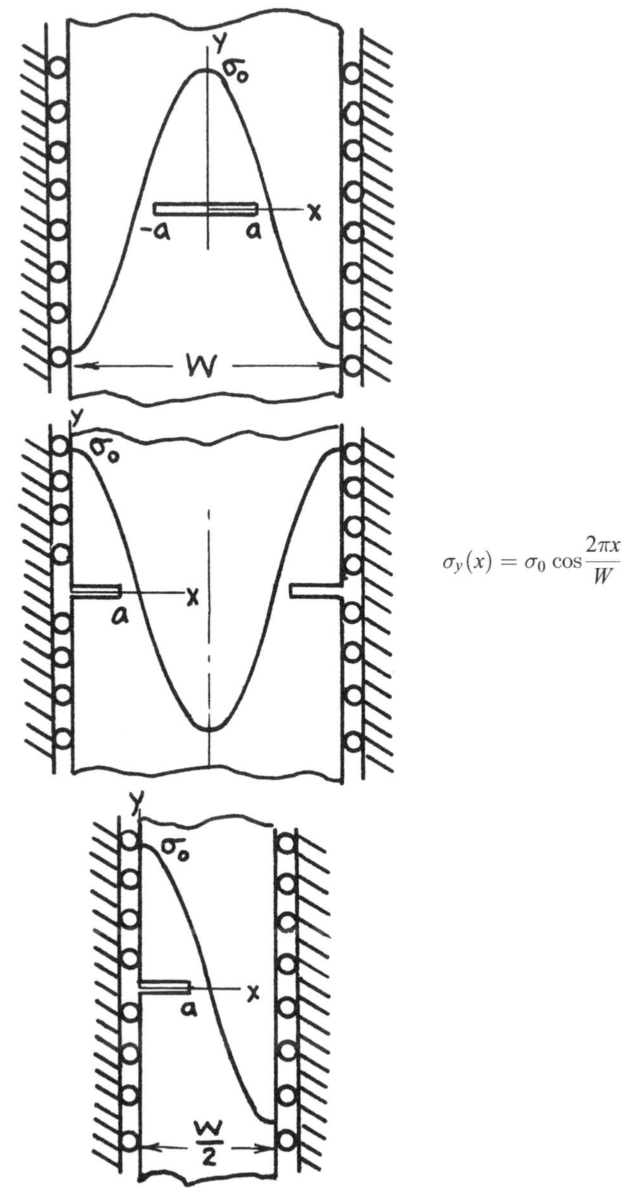

$$\sigma_y(x) = \sigma_0 \cos\frac{2\pi x}{W}$$

See **page F.6**.

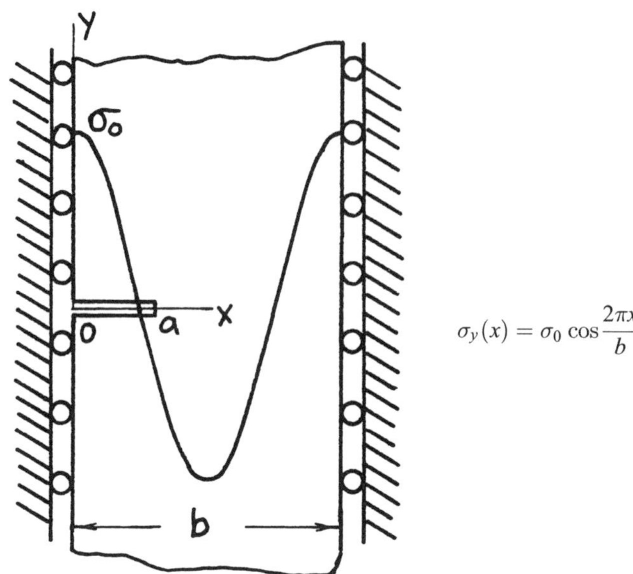

$$\sigma_y(x) = \sigma_0 \cos \frac{2\pi x}{b}$$

See **page F.7** with $W = 2b$.

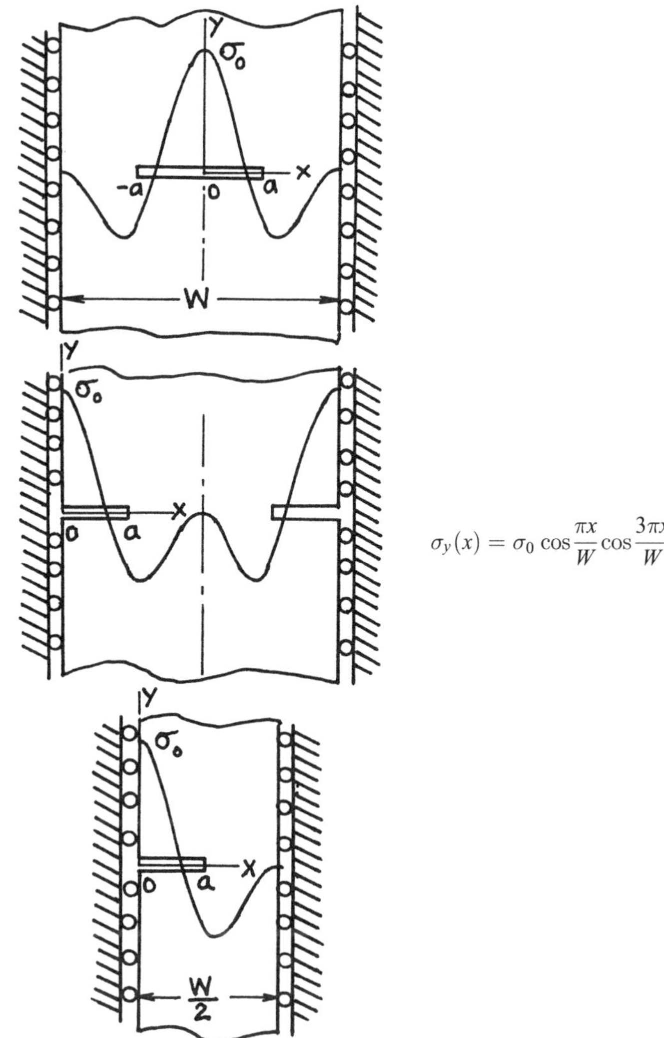

$$\sigma_y(x) = \sigma_0 \cos\frac{\pi x}{W} \cos\frac{3\pi x}{W}$$

See **page F.8**.

$$(K_I)_{\max} = K_{IA} = \frac{2}{\pi}\sigma_0\sqrt{\pi a}\cdot F_A\left(\frac{a}{c}\right)$$

$$(K_I)_{\min} = K_{1B} = \frac{2}{\pi}\sigma_0\sqrt{\pi a}\cdot F_B\left(\frac{a}{c}\right)$$

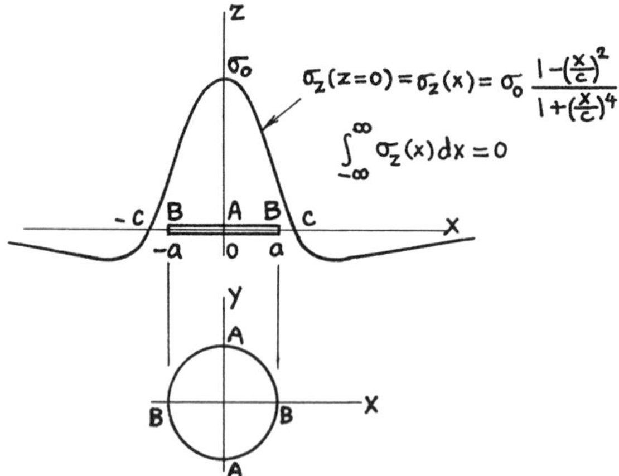

Volume of Crack:

$$V = \frac{16\left(1-\nu^2\right)}{3E}\sigma_0 a^3\cdot G\left(\frac{a}{c}\right)$$

Opening at Center:

$$\delta = \frac{8\left(1-\nu^2\right)}{\pi E}\sigma_0 a\cdot H\left(\frac{a}{c}\right)$$

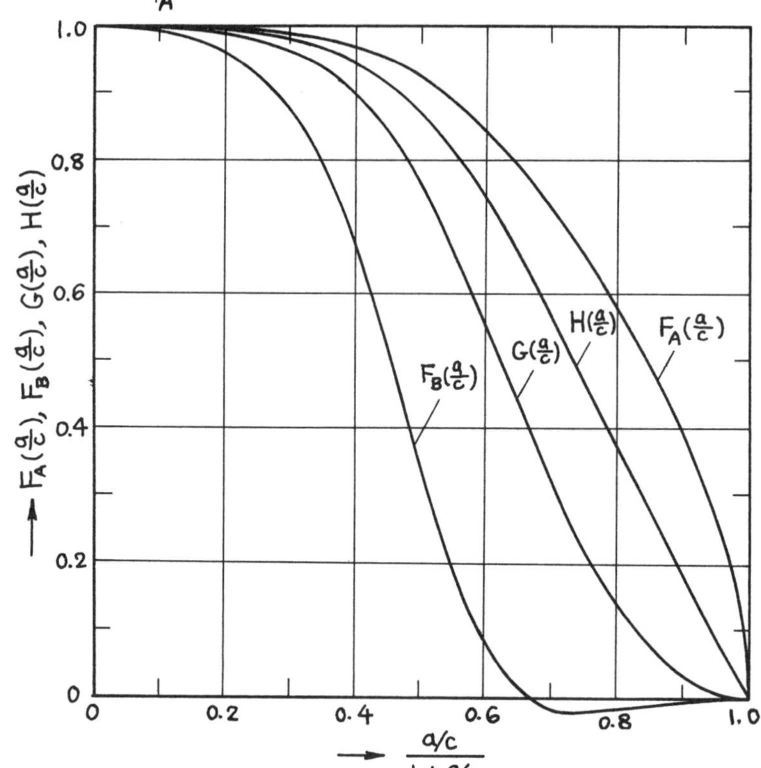

Method: Integration of **page 24.11**
Accuracy: Curves are based on numerical values with 0.1% accuracy.
Reference: **Tada 1985**

$$\sigma_z(z=0) = \sigma_z(x) = \sigma_0 \frac{1-\left(\frac{x}{c}\right)^2}{\left[1+\left(\frac{x}{c}\right)^2\right]^2}$$

$$\int_{-\infty}^{\infty} \sigma_z(x)\,dx = 0$$

$$(K_I)_{\max} = K_{IA} = \frac{2}{\pi}\sigma_0\sqrt{\pi a}\cdot F_A\left(\frac{a}{c}\right)$$

$$(K_I)_{\min} = K_{IB} = \frac{2}{\pi}\sigma_0\sqrt{\pi a}\cdot F_B\left(\frac{a}{c}\right)$$

Volume of Crack:

$$V = \frac{16\left(1-\nu^2\right)}{3E}\sigma_0 a^3\cdot G\left(\frac{a}{c}\right)$$

Opening at Center:

$$\delta = \frac{8\left(1-\nu^2\right)}{\pi E}\sigma_0 a\cdot H\left(\frac{a}{c}\right)$$

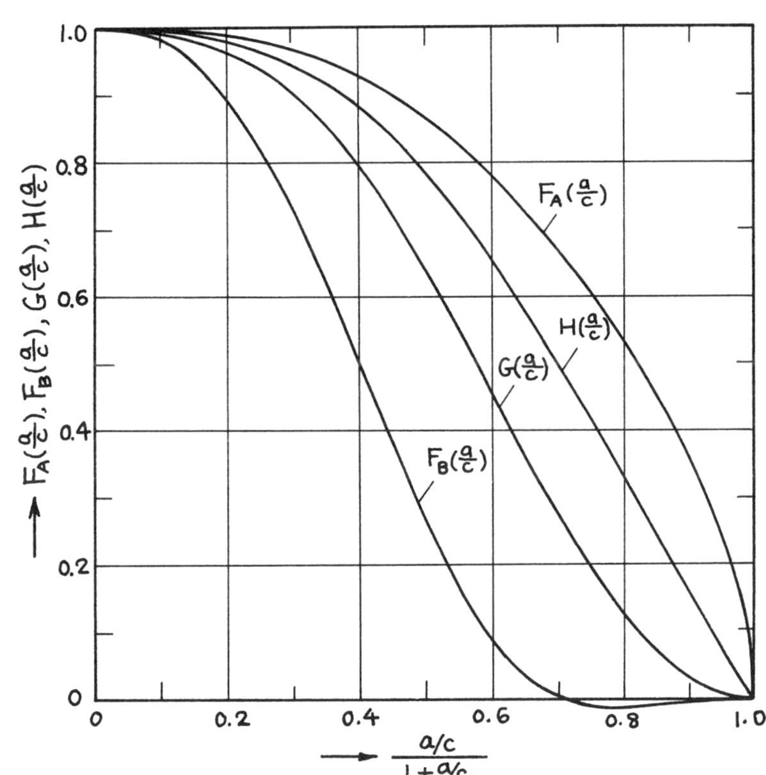

Method: Integration of **page 24.11**
Accuracy: Curves are based on numerical values with 0.1% accuracy.
Reference: **Tada 1985**

$$(K_I)_{\max} = K_{IA} = \frac{2}{\pi}\sigma_0\sqrt{\pi a}\cdot F_A\left(\frac{a}{c}\right)$$

$$(K_I)_{\min} = K_{IB} = \frac{2}{\pi}\sigma_0\sqrt{\pi a}\cdot F_B\left(\frac{a}{c}\right)$$

Volume of Crack:

$$V = \frac{16\left(1-\nu^2\right)}{3E}\sigma_0 a^3\cdot G\left(\frac{a}{c}\right)$$

Opening at Center:

$$\delta = \frac{8\left(1-\nu^2\right)}{\pi E}\sigma_0 a\cdot H\left(\frac{a}{c}\right)$$

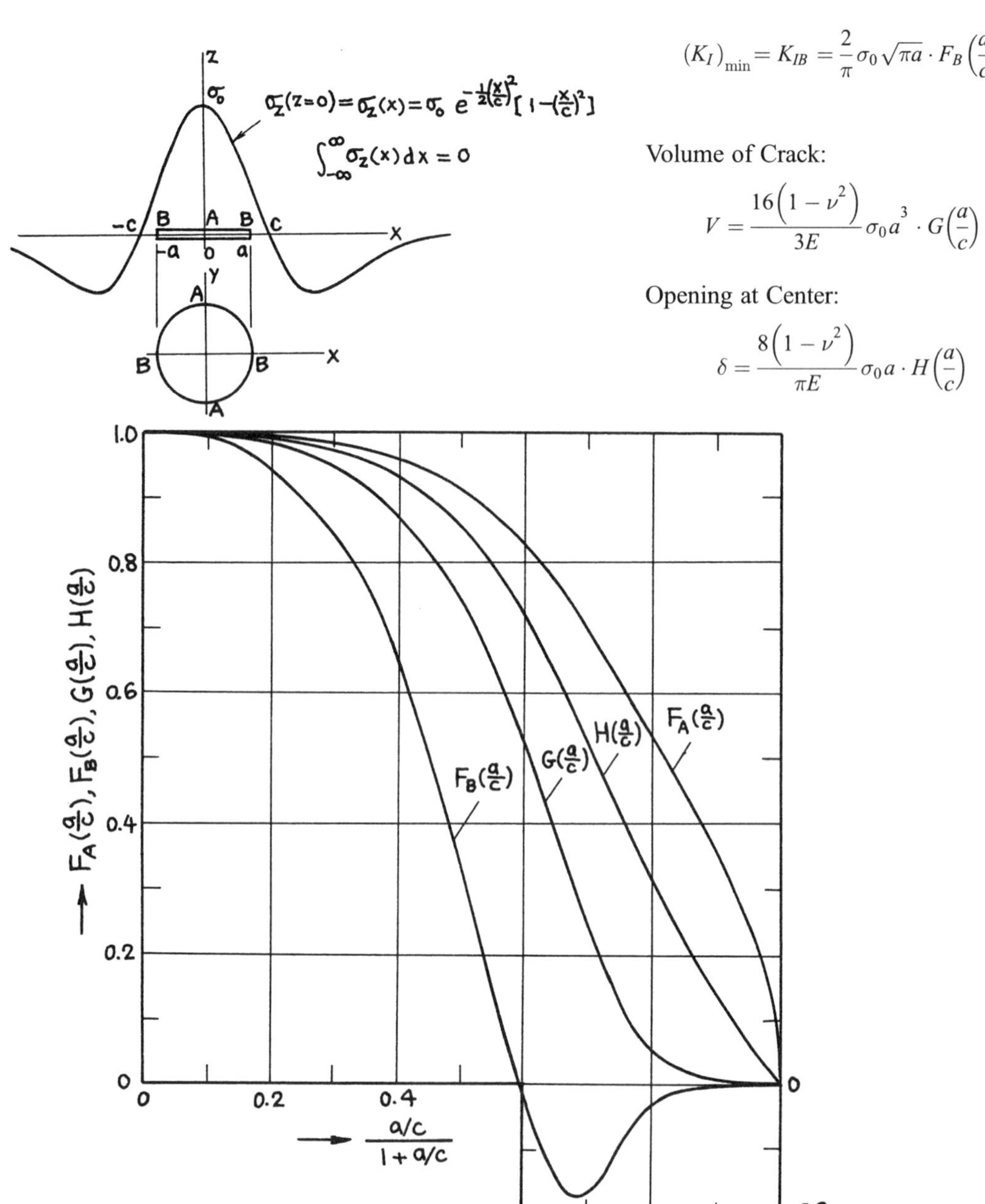

Method: Integration of **page 24.11**
Accuracy: Curves are based on numerical values with 0.1% accuracy.
Reference: **Tada 1985**

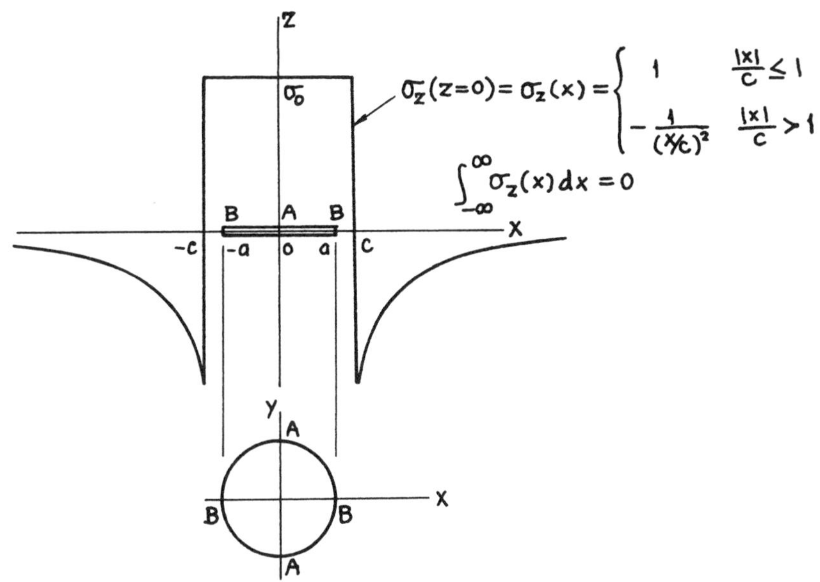

$$(K_I)_{\max} = K_{IA} = \frac{2}{\pi}\sigma_0\sqrt{\pi a}\cdot\begin{cases} 1 & \frac{a}{c}\le 1 \\ \frac{1}{3}\left[4\sqrt{\frac{c}{a}}-\left(\frac{c}{a}\right)^2\right] & \frac{a}{c}>1 \end{cases}$$

$$(K_I)_{\min} = K_{IB} = \frac{2}{\pi}\sigma_0\sqrt{\pi a}\cdot\begin{cases} 1 & \frac{a}{c}\le 1 \\ F_B\left(\frac{a}{c}\right) & \frac{a}{c}>1 \end{cases}$$

where

$$F_B\left(\frac{a}{c}\right) = \frac{1}{\sqrt{2}}\left[\left(2-\frac{c}{a}\right)\sqrt{1+\frac{c}{a}}-\left(2+\frac{c}{a}\right)\sqrt{1-\frac{c}{a}}-\left(\frac{c}{a}\right)^2\left(\coth^{-1}\sqrt{2}-\coth^{-1}\sqrt{1+\frac{c}{a}}+\tanh^{-1}\sqrt{1-\frac{c}{a}}\right)\right]$$

Volume of Crack:

$$V = \frac{16\left(1-\nu^2\right)}{3E}\sigma_0 a^3\cdot\begin{cases} 1 & \frac{a}{c}\le 1 \\ \left(\frac{c}{a}\right)^2\left(3-2\frac{c}{a}\right) & \frac{a}{c}>1 \end{cases}$$

Opening at Center:

$$\delta = \frac{8\left(1-\nu^2\right)}{\pi E}\sigma_0 a\cdot\begin{cases} 1 & \frac{a}{c}\le 1 \\ \frac{c}{a}\left(2-\frac{c}{a}\right) & \frac{a}{c}>1 \end{cases}$$

Method: Integration of **page 24.11**
Accuracy: Exact
Reference: **Tada 1985**

Appendix G

Westergaard Stress Functions for Dislocations and Cracks[1]

INTRODUCTION

The single-stress function approach of **Westergaard (1939)** is effective for a certain class of crack problems in which cracks occupy collinear line segments (e.g., along the x-axis). Considerable effort has been expended on finding Westergaard stress functions for various load-geometry configurations, and solutions have been obtained for many crack problems (**Westergaard, 1939**). Few Westergaard functions, however, are available for displacement-prescribed problems because of the presumably inherent mathematical difficulties. As noted in a few examples that were solved by other methods, the so-called "rigid wedge" problems or dislocation problems, where displacements of the crack surfaces are partially or wholly prescribed, seem to entail a great deal of mathematical complication involving, in general, single or multiple integral equations (**Barenblatt, 1962; Sneddon, 1969a; Tweed, 1970**).

In this appendix, for ease of subsequent descriptions, the term "dislocations" is generally used in the following sense to distinguish it from the term "cracks": "dislocations" represent macromechanical "cracks with wholly prescribed surface displacements"; whereas "cracks" retain their ordinary interpretation as "cracks with wholly prescribed surface tractions." The micromechanical analogy of "dislocations" will prove to be physically and analytically evident.

The objective of this appendix is to present a different approach to such dislocation problems and then extend it to combined problems of dislocations and cracks using the Westergaard stress function method. It will be shown that the stress functions for dislocations can be more easily obtained than those for cracks and also that the combined problems of dislocations and cracks are nothing more than the ordinary "crack" problems.

The nature of displacement discontinuity and the resulting internal stress field due to a Volterra dislocation in an elastic medium are well known (**Love, 1944; Cottrell, 1953**). The three fundamental displacement discontinuities of this kind are of the form

$$[u] = \Delta_{II}, \quad [v] = \Delta_I, \quad \text{and } [w] = \Delta_{III} \quad \text{on } x < 0, \; y = 0$$

where u, v, and w are, respectively, displacement components in x, y, and z directions, and Δ_{II}, Δ_I, and Δ_{III} are constants. These two-dimensional displacement fields correspond to the Mode II, Mode I, and Mode III displacements, respectively, in fracture mechanics. An analysis based on the theory of dislocation in crystals using distributed screw (Mode III) or edge (Modes I and II) dislocations has also received previous attention (**Cottrell, 1953; Bilby, 1968**). In this appendix, our attention is focused on Mode I problems.

---

[1] **Published in two parts with some modifications: Tada 1993, and 1994.**

A crack that is filled with a wedge of uniform thickness, the classical dislocation problem, leads to a Westergaard stress function having a $^1\!/_z$ singularity at the crack tip or at the end of the wedge. This observation readily leads to the Green's function (i.e., the solution for the "concentrated displacement" problem) for dislocation problems. It is then only necessary to integrate the Green's function to obtain Westergaard stress functions for dislocations with arbitrarily prescribed displacements. Once this method is established, the combined problems of dislocations and cracks are, in effect, reduced to the ordinary crack problems. The present method thus requires no more than integrations, even for the combined problems, and significantly simplifies the analysis.

An analysis of dislocations is presented in the first half of this appendix, which is followed by an analysis of combined problems in the second half. Many examples follow both analyses, and some of them are discussed in detail.

STRESS FUNCTIONS FOR DISLOCATIONS WITH UNIFORM THICKNESS

We define the displacement among the x-axis by

$$D(x) = 2v(x,0) = \overset{+}{v}(x,\ 0) - \overset{-}{v}(x,\ 0) \tag{G1}$$

where the $\pm$ signs indicate the upper and lower surfaces along $y = 0$.

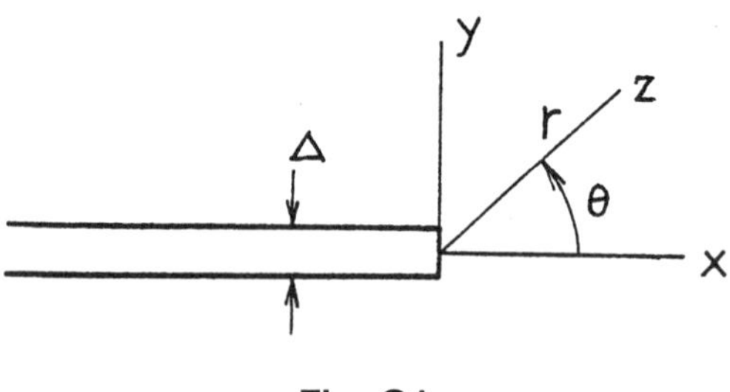

Fig. G1

First we consider the Volterra dislocation with discontinuity along the negative x-axis, as shown in **Fig. G1**. The discontinuous displacement is given by

$$\begin{aligned} D_0(x) &= \Delta\,[1 - H(x)] \\ \text{or}\quad &= \Delta\, H(-x) \end{aligned} \tag{G2}$$

where $H(x)$ denotes the unit step function.

For Westergaard stress functions, $Z(z)$, $z = x + iy$, the displacement along $y = 0$ is given by **Eq. (39)**:

$$2v(x,0) = \frac{4}{E'}\ \text{Im}\{\,\overline{Z}(z)\}_{y=0} \tag{G3}$$

where the notation Im{ } and, for subsequent use, Re{ } denote the imaginary and real part of { }, respectively, and $\overline{Z}(z) = \int Z(z)dz$, $E' = E$ for plane stress condition, and $E' = E/(1-\nu^2)$ for plane strain condition ($E' = 4\alpha G$). Noting that for polar coordinates (r,θ)

$$\ell n\ z = \ell n\ r + i\theta$$

and that $x > 0,\ y = 0$ and $x < 0,\ y = 0$ give $\theta = 0$ and $\theta = \pm\pi$, respectively, $\overline{Z}(z)$, which is in agreement with the displacement condition of **Eq. (G2)** is given by

$$\overline{Z}_0(z) = \frac{E'\Delta}{4\pi}\ell n\ z \tag{G4}$$

The corresponding Westergaard stress function, $Z(z)$, is

$$Z_0(z) = \frac{E'\Delta}{4\pi}\frac{1}{z} \tag{G4a}$$

Eqs. (G4) and (G4a) are well-known functions for the Volterra dislocation, or for an edge dislocation with the Burgers vector Δ in the y-direction.

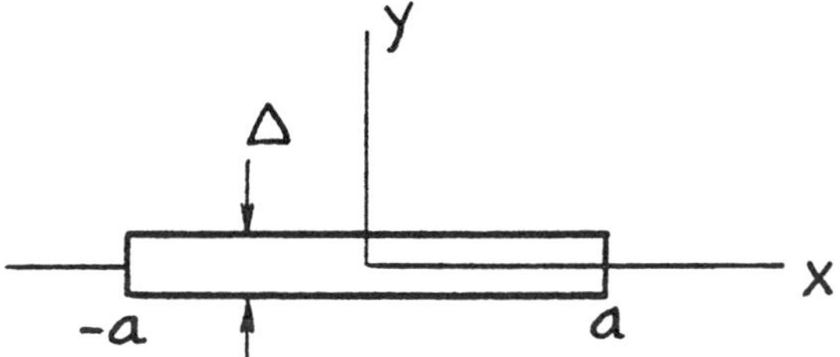

Fig. G2

Next consider a few simple examples that are directly derived from the previous solution. The first example is a uniform dislocation of finite length, as illustrated by **Fig. G2**. Since the displacement along the x-axis is given by

$$\begin{aligned} D(x) &= \Delta\left[H\{-(x-a)\} - H\{-(x+a)\}\right] \\ \text{or} \quad &= D_0(x-a) - D_0(x+a) \end{aligned} \tag{G5}$$

we can readily write the stress function $Z(z)$ and $\overline{Z}(z)$ by superposition of two dislocations with opposite signs at $(\pm a, 0)$:

$$\begin{aligned} Z(z) &= Z_0(z-a) - Z_0(z+a) \\ &= \frac{E'\Delta}{4\pi}\left(\frac{1}{z-a} - \frac{1}{z+a}\right) \\ \text{or} \quad &= \frac{E'\Delta}{4\pi}\left(\frac{2a}{z^2 - a^2}\right) \end{aligned} \tag{G6a}$$

and

$$\overline{Z}(z) = \frac{E'\Delta}{4\pi}\ \ell n\left(\frac{z-a}{z+a}\right) \tag{G6b}$$

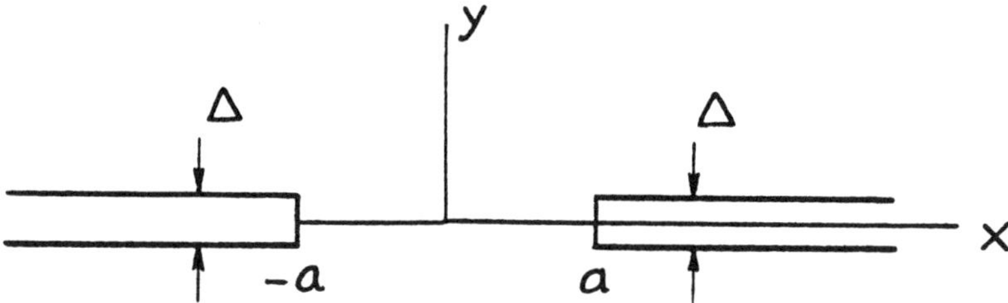

Fig. G3

Similarly, for two semi-infinite dislocations, as shown in Fig. G3,

$$\begin{aligned} D(x) &= \Delta[H\{-(x+a)\} - H(x-a)] \\ \text{or} \quad &= D_0(x+a) - D_0(a-x) \end{aligned} \tag{G7}$$

thus

$$Z(z) = \frac{E'\Delta}{4\pi}\left(\frac{2a}{a^2 - z^2}\right) \tag{G8a}$$

and

$$\overline{Z}(z) = \frac{E'\Delta}{4\pi}\,\ell\text{n}\left(\frac{a-z}{a+z}\right) \tag{G8b}$$

For special dislocation problems under consideration, unlike usual stress-prescribed crack problems, there are no interactions (couplings) among stress functions and the stress functions are obtained by direct superposition of individual stress functions for each dislocation that exists independently. Thus for collinearly located dislocations, as shown in **Fig. G4**, since displacement discontinuities are given by

$$D(x) = \sum_{k=1}^{n} \Delta_k[H\{-(x-\beta_k)\} - H\{-(x-\alpha_k)\}] \tag{G9}$$

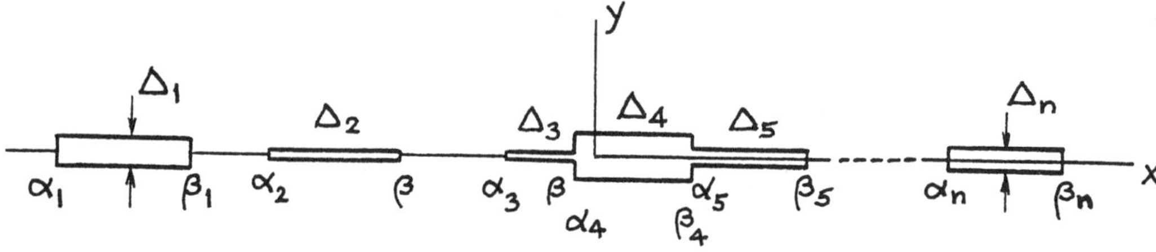

Fig. G4

we can write the stress function $Z(z)$ and $\overline{Z}(z)$ as

$$Z(z) = \frac{E'}{4\pi}\sum_{k=1}^{n}\Delta_k\left[\frac{1}{z-\beta_k} - \frac{1}{z-\alpha_k}\right] \quad \text{(G10a)}$$

and

$$\overline{Z}(z) = \frac{E'}{4\pi}\sum_{k=1}^{n}\Delta_k\,\ell\text{n}\left(\frac{z-\beta_k}{z-\alpha_k}\right) \quad \text{(G10b)}$$

The situations shown in **Figs. G1, G2, and G3** are special cases of **Fig. G4**. In **Fig. G4**, if we put, for example, $\Delta_1 = \Delta$, $\alpha_1 = -\infty$, $\beta_1 = 0$ with $\Delta_k = 0 (k = 2, \cdots, n)$, $\Delta_1 = \Delta$, $\alpha_1 = -a$, $\beta_1 = a$ with $\Delta_k = 0$ $(k = 2, \cdots, n)$, and $\Delta_1 = \Delta_n = \Delta$, $\alpha_1 = -\infty$, $\beta_1 = -a$, $\alpha_n = a$, $\beta_n = +\infty$ with $\Delta_k = 0$ $(k = 2, \cdots, n-1)$, we have the situations of **Figs. G1, G2, and G3**, respectively.

Another example is the periodic array of identical dislocations with discontinuity Δ, length $2a$, and period $2b$, as shown in **Fig. G5**.

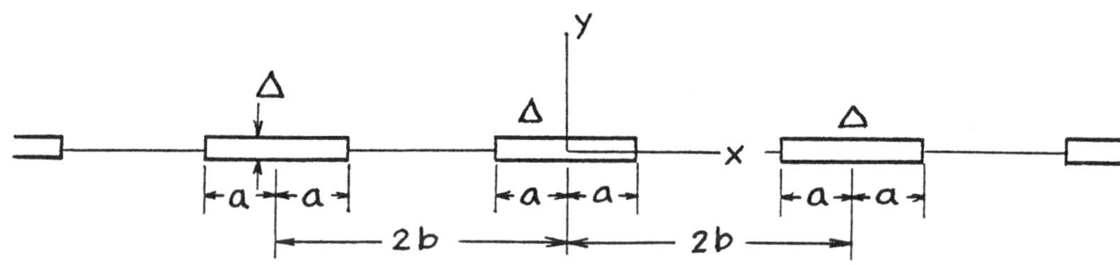

Fig. G5

Substituting $\Delta_k = \Delta$, $\alpha_k = 2kb - a$, $\beta_k = 2kb + a$ into **Eqs. (G10a) and (G10b)**, we can write $Z(z)$ and $\overline{Z}(z)$ as

$$Z(z) = \frac{E'\Delta}{4\pi}\sum_{k=-\infty}^{\infty}\left(\frac{1}{z-a-2kb} - \frac{1}{z+a-2kb}\right) \quad \text{(G11a)}$$

$$\overline{Z}(z) = \frac{E'\Delta}{4\pi}\sum_{k=-\infty}^{\infty}\ell\text{n}\left(\frac{z-a-2kb}{z+a-2kb}\right) \quad \text{(G11b)}$$

After some manipulations, we apply the identities given by **Eqs. (G12a) and (G12b)** to **Eqs. (G11a) and (G11b)**, respectively.

$$\sum_{k=0}^{\infty}\frac{1}{k^2-\alpha^2} = -\frac{\pi}{2\alpha}\left[\frac{1}{\pi\alpha} + \cot(\pi\alpha)\right] \quad \text{(G12a)}$$

$$\prod_{k=0}^{\infty}\left(1 - \frac{\alpha^2}{k^2}\right) = \frac{\sin\,\pi\alpha}{\pi\alpha} \quad \text{(G12b)}$$

Then these functions are reduced to the following simple, closed-form functions:

$$Z(z) = \frac{E'\Delta}{4\pi}\frac{\pi}{2b}\left[\cot\frac{\pi(z-a)}{2b} - \cot\frac{\pi(z+a)}{2b}\right] \tag{G13a}$$

$$\overline{Z}(z) = \frac{E'\Delta}{4\pi}\,\ell\text{n}\,\frac{\sin\frac{\pi(z-a)}{2b}}{\sin\frac{\pi(z+a)}{2b}} \tag{G13b}$$

Note that either **Eq. (G13a) or (G13b)** can be derived directly from the other by the relationship $\overline{Z}(z) = \int Z(z)dz$ or $Z(z) = d\overline{Z}(z)/dz$.

GREEN'S FUNCTIONS FOR DISLOCATIONS OF ARBITRARY SHAPE

The previous results suggest the possibility of constructing stress functions for dislocations of arbitrary shape. We have considered so far the elastic response to dislocations whose discontinuous displacements are represented by the sum of the unit step functions. In order to obtain elastic solutions for dislocation problems of arbitrary displacement, it is more convenient to consider the elastic response to the dislocation displacement in the form of the unit impulse or the delta function. The delta function, $\delta(x)$, is related to the unit step function, $H(x)$, as follows:

$$\delta(x) = \frac{dH(x)}{dx} = -\frac{dH(-x)}{dx} \tag{G14}$$

The same relationship is satisfied between the elastic responses to the unit step dislocation, $H(x): Z_H(z)$, $\overline{Z}_H(z)$, and those to the unit impulse dislocation, $\delta(x): Z_\delta(z)$, $\overline{Z}_\delta(z)$.

$$Z_\delta(z) = \frac{dZ_H(z)}{dx} = \frac{dZ_H(z)}{dz} \tag{G15a}$$

$$\overline{Z}_\delta(z) = \frac{d\overline{Z}_H(z)}{dz} = Z_H(z) \tag{G15b}$$

Comparison of **Eqs. (G14), (G15a), and (G15b)** with **Eqs. (G2), (G4a), and (G4)** leads to the stress functions for the dislocation with displacement density $d \cdot \delta(x)$. This nucleus of displacement may be construed as a "concentrated displacement". Noting that $Z_H(z) = -\frac{1}{\Delta}Z_0(z)$, these functions are written as follows:

$$Z(z) = -\frac{d}{\Delta}Z_0'(z) = \frac{E'd}{4\pi}\frac{1}{z^2} \tag{G16a}$$

$$\overline{Z}(z) = -\frac{d}{\Delta}Z_0(z) = -\frac{E'd}{4\pi}\frac{1}{z} \tag{G16b}$$

These functions serve as the Green's functions for dislocations of arbitrary shape.

The distinct advantage of this method is that the desired displacements can be built up by putting together elementary "concentrated displacements" with no cracks to begin with, thus avoiding the interactions (couplings) among the solutions for individual dislocations. That is, these solutions can be directly superimposed; consequently, Westergaard stress functions are, in general, actually more readily obtained for dislocations than for cracks.

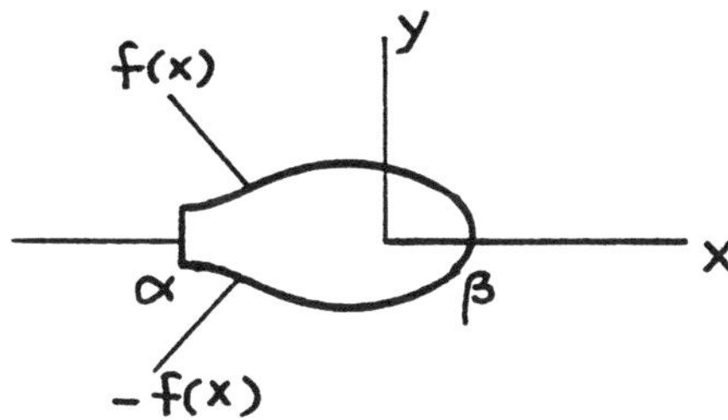

Fig. G6

Let a dislocation displacement on the x-axis, $D(x)$, defined by **Eq. (G1)**, be given by a piecewise continuous function $f(x)$ that is nonnegative[2] in a region $\alpha \leq x \leq \beta$ and zero in regions $x < \alpha$ and $x > \beta$, as shown in **Fig. G6**. That is,

$$\begin{aligned} D(x) &= 2f(x), \qquad \alpha \leq x \leq \beta \\ D(x) &= 0 \qquad\qquad x < \alpha,\ x > \beta \end{aligned} \tag{G17}$$

Then, using the Green's functions given by **Eqs. (G16a) and (G16b)**, the Westergaard stress functions $Z(z)$ and $\overline{Z}(z)$ are obtained by simple integrations.

$$Z(z) = \frac{E'}{2\pi}\int_{\alpha}^{\beta} \frac{f(x')}{(z-x')^2}\,dx' \tag{G18a}$$

$$\overline{Z}(z) = -\frac{E'}{2\pi}\int_{\alpha}^{\beta} \frac{f(x')}{z-x'}\,dx' \tag{G18b}$$

Again, note that either $Z(z)$ or $\overline{Z}(z)$ of **Eq. (G18)** is also derived directly from the other.

[2] This requirement is only for physical purposes.

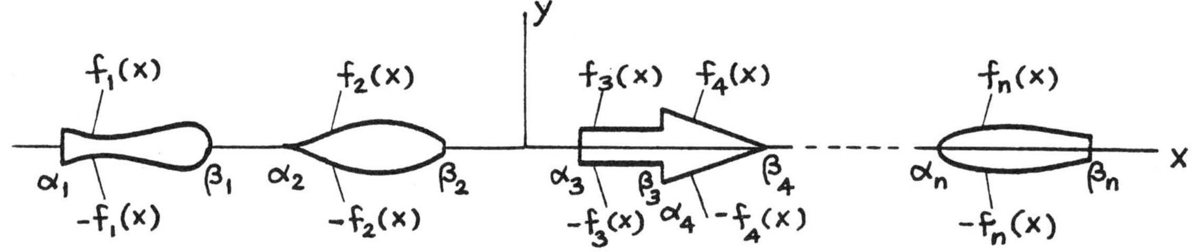

Fig. G7

When the dislocation displacement, $D(x)$, consists of more than one piecewise continuous functions, $2f_i(x)$, which are nonnegative in $\alpha_i \leq x \leq \beta_i$ and zero in $\beta_{i-1} < x < \alpha_i$ and $\beta_i < x < \alpha_{i+1}$, as shown in **Fig. G7**, by simply applying the direct superposition of stress functions, we can write the stress functions as

$$Z(z) = \frac{E'}{2\pi} \sum_{i=1}^{n} \int_{\alpha_i}^{\beta_i} \frac{f_i(x')}{(z-x')^2} dx' \tag{G19a}$$

$$\overline{Z}(z) = -\frac{E'}{2\pi} \sum_{i=1}^{n} \int_{\alpha_i}^{\beta_i} \frac{f_i(x')}{z-x'} dx' \tag{G19b}$$

Note that integrating **Eq. (G18a)** by parts we have its alternative form

$$Z(z) = \frac{E'}{2\pi} \left[\frac{f(x')}{z-x'} \bigg|_{\alpha}^{\beta} - \int_{\alpha}^{\beta} \frac{f'(x')}{z-x'} dx' \right] \tag{G18a$'$}$$

where $f'(x) = df(x)/dx$. The choice between **Eqs. (G18a) and (G18a)$'$** can be made on the simplicity of the integration.

WESTERGAARD STRESS FUNCTIONS FOR SEVERAL DISLOCATION PROBLEMS

Some examples are given to illustrate the possibilities for easily obtaining solutions.

1. Periodically located identical dislocations whose displacements densities are given by $d \cdot \delta(x \pm 2kb)$, $k = 0, \ldots, \infty$ **(Fig. G8)**.

Applying the relations given by the first halves of **Eqs. (G16a) and (G16b)** to **Eqs. (G13a) and (G13b)**, respectively, we readily have

$$Z(z) = \frac{E'd}{4\pi} \left(\frac{\pi}{2b} \operatorname{cosec} \frac{\pi z}{2b} \right)^2 \tag{G20a}$$

$$\overline{Z}(z) = \frac{E'd}{4\pi} \left(-\frac{\pi}{2b} \cot \frac{\pi z}{2b} \right) \tag{G20b}$$

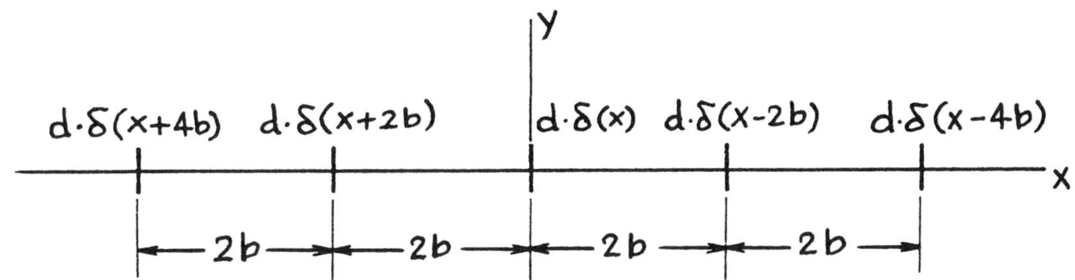

Fig. G8

These functions can also be obtained from the Green's functions, **Eqs. (G16a) and (G16b)**, by the direct superposition

$$Z(z) = \frac{E'd}{4\pi} \sum_{k=-\infty}^{\infty} \frac{1}{(z-2kb)^2}$$

$$\overline{Z}(z) = -\frac{E'd}{4\pi} \sum_{k=-\infty}^{\infty} \frac{1}{z-2kb}$$

with the aid of the following identity [or the relation $\overline{Z}(z) = \int Z(z)dz$] in addition to **Eq. (G12a)**

$$\sum_{k=0}^{\infty} \frac{a^2+k^2}{\left(k^2-a^2\right)^2} = \frac{1}{2}\left[\frac{1}{a^2} + (\pi \operatorname{cosec} \pi a)^2\right] \tag{G12c}$$

Eqs. (G20a) and (G20b) can serve as the Green's functions for periodically repeated dislocations with arbitrary shape.

2a. Diamond-shaped dislocation (**Fig. G9**).
$f(x)$ is given by

$$f(x) = \frac{\Delta_0}{2}\left(1 - \frac{|x|}{2}\right)$$

or

$$= \alpha(a - |x|) \qquad |x| \leq a$$

$$f(x) = 0 \qquad |x| > a$$

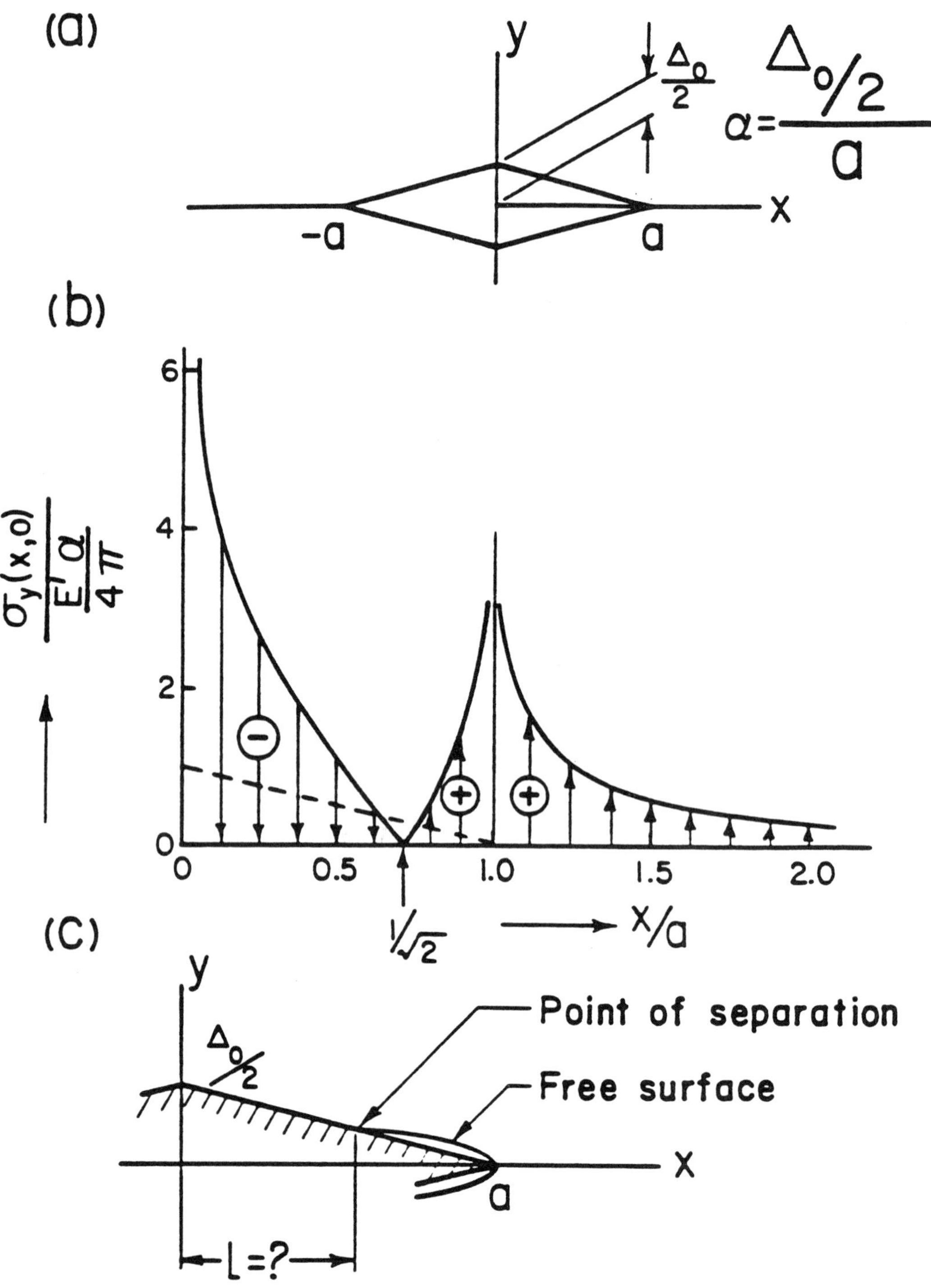
(a)
y
$\Delta_0/2$
$\alpha = \frac{\Delta_0/2}{a}$
x
-a
a
(b)
$\frac{\sigma_y(x,0)}{\frac{E'\alpha}{4\pi}}$
6
4
2
0
0
0.5
1.0
1.5
2.0
$1/\sqrt{2}$
x/a
(c)
y
$\Delta_0/2$
Point of separation
Free surface
x
a
L=?

Fig. G9

Using **Eq. (G18a)**, we have the following Westergaard stress function:

$$Z(z) = \frac{E'\alpha}{4\pi}\,\ell\text{n}\,\frac{z^2}{(z+a)(z-a)} \tag{G21a}$$

and

$$\begin{aligned}\bar{Z}(z) &= \int Z(z)dz \\ &= \frac{E'\alpha}{4\pi}\,\ell\text{n}\,\frac{\left(z^2\right)^z}{(z+a)^{z+a}(z-a)^{z-a}}\end{aligned} \tag{G21b}$$

The following observation supports visualization of our intuition that we cannot always fill a crack with a rigid wedge of arbitrary shape without welding the surfaces together.

The normal stress distribution, $\sigma_y(x,0)$, is calculated from **Eq. (G21a)** using **Eq. (38)**

$$\begin{aligned}\sigma_y(x,0) &= \text{Re}\{Z(z)\}_{y=0} \\ &= \frac{E'\alpha}{4\pi}\,\ell\text{n}\,\frac{x^2}{(x+a)(x-a)}, \quad |x|> a \\ &= \frac{E'\alpha}{4\pi}\,\ell\text{n}\,\frac{x^2}{(a+x)(a-x)}, \quad |x|< a\end{aligned} \tag{G21a$'$}$$

The stress distribution given by **Eq. (G21a)′** is plotted in **Fig. G9b**. As readily seen, to maintain the diamond shape we have to apply the tensile surface stress over the region $a/\sqrt{2} < x < a$. This implies that when a diamond-shaped wedge is inserted into a crack with equal length, the contact between the surfaces is lost and free surfaces are formed near the tips of the wedge (see **Fig. G9c**). The location of the point of separation is, however, not known beforehand. The method for determining the position of this point is discussed subsequently.

2b. Dislocation shown in **Fig. G10**.

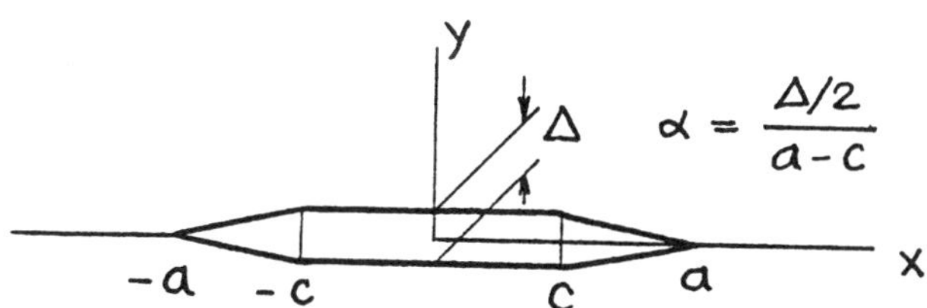

Fig. G10

$$f(x) = \Delta/2 = \alpha(a-c) \qquad |x| \le c$$

$$f(x) = \alpha(a-|x|) \qquad c \le |x| \le a$$

$$f(x) = 0 \qquad |x| > a$$

By direct superpositions of **Eqs. (G21a) and (G21b)**, we immediately have the following solutions:

$$Z(z) = \frac{E'\alpha}{4\pi} \ell n \frac{(z+c)(z-c)}{(z+a)(z-a)} \tag{G22a}$$

$$\overline{Z}(z) = \frac{E'\alpha}{4\pi} \ell n \frac{(z+c)^{z+c}(z-c)^{z-c}}{(z+a)^{z+a}(z-a)^{z-a}} \tag{G22b}$$

3a. Periodically located diamond-shaped dislocations (**Fig. G11**).

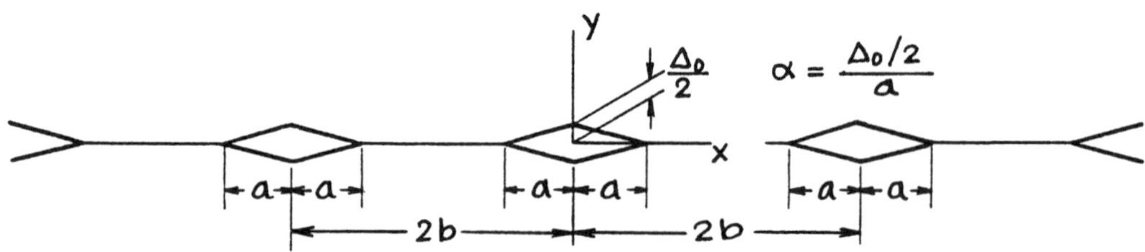

Fig. G11

By superposition of **Eq. (G21a)**, $Z(z)$ is given by

$$Z(z) = \frac{E'\alpha}{4\pi} \sum_{k=-\infty}^{\infty} F(z-2kb)$$

where

$$F(z) = \ell n \frac{z^2}{(z+a)(z-a)}$$

Using the identity given by **Eq. (G12b)**, we have $Z(z)$ in the following simple closed form:

$$Z(z) = \frac{E'\alpha}{4\pi} \ell n \frac{\left(\sin\frac{\pi z}{2b}\right)^2}{\sin\frac{\pi(z+a)}{2b}\sin\frac{\pi(z-a)}{2b}} \tag{G23}$$

3b. Periodic dislocations shown in **Fig. G12**.

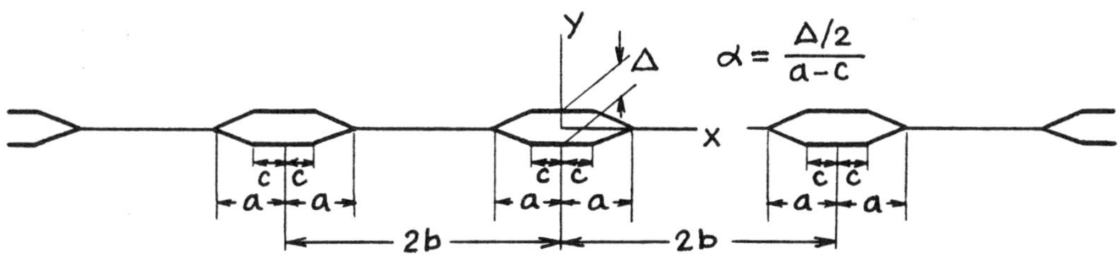

Fig. G12

Directly from **Eq. (G23)**, we have

$$Z(z) = \frac{E'\alpha}{4\pi}\,\ell\text{n}\,\frac{\sin\frac{\pi(z+c)}{2b}\ \sin\frac{\pi(z-c)}{2b}}{\sin\frac{\pi(z+a)}{2b}\ \sin\frac{\pi(z-a)}{2b}}$$

(G24)

or

$$= \frac{E'\alpha}{4\pi}\,\ell\text{n}\,\frac{\left(\sin\frac{\pi z}{2b}\right)^2 - \left(\sin\frac{\pi c}{2b}\right)^2}{\left(\sin\frac{\pi z}{2b}\right)^2 - \left(\sin\frac{\pi a}{2b}\right)^2}$$

4. Elliptical-shaped dislocation (**Fig. G13**).

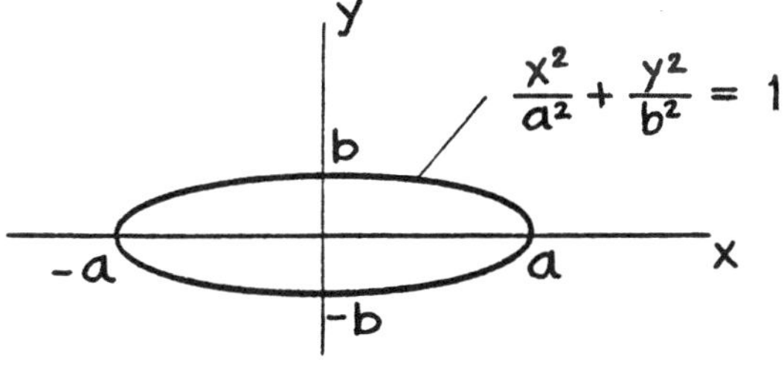

Fig. G13

$f(x)$ for this case is given by

$$f(x) = b\sqrt{1-\left(\frac{x}{a}\right)^2} \qquad |x| \le a$$

$$f(x) = 0 \qquad |x| > a$$

We can immediately write the stress functions $Z(z)$ and $\overline{Z}(z)$ for this problem by comparison with the well-known solution of the Griffith crack under uniform internal pressure as follows [see **Eq. (71)**].

$$Z(z) = \frac{E'b}{2a}\left[\frac{1}{\sqrt{1-(a/z)^2}} - 1\right] \tag{G25a}$$

$$\overline{Z}(z) = \frac{E'b}{2a}\left[\sqrt{z^2 - a^2} - z\right] \tag{G25b}$$

In fact, **Eqs. (G25a) and (G25b)** are obtained by the direct application of **Eqs. (G18a) and (G18b)**, respectively.

The solution shows that, when an internal crack in an infinite plate is filled with a smooth rigid wedge with elliptical shape, the traction between the surfaces becomes uniform compression. If uniform remote tensile stress, $E'b/2a$, is applied, these surfaces become traction free.

5. Parabolic-shaped dislocation (**Fig. G14**).

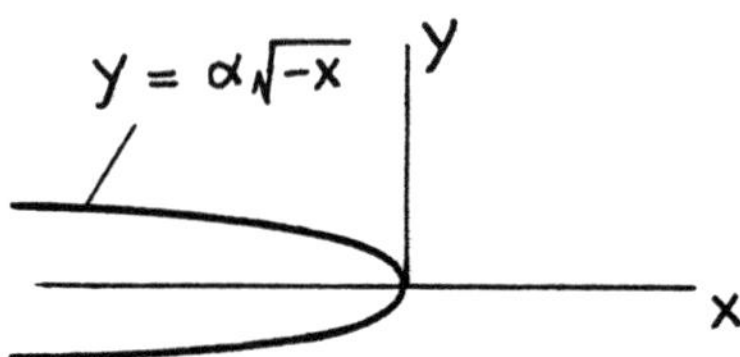

Fig. G14

$f(x)$ for this example is

$$f(x) = \alpha\sqrt{-x} \qquad x \leq 0$$

$$f(x) = 0 \qquad x > 0$$

Again we can immediately expect the functions $Z(z)$ and $\overline{Z}(z)$ to be identical to those for a semi-infinite crack with stress-free surfaces along the negative x-axis (see **page 3.1**). In fact, by **Eq. (G18a)**,

$$\begin{aligned} Z(z) &= \frac{E'\alpha}{2\pi}\int_{-\infty}^{0} \frac{\sqrt{-x'}}{(z-x')^2}\,dx' = \frac{E'\alpha}{2\pi}\int_{0}^{\infty} \frac{\sqrt{\xi}}{(z+\xi)^2}\,d\xi \\ &= \frac{E'\alpha}{4}\frac{1}{\sqrt{z}} \end{aligned} \tag{G26a}$$

and

$$\overline{Z}(z) = \int Z(z)\,dz = \frac{E'\alpha}{2}\sqrt{z} \tag{G26b}$$

Eq. (G26) gives the elastic field near the tip of a crack with free surfaces on $x < 0,\ y = 0$, where the remote loads cause the stress intensity factor $K = \sqrt{\pi}E'\alpha/(2\sqrt{2})$, which is readily obtained by comparing (G26a) with $Z(z) = K/\sqrt{2\pi z}$ on **page 3.1**. The crack-tip field for the preceding example (4), since $K = (E'b/2a)\ \sqrt{\pi a}$, corresponds to the case where $\alpha = \sqrt{2}\ b/\sqrt{a}$.

6. A Dislocation whose shape is given by **Eq. (G27)** **(Fig. G15)**.

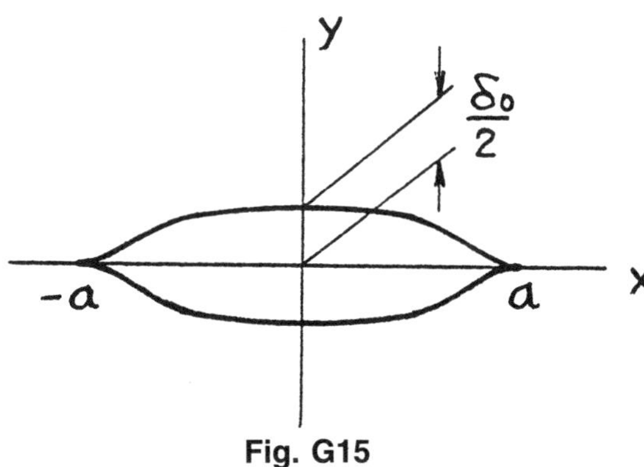

Fig. G15

$$f(x) = \frac{\delta_0}{2}\left\{1 - \left(\frac{x}{a}\right)^2\right\}^{\alpha}, \qquad \alpha \geq 0 \qquad |x| \leq a$$
$$f(x) = 0 \qquad |x| > a \tag{G27}$$

By **Eq. (G18a)**, $Z(z)$ can be obtained from the following integral:

$$Z(z) = \frac{E'\delta_0}{4\pi}\int_{-a}^{a}\frac{\left\{1 - \left(\frac{x'}{a}\right)^2\right\}^{\alpha}}{(z - x')^2}dx'$$

After some manipulations, we arrive at

$$Z(z) = \frac{E'\delta_0}{2\sqrt{\pi}a}\frac{\Gamma(\alpha + 1)}{\Gamma\left(\alpha + \frac{1}{2}\right)}{}_2F_1\left(1,\ \frac{1}{2} - \alpha;\ \frac{1}{2};\ \left(\frac{z}{a}\right)^2\right) \tag{G28}$$

where ${}_2F_1(\alpha,\ \beta;\ \gamma;\ z)$ is the usual notation for the hypergeometric series.

Note that special cases $\alpha = 0$ and $\alpha = \frac{1}{2}$ correspond to the cases of **Fig. G2** and **Fig. G13**, respectively.

Sneddon (1969a) solved this problem as an example of his general solution which was derived by the method of Fourier transforms and dual integral equations. His interpretation of the problem is to find the surface stress distribution that is necessary to maintain the Griffith crack in the prescribed shape. The general solution was given in the form of two relatively simple integrations. The present method requires only one simple integral and seems to provide a simpler and more direct approach to the problem.

The stress distribution, $\sigma_y(x, 0)$, calculated from these stress functions on the dislocations (i.e., the displacement-prescribed segments) is interpreted as the surface stress necessary to maintain the crack surfaces in the prescribed shape. Therefore, when this stress is compressive, that is, $\sigma_y(x, 0) < 0$, over the entire portion of the crack, the problem is equivalent to the insertion of a smooth rigid wedge with prescribed shape into the crack. Otherwise, the stress distribution is regarded as the traction between the surfaces that would result if the surfaces of the wedge and crack are welded together. When the surfaces are not welded, free surfaces form over certain portions of the crack, as discussed in the example (4) above (**Fig. G9**). The positions of the points of separation of the surfaces are determined by the method discussed in the following sections.

COMBINED PROBLEMS OF DISLOCATIONS AND CRACKS

When dislocations (or cracks with fully prescribed displacements) and cracks in the ordinary sense (cracks with prescribed surface stress distributions) are present collinearly on the x-axis, the Westergaard stress functions can be obtained relatively easily. The method is simply successive applications of the preceding method of dislocation analysis and the usual method of stress analysis of cracks. The procedure of the analysis is as follows.

First, obtain the stress function for the dislocation only, $Z_D(z)$, assuming that cracks are not present, by use of the method described in the preceding sections. The normal stress distribution on the x-axis due to the dislocation, $\sigma_{yD}(x, 0)$, is calculated by **Eq. (38)**

$$\sigma_{yD}(x,0) = \mathrm{Re}\{Z_D(z)\}_{y=0} \tag{G29}$$

This is the "crack-absent" stress distribution to be removed along the cracks. Thus the problem is now reduced to the ordinary crack problem where the crack surfaces are assumed to be stress free. Green's functions for the Westergaard stress function, $Z_G(z,x')$, are available in this handbook for various crack geometries. The stress function for the crack, $Z_C(z)$, under the stress given by **Eq. (G29)**, is calculated by the integral

$$Z_C(z) = \int_C \sigma_{yD}(x',0) Z_G(z,x')dx' \tag{G30}$$

where the integral $\int_C (\ldots)dx'$ is a definite integral over the entire portion of the crack. Since the stress function, $Z_C(z)$, generates no vertical displacement, $v(x,0)$, outside the crack, the prescribed shape of dislocation is maintained. Thus the combined stress function for the dislocation and the crack is obtained by the superposition of $Z_D(z)$ and $Z_C(z)$, that is,

$$Z(z) = Z_D(z) + Z_C(z) \tag{G31}$$

As discussed earlier, the normal stress, $\sigma_y(x,\ 0)$, on the dislocations, calculated from the total stress function $Z(z)$, **Eq. (G31)**, is the stress distribution necessary to maintain the dislocations in the prescribed shapes. When this stress is compressive over the entire segments of dislocations, the problem is equivalent to plugging the thin rigid wedges with the same prescribed shapes into the corresponding cracks. When the stress is tensile over any portion of the segments, the equivalency is retained only if surfaces of the cracks and the wedges are welded. Otherwise, the contact between the surfaces will be lost and free surfaces will form. Then, the portions of the free surface should be treated as the usual stress-prescribed cracks, that is, $Z_C(z)$ should incorporate the analysis of these portions while $Z_D(z)$ remains unchanged. The points of separation of the surfaces, however, are generally not known beforehand. The method for determining the positions of the separation points is also discussed subsequently.

In the subsequent discussion, the method is applied to several very simple examples to illustrate the possibilities for obtaining the stress function for more complicated problems without great mathematical difficulties.

A CRACK PARTIALLY FILLED WITH A RIGID WEDGE

As a simple illustrative problem of this type, consider a semi-infinite, thin rigid wedge with uniform thickness, Δ, inserted into a semi-infinite crack leaving free surfaces of length $2a$ ahead of the wedge, as shown in **Fig. G16**. Note that **Fig. G16** is a limiting case of **Fig. G23**. For convenience, we take the origin of the coordinates at the center of the stress-free portion.

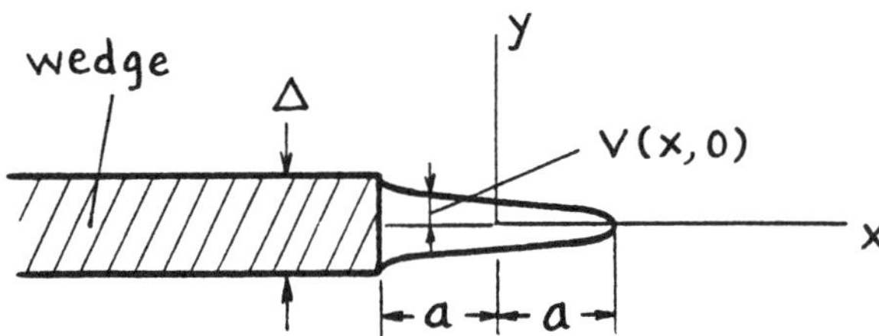

Fig. G16

When there is no free surface ahead of the wedge (or when the crack is filled with the wedge), the stress function is given by **Eq. (G4a)**. That is, the stress function, $Z_D(z)$, in this case is

$$Z_D(z) = \frac{E'\Delta}{4\pi}\frac{1}{z+a} \tag{G32}$$

The corresponding crack-absent stress distribution, $\sigma_{yD}(x, 0)$, is calculated from **Eq. (G29)**

$$\sigma_{yD}(x, 0) = \ \mathrm{Re}\{Z_D(z)\}_{y=0} = \frac{E'\Delta}{4\pi}\frac{1}{x+a} \tag{G33}$$

The Green's function, $Z_G(z, x')$, for a crack in an infinite plane is well known as (see **page 5.10**)

$$Z_G(z, x') = \frac{1}{\pi}\frac{\sqrt{a^2 - x'^2}}{(z - x')\sqrt{z^2 - a^2}} \tag{G34}$$

The additional stress function due to crack, $Z_C(z)$, is calculated by performing the integration of **Eq. (G30)**

$$\begin{aligned} Z_C(z) &= \int_{-a}^{a} \sigma_{yD}(x', 0) Z_G(z, x')dx' \\ &= \frac{E'\Delta}{4\pi^2}\frac{1}{\sqrt{z^2 - a^2}}\int_{-a}^{a}\frac{1}{z - x'}\sqrt{\frac{a - x'}{a + x'}}dx' \end{aligned}$$

The final expression of $Z_C(z)$ has the following simple form.

$$Z_C(z) = \frac{E'\Delta}{4\pi}\left\{\frac{1}{\sqrt{z^2 - a^2}} - \frac{1}{z + a}\right\} \tag{G35}$$

The total stress function of the problem is then, by **Eq. (G31)**, written as

$$Z(z) = Z_D(z) + Z_C(z) = \frac{E'\Delta}{4\pi}\frac{1}{\sqrt{z^2 - a^2}} \tag{G36a}$$

and

$$\overline{Z}(z) = \int Z(z)dz = \frac{E'\Delta}{4\pi}\cosh^{-1}\frac{z}{a} \tag{G36b}$$

These functions are among the few Westergaard functions known for displacement-prescribed crack problems.

The stress intensity factor, $K_{\pm a}$, at the ends of the free surface, $x = \pm a$, is directly obtained from **Eq. (G36a)** as

$$\begin{aligned} K_{+a} &= \lim_{x \to a+0}\left\{\sqrt{2\pi(x-a)}\ \sigma_y(x,0)\right\} \\ &= \lim_{x \to a+0}\left[\sqrt{2\pi(x-a)}\ \mathrm{Re}\{Z(z)\}_{y=0}\right] \\ K_{+a} &= \frac{E'\Delta}{4\sqrt{\pi a}} \end{aligned} \tag{G37a}$$

Similarly,

$$K_{-a} = -\frac{E'\Delta}{4\sqrt{\pi a}} \tag{G37b}$$

The profile of the free surface (see **Fig. G16**), $v(x,0)$; $|x| \leq a$, is calculated from **Eq. (G36b)** by the formula given by **Eq. (G3)**.

$$\begin{aligned} \underset{|x|\leq a}{v(x,0)} &= \frac{2}{E'}\ \mathrm{Im}\{\overline{Z}(z)\}_{y=0} \\ &= \frac{\Delta}{2\pi}\cos^{-1}\frac{x}{a} \end{aligned} \tag{G38}$$

Eqs. (G37) and (G38) agree with the known solution **(Barenblatt, 1962)** (see **page 3.11** and the dashed curve in **Fig. G24**).

In the preceding example, we discussed the case where the position of the point of separation of the crack surface and the wedge is definite and known beforehand. For the case where the position of the separating point is not known, we can still apply the present method by simply imposing an additional condition at the point of separation. Some discussion on a problem of this type is found in Barenblatt's review **(Barenblatt, 1962)**. The following simple examples illustrate the applicability of the method.

Consider a semi-infinite crack whose elastic field is characterized by the stress intensity factor K_{appl}. When a thin, smooth, rigid wedge with uniform thickness, Δ, is inserted into the crack up to the leading edge, the contact between the crack surface and the wedge will occur over a portion, ℓ, near the crack tip, as shown in **Fig. G17**, and the surfaces will separate outside this portion since the crack opening profile is parabolic before the insertion of the wedge.

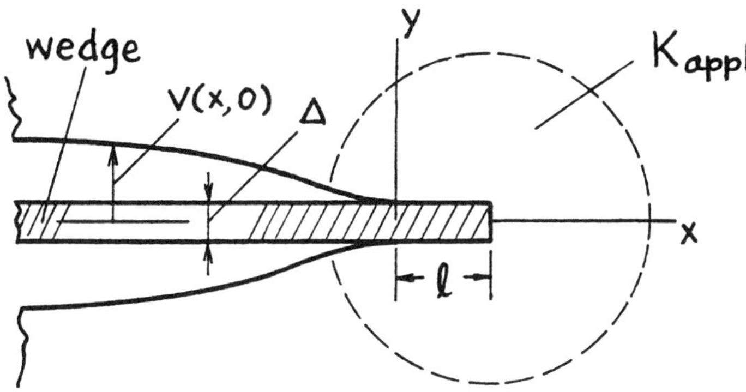

Fig. G17

To apply the present method to this example, we interpret the problem in the following way. First, consider a semi-infinite crack filled with the wedge. We apply a remote external load that would generate the stress intensity factor K_{appl} if the wedge is not inserted. The separation of the contact surfaces will occur at a certain distance, ℓ, from the tip. For convenience, the coordinates are taken as shown in **Fig. G17**. $Z_D(z)$ and $\sigma_{yD}(x,0)$ for this case are

$$Z_D(z) = \frac{E'\Delta}{4\pi}\frac{1}{z-\ell} \tag{G39}$$

and

$$\sigma_{yD}(x,0) = \frac{E'\Delta}{4\pi}\frac{1}{x-\ell} \tag{G40}$$

$Z_G(z,x')$ for a semi-infinite crack is known as (see **page 3.6**)

$$Z_G(z,x') = \frac{1}{\pi}\sqrt{\frac{-x'}{z}}\cdot\frac{1}{z-x'} \tag{G41}$$

Thus $Z_C(z)$ is calculated as

$$\begin{aligned} Z_C(z) &= \frac{E'\Delta}{4\pi^2}\frac{1}{\sqrt{z}}\int_{-\infty}^{0}\frac{1}{x'-\ell}\cdot\frac{\sqrt{-x'}}{z-x'}\,dx' = \frac{E'\Delta}{4\pi^2}\frac{1}{\sqrt{z}}\int_{0}^{\infty}\frac{1}{\ell+x'}\cdot\frac{\sqrt{x'}}{z+x'}\,dx' \\ &= -\frac{E'\Delta}{4\pi}\left[\sqrt{\frac{\ell}{z}}\frac{1}{\ell-z}-\frac{1}{\ell-z}\right] \end{aligned} \tag{G42}$$

Note that the function

$$Z_D(z)+Z_C(z) = \frac{E'\Delta}{4\pi}\frac{1}{z-\ell}\sqrt{\frac{\ell}{z}} \tag{G43}$$

is the solution to the problem where the portion of the wedge that is located $x<0$ is removed from the crack (**Fig. G18**). This solution is also obtained using the finite wedge which fills a single crack, as follows (**Fig. G19**). For this case (see **Eq. (G6a)**)

$$Z_D(z) = \frac{E'\Delta}{4\pi}\left(\frac{1}{z-\ell}-\frac{1}{z}\right) \tag{G44}$$

$$\sigma_{yD}(x,0) = \frac{E'\Delta}{4\pi}\left(\frac{1}{x-\ell}-\frac{1}{x}\right) \tag{G45}$$

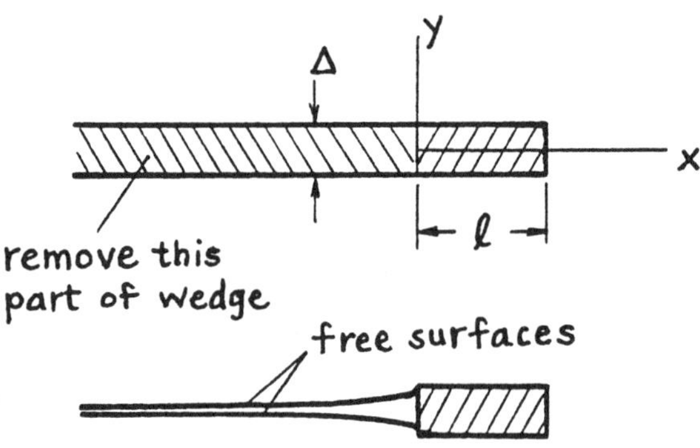

Fig. G18

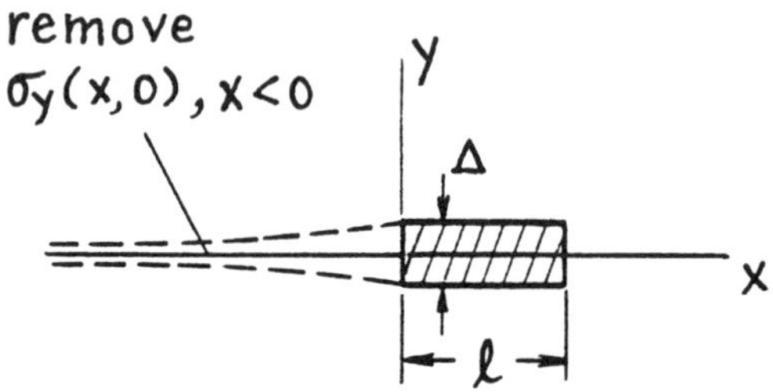

Fig. G19

$Z_G(z, x')$ is the same as **Eq. (G41)**. Thus

$$\begin{aligned} Z_C(z) &= \int_{-\infty}^{0} \sigma_{yD}(x', 0) Z_G(z, x') dx \\ &= \frac{E'\Delta}{4\pi} \frac{1}{z-\ell} \sqrt{\frac{\ell}{z}} - Z_D(z) \end{aligned}$$

$$Z_D(z) + Z_C(z) = \frac{E'\Delta}{4\pi} \frac{1}{z-\ell} \sqrt{\frac{\ell}{z}} \quad \text{(G46)}$$

Eq. (G46) agrees with **Eq. (G43)**.

Since we have an additional crack-stress field, $Z_{C\,appl}(z)$, due to applied external load, which is given by (see **page 3.1**)

$$Z_{C\,appl}(z) = \frac{K_{appl}}{\sqrt{2\pi z}} \quad \text{(G47)}$$

the total stress function, $Z(z)$, is given by

$$\begin{aligned} Z(z) &= Z_D(z) + Z_C(z) + Z_{C\,appl}(z) \\ &= \frac{E'\Delta}{4\pi}\frac{1}{z-\ell}\sqrt{\frac{\ell}{z}} + \frac{K_{appl}}{\sqrt{2\pi z}} \end{aligned} \tag{G48}$$

The unknown length of contact, ℓ, is obtained in terms of Δ and K_{appl} from the boundary condition at the point of separation. The condition is that the separation of surfaces occurs smoothly, or that the stress singularity at the separation point, $K_{x=0}$, is zero. That is, from **Eq. (G48)**,

$$-\frac{E'\Delta}{2\sqrt{2\pi\ell}} + K_{appl} = 0$$

or

$$\ell = \frac{1}{8\pi}\left(\frac{E'\Delta}{K_{appl}}\right)^2 \tag{G49}$$

Eq. (G48) is then simplified as

$$Z(z) = \frac{E'\Delta}{4\pi}\frac{1}{z-\ell}\sqrt{\frac{z}{\ell}}$$

or

$$= \frac{K_{appl}}{\sqrt{2\pi}}\frac{\sqrt{z}}{z-\ell} \tag{G50}$$

The function $\overline{Z}(z) = \int Z(z)dz$ is also written in a simple form as

$$\overline{Z}(z) = \frac{E'\Delta}{2\pi}\left\{\sqrt{\frac{z}{\ell}} + \frac{1}{2}\ell\text{n}\left(\frac{\sqrt{z/\ell}-1}{\sqrt{z/\ell}+1}\right)\right\} \tag{G51}$$

The stress distribution, $\sigma_y(x,0)$, and the displacement, $v(x,0)$, along the x-axis are calculated from **Eqs. (G50) and (G51)**, respectively, as

$$\sigma_y(x,0) = \text{Re}\{Z(z)\}_{y=0} = \begin{cases} 0 & x < 0 \\ \dfrac{E'\Delta}{4\pi\sqrt{\ell}}\dfrac{\sqrt{x}}{x-\ell}\left(=\dfrac{K_{appl}}{\sqrt{2\pi}}\dfrac{\sqrt{x}}{x-\ell}\right) & x \geq 0 \end{cases} \tag{G52}$$

$$v(x,0) = \frac{2}{E'}\,\text{Im}\{\overline{Z}(z)\}_{y=0} = \begin{cases} \dfrac{\Delta}{\pi}\left(\sqrt{-\dfrac{x}{\ell}} + \cot^{-1}\sqrt{-\dfrac{x}{\ell}}\right) & x \leq 0 \\ \Delta/2 & 0 \leq x \leq \ell \\ 0 & x > \ell \end{cases} \tag{G53}$$

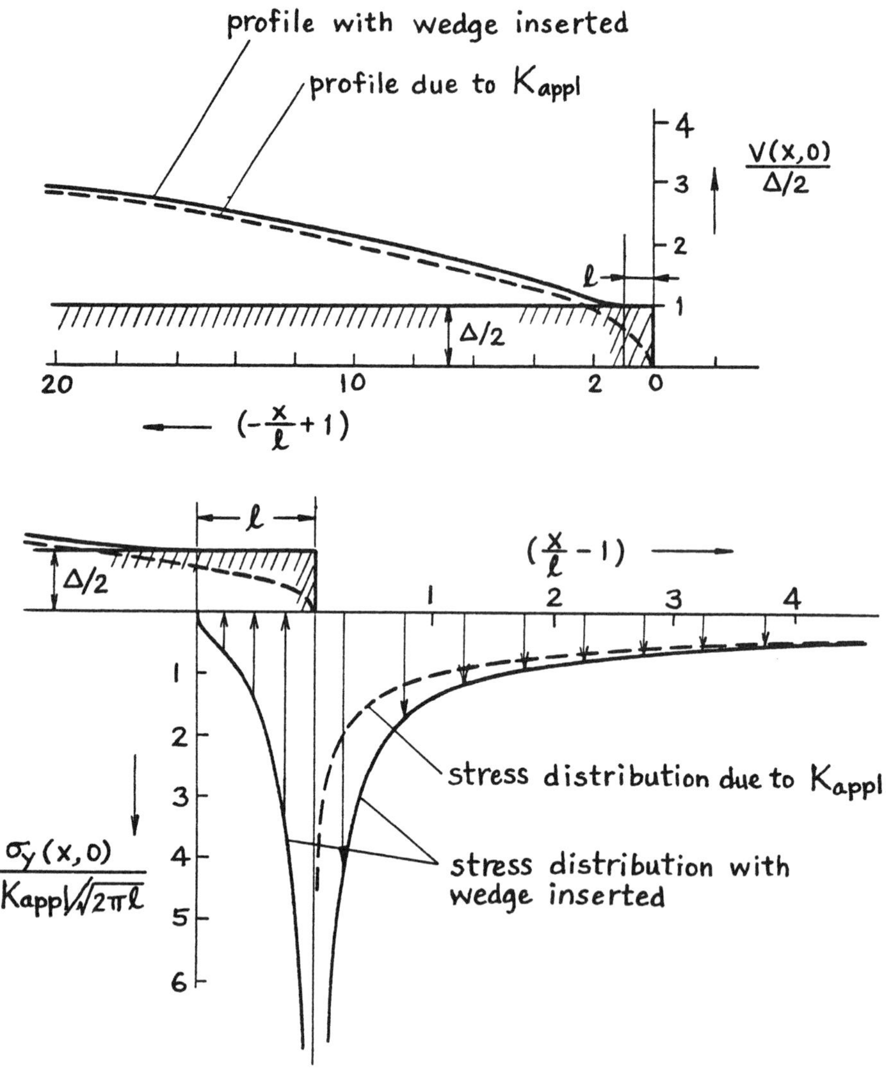

Fig. G20. Crack profile and stress distribution given by Eqs. (G53) and (G52), respectively.

The stress distribution and displacement given by **Eqs. (G52) and (G53)** are shown in **Fig. G20**, where the stress distribution and crack profile due to the applied force (K_{appl}) are also shown by dashed lines for comparison.

It is noted from **Eq. (G49)** that the contact between the wedge and the crack surfaces always exists ($\ell > 0$), as expected, regardless of the magnitude of applied load.

As another simple example of this type, we again consider the situation given by **Fig. G16**. In addition to the insertion of the wedge, the crack is now assumed to be subjected to a remote uniform compressive stress, p, as shown in **Fig. G21**.

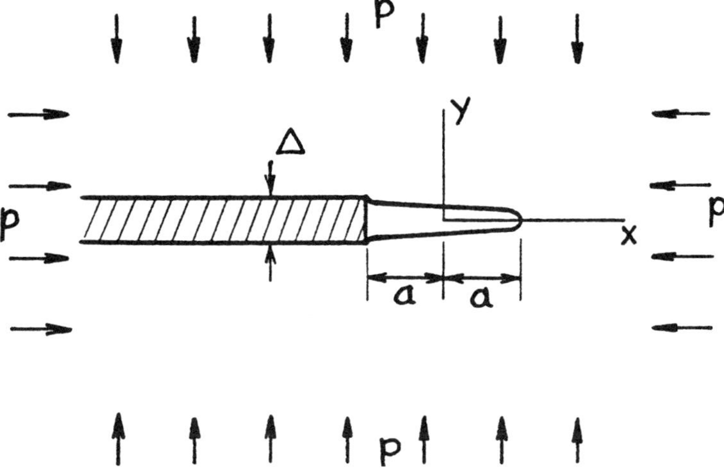

Fig. G21

The stress function is readily obtained by the superposition of **Eq. (G36)** and the well-known solution for the Griffith crack **(page 5.1)**. Thus the total stress functions are written as

$$Z(z) = \frac{E'\Delta}{4\pi}\frac{1}{\sqrt{z^2 - a^2}} - \frac{p}{\sqrt{1-(a/z)^2}} \tag{G54a}$$

$$\overline{Z}(z) = \frac{E'\Delta}{4\pi}\cosh^{-1}\frac{z}{a} - p\sqrt{z^2 - a^2} \tag{G54b}$$

The stress intensity factors at $x = \pm a$ are

$$K_{+a} = \frac{E'\Delta}{4\sqrt{\pi a}} - p\sqrt{\pi a} \tag{G55a}$$

$$K_{-a} = -\frac{E'\Delta}{4\sqrt{\pi a}} - p\sqrt{\pi a} \tag{G55b}$$

While $K_{-a} < 0$ always holds, from **Eq. (G55a)**,

$$K_{+a} \leq 0 \quad \text{when } p \geq p_0 = \frac{E'\Delta}{4\pi a} \tag{G56}$$

Therefore, **Eqs. (G54) and (G55)** are valid as they are when $p \leq p_0 = E'\Delta/(4\pi a)$. When the compressive stress p exceeds p_0, the crack closure will occur over a certain length, ℓ, from the right end of the crack, as shown in **Fig. G22b**. The length, ℓ, and the resulting elastic field are easily obtained by imposing the condition for smooth contact between the closed surfaces.

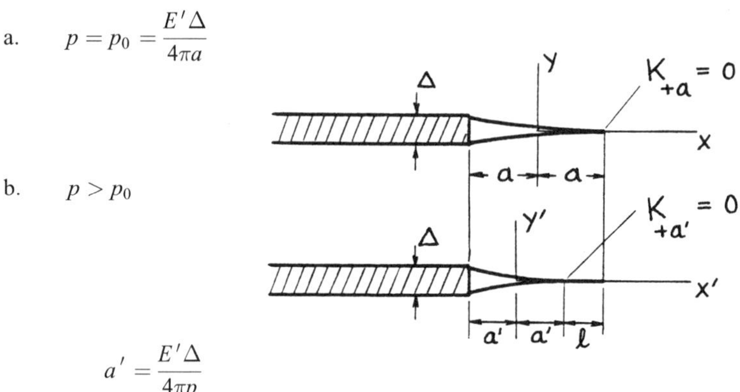

Fig. G22

We first examine the situation shown in **Fig. G22a**, where $p = p_0$. We can immediately write stress functions and $K_{\pm a}$ from **Eqs. (G54) and (G55)**.

$$Z(z) = -p_0\sqrt{\frac{z-a}{z+a}} \tag{G57a}$$

$$\overline{Z}(z) = p_0 a\left(\cosh^{-1}\frac{z}{a} - \sqrt{\left(\frac{z}{a}\right)^2 - 1}\right) \tag{G57b}$$

$$K_{+a} = 0 \tag{G58a}$$

$$K_{-a} = -2p_0\sqrt{\pi a} = -\frac{E'\Delta}{2\sqrt{\pi a}}\left(\text{or} = -\sqrt{E'\Delta p_0}\right) \tag{G58b}$$

Now the complete solution for $p > p_0$ (**Fig. G22b**) can be expressed in the same form simply replacing z, a and p_0 by z', a' and p, respectively, where $z' = z + \ell/2$, $a' = E'\Delta/(4\pi p)$, and $\ell = 2(a - a')$. Thus

$$Z(z') = -p\sqrt{\frac{z'-a'}{z'+a'}} \tag{G59a}$$

$$\overline{Z}(z') = \frac{E'\Delta}{4\pi}\left(\cosh^{-1}\frac{z'}{a'} - \sqrt{\left(\frac{z'}{a'}\right)^2 - 1}\right) \tag{G59b}$$

$$K_{+a'} = 0 \tag{G60a}$$

$$K_{-a'} = -\sqrt{E'p\Delta}\left(= -2p\sqrt{\pi a'} = -\frac{E'\Delta}{2\sqrt{\pi a'}}\right) \tag{G60b}$$

where

$$p = \frac{E'\Delta}{4\pi a'}$$

The length of closure, ℓ, is given by

$$\ell = 2(a - a') = 2\left(a - \frac{E'\Delta}{4\pi p}\right) = \frac{E'\Delta}{4\pi}\left(\frac{1}{p_0} - \frac{1}{p}\right) \tag{G61}$$

When an internal pressure, p, is applied inside the crack instead of a remote compression, it is readily shown that the contact between the surfaces along $x < -a$ is lost completely when the internal pressure p exceeds $p_0 = E'\Delta/(4\pi a)$.

CRACKS COLLINEARLY LOCATED WITH DISLOCATIONS

(Combinations of Displacement-Specified Cracks and Stress-Specified Cracks)

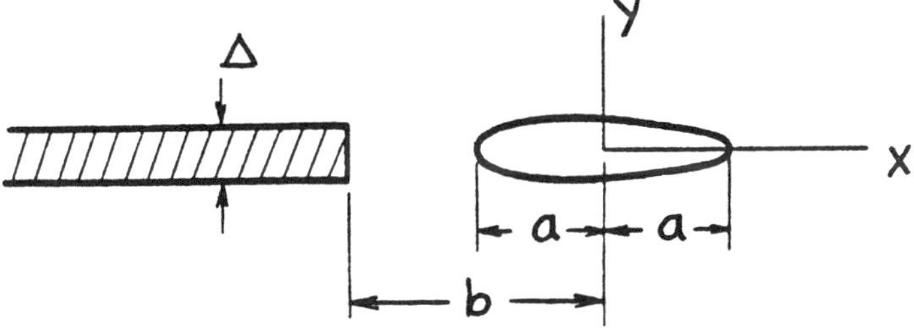

Fig. G23

In the very simple example shown in **Fig. G23**, a uniform dislocation with thickness Δ is located along $x \leq -b,\ y = 0$ and a single crack is located at $|x| \leq a,\ y = 0$. This situation is realized when a semi-infinite crack located at $x \leq -b,\ y = 0$ is filled with a thin, smooth, rigid wedge with uniform thickness Δ, leaving another crack stress free.

$Z_D(z)$ for this case is given, again from **Eq. (G4a)**, by

$$Z_D(z) = \frac{E'\Delta}{4\pi}\frac{1}{z+b} \tag{G62}$$

Then $\sigma_{yD}(x,0)$ is readily given by

$$\sigma_{yD}(x,0) = \frac{E'\Delta}{4\pi}\frac{1}{x+b} \tag{G63}$$

Since the Green's function, $Z_G(z,x')$, given by **Eq. (G34)** is again valid,

$$Z_C(z) = \int_C \sigma_{yD}(x',0)Z_G(z,x')dx' = \frac{E'\Delta}{4\pi^2}\frac{1}{\sqrt{z^2-a^2}}\int_{-a}^{a}\frac{\sqrt{a^2-x'^2}}{(z-x')(x'-b)}dx'$$

Performing the integration, we can obtain the following final expression for $Z_C(z)$:

$$Z_C(z) = \frac{E'\Delta}{4\pi}\frac{1}{\sqrt{z^2-a^2}}\left(1-\frac{\sqrt{b^2-a^2}}{z+b}\right) - \frac{E'\Delta}{4\pi}\frac{1}{z+b} \tag{G64}$$

Thus, the total stress function is

$$Z(z) = Z_D(z) + Z_C(z) = \frac{E'\Delta}{4\pi}\frac{1}{\sqrt{z^2-a^2}}\left(1-\frac{\sqrt{b^2-a^2}}{z+b}\right) \tag{G65a}$$

and

$$\overline{Z}(z) = \int Z(z)dz = \frac{E'\Delta}{4\pi}\left\{\cosh^{-1}\frac{z}{a} - i\cos^{-1}\frac{a^2+bz}{a(b+z)}\right\} \tag{G65b}$$

The stress intensity factors at $x = \pm a$ are

$$K_{+a} = \frac{E'\Delta}{4\sqrt{\pi a}}\left(1-\sqrt{\frac{b-a}{b+a}}\right) \tag{G66a}$$

and

$$K_{-a} = -\frac{E'\Delta}{4\sqrt{\pi a}}\left(1-\sqrt{\frac{b+a}{b-a}}\right) \tag{G66b}$$

When $a = b$, **Eq. (G65)** is reduced to **Eq. (G36)**, that is, the solution for the situation of **Fig. G16**. Therefore, it is seems interesting to examine how the profile of the crack changes when the ratio a/b varies.

The profile is calculated from **Eq. (G65b)** using **Eq. (G3)**.

$$v(x,0) = \frac{2}{E'}\,\mathrm{Im}\{\overline{Z}(z)\}_{y=0} = \frac{\Delta}{2\pi}\left\{\cos^{-1}\frac{x}{a} - \cos^{-1}\frac{a^2+bx}{a(b+x)}\right\} \tag{G67}$$

The shape of the crack given by **Eq. (G67)** is shown in **Fig. G24** for various values of a/b while the distance b is fixed for convenience.

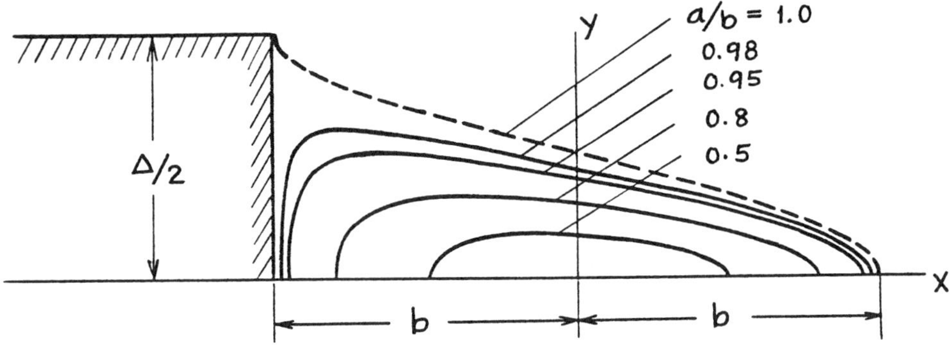

Fig. G24

When a is very small compared with b, the elastic field near the crack should correspond to a Griffith crack under uniform tension $\sigma_y(0,0) = E'\Delta/(4\pi b)$. In fact, when $^a/_b \ll 1$ and $^z/_b \ll 1$, **Eq. (G65a)** is reduced to the following well-known solution:

$$Z(z) = \frac{E'\Delta}{4\pi b}\frac{1}{\sqrt{1-\left(\frac{a}{z}\right)^2}} = \frac{\sigma_y(0,0)}{\sqrt{1-\left(\frac{a}{z}\right)^2}} \qquad \text{(G68)}$$

It should be noted that the solution, **Eq. (G65)**, is very useful in constructing solutions for various problems associated with a single stress-free (or stress-specified) crack and dislocations with uniform thickness. It is possible to derive solutions by superposition of **Eq. (G65)** and the limiting procedure without performing the integration of **Eq. (G30)**. Westergaard functions, $Z(z)$, for a few simple examples, which are directly obtained from **Eq. (G65)**, are illustrated next.

EXAMPLES OF WESTERGAARD FUNCTIONS ASSOCIATED WITH UNIFORM DISLOCATIONS AND A STRESS-FREE CRACK

a. Two opposing cracks, one of which is filled with an infinite uniform wedge, **Fig. G25(a)**.

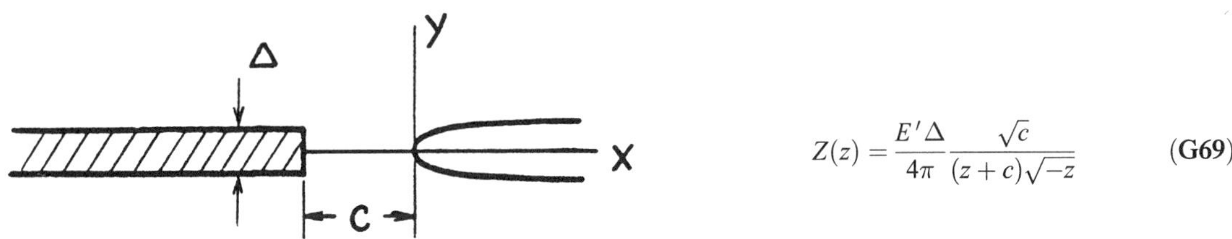

$$Z(z) = \frac{E'\Delta}{4\pi}\frac{\sqrt{c}}{(z+c)\sqrt{-z}} \qquad \text{(G69)}$$

Fig. G25(a)

b. A finite crack filled with uniform wedge and a semi-infinite crack, **Fig. G25(b)**.

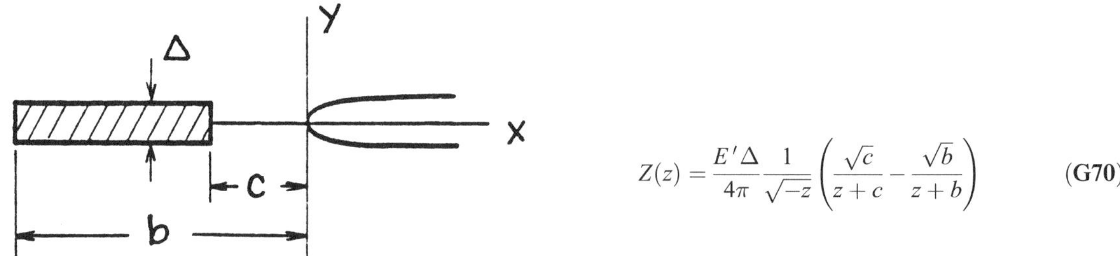

$$Z(z) = \frac{E'\Delta}{4\pi}\frac{1}{\sqrt{-z}}\left(\frac{\sqrt{c}}{z+c} - \frac{\sqrt{b}}{z+b}\right) \tag{G70}$$

Fig. G25(b)

c. A semi-infinite crack, the tip of which is filled with a uniform wedge with finite length, **Fig. G25(c)**.

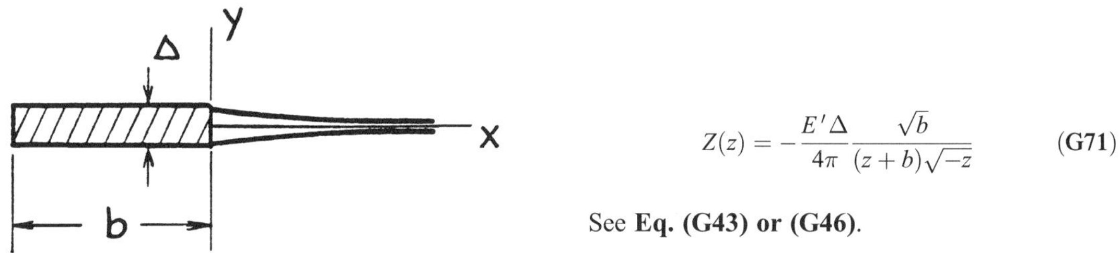

$$Z(z) = -\frac{E'\Delta}{4\pi}\frac{\sqrt{b}}{(z+b)\sqrt{-z}} \tag{G71}$$

See **Eq. (G43) or (G46)**.

Fig. G25(c)

d. Two finite cracks, one of which is filled with a uniform wedge, **Fig. G25(d)**.

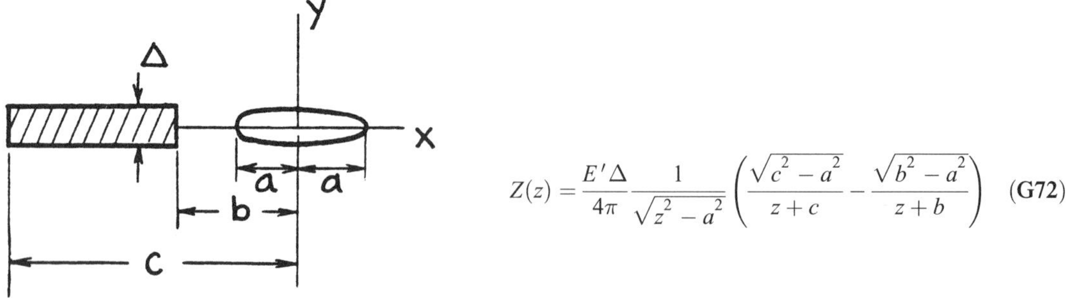

$$Z(z) = \frac{E'\Delta}{4\pi}\frac{1}{\sqrt{z^2-a^2}}\left(\frac{\sqrt{c^2-a^2}}{z+c} - \frac{\sqrt{b^2-a^2}}{z+b}\right) \tag{G72}$$

Fig. G25(d)

e. A finite crack partially filled with a uniform wedge at one end, **Fig. G25(e)**.

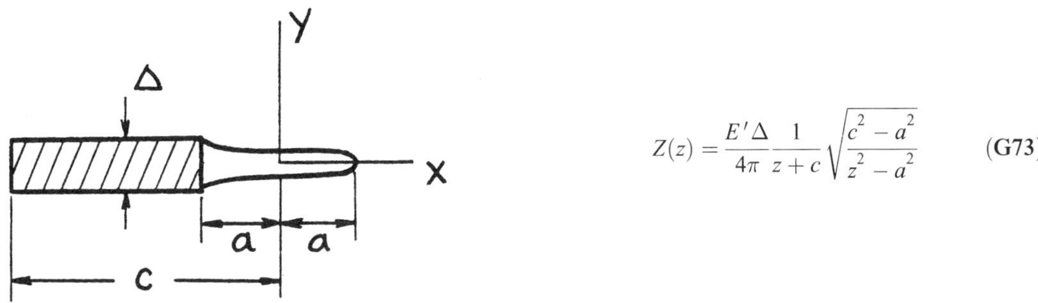

$$Z(z) = \frac{E'\Delta}{4\pi}\frac{1}{z+c}\sqrt{\frac{c^2-a^2}{z^2-a^2}} \tag{G73}$$

Fig. G25(e)

f. A finite crack partially filled with two identical uniform wedges at both ends, **Fig. G25(f)**.

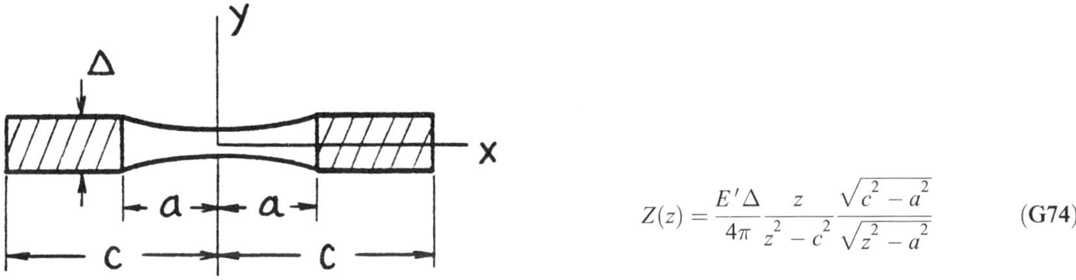

$$Z(z) = \frac{E'\Delta}{4\pi}\frac{z}{z^2-c^2}\frac{\sqrt{c^2-a^2}}{\sqrt{z^2-a^2}} \tag{G74}$$

Fig. G25(f)

g. A finite crack filled with a uniform wedge subjected to remote tension, **Fig. G25(g)**.

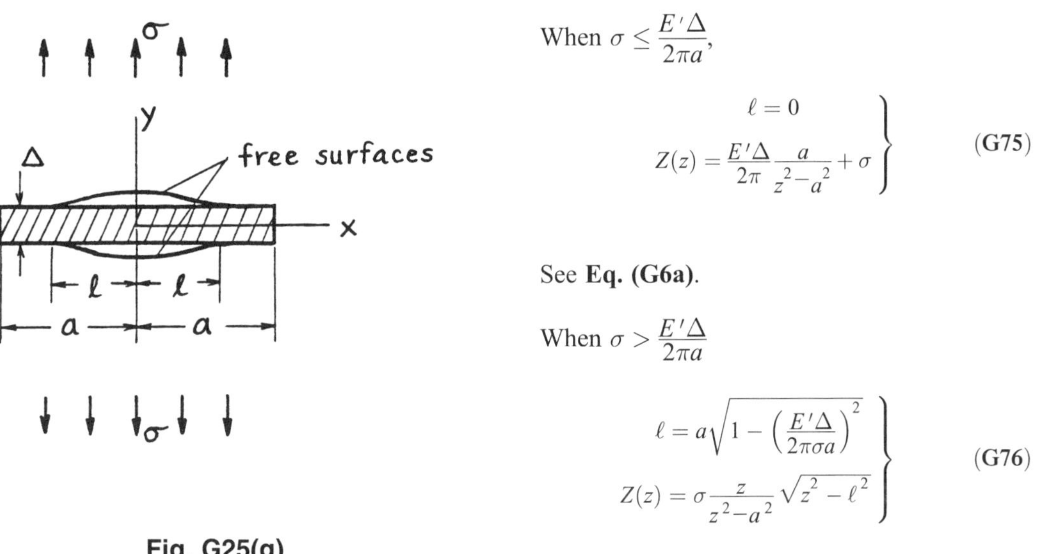

When $\sigma \leq \dfrac{E'\Delta}{2\pi a}$,

$$\left.\begin{aligned} \ell &= 0 \\ Z(z) &= \frac{E'\Delta}{2\pi}\frac{a}{z^2-a^2} + \sigma \end{aligned}\right\} \tag{G75}$$

See **Eq. (G6a)**.

When $\sigma > \dfrac{E'\Delta}{2\pi a}$

$$\left.\begin{aligned} \ell &= a\sqrt{1-\left(\frac{E'\Delta}{2\pi\sigma a}\right)^2} \\ Z(z) &= \sigma\frac{z}{z^2-a^2}\sqrt{z^2-\ell^2} \end{aligned}\right\} \tag{G76}$$

Fig. G25(g)

where 2ℓ is the length of separated (free) surfaces.

THE GREEN'S FUNCTION FOR A SINGLE STRESS-FREE CRACK AND DISLOCATIONS WITH ARBITRARY SHAPE

Solutions are easily obtained from **Eq. (G65)** by superposition for many other problems in which there is only one stress-free crack or only one stress-free portion along the x-axis, such as shown in **Fig. G25, (c), (e), and (f)**. This suggests the use of **Eq. (G65)** as the Green's function for such problems. When the same portion is subjected to a specified stress instead of stress free, the ordinary method of crack stress analysis is separately applied. Then the result must be superimposed on the solution obtained under the stress-free condition. Note that the cases illustrated by **Fig. G25, (c), (e), and (f)** are simply treated as special cases of the single stress-free crack problem. Therefore, we address only the Green's function for the problem associated with a single stress-free crack and dislocations (or cracks) with arbitrarily prescribed shape.

For convenience, replacing b by $-x'$ we rewrite **Eq. (G65a)** as

$$Z_0(z,x') = \frac{E'\Delta}{4\pi}\frac{1}{\sqrt{z^2-a^2}}\left(1-\frac{\sqrt{x'^2-a^2}}{z-x'}\right) \tag{G77}$$

Recalling the previous discussion leading to **Eq. (G16)**, the Green's function is readily derived as

$$\begin{aligned} Z_G(z,x') &= \lim_{\substack{\varepsilon\to 0\\ \Delta\varepsilon=d}}\{Z_0(z,x')-Z_0(z,x'-\varepsilon)\} \\ &= \frac{d}{\Delta}\frac{d}{dx'}\{Z_0(z,x')\} \end{aligned}$$

that is,

$$Z_G(z,x') = -\frac{E'd}{4\pi}\frac{1}{\sqrt{z^2-a^2}}\frac{d}{dx'}\left(\frac{\sqrt{x'^2-a^2}}{z-x'}\right) \tag{G78}$$

or

$$\begin{aligned} Z_G(z,x') &= -\frac{E'd}{4\pi}\frac{1}{\sqrt{z^2-a^2}}\left\{\frac{x'}{\sqrt{x'^2-a^2}}\frac{1}{z-x'}+\frac{\sqrt{x'^2-a^2}}{(z-x')^2}\right\} \\ &= -\frac{E'd}{4\pi}\frac{1}{\sqrt{z^2-a^2}}\cdot\frac{zx'-a^2}{(z-x')^2\sqrt{x'^2-a^2}} \end{aligned} \tag{G79}$$

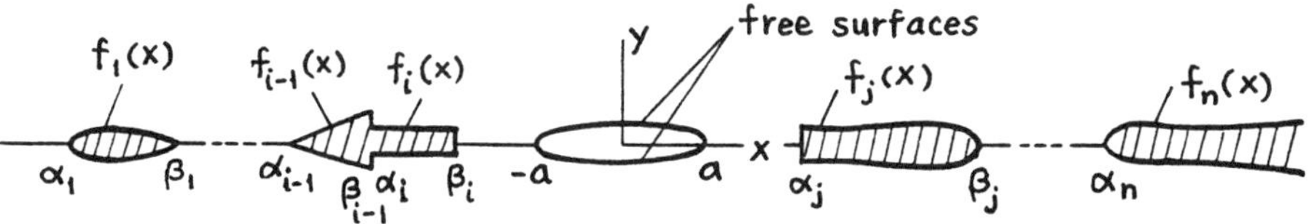

Fig. G26

When symmetric vertical displacement, $v(x,0)$, is specified along the x-axis, except in the region $|x|\leq a$ where a stress-free condition is given (**Fig. G26**), $Z(z)$ is obtained by the following integration. Note that in the region $|x|\leq a$ we can simply assume $v(x,0)=0$.

$$Z(z)=\frac{1}{d}\int_{-\infty}^{\infty}v(x',0)Z_G(z,x')dx' \tag{G80}$$

or using the same notations used in **Fig. G7** and **Eq. (G19)**

$$Z(z)=\frac{1}{d}\sum_{i=1}^{n}\int_{\alpha_i}^{\beta_i}f_i(x')Z_G(z,x')dx' \tag{G81}$$

When **Eq. (G79)** is used for $Z_G(z,x')$, **Eq. (G80)** becomes

$$Z(z)=-\frac{E'}{4\pi}\frac{1}{\sqrt{z^2-a^2}}\int_{-\infty}^{\infty}v(x',0)\cdot\frac{zx'-a^2}{(z-x')^2\sqrt{x'^2-a^2}}dx' \tag{G82}$$

and when **Eq. (G78)** is used, **Eq. (G80)** is integrated by parts,

$$\begin{aligned}Z(z)&=-\frac{E'}{4\pi}\frac{1}{\sqrt{z^2-a^2}}\int_{-\infty}^{\infty}v(x',0)\frac{d}{dx'}\left(\frac{\sqrt{x'^2-a^2}}{z-x'}\right)dx'\\&=-\frac{E'}{4\pi}\frac{1}{\sqrt{z^2-a^2}}\left[v(x',0)\frac{\sqrt{x'^2-a^2}}{z-x'}\Bigg|_{-\infty}^{\infty}-\int_{-\infty}^{\infty}v'(x',0)\frac{\sqrt{x'^2-a^2}}{z-x'}dx'\right]\end{aligned} \tag{G83}$$

where $v'(x',0)=\frac{d}{dx'}v(x',0)$. The choice between **Eqs. (G82) and (G83)** can be made for simplicity of integration. Note that this method can be used even when $\beta_i=-a$ and/or $\alpha_j=a$ in **Fig. G26**.

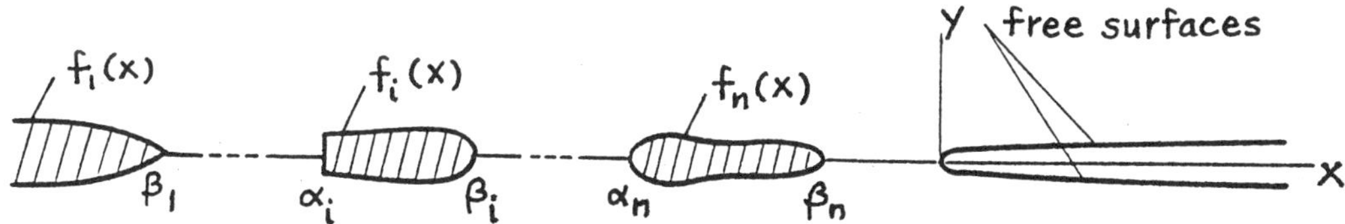

Fig. G27

When this stress-free crack is a semi-infinite crack located along the positive x-axis (**Fig. G27**), in a similar manner, starting with **Eq. (G69)**, or as a limiting case of the above, the Green's function is obtained as follows. $Z_0(z,x')$ in this case is

$$Z_0(z,x')=\frac{E'\Delta}{4\pi}\frac{\sqrt{-x'}}{(z-x')\sqrt{-z}} \tag{G84}$$

The Green's function is

$$Z_G(z,x') = \frac{d}{\Delta}\frac{d}{dx'}\{Z_0(z,x')\}$$
$$= \frac{E'd}{4\pi}\frac{1}{\sqrt{-z}}\frac{1}{dx'}\left(\frac{\sqrt{-x'}}{z-x'}\right) \tag{G85}$$

or

$$= -\frac{E'd}{4\pi}\frac{1}{\sqrt{-z}}\left\{\frac{1}{2}\frac{1}{\sqrt{-x'}(z-x')} - \frac{\sqrt{-x'}}{(z-x')^2}\right\}$$
$$= -\frac{E'd}{4\pi}\frac{1}{\sqrt{-z}}\cdot\frac{z+x'}{(z-x')^2\sqrt{-z}} \tag{G86}$$

Then, $Z(z)$ for the situation shown in **Fig. G27**, again using the same notation, is given by the integral

$$Z(z) = \frac{1}{d}\int_{-\infty}^{0} v(x',0)Z_G(z,x')dx' \tag{G87}$$

or

$$Z(z) = \frac{1}{d}\sum_{i=1}^{n}\int_{\alpha_i}^{\beta_i} f_i(x')Z_G(z,x')dx' \tag{G88}$$

Using **Eq. (G86)**

$$Z(z) = -\frac{E'}{4\pi}\frac{1}{\sqrt{-z}}\int_{-\infty}^{0} v(x',0)\cdot\frac{z+x'}{(z-x')^2\sqrt{-x'}}dx' \tag{G89}$$

or using **Eq. (G85)**,

$$Z(z) = \frac{E'}{4\pi}\frac{1}{\sqrt{-z}}\int_{-\infty}^{0} v(x',0)\frac{d}{dx'}\left(\frac{\sqrt{-x'}}{z-x'}\right)dx'$$
$$= \frac{E'}{4\pi}\frac{1}{\sqrt{-z}}\left[v(x',0)\frac{\sqrt{-x'}}{z-x'}\bigg|_{-\infty}^{0} - \int_{-\infty}^{0} v'(x',0)\frac{\sqrt{-x'}}{z-x'}dx'\right] \tag{G90}$$

where $v'(x',0) = \frac{d}{dx'}v(x',0)$.

The choice between **Eqs. (G89) and (G90)** is again based on simplicity of integration. This method applies to a special case of $\beta_n = 0$ in **Fig. G27**.

The method discussed in this section combines the two integrals, **Eq. (G18) or (G19) and Eq. (G30)**, into a single integral. **Equation (G18) or (G19)** is needed to calculate $\sigma_y(x,0)$, which is then used in **Eq. (G30)**. Thus the general method given by **Eq. (G30)** was further simplified for problems involving a single stress-free crack and cracks or dislocations with arbitrary shape.

The method is easily extended to problems where there are more than one stress-specified cracks, along with cracks or dislocations with arbitrarily specified shape.

SUMMARY

This appendix presents a simple method for obtaining the Westergaard stress function for dislocations and cracks.

First, problems associated with dislocations with uniform height (or cracks with uniform displacement) were discussed. Then, to extend the method's application to dislocations (or cracks) with arbitrarily prescribed shape, the Green's function was obtained. Thus no more than integrations were required to obtain the Westergaard stress function. Examples of $Z(z)$ and $\overline{Z}(z)$ for various dislocations were given.

The method was then further extended to problems with a combination of displacement-specified cracks (or dislocations) and stress-specified cracks. The general approach was presented and its applications to a few problems were discussed in detail. Many examples of $Z(z)$ were compiled.

Finally, the Green's functions for problems involving both displacement-specified and stress-specified cracks were established for a few simple configurations, further simplifying the general method.

The method presented in this appendix requires only integrations, and thus provides a simple way of treating problems associated with collinear dislocations and cracks.

APPENDIX H

THE PLASTIC ZONE INSTABILITY CONCEPT APPLIED TO ANALYSIS OF PRESSURE VESSEL FAILURE[1]

This appendix reviews the plastic zone instability concept and refines the procedures for determining the conditions for instability. The plastic zone instability failure criterion is applied to several crack configurations in a specific thin-walled cylindrical pressure vessel with semi-spherical end caps. The cracks analyzed are a longitudinal and a circumferential through-wall cracks in a cylindrical shell, and a through-wall crack in a spherical shell. The results predict the possibility that for ductile materials in certain combinations of geometry and loading the failure controlled by the plastic zone instability may precede the failure controlled by the plastic collapse load. Perhaps the results suggest the necessity, in the presence of cracks, for consideration of the plastic zone instability along with the failure criterion based on the plastic limit load.

INTRODUCTION

In the evaluation of mechanical integrity of pressure vessels, the stability analysis of through-wall cracks is documented in various international sources [2]. In these sources, the dual approach for the following two conditions is generally regarded as reasonable.

1. Avoidance of conditions for unstable running crack growth
2. Avoidance of conditions for unrestricted plastic flow (or for plastic collapse)

The plastic zone instability analysis discussed in this appendix addresses the second condition.

The concept of the plastic zone instability (PZI) was proposed by Vazquez in 1971 **(Vazquez, 1971)**, but it has not received much attention. It seems that the only extensive application of this method is found in the work of Tada and Paris **(Tada, 1983c)**. Some of the results obtained in **Tada (1983c)** are included in this discussion.

The PZI concept is based on the plastic zone size (r_Y) corrected linear-elastic fracture mechanics (LEFM) analysis, and therefore only LEFM solutions are required. In what follows, the LEFM solutions are presented first for the crack configurations of interest for convenience of subsequent discussions. Then the r_Y-correction

[1] **Tada 1996.**

[2] **For example: British Standards Institute PD 64-93-91; U.K. Central Electricity Generating Board R-6; International Insitute of Welding 11S-SST 1157-90; American Petroleum Institute RP579; and Materials Properites Council (USA), Fitness for Service Report.**

method and the PZI concept are described in detail. The PZI analysis is made for cracks in a specific pressure vessel and the results are compared with those obtained by formulas based on yield load or limit load solutions.

LINEAR-ELASTIC SOLUTIONS

The following stress intensity factor (K) formulas are taken directly from **pages 33.6, 35.1, and 36.1**. For through-wall cracks of length, $2a$, in pressure vessels of mean radius, R, and wall thickness, t, subjected to an internal pressure, p, K values are expressed in the following form using a function of a single, dimensionless geometric shell parameter λ.

$$K = \sigma\sqrt{\pi a}F(\lambda) \quad \textbf{(H1)}$$

where the parameter λ is defined as

$$\lambda = \frac{a}{\sqrt{Rt}} \quad \text{(H2)}$$

and the definition of σ and the expression of $F(\lambda)$ are given below for each crack configuration.

a. For a longitudinal through-wall crack in a cylinder,

$$\begin{aligned} \sigma &= pR/t \\ F(\lambda) &= \left(1 + 1.25\lambda^2\right)^{1/2} \qquad 0 \le \lambda \le 1 \\ &= 0.6 + 0.9\lambda \qquad 1 \le \lambda \le 5 \end{aligned} \quad \text{(H3)}$$

b. For a circumferential through-wall crack in a cylinder,

$$\begin{aligned} \sigma &= pR/(2t) \\ F(\lambda) &= \left(1 + 0.3225\lambda^2\right)^{1/2} \qquad 0 \le \lambda \le 1 \\ &= 0.9 + 0.25\lambda \qquad 1 \le \lambda \le 5 \end{aligned} \quad \text{(H4)}$$

c. For a circumferential through-wall crack in a spherical shell,

$$\begin{aligned} \sigma &= pR/(2t) \\ F(\lambda) &= \left(1 + 1.41\,\lambda^2 + 0.04\,\lambda^3\right)^{1/2} \qquad 0 \le \lambda \le 3 \end{aligned} \quad \text{(H5)}$$

These formulas are presented here for subsequent use in the plastic zone instability analysis.

PLASTIC ZONE SIZE (r_Y) CORRECTION METHOD

A localized yielding near the crack tip begins immediately with the load application because of the presence of stress singularity and causes deviation from the behavior based on the purely elastic analysis. Therefore, rigorously speaking, there is no linear portion in the load-displacement relation for a cracked body. Interestingly, as shown later, this nonlinearity manifests itself as the linear portion of the material J (or $\mathcal{G}$)-resistance curve prior to the actual extension of the crack.

It is the plastic zone size (r_Y) correction method that is commonly used to account for the effect of such local yielding. The plastic zone size r_Y given in the form of **Eq. (H7)** with $\alpha = 2$, was initially pointed out by Paris in 1957 **(Paris, 1957)**, and subsequently theoretical grounds for the use of the r_Y-corrected effective crack size in LEFM analysis was provided by Irwin and Koskinen **(Irwin, 1963)**, The method was developed for evaluating material toughness in small-scale yielding conditions where the yielding zone near the crack tip is well contained within the surrounding elastic field.

In this analysis model, the actual crack size, a, in the elastic solution is replaced by the r_Y-corrected effective size; that is, the effective elastic solution is used. The effective crack size, a_{eff}, is given by

$$a_{eff} = a + r_Y \tag{H6}$$

where

$$r_Y = \frac{1}{\alpha\pi}\left(\frac{K}{\sigma_Y}\right)^2 \tag{H7}$$

σ_Y is the yield strength of the material; the constant α is customarily taken as $\alpha = 2$ (plane stress) or $\alpha = 6$ (plane strain) depending on the constraining conditions (triaxiality) near the crack tip. It is sometimes convenient to take α as a geometry-dependent constant to extend the range of applicability of the method **(Tada, 1972a)**.

Because K on the right-hand side of **Eq. (H7)** is a function of crack size, to obtain the plastic zone size adjustment, r_Y, in a consistent manner, an iterative procedure or a graphical method is required. When K is given in a function form, as is the case of the shell crack problems of interest, the graphical method is conveniently used on the principle of "a single curve and straight lines."

The crack size is represented by the dimensionless parameter λ defined by **Eq. (H2)**. For convenience, writing $\lambda = {}^{a}/_{b}$ with $b = \sqrt{Rt}$, the following are equivalent to **Eqs. (H6) and (H7)**, respectively,

$$\lambda_{eff} = \lambda + \lambda_Y \tag{H8}$$

where

$$\lambda_Y = \frac{r_Y}{b} = \frac{1}{\alpha\pi b}\left(\frac{K}{\sigma_Y}\right)^2 \tag{H9}$$

Substituting **Eq. (H1) into Eq. (H9)** for K,

$$\lambda_Y = \frac{1}{\alpha}\cdot s^2 \cdot Y(\lambda_{eff}) \tag{H10}$$

where

$$\begin{aligned} s &= \sigma/\sigma_Y \\ Y(\lambda) &= \lambda\{F(\lambda)\}^2 \end{aligned} \tag{H11}$$

For a graphical solution for the unknown, λ_{eff}, it may be more convenient to write **Eq. (H10)** in the following form in terms of λ_{eff} eliminating λ_Y:

$$\lambda_{eff} - \lambda = \frac{1}{\alpha} \cdot s^2 \cdot Y(\lambda_{eff})$$

or further,

$$Y(\lambda_{eff}) = \alpha \cdot \frac{1}{s^2} (\lambda_{eff} - \lambda) \quad \textbf{(H12)}$$

The form of **Eq. (H12)** allows a simple graphical solution for λ_{eff} for any specified crack size parameter $\lambda = \lambda_0$ and various load levels represented by $s = \sigma/\sigma_Y$. That is, λ_{eff} is determined from the intersection points of the curve

$$y = Y(\lambda) \quad \textbf{(H13)}$$

and a series of straight lines

$$y = g(\lambda; \lambda_0, s) = \alpha \cdot \frac{1}{s^2} (\lambda - \lambda_0) \quad \textbf{(H14)}$$

The method is schematically presented in **Fig. H.1**. The λ_{eff} illustrated in the figure is $\lambda_{eff} = \lambda_{eff}(\lambda_0, s_2)$, corresponding to the given crack size λ_0 and the load level $s = s_2$.

PLASTIC ZONE INSTABILITY CONCEPT

Before discussing this concept, let us make some useful observations regarding **Fig. H.1**. From **Fig. H.1** we see that the plastic zone size, $r_Y = b\lambda_Y = \sqrt{Rt}\lambda_Y$, increases as the load level, $s = \sigma/\sigma_Y$, increases until the straight line given by **Eq. (H14)** becomes the tangent to the curve $y = Y(\lambda)$, **Eq. (H13)**, at $s = s_{\tan}$. That is, the solution of the iterative equation, **Eq. (H7)**, does not converge beyond this load level, $s > s_{\tan}$. The values of λ_Y and λ_{eff} corresponding to $s = s_{\tan}$ are also shown in **Fig. H.1** as $(\lambda_Y)_{\tan}$ and $(\lambda_{eff})_{\tan}$, respectively, along with the corresponding slope $(\alpha/s^2)_{\tan}$ of **Eq. (H14)**. Note that, although the point of tangency, A, may not be determined very accurately, the slope $\alpha/s^2_{\tan}$ is determined accurately. It is interesting to note also that the point of tangency A is unique for a given crack size λ_0 irrespective of the value of constant α. In other words, although $s_{\tan}$ depends on the value of α, the slope of the tangent, $(\alpha/s^2)_{\tan} = \alpha/s^2_{\tan}$, and the corresponding value of $(\lambda_Y)_{\tan}$ are uniquely determined for each given crack size λ_0.

Also noted is that **Fig. H.1** is an alternative form of **Fig. H.2**, which was used by **Vazquez (1971)** in his analysis of the plastic zone instability (PZI). **Figure H.1** was chosen for simplicity of graphical solution, where the load parameter $s = \sigma/\sigma_Y$ is separated from the function (λ_Y) to permit the use of a single curve and a series of straight lines instead of a series of curves and straight lines used in the method of **Fig. H.2**. However, since **Eq. (H12) and Figure H.1** are not designed for discussing the PZI concept, the alternative form, **Eq. (H17)**, and the corresponding graphical interpretation, **Fig. H.2**, are more convenient for directly following the discussion by Vazquez.

Eq. (H7) in terms of J(or $\mathcal{G}$) $= K^2/E$ is

$$r_Y = \frac{1}{2\pi} \frac{E}{\sigma_Y^2} J \quad \textbf{(H15)}$$

where E is the Young's modulus. Let us introduce the normalized form of J,

$$\bar{J} = \frac{J}{\left(\frac{\sigma_Y^2 b}{E}\right)} = \frac{J}{\left(\frac{\sigma_Y^2 \sqrt{Rt}}{E}\right)} \quad \textbf{(H16)}$$

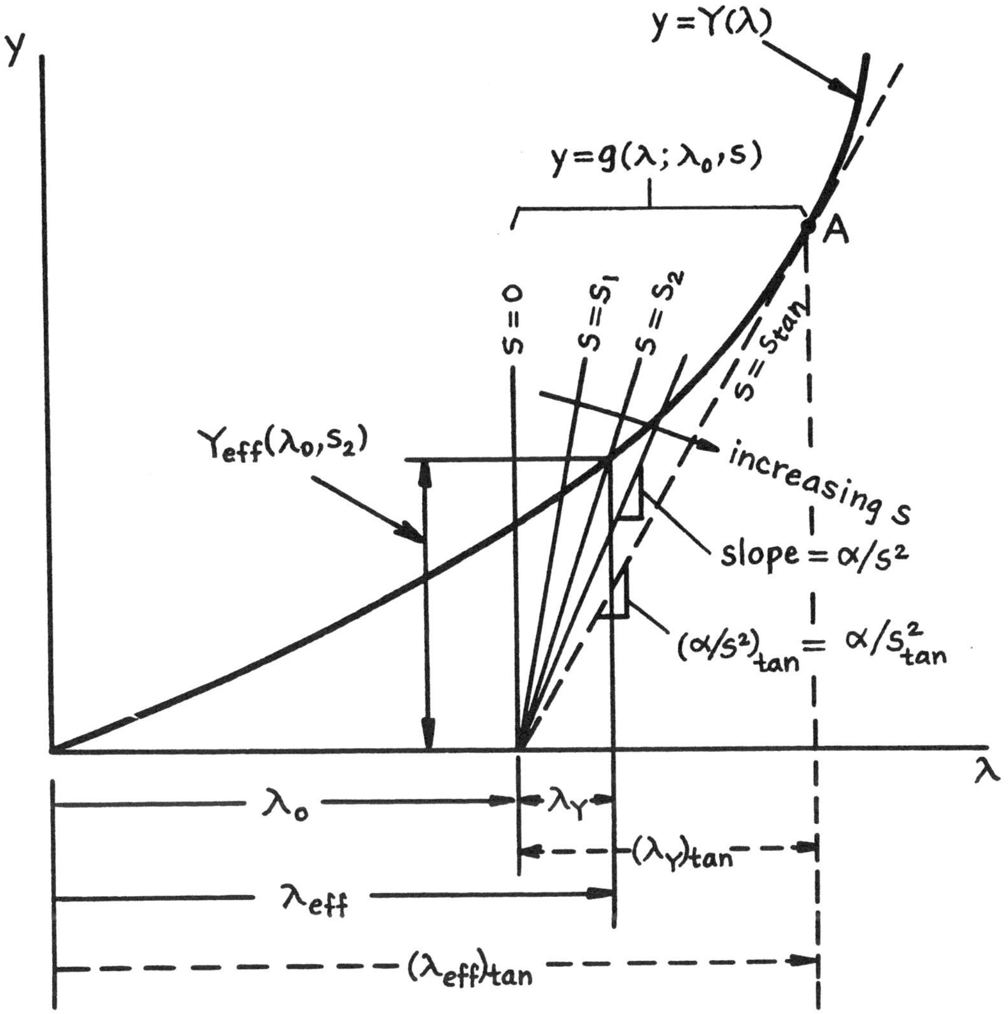

Fig. H.1. Graphical solution of Eq. (H12).

and rewrite **Eq. (H15) or (H11)** as

$$\alpha\pi\lambda_Y = \bar{J}(\lambda_{eff}, s)$$

or

$$\alpha\pi(\lambda_{eff} - \lambda) = \bar{J}(\lambda_{eff}, s) \quad \textbf{(H17)}$$

where

$$\bar{J}(\lambda, s) = \pi s^2 Y(\lambda) \quad \textbf{(H18)}$$

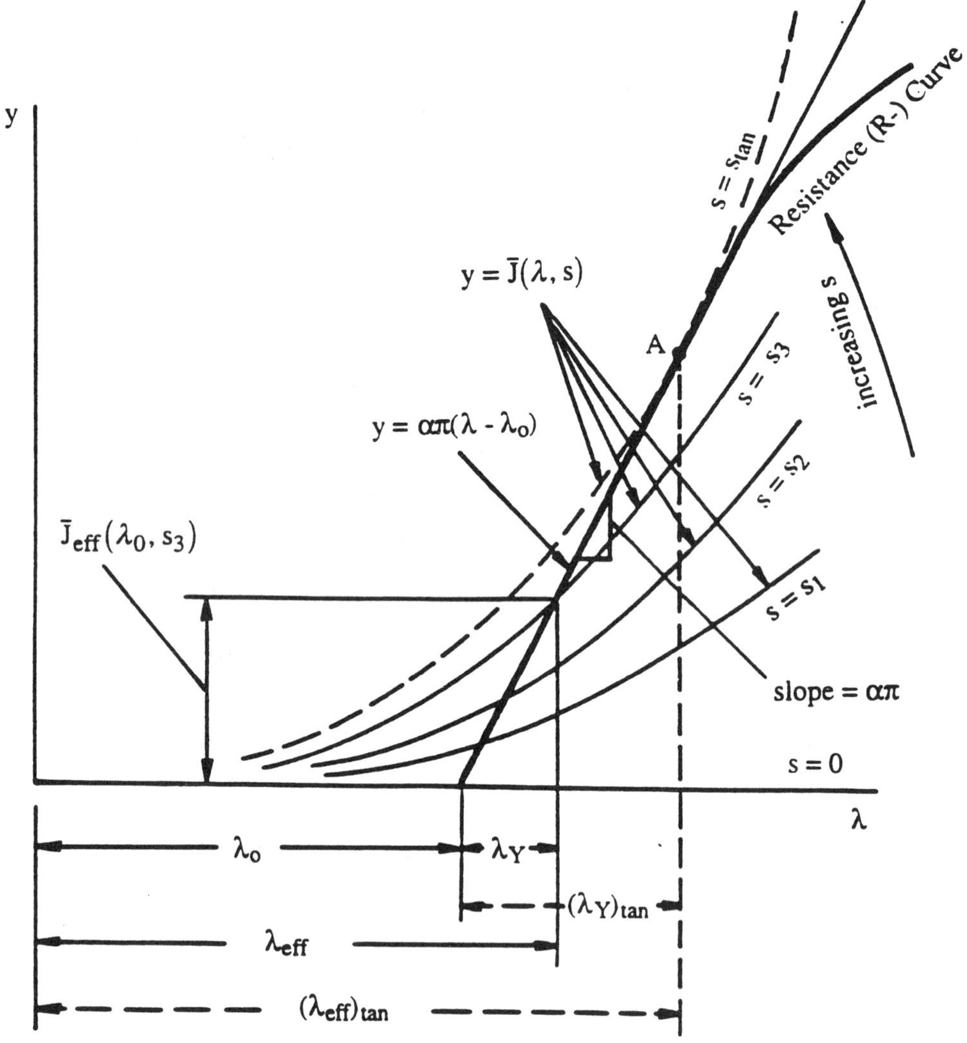

Fig. H.2. Schematic of plastic zone instability.

In **Fig. H.2**, a series of curves $y = \bar{J}(\lambda, s)$ and a straight line $y = \alpha\pi(\lambda - \lambda_0)$ are shown. The PZI concept is discussed on the basis of **Fig. H.2**, which is equivalent to **Fig. H.1**.

The possibility of extending the r_Y-corrected LEFM approach to non-small-scale yielding situations was explored and some justifications were given by **Tada (1983c, 1972a)**. **Vazquez (1971)** examined the static equilibrium between the plastic zone near the crack tip and the surrounding elastic field and proposed the concept of plastic zone instability (instability without crack growth). Detailed discussions on the mechanism and the physical interpretations of this concept are found in **Vazquez (1971)**, and some similar considerations are also found in **Tada (1972a)**. Vazquez's analysis may be summarized as follows.

Let us consider a crack extension resistance (R-Δa) curve and the interpretations of crack equilibrium and crack stability associated with it. It is noted that, as long as the crack does not extend ($\Delta a = 0$ or $\Delta a_{eff} = r_Y$), the plastic zone extension resistance $(R\text{-}r_Y)$ curve, **Eq. (H15)**, represented in **Fig. H.2** by the straight line, $y = \alpha\pi(\lambda - \lambda_0)$, constitutes the initial portion of the material resistance curve. Thus, when $F(\lambda)$ in **Eq. (H1)** is a rapidly increasing function, as in the case of through-wall cracks in shells, two distinct types of instability behaviors may be expected depending on the material's capability of absorbing plastic deformation; that is, the instability controlled by crack propagation – fracture toughness instability, and the instability controlled by plastic zone development – plastic zone instability.

It is observed from **Fig. H.2** that there is an equilibrium plastic zone size at each load level determined by the intersection point $(\lambda_{eff}, \bar{J}_{eff})$ between the curve $y = \bar{J}(\lambda, s)$ and the straight r_Y-line, $y = \alpha\pi(\lambda - \lambda_0)$, in the absence of crack extension. When the applied load level reaches $s = s_{\tan}$, the curve becomes tangent to the straight line at the point A and no further equilibrium plastic zone is determined. That is, for $s > s_{\tan}$ the solution for the iterative equation, **Eq. (H17)**, no longer converges. The interpretation of Vazquez is that for $s > s_{\tan}$ the plastic zone can not maintain stable equilibrium with the surrounding elastic field. If this tangency occurs before an appreciable crack extension (or deviation of the resistance curve from the straight r_Y-line) takes place, the plastic zone becomes unstable and spreads across the body. This is referred to as the plastic zone instability. The load level at instability is given by $s_{inst} = s_{\tan}$. When unstable crack propagation precedes the PZI, the fracture toughness analysis is to be followed. Note that the only material property involved in the PZI analysis is the yield strength σ_Y. Therefore, the PZI analysis alone does not determine which mode of failure occurs first.

Vazquez (1971) experimentally confirmed the prediction for existence of the PZI for a longitudinal through-wall crack in a pressurized cylinder in plane stress conditions ($\alpha = 2$) and concluded that the PZI analysis provides a conservative estimate of failure conditions.

r_Y AT PLASTIC ZONE INSTABILITY AND CONDITIONS FOR UNRESTRICTED PLASTIC FLOW

As previously observed, the values of r_Y at the occurrence of plastic instability $(r_Y)_{inst} = (r_Y)_{\tan}$ are uniquely determined by the geometry irrespective of the value of α. However, the values of r_Y at PZI do not generally represent the fully yielded conditions geometrically. The use of the LEFM analysis with the r_Y-adjusted effective crack size in such large-scale yielding conditions may seem unjustified.

Vazquez (1971) suggested that the quantity r_Y may be regarded as an index representing the development of the plastic yield zone rather than the physical plastic zone size itself in large-scale yielding conditions. Then it may be reasonable to assume the correspondence between the conditions for plastic zone instability and the attainment of the conditions for unrestricted plastic flow.

Interestingly, such correspondence is what exactly occurs in a few extreme two-dimensional crack configurations with the interpretation of r_Y as the physical size of plastic zone. Referring to **Fig. H.3**, consider a few deeply cracked bodies where the sole dimension relevant to the analysis is b, that is, the net ligament size. For the tensile configurations, **Fig. H.3, a and b**, the r_Y at the plastic zone instability, $(r_Y)_{\tan}$, is given by

$$(r_Y)_{\tan} = \frac{b}{2} \tag{H.19}$$

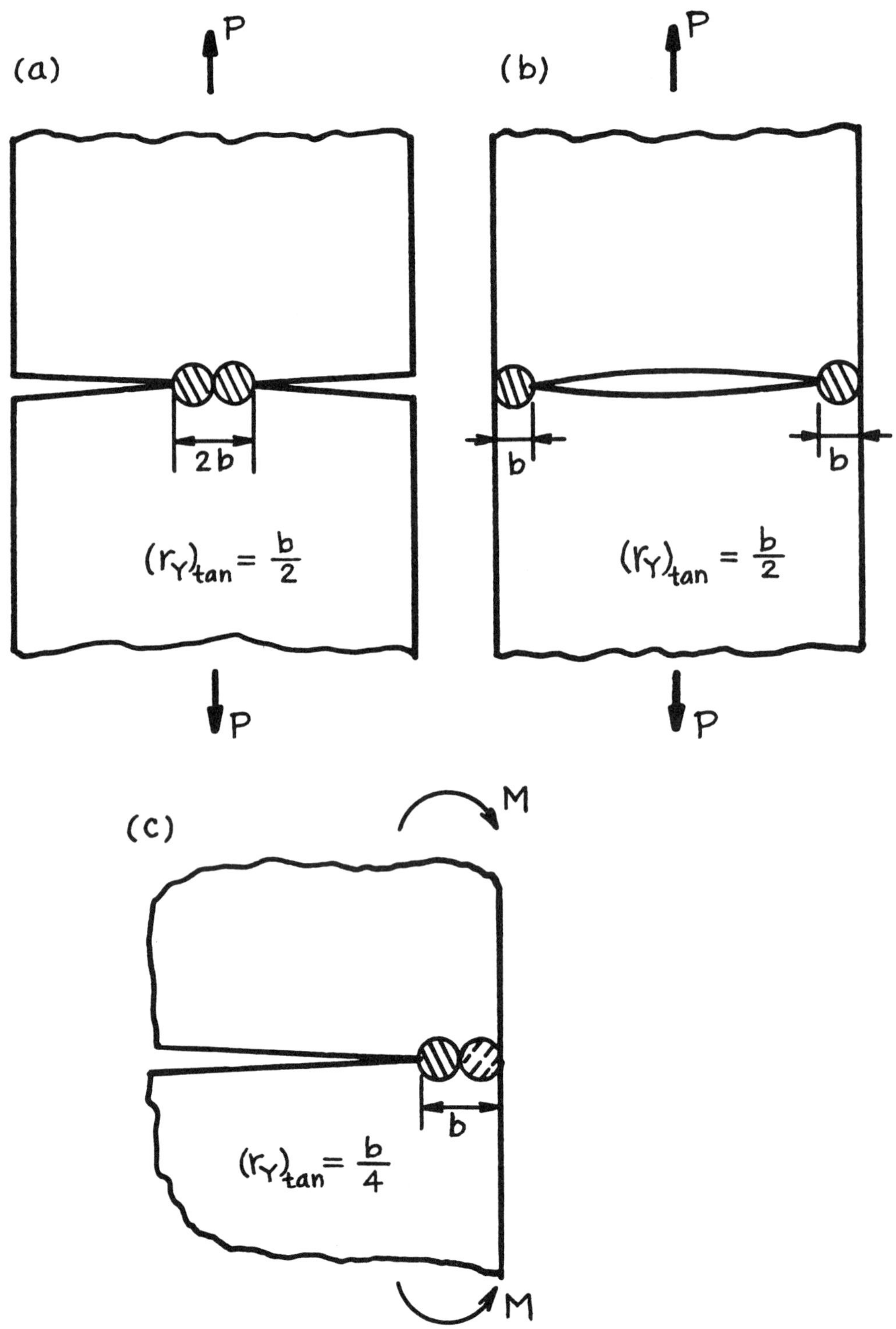

Fig. H.3. r_Y **at plastic zone instability for several extreme configurations.**

and for the bending configuration, **Fig. H.3c**, it is given by

$$\left(r_Y\right)_{\tan} = \frac{b}{4} \tag{H.20}$$

regardless of the value of α in **Eq. (H7)**. These conditions are readily obtained algebraically by noting that the stress intensity factors for these configurations have the following forms. For the tension configurations, **Fig. H.3, a and b,**

$$K = A_t\left(\frac{P}{B}\right)\frac{1}{b^{1/2}} \tag{H21}$$

and for the bending configuration, **Fig. H.3c,**

$$K = A_b\left(\frac{M}{B}\right)\frac{1}{b^{3/2}} \tag{H22}$$

where B is the thickness of the plates. The values of proportionality constants, A_t and A_b, are found on **pages 2.6, 2.1, and 9.1**. The values of A_t and A_b are immaterial for the validity of **Eqs. (H19) and (H20)**.

The use of the r_Y-adjusted effective LEFM solution may be reasonably extended to cover a wide range of elastic–plastic loading leading to the plastic zone instability. Some discussions on the use of the geometry-adjusted α are found in **Tada (1983c) and Tada (1972a)**.

APPLICATION TO A PRESSURE VESSEL

The plastic zone instability analysis was applied to determine the conditions for unrestricted plastic flow for through-wall cracks in an actual pressure vessel for a chemical reactor made of a ferritic steel. Its actual dimensions are

R = 125" (3160 mm)
t = 2.52" (64 mm) for cylindrical part
t = 1.26" (32 mm) for spherical end caps

The iterative equation, **Eq. (H7)**, was solved graphically (**Fig. H.1**) for the values of σ/σ_Y at the plastic zone instability, that is, $\sigma/\sigma_Y = s_{\tan}$, for the three crack configurations listed in the section: Linear-Elastic Solutions. The functions $F(\lambda)$ used here are **Eqs. (H3), (H4), and (H5)**. The results are presented in **Table H.1** in PZI columns. A report for the Materials Properties Council (MPC) by **Anderson (1993)** gives

$$F(\lambda) = \left(1 + 1.6\lambda^2\right)^{1/2} \tag{H23}$$

for both circumferential cracks and cracks in spheres without limits on λ. As observed from the comparison of **Eq. (H23)** with **Eqs. (H4) and (H5), Eq. (H23)** is an overestimate, especially for circumferential cracks. Therefore, the use of **Eq. (H23)** is not recommended here.

In **Table H.1**, these results are contrasted with the results in plastic collapse load (PCL) columns obtained through the following formulas of the critical stress levels, σ/σ_Y, for unrestricted plastic flow. These formulas were taken from the MPC report **(Anderson, 1993)**.

a. Longitudinal through-wall cracks in cylinders

$$\frac{\sigma}{\sigma_Y} = \left[1 + 1.6\lambda^2\right]^{-1/2} \tag{H24}$$

where $\sigma = {}^{PR}/_{t}$.

b. Circumferential through-wall cracks in cylinders

$$\frac{\sigma}{\sigma_Y} = \frac{\pi - \theta}{\pi}\left[1 + \frac{2\sin\theta\left(\cos\theta + \frac{\sin\theta}{\pi-\theta}\right)}{\pi - \theta - \frac{2\sin^2\theta}{\pi-\theta} - \frac{\sin^2\theta}{2}}\right]^{-1} \tag{H25}$$

where $\theta = {}^{a}/_{R}$, and $\sigma = pR/(2t)$. This is the form found in MPC **(Anderson, 1993)**, which is somewhat simplified to

$$\frac{\sigma}{\sigma_Y} = \frac{\pi - \theta}{\pi}\left[1 - \frac{\sin 2\theta + \frac{1-\cos 2\theta}{\pi-\theta}}{\pi - \theta + \frac{1}{2}\sin 2\theta}\right] \tag{H25a}$$

c. Through-wall cracks in spheres

$$\frac{\sigma}{\sigma_Y} = 2\left[1 + \sqrt{1 + \frac{8\lambda^2}{(\cos\theta)^2}}\right]^{-1} \tag{H26}$$

where again $\theta = {}^{a}/_{R}$ and $\sigma = pR/(2t)$.

Table H.1. Conditions for Unrestricted Plastic Flow for Through-Wall Cracks in Pressure Vessels (Plastic Zone Instability vs. Plastic Collapse Load)

	Cylinder Longitudinal Crack (σ/σ_Y)		Cylinder Circumferential Crack (σ/σ_Y)		Sphere Spherical Crack (σ/σ_Y)	
λ	P.Z.I.	P.C.L.	P.Z.I.	P.C.L.	P.Z.I.	P.C.L.
0	$\sqrt{2}$	1.0	$\sqrt{2}$	1.0	$\sqrt{2}$	1.0
0.25	.900	.954	1.07	.965	.878	.899
0.50	.669	.845	.894	.932	.642	.730
0.75	.523	.726	.764	.899	.494	.596
1.0	.423	.620	.661	.866	.396	.494
1.5	.314	.446	.605	.800	.279	.365
2.0	.250	.368	.526	.737	.213	.285
2.5	.206	.302	.460	.677	.170	.229
3.0	.170	.255	.418	.619	.142	.191
	$\sigma = \frac{pR}{t}$		$\sigma = \frac{pR}{2t}$		$\sigma = \frac{pR}{2t_{sp}}$	

The actual dimensions of the pressure vessel analyzed are R = 125" (3160 mm), t=2.52" (64 mm), and t_{sp} = 1.26" (32mm).

DISCUSSIONS

Some observations can be made from the results presented in **Table H.1**.

The critical stress values σ/σ_Y for unrestricted plastic flow predicted by the plastic zone instability analysis are generally lower than those predicted by the formulas based on the plastic collapse (the yield or limit) loads **(Anderson, 1993)** except for very small cracks. That is, the PZI results are more conservative for most crack sizes than those from **Eqs. (H24) (H25), and (H26)**, recommended by MPC **(Anderson, 1993)** and other documents [2].

In MPC **(Anderson, 1993)**, these formulas are assumed to be applicable to both cracks and grooves, and are claimed to be conservative. Such an approach, however, may not yield conservative results in the presence of cracks, as observed from the comparison. Consideration of the possibility of plastic zone instability may be necessary.

As discussed earlier, the r_Y at PZI, $(r_Y)_{\tan}$, and, correspondingly, $(\alpha/s^2)_{\tan}$, are uniquely determined for a given crack geometry. That is, the effect of α on σ/σ_Y at PZI, $s_{\tan}$, is proportional to $\sqrt{\alpha}$; $s_{\tan} \propto \sqrt{\alpha}$. In other words, the plane stress situations $(\alpha = 2)$ yield the most conservative results. For the thin-walled shell analyzed here, the plane stress conditions are expected to prevail except perhaps for very short cracks. As long as the same flow stress is used for σ_Y in both analyses, the results in **Table H.1** are good for direct comparison in plane stress conditions. Also note that the values of α have the opposite effect on the plastic zone instability and the fracture toughness.

The plastic zone instability analysis is based on r_Y-corrected effective linear-elastic fracture mechanics (and the material crack extension resistance curve). Therefore, PZI analysis requires only LEFM solutions. Accurate LEFM solutions are readily available for many crack configurations, in handbooks, such as this handbook, or are otherwise easily obtained.

The methods for determining the progressive increases of r_Y leading to the conditions for PZI are described in sufficient detail so that these procedures can be easily followed in the analysis of other crack problems.

APPENDIX I

APPROXIMATIONS AND ENGINEERING ESTIMATES OF STRESS INTENSITY FACTORS

The uses of stress intensity factor values in engineering applications of fracture mechanics do not require values computed with absolute precision. Often estimates within 5% or better are sufficient for the purpose, as loads and material properties are seldom known with great accuracy. Therefore, the techniques of developing approximations for stress intensity factors for applications are discussed here. The solutions and formulas found earlier in this handbook for idealized crack and body configurations are explored for estimating stress intensity factors for the nonideal configurations often confronted in applications. Intuitive methods for estimating the magnitude of errors in such estimates will be emphasized. In addition, methods of developing bounds on values will be discussed. Indeed, the art of developing useful stress intensity factor estimates with the intuitive tools of linear elasticity and "strength of materials" approximations can be developed with practice for many advantages in applications.

DIMENSIONAL ANALYSIS AND OTHER ASPECTS OF STRESS INTENSITY FACTORS

The dimensional nature of stress intensity factors, K, is defined from their definition as expressed in the equation in Part I of this hardbook. All formulas for K have units of force over length to the three-halves power. Therefore, if a remote stress, σ, is the applied load for a member containing a crack, then the K-formula for that situation will have the form

$$K = \sigma\sqrt{\pi a} f\left(\frac{a}{W}, \frac{B}{W}, \frac{L}{W}, \textit{etc.}\right) \tag{I1}$$

whereas if a force per unit thickness, P, is the applied load, the form is

$$K = \frac{P}{\sqrt{\pi a}} g\left(\frac{a}{W}, \frac{B}{W}, \frac{L}{W}, \textit{etc.}\right) \tag{I2}$$

where for simplicity only Mode I K is considered in both of these formulas and a, B, L, W, etc. are the crack size and other characteristic dimensions of each configuration considered. A quick review of a few solution pages will reveal these formats and alternatives for other types of applied loading, such as moment per unit thickness, M. The nondimensional "configuration functions," f ( ) and g ( ), also have special characteristic behavior depending on the type of configuration being considered. In particular, the finiteness of the body containing the crack, as described by the characteristic dimensions, often affects these formulas through separate factors, giving the forms

$$f\left(\frac{a}{W}, \frac{B}{W}, \frac{L}{W}, \text{ etc.}\right) = f_1\left(\frac{a}{W}\right) \cdot f_2\left(\frac{B}{W}\right) \cdot f_3\left(\frac{L}{W}\right) \cdot \text{ etc.} \tag{I3}$$

and

$$g\left(\frac{a}{W}, \frac{B}{W}, \frac{L}{W}, \text{ etc.}\right) = g_1\left(\frac{a}{W}\right) \cdot g_2\left(\frac{B}{W}\right) \cdot g_3\left(\frac{L}{W}\right) \cdot \text{ etc.} \tag{I4}$$

Reviewing a few solution pages also shows this to be characteristic of most of the K formulas. It is the intent here to exploit this tendency to make estimates of stress intensity factors.

EDGE CRACK

We begin by considering the configurations on solution **pages 5.1 and 8.1**. The discussion is restricted to normal stress, σ, on the boundaries. Note that the configuration of **page 8.1** can be formed from that of **page 5.1** by cutting the latter along its vertical axis of symmetry. However, before cutting, uniform stress, σ, on the boundary must be removed so that no (net) horizontal force is present on the free edge of configuration 8.1. This removal is illustrated by superposition in **Fig. I.1**, which shows that although no net force remains, a distribution of self-equilibrating normal stress is left on the vertical cut surface. The maximum intensity of this

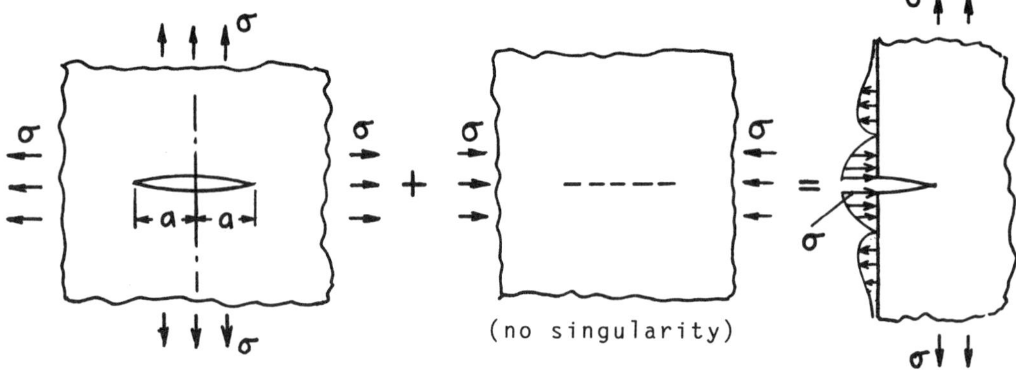

Fig. I.1.

stress is compression, - σ, near the emanation of the crack from the edge and it dies away to zero at locations far from the crack as compared with the crack size, a. These remaining stresses on the cut in **Fig. I.1** must also be removed to make it into the free edge of configuration 8.1. In order to remove these stresses on the edge, tension is required near the crack site and compression farther away, to net zero everywhere on the cut. Note that these added stresses required to make the cut a free surface would tend to open the crack more than before cutting. This implies that the K values should be higher for configuration 8.1 than for configuration 5.1.

Indeed, the factor is 1.1215 from the formula on solution **page 8.1**, or the difference in values is about a 12% increase in K due to the free-surface effect. Also notice that the introduction of the free surface caused a factor, f, to be incorporated in the K formula of the form

$$K = \sigma\sqrt{\pi a} \cdot f \qquad f = 1.1215 \tag{I5}$$

Although this result was known from the solution pages, it illustrates the idea of using the 5.1 solution with a modifying factor to approximate solution 8.1. Note that a stress distribution on an edge perpendicular to the crack of maximum intensity equal to the applied stress caused a error of only 12%.

Next, consider the center cracked strip configuration on solution **page 2.1.** Again the solution for K is given, but it is informative to consider how it might be approximated if the solution was not known. On solution **page 7.1**, the closed-form solution is given for the repeated crack problem, which could be cut on successive axes of symmetry between cracks to form a strip. The stresses along the cuts can be computed from the stress function, Z, after again removing the uniform lateral stress, σ, as in the previous example. The final remaining self-equilibrating stresses on the cuts are again maximum at the crack line and of intensity given by

$$\sigma_x\left(\frac{W}{2}, 0\right) = \sigma\left[\frac{1}{\cos\frac{\pi a}{W}} - 1\right] \tag{I6}$$

Thus as the crack size a goes to zero compared with the strip width, W, the stresses to be removed tend to zero. Moreover, for a crack length of one-half of the strip width, the maximum stress to be removed is $\sigma_x = 0.414\sigma$. Again note that removing these stresses would tend to open the crack further. Noting the magnitude of stresses to be removed compared to the edge crack example, up to a crack size of half the strip width the estimated error in K would be 12% times 0.414 or about a 5% underestimate. Therefore, historically, the K formula on solution **page 7.1** was considered a good approximation for the center cracked strip test configuration until a better formula became available. Indeed, we can improve it here by writing it as

$$K = \sigma\sqrt{\pi a} \cdot f_1 \cdot f_2 \tag{I7}$$

where

$$f_1 = \sqrt{\frac{W}{\pi a}\tan\frac{\pi a}{W}} \tag{I8}$$

and knowing that f_2 should be asymptotic to 1 for a diminishing crack, then we estimate

$$f_2 = 1 + 0.05\left(\frac{4a}{W}\right)^2 \quad \text{(compensates for 5\% error for } 4a = \text{W)} \tag{I9}$$

as an approximation. Now this result may not be as good as some of the better formulas given on solution **page 2.1**, but we can see that it is probably accurate within 1% for crack sizes up to more than half the strip width. Hence, without the benefit of other solution techniques, a very good "engineering estimate" is illustrated here. The purpose is simply to develop the method of forming good estimates.

In addition, this example also illustrates the method of using factors for various configuration effects. The factor, f_1, is the exact factor for the nearness of periodic cracks in an infinite sheet, and f_2 is the effect of cutting on the axes of symmetry between the cracks.

DOUBLE EDGE CRACKED STRIP

This configuration is also treated in Part II and presents an interesting example for estimation purposes. Notice that the configuration on solution **page 5.1** can be cut on the vertical axes of symmetry centering on two adjacent cracks to form the configuration. Again removing the uniform horizontal stress field prior to cutting, the remaining self-equilibrating normal stresses on the cut surfaces are a maximum of - σ compression at the locations of the cracks. For small cracks compared to the strip width, it is thus appropriate to use the 1.1215 factor for edge cracks. However, as the cracks deepen and the tips approach each other, the force transmitted from top to bottom of the strip is transmitted through a narrowing neck, which is already adequately represented in the solution without cutting. Therefore, the 1.1215 factor should diminish to 1 asymptotically as the crack tips approach each other. Consequently, an estimate of the appropriate effects is

$$K = \sigma\sqrt{\pi a} \cdot f_1 \cdot f_2 \qquad \textbf{(I10)}$$

where

$$f_1 = \sqrt{\frac{2b}{\pi a}\tan\frac{\pi a}{2b}} \quad (2b \text{ is the strip width } W) \qquad \textbf{(I11)}$$

and

$$f_2 = 1 + 0.1215\cos^n\frac{\pi a}{2b} \qquad \textbf{(I12)}$$

where n is guessed to be 2 or greater. In Part II, Tada found, taking n = 4, results in less than 0.5% error over the full range of a/b. However, if we simply take the straight line approximation given by

$$f_2 = 1 + 0.1215\left(1 - \frac{a}{b}\right) \qquad \textbf{(I13)}$$

the results are within 3%. Consequently, this example provides good approximations (even where the asymptotic behavior has been ignored). In the text of Part II for this configuration, other asymptotic approximations are given as examples of the possibilities.

As a further note, for deep cracks in the double edge cracked strip, that is, a → b, the above solution can be shown to agree with the closed-form solution of **page 4.9**. This fact adds to confidence through agreement with an exact limiting case. When developing approximate formulas, checking with limiting cases should be done whenever possible.

WEDGE FORCE SOLUTIONS

Although configurations with concentrated forces applied to the surface of a crack are seldom of direct practical interest, they are useful through superposition techniques in forming the solutions to many other configurations and in developing estimates of K. The technique involves first solving for the stresses present on the crack surface with the crack absent and then using the wedge force solutions to apply distributed forces (or effectively opposite stresses) to wipe out these stresses on the crack surfaces. This technique is shown schematically in **Fig. I.2**. The geometric configuration must be the same as the original configuration for the companion wedge force solution for superposition to apply exactly. However, for estimates, the companion wedge force configuration may differ slightly and with good judgment provide suitable approximations. This judgment can be developed through an understanding of "St. Venant's Principle" and like methods of intuitive stress analysis.

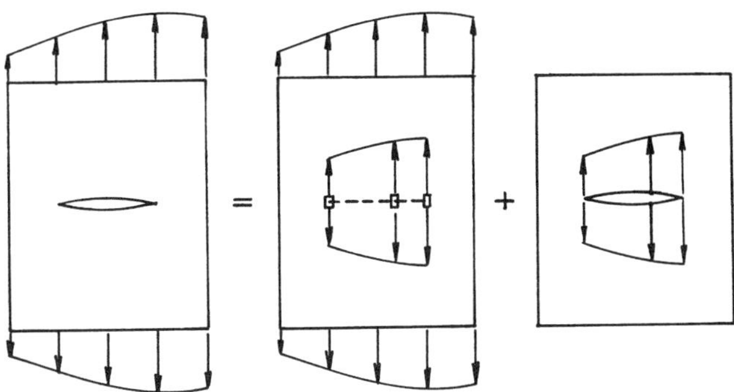

Fig. I.2

WEDGE FORCES NEAR THE TIP OF A CRACK

The wedge force solution for a semi-infinite crack with forces at distance, b, from the tip is given on page 3.6 and is

$$K = \frac{P}{\sqrt{\pi b}} \tag{I14}$$

where P is force per unit thickness. It is especially noted that forces nearer the crack tip are of greater influence due to the inverse $\sqrt{b}$ in the formula. Taking limiting cases of all other wedge force solutions will show this same influence as b becomes small compared with other dimensions. Therefore the stresses on the crack surface near the crack tip with the crack absent are of greatest influence. This fact is very useful in making estimates.

GREEN'S FUNCTIONS FOR STRESS INTENSITY FACTORS FROM WEDGE FORCE SOLUTIONS

For internally cracked infinite sheets, the method illustrated by **Fig. I.2** can be used with the wedge force solutions on **pages 5.10 and 5.11** to write forms that can be integrated to obtain K. In each of these solutions, the load P may be replaced by $\sigma_y dx$, and b with x for the integration, where $\sigma_y = \sigma_y(x, 0)$ is the normal stress on the crack surface with the crack absent. For stress distributions, σ_y, which are not symmetric with respect to the y-axis, from **page 5.10**:

$$K = \frac{1}{\sqrt{\pi a}} \int_{-a}^{a} \sigma_y \sqrt{\frac{a+x}{a-x}}\, dx \tag{I15}$$

For stress distributions symmetric with respect to the y-axis, from **page 5.11**:

$$K = 2\sqrt{\frac{a}{\pi}} \int_{0}^{a} \frac{\sigma_y}{\sqrt{a^2 - x^2}}\, dx \tag{I16}$$

Note that for uniform stress, σ, applied to the sheet, the crack absent stress on the crack surface is $\sigma_y = \sigma$ (constant), which may then be moved outside the integral sign, whereupon integration gives $\frac{\pi}{2}$. The net result is the familiar form

$$K = \sigma\sqrt{\pi a} \tag{I17}$$

Although this result is trivial it verifies the method. More important is to note that for any crack absent stress distributions, σ_y, the integrals may be evaluated numerically, if not analytically. This is the reason that many wedge force solutions are given in this handbook for other configurations. They will be used later in this discussion on estimating.

Wedge force solutions are also given for many three-dimensional problems (see solution **pages 23.1, 24.3, 24.24, 24.25, 25.1**, etc.). In many cases, other solutions have been generated from these solutions (see adjacent solution pages) using the Green's function method for the same geometrical configuration with different loadings. These solutions are of assistance in estimating as well.

PRELIMINARIES FOR ESTIMATING SOLUTIONS FOR THREE-DIMENSIONAL CRACKS

The plane problems presented here are stress solutions that are independent of the thickness of the sheet. Consequently, they are really three-dimensional solutions which simply do not vary in the out-of-plane direction. However, for truly nonvarying conditions in the out-of-plane direction, the "constraint" conditions must be either plane stress or plane strain. With sheet thicknesses of the order of the crack size, a (or other significant planar dimensions), neither plane stress nor plane strain fully applies. Indeed, in regions of high planar stress gradients, the inner sheet conditions may tend toward plane strain with plane stress near the surfaces. The crack tip is just such a high-stress gradient region. Therefore, the K formulas have inherent error due to these constraint effects. However, if we view the in-plane displacements as being compatible (approximately equal through the sheet), then the stress differences are at most by the factor $1 - \nu^2$, where ν is Poisson's ratio. Therefore, for a perfectly straight crack front through a sheet of finite thickness, at some portions of that crack front our K formulas may be incorrect by a maximum equal to that factor (i.e., less than 10% error maximum) and for the largest errors only a very small portion of the crack front would be involved. Therefore, the application of our "exact solutions" for K inherently involve some error. Moreover, crack fronts though the thickness of a sheet are never perfectly straight. Nevertheless it is the final objective to make "estimates" of reasonable engineering precision and feasible accuracy.

INTERIOR "PENNY-SHAPED" CRACK

The exact solution for the penny-shaped (circular), crack is found on solution **page 24.1**. Suppose that the solution for K is not known and we wish to estimate it for practice in forming estimates. Referring to **Fig. I.3**, the K formula must contain a factor proportional to the applied stress, σ, and also to the square root of the only characteristic dimension, a. Then the only remaining element is to find the proper numerical factor, f, in the following form

$$K = \sigma\sqrt{\pi a} \cdot f \tag{I18}$$

Now if the crack were a straight through-the-body "tunnel crack" of width $2a$, then f would be 1. Viewing the circular crack as made from the tunnel crack but with portions of the crack surface pulled closed, we can see that f must be less than 1 for the circular crack. In considering the technique of wiping out the stresses on the crack surface with the crack absent described in the previous paragraphs, we note that uniform stresses would be present. For the ratio of the area of crack surface to the length of crack front (about which the load transmission interrupted by the crack must be carried), note that the circular flaw has a ratio of one-half that of

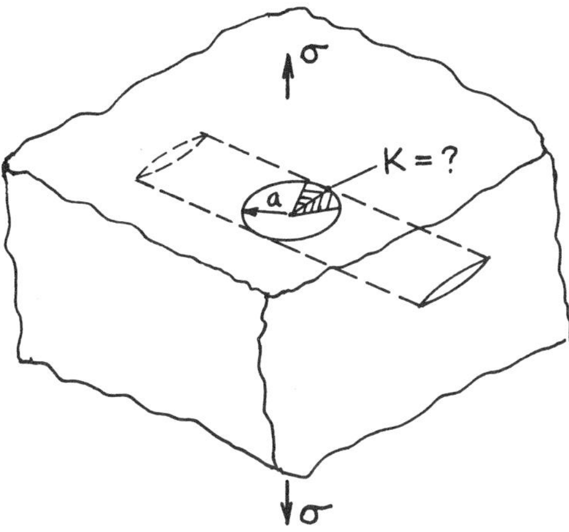

Fig. I.3

the tunnel crack. Based on that ratio, our first estimate of the appropriate f might be $^1/_2$. However, we note that the area of the circular crack is more closely gathered about its crack front. Our discussion of wedge forces noted that interruption of forces closer to the crack front has greater influence on K. Therefore our estimate of f should obviously be more than $^1/_2$; so a first guess might be about $^2/_3$. The exact solution for f shows this intuitive estimate to be quite good. Comparing shows

$$f = \frac{2}{\pi} = 0.6366 \cong \frac{2}{3} \quad \text{(within 5\% error)} \tag{I19}$$

With a little practice in guessing we would conclude that 0.6 would be too small and $^3/_4$ would be too high and conclude that the guess of $^2/_3$ is a reasonable estimate between these "bounds," even if the exact solution is unknown. (If you agree, you are ready for further developments of techniques of estimating.)

INTERNAL ELLIPTICAL CRACK (ESTIMATED SOLUTION)

Consider a flat elliptical internal crack subject to uniform tension normal to the crack. Let the smaller of the semi-axes of the ellipse be described by a and the greater semi-axis by b. If $a = b$ it would be the circular crack previously discussed, and $b \to \infty$, then it becomes the tunnel crack. The K along the crack front of the ellipse will vary but is obviously a maximum at the semi-minor-axis location since more of the area of crack surface is nearby per increment of crack front length. Suppose for practical reasons the maximum K is desired. The previous formula for the circular crack is appropriate with f assumed to be a nondimensional function of the ratio $^a/_b$. A first estimate might be to assume that f varies approximately linearly from 1 for $^a/_b = 0$ to $^2/_\pi$ for $^a/_b = 1$. This can be expressed by

$$f = 1 - \left(1 - \frac{2}{\pi}\right)\frac{a}{b} = 1 - 0.363\frac{a}{b} \tag{I20}$$

Now for $^a/_b = {}^1/_2$ this estimate gives $f = 0.818$, whereas the exact solution gives $f = 0.826$, which indicates about a 1% error (not only a good guess but a very lucky guess!). Perhaps over the full range, it can be simply said that the error is expected to be within 5%.

INTERNAL ELLIPTICAL CRACK (EXACT SOLUTION)

The full, exact solution for the internal elliptical crack subject to uniform remotely applied stresses is given on solution **pages 26.1 through 26.4**. The variation of K for all modes is given for points on the crack front around the ellipse. For practicality, the discussion here is restricted to applied normal stress perpendicular to the crack, which results in first mode K only. For this configuration, the result from **page 26.2** with a and b interchanged is

$$K = \frac{\sigma\sqrt{\pi a}}{E(k)}\left(\sin^2\theta + \frac{a^2}{b^2}\cos^2\theta\right)^{1/4} \tag{I21}$$

where $E(k)$ is a complete elliptic integral of the second kind depending only on the a/b ratio, which is found in **Appendix L**, and θ is the parametric angle for the ellipse as on **page 26.1**. Of special interest in estimating K for imperfect elliptical shaped cracks is l, which is the length of the normal (from the ellipse itself to an intersection with its major axes). This length is also illustrated on **page 26.1**. As noted, it can be expressed by

$$l = a\left(\sin^2\theta + \frac{a^2}{b^2}\cos^2\theta\right)^{1/2} \tag{I22}$$

The expression for K then simplifies to give

$$K = \frac{\sigma\sqrt{\pi l}}{E(k)} \tag{I23}$$

This formula demonstrates that K values for irregular internal flaws are dominated by the local geometry near the crack front region being considered through the strong dependence on l. On the other hand, the dependence on the overall crack shape through $E(k)$ is quite limited since its values only vary from 1 to 1.57 over extreme changes in shape. This concept is useful in making K estimates for both irregular internal cracks and surface cracks of various kinds.

SEMI-ELLIPTICAL SURFACE CRACK

Figure I.4 shows a semi-elliptical surface crack perpendicular to a flat surface with remotely applied normal stress parallel to the surface and normal to the crack. It is chosen as a configuration that is typical of many flaws found in practice that are critical in impairing the strength of structures. These flaws are almost always two or more times as long on the surface from which they emanate as they are deep. Although no completely closed-form solution determined analytically is available for this type of surface crack, it has been discussed extensively in the literature by complicated semianalytical techniques and numerical methods, which almost always have hidden assumptions not accessible to the user. For this reason semi-elliptical surface cracks are good candidates for checking by K estimating procedures that are convenient, quick, and have no hidden assumptions. Therefore this problem is discussed as a first example of estimating for surface cracks.

The standard approach here is to apply the usual form of

$$K = \sigma\sqrt{\pi a}\cdot f_1 \cdot f_2 \tag{I24}$$

where the geometrical factors, f_1 and f_2, are the effect of the elliptical shape, a/b, and the effect of introducing the front free surface, respectively. The first factor is taken from the exact elliptic solution as

$$f_1 = \frac{1}{E(k)} \quad \text{(exact)} \tag{I25}$$

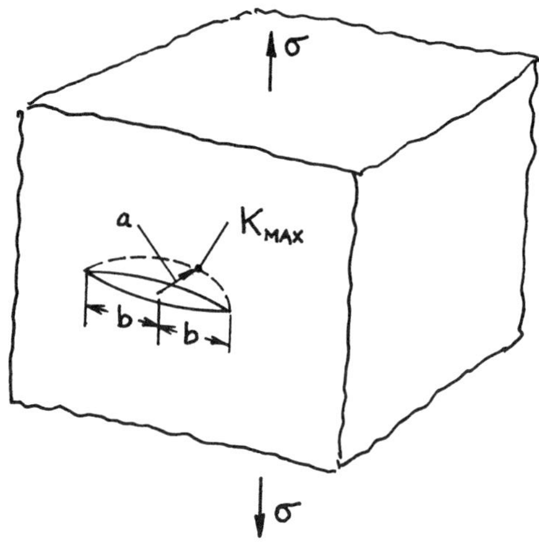

Fig. I.4

The second factor is that the free surface, which tends to 1.1215 for long surface cracks, that is, $a/b \to 0$, and would tend toward 1 for semicircular shapes, $a/b \to 1$, considering that the tendency for a surface crack to open further than an internal crack would be almost entirely suppressed for the semicircular crack. Therefore, it seems reasonable to estimate the second factor as

$$f_2 = 1 + 0.1215\left(1 - \frac{a}{b}\right) \quad \text{(estimate)} \tag{I26}$$

This estimate is surely within 5% error and probably much better than that if good intuition prevails. Furthermore, there have been no hidden errors, numerical or otherwise, in our estimate; this fact instills confidence.

SEMI-ELLIPTICAL SURFACE CRACK IN A FINITE THICKNESS PLATE

The ASME Nuclear Pressure Vessel Code requires that postulated semi-elliptical surface cracks in the wall of a vessel, of a depth of $1/4$ of the wall thickness and of an aspect ratio a/b, are to be analyzed. Therefore our interest in semi-elliptic surface cracks in finite thickness plates is evident. **Figure I.5** shows the configuration, including the finite thickness, t, the effect of which is another factor $f_3(a/t)$, beyond those in the previous analysis of semi-elliptic surface cracks. Indeed, f_2 and f_3 should be considered together in the analysis, since their effects are physically coupled. As an alternative, writing separate factors, their combined error effects should be considered together to form a total error for the separated functions. Now for a surface flaw in a finite thickness plate, f_2 may be noted to be too large a correction by perhaps 5% at most. On the other hand, for the f_3 correction factor for a crack tip approaching a back free surface, the "square root of tangent" correction previously applied to the central crack strip estimate was seen by itself to undercorrect by about 5%. Therefore, by using these two corrections together in this analysis, we create compensating errors well within 5%. For this reason, we choose f_3 as (f_1 in the strip example)

$$f_3 = \sqrt{\frac{2t}{\pi a}\tan\frac{\pi a}{2t}} \tag{I27}$$

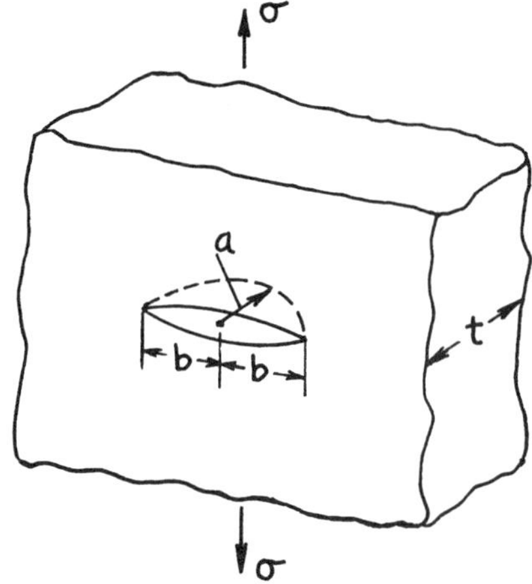

Fig. I.5

This combined with the immediately preceding analysis leads to

$$K = \frac{\sigma\sqrt{\pi a}}{E(k)}\left[1 + 0.12\left(1 - \frac{a}{b}\right)\right]\sqrt{\frac{2t}{\pi a}\tan\frac{\pi a}{2t}} \qquad \text{(I28)}$$

(estimated error $< 5\%$ for $a/t < 1/2$)

This estimating formula for semi-elliptic surface cracks in finite thickness plates was first proposed by **Paris (1965)**. **Newman (1979b)** used finite element analysis to compare Paris' formula with about 10 alternative formulas. He found Paris' formula to be the second most accurate of all of the formulas, many of which had complicated mathematical justifications associated with them. The simplicity of the logic and the freedom from hidden assumptions for the this formula makes it even more appealing.

IRREGULARITIES IN SHAPE OF ALMOST SEMI-ELLIPTICAL SURFACE CRACKS

Real surface cracks invariably are not of a perfect semi-elliptical shape, but almost all analyses of them assume they are perfect without special justification or consideration. It is most noticeable that the ends of these cracks, where they meet the surface, are almost never perpendicular to that surface. At other places along such a crack front the curvature also rarely perfectly matches the ellipse chosen. To better evaluate K along crack fronts of such irregularly shaped cracks it is desirable to discuss some possibilities for improvements.

A first example is based on the actual shape of a preexisting crack in an $F - 111$ wing box that led to the loss of this aircraft in late 1969 and the grounding of this type of aircraft for more than 6 months. The crack is carefully drawn in **Fig. I.6** to illustrate its actual shape. In addition, the figure shows two perfect ellipses, one matching the length and depth of the real flaw, and the other matching the depth and curvature of the real flaw at the deepest point where K is anticipated to be maximum. The reason for selecting these two ellipses is that the first will give an "upper bound" on the K value and the latter will give a "lower bound" of the value and a better estimate as well. The second ellipse is a closer estimate because it better approximates the local geometric conditions at the point where K is desired.

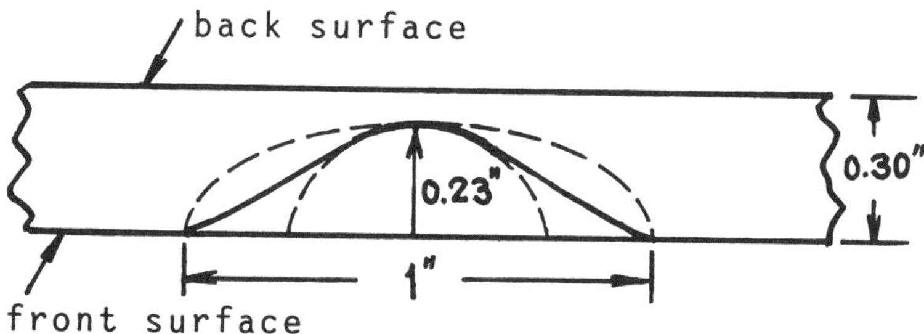

Fig. I.6

Taking the approximate dimensions from the ellipses and computing the correction f for both results in

$$K = (1.18 \text{ to } 1.34)\sigma\sqrt{\pi a} \tag{I29}$$

This shows a difference of about 15% in the coefficient f. The correct result is undoubtedly closer to the lower value but also surely above it (rather than the upper value, which most often is blindly adopted). Thus if we elect to estimate the proper value at 1.24 it would be 5% above the lower bound and 9% below the upper bound, and likely within $\pm$ 3% or probably better for the geometrical correction factor f. Now the basis of the formula was no better than $\pm$ 5% in the first place, so the maximum combined error would be $\pm$ 8%. Knowing these bounds, the 1.24 factor could be intelligently adjusted up or down for a particular application depending on whether a low or high estimate is dictated by that application. Although engineering practice often simply computes a single "answer," the bounding technique and error estimates provided here have obvious advantages. Furthermore, in a real analysis of strength impairment of a structural component, the precision of the values of applied stresses and material properties should also be considered in a consistent fashion. Such a practice was actually followed in assessing the $F - 111$ failure.

ESTIMATES OF K NEAR THE INTERSECTION OF A SURFACE CRACK WITH THE SURFACE

Two difficulties in estimating K occur near the intersection of a surface crack with a surface. First, the crack tip stress field and its constraint parallel to the crack front change from plane strain in the interior to plane stress at the surface. Moreover, plastic flow near a crack tip will modify this constraint so that detailed solutions by purely elastic analysis for constraint effects will themselves be inaccurate. However, the difference between plane stress and strain is obviously a maximum factor in stress or K results approaching $1 - \nu^2$. That is to say, at the plane stress region near the surface, K formulas may give up to about a 9% overestimate of K. Notice that near surfaces, fatigue cracks frequently lag behind interior growth, which is explained at least in part by this effect. The second difficulty is simply that flaws almost never intersect the surface with their fronts perpendicular to the surface. Even when almost semi-elliptic surface flaws intersect a surface, near that intersection, the curvature of the crack front does not match that of a perfect semi-ellipse. This difficulty in assessing the local K can produce large errors if the formulas for perfect ellipses are blindly applied. This second difficulty can be remedied by suitable estimating procedures for K.

It is noted from solution **pages 26.1 and 26.2** that the formula for K near the narrow end of an ellipse is

$$K = \frac{\sigma\sqrt{\pi\rho}}{E(k)} \tag{I30}$$

where ρ is the end radius of the ellipse. For almost semi-elliptic surface cracks where they do meet the surface in a perpendicular manner this formula shows that K is strongly dependent on the local crack front radius ρ and more weakly dependent on the gross proportions of the ellipse through $E(k)$, since it only varies from 1 to 1.57 for extreme changes in proportions from infinitely long to circular. Consequently, the above formula for K should be used with the local radius of the actual elliptical shape. This will normally give reasonable results when the crack front varies somewhat from perpendicular to the surface.

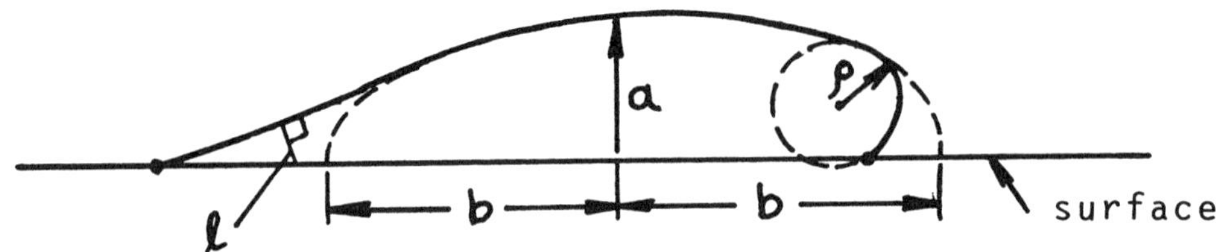

Fig. I.7

For more extreme cases of lack of perpendicularity of the crack front to the surface, further considerations are appropriate. **Figure I.7** illustrates cases of interest. The left and right intersections with the surface (solid curve) are the extreme cases. The theoretical elastic stress intensity values, K, for the intersection points are actually zero for the left nonperpendicular type of intersection and infinite for the right one. However, practical treatment and judgment of the relevant values would probably dictate using the local radius ρ in the above formula for the right crack end, along with an $E(k)$ using the proportion of the dashed ellipse. On the other hand for the left crack end, it would be appropriate to substitute l, the length of the perpendicular from a local portion of the crack front, for ρ in the above formula. Solution **pages 26.1, 26.2, and 26.6** support this suggestion. Of course, corrections for front free surface and back free surface, as with previous estimates, should be added if finite thickness plate is involved. However, the back free surface correction would be less influential at the crack ends, so l or ρ should be used in the tangent correction rather than a.

It needs to be emphasized here that, in most treatments of irregular semi-elliptical surface cracks, and in many computational programs, the effects of the irregularities discussed here have been simply ignored. As noted, these effects can be significant even for moderately irregular cracks.

GROSSLY IRREGULAR SURFACE CRACKS

For grossly irregular surface cracks, it is reasonable to use the intuitive knowledge gained from the preceding discussion to make estimates of stress intensity, K, at points along the crack front. **Figure I.8** is an example of such an irregular surface crack. For the initial discussion of this example, it is assumed that uniform normal stress, σ, normal to the crack is applied remotely to this region of the body. The objective here is to estimate K on each segment of the crack front between points labeled A through H. With each segment, the logic of the estimate is explained to assist the reader in understanding estimating.

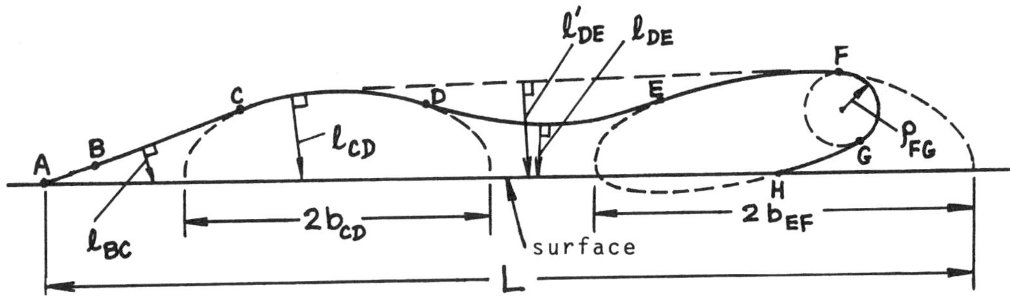

Fig. I.8

$A-B$: Along this segment, the K tends to zero at A and the relatively large crack to the right may be regarded as infinitely large. The K is then estimated as if it were an infinitely long surface (edge) crack as

$$K_{AB} = 1.1215\sigma\sqrt{\pi l} \tag{I31}$$

where the crack depth l is most appropriately taken as locally normal to the crack front for the point of interest along $A-B$.

$B-C$: By the time it reaches point C, the crack is deep enough that the total length, L, is finite by relative proportions so that a correction for finite length such as $E(k)$ is appropriate. Since $E(k)$ itself only varies between 1 and 1.57 for extreme changes in proportions, a/b, the estimation of the proportion to use is not critical. As the proportions are varied from fitting the curve of segment $C-D$ to using the full length of the surface crack, L, the $E(k)$ varies from about 1.21 to 1.07. Noting that the effective length is somewhere midway between these extremes, selecting approximately 1.14 is surely within 3% of correct. The 1.1215 surface correction is also reduced by the finite length toward a minimum of 1 for a semi-circular crack. Obviously, this correction should be quite close to 1.1215, so we choose 1.09 as surely within 2% of correct. The overall correction is thus

$$\frac{1.09}{E(k)} = \frac{1.09}{1.14} = 0.96 \tag{I32}$$

within a maximum error of 5% (really believed to be within 3%). Therefore as the point of interest moves from B to C, the K formula would appropriately change gradually (respectively) as

$$K_{BC} = (1.1215 \textit{ to } 0.96)\sigma\sqrt{\pi l_{BC}} \tag{I33}$$

This should allow estimating K on this segment within 5%.

$C-D$: For this segment, the K values would be best approximated by the dashed ellipse fitting the segment, with the following modifications. The front free surface correction should be reduced to the 1.09 value previously elected, and the $E(k)$ value should be modestly reduced from the 1.21 value for this ellipse. Electing a reduction to about 1.15 is surely within 3% of correct. The result is

$$K_{CD} = 0.95\sigma\sqrt{\pi l_{CD}} \tag{I34}$$

For this segment the values of K are surely well within 4% of correct with the worst error likely approaching D. Since the values that will result from this formula are considerably larger than those from the formulas for earlier segments due to larger values of l, its precision would probably be more important in a real application. Indeed, for this configuration of crack, the maximum K probably occurs on this segment and precision is therefore most important here.

$D-E$: Estimating K over this segment poses special problems due to the reversed curvature of the crack front. In particular the surface correction would gradually exceed the 1.1215 value with its maximum at the nearest point of the crack front to the surface, because the crack is deeper on both sides of this point. Accounting for the finite length, L, for this configuration would reduce the 1.1215 factor to a minimum of about 1.09 using previous forms. Therefore, at the portion of this segment nearest the surface it can be stated that

$$K_{DE}(\text{min}) \geq 1.09\sigma\sqrt{\pi l_{DE(\text{min})}} \tag{I35}$$

On the other hand, taking the dashed straight line joining the peaks in the crack depth and using l' at that same location would indicate a maximum K, again adjusting for finite length, as

$$K_{DE}(\text{max}) \leq 1.03\sigma\sqrt{\pi l'_{DE}} \tag{I36}$$

At this location l' is noted to be about twice the size of l. Accounting for the numerical coefficients in these expressions, the second exceeds the first by about 33%. Intuitively, the correct result should be closer to the minimum than the maximum, since it clearly is a closer representation of the real configuration. Hence, the estimate should increase the minimum by considerably less then half of 33%, so 10% is taken. Based on this logic, the K at the nearest point on $D-E$ to the surface is estimated as

$$K_{DE} = 1.20\sigma\sqrt{\pi l_{DE(\text{min})}} \quad (\text{nearest point}) \tag{I37}$$

This estimate is also believed to be within 5% of the correct result.

As a check on this result the notched round bar solution on **page 27.1** may be consulted. It has a crack front with reversed curvature as is the case for the nearest location on segment $D-E$. Matching this crack front curvature in a notched round bar and taking $l_{DE(\text{min})}$ as the crack depth in the notched round bar (estimated as $a/b = 0.75$ for the round bar) results in a coefficient of 1.21 instead of the 1.20 estimated above. This degree of agreement is somewhat fortuitous, since the notched round bar does not have increasing crack depth as the point is moved away from the nearest to surface location and neither does it have a finite length crack, which would cause errors to be somewhat compensating. However, the agreement is acceptable and helpful in giving confidence in the claim of 5% accuracy.

$E-F$: For this segment the suggested procedure is to consider both the dashed ellipse best fitting points $E-F-G-H$, with emphasis on fitting the curvature of $E-F$, and also taking the dashed semi-ellipse fitting $E-F$. For the former, the free surface correction should acknowledge more effect at E diminishing toward F, from a value exceeding 1.1215 to 1, respectively. On the other hand, for the latter semi-ellipse, the results at E should be modestly reduced considering the gross deviation in shape at the right end and this reduction should be increased significantly approaching F. Indeed, upon reaching F the results of the first complete ellipse should be considered as the relevant estimate, whereas the estimates of the semi-ellipse are most relevant at E. Estimating the free surface correction for the semi-ellipse at E to be about 1.08 should give results within 5%. Estimating the full dashed ellipse's surface correction to be entirely absent or 1 should give results within 5% at F. It then remains to simply interpolate between the two for points E and F smoothly. This is left to the reader.

$F-G$: Between F and G the curve should be best fit with a radius, ρ_{FG}, which, using the full dashed ellipse, leads to

$$K_{FG} = \frac{\sigma\sqrt{\pi\rho_{FG}}}{E(k)} \tag{I38}$$

Because this segment is away from the influence of the free surface, it has not been corrected for it. However, since there is some remote free surface, it is a lower bound of the K estimate but certainly well within a 5% error of the actual value. A factor of 1.02 could be added to this expression, which would imply the accuracy then to be within 1% (if the inherent error is ever that small).

$G-H$: Before even getting to G along $F-G$, a correction for a crack approaching a back free surface might be considered. However, it was omitted here because conditions along $G-H$ become exceedingly difficult for simple estimating. First, it should be noted that K tends to infinity at reversed intersections with the surface, such as at H. In any analysis of a real problem, the ligament between the crack and the surface would be subject to stresses causing yielding to spread away from H, perhaps beyond G, depending on nominal stress levels. Therefore, it may be dangerously irrelevant to attempt an estimate. With that in mind, we proceed assuming that elastic action prevails. The opened part of the crack in the region between $E-F-G-H$ would shed load onto the uncracked ligament adjacent to $G-H$. Crudely, the stress interrupted by half of the width of the dashed full ellipse would be dumped through that ligament. The net section stresses there would thereby increase from approximately 3σ at G toward infinity at H. These increased ligament stresses could then be used in a solution, such as on **page 9.2**, to estimate K. Such a procedure is omitted here, as large errors would be present and the relevance is doubtful.

It might be more relevant to expect cracking of such a ligament in any application and anticipate the crack to grow in the end shape of the semi-ellipse shown. That is, we could start our estimates from some intermediate shape that would be easier to analyze with less loss of relevance. This also is left to the reader.

ADJUSTMENTS FOR STRESS GRADIENTS IN ESTIMATES OF K FOR SURFACE CRACKS

Generally with cracks absent, the stresses in bodies containing cracks vary gradually and continuously throughout the region where a crack appears. Exceptions are left for more advanced individual treatment. Restricting the discussion to normal stresses perpendicular to a crack, first consider the variation in stress, σ, along a surface in the direction of its intersection with the crack. **Figure I.9** shows the crack under consideration.

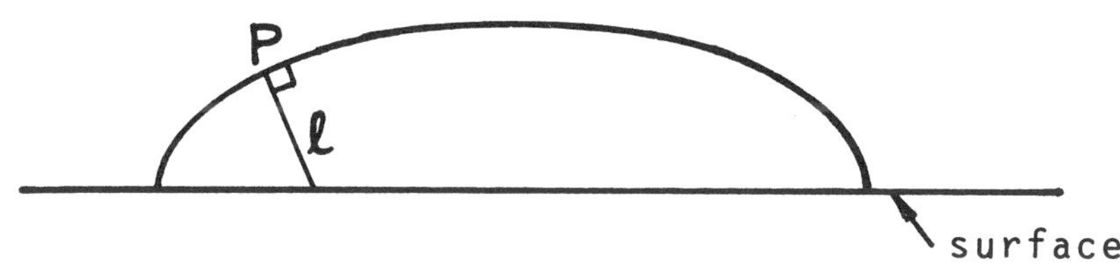

Fig. I.9

The analysis of the behavior of K along an elliptical crack shows a strong dependence on local stress conditions. For a point P, as on **Fig. I.9**, the stresses are on and near the normal to the crack front, l. This local stress dependence is also borne out by the character of K caused by wedge force solutions in both two and three dimensions as noted on solution **pages 3.6** and especially **page 23.1**, as well as others. Therefore, the simple rule to accommodate gradients of stress along the surface is simply to take the average stress, σ, on l with the crack absent to evaluate K at point P.

Furthermore, if stress gradients occur normal to the surface, additional consideration should be superimposed. **Fig. I.10** illustrates a linear gradient normal to the surface that is divided into the constant stress occurring at the crack front point, P, of interest and the additional stress away from that point. Now the K due to each of these distributions should simply be added. The K for the constant stress is as treated earlier. For the additional linearly varying stress normal to the surface with zero at the crack front, it is helpful to review the solutions on **pages 8.1 and 8.6, and 28.1 and 28.2** and to note the effect of stress gradients normal

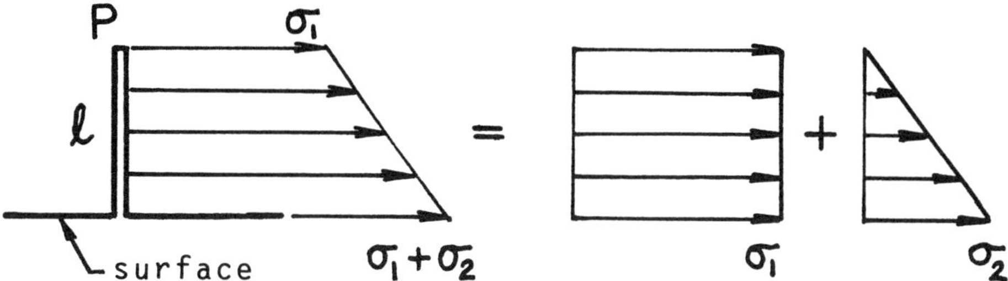

Fig. I.10

to the surface for both infinitely long surface cracks (edge cracks) and the other extreme of semi-elliptical cracks (i.e., semicircular cracks). The results are that an equivalent stress σ can be found for insertion into formulas for uniform stress, where for long cracks it gives

$$\sigma = \sigma_1 + 0.39\sigma_2 \tag{I39}$$

and for semicircular cracks it gives

$$\sigma = \sigma_1 + 0.30\sigma_2 \tag{I40}$$

Now these forms are so similar that it is relevant to suggest for elliptical cracks that a suitable approximation is

$$\sigma = \sigma_1 + \left[0.30 + 0.09\left(1 - \frac{a}{b}\right)\right]\sigma_2 \tag{I41}$$

This form is obviously quite accurate since it is exact at both extremes of elliptical shape and the adjustable term containing a/b is very small for stress distributions with σ_2 of the same magnitude or smaller than σ_1. Thus any error from this factor would normally be within 2 or 3%. Moreover, using the stress distribution along l in **Fig. I.9** is entirely appropriate within the stated degree of accuracy. Finally, if the stress distribution with the crack absent varies in a nonlinear manner, as in **Fig. I.11**, then taking the extreme linear approximations using σ_2 or σ'_2 will give error extremes between which suitable judgments can be made with assured accuracy. This method accommodates having continuous stress gradients both along and into the surface. However, if the stress distributions are sharply peaked at the surface, such as near a notch, or otherwise not a smooth stress distribution, then estimating should be approached with extra caution. An example follows.

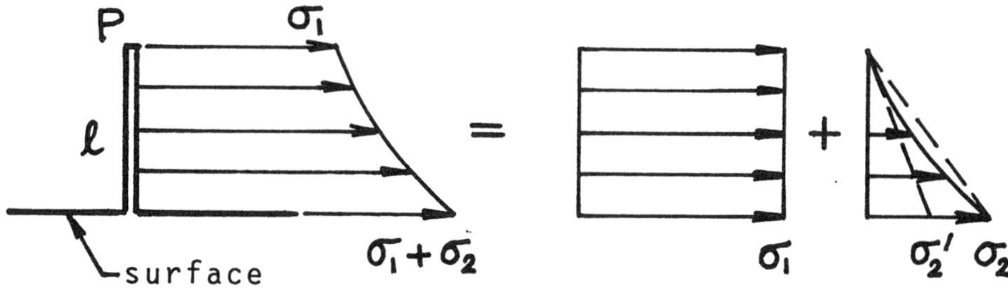

Fig. I.11

CRACKS EMANATING FROM A ROUND HOLE IN A PLATE IN TENSION

Figure. I.12 shows a circular hole in a plate subject to uniform uniaxial tension, σ, remotely applied. Assuming symmetrical through-thickness cracks exist at the maximum stress concentration points on opposite sides of the hole, it is desired to determine the K values for this case. Although a solution appears on **page 19.1** for this configuration, the estimating method as developed in the previous paragraphs here will be compared with this solution to evaluate the estimating method with nonlinear stress gradients.

With the crack absent, the stress distribution for normal stress on the crack plane is described by

$$\sigma_y(r) = \sigma\left[1 + \frac{1}{2}\left(\frac{R}{r}\right)^2 + \frac{3}{2}\left(\frac{R}{r}\right)^4\right] \tag{I42}$$

From this expression it is noted that a sharp peak in stress, of a value 3σ, exists next to the hole, which nonlinearly diminishes to a constant value, σ, away from the hole. The crack emanates from the surface of the hole that is not flat so that the 1.1215 surface factor should diminish rapidly away from the hole but in an unknown manner. Therefore, using that factor without diminution would cause an overestimate in K as the crack lengthens. Consequently, the previous method, using σ'_2, which would cause underestimates, is used with that factor for the evaluation. The resulting form is

$$K = 1.1215(\sigma_1 + 0.39\sigma'_2)\sqrt{\pi a} \quad ({}^{a}/_{R} \ll 1) \tag{I43}$$

On the other hand, for larger cracks, the hole itself may be regarded as part of a longer crack of length $2a + 2R$ in a sheet with uniform tension. This can be evaluated with the familiar form

$$K = \sigma\sqrt{\pi(a+R)} = \sigma\sqrt{\pi a}\cdot\sqrt{1+\frac{R}{a}} \quad ({}^{a}/_{R} \gg 0) \tag{I44}$$

Now both of these formulas for K will tend to give exact results, the first as the crack size approaches zero and the second for very large cracks (i.e., a large compared with R). In the intermediate region, both tend to give overestimates of the real K. Therefore, electing to take the smaller value of the two would tend to give the most reasonable results. The results are tabulated in **Table I.1** and are compared with Tada's best fit form from **page 19.1**.

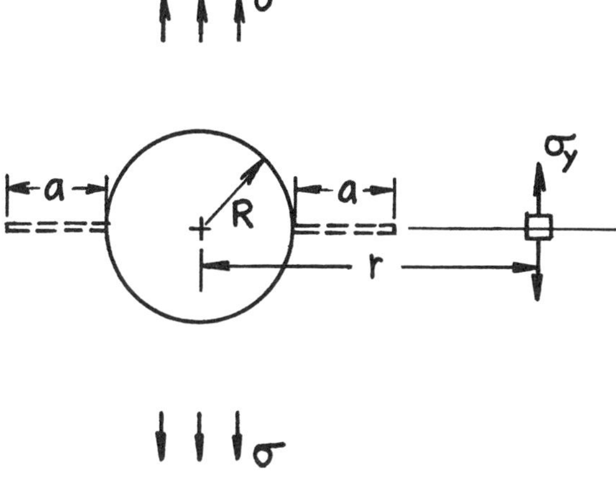

Fig. I.12

Table I.1

$\frac{a}{R}$	$1.1215\left[\frac{\sigma_1}{\sigma}+0.39\frac{\sigma'_2}{\sigma}\right]$	$\sqrt{1+\frac{R}{a}}$	Tada p. 19.1	% Error	Green's Function
0	3.3645	(infinite)	3.3645	0%	
0.02	3.2804		3.2357	+1.4%	
0.05	3.1422	(4.5826)	3.0614	+3.0%	
0.10	2.9494	(3.3166)	2.8131	+4.8%	
0.15	(2.7822)	2.7689	2.6079	+6.2%	(2.2397)
0.20	(2.6359)	2.4495	2.4357	+0.6%	(2.1305)
0.30		2.0817	2.1679	-4.1%	1.9554
0.50		1.7321	1.8246	-5.3%	1.7189
0.70		1.5584	1.621	-4.0%	1.5697
1.00		1.4142	1.444	-2.1%	1.4301
Infinite		1.0000	1.000	None	1.0000

These results speak for themselves for confirming the method of estimating in this difficult case. Other improvements could be attempted, but this seems sufficient to demonstrate the method. For example, the second integral **(I16)** can also be applied by regarding the hole as part of the crack. It will give even better values at intermediate crack sizes, that is, for $^a/_R$ of about 0.50 to 3.0 or more. Defining the apparent half-crack length as $l = a + R$, the appropriate form is

$$K = 2\sqrt{\frac{l}{\pi}}\int_R^l \frac{\sigma_y(r)dr}{\sqrt{l^2 - r^2}} \tag{I45}$$

where $\sigma_y(r)$ is given by **(I42)**.
Integrating, the numerical values are the final column in **Table I.1**.

Appendix J

Rice's J-Integral as an Analytical Tool in Stress Analysis

The J-Integral, which is most often associated with elastic–plastic fracture mechanics, is also a useful tool for linear and nonlinear elastic stress analysis of cracks. It is therefore discussed here, mostly in the latter elastic context, but with some comments on elastic–plastic analysis as well. It has also often been used as a powerful analytical tool in combination with finite element analysis, which shall get only minimal comment here.

The path-independent integral form of J, as illustrated by **Fig. J.1**, is given by **Rice (1968a)** as

$$J = \oint_{\Gamma} \left(W dy - T_i \frac{\partial u_i}{dx} ds \right) \tag{J1}$$

where

$$W = \int_0^{\varepsilon_{ij}} \sigma_{ij} d\varepsilon_{ij} = \text{ elastic strain energy per unit volume or work per unit volume if plastic}$$

$$T_i = \sigma_{ij} n_j = \text{ traction on the contour } \Gamma$$

and

$$n_j = \text{ components of the unit vector normal to contour}$$

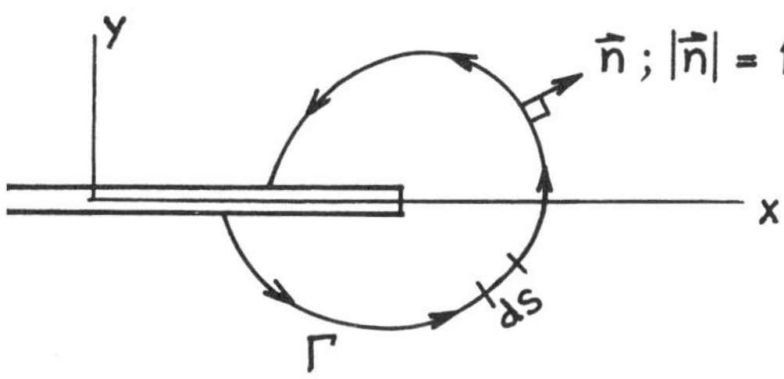

Fig. J.1

With the usual assumptions of equilibrium, small strains and rotations, and assuming that elastic behavior or deformation theory of plasticity applies, then

$$W = W(\varepsilon_{ij}) \quad \text{or} \quad \frac{\partial W}{\partial \varepsilon_{ij}} = \sigma_{ij} \tag{J2}$$

These assumptions, along with the Green-Gauss Theorem, allow the proof that the J-Integral is path independent when integrating on any contour, Γ, from the lower to upper crack surfaces, provided that the material is a continuum and that no singularities or loads are enclosed within Γ, except the crack tip itself.

PHYSICAL INTERPRETATION OF *J* FOR NONLINEAR ELASTIC CONDITIONS

The crack tip with the contour attached may be advanced by an increment, *da*. The terms in the J-Integral interpreted as a result of that advance are illustrated in **Fig. J.2**. **Rice (1968a)** notes that the value of

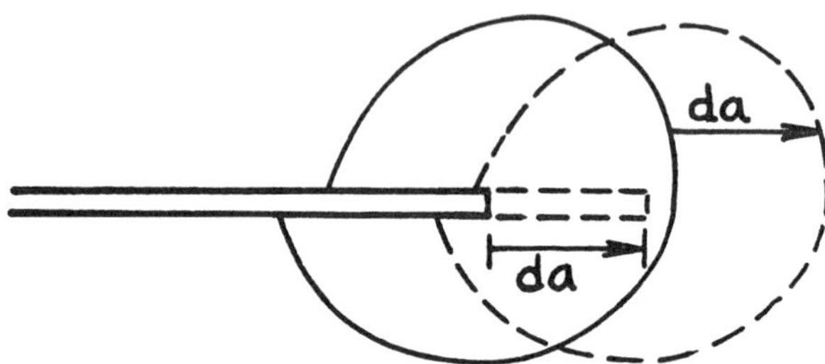

Fig. J.2

the J-Integral is the amount of energy pouring through the contour per unit increase in crack area, as characterized by *da.* This means that for nonlinear or linear elastic conditions, J gives the Griffith energy rate per unit increase in crack area, or

$$J = G = \frac{K^2}{E'} \text{ (elastic: linear or nonlinear)} \tag{J3}$$

Shrinking the contour to zero then shows that J is the energy made available for crack extension under elastic conditions. However, under conditions of plasticity, since W is not recoverable energy, J is <u>not</u> the energy made available for separation. Nevertheless, for the deformation theory of plasticity J remains a path-independent integral, subject to the other interpretations to follow. However, the above relationship to K allows the J-Integral to be used to evaluate K directly for various analytical and finite element numerical analyses.

EXAMPLE OF THE USE OF THE *J*-INTEGRAL IN EVALUATING *G* AND *K*

Consider the mixed boundary value problem for the configuration shown in **Fig. J.2a**. It shows a sheet stretched uniformly prior to clamping rigid, smooth parallel boundaries perpendicular to the stretch direction, and with a centrally introduced crack from one edge subsequent to clamping. The sheet is considered to be

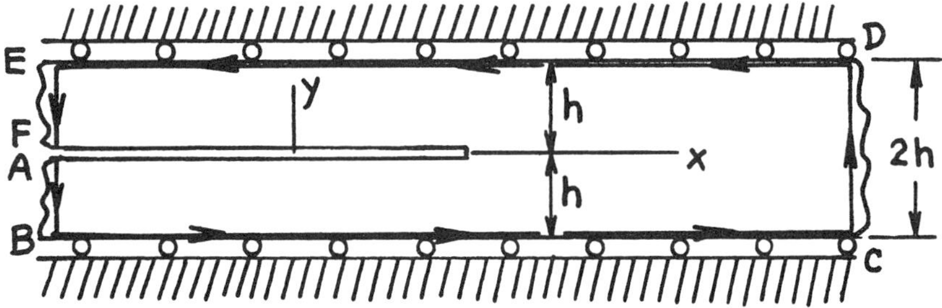

Fig. J.2a

elastic but not necessarily linear. This problem is analogous to the vertical periodic array of cracks problem of **page 12.1**. The application of the J-Integral follows by integrating on the contour shown. On segments $A-B$ and $E-F$, where all stresses are relieved, there are no tractions, nor any strain energy, W, so that portion of the integral gives zero. Further on segments $B-C$ and $D-E$, dy, T_x, and $\frac{\partial u_y}{\partial x}$ are zero so the integration there gives zero. On the final segment $C-D$, the tractions are zero but the strain energy, W, remains undisturbed by the crack and is constant with y.

Therefore, integration over the full contour, Γ, is reduced to

$$J = G = \int_{-h}^{h} W dy = 2Wh \quad \text{(nonlinear elastic)} \tag{J4}$$

However, if the sheet is linear elastic, then

$$J = G = \frac{\sigma^2}{E} h = \frac{K^2}{E} \quad \text{(linear elastic)} \tag{J5}$$

This otherwise difficult problem for finding G or K is made almost trivial by the J-Integral method. For such special problems, the method can be very useful, including use with numerical methods such as finite element methods into the elastic–plastic regime.

RELATIONSHIP OF *J* TO CRACK OPENING STRETCH, δ_T, FOR THE STRIP YIELD MODEL

The strip yield model, as discussed in **Section 30** and its solution pages, has a crack opening stretch at the crack tip that is directly related to J as follows. **Figure J.2b** shows any of the crack tip situations for all configurations. The shaded plastic zone is assumed to be perfectly plastic with a yield strength, σ_0, applied to

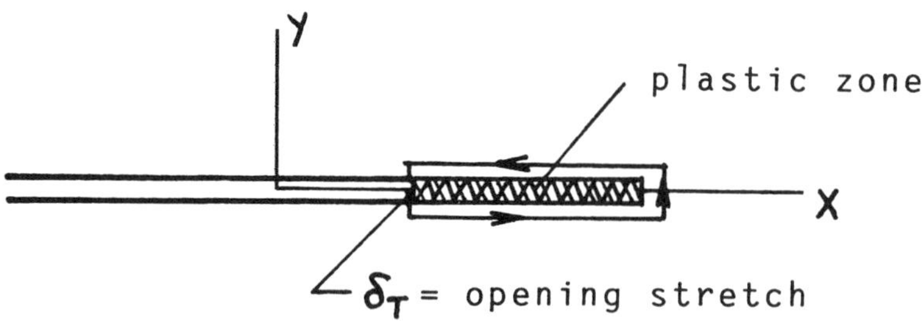

Fig. J.2b

the zone as a constant traction, $T_y = \sigma_0$, top and bottom. Then, since dy is zero, and ds is simply dx or $-dx$, integrating around the contour gives

$$J = -\sigma_0 \int \frac{\partial U_y}{\partial x} ds = \sigma_0 \delta_T \qquad \text{(J6)}$$

This simple result allows conversion to J for the strip yield results in Section 30. However, the idealization of strip yielding can simply be dropped to allow the yielding to spread, in which case it is still found that the relationship is given by

$$\delta_T = \gamma \frac{J}{\sigma_0} \qquad \text{(J7)}$$

where γ is a constant of the order of one, depending on plane stress or plane strain yield zone conditions and the hardening coefficient of the material. In any event, this J-Integral calculation has demonstrated the appropriate type of relationship in a direct fashion.

ALTERNATE "NONLINEAR COMPLIANCE" DEFINITION OF *J*

The preceding nonlinear elastic energy rate interpretation of J itself suggests further alternative forms for the definition of J for both elastic and elastic–plastic circumstances. Consider, a nonlinear elastic body containing a crack and loaded as shown in **Fig. J.3**. Suppose the body is loaded by a force, P, whose work-producing component of displacement is denoted δ. For elastic conditions the load displacement diagram might look as shown in **Fig. J.3** up to point A for constant crack length, a, where the value of J is desired. At that point, if the displacement is fixed, and the crack is advanced by an increment, da, the load will drop to point B with no further work of loading induced. Then at point B, with crack size now $a + da$, unloading will result in the curve from B to O. The region $O - A - B - O$ on that load displacement curve is the energy

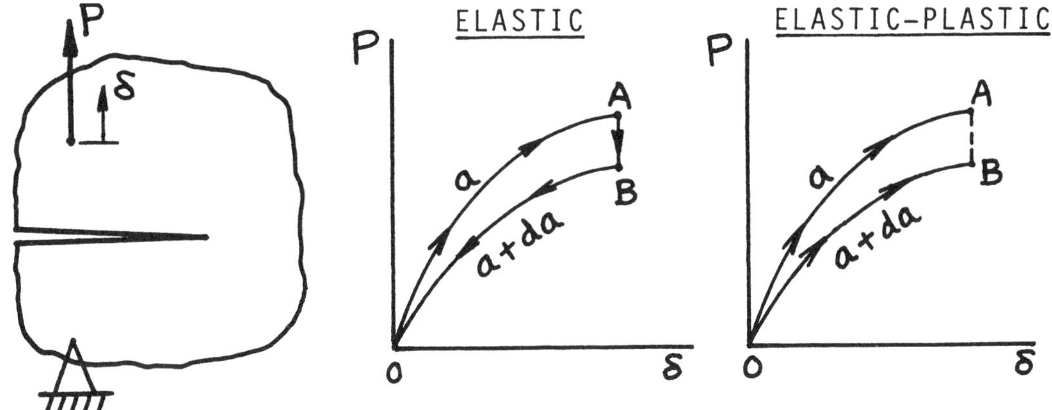

Fig. J.3

given up for the crack extension da, and is therefore equal to Jda. Considering that fixity of the displacement in the preceding argument would change the result only by a higher-order double differential, it can be seen that fixity does not influence the result. Therefore, J may be determined from

$$J = -\int_0^\delta \frac{\partial P}{\partial a} d\delta = \int_0^P \frac{\partial \delta}{\partial a} dP \tag{J8}$$

These two integral forms in terms of load displacement relations are equally valid definitions of J, consistent with the original contour integral definition given by **(J1)**. Indeed, the elastic–plastic load displacement curves for two samples of crack sizes a and $a + da$ on the final load displacement curve in **Fig. J.3** would be identical to those in the nonlinear elastic case under deformation theory of plasticity assumptions, if the loading stress–strain curves of the elastic and elastic–plastic materials were identical. Therefore, these alternate definitions of J, by the above integrals, also equally apply under deformation theory restrictions for the evaluation of J.

These alternate definitions of J made possible the experimental technique used in the famous works of **Landes (1972)** and **Begley (1972)** for measurement of J under elastic–plastic conditions. Further, for purposes of computing or estimating J in practice, the above "nonlinear compliance" form of J leads to the following consideration. For linear–elastic plastic conditions the displacement can always be exactly divided into its elastic and plastic parts. That is to say:

$$\delta = \delta_{el} + \delta_{pl} \tag{J9}$$

The second integral form then becomes

$$J = \int_0^P \frac{\partial \delta_{el}}{\partial a} dP + \int_0^P \frac{\partial \delta_{pl}}{\partial a} dP \tag{J10}$$

Now the first of these two integrals is exactly the linear–elastic energy rate, G, and the second can be converted by analysis of load versus plastic displacement (plastic part only) to result in

$$J = J_{el} + J_{pl} = G - \int_{0}^{\delta_{pl}} \frac{\partial P}{\partial a} d\delta_{pl} \tag{J11}$$

This form has a great advantage when used to estimate J under elastic–perfectly plastic conditions (no hardening), because for large plastic deformations it gives

$$J \cong G - \frac{\partial P_{\lim}}{\partial a} \delta_{pl} \tag{J12}$$

where $P_{\lim}$ is the fully plastic limit load. Actually, since the first term, G, in this expression is small for large plastic deformation, the second term by itself is often an adequate estimate of J.

THE HUTCHINSON–RICE–ROSENGREN (HRR) CRACK TIP FIELD EQUATION INTERPRETATION OF J

Hutchinson (1968) and **Rice (1968b)** independently demonstrated that the near-tip stress and strain fields within the plastic region can be found for materials, if they are reasonably represented as purely power hardening well beyond their elastic range, which are the conditions near a crack tip in elastic–plastic materials even with moderate loading. A power-hardening material is represented by the stress–strain relation

$$\frac{\varepsilon}{\varepsilon_0} = \left(\frac{\sigma}{\sigma_0}\right)^n \tag{J13}$$

For such a material, in the intense plastic enclave at a crack tip, the field equations have a dominant term (the only singular term) of the forms

$$\sigma_{ij} = \sigma_0 \left(\frac{J}{\sigma_0 \varepsilon_0 r}\right)^{\frac{1}{n+1}} \Sigma_{ij}(\theta, n) \tag{J14}$$

$$\varepsilon_{ij} = \varepsilon_0 \left(\frac{J}{\sigma_0 \varepsilon_0 r}\right)^{\frac{n}{n+1}} E_{ij}(\theta, n) \tag{J15}$$

Again these equations were developed assuming deformation theory of plasticity with small strains and rotations. They are therefore "valid" in a region surrounding a crack tip, away from the tip by an amount measured by the crack opening stretch, but near to the tip compared with planar body dimensions and well within the crack tip plastic zone. Therefore, J may be interpreted as the strength of the plastic field surrounding the crack tip, just as K has been interpreted as the strength of the elastic field for small-scale yielding conditions, However, the J field strength does not depend on the presence of small scale yielding conditions! Further, it is clear that, for any monotonicaly increasing true stress–strain behavior, the power law assumed is not necessary to have J as the relevant intensity parameter. Indeed, J is the parameter that reflects the effects of body configuration and loading on the surrounding crack tip plastic field.

J-CONTROLLED CRACK GROWTH

In various applications, it is desirable to use a parameter such as J to measure the effects of loading and configuration on crack tip deformation and stressing. To justify using J, the assumptions of deformation theory of plasticity must reasonably prevail in the surrounding crack tip field. The "no unloading" and "proportional straining" assumptions are of concern if the crack is extending, da, as loading or imposed deformation, as reflected by dJ, is occurring. To consider the proportionality of straining, **Hutchinson (1979)** formed the differential of strain $d\varepsilon_{ij}$ to compare it with the preceding strain field equation. The result was of the form

$$d\varepsilon_{ij} = \varepsilon_{ij}\left(\frac{n}{n+1}\right)\frac{dJ}{J} + \left(\frac{\partial\varepsilon_{ij}}{\partial r}\cdot\frac{\partial r}{\partial a} + \frac{\partial\varepsilon_{ij}}{\partial\theta}\cdot\frac{\partial\theta}{\partial a}\right)da \tag{J16}$$

where the parentheses of the second term, if multiplied by a characteristic planar dimension, is of the order of the first term excluding the ${}^{dJ}/_{J}$. However, the second term results in an r, θ distribution of strain that is nonproportional to the original strains. Hence proportional straining only occurs for situations where the complete first term dominates the second. This is the case if

$$\frac{dJ}{J} \gg \frac{da}{b} \tag{J17}$$

where b is the relevant characteristic dimension for the field, which is normally the remaining uncracked ligament ahead of the crack. On this basis the conditions for valid "J - controlled crack growth" are stated as

$$\omega = \frac{dJ}{da}\cdot\frac{b}{J} \gg 1 \quad (\text{and } \Delta a_{total} \ll b) \tag{J18}$$

For example, the "validity" of $J - R$ curve data is limited to portions of the curve where its current slope $\frac{dJ}{da}$ divided by current J is large enough for the specimen size, as measured by b to keep $\omega \gg 1$. Valid conditions for crack growth must also apply in any proper structural application.

FURTHER DIMENSIONAL CONSIDERATION OF THE HRR FIELD EQUATIONS

Note that the preceding HRR field equations imply that J is characterized by stress times strain times a characteristic length. In the analysis of a cracked body with a single characteristic dimension (the crack size) subject to a uniform remote field of stress (and thereby strain), the form of the applied J must be

$$J = C\sigma\varepsilon a \tag{J19}$$

where C is a constant for the particular configuration (an infinite plate with a crack of length $2a$ perpendicular to the stress direction, or a semi-infinite plate with an edge crack or an internal circular crack or semi-circular surface flaw, etc.). For example, the familiar Griffith configuration gives

$$J = G = \frac{\pi\sigma^2 a}{E} = \pi\sigma\varepsilon a \tag{J20}$$

Now, for example, applying this result to a material that is represented appropriately by the Ramberg-Osgood relation

$$\frac{\varepsilon}{\varepsilon_0} = \frac{\sigma}{\sigma_0} + \alpha\left(\frac{\sigma}{\sigma_0}\right)^n \tag{J21}$$

then the relationship for J for this particular configuration becomes

$$J = \pi\left\{\left(\frac{\sigma}{\sigma_0}\right)^2 + \alpha\left(\frac{\sigma}{\sigma_0}\right)^{n+1}\right\}\sigma_0\varepsilon_0 a \qquad \text{(J22)}$$

This is an excellent form for approximating J when the applied stress, σ, exceeds σ_0 for the Griffith configuration, and for others with the single characteristic dimension of the crack, by appropriately adjusting the constant π in this expression. The first change is for the constant adjustment in the elastic solution, for example, the π should be replaced by $^4/_\pi$ for the penny-shaped circular crack. Second, the coefficient, α, should also be increased because of hardening, approximately by the factor $(1 + 0.26n)$ to accommodate fully adjustments for the nonlinear material behavior term according to **Paris (1983a)**.

A reasonable approximation for configurations with more than one characteristic dimension is to use the linear elastic configuration correction factors for stress levels, σ, at or not far above σ_0. The final form for such approximations uses both the above stress correction bracket and a geometry correction bracket as well, in the form

$$J = \sigma_0\varepsilon_0 a\{Stress\}[Geometry] \qquad \text{(J23)}$$

The separated nondimensional brackets are convenient in applying Tearing Instability Theory for the stability of growing cracks using R-curve concepts, as well as other applications. (It should be noted that the EPRI Elastic-Plastic Fracture Analysis Handbook method is based on similar approximations from dimensional considerations.)

RICE'S ANALYSIS FOR PURE BENDING OF A REMAINING LIGAMENT

For a semi-infinite plate with a semi-infinite crack approaching the free edge leaving a ligament subject to pure bending, as shown in **Fig. J.4**, dimensional analysis leads to a very useful result. For this configuration, the relative angle change, θ, between the applied moments (per unit thickness), M, must depend only on that moment, M, and the only characteristic dimension, b, the ligament size, and the stress–strain properties of the material (whose dimensions are those of stress, $^F/_{L^2}$, and strain, nondimensional). Therefore, M and b must appear in a combination to give $^F/_{L^2}$ to allow θ to be nondimensional. Formally stated, θ must be

$$\theta = f\left(\frac{M}{b^2}, \textit{stress-strain}\right) \qquad \text{(J24)}$$

This functional form is true for both elastic and plastic behavior. The form may be inverted to give

$$M = b^2 F(\theta) \qquad \text{(J25)}$$

where $F(\ )$ includes the materials properties (constant). Considering that $da = -db$, then this result may be put into the alternative nonlinear compliance form for the J-Integral to give

$$J = -\int_0^\theta \frac{\partial M}{\partial a}\bigg|_\theta d\theta = +\int_0^\theta 2bF(\theta)d\theta = \frac{2}{b}\int_0^\theta M d\theta \qquad \text{(J26)}$$

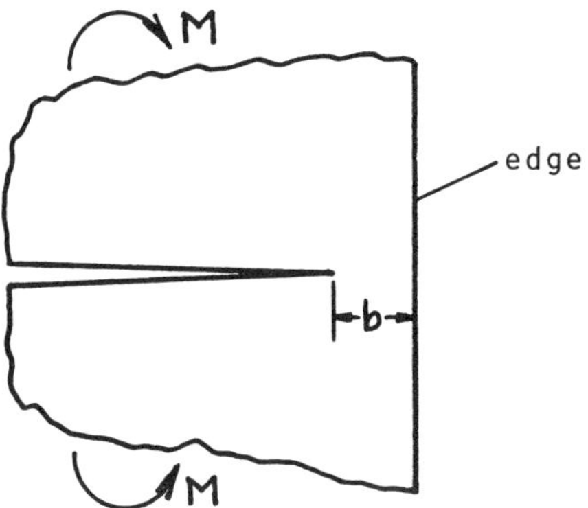

Fig. J.4

where the crack size, b, remains constant and the final integral of $Md\theta$ is the work done in loading. This result is independent of the actual elastic–plastic stress–strain relation as long as no crack growth occurs. Again, for pure bending, it is

$$J = \frac{2}{b}(Work) \quad \text{(no crack growth)} \tag{J27}$$

where $(Work)$ is per unit thickness of plate. For other configurations with a single characteristic dimension similar simple forms are possible, (see **Rice 1973**).

Now, when crack growth is occurring under J-controlled growth conditions so that deformation theory of plasticity applies, J may be considered to be a function of deformation, θ, and crack size, a, (or b): $J = J(\theta, a)$ and an increment in J must be computed by

$$dJ = \frac{\partial J}{\partial \theta} d\theta + \frac{\partial J}{\partial a} da \tag{J28}$$

Forming dJ for pure bending of a ligament from this result and **(J26)** gives

$$dJ = \frac{2}{b} d(Work) - \frac{J}{b} da \quad \text{(for } J\text{-controlled growth)} \tag{J29}$$

Consequently, J may be computed as an increment-by-increment accumulation under J-controlled crack growth conditions. For other configurations similar forms may be written for dJ, including those with more than one characteristic dimension (see **Ernst 1979**). However, the principles have been adequately demonstrated here.

As a final example, showing the results of the preceding dimensional arguments, the formula developed by these means for the determination of J for ASTM standard Compact Specimen tests is

$$J + \Delta J = \left[J + \frac{\eta}{b}(\Delta\ Work)\right]\left[1 - \frac{\gamma}{b}\Delta a\right] \tag{J30}$$

where

$$\begin{aligned} \eta &= 2 + 0.522\left({}^{b}/_{W}\right) \\ \gamma &= 1 + 0.76\left({}^{b}/_{W}\right) \end{aligned} \tag{J31}$$

and the current values of each parameter here should be used in evaluating the next increment, ΔJ. Here the 2 in η, and the 1 in γ are limiting factors agreeing with the preceding analysis where W is large compared to b, tending toward pure bending of the remaining ligament.

The result that the work in pure bending of a remaining ligament is directly related to J, the strength of the crack tip field of deformation, undoubtedly gives a sound basis for the use of bending tests, such as the Charpy impact test, to evaluate fracture toughness. Indeed, that result, which does not rely on the particular stress–strain curve of the material but only on the consequent work of the loading, is surprising in its generality. Another example of a surprising result follows here to demonstrate the unique qualities of the J-Integral.

J FOR THE DOUBLE CANTILEVER BEAM CONFIGURATION

The configuration of a double cantilever beam is shown in **Fig. J.5**, with opposing concentrated loads

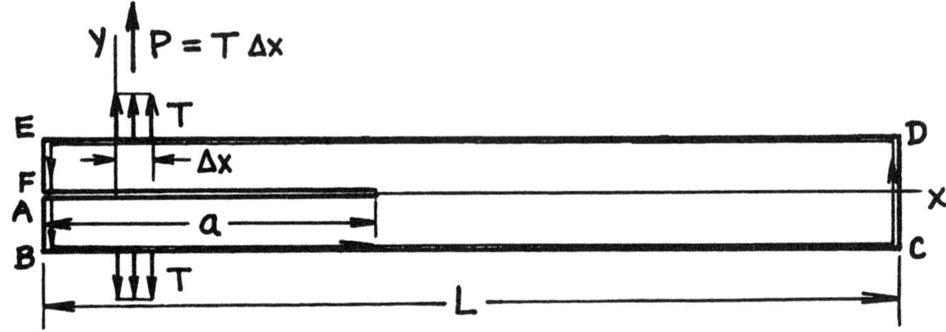

Figure J.5

modeled by uniform tractions over short segments (**A.J. Paris 1988**). The original contour J-Integral form, **(J1)**, will be used to evaluate J over the contour $A - B - \cdots - F$. Note that Wdy is zero over the whole contour and tractions are zero everywhere except at the load points where $T_y = T$. Therefore the integral becomes

$$J = \oint \left(Wdy - T_i \frac{\partial u_i}{\partial x} ds \right) = - \int_0^{\Delta x} T \frac{\partial u_y}{\partial x} dx + \int_{\Delta x}^{0} T \frac{\partial u_y}{\partial x} dx \tag{J32}$$

where the items in the final integrands are constant. The partial derivatives of the displacements at the load points are the relative rotations of the arms with respect to each other, totaling θ, which gives

$$J = P\theta \tag{J33}$$

This result gives J in terms of the load, P, and relative arm rotations, θ, independent of the elastic–plastic stress–strain properties, in terms of the final values of load and rotation, without requiring values during loading. This is a most amazingly simple result, showing the power of the J-Integral in analyzing the effects of loading on the intensity of the stress–strain singularity at a crack tip.

ELASTIC–PERFECTLY PLASTIC ESTIMATES OF J

For elastic–perfectly plastic behavior (no hardening) of the load-displacement relationship for a crack body, J can be estimated with good accuracy in terms of linear-elastic results and limit load behavior, as follows. The alternate nonlinear compliance form of J is given by **(J11)** and **(J12)**

$$J = G - \int_0^{\delta_{pl}} \frac{\partial P}{\partial a} d\delta_{pl} = G - \frac{\partial P_{\lim}}{\partial a}\delta_{pl} \tag{J34}$$

where the rate of change of limit load, $P_{\lim}$, with crack size is often known (always negative) or easy to compute. Especially for large plastic deformations, this is an accurate and practical way to determine J. In addition to many other potential applications, this method provides a direct way to evaluate J in bending of beams, tubes (pipes), and the like where plastic hinges form at cracked sections in statically determinate or indeterminate structures. Notice that this method can be easily and consistently incorporated into "plastic design" approaches, including collapse mechanism analysis. It is also effective in analyses of crack stability in beams under J-controlled growth of cracks (see **Paris 1983b**).

CRACK STABILITY UNDER J-CONTROLLED GROWTH CONDITIONS

Tearing instability theory, as initially developed by **Paris (1979)** and others, such as **Hutchinson (1979)** identifying conditions for applying J-controlled growth, allows evaluations of crack stability for linear and nonlinear elastic and elastic–plastic conditions where deformation theory of plasticity applies, that is, for all J-controlled growth conditions. It argues that equilibrium of a growing crack is expressed by the first derivative of energy, U, with respect to crack size, which is expressed as

$$J = \frac{dU}{da} = R \quad \text{(equilibrium)} \tag{J35}$$

where R is the material's resistance to crack growth in terms of energy or work required per unit increase in crack area. R is frequently presented as an "R-curve" of a material in terms of resistance, R, for a given crack extension, Δa, from its initial size. The J here is considered to be the applied J reflecting the loading and configuration of the deformed body. Crack stability then depends on the second derivative of energy, or

$$\frac{dJ}{da} = \frac{d^2U}{da^2} (< \text{or} \geq) \frac{dR}{da} \quad (< \text{stable}; \geq \text{unstable}) \tag{J36}$$

where $\frac{dR}{da}$ is the slope of a J-Integral R-curve.

Defining the nondimensional parameters

$$T_{appl} = \frac{dJ}{da} \cdot \frac{E}{\sigma_0^2} \quad \text{and} \quad T_{mat} = \frac{dR}{da} \cdot \frac{E}{\sigma_0^2} \tag{J37}$$

stability is judged by

$$T_{appl} \; (< \text{or} \geq) \; T_{mat} \quad (< \text{stable}; \geq \text{unstable}) \tag{J38}$$

It is convenient to plot diagrams of T vs. J to analyze stability continuously as J increases with loading or deformation. These diagrams identify first instability somewhat more conveniently than techniques employing using tangency with R-curves (see **Paris 1983a**).

APPLICATION OF C^* (TIME DEPENDENT J) TO CREEP CRACK GROWTH

Begley (1976) showed that a modified J-Integral, denoted C^*, replacing strain and displacement with their time derivatives, is a controlling parameter for certain creep crack growth behaviors. In such cases the integral is defined as

$$C^* = \oint \left(W' dy - T_i \frac{\partial \dot{u}_i}{\partial x} ds \right) \quad \text{(J39)}$$

where

$$W' = \int \sigma_{ij} d\dot{\varepsilon}_{ij} \quad \text{(J40)}$$

and the dot over them indicates time derivatives of ϵ and u. Later discussions of this time rate-modified J-Integral have been provided by **Riedel (1980, 1989)**, in which its further relevance to applications has been given.

APPLICATION OF ΔJ TO FATIGUE CRACK GROWTH

Dowling (1976) suggested computing the range of J or ΔJ added during a load cycle in fatigue and correlated that parameter with the cyclic rate of crack growth. The ΔJ in this work was defined in cyclic plastic bending of a remaining ligament as

$$\Delta J = \frac{2}{b} \Delta(Work) \quad \text{(J41)}$$

where Δ $Work$ was defined as the area under the crack-opened portion of the loading cycle. This parameter correlated fatigue crack growth rates under cyclic plastic conditions, and also related these rates to other fatigue cracking rates for tests in the small-scale yielding regime by

$$\Delta K = \sqrt{E' \Delta J} \quad \text{(J42)}$$

Although the assumptions of deformation theory and thus J normally do not allow unloading, the success of this method may be taken to imply no effects of the prior unloading cycles, that is, no history dependence. Indeed, these last two examples of applying J to time-dependent and cyclic loaded subcritical crack growth demonstrate the power of J as a relevant analysis parameter.

In conclusion J has been demonstrated here to be a powerful analytical tool for a wide range of applications, from providing a method to evaluate K and G under elastic conditions to nonlinear elastic and elastic–plastic applications of correlating cracking behavior under static, time-dependent, and even cyclic deformation and loading. It generalizes analysis under a broader scope, including as a special case, linear-elastic method employing K and G. Its arguments and assumptions go beyond but are completely consistent with linear-elastic fracture mechanics methods.

Appendix K

Elasto-Plastic Pure Shear Stress-Strain Analysis (Mode III)

This appendix presents a method of developing "exact solutions" to elastic-perfectly plastic problems in pure shear for the stresses and strains in both the elastic and plastic regions. Since no such analytical method is available for any other (Mode I or II) crack problems, this Mode III method is given to illustrate the nature of plasticity in this rather special case as it may aid intuitively in more general problems. The method to be presented was fully developed in **Rice (1966, 1968c, 1984)** and similar problems were previously investigated by F. A. McClintock in the 1950s.

Pure shear (or anti-plane shear) results from problems where the deformations can be described in or are restricted to the following form of the induced displacements:

$$u(x, y, z) = 0, \quad v(x, y, z) = 0, \quad w(x, y, z) = w(x, y)$$

This form for the displacements leads to strains of the form

$$\varepsilon_x = \varepsilon_y = \varepsilon_z = 0$$

$$\gamma_{xy} = \frac{\partial u}{\partial y} + \frac{\partial v}{\partial x} = 0, \quad \gamma_{yz} = \frac{\partial w}{\partial y}, \quad \gamma_{xz} = \frac{\partial w}{\partial x}$$

Differentiating each of these two nonzero strains with respect to the opposite variable and equating leads to the compatibility equation

$$\frac{\partial \gamma_{xz}}{\partial y} - \frac{\partial \gamma_{yz}}{\partial x} = 0 \qquad \textbf{(K1)}$$

The absence of the three extensional strains as well as the *xy* shear strain implies the absence of the following stress components:

$$\sigma_x = \sigma_y = \sigma_z = \tau_{xy} = 0$$

Therefore the only nontrivial equilibrium equation is

$$\frac{\partial \tau_{xz}}{\partial x} + \frac{\partial \tau_{yz}}{\partial y} = 0 \tag{K2}$$

Equations (K1) and (K2) are usually solved directly for elastic problems, also making use of the stress-strain relations

$$\gamma_{xz} = \frac{\tau_{xz}}{G} \quad \text{and} \quad \gamma_{yz} = \frac{\tau_{yz}}{G} \tag{K3}$$

The solution to this set of equations, **Eqs. (K1), (K2), and (K3)**, would normally take the form

$$\tau_{xz} = \tau_{xz}\ (x, y) \quad \text{and} \quad \tau_{yz} = \tau_{yz}\ (x, y) \tag{K4}$$

However, for elastic–plastic problems, advantages occur upon inverting the variables and solving the problem on the stress plane, the τ_{xz} vs. τ_{yz} plane, rather than on the physical plane, the x vs. y plane. The solutions will then involve mapping points on the physical plane onto the stress plane, which is sometimes called the "Hodograph Method" in other applications.

INVERTING TO THE STRESS PLANE

The stress equations. **Eq. (K4)**, can be inverted to express the coordinates as functions of the stresses, or

$$x = x\ (\tau_{xz},\ \tau_{yz}) \quad \text{and} \quad y = y\ (\tau_{xz},\ \tau_{yz}) \tag{K5}$$

Taking the differentials of **Eqs. (K5)** and then substituting for the differentials of stress from **Eqs. (K4)** leads to two equations, which, upon noting that dx and dy are independent, become four equations. Performing some direct algebraic manipulation one can then show that **Eq. (K2)** becomes

$$\frac{\partial x}{\partial \tau_{xz}} + \frac{\partial y}{\partial \tau_{yz}} = 0 \tag{K6}$$

Similarly, **Eqs. (K1) and (K3)** may be combined and inverted to give

$$\frac{\partial x}{\partial \tau_{yz}} - \frac{\partial y}{\partial \tau_{xz}} = 0 \tag{K7}$$

which applies only to the elastic region of any problem.

STRESS FUNCTION SOLUTION FOR THE ELASTIC REGION

Figure K.1 shows the stress plane where the circle represents the yield surface for homogeneous isotropic yielding and points within the circle are the elastic region on this plane. For the solution within this elastic region, a stress function, Φ, is elected where

$$\Phi = \Phi(\tau_{xz},\ \tau_{yz})$$

and

$$x = \frac{\partial \Phi}{\partial \tau_{yz}}, \quad y = -\frac{\partial \Phi}{\partial \tau_{xz}} \tag{K8}$$

so that the equilibrium equation, **Eq. (K6)**, is automatically satisfied and the elastic region compatibility equation, **Eq. (K7)**, becomes

$$\frac{\partial^2 \Phi}{\partial \tau_{xz}^2} + \frac{\partial^2 \Phi}{\partial \tau_{yz}^2} = \nabla^2 \Phi = 0 \tag{K9}$$

which is the well-known Laplace's Equation.

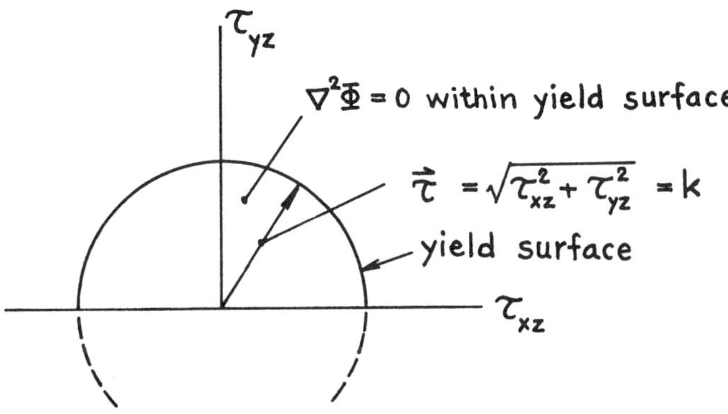

Fig. K.1

BOUNDARY CONDITIONS

The boundary conditions for the solution of **Eq. (K9)** in the elastic region of the stress plane for typical crack problems, with stress boundary conditions on external surfaces, are dictated by the mapping of those external surfaces from the physical plane onto the stress plane, as will be illustrated. However, the boundary conditions for Φ at the elastic–plastic boundary are more subtle. **Figure K.2** shows an anisotropic yield surface (other than circular) with ds defined as an increment of length along that surface at the point indicated by the stress vector, τ. Further, the normal unit vector to the yield surface, $\vec{n}$, and the tangent unit vector, $\vec{s}$, and their components are shown on the stress plane. In addition the corresponding crack tip is shown on the physical plane in **Fig. K.2** with the plastic zone at its tip. The plastic slips must begin at the singularity of the crack tip forming a staircase of slips emanating at that tip. For elastic–perfectly plastic material (no hardening) the emphasized slip at an angle θ must correspond to the point on the yield surface with the corresponding angle θ because, as is well known in plasticity theory, plastic strains must be normal to the yield surface.

Stress Plane

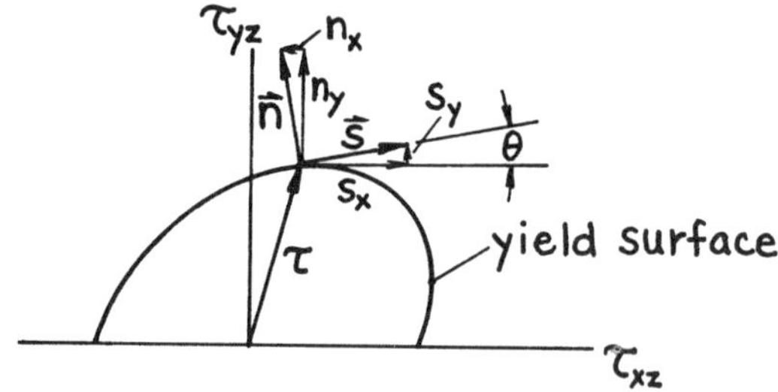

Physical Plane

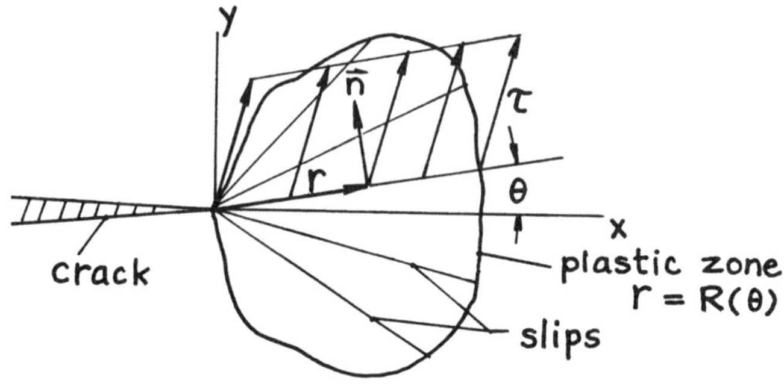

Fig. K.2

From conditions on the yield surface it is noted that

$$s_x = n_y = \frac{\partial \tau_{xz}}{\partial s}, \quad s_y = -n_x = \frac{\partial \tau_{yz}}{\partial s}$$
$$\text{and } \vec{n} = n_x\vec{i} + n_y\vec{j} \tag{K10}$$

Also on the physical plane

$$\vec{r} = x\vec{i} + y\vec{j} \tag{K11}$$

Now the rate of change of the stress function along the yield surface may be written as

$$\frac{d\Phi}{ds} = \frac{\partial \Phi}{\partial \tau_{xz}}\frac{\partial \tau_{xz}}{\partial s} + \frac{\partial \Phi}{\partial \tau_{yz}}\frac{\partial \tau_{yz}}{\partial s}$$

which upon relevant substitutions from **Eqs. (K8), (K10), and (K11)** give

$$\frac{d\Phi}{ds} = -(xn_x + yn_y) = -\vec{r}\cdot\vec{n} = 0 \tag{K12}$$

since the vectors $\vec{r}$ and $\vec{n}$ are perpendicular. Therefore the boundary condition along the yield surface on the stress plane is that Φ is a constant, which may be arbitrarily chosen to be zero, thus

$$\Phi = \underline{\text{constant} = 0} \text{ (on the yield surface)} \tag{K13}$$

In addition, the radial coordinate on the physical plane from the crack tip to the elastic–plastic boundary may be denoted

$$R(\theta) = \sqrt{x_\theta^2 + y_\theta^2} = \sqrt{\left(-\frac{\partial\Phi}{\partial\tau_{xz}}\right)^2 + \left(\frac{\partial\Phi}{\partial\tau_{yz}}\right)^2}$$

or

$$R(\theta) = \sqrt{\left(\frac{\partial\Phi}{\partial n}\right)^2 + \left(\frac{\partial\Phi}{\partial s}(=0)\right)^2} = \frac{\partial\Phi}{\partial n} \tag{K14}$$

From this expression, the physical radius of the plastic zone is determined along any radial slip line by the gradient of the stress function on the stress plane.

Let us further invoke Neuber's observation that at a crack tip the stress singularity times the strain singularity of the field is always $^1/_r$. For no hardening, the stress is a nonsingular, constant, τ, along radial lines within the plastic zone. Therefore the strain along these radial lines must be determined by

$$\gamma_{\theta z} = \gamma_{r=R}\left(\frac{R(\theta)}{r}\right) = \frac{\partial w}{r\partial\theta}$$

$$\gamma_{rz} = \frac{\partial w}{\partial r} = 0$$

Consequently, the stresses and strains are completely known within the plastic zone once the stress function, Φ, is determined for the adjacent elastic region. The discussion now proceeds to the determination of Φ.

MAPPING INDIVIDUAL PROBLEMS AND FULL BOUNDARY CONDITION DETERMINATION

Figure K.3 shows the example of a half plane with a crack of length, a, perpendicular to its edge. The crack is along the x-axis with its tip at the origin making the edge parallel to the y-axis. The body is loaded with uniform shear stress, $\tau_{yz} = \tau_0$, at $y = \pm\infty$.

Physical Plane

Stress Plane

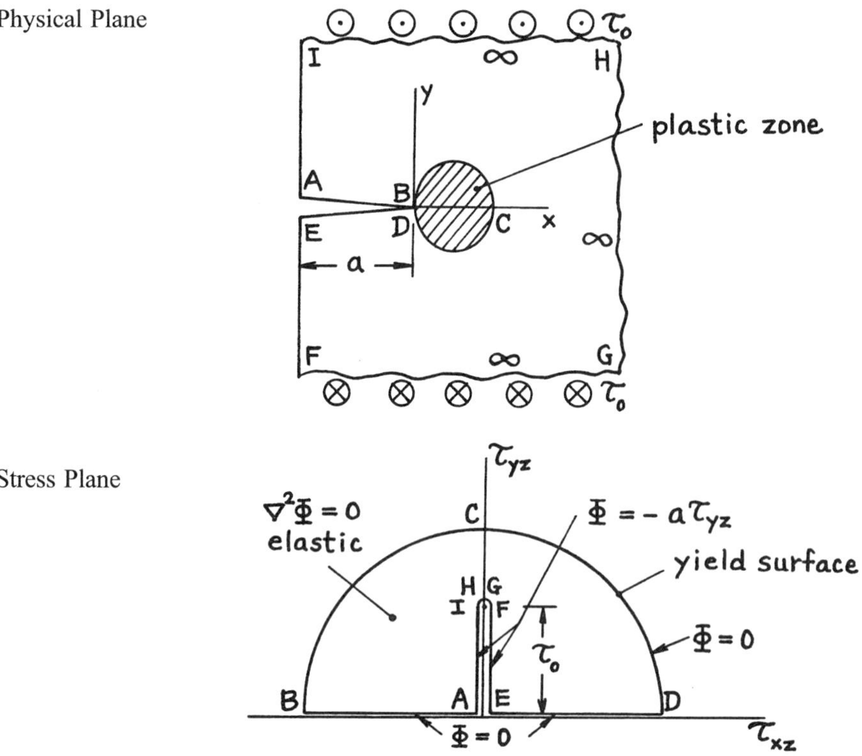

Fig. K.3

In addition, the stress plane is shown with a circular yield surface corresponding to isotropic yield strength, k, or the arc $\tau^2 = \tau_{xz}^2 + \tau_{yz}^2 = k^2$. To complete the mapping, successive points around the boundary of the elastic region on the physical plane are labeled $A, B, C, ..., I$, which correspond to similarly labeled points on the stress plane. For example, all along the upper crack surface is labeled AB, and $\tau_{yz} = 0$ as a load free boundary condition. At the corner at A it is entirely stress free, or $\tau_{xz} = 0$, whereas the magnitude of this stress component gradually increases along AB, meeting the yield surface, $\tau_{xz} = -k$, so that B is at the origin on the physical plane but at the yield surface on the stress plane. Note that where the plastic zone crosses the x-axis remote from the origin, labeled C on the physical plane, because of symmetry, $\tau_{xz} = 0$ so that on the stress plane $\tau_{yz} = +k$. Similar to AB on the stress plane, DE occupies the $\tau_{xz}(= +)$ axis from the origin to the yield surface. Further, for all boundaries at infinity on the physical plane corresponding to $FGHI$ the stresses are $\tau_{yz} = +\tau_0$ and $\tau_{xz} = 0$, that is, all at the same point on the stress plane. It is also noted that points along EF and IA on the physical plane are on the vertical axis on the stress plane. The map is then completed and the elastic region is within the boundary, as described on the stress plane. Within that boundary the stress function, Φ, obeys **Eq. (K9)** or $\nabla^2\Phi = 0$ as indicated. It remains to express the boundary conditions on Φ on the stress plane. They are:

a.) On the elastic–plastic boundary, BCD, **Eq. (K13)** specifies

$$\Phi_{BCD} = 0$$

b.) On AB and DE it is noted that

$$y = 0 = -\frac{\partial\Phi}{\partial\tau_{xz}} \quad \text{along} \quad \tau_{yz} = 0$$

Therefore integrating gives

$$\Phi_{AB} = \Phi_{DE} = \quad \text{constants} \quad = 0$$

where the zero values are required to avoid a discontinuity in the boundary condition as it passes through B and D on the stress plane.

c.) On EF and IA it is noted that

$$x = -a = \frac{\partial\Phi}{\partial\tau_{yz}} \quad \text{along} \quad \tau_{xz} = 0$$

Again integrating this result gives

$$\Phi_{IA} = \Phi_{CF} = -a\tau_{yz} + 0$$

where the constants of integration are again both taken to be zero to avoid a discontinuity passing through A and C on the stress plane.

These boundary conditions on Φ are noted on the stress plane map in **Fig. K.3** and it is left to solve for it within these boundaries by using Laplace's **Eq. (K9)**. The final solution is left undone, as the object of this example is to illustrate the setting up of the problem for solution. The exact analytical solution to a similar problem shall now be given because of its special significance to small-scale yielding fracture mechanics assumptions.

THE COMPLETE ANALYTICAL SOLUTION FOR SMALL-SCALE YIELDING

A significant special case of the preceding example is that for which the free edge is removed to infinity giving an infinite region on the physical plane with a semi-infinite crack with its tip at the origin, as shown in **Fig. K.4**. The loading on the boundaries as shown on **Fig. K.3** must approach zero but not without leaving significant finite stresses and strains, causing the usual singular effects at the crack tip. The physical plane and mapping to the stress plane are shown in **Fig. K.4** for this case. Note that the boundary condition for Φ exhibits singular behavior at the origin of the stress plane.

The stress function for this problem is

$$\Phi(\tau_{xz}, \tau_{yz}) = \frac{A^2}{2\pi}\left[\frac{1}{\tau_{xz}^2 + \tau_{yz}^2} - \frac{1}{k^2}\right]\tau_{yz} \tag{K15}$$

Note that this stress function satisfies Laplace's Equation and gives zero values on the yield surface and horizontal axis of the stress plane except at the origin where a singular spike in value occurs. Differentiating this stress function using **Eqs. (K8)**, the elastic–plastic boundary on the physical plane is described by

$$y^2 + \left(x - \frac{A^2}{2\pi k^2}\right)^2 = \left(\frac{A^2}{2\pi k^2}\right)^2 \tag{K16}$$

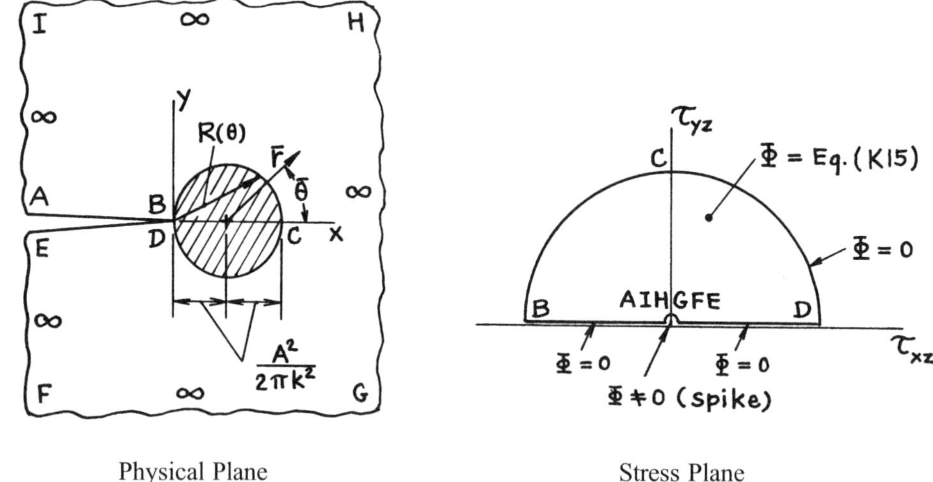

Physical Plane

Stress Plane

Figure K.4

which is a circle directly ahead of and just touched by the crack tip. Further, it is found that larger circles with the same center on the physical plane correspond to semi-circles centered at the origin within the elastic region on the stress plane. For mapping corresponding points from the physical plane to the stress plane on these circles, the angular measure from their centers differs by the factor 2. Consequently, stresses in the elastic region may be most simply expressed in terms of new coordinates, $\bar{r}$, $\bar{\theta}$, measured on the physical plane from the center of the plastic zone given by **Eq. (K16)**, and they are

$$\tau_{xz} = -\frac{A}{\sqrt{2\pi\bar{r}}}\sin\frac{\bar{\theta}}{2}$$

$$\tau_{yz} = \frac{A}{\sqrt{2\pi\bar{r}}}\cos\frac{\bar{\theta}}{2} \tag{K17}$$

Comparing these equations with **Eq. (3)** in Part I of this handbook, it is noted that exactly the same form of elastic stress field is found for the purely elastic analysis except that here the coordinates are taken at the center of the plastic zone instead of at the origin. That means that the plastic zone embeds itself within the same elastic stress field as that for the purely elastic case, except that the "effective (or equivalent) elastic crack length" is the physical crack length plus the plastic zone radius. These same "plastic zone corrections" to crack size, as suggested by **Paris 1957** and later developed by **Irwin (1960b)**, have been used without full analytical justification for many years for Mode I linear-elastic fracture mechanics applications. Although this analytic justification is shown only for a Mode III case here, it adds significant credibility for the concept applied to other modes. Indeed, it shows that a small-scale plastic zone embeds itself within the elastic crack tip stress field without significantly disturbing the form of that field but with an effective crack size including half the plastic zone.

EXAMPLES OF OTHER MAPS

Physical Plane

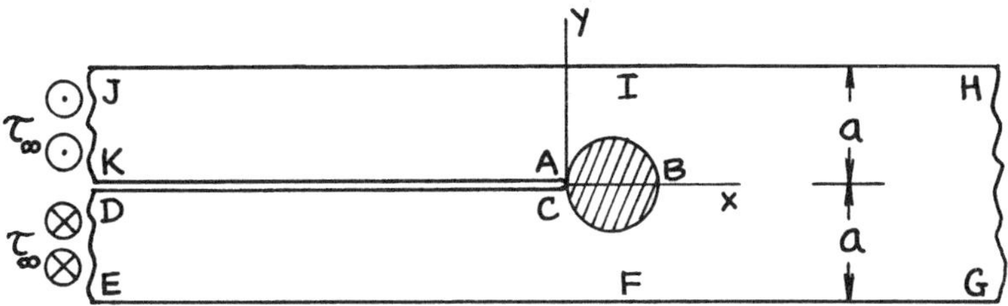

Stress Plane

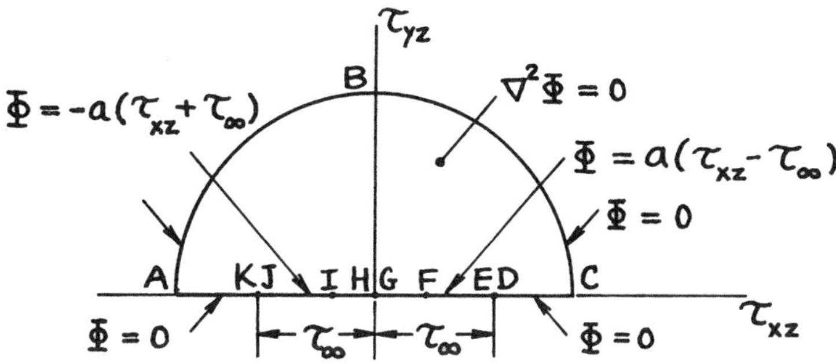

Fig. K.5

For the example shown in **Fig. K.5**, it is noted that
a.) On CD and AK:

$$y = 0 = -\frac{\partial \Phi}{\partial \tau_{xz}} \quad \text{or} \quad \Phi_{CD} = \Phi_{AK} = C_1 = 0$$

b.) On HIJ and EFG:

$$y = \pm a = -\frac{\partial \Phi}{\partial \tau_{xz}} \quad \text{or} \quad \begin{aligned} \Phi_{EFG} &= a(\tau_{xz} - \tau_\infty) \\ \Phi_{HIJ} &= -a(\tau_{xz} + \tau_\infty) \end{aligned}$$

with simultaneous points DE and KJ at $\tau_{yz} = 0$, $\tau_{xz} = \pm\tau_\infty$.

Physical Plane

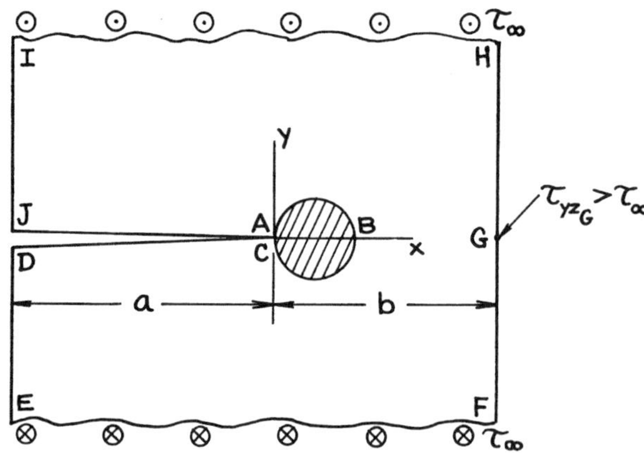

Stress Plane

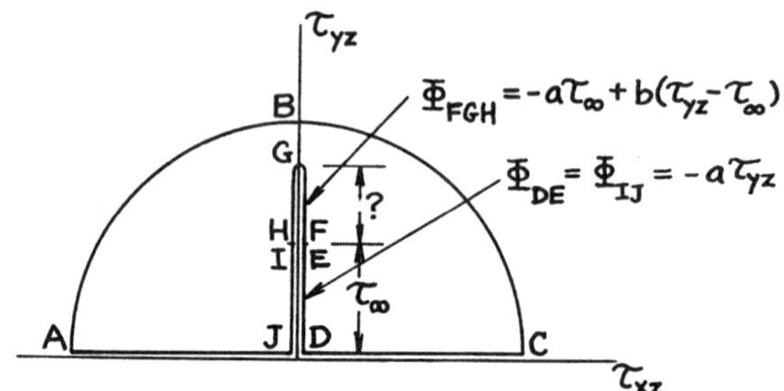

Sketch of Stress Function Surface

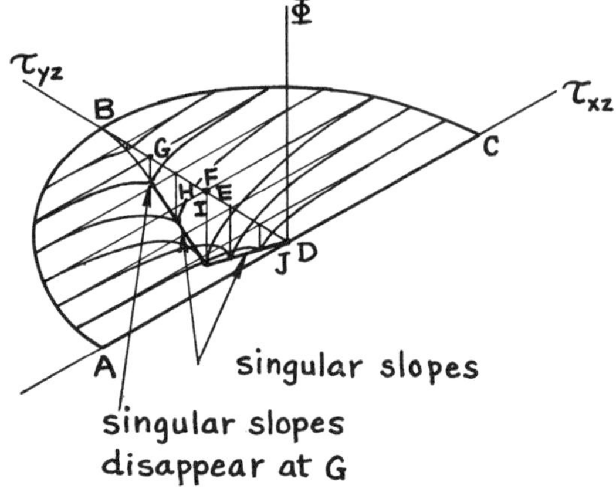

Fig. K.6

Effect of Angled Sharp Notches:

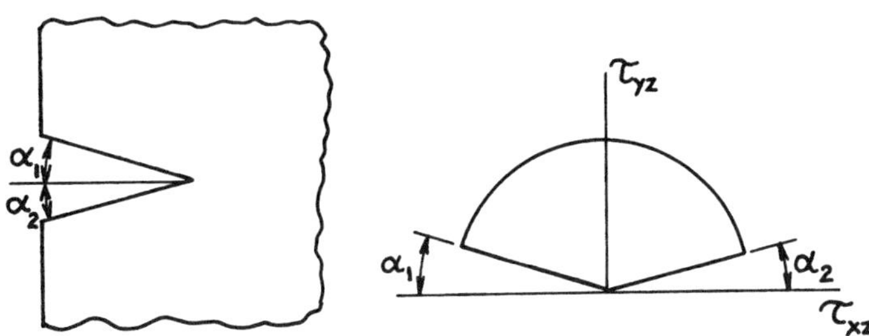

Effect of Square Ended Notches:

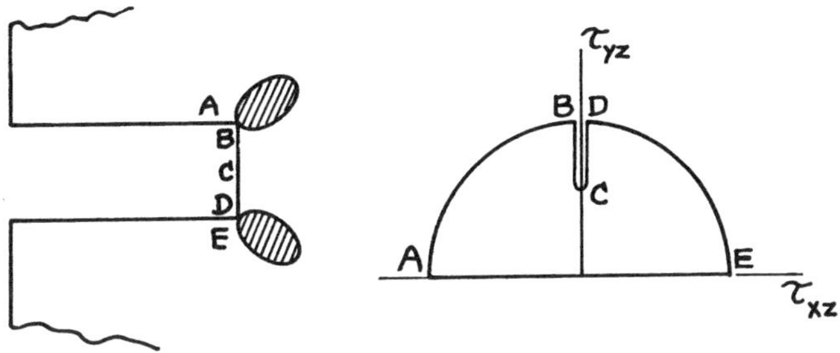

Effect of Discrete Limited Slip Directions:

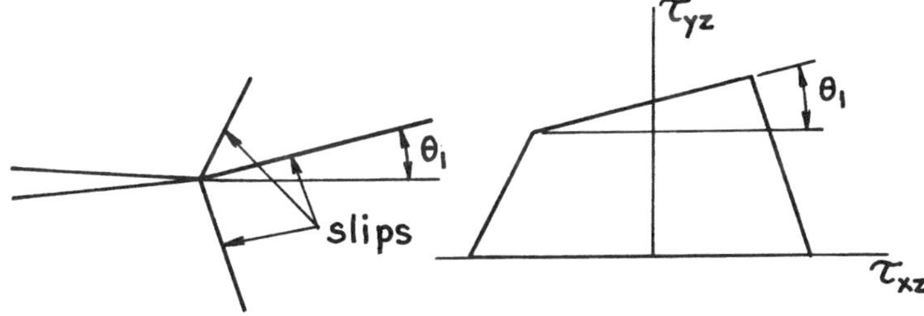

For Plane Anisotropic Elastic Behavior:

$$\tau_{xz} = A_{xx}\gamma_{xz} + A_{xy}\gamma_{yz}$$

$$\tau_{yz} = A_{yx}\gamma_{xz} + A_{yy}\gamma_{yz} \quad \text{with} \quad A_{yx} = A_{xy}$$

results in replacing Laplace's Equation for the elastic region by

$$A_{xx}\frac{\partial^2\Phi}{\partial\tau_{xz}^2}+2A_{xy}\frac{\partial^2\Phi}{\partial\tau_{xz}\partial\tau_{yz}}+A_{yy}\frac{\partial^2\Phi}{\partial\tau_{yz}^2}=0$$

For Power Hardening Plasticity Beyond an Elastic Range:

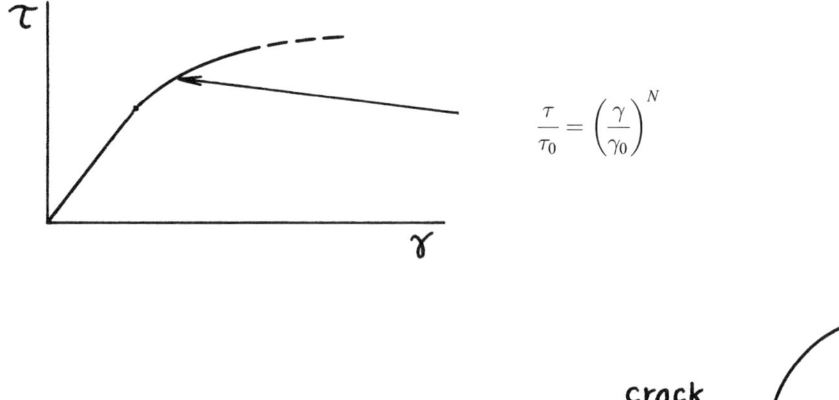

$$\frac{\tau}{\tau_0}=\left(\frac{\gamma}{\gamma_0}\right)^N$$

$$x_0=\frac{1-N}{1+N}\overline{R}$$

(See **Rice 1968c** for details and numerical solutions.)

Appendix L

Table of Complete Elliptic Integrals, $K(k)$ and $E(k)$

$k = \sin \alpha$

α°	K	E	α°	K	E	α°	K	E
0	1.5708	1.5708	30	1.6858	1.4675	60	2.1565	1.2111
1	1.5709	1.5707	31	1.6941	1.4608	61	2.1842	1.2015
2	1.5713	1.5703	32	1.7028	1.4539	62	2.2132	1.1920
3	1.5719	1.5697	33	1.7119	1.4469	63	2.2435	1.1826
4	1.5727	1.5689	34	1.7214	1.4397	64	2.2754	1.1732
5	1.5738	1.5678	35	1.7312	1.4323	65	2.3088	1.1638
6	1.5751	1.5665	36	1.7415	1.4248	66	2.3439	1.1545
7	1.5767	1.5649	37	1.7522	1.4171	67	2.3809	1.1453
8	1.5785	1.5632	38	1.7633	1.4092	68	2.4198	1.1362
9	1.5805	1.5611	39	1.7748	1.4013	69	2.4610	1.1272
10	1.5828	1.5589	40	1.7868	1.3931	70	2.5046	1.1184
11	1.5854	1.5564	41	1.7992	1.3849	71	2.5507	1.1096
12	1.5882	1.5537	42	1.8122	1.3765	72	2.5998	1.1011
13	1.5913	1.5507	43	1.8256	1.3680	73	2.6521	1.0927
14	1.5946	1.5476	44	1.8396	1.3594	74	2.7081	1.0844
15	1.5981	1.5442	45	1.8541	1.3506	75	2.7681	1.0764
16	1.6020	1.5405	46	1.8691	1.3418	76	2.8327	1.0686
17	1.6061	1.5367	47	1.8848	1.3329	77	2.9026	1.0611
18	1.6105	1.5326	48	1.9011	1.3238	78	2.9786	1.0538
19	1.6151	1.5283	49	1.9180	1.3147	79	3.0617	1.0468
20	1.6200	1.5238	50	1.9356	1.3055	80	3.1534	1.0401
21	1.6252	1.5191	51	1.9539	1.2963	81	3.2553	1.0338
22	1.6307	1.5141	52	1.9729	1.2870	82	3.3699	1.0278
23	1.6365	1.5090	53	1.9927	1.2776	83	3.5004	1.0223
24	1.6426	1.5037	54	2.0133	1.2681	84	3.6519	1.0172
25	1.6490	1.4981	55	2.0347	1.2587	85	3.8317	1.0127
26	1.6557	1.4924	56	2.0571	1.2492	86	4.0528	1.0086
27	1.6627	1.4864	57	2.0804	1.2397	87	4.3387	1.0053
28	1.6701	1.4803	58	2.1047	1.2301	88	4.7427	1.0026
29	1.6777	1.4740	59	2.1300	1.2206	89	5.4349	1.0008
30	1.6858	1.4675	60	2.1565	1.2111	90	∞	1.0000

The definitions of the complete elliptic integrals of the first kind, $K(k)$, and the second kind, $E(k)$, are as follows:

$$K(k) = \int_0^{\pi/2} \frac{d\varphi}{\sqrt{1 - k^2 \sin^2 \varphi}}$$

and

$$E(k) = \int_0^{\pi/2} \sqrt{1 - k^2 \sin^2 \varphi} \quad d\varphi$$

Approximate Formulas for $E(k); k = \sin \alpha$

$$E(k) = \left[1 + 1.464 \left(\cos \alpha\right)^{1.65}\right]^{1/2}$$

Accuracy: 0.1%
References: **Rawe 1970**; see also **Merkle 1973, Raju 1979**

$$E(k) = \frac{\pi}{2} \cdot \frac{1}{1 + \lambda} \cdot \frac{4 - 0.18\lambda^4}{4 - \lambda^2}$$

where $\lambda = \tan^2 \frac{\alpha}{2}$

Accuracy: 0.05%
Reference: **Tada 2000**

Appendix M

A Table of Gamma Function $\Gamma(x)$

x	$\Gamma(x)$	x	$\Gamma(x)$
1.000	1.00000	1.500	.88623
1.025	.98617	1.525	.88729
1.050	.97350	1.550	.88887
1.075	.96191	1.575	.89094
1.100	.95135	1.600	.89351
1.125	.94174	1.625	.89657
1.150	.93304	1.650	.90012
1.175	.92520	1.675	.90414
1.200	.91817	1.700	.90864
1.225	.91192	1.725	.91361
1.250	.90640	1.750	.91906
1.275	.90160	1.775	.92499
1.300	.89747	1.800	.93138
1.325	.89400	1.825	.93825
1.350	.89115	1.850	.94561
1.375	.88891	1.875	.95345
1.400	.88726	1.900	.96177
1.425	.88618	1.925	.97058
1.450	.88566	1.950	.97988
1.475	.88567	1.975	.98969
1.500	.88623	2.000	1.00000

Values of $\Gamma(x)$ for $x<1$ and $x>2$ are calculated by the following formula:

$$\Gamma(x) = \frac{\Gamma(x+1)}{x} \text{ or } \Gamma(x) = (x-1)\Gamma(x-1)$$

Examples:

1) $$\Gamma(0.65) = \frac{\Gamma(1.65)}{0.65} = \frac{0.90012}{0.65} = 1.38480$$

2) $$\begin{aligned}\Gamma(3.65) &= 2.65 \text{ x } \Gamma\ (2.65)\\ &= 2.65 \text{ x } 1.65 \text{ x } \Gamma\ (1.65)\\ &= 2.65 \text{ x } 1.65 \text{ x } 0.90012 = 3.93577\end{aligned}$$

PROPERTIES OF GAMMA FUNCTION $\Gamma(x)$

Definition:

$$\Gamma(x) = \begin{cases} \displaystyle\int_0^\infty e^{-t} t^{x-1} dt & \text{for } x \geq 0^{*} \\ \displaystyle\lim_{n\to\infty} \frac{n! n^{x-1}}{x(x+1)(x+2)\cdots\cdots(x+n-1)} & \text{for any } x^{**} \end{cases}$$

* when x is complex; $Re(x)>0$
** real or complex

Basic Formulas:

$$\Gamma(x+1) = x\Gamma(x)$$

$$\Gamma(n) = (n-1)! \quad (\text{n = positive integer})$$

$$\Gamma(x)\Gamma(1-x) = \frac{\pi}{\sin \pi x}$$

$$\Gamma(x)\Gamma\left(x+\frac{1}{2}\right) = \frac{\sqrt{\pi}}{2^{2x-1}}\Gamma(2x)$$

$$\Gamma\left(\frac{1}{2}\right) = \sqrt{\pi}, \ \Gamma(1) = 1, \ \Gamma\left(\frac{3}{2}\right) = \frac{\sqrt{\pi}}{2}$$

For crack problems, the following formulas are useful.

(1)

$$\int_0^1 x^{2\alpha+1}\left(1-x^2\right)^\beta dx = \left(\frac{1}{2}\int_0^1 x^\alpha (1-x)^\beta dx\right)$$

$$= \int_0^{\pi/2} (\sin\theta)^{2\alpha+1}(\cos\theta)^{2\beta+1} d\theta$$

$$= \frac{\Gamma(\alpha+1)\Gamma(\beta+1)}{2\Gamma(\alpha+\beta+2)} \left(= \frac{1}{2}B(\alpha+1, \beta+1)\right)$$

$$\left(= \frac{\alpha!\beta!}{2(\alpha+\beta+1)!} \text{ when } \alpha, \beta = \text{ positive integers}\right)$$

(2) Special cases: $2\alpha + 1 = \gamma,\ \beta = -\frac{1}{2}$

or $\alpha = -\frac{1}{2},\ 2\beta + 1 = \gamma$

$$\int_0^{\pi/2} (\sin\theta)^\gamma\, d\theta = \int_0^{\pi/2} (\cos\theta)^\gamma\, d\theta = \frac{\sqrt{\pi}}{2}\,\frac{\Gamma\left(\frac{\gamma+1}{2}\right)}{\Gamma\left(\frac{\gamma}{2}+1\right)}$$

when $\gamma = n$ (n = positive integer)

$$\begin{cases} = \dfrac{(n-1)(n-3)(n-5)\cdots\cdots 3.1}{n\,(n-2)(n-4)\cdots\cdots 4.2}\cdot\dfrac{\pi}{2} & (n = \text{ even}) \\[2ex] = \dfrac{(n-1)(n-3)(n-5)\cdots\cdots 4.2}{n\,(n-2)(n-4)\cdots\cdots 3.1} & (n = \text{ odd}) \end{cases}$$

REFERENCES

Akao 1967
H. T. Akao and A. S. Kobayashi, "Stress Intensity Factor for a Short Edge Notched Specimen Subjected to Three-Point Loading," Trans. ASME, *Journal of Basic Engineering*, Vol. 89, p. 7, 1967.

Anderson 1993
T. L. Anderson et al., "Fitness-for-Service Evaluation Procedures for Operating Pressure Vessels, Tanks, and Piping in Refinery and Chemical Service," Consultants' Report for the Materials Properties Council, Draft No. 3, Sept., 1993.

ASME 1972
"The Surface Flaw," Proceedings, ASME ComCam Symposium, J. Swedlow, ed., November, 1972.

ASTM 1972
ASTM Standards E399-72, "Standard Method of Test for Plane-Strain Fracture Toughness of Metallic Materials," *Annual Book of ASTM Standards*, 1972.

Barenblatt 1962
G. I. Barenblatt, "Mathematical Theory of Equilibrium Cracks in Brittle Fracture," *Advances in Applied Mechanics*, Vol. VII, p. 55, Academic Press, New York, 1962.

Begley 1972
J. A. Begley and J. D. Landes, "The J Integral as a Fracture Criterion," ASTM STP 514, pp. 1–24, 1972.

Begley 1976
J. A. Begley and J. D. Landes, "A Fracture Mechanics Approach to Creep Crack Growth," ASTM STP 590, pp. 128–148, 1976.

Benthem 1972
J. P. Benthem and W. T. Koiter, "Asymptotic Approximations to Crack Problems," Chapter 3 in *Methods of Analysis of Crack Problems,* edited by G. C. Sih, Noordhoff International Publishing, Jan., 1972.

Berezhnitskii 1966
L. T. Berezhnitskii, "Propagation of Cracks Terminating at the Edge of a Curvilinear Hole in a Plate," *Soviet Materials Science*, Vol. 2, pp. 16–23, 1966.

Bilby 1963
B. A. Bilby, A. H. Cottrell and K. H. Swinden, "The Spread of Plastic Strip from a Notch," *Proc. Royal Soc. London*, Ser. A, Vol. 272, pp. 304–314, 1963.

Bilby 1968
B. A. Bilby and J. D. Eshelby, "Dislocations and the Theory of Fracture," *Fracture*, Academic Press, Vol. 1, 1968.

Bowie 1956
O. L. Bowie, "Analysis of an Infinite Plate Containing Radial Cracks, Originating at the Boundaries of an Internal Circular Hole," *Journal of Mathematics and Physics*, Vol. 35, p. 60, 1956.

Bowie 1964a
O. L. Bowie, "Rectangular Tensile Sheets with Symmetric Edge Cracks," Trans. ASME, Ser. E, *Journal of Applied Mechanics*, Vol. 31, p. 208, 1964.

Bowie 1964b
O. L. Bowie, "Symmetric Edge Cracks in Tensile Sheet with Constrained Ends," Trans. ASME, Ser. E, *Journal of Applied Mechanics*, Vol. 31, p. 726, 1964.

Bowie 1965
O. L. Bowie, "Single Edge Crack in Rectangular Tensile Sheet," Trans. ASME, Ser. E, *Journal of Applied Mechanics*, Vol. 32, p. 708, 1965.

Bowie 1970a
O. L. Bowie and D. M. Neal, "A Note on the Central Crack in a Uniformly Stressed Strip," *Engineering Fracture Mechanics*, Vol. 2, p. 181, 1970.

Bowie 1970b
O. L. Bowie and D. M. Neal, "A Modified Mapping Collocation Technique for Accurate Calculation of Stress Intensity Factors," *International Journal of Fracture Mechanics*, Vol. 6, pp. 199–206, 1970.

Bowie 1973
O. L. Bowie, "Solutions of Plane Crack Problems by Mapping Techniques," *Mechanics of Fracture*, Vol. 1, *Methods of Analysis and Solutions of Crack Problems*, edited by G. C. Sih, Chapter 1, pp. 1–55, Noordhoff, 1973.

Bowie 1976a
O. L. Bowie and C. E. Freese, "On the 'Overlapping' Problem in Crack Analysis," *Engineering Fracture Mechanics*, Vol. 8, pp. 373–379, 1976.

Bowie 1976b
O. L. Bowie and C. E. Freese, "A Note on the Bending of a Cracked Strip Including Crack Surface Interference," *International Journal of Fracture*, Vol. 12, pp. 457–459, 1976.

Brown 1965
W. F. Brown, Jr. and J. E. Srawley, "Fracture Toughness Testing," *Fracture Toughness Testing and Its Applications*, ASTM STP 381, 1965.

Brown 1966
W. F. Brown, Jr. and J. E. Srawley, "Plane Strain Crack Toughness Testing of High Strength Metallic Materials," ASTM STP 410, 1966.

Bubsey 1973
R. Bubsey, D. Fisher, M. Jones, and J. E. Srawley, "Compliance Measurements," Chapter IV in *Experimental Techniques in Fracture Mechanics,* SESA Monograph, ed. by A. S. Kobayashi, 1973.

Bueckner 1958
H. F. Bueckner, "The Propagation of Cracks and Energy of Elastic Deformation," Trans. ASME, *Journal of Basic Engineering*, Vol. 80, pp. 1225–1229, 1958.

Bueckner 1960
H. F. Bueckner, "Some Stress Singularities and Their Computation by Means of Integral Equations," in *Boundary Value Problems in Differential Equations*, edited by R. E. Langer, University of Wisconsin Press, Madison, WI, pp. 215–230, 1960.

Bueckner 1965
H. F. Bueckner, discussion to **Paris 1965**, ASTM, STP. 381, p. 82, 1965.

Bueckner 1966
H. F. Bueckner and L. Giaever, "The Stress Concentration of a Notched Rotor Subjected to Centrifugal Forces," *Zeitschrift fur Angewandte Mathematik und Mechanik (ZAMM)*, Vol. 46, pp. 265–273, 1966.

Bueckner 1970
H. F. Bueckner, "A Novel Principle for the Computation of Stress Intensity Factors," *ZAMM*, Vol. 50, pp. 529–545, 1970.

Bueckner 1971
H. F. Bueckner, "Weight Functions for the Notched Bar," *ZAMM*, Vol. 51, pp. 97–109, 1971.

Bueckner 1972
H. F. Bueckner, "Field Singularities and Related Integral Representations," Chapter 5 in *Methods of Analysis and Solutions of Crack Problems*, edited by G. C. Sih, Noordhoff International Publishing, 1972.

Collins 1962
W. D. Collins, "Some Axially Symmetric Stress Distributions in Elastic Solids Containing Penny-Shaped Cracks," Part I, Proc. Roy. Soc. London, Series A, Vol. 266, pp. 359 – 386, 1962; Part II, Mathematika, Vol. 9, pp. 25–37, 1962; Part III, Proc. Edinburgh Math. Soc., Vol. 13, pp. 69–78, 1962.

Cottrell 1953
A. H. Cottrell, *Dislocations and Plastic Flow in Crystals.* Clarendon Press, Oxford, England, 1953.

Creager 1967
M. Creager and P. C. Paris, "Elastic Field Equations for Blunt Cracks with Reference to Stress Corrosion Cracking," *International Journal of Fracture Mechanics*, Vol. 3, pp. 247–252, 1967.

Doran 1969
H. E. Doran and V. T. Buchwald, "The Half-Plane with an Edge Crack in Plane Elastostatics," *Journal of the Institute of Mathematics and Its Applications*, Vol. 5, pp. 91–112, 1969.

Dowling 1976
N. E. Dowling and J. A. Begley, "Fatigue Crack Growth During Gross Plasticity and the J-Integral," ASTM STP 590, pp. 82–103, 1976.

Dugdale 1960
D. S. Dugdale, "Yielding of Steel Sheets Containing Slits," *Journal of Mech. Phys. Solids*, Vol. 8, pp. 100–104, 1960.

Emery 1969
A. F. Emery, G. E. Walker, Jr., and J. A. Williams, "A Green's Function for the Stress-Intensity Factors of Edge Cracks and Its Application to Thermal Stresses," Trans. ASME, Series D., *Journal of Basic Engineering*, Vol. 91, pp. 618–624, 1969.

Emery 1972
A. F. Emery and C. M. Segedin, "The Evaluation of the Stress Intensity Factors for Cracks Subjected to Tension, Torsion, and Flexure by an Efficient Numerical Technique," Trans. ASME, Series D., *Journal of Basic Engineering*, Paper No. 72-Met-B.

Ernst 1979
H. Ernst, P. C. Paris, M. Rossow, and J. W. Hutchinson, "Analysis of Load-Displacement Relationship to Determine J-R Curve and Tearing Instability Material Properties," ASTM STP 677, pp. 589–599, 1979.

Erdogan 1962
F. Erdogan, "On the Stress Distribution on Plates with Collinear Cuts under Arbitrary Loads," Proceedings, Fourth U. S. National Congress of Applied Mechanics, p. 547, 1962.

Erdogan 1963
F. Erdogan and G. C. Sih, "On the Crack Extension Under Plane Loading and Transverse Shear," Trans. ASME, Series D, *Journal of Basic Engineering*, Vol. 85, pp. 519–527, 1963.

Erdogan 1966
F. Erdogan, "Elastic-Plastic Anti-Plane Problems for Bonded Dissimilar Media Containing Cracks and Cavities," *Int. Journal of Solids and Structures*, Vol. 2, pp. 447–465, 1966.

Erdogan 1969
F. Erdogan and J. J. Kibler, "Cylindrical and Spherical Shells with Cracks," *Int. Journal of Fracture Mechanics*, Vol. 5, pp. 229–237, 1969.

Erdogan 1973
F. Erdogan, G. D. Gupta, and T. S. Cook, "Numerical Solutions of Singular Integral Equations," *Mechanics of Fracture*, Vol. 1, *Methods of Analysis and Solutions of Crack Problems*, edited by G. C. Sih, Chapter 7, pp. 368–425, Noordhoff, 1973.

Erdogan 1982
F. Erdogan, "Theoretical and Experimental Study of Fracture in Pipelines containing Circumferential Flaws," U. S. Department of Transportation, DOT-RSPA-DMA, 50/83/3, September, 1982.

Ernst 1979
H. Ernst, P. C. Paris, M. Rossow, and J. W. Hutchinson, "Analysis of Load-Displacement Relationships to Determine J-R Curve and Tearing Instability Material Properties," ASTM STP 677, pp. 581–599, 1979.

Fama 1972
M. E. Duncan-Fama and J. L. Sanders, Jr., "A Circumferential Crack in a Cylindrical Shell Under Tension," *Int. Journal of Fracture Mechanics*, Vol. 6, pp. 379–392, 1970.

Feddersen 1966
C. E. Fedderson, discussion to "Plane Strain Crack Toughness Testing," ASTM, Special Technical Publication No. 410, p. 77, 1966.

Fichter 1967
W. B. Fichter, "Stresses at the Tip of a Longitudinal Crack in a Plate Strip," NASA Technical Report, TR-R-265, NASA Langley, 1967.

Folias 1965
E. S. Folias, "An Axial Crack in a Pressurized Cylindrical Shell," *Int. Journal of Fracture Mechanics*, Vol. 1, pp. 104–113, 1965.

Folias 1967
E. S. Folias, "A Circumferential Crack in a Pressurized Cylindrical Shell," *Int. Journal of Fracture Mechanics*, Vol. 3, pp. 1–11, 1967.

Forman 1964
R. G. Forman and A. S. Kobayashi, "On the Axial Rigidity of a Perforated Strip and Strain Energy Release Rate in a Centrally Notched Strip Subjected to Uniaxial Tension," Trans. ASME, *Journal of Basic Engineering*, Vol. 86, p. 693, 1964.

Galin 1953
L. A. Galin, *Contract Problems in the Theory of Elasticity*, GITTL, M., 1953 (in Russian); translation: North Carolina State College Publications, 1961.

Gilman 1959
J. J. Gilman, "Cleavage, Ductility, and Tenacity in Crystals," in *Fracture*, edited by B. L. Auerbach et. al., pp. 193–221, John Wiley and Sons, Inc., New York, 1959.

Green 1950
A. E. Green and I. N. Sneddon, "The Distribution of Stress in the Neighborhood of a Flat Elliptical Crack in an Elastic Solid," Proceedings, Cambridge Philosophical Society, Vol. 46, p. 159, 1950.

Griffith 1920
A. A. Griffith, "The Phenomenon of Rupture and Flow in Solids," Phil. Trans. Roy. Soc., London, Series A. Vol. 221, p. 163, 1920.

Gross 1964
B. Gross and J. E. Srawley, "Stress-Intensity Factors for a Single-Edge-Notch Tension Specimen by Boundary Collocation of a Stress Function," NASA TN D-2395, 1964.

Gross 1965a
B. Gross and J. E. Srawley, "Stress-Intensity Factors for Single-Edge-Notch Specimens in Bending or Combined Bending and Tension by Boundary Collocation of a Stress Function," NASA TN D-2603, 1965.

Gross 1965b
B. Gross and J. E. Srawley, "Stress-Intensity Factors for Three-Point Bend Specimens by Boundary Collocation," NASA TN D-3092, 1965.

Gross 1966
B. Gross and J. E. Srawley, "Stress-Intensity Factors by Boundary Collocation for Single-Edge-Notch Specimens Subject to Splitting Forces," NASA TN D-3295, 1966.

Gross 1967
B. Gross, E. Roberts, Jr., and J. E. Srawley, "Elastic Displacements for Various Edge-Cracked Plate Specimens," *International Journal of Fracture Mechanics*, Vol. 4, p. 267, 1968; correction: Vol. 6, p. 87, 1970 and also NASA TN D-4232, 1967.

Gross 1970
B. Gross, "Some Plane Elastostatic Solutions for Plates Having a V-Notch," Ph. D. Dissertation, Case Western Reserve University, 1970.

Gustafson 1976
C. -G. Gustafson, discussion to Paris 1975b, *International Journal of Fracture*, Vol. 12, pp. 460–462, 1976.

Harris 1967
D. O. Harris, "Stress Intensity Factors for Hollow Circumferentially Notched Round Bars," Trans. ASME, *Journal Basic Engineering*, Vol. 89, p. 49, 1967.

Harris 1972
D. O. Harris, "Stress Intensity Factors for Transversely Loaded Elastic Plates and Their Application to Predictions of Crack Arrest," *Engineering Fracture Mechanics*, Vol. 4, pp. 277–294, 1972.

Harris 1997
D. O. Harris and P. J. Woytowitz, "Fully Plastic J-Integrals for Through-Wall Axial Cracks in Pipes," presented at ASTM 29th Fracture Mechanics Symposium, Stanford California, June, 1997.

Hartranft 1973
R. J. Hartranft and G. C. Sih, "Alternating Method applied to Edge and Surface Crack Problems," *Mechanics of Fracture*, Vol. 1, *Methods of Analysis and Solutions of Crack Problems*, edited by G. C. Sih, Chapter 4, pp. 179–238, Noordhoff, 1973.

Hasebe 1978
N. Hasebe and J. Iida, "A Crack Originating from a Triangular Notch on a Rim of a Semi-Finite Plate," *Engineering Fracture Mechanics*, Vol. 10, pp. 773–782, 1978.

Hasebe 1980
N. Hasebe and M. Ueda, "A Crack Originating from an Angular Corner of a Semi-Infinite Plate with a Step," *Trans. Japan Soc. Mech. Engrs.*, Vol. 46, No. 407, pp. 739–744, 1980.

Hasebe 1981
N. Hasebe and M. Takemura, "Cracks at Joint of Strip and Semi-Infinite Plate," *Engineering Fracture Mechanics*, Vol. 15, pp. 45–53, 1981.

Hasebe 1987
N. Hasebe and M. Taira, "A Crack at a Juncture of a Strip and a Semi-Infinite Plate," *Stress Intensity Factors Handbook*, edited by Y. Murakami, Pergamon Press, 1987.

Hutchinson 1968
J. W. Hutchinson, "Singular Behavior at the End of Tensile Cracks in Hardening Material," *Journal of Mechanics and Physics of Solids*, Vol. 16, pp. 13–31, 1968.

Hutchinson 1979
J. W. Hutchinson and P. C. Paris, "Stability Analysis of J-Controlled Crack Growth," ASTM STP 668, pp. 37–64, 1979.

Hutchinson 1992
J. W. Hutchinson and Z. Suo, "Mixed Mode Cracking in Layered Materials," *Advances in Applied Mechanics*, Vol. 29, edited by J. W. Hutchinson and T. Y. Wu, pp. 63–191, Academic Press, 1992.

Inglis 1913
C. E. Inglis, "Stresses in a Plate Due to the Presence of Cracks and Sharp Corners," Proceedings, Inst. Naval Architects, Vol. 60, 1913.

Irwin 1948
G. R. Irwin, "Fracture Dynamics," in *Fracturing of Metals*, edited by F. Jonassen et al., pp. 147–166, American Soc. of Metals, Cleveland, OH, 1948.

Irwin 1952
G. R. Irwin and J. A. Kies, "Fracturing and Fracture Dynamics," *The Welding Journal*, Res. Suppl., Vol. 31, p. 95, 1952.

Irwin 1954
G. R. Irwin and J. A. Kies, "Critical Strain Rate Analysis of Fracture Strength of Large Welded Structures," *The Welding Journal*, Res. Suppl., Vol. 33, pp. 193–198, 1954.

Irwin 1957
G. R. Irwin, "Analysis of Stresses and Strains Near the End of a Crack Transversing a Plate," Transactions, ASME, *Journal of Applied Mechanics*, Vol. 24, p. 361, 1957.

Irwin 1958a
G. R. Irwin, *Fracture*, Handbuch der Physik, Vol. VI, pp. 551–590, Springer, Berlin, 1958.

Irwin 1958b
G. R. Irwin, "The Crack Extension Force for a Crack at a Free Surface Boundary," Report No. 5120, Naval Research Lab., 1958.

Irwin 1958c
G. R. Irwin, J. A. Kies, and H. L. Smith, "Fracture Strengths Relative to Onset and Arrest of Crack Propagation," Proc. ASTM, Vol. 58, pp. 640–657, 1958.

Irwin 1960a
G. R. Irwin, "Fracture Mechanics," in *Structural Mechanics,* Proceedings of the 1st Symposium on Naval Structural Mechanics, 1958, edited by J. N. Goodier and N. J. Hoff, pp. 557–591, Pergamon Press, New York, 1960.

Irwin 1960b
G. R. Irwin, "Plastic Zone Near a Crack and Fracture Toughness," Proceedings of Seventh Sagamore Ordinance Material Research Conference, Report No. MeTE 661-611/F, Syracuse University Research Inst., p. IV–63, Aug., 1960.

Irwin 1961
G. R. Irwin, Supplement to Notes for May, 1961 meeting of ASTM Committee for Fracture Testing of High-Strength Metallic Materials, 1961.

Irwin 1962a
G. R. Irwin, "Analytical Aspects of Crack Stress Field Problems," T&AM Report No. 213, University of Illinois, March, 1962.

Irwin 1962b
G. R. Irwin, "The Crack Extension Force for a Part Through Crack in a Plate," Trans. ASME, *Journal of Applied Mechanics*, Vol. 29, pp. 651–654, 1962.

Irwin 1963
G. R. Irwin and M. F. Koskinen, discussion and author's closure to "Elastic-Plastic Deformation of a Single Grooved Flat Plate Under Longitudinal Shear," by J. F. Koskinen, *Trans. ASME*, Vol. 85, D, p. 585; p. 590, 1963.

Irwin 1968
G. R. Irwin, H. Liebowitz, and P. C. Paris, "A Mystery of Fracture Mechanics," *Engineering Fracture Mechanics*, Vol. 1, p. 235, 1968.

Irwin 1969
G. R. Irwin, B. Lingaraju, and H. Tada, "Interpretation of the Crack Opening Dislocation Concept," Fritz Engr. Lab. Rep. No. 358-2, Lehigh University, June, 1969.

Irwin 1971
G. R. Irwin and P. C. Paris, "Fundamental Aspects of Crack Growth and Fracture," Chapter 1 in *Fracture*, Vol. III, edited by H. Liebowitz, Pergamon Press, pp. 1–46, 1971.

Isida 1955
M. Isida, "On the Tension of a Strip with a Central Elliptical Hole," *Trans. Japanese Soc. of Mech. Engineers*, Vol. 21, p. 511, 1955.

Isida 1956
M. Isida, "On the In-Plate Bending of a Strip with a Central Elliptical Hole," *Trans. Japanese Soc. for Mech. Engineers*, Vol. 22, p. 809, 1956.

Isida 1962
M. Isida and Y. Itagaki, "Stress Concentration at the Tip of a Central Transverse Crack in a Stiffened Plate Subjected to Tension," Proceedings of the Fourth U. S. National Congress of Applied Mechanics, p. 955, 1962.

Isida 1965a
M. Isida, "Stress-Intensity Factors for the Tension of an Eccentrically Cracked Strip," Trans. ASME, Series E, *Journal of Applied Mechanics*, Vol. 33, p. 674, 1965.

Isida 1965b
M. Isida, Y. Itagaki, and S. Iida, "On the Crack Tip Stress Intensity Factor for the Tension of a Centrally Cracked Strip with Reinforced Edges," Dept. of Mech., Lehigh University, Bethlehem, PA, Sept., 1965.

Isida 1970a
M. Isida, "On the Determination of Stress Intensity Factors for Some Common Structural Problems," *Engineering Fracture Mechanics*, Vol. 2, p. 61, 1970.

Isida 1970b
M. Isida, "Analysis of Stress Intensity Factors for Plates Containing Random Array of Cracks," *Bulletin of Japan Soc. Mech. Engrs.*, Vol 13, p. 635, 1970.

Isida 1971a
M. Isida, "Effect of Width and Length on Stress Intensity Factors of Internally Cracked Plates under Various Boundary Conditions," *International Journal of Fracture Mechanics*, Vol. 7, p. 301, 1971.

Isida 1971b
M. Isida, "Stress Intensity Factors and Dugdale Type Plastic Zones for a Finite Plate with an Internal Crack," Proceedings, Symposium on Applications and Extensions of Fracture Mechanics, Japanese Soc. Mech. Engrs., No. 710-7, p. 119, 1971 (in Japanese).

Isida 1972
M. Isida, "Method of Laurent Series Expansion for Internal Crack Problems," Chapter 2 in *Methods of Analysis and Solutions of Crack Problems*, edited by G. C. Sih, Noordhoff, 1972.

Isida 1973
M. Isida, "Analysis of Stress Intensity Factors for the Tension of a Centrally Cracked Strip with Stiffened Edges," *Engineering Fracture Mechanics*, Vol. 5, pp. 647–655, 1973.

Isida 1974
M. Isida and H. Terada, *Journal of Strength and Fracture of Materials*, Vol. 9, No. 4, pp. 10–18, 1974 (in Japanese).

Isida 1975
M. Isida, "Arbitrary Loading Problems of Doubly Symmetric Regions Containing a Central Crack," *Engineering Fracture Mechanics*, Vol. 7, pp. 505–514, 1975.

Johnson 1965
H. H. Johnson, "Calibrating the Electric Potential Method for Studying Slow Crack Growth," *Materials Research and Standards*, Vol. 5, No. 9, pp. 442–445, 1965.

Kamei 1974
A. Kamei and T. Yokobori, "Some Results on Stress Intensity Factors of the Cracks and/or Slip Bands System," *Report of Research Institute for Strength and Fracture of Materials*, Special Issue, Tohoku University, Sendai, Japan, Vol. 10, No. 2, 1974.

Kapp 1980
J. A. Kapp, J. C. Newman, Jr., and J. H. Underwood, "A Wide Range Stress Intensity Factor Expression for the C-Shaped Specimen," *J. of Test. and Eval.*, Vol. 8, pp. 314–317, 1980.

Kassir 1966
M. K. Kassir and G. C. Sih, "Three Dimensional Stress Distribution Around an Elliptical Crack under Arbitrary Loadings," Trans. ASME, *Journal of Applied Mechanics*, Vol. 33, pp. 601–611, 1966.

Kassir 1973
M. K. Kassir and G. C. Sih, "Application of Papkovich-Neuber Potentials to a Crack Problem," *International Journal of Solids and Structures*, Vol. 9, pp. 643–645, 1973.

Kassir 1975
M. K. Kassir and G. C. Sih, "Three Dimensional Crack Problems," *Mechanics of Fracture*, Vol. II, edited by G. C. Sih, Noordhoff, 1975.

Kaya 1980
A. C. Kaya and F. Erdogan, "Stress Intensity Factors and COD in an Orthotropic Strip," *International Journal of Fracture*, Vol. 16, pp. 171–190, 1980.

Keer 1974
L. M. Keer, "Stress Analysis for Bonded Layers," *Journal of Applied Mechanics*, Vol. 41, pp. 679–683, 1974.

Keer 1989
L. M. Keer and Q. Guo, "Stress Analysis for Thin Bonded Layers," *Advances in Fracture Research*, Vol. 4, pp. 3073–3080, 1989.

Keer 1990
L. M. Keer and Q. Guo, "Stress Analysis for Symmetrically Loaded Bonded Layers," *International Journal of Fracture*, Vol. 43, pp. 69–81, 1990.

Kitagawa 1975
H. Kitagawa and R. Yuki, *Trans. Japan Soc. Mech. Engineers*, Vol. 41, No. 346, pp. 1641–1649, 1975.

Kobayashi 1964
A. S. Kobayashi, R. D. Cherepy, and W. C. Kinsel, "A Numerical Procedure for Estimating the Stress Intensity Factors of a Crack in a Finite Plate," Trans. ASME, Ser. D, *Journal of Basic Engineering*, Vol. 86, p. 681, 1964.

Kobayashi 1973
A. S. Kobayashi, *Experimental Techniques in Fracture Mechanics*, SESA Monograph, edited by A. S. Kobayashi, p. 18, 1973.

Kobayashi 1976
A. S. Kobayashi and A. N. Enetanya, "Stress Intensity Factor of a Surface Crack," 8th National Symposium on Fracture Mechanics, Brown University, August, 1974, ASTM STP 590, pp. 477–495, 1976.

Koiter 1956
W. T. Koiter, "On the Flexural Rigidity of a Beam Weakened by Transverse Saw Cuts II," Koninkl. Nederl. Akademic Van Wetenschappen, Amsterdam, Proceedings Series B. Vol. 59, p. 354, 1956.

Koiter 1959
W. T. Koiter, "An Infinite Row of Collinear Cracks in an Infinite Elastic Sheet," Ingenieur-Archiv, Vol. 28, p. 168, 1959.

Koiter 1961
W. T. Koiter, "An Infinite Row of Parallel Cracks in an Infinite Elastic Sheet," Problems of Contiuum Mechanics (Muskhelishvili Ann. Vol.), p. 246, Philadelphia, PA, 1961.

Koiter 1965a
W. T. Koiter, "Rectangular Tensile Sheet with Symmetric Edge Cracks," Trans. ASME, *Journal of Applied Mechanics*, Vol. 32. p. 237, 1965 (Discussion to Bowie 1964a).

Koiter 1965b
W. T. Koiter, "Note on the Stress Intensity Factor for Sheet Strips with Cracks under Tensile Loads," University of Technology, Laboratory of Engineering Mechanics, Report No. 314, Delft, Netherlands, 1965.

Lachenbruch 1961
A. H. Lachenbruch, "Depth and Spacing of Tension Cracks," *Journal of Geophysical Research*, Vol. 66, p. 4273, 1967.

Landes 1972
J. D. Landes and J. A. Begley, "The Effect of Spencimen Geometry on J_{Ic}," ASTM STP 514, pp. 24–39, 1972.

Libatskii 1965
L. L. Libatskii, "Application of Singular Integral Equations to Determine the Critical Forces in Cracked Plates," *Soviet Materials Science*, Vol. 1, p. 281–288, 1965.

Libatskii 1967
L. L. Libatskii and S. E. Kovchik, "Fracture of Discs Containing Cracks," *Soviet Materials Science*, Vol. 3, pp. 334–339, 1967.

Love 1944
A. E. H. Love, *A Treatise on the Mathematical Theory of Elasticity*, Dover, New York, 1944.

Lowengrub 1965
M. Lowengrub and I. N. Sneddon, "The Distribution of Stress in the Vicinity of an External Crack in an Infinite Elastic Solid," *International Journal of Engineering Science*, Vol. 3, pp. 451–460, 1965.

Lowengrub 1966
M. Lowengrub, "A Two-Dimensional Crack Problem," *Int. Journal of Engineering Science*, Vol. 4. pp. 289–299, 1966.

Lubahn 1959
J. D. Lubahn, "Experimental Determination of Energy Release Rate for Notch Bending and Notch Tension," Proc. ASTM. Vol. 59, pp. 885–915, 1959.

Markuzon 1961
I. A. Markuzon, "On the Wedging of a Brittle Body by a Wedge of Finite Length," Prikl. Matem. i Mekhau, Vol. 25, pp. 356–361, 1961.

Masubuchi 1960
K. Masubuchi, "Analytical Investigation of Residual Stresses and Distortions due to Welding," *The Welding Journal*, Vol. 39, Research Supplement, p. 5255, 1960.

Masubuchi 1961
K. Masubuchi and D. C. Martin, "Investigation of Residual Stresses by Use of Hydrogen Cracking," *The Welding Journal*, Vol. 40, Research Supplement, p. 5535, 1961.

Mendelson 1972
A. Mendelson, B. Gross, and J. E. Srawley, "Evaluation of the Use of Singularity Element in Finite-Element Analysis of Center-Cracked Plates," NASA Technical Note, NASA, TN D-6703, 1972.

Merkle 1973
J. G. Merkle, "A Review of Some of the Existing Stress Intensity Factor Solutions for Part-Through Surface Cracks," Oak Ridge National Laboratory, ORNL-TM-3983, Oak Ridge, TN, January, 1973.

Moss 1971
L. W. Moss and A. S. Kobayashi, "Approximate Analysis of Axisymmetric Problems in Fracture Mechanics with Application to a Flat Toroidal Crack," *Int. Journal of Fracture Mechanics*, Vol. 7, pp. 89–99, 1971.

Mostovoy 1966
S. Mostovoy, P. G. Crosley, and E. J. Ripling, "Use of Crackline Loaded Specimens for Measuring Plane Strain Fracture Toughness," *Materials Research Labs., Inc.*, June, 1966.

Mowbray 1970
D. F. Mowbray, "A Note on the Finite Element Method in Linear Fracture Mechanics," *Compendium Engineering Fracture Mechanics*, Vol. 2. pp. 173–176, 1970.

Muskhelishvili 1933
N. I. Muskhelishvili, *Some Basic Problems of Mathematical Theory of Elasticity*, published in Russian in 1933; English translation: P. Noordhoff and Co., 1953.

Muskhelishvili 1953
N. I. Muskhelishvili, *Singular Integral Equations,* P. Noordhoff Limited, Amsterdam, The Netherlands, 1953.

Narendran 1982
V. M. Narendran and M. P. Cleary, "Elastostatic Interaction of Multiple Arbitrarily Shaped Cracks in Plane Inhomogeneous Regions," Report of Research in Mechanics and Materials, REL-82-6, Dept. of Mech. Eng., M. I. T., July, 1982.

Neal 1970
D. M. Neal, "Stress Intensity Factors for Cracks Emanating from Rectangular Cutouts," *Int. Journal of Fracture Mechanics,* Vol. 6, pp. 393–400, 1970.

Neuber 1937, 1958
H. Neuber, *Kerpspannungslehre*, Springer, Berlin, 1937 and 1958, English Translation available from Edwards Bros., Ann Arbor, Mich.

Newman 1971
J. C. Newman, Jr., "An Improved Method of Collocation for the Stress Analysis of Cracked Plates with Various Shaped Boundaries," NASA Technical Note, NASA TN D-6376, Aug., 1971.

Newman 1979a
J. C. Newman, Jr., "Stress Intensity Factors and Crack Opening Displacements for Round Compact Specimens," NASA Technical Note, TM-80174, 1979.

Newman 1979b
J. C. Newman, Jr., " A Review of Stress Intensity Factors for the Surface Crack," ASTM STP 687, pp. 16 – 42, 1979.

Newman 1981a
J. C. Newman, Jr. and I. S. Raju, "Stress Intensity Factor Equations for Cracks in Three-Dimensional Finite Bodies," NASA TM-83200, NASA Langley Research Center, Hampton, Virginia, August, 1981.

Newman 1981b
"Stress Intensity Factors and Crack Opening Displacements for Round Compact Specimens," *International Journal of Fracture*, Vol. 17, pp. 567–578, 1981.

Nishitani 1969
H. Nishitani and M. Isida, "Stress Intensity Factor for the Tension of an Infinite Plate Containing an Ellipitical Hole with Two Symmetrical Edge Cracks," *Trans. Japanese Soc. for Mechanical Engineers*, No. 212, p. 131, 1969.

Nishitani 1971a
H. Nishitani and Y. Murakami, "Interaction of Elasto-Plastic Cracks Subjected to a Uniform Tensile Stress in an Infinite or a Semi-Infinite Plate," Proc. International Conference on Mechanical Behavior of Metals, Kyoto, 1971.

Nishitani 1971b
H. Nishitani and T. Murakami, "Elastic-Plastic Stresses in Plates under Tension: An Infinite Plate with Collinear Periodic Cracks and a Semi-Infinite Plate with an Edge Crack," Proceedings, Symposium on Applications and Extension of Fracture Mechanics, *Japanese Soc. Mech. Engrs.*, No. 710-7, pp. 111–118, 1971 (in Japanese).

Nishitani 1973
H. Nishitani, "Elastic-Plastic Stress in a Semi-Infinite Plate Having an Elliptical Arc Notch with an Edge Crack under Tension or Longitudinal Shear," Proceedings, Third International Conference on Fracture, Munchen, Vol. 5, 1-513, April, 1973.

Nishitani 1976
H. Nishitani, "A Method For Calculating Stress Concentrations and Its Applications," *Strength and Structure of Solid Materials*, ed. by H. Miyamoto et. al., Noordhoff International Publishing, Leyden, pp. 53–67, 1976.

Okamura 1976
H. Okamura, *Introduction to Linear Elastic Fracture Mechanics*, *Fracture Mechanics and Materials' Strength Series*, edited by H. Kihara, Baifu-Kan, Tokyo, 1976 (in Japanese).

Orowan 1949
E. Orowan, "Fracture and Strength of Solids," *Report on Progress in Physics*, Physics Soc., London, Vol. 12, pp. 185–232, 1949.

Orowan 1952
E. Orowan, "Fundamentals of Brittle Behavior of Metals," in *Fatigue and Fracture of Metals*, John Wiley and Sons, Inc., New York, 1952.

Ouchterlony 1975
F. Ouchterlony, "Concentrated Loads Applied to the Tips of a Symmetrically Cracked Wedge," DS 1975: 3, Swedish Detonic Research Foundation, 1975.

Ouchterlony 1976
F. Ouchterlony, "Stress Intensity Factors for the Expansion Loaded Star Crack," *Engineering Fracture Mechanics*, Vol. 8, pp. 447–448, 1976.

Paris 1955
P. C. Paris, "The Mechanics of Fracture Initiation and Propagation," Document D-17867, The Boeing Company, 1955.

Paris 1957
P. C. Paris, "The Mechanics of Fracture Propagation and Solutions to Fracture Arrester Problems," Document D2-2195, The Boeing Company, 1957.

Paris 1960
P. C. Paris, A Short Course in Fracture Mechanics, University of Washington Press, 1960.

Paris 1961
P. C. Paris, "Stress-Intensity-Factors by Dimensional Analysis," Institute of Research Report, Lehigh University, 1961.

Paris 1965
P. C. Paris and G. C. Sih, "Stress Analysis of Cracks," *Fracture Toughness Testing and Its Applications*, *Am. Soc. for Testing and Materials*, Special Technical Publication 381, pp. 30 – 83, 1965.

Paris 1975a
P. C. Paris and R. M. McMeeking, "Efficient Finite Element Methods for Stress Intensity Factors Using Weight Functions," *Int. Journal of Fracture Mechanics*, April, 1975.

Paris 1975b
P. C. Paris and H. Tada, "The Stress Intensity Factors for Cyclic Reversed Bending of a Single Edge Cracked Strip Including Crack Surface Interference," *International Journal of Fracture*, Vol. 11, pp. 1070 – 1072, 1975.

Paris 1976
P. C. Paris, R. M. McMeeking, and H. Tada, "The Weight Function Method for Determining Stress Intensity Factors," ASTM STP 601, pp. 471 – 489, 1976.

Paris 1979
P. C. Paris, H. Tada, A. Zahoor, and H. Ernst, "The Theory of Instability of the Tearing Mode of Elastic-Plastic Crack Growth," ASTM STP 668, pp. 37–64, 1979.

Paris 1983a
P. C. Paris and R. E. Johnson, "A Method of Application of Elastic-Plastic Fracture Mechanics to Nuclear Vessel Applications," ASTM STP 803, pp. 5–40, 1983.

Paris 1983b
P. C. Paris and H. Tada, "The Application of Fracture Proof Design Methods Using Tearing Instability Theory to Nuclear Piping Postulating Circumferential Through Wall Cracks," U. S. Nuclear Regulatory Commission, NUREG/CR-3464, Sept., 1983.

A. J. Paris 1988
A. J. Paris and P. C. Paris, "Instantaneous Evaluation of J and C*," *International Journal of Fracture*, Vol. 38, pp. R19–R21, 1988.

Peterson 1953
R. E. Peterson, *Stress Concentration Design Factors*, John Wiley and Sons, Inc., New York, 1953.

Raju 1979
I. S. Raju and J.C. Newman, Jr., "Stress Intensity Factors for a Wide Range of Semi-Elliptical Surface Cracks in Finite Thickness Plates," *Engineering Fracture Mechanics*, Vol. 11, pp. 817–829, 1979.

Rawe 1970
R. E. Rawe, "Fracture Mechanics and Safe Life Design," Douglas Aircraft Corporation, DAC 59591, 1970.

Rice 1964
J. R. Rice, private communication, 1964.

Rice 1966
J. R. Rice, "Contained Plastic Deformation Near Cracks and Notches Under Longitudinal Shear," *International Journal of Fracture Mechanics*, Vol. 2, pp. 426–447, 1966.

Rice 1967
J. R. Rice, "Stresses in an Infinite Strip Containing a Semi-Infinite Crack," (Discussion to W. G. Knauss, Vol. 33, p. 356, 1966), Trans. ASME, Ser. E, *Journal of Applied Mechanics*, Vol. 34, p. 248, 1967.

Rice 1968a
J. R. Rice, "A Path Independent Integral and the Approximate Analysis of Strain Concentration by Notches and Cracks," Trans. ASME, *Journal of Applied Mechanics*, Vol. 35, pp. 379–386, 1968.

Rice 1968b
J. R. Rice and G. F. Rosengren, "Plane Strain Deformation Near a Crack Tip in Power-Law Hardening Material," *Journal of Mechanics and Physics of Solids*, Vol. 16, pp. 1–12, 1968.

Rice 1968c
J. R. Rice, "Stresses Due to a Sharp Notch in a Work-Hardening Elastic-Plastic Material Loaded by Longitudinal Shear," Trans. ASME, Ser. E, *Journal of Applied Mechanics*, Vol. 33, pp. 287–298, 1967.

Rice 1972
J. R. Rice, "Some Remarks on Elastic Crack-Tip Stress Fields," *Int. Journal of Solids and Structures*, Vol. 8, pp. 751–758, 1972.

Rice 1973
J. R. Rice, P. C. Paris, and J. G. Merkle, "Some Further Results of J-Integral Analysis and Estimates," ASTM STP 536, pp. 231–245, 1973.

Rice 1984
J. R. Rice, "On the Theory of Perfectly Plastic Anti-Plane Straining," *Mechanics of Materials*, Vol. 3, pp. 55–80, 1984.

Rich 1968
T. Rich and R. Roberts, "Plastic Enclave Sizes for Internal Cracks Emanating from Circular Cavities within Elastic Plates," *Engineering Fracture Mechanics*, Vol. 1, pp. 167–173, 1968.

Riedel 1980
H. Riedel and J. R. Rice, "Tensile Cracks in Creeping Solids," ASTM STP 700, pp. 112–130, 1980.

Riedel 1989
H. Riedel, "Creep Crack Growth," ASTM STP 1020, pp. 101–126, 1989.

Ripling 1964
E. J. Ripling, S. Mostovoy, and R. L. Patrick, "Measuring Fracture Toughness of Adhesive Joints," *Materials Research and Standard*, Vol. 4, No. 3, pp. 129–134, 1964.

Roberts 1969
E. Roberts, Jr., "Elastic Crack-Edge Displacements for the Compact Tension Specimen," *Materials Research and Standards*, Vol. 9, No. 2, p. 27, 1969.

Romualdi 1957
J. P. Romualdi, J. T. Fraiser, and G. R. Irwin, "Crack Extension Force Near a Riveted Stringer," Naval Research Lab. Report 4956, 1957.

Rooke 1969
D. P. Rooke and I. N. Sneddon, "The Crack Energy and the Stress Intensity Factor for a Cruciform Crack Deformed by Internal Pressure," *Int. Journal of Engineering Science*, Vol. 7, pp. 1079–1089, 1969.

Rooke 1972
D. P. Rooke and J. Tweed, "The Stress Intensity Factors of a Radial Crack in a Finite Rotating Disc," *Int. Journal of Engineering Science*, Vol. 10, pp. 709–714, 1972.

Rooke 1973a
D. P. Rooke and J. Tweed, "The Stress Intensity Factors of an Edge Crack in a Finite Rotation Disc," *Int. Journal of Engineering Science*, Vol. 11, pp. 279–283, 1973.

Rooke 1973b
D. P. Rooke and J. Tweed, "The Stress Intensity Factors of a Radial Crack in a Point Loaded Disc," *Int. Journal of Engineering Science*, Vol. 11, pp. 285–290, 1973.

Sadowsky 1949
M. A. Sadowsky and E. G. Sternberg, "Stress Concentration Around a Triaxial Ellipsoidal Cavity," Trans. ASME, *Journal of Applied Mechanics*, Vol. 16, pp. 149–157, 1949.

Sanders 1959
J. L. Sanders, Jr., "Effect of a Stringer on the Stress Concentration Due to a Crack in a Thin Sheet," NASA Technical Report R-13, 1959.

Sanders 1960
J. L. Sanders, Jr., "On the Griffith-Irwin Fracture Theory," Trans. ASME, Series E., *Journal of Applied Mechanics*, Vol. 27, pp. 352–353, 1960.

Sanders 1982
J. L. Sanders, Jr., "Circumferential Through-Cracks in a Cylindrical Shell Under Tension," Trans. ASME, *Journal Applied Mechanics*, Vol. 49, pp. 103–107, 1982.

Sanders 1983
J. L. Sanders, Jr., "Circumferential Through-Crack in a Cylindrical Shell Under Combined Bending and Tension," Trans. ASME, *Journal of Applied Mechanics*, Brief Note, Vol. 50, p. 221, 1983.

Savin 1961
G. N. Savin, *Stress Concentrations Around Holes*, Pergamon Press, New York, 1961.

Savin 1968
G. N. Savin, *Stress Distribution Around Holes,* Naukova Dumke Press, Kiev, 1968 (in Russian); Translation: NASA Technical Translation, NASA TT F607, 1970.

Saxena 1978
A. Saxena and S. J. Hudak, Jr., "Review and Extension of Compliance Information for Common Crack Growth Specimens," *International Journal of Fracture*, Vol. 14, pp. 453–467, 1978.

Seeger 1973
T. Seeger, "Ein Beitrag zur Berechung von statisch und zyklisch belasteten Rissscheiben nach dem Dugdale-Barenblatt-Modell (A Calculation Method Based on the Dugdale-Barenblatt-Model for Static and Fatigue Loaded Cracked Plates)," Publications of the Institut fur Statik und Stahlbau Technische Hochschule Darmstadt, Germany, Heft 21, 1973.

Segedin 1950
C. M. Segedin, "Note on a Penny-Shaped Crack Under Shear," Proc. Cambridge Phil. Soc., Vol. 47, pp. 396–400, 1950.

Shah 1968
R. C. Shah and A. S. Kobayashi, "On the Parabolic Crack in an Elastic Solid," *Engineering Fracture Mechanics*, Vol. 1, pp. 309–325, 1968.

Sih 1962a
G. C. Sih, "Application of Muskhelishvili's Method to Fracture Mechanics," Transactions, The Chinese Assn. for Advanced Studies, p. 25, November, 1962.

Sih 1962b
G. C. Sih, P. C. Paris, and F. Erodogan, "Crack Tip Stress Intensity Factors for Plane Extension and Plate Bending Problems," Trans. ASME, *Journal of Applied Mechanics*, Vol. 29, pp. 306–312, 1962.

Sih 1963
G. C. Sih, "Strength of Stress Singularities at Crack Tips for Flexural and Torsional Problems," Trans. ASME, *Journal of Applied Mechanics*, Vol. 30, p. 419, 1963.

Sih 1964
G. C. Sih, "Boundary Problems for Longitudinal Shear Cracks," Proceedings, Second Conference on Theoretical and Applied Mechanics, Pergamon Press, p. 117, New York, 1964.

Sih 1965a
G. C. Sih, "Stress Distribution Near Internal Crack Tips for Longitudinal Shear Problems," Trans. ASME, *Journal of Applied Mechanics*, Vol. 32, p. 51, 1965.

Sih 1965b
G. C. Sih, P. C. Paris, and G. R. Irwin, "On Cracks in Rectilinearly Anisotropic Bodies," *International Journal of Fracture Mechanics*, Vol. 1, pp. 189–203, 1965.

Sih 1968
G. C. Sih and H. Liebowitz, "Mathematical Theories of Brittle Fracture," *Treatise on Fracture*, Vol. II, edited by H. Liebowitz, p. 67, Academic Press, New York, 1968.

Sire 1989
R. A. Sire, D. O. Harris, and E. D. Eason, "Automated Generation of Influence Functions for Planar Crack Problems," ASTM STP 1020, pp. 351–365, 1989.

Smetanin 1968
B. I. Smetanin, "Problem of Extension of an Elastic Space Containing a Plane Annular Slit," (translation from PMM Vol 32, pp. 458–462, 1968), *Applied Mathematics and Mechanics*, Vol. 32, pp. 461–466, 1968.

Smith 1967
F. W. Smith, A. F. Emery, and A. S. Kobayashi, "Stress Intensity Factors for Semi-Circular Cracks, Part 2: Semi-Infinite Solid," *Journal of Applied Mechanics*, Vol. 34, pp. 953–959, 1967.

Sneddon 1946
I. N. Sneddon, "The Distribution of Stress in the Neighborhood of a Crack in an Elastic Solid," Proceedings, Royal Soc. London, Series A, Vol. 187, p. 229, 1946.

Sneddon 1951
I. N. Sneddon, *Fourier Transforms,* McGraw Hill, New York, 1951.

Sneddon 1963a
I. N. Sneddon and R. J. Tait, "The Effect of a Penny-Shaped Crack on the Distribution of Stress in a Long Circular Cylinder," *Int. Journal of Engineering Science*, Vol. 1, pp. 391–409, 1963.

Sneddon 1963b
I. N. Sneddon and J. T. Welch, "A Note on the Distribution of Stress in a Cylinder Containing a Penny-Shaped Crack," *Int. Journal of Engineering Science*, Vol. 1, pp. 411–419, 1963.

Sneddon 1967a
I. N. Sneddon and J. Tweed, "The Stress Intensity Factor for a Penny-Shaped Crack in an Elastic Body under the Action of Symmetric Body Forces," *Int. Journal of Fracture Mechanics*, Vol. 3, p. 291, 1967.

Sneddon 1967b
I. N. Sneddon and J. Tweed, "The Stress Intensity Factor for a Griffith Crack in an Elastic Body in which Body Forces are Acting," *Int. Journal of Fracture Mechanics*, Vol. 3, p. 317, 1967.

Sneddon 1969a
I. N. Sneddon, "The Distribution of Surface Stress Necessary to Produce a Griffith Crack of Prescribed Shape," in Two Lectures Concerning Pressurized Cracks, North Carolina State University, Dept. of Mathematics Report, February, 1969.

Sneddon 1969b
I. N. Sneddon and M. Lowengrub, *Crack Problems in Classical Theory of Elasticity*, The SIAM Series in Applied Mathemetics, John Wiley and Sons, Inc., New York, 1969.

Sneddon 1971a
I. N. Sneddon and S. C. Das, "The Stress Intensity Factor at the Tip of an Edge Crack in an Elastic Half-Plane," *Int. J. of Engineering Science*, Vol. 9, p. 25, 1971.

Sneddon 1971b
I. N. Sneddon and R. P. Srivastava, "The Stress Field in the Vicinity of Griffith Crack in a Strip of Finite Width," *Int. J. of Engineering Science*, Vol. 9, p. 479, 1971.

Srawley 1964
J. E. Srawley, M. H. Jones, and B. Gross, "Experimental Determination of the Dependence of Crack Extension Force on Crack Length for a Single-Edge-Notch Tension Specimen," NASA TN D-2396, 1964.

Srawley 1965
J. E. Srawley and W. F. Brown, "Fracture Toughness Testing Methods," ASTM STP 381, pp. 139–196, 1965.

Srawley 1967
J. E. Srawley and B. Gross, "Stress Intensity Factors for Crackline-Loaded Edge-Crack Specimens," NASA TN D-3820, 1967.

Srawley 1969
J. E. Srawley, "Plane Strain Fracture Toughness," in *Fracture*, edited by H. Liebowitz, Vol. IV, pp. 45–69, Academic Press, New York and London, 1969.

Srawley 1972
J. E. Srawley and B. Gross, "Stress Intensity Factors for Bend and Compact Specimens," *Engineering Fracture Mechanics*, Vol. 4, p. 587, 1972.

Srawley 1976
J. E. Srawley, "Wide Range Stress Intensity Factor Expressions for ASTM E399 Standard Fracture Toughness Specimens," *Int. Journal of Fracture*, Vol. 12, pp. 475–476, June, 1976.

Srivastava 1971
K. N. Srivastava and J. P. Durvedi, "The Effect of a Penny-Shaped Crack on the Distribution of Stress in an Elastic Sphere," *Int. Journal of Engineering Science*, Vol. 9, pp. 399–420, 1971.

Stallybrass 1969
M. P. Stallybrass, "A Cruciform Crack Deformed by an Arbitrary Internal Pressure," *Int. Journal of Engineering Science* Vol. 7, pp. 1103–1116, 1969.

Stallybrass 1970
M. P. Stallybrass, "A Crack Perpendicular to an Elastic Half-Plane," *Int. Journal of Engineering Science*, Vol. 8, pp. 351–362, 1970.

Stallybrass 1971
M. P. Stallybrass, "A Semi-Infinite Crack Perpendicular to the Surface of an Elastic Half-Plane," *Int. Journal of Engineering Science*, Vol. 9, p. 133, 1971.

Sullivan 1964
A. M. Sullivan, "New Specimen Design for Plane Strain Fracture Toughness Tests," *Materials Research and Standards*, Vol. 4, pp. 20–24, 1964.

Suo 1990a
Z. Suo and J. W. Hutchinson, "Interface Crack Between Two Elastic Layers," *International Journal of Fracture*, Vol. 43, pp. 1–18, 1990.

Suo 1990b
Z. Suo, "Failure of Brittle Adhesive Joints," *Applied Mechanics Review*, Vol. 43, pp. S276–S279, 1990.

Swedlow 1972
J. L. Swedlow (ed.), *Surface Crack: Physical Problems and Computational Solutions*, ASME, New York, 1972.

Tada 1970
H. Tada, "Westergaard Stress Functions for Several Periodic Crack Problems," *Engineering Fracture Mechanics*, Vol. 2, pp. 177–180, 1970.

Tada 1971
H. Tada, "A Note on the Finite Width Corrections to the Stress Intensity Factor," *Engineering Fracture Mechanics*, Vol. 3, pp. 345–347, 1971.

Tada 1972a
H. Tada, "Studies of the Crack Opening Stretch Concept in Application to Several Fracture Problems," Ph. D. Dissertation, Lehigh University, June, 1972.

Tada 1972b
H. Tada, "A Note on a Central Crack in a Strip Under Uniaxial Tension, *Engineering Fracture Mechanics*, Vol. 4, p. 585, 1972.

Tada 1973
H. Tada, P. C. Paris, and G. R. Irwin, *The Stress Analysis of Cracks Handbook*, 1st Ed. Del Research Corporation, Hellertown, PA, 1973.

Tada 1974
H. Tada, additions to **Tada 1973**.

Tada 1975
H. Tada and G. R. Irwin, "K-Value Analysis for Cracks in Bridge Structures," Fritz Engineering Laboratory Report, Lehigh University, June, 1975.

Tada 1983a
H. Tada, "The Effects of Shell Corrections on Stress Intensity Factors and the Crack Opening Areas of a Circumferential and a Longitudinal Through-Crack in a Pipe," Section III-1 of *The Application of Fracture Proof Design Methods Using Tearing Instability Theory to Nuclear Piping Postulating Circumferential Through Wall Cracks*, by P. C. Paris and H. Tada, U.S. Nuclear Regulatory Commission, NUREG/CR-3464, Sept., 1983.

Tada 1983b
H. Tada, "The Stress Intensity Factor for a Crack Perpendicular to the Welding Bead," *International Journal of Fracture*, Vol. 21, pp. 279–284, 1983.

Tada 1983c
H. Tada and P. C. Paris, "Estimation Procedures for Load-Displacement Relation and J-Integral for Entire Range of Elastic-Plastic Loading of Circumferentially Cracked Pipes," and other related articles in *The Application of Fracture Proof Design Methods Using Tearing Instability Theory to Nuclear Piping Postulating Circumferential Through Wall Cracks*, U.S. NRC, NUREG/CR-3464, Sept., 1983.

Tada 1985
H. Tada, P. C. Paris, and G. R. Irwin, *The Stress Analysis of Cracks Handbook*, 2nd ed., Paris Productions Incorporated (and Del Research Corporation), St. Louis, MO, 1985.

Tada 1993
H. Tada, H. A. Ernst, and P. C. Paris, "Westergaard Stress Functions for Displacement-Prescribed Crack Problems: I," *International Journal of Fracture*, Vol. 61, pp. 39–53, 1993.

Tada 1994
H. Tada, H. A. Ernst, and P. C. Paris, "Westergaard Stress Functions for Displacement-Prescribed Cracks: II: Extension to Mixed Problems," *International Journal of Fracture*, Vol. 67, pp. 151–167, 1994.

Tada 1996
H. Tada, P. C. Paris, and L. L. Loushin, "The Plastic Zone Instability Concept in Application to Analysis of Pressure Vessel Failure," Presented at the ASME Pressure Vessel and Piping Conference, Montreal, Canada, July 21-26, 1996.

Tada 2000
H. Tada, P. C. Paris, and G. R. Irwin, 2000, The Stress Analysis of Cracks Handbook, 3rd ed., ASME, New York, 2000 (this handbook).

Terada 1976
H. Terada, "An Analysis of the Stress Intensity Factor of a Crack Perpendicular to the Welding Bead," *Engineering Fracture Mechanics*, Vol. 8, pp. 441–444, 1976.

Tracey 1973
D. M. Tracey, "Three-Dimensional Elastic Singularity Element for Evaluation of K along an Arbitrary Crack Front," *International Journal of Fracture*, Vol. 9, pp. 340–343, 1973.

Tweed 1969a
J. Tweed, "The Distribution of Stresses in the Vicinity of a Penny-Shaped Crack in an Elastic Solid Under the Action of Symmetric Body Forces," *Int. Journal of Engineering Science*, Vol. 7, pp. 723–735, 1969.

Tweed 1969b
J. Tweed, "The Distribution of Stresses in the Vicinity of a Griffith Crack in an Elastic Solid in which Body Forces are Acting," *Int. Journal of Engineering Science*, Vol. 7, pp. 815–842, 1969.

Tweed 1970
J. Tweed, "The Stress Intensity Factor of a Griffith Crack which is Opened by a Thin Symmetric Wedge," *Applied Mathematics Research Group Report*, North Carolina State University, June, 1970.

Tweed 1971
J. Tweed, "The Determination of the Stress Intensity Factors, of a Pair of Coplanar Griffith Cracks whose Surfaces are Loaded Asymmetrically," *Engineering Fracture Mechanics*, Vol. 3, p. 381, 1971.

Tweed 1972a
J. Tweed, S. C. Das, and D. P. Rooke, "The Stress Intensity Factors of a Radial Crack in a Finite Elastic Disc," *Int. Journal of Engineering Science*, Vol. 10, pp. 325–335, 1972.

Tweed 1972b
J. Tweed and D. P. Rooke, "The Torsion of a Circular Cylinder Containing a Symmetric Array of Edge Cracks," *Int. Journal of Engineering Science*, Vol. 10, pp. 801–812, 1972.

Tweed 1973
J. Tweed and D. P. Rooke, "The Stress Intensity Factors of an Edge Crack in a Finite Elastic Disc," *Int. Journal of Engineering Science*, Vol. 11, pp. 65–73, 1973.

Uflyand 1965
Y. S. Uflyand, "Survey of Articles on the Application of Integral Transforms in the Theory of Elasticity," Dept. of Mathematics Report, North Carolina State University, Raleigh, NC, 1965.

Vazquez 1971
J. A. Vazquez, "Rol De La Deformacion Plastica En El Fenomeno De La Propagacion De Una Rajadura," Ph.D. Dissertation, Universidad Nacional De Cuyo, Nov., 1971 (in Spanish).

Vazquez 1979
J. A. Vazquez and P. C. Paris, "Plastic Zone Instability Phenomenon Leading to Crack Propagation," and "The Application of the Plastic Zone Instability Criterion to Pressure Vessel Failure," Proceedings of CSNI Specialist Meeting on Plastic Tearing Instability, St. Louis, Missouri, September 1979, US NRC NUREG/CP-0010 and OECD Nuclear Energy Agency, CSNI Report No. 39. pp. 601–631 and pp. 632–654, 1979.

Vooren 1967
J. V. Vooren, "Remarks on an Existing Numerical Method to Estimate the Stress Intensity Factor of a Straight Crack in a Finite Plate," Trans. ASME, Ser. D., *Journal of Basic Engineering*, Vol. 89, 1967.

Watwood 1969
V. B. Watwood, "The Finite Element Method for Prediction of Crack Behavior," *Nuclear Engineering and Designs*, Vol. 1, pp. 323–332, 1969.

Weinstein 1952
A. Weinstein, "On Cracks and Dislocations in Shafts under Tension," *Quart. of Appl. Math.*, Vol. X, p. 77, 1952.

Wessel 1968
E. T. Wessel, "State of the Art of the WOL Specimen for K_{IC} Fracture Toughness Testing," *Engineering Fracture Mechanics*, Vol. 7, p. 77, 1968.

Westergaard 1939
H. M. Westergaard, "Bearing Pressures and Cracks," Trans. ASME, *Journal of Applied Mechanics*, Series A, Vol. 66, p. 49, 1939.

Westmann 1964
R. A. Westmann, "Pressurized Star Crack," *Journal of Mat. and Physics*, Vol. 43, pp. 191–198, 1964.

Westmann 1966
R. A. Westmann, "Notes on Estimating Critical Stress for Irregularly-Shaped Planar Cracks," *International Journal of Fracture Mechanics*, Vol. 2, p. 561, 1966.

Wigglesworth 1957
L. A. Wigglesworth, "Stress Distribution in a Notched Plate," *Mathematica*, Vol. 5, p. 67, 1957.

Williams 1957
M. L. Williams, "On the Stress Distribution at the Base of a Stationary Crack," Trans. ASME, *Journal of Applied Mechanics*, Vol. 24, No. 1, pp. 109–114, 1957.

Williams 1961
M. L. Williams, "The Bending Stress Distribution at the Base of a Stationary Crack," Trans. ASME, Ser. E., *Journal of Applied Mechanics*, Vol. 28, pp. 78–82, 1961.

Williams 1971
W. E. Williams, "A Star-Shaped Crack Deformed by an Arbitrary Internal Pressure," *Int. Journal of Engineering Science*, Vol. 9, pp. 705–712, 1971.

Wilson 1969
W. K. Wilson, "On Combined Mode Fracture Mechanics," Report 69-1E7-FMECH-R1, Westinghouse Research Laboratories, June, 1969.

Wilson 1970
W. K. Wilson, "Stress Intensity Factors for Deep Cracks in Bending and Compact Tension Specimens," *Engineering Fracture Mechanics*, Vol. 2, p. 169, 1970.

Wilson 1972
W. K. Wilson, "Some Crack Tip Finite Elements for Plane Elasticity," ASTM STP 513, pp. 90–105, 1972.

Winne 1958
E. H. Winne and B. M. Wundt, "Application of the Griffith-Irwin Theory of Crack Propagation to Bursting Behavior of Disks," Trans. ASME, *Journal of Basic Engineering*, Vol. 80, p. 1643, 1958.

Wu 1970
T. S. Wu, Y. C. Pao, and Y. P. Chin, "Analysis of a Finite Elastic Layer Containing a Griffith Crack," *Int. Journal of Engineering Science*, Vol. 8, pp. 575–582, 1970.

Wundt 1959
B. M. Wundt, "A Unified Interpretation of Room Temperature Strength of Notch Specimens as Influenced by Size," ASME Paper No. 59, Met 9, 1959.

Yamamoto 1972
Y. Yamamoto and N. Tokuda, "Determination of Stress Intensity Factors in Cracked Plates by the Finite Element Method," NAUT Report No. 4003, Dept. of Naval Architecture, University of Tokyo, 1972.

Yamamoto 1973
Y. Yamamoto and Y. Sumi, "Stress Intensity Factors in a Cracked Axisymmetric Body Calculated by the Finite Element Method," *J. Soc. of Naval Arch. Japan*, Vol. 121, 1967.

Yamera 1966
S. Yamera and G. S. Krestin, "Determination of the Modulus of Cohesion of Brittle Materials by Compressive Tests on Disc Specimen Containing Cracks," *Soviet Materials Science*, Vol. 2, pp. 710, 1966.

Yokobori 1965
T. Yokobori, M. Ohashi and M. Ichikawa, Report of the Research Institute for Strength and Fracture of Materials, Tohoku University, Sendai, Japan, Vol. 7, No. 2, 1965.

Yokobori 1972
T. Yokobori, M. Ichikawa, S. Konosu, and R. Takahashi, Report of the Research Institute for Strength and Fracture of Materials, Tohoku University, Sendai, Japan, Vol. 8, No. 1, 1972.

Zahn 1965
J. J. Zahn, "Stress Intensity Factors for a Sharply Notched Cylinder under Tension," *Developments in Mechanics*, Vol. 3, Part 1, p. 91, John Wiley and Sons, Inc., New York, 1965.

Reference Index

U

V

W

Y

Z

Subject Index

T

U

FREE SOFTWARE (SMARTCRACK-LITE)

Free software (SmartCrack-Lite) by Engineering Mechanics Technology, Inc. is available to purchasers of this book as a supplement to this Handbook upon request. A description of this software can be found on the following page.

This software is not an ASME product, but by special arrangement with the developers, purchasers of this book can request free copies by contacting:

Engineering Mechanics Technology, Inc.
4340 Stevens Creek Blvd., Suite 166
San Jose, CA 95129
(408) 247-9274 (PHONE)
(408) 247-9272 (FAX)
dharris@emtinc.com (E-MAIL)
www.emtinc.com (WEB ADDRESS)

Proof of purchase, a copy of the invoice or packing slip that comes with the book, will be required to obtain the software. This software is owned by Engineering Mechanics Technology, Inc.

(This is not an endorsement by ASME, implied, deliberate or otherwise, of the software (SmartCrack-Lite) or of Engineering Mechanics Technology, Inc. and should not be construed as such.)

SOFTWARE FOR EVALUATION OF STRESS INTENSITY FACTORS BY USE OF FORMULAS IN THE STRESS ANALYSIS OF CRACKS HANDBOOK

The third edition of *The Stress Analysis of Cracks Handbook* contains many formulas and plots for evaluation of stress intensity factors. The use of such results is made more convenient by the software that is included in the accompanying computer diskette. The software is called SmartCrack-Lite, is in Windows, and contains the following 21 K-solutions included in the Handbook.

Standard Specimens	**Infinite & Semi-Infinite**	**Round Bars**
center cracked strip	crack from V-notch	exterior circum. crack
double edge crack	crack near edge	interior crack
single edge crack	crack from elliptical notch	**Hollow Round Bars**
3-point bend	crack from elliptical hole	interior crack
compact tension	2 cracks from ell. hole	exterior crack
arc shaped	**Disks**	**Shells**
round compact	edge loaded center crack	axial crack cylinder
	edge crack	circum. crack cylinder
		sphere

Many of the K-solutions include the influence function, which normally requires numerical integration for their use. The numerical integration procedure is included in the software and is transparent to the user. This makes application of the influence functions much easier. Further description of the software is included on the accompanying diskette, which also describes the installation procedure for SmartCrack-Lite. The following are the minimum system requirements to run SmartCrack-Lite: any IBM-compatible running Windows 3.1, 95, 98 or NT 4.0, a hard disk, and a 3 1/2" floppy disk.

The dialog box for the selection of cracks in the "infinite and semi-infinite planes" category is shown below as an example.

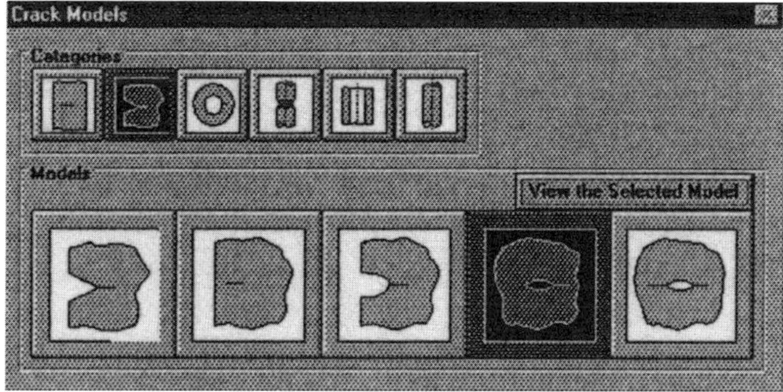

The K-solutions available in the software are a subset of the solutions included in a more comprehensive software package called SmartCrack, hence the name SmartCrack-Lite. The full SmartCrack software package, which is available from Engineering Mechanics Technology, Inc., in San Jose, California, contains many additional K solutions, and is able to perform fatigue and stress corrosion crack growth calculations, evaluate critical crack lengths, and evaluate crack opening areas. An extensive compilation of material fatigue crack growth characteristics is included.